Pharmaceutical Biotechnology

Pharmaceutical Biotechnology

A Focus on Industrial Application

Edited by

Adalberto Pessoa Jr., Michele Vitolo and Paul Frederick Long

CRC Press
Taylor & Francis Group
Boca Raton London New York

CRC Press is an imprint of the
Taylor & Francis Group, an **informa** business

Cover design courtesy of Leonardo Schultz da Silva.

First edition published 2022
by CRC Press
6000 Broken Sound Parkway NW, Suite 300, Boca Raton, FL 33487-2742

and by CRC Press
2 Park Square, Milton Park, Abingdon, Oxon, OX14 4RN

ISBN: 978-1-032-01374-9 (hbk)
ISBN: 978-1-032-00606-2 (pbk)
ISBN: 978-1-003-17836-1 (ebk)

Typeset in Times
by codeMantra

Contents

Preface

Biotechnology can be defined as a group of technologies that use cells, organelles and biomolecules to find solutions to scientific problems in general as well as for developing and/or improving biological drugs of economical interest. In this book, the terms 'biopharmaceuticals' and 'biological drugs' are being considered as synonyms.

Undoubtedly, biotechnology impacts areas of great economic and social interests such as agriculture, cattle raising, human and animal health, and the environment.

Modern biotechnology has rapidly advanced as a scientific discipline with the advent of recombinant DNA and cell fusion techniques that have allowed specific modifications to be introduced into the genome of living beings. Conversely, the term 'classic biotechnology' refers to all technologies developed since pre-history in which unspecific genomic modifications in living beings were made – leading to consolidation of methodologies and applications into either **industrial biotechnology** (focused on the use of techniques and processes mediated by biological agents) or **pharmaceutical biotechnology**.

Pharmaceutical biotechnology employs laboratory techniques and industrial processes to produce biopharmaceutical products (hormones, vaccines, antibiotics, monoclonal antibodies), develop tools for prognosis and diagnosis of diseases (biosensors, optogenetics, plasmonic, chips associated with macromolecules), administration of biopharmaceuticals, personalizing therapy and to develop synthetic biomolecules (for example, peptide-nucleic acids).

Undoubtedly, pharmaceutical biotechnology is the cornerstone of several scientific areas (Figure 0.1). The multidisciplinary nature of this science creates challenges in teaching undergraduate and graduate students across the health sciences. To face this challenge, we propose this book titled *Pharmaceutical Biotechnology: A Focus on Industrial Application*.

This book was conceived by professors, who since 2004 have been teaching pharmaceutical biotechnology to undergraduate students reading for degrees in Pharmacy and Biochemistry offered by the School of Pharmaceutical Sciences of USP. Colleagues from other Brazilian teaching and research institutions were invited as co-authors of several chapters and contributed significantly to the selection and consolidation of topics discussed throughout this book.

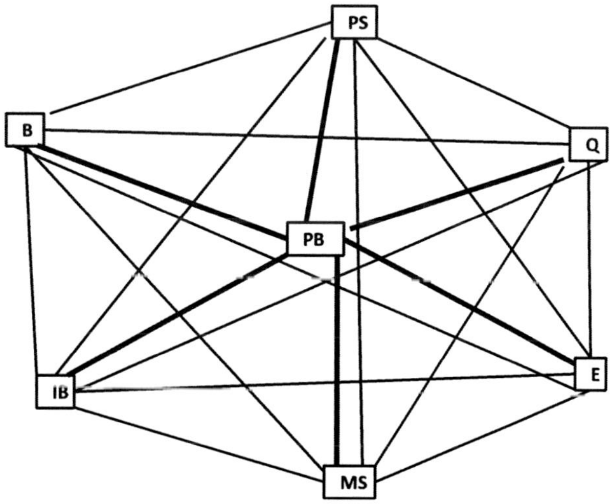

FIGURE 0.1. Simplified sketch of the multidisciplinary network of disciplines which mould Pharmaceutical Biotechnology (PB). B, Biology; E, Engineering; Q, Chemistry; IB, Industrial Biotechnology; PS, Pharmaceutical Sciences; MS, Medical Sciences.

Editors

Dr. Adalberto Pessoa Jr., PhD, is Full Professor at the School of Pharmaceutical Sciences of the Universidade de São Paulo and Visiting Senior Professor at King's College London. He has experience in enzyme and fermentation technology and in the purification processes of biotechnological products such as liquid–liquid extraction, cross-flow filtration and chromatography of interest to the pharmaceutical and food industries.

Dr. Michele Vitolo, PhD, is Full Professor at the School of Pharmaceutical Sciences of the Universidade de São Paulo. He has experience in enzyme technology, in immobilization techniques (aiming the reuse of the biocatalyst) and in the operation of membrane reactors for obtaining biotechnological products of interest to the pharmaceutical, chemical and food industries.

Dr. Paul Frederick Long, PhD, is Professor of Biotechnology at King's College London and Visiting International Research Professor at the Universidade de São Paulo. Prof Long is a microbiologist by training, and his research uses a combination of bioinformatics, laboratory and field studies to discover new medicines from nature, particularly from the marine environment.

Contributors

Alexsandra Conceição Apolinário
Department of Pharmacology
Institute of Biomedical Sciences
Universidade de São Paulo
São Paulo, Brazil

Adriano Rodrigues Azzoni
Departamento de Engenharia Química
Escola Politécnica
Universidade de São Paulo
São Paulo, Brazil

Thiago Olitta Basso
Departamento de Engenharia Química
Escola Politécnica
Universidade de São Paulo
São Paulo, Brazil

Antônio Sergio Kimus Braz
Centro de Ciências Naturais e Humanas – CCNH
Universidade Federal do ABC - Câmpus de Santo André
Santo André, Brazil

Carlos Alexandre Breyer
Biobreyer Pesquisa e Desenvolvimento Científico LTDA
Santos, Brazil

Larissa Pereira Brumano
Departamento de Tecnologia Bioquímico-Farmacêutica
Faculdade de Ciências Farmacêuticas
Universidade de São Paulo
São Paulo, Brazil

Celso Pereira Caricati
Instituto Butantan
São Paulo, Brazil

Attilio Converti
Department of Civil, Chemical and Environmental Engineering
Genoa University
Genoa, Italy

Camila Manoel Crnkovic
Departamento de Tecnologia Bioquímico-Farmacêutica
Faculdade de Ciências Farmacêuticas
Universidade de São Paulo
São Paulo, Brazil

Francisco Vitor Santos da Silva
Department of Biochemical-Pharmaceutical Technology
School of Pharmaceutical Sciences
Pharmaceutical Biotechnology Laboratory
Universidade de São Paulo
São Paulo, Brazil

Leonardo Schultz da Silva
Instituto de Biociências
Universidade Estadual Paulista (UNESP) – Campus do Litoral Paulista (CLP)
São Paulo, Brazil

Jônatas de Carvalho Silva
Unidade Acadêmica de Garanhuns
Rural Federal University of Pernambuco
Garanhuns, Brazil

Fernanda de Oliveira
Department of Engineering Bioprocesses and Biotechnology
School of Pharmaceutical Sciences
São Paulo State University (UNESP)
São Paulo, Brazil

Rodrigo Lira de Oliveira
Unidade Acadêmica de Garanhuns
Rural Federal University of Pernambuco
Garanhuns, Brazil

Marcos Antonio de Oliveira
Departamento de Ciências Biológicas e Ambientais
Instituto de Biociências
Universidade Estadual Paulista (UNESP) – Campus do Litoral Paulista (CLP)
São Paulo, Brazil

Ricardo Pinheiro de Souza Oliveira
Departamento de Tecnologia Bioquímico-Farmacêutica
Faculdade de Ciências Farmacêuticas
Universidade de São Paulo
São Paulo, Brazil

Nathalia V. Veríssimo
Department of Engineering Bioprocesses and Biotechnology
School of Pharmaceutical Sciences
São Paulo State University (UNESP)
São Paulo, Brazil

Patrícia Barros dos Santos
Serviço de Transferência de Tecnologia
Empresa Brasileira de Hemoderivados e Biotecnologia (Hemobras)
Goiana, Brazil

Valéria de Carvalho Santos Ebinuma
Department of Engineering Bioprocesses and Biotechnology
School of Pharmaceutical Sciences
São Paulo State University (UNESP)
São Paulo, Brazil

Valker Araujo Feitosa
Departamento de Tecnologia Bioquímico-Farmacêutica
Faculdade de Ciências Farmacêuticas
Universidade de São Paulo/Institute for Technological Research

Renata Bannitz Fernandes
Departamento de Tecnologia Bioquímico-Farmacêutica
Faculdade de Ciências Farmacêuticas
Universidade de São Paulo
São Paulo, Brazil

Rafael da Gama Ferreira
Intelligen, Inc.
Scotch Plains, New Jersey

Marina Gabriel Fontes
Department of Biochemical-Pharmaceutical Technology
School of Pharmaceutical Sciences
Universidade de São Paulo
Sao Paulo, Brazil

Sindelia Freitas
Centro de Biologia Molecular e Engenharia Genética
Universidade Estadual de Campinas (UNICAMP)
Campinas, Brazil
Instituto SENAI de Inovação Biotecnologia
SENAI SP
São Paulo, SP

Nicola Sgrignoli Fuganholi
Departamento de Tecnologia Bioquímico-Farmacêutica
Faculdade de Ciências Farmacêuticas
Universidade de São Paulo
São Paulo, Brazil

Francislene Andréia Hasmann
PPG Mestrado e Doutorado em Desenvolvimento e Meio Ambiente Urbano
Universidade da Amazônia
Belém, Brazil

Beatriz Vahan Kilikian
Departamento de Engenharia Química
Escola Politécnica
Universidade de São Paulo
São Paulo, Brazil

Patrícia Léo
Laboratório de Biotecnologia do Núcleo de Bionanomanufatura
Instituto de Pesquisas Tecnológicas do Estado de São Paulo
São Paulo, Brazil

André Moreni Lopes
Faculdade de Ciências Farmacêuticas
Universidade Estadual de Campinas
Campinas, Brazil

Felipe Rebello Lourenço
Departamento de Farmácia
Faculdade de Ciências Farmacêuticas
Universidade de São Paulo
São Paulo, Brazil

Carlos Miguel Nóbrega Mendonça
Departamento de Tecnologia Bioquímico-Farmacêutica
Faculdade de Ciências Farmacêuticas
University of São Paulo
Sao Paulo, Brazil

Amanda Mizukami
Hemotherapy Center of Ribeirao Preto
Universidade de São Paulo
São Paulo, Brazil

João Vitor Dutra Molino
Department of Chemistry
University of Zürich
Zürich, Switzerland

Gisele Monteiro
Departamento de Tecnologia Bioquímico-Farmacêutica
Faculdade de Ciências Farmacêuticas
Universidade de São Paulo
São Paulo, Brazil

Laura de Oliveira Nascimento
Faculty of Pharmaceutical Sciences
University of Campinas
Campinas, Brazil

Pedro Henrique Camargo Penna
NAIAD Drug Design LTDA
Universidade Federal do ABC
São Paulo, Brazil

Jorge Fernando Brandão Pereira
Department of Chemical Engineering
University of Coimbra
Coimbra, Portugal

Adalberto Pessoa Jr.
Departamento de Tecnologia Bioquímico-Farmacêutica
Faculdade de Ciências Farmacêuticas
University of São Paulo
São Paulo, Brazil

Tatiana Souza Porto
Unidade Acadêmica de Garanhuns
Rural Federal University of Pernambuco
Garanhuns, Brazil

Inês Conceição Roberto
Departamento de Biotecnologia
Escola de Engenharia de Lorena
Universidade de São Paulo
São Paulo, Brazil

João Henrique Picado Madalena Santos
Biochemical-Pharmaceutical Technology Department
Faculty of Pharmaceutical Sciences
University of São Paulo
São Paulo, Brazil

Marco Antonio Stephano
Departamento de Tecnologia Bioquímico-Farmacêutica
Faculdade de Ciências Farmacêuticas
Universidade de São Paulo
São Paulo, Brazil

Kamilla Swiech
Departamento de Ciências Farmacêuticas
Faculdade de Ciências Farmacêuticas de Ribeirão Preto
Universidade de São Paulo
São Paulo, Brazil

Aldo Tonso
Departamento de Engenharia Química
Escola Politécnica
Universidade de São Paulo
São Paulo, Brazil

Michele Vitolo
Department of Biochemical-Pharmaceutical Technology
School of Pharmaceutical Sciences
Universidade de São Paulo
São Paulo, Brazil

Carlota de Oliveira Rangel-Yagui
Department of Biochemical-Pharmaceutical Technology
School of Pharmaceutical Sciences
Universidade de São Paulo
São Paulo, Brazil

1 Fundamentals of Biotechnology

Michele Vitolo
Universidade de São Paulo

CONTENTS

1.1 INTRODUCTION

Biotechnology can be defined as an applied science that exploits biological systems (e.g. molecules, organelles, cells, tissues or whole animals or plants) to find solutions of economic value that benefit humankind. Within the context, we consider the very latest uses of biomolecules for therapeutic purposes, from antibiotics to more complex macromolecules and tissue systems defined as biological drugs. The definition of biological drug or biological product vary among regulatory agencies; they can also be known as 'biologics', 'biological medicines', 'biological medicinal products' or 'biopharmaceuticals'. Most regulatory agencies, such as the American Food and Drug Administration (FDA) and the European Medicines Agency (EMA), consider biological products as a wide range of items including vaccines, blood and its components, allergenics, somatic cells, gene therapy, tissues and therapeutic proteins. As described by the FDA, 'biologics can be composed of sugars, proteins, nucleic acids or complex combinations of these substances, or may be living entities such as cells and tissues'. In this book, the terms 'biopharmaceutical' and 'biological drug' are used interchangeably.

In spite of biotechnology becoming of immense scientific and industrial interest over the last three decades, biotechnological processes have been used by humans for millennia for beer, bread, cheese and wine production.

These food products were produced empirically without any theoretical knowledge about the processes involved. But, following the invention of the microscope by Antonie van Leeuwenhoek in 1667, the microbial world was discovered and could be studied. Knowledge about the characteristics of microorganisms (growth, metabolism, cell structure, reproduction, etc.) accumulated throughout subsequent centuries, culminating with Louis Pasteur's (1876) conclusion that the fermentative process was a microbial process. As a consequence, fermentation technology carried out under no aseptic conditions grew so fast that several bioproducts such as ethanol, acetic acid, butanol and acetone quickly appeared on the market and in huge amounts.

Fermentations carried out under aseptic conditions and with pure microbial strains started around the 1940s and led to an explosion in the variety of new commercial bioproducts (antibiotics, steroids, amino acids, vaccines, enzymes, among others).

Pure microbial strains that could produce a specific bioproduct in high yield were obtained by the genetic improvement of wild strains. The main techniques employed were random mutations induced in the wild strains by exposure of the microbes to physical (UV radiation, for instance) and/or chemical agents or by mating between strains belonging to closely related species, followed by screening for more productive mutants using auxotrophic culture media. However, a great technical advance for modifying the metabolic characteristics of prokaryotic and eukaryotic cells occurred in 1972, when the techniques of cell fusion (hybridoma) and recombinant DNA were introduced. Both techniques enhanced the development of so-called genetic engineering, insofar as they allowed introducing specific and planned changes directly into cellular DNA. The era of empirical biotechnology was over.

1.2 BIOLOGICAL MOLECULES

1.2.1 Introduction

Living cells are composed of a large variety of macromolecules (labelled generically as biomolecules) such as carbohydrates (polymers of low MW sugars), proteins (polymers of amino acids), nucleic acids (polymers of nucleotides; DNA and RNA) and lipids (a heterogeneous group of compounds having high solubility in organic solvents in common).

Although carbohydrates and lipids are essential constituents of cells, from the viewpoint of biotechnology, proteins and nucleic acids are the main biomolecules of interest. Carbohydrates have a variety of roles inside the cell, such as energy storage, maintaining structure and stimulating the immune response (antigen–antibody interaction, amongst others). Lipids are also important energy stores within adipose tissue, but the most crucial role of lipids is as components of cell membranes, promoting fluidity of cell membranes as well as effector molecules in their own right such as PAF (Platelet Activator Factor). PAF is responsible for platelet aggregation during blood coagulation, for dilatation of blood vessels which mediates the inflammatory process and allergic response and for implantation of a zygote into the wall of the uterus.

1.2.2 Proteins

Proteins – from the Greek term meaning 'first class' – are unbranched polymers of amino acids joined head to tail, from a carboxyl group (called the C-terminus) to an amino group (called the N-terminus), through the formation of covalent peptide bonds, which are a type of amide linkage. Except for the amino acid glycine, all other amino acids have an asymmetrical C atom leading to enantiomers designed as L and D. Only L-amino acids are constituents of active proteins in living cells.

Peptide bonds are rigid and planar structures that favour folding of the primary structure of protein (the primary structure is the sequential order of amino acids in the chain). As a consequence, two main structural folds can appear along the protein chain, i.e., α-helix and β-pleated sheet. Often both types of folds are found in proteins intermixed with parts of the protein where the conformation doesn't fit any regular pattern.

An α-helix can be visualized like a stick. It is stabilized by hydrogen bonds established between a peptide carbonyl group (C=O) of one amino acid and the peptide NH group of another amino acid, four residues farther up the chain. The α-helix can be oriented to the right (clockwise) or left (anti-clockwise), but the clockwise rotation is the most common pattern. Myoglobin is an example of a protein rich in α-helixes.

The β-pleated sheets resemble a concertinaed piece of paper sheet. It is stabilized by hydrogen bonds established between NH and C=O groups belonging to amino acids of adjacent peptide chains. The pleated sheet can exist in both parallel and antiparallel forms. In the parallel β-pleated sheet, adjacent chains run in the same direction (N → C or C → N). In the antiparallel β-pleated sheet, adjacent strands run in opposite directions. Ribonuclease is an example of protein rich in β-pleated sheets.

Regarding the architecture of protein molecules, it is by convention that we can distinguish four structural patterns: **primary structure** (corresponds to the sequence of amino acids along the chain and the position of disulphide bridges (S-S), if they exist); **secondary structure** (represented by α-helix and β-pleated sheets, both secondary structures stabilized by hydrogen bonds); **tertiary structure** (the polypeptide chains of protein molecules bend and fold in order to assume a more compact three-dimensional shape); **quaternary structure** (occurs when a protein is composed of two or more interacting polypeptide chains with characteristic tertiary structure, each polypeptide is commonly referred to as a subunit of the protein). It must be borne out that all the information necessary for a protein molecule to achieve its intricate architecture is contained within the amino acid sequence of its polypeptide chain(s). Whereas the primary structure of a protein is determined by the covalently linked amino acids in the polypeptide backbone, secondary and higher orders of structure are determined mainly by non-covalent forces such as hydrogen bonds, ionic bonds, and Van der Walls and hydrophobic interactions. In addition to the four structural levels cited, there

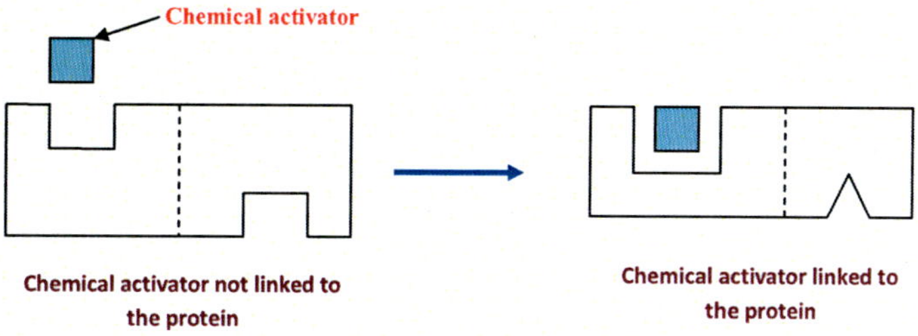

FIGURE 1.1 Sketch of a conformational change induced within a protein following interaction with a specific chemical activator called the ligand.

are also the so-called **super-secondary structures** (which consist of aggregates of secondary structures) and **domains** (compact folding of amino acids – generally composed of 100–400 amino acids – found in a specific location of the protein molecule; the active site of an enzyme is an example of a catalytic domain).

In living cells, proteins have a wide variety of different functions: **enzymes** (e.g. catalase and trypsin), **regulatory proteins** (insulin, somatotropin), **transport proteins** (haemoglobin and serum albumin), **storage proteins** (ovalbumin, casein and ferritin), **contractile and motile proteins** (actin, myosin, tubulin), **structural proteins** (collagen, elastin, fibroin), **protective proteins** (immunoglobulins, fibrinogen and ricin) and **exotic proteins** (antifreeze proteins, monellin and glue proteins).

Simply described, the function of a protein is mediated by an interaction between the protein and a specific chemical activator (usually called a ligand) that results in a conformational change throughout the protein structure (Figure 1.1).

1.2.3 Nucleic Acids

Nucleic acids – deoxyribonucleic acid (DNA) and ribonucleic acid (RNA) – are involved in the flux of genetic information, the control of gene expression and the control of cell metabolism.

DNA consists of two polynucleotide strands wound together to a form a long, slender, helical molecule (DNA double helix). Each strand is formed by nucleotides linked 3′–5′ by phosphodiester bridges. The nucleotides, in turn, are composed of deoxyribose, phosphate group and nitrogenous base (guanine [**G**], adenine [**A**], thymine [**T**] or cytosine [**C**]). The nitrogenous bases of one strand are complementary to the bases of the other strand, i.e., **A** pairs **T** and **G** pairs **C**. The pairing between bases is stabilized by hydrogen bonds (a type of weak chemical bond widely found throughout the biological world). In the double helix, the pair of bases is located inside the helix whereas the phosphodiester bridges and deoxyribose are outside the helix. The structure of DNA is essentially the same in all living cells, differing only in the number and sequence of nitrogenous bases **(A-C-G ≠ G-C-A).** DNA has the capacity to self-replicate due to the action of a specific enzyme called DNA polymerase. It is estimated that human DNA is composed of about 3 billion nitrogenous bases (Felsenfeld and Groudinein, 2003).

The summation of all hereditary information stored within chromosomes is referred to as the genome. So it is the genome that orchestrates the development and life of an organism. A chromosome is not a simple structure of weaving of DNA but is a complex of DNA, proteins (histones) and other molecules (methyl, acetyl, etc.). This complex is called chromatin. Normally, chromatin is wrapped in aggregates of histones, so that it is possible to insert long DNA chains inside small organelles such as the nucleus of eukaryotic cells and mitochondria. Moreover, parts of chromatin can condense –; hiding long stretches of a DNA molecule – or can expand in order to expose parts of DNA sequence to be a transcript by RNA polymerase. The genome is a complex non-stop biochemical machine that constantly suffers intense and unspecific modifications, so even homozygote twins are not truly genetically identical but can present different gene copies. In this way, one twin can suffer from diabetes whereas the other twin is healthy, for instance.

Ribonucleic acid (RNA) is also a polymer of nucleotides differing from DNA in the sugar type (**ribose**) and one of the nitrogenous bases, i.e., **uracil** instead of thymine. RNA can be linear or cyclic as well as single or double stranded. There are several types of RNA (Figure 1.2), the best known being messenger (${}_{m}$RNA), ribosomal (${}_{r}$RNA) and transfer (${}_{t}$RNA). Each type of RNA has a specific role on the flux of genetic information. This occurs in two sequential steps named, respectively, transcription (in which the ${}_{m}$RNA is synthesized by RNA polymerase that copies a part of the DNA strand taken as the template) and translation (the ${}_{m}$RNA links to the ribosome followed by ${}_{t}$RNA that brings amino acids to the surface of the ribosome which are sequentially bonded leading to the formation of a specific polypeptide, the final product).

Stretches of chromosomal DNA which encode for an entire polypeptide chain are called **exons**, whereas those stretches that are not protein-coding are called **introns** (also named **intervening sequences** because they intervene between exons). Thereby, genes (delimited regions of DNA molecule involved in the flux of genetic information) are formed only by exons, as observed in prokaryotes, or by exons and introns intercalated as occur in eukaryotes. It must be borne out that introns do encode RNA, but this RNA is not translated into protein and is thought to be involved in various aspects of genome regulation. Moreover, it seems that the number of introns a gene contains increases with the complexity of the organism (Sorz, 2002; Fu, 2014).

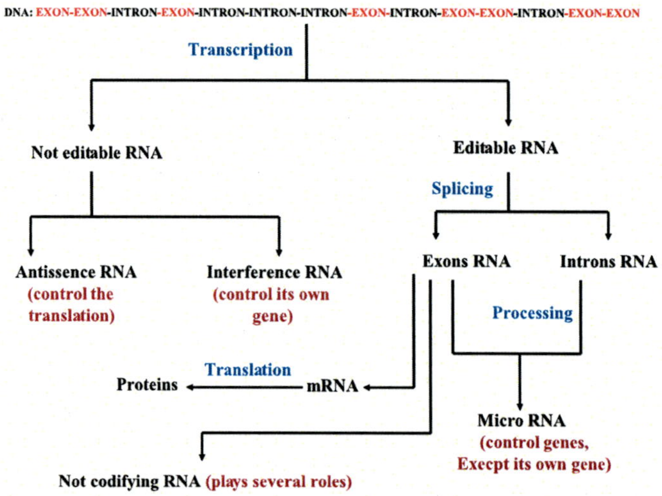

FIGURE 1.2 Flux of genetic information in eukaryotes.

As more genomes of complex organisms, including man (*Homo sapiens*), are sequenced and become publicly available, so our understanding of the complexity of the flux of genetic information increases (Figure 1.2). For instance, the human genome is composed of about 25,000 genes, which are responsible for the synthesis of some 90,000 different proteins. Scrutinizing these numbers, it is clear that the concept of 'one gene–one protein' is not valid (Ast, 2004; Park et al., 2018).

In prokaryotes, a single $_{m}$RNA encodes all of the information for the synthesis of a single polypeptide chain within its nucleotide sequence (or the synthesis of several polypeptide chains in the case of operons). In contrast, eukaryotic $_{m}$RNAs encode only one polypeptide but are more complex in that they are synthesized in the nucleus in the form of much larger precursor molecules called **heterogeneous nuclear RNA (hnRNA)**. The hnRNA is composed of introns and exons copied from the gene and is inadequate for the protein synthesis whilst the introns are present. So the introns must be spliced out for the formation of functional $_{m}$RNA.

The splicing occurs when a nuclear protein, the so-called protein regulator of splicing (PRS), interacts with exons belonging to the hnRNA creating points in which low MW ribonucleic acids molecules ($_{sn}$RNA) can aggregate. The exon-PRS point of contact is named the **splicing exon activator (SEA)**. To date, six different types of $_{sn}$RNA have been identified (Ast, 2004). The $_{sn}$RNAs bind at specific points of the hnRNA in order to delimit the introns to be discharged. At these points, several nuclear proteins form a complex called the **spliceosome**, which cuts the introns at the connections with the exons, leading to the formation of mature $_{m}$RNA. However, a PRS – the so-called **splicing exon inhibitor (SEI)** – can be linked at a different site to the SEA, inhibiting the binding of the $_{sn}$RNA at the points required to remove the introns. As consequence, exons can also be erased from the hnRNA sequence. This process of hnRNA splicing is termed **alternative splicing**. According to Ast (2004), the deletion of exons through alternative splicing is quite common in mammalian cells and in humans, for example, accounts for about 38% of all genetic processing. A human gene is on average composed of 28,000 nucleotides with 8 exons separated by 7 introns. In general, each exon is composed of about 120 nucleotides, whereas introns are composed of about 100,000 nucleotides. It is worth noting at this point that the human genome has the highest number of introns per gene than any other organism on Earth.

The most likely explanation as to why the human genome has the potential to produce 90,000 proteins without needing to have the same number of genes is due to alternative splicing. On average, each human gene can transcribe three $_{m}$RNA through the process of alternative splicing (Ast, 2004). However, such a mechanism does not explain why the human genome is composed of about 97% and 3% of introns and exons, respectively. Besides, the complexity of the splicing mechanisms, both demand high-energy consumption as run the risk that errors could often occur during the cut and repair of the transcript RNA. An error during the cut–repair mechanism can lead to the appearance of a hereditary disease named Riley–Day syndrome, which causes the malfunction of the body's neural web. This syndrome is caused by the insertion or deletion of a single nucleotide in the IKBKAP gene (located at chromosome 9), leading to undesired alternative splicing in nervous central system tissues. It is estimated that at least 15% of mutations, which cause hereditary diseases as well as some types of cancer, result from the incorrect splicing of transcript RNA. The occurrence of high numbers of introns within the human genome can reasonably be understood if it is considered that introns occur at loci in which small fragments of DNA – called transposons – can insert. Transposons can be moved enzymatically from place to place in the genome, that is, their location within the DNA is unstable. Transposons contain a gene encoding the enzyme necessary for insertion into a chromosome and for the remobilization of the transposon to different locations. These movements are termed **transposition events**. The smallest transposons are called **insertion sequences**, meaning that they are able to insert apparently at random anywhere in the genome. Although certain transposons may undergo transposition once per cell generation – as observed generally in prokaryotes –, most transposition events are rare occurring only once every 10^4–10^7 generations (Garrett and Grisham, 1995).

At first glance, it seems that the function of transposons would be the generation of self-copies followed by random reinsertion amongst the introns of the genome. However, if the transposon-intron system suffers a mutation – say, during mitosis or meiosis – from which a favourable point for anchoring $_{sn}$RNAs appears, a spliceosome would be formed resulting in an intron being expressed, i.e., it becomes an exon. As a consequence, eukaryotic cells can acquire the capacity to synthesize a new protein. If the mutation introduces some metabolic advantage, it will be integrated into the cell's metabolism by natural selection. It must be emphasized that the capability to produce a new protein does not interfere with the normal function of the gene because the previous types of $_m$RNA continue to be synthesized, since the new exon is left out of the splicing. However, when the new exon becomes constitutive, i.e., the exon is included in all $_m$RNA formed, some diseases can occur such as Alport's syndrome (hereditary miscarriage linked to the chromosome X that causes kidneys malfunction, deafness and visual deficiency) and Sly's syndrome (the biosynthesis of mucopolysaccharides is affected leading to weakness of bones, cartilages and tendons as well as any other body structures dependent on these tissues). Some evidence suggests that transpositions into the genome of tissue-specific stem cells of the hippocampus (a region of the brain related to memory and attention) might lead to cognition miscarriage and schizophrenic symptoms in humans, even in identical twins (Coufal, 2009).

According to a conventional paradigm, genes that encode proteins correspond to only 3% of the overall DNA in a cell. However, 97% of 'non-protein coding' DNA does transcribe several distinct types of quite biochemically active RNA molecules, some of which are capable of regulating the function of genes. Four different types of RNA have been identified, i.e., antisense RNA ($_{ant}$RNA), interference RNA ($_i$RNA), micro RNA ($_{mic}$RNA) and riboswitch RNA ($_{rib}$RNA).

There are two possible genetic origins for $_{ant}$RNA: this RNA can be transcribed from a 'pseudogene' that encodes only RNA, or from the DNA strand complementary to that transcript of $_m$RNA. So far, 1,600 genes have been identified where complementary strands can transcribe $_{ant}$RNA (Sorz, 2002). The pseudogene and the normal protein-encoding gene are located on different chromosomes. When levels of $_{ant}$RNA transcripts reach a determined concentration inside the cell nucleus, the transcript becomes linked to the complementary $_m$RNA to form a double-stranded RNA molecule which cannot be translated on a ribosome to biosynthesize a protein. Additionally, the $_{ant}$RNA can also bind itself not only to a $_m$RNA but also to DNA, proteins and low MW compounds (metabolic inhibitors or stimulators).

Transcripts of $_{mic}$RNA are formed from the removal of introns during splicing of the primary RNA transcript – in this case, we have a precursor of $_{mic}$RNA because its MW is higher than that of a functional $_{mic}$RNA – and/or an intermediate in the mechanism that generates $_i$RNA. In this case, $_{mic}$RNA are formed by the transcription of transposons from anomalous genes or by retroviruses infecting the cell. $_{mic}$RNA is a short double-stranded RNA molecule that forms a folded secondary structure. Normally, $_{mic}$RNAs are found moving freely throughout the cell cytoplasm but unfold as soon as they encounter $_m$RNA to be silenced. The unfolding mechanism has yet to be elucidated, although there are about 150 $_{mic}$RNAs that have been described (Sorz, 2002).

The biosynthesis of $_i$RNA has been found in all animal and plant cells examined to date. $_i$RNA acts as a sensor of cell gene activity so that if at any time an anomalous gene is expressed – which may jeopardize the cell – the $_i$RNA mechanism is activated. This mechanism for regulating gene expression has also been found to be important for regulating normal gene transcription, as it prevents genes from becoming constitutively expressed which would otherwise lead to 'protein intoxication'. Moreover, it seems that $_i$RNA plays an important role in the development of an organism from the embryo to the adulthood. From one embryonic cell, thousands of different types of cells originate and form what constitutes all body tissues. Many genes that are expressed during the embryonic phase must be silenced during maturation, and conversely, genes that are silenced during embryogenesis must be expressed at other times during the life course of the organism. The role of $_i$RNA is not only to regulate the genome from aberrant gene expression but also to regulate differentiation of cells (neurons, hepatocytes, etc.) and different organs (brain, liver, kidneys, etc).

A simplified description of $_i$RNA mechanism would be inside the cell double-stranded RNA interacts with an enzyme (called the **dicer**). The dicer hydrolyses the RNA into short interference RNA ($_{si}$RNA), no more than 22 ribonucleotides in length. The dicer does not cut each strand of RNA uniformly, resulting in double-stranded $_{si}$RNA fragments with two unpaired nucleotides at each end. Then, the double-stranded $_{si}$RNA unfolds and one of the strands interacts with specific proteins forming the **RNA inducer silencing complex (Risc-RNA)**. Inside the Risc-RNA, the $_{si}$RNA molecule is positioned in order to pair with the $_m$RNA to be silenced. However, the Risc-RNA will encounter many $_m$RNA molecules in the cytoplasm, but pairs only to those RNA molecules with a nucleotide sequence that is complementary to the $_{si}$RNA. Once the $_m$RNA that is targeted for silencing has become paired to the Risc-RNA complex, then an enzyme called the **slicer** breaks the $_m$RNA molecule into two parts so that each becomes incapable of being translated into protein. It must be said that the Risc-RNA complex remains intact at the end of the process, allowing the complex to be available for silencing another $_m$RNA molecule (MacManus and Sharp, 2002; Chalbatani et al., 2019).

Understanding the mechanism of $_{si}$RNA gene silencing has become a useful tool for research. The reason being that, in principle, any gene in any cell can be silenced in a matter of hours, allowing sufficient time when the gene is silenced for an experiment to study the function of the gene product to be performed. Such approaches would be useful in screening new drugs that could interact with particular genes believed to be involved in cancer cell development, in the control of genetic disorders and in new treatments for viral infections. Following this philosophy, Ribopharma is a company that is developing a new drug (a $_{si}$RNA molecule modified chemically) for treating a type of brain tumour called glioblastome which is already in clinical trials. Nonetheless, the greatest difficulty in using $_{si}$RNA as potential new drugs is drug delivery and subsequent drug targeting to the tumour in sufficient concentration to have a positive therapeutic effect. The high MW of a $_{si}$RNA molecule is a handicap, preventing passage across membranes and other body barriers, especially the blood–brain barrier. The most likely route for administration would be the development of liposome vehicles that could deliver the $_{si}$RNA by transdermal or respiratory routes, thus avoiding oral administration and likely degradation by digestive enzymes.

At this point, it would be useful to distinguish between $_{mic}$RNA and $_{si}$RNA, since both are small fragments of ribonucleic acid and are involved in regulating the global gene expression. Essentially, both differ according to the biosynthetic origin, while $_{si}$RNA(s) silence the same genes from which they were transcripted, $_{mic}$RNA(s) result from genes specifically designed for their transcription.

It must be remembered that there is a class of low MW proteins (15–35 kDa), called interferons, which can also regulate gene expression. However, the so-called interferon-response occurs only when a cell is infected by a virus. There are three types of interferons, i.e., α-interferon (produced by leucocytes), β-interferon (produced by fibroblasts) and γ-interferon (produced by T-lymphocytes and NK cells of the immune system). Their mechanism of action consists of binding to receptors on the cytoplasmic membranes of cells not infected by viruses, inducing the so-called viral condition, which means that the cells become resistant against infection by a large variety of different viruses. The 'viral condition' stimulates the production of three intracellular enzymes called protein kinase (PKR), oligoadenylate synthetase and endoribonuclease (RNAse-L). The PKR inhibits the action of eIF2 (an initiation factor for protein synthesis) – whose function is to bind $_{t}$RNA onto the 40S subunit of the ribosome – by introducing into its structure a phosphate group leading to inhibition of the translation of $_{m}$RNA to protein. The oligoadenylate synthetase catalyses the formation of oligoadenylates (composed of 2–15 nucleotides), which activate the RNAse-L that will hydrolyse either $_{m}$RNA (viral or not) or $_{r}$RNA thereby preventing protein synthesis. In other words, interferons are molecules of the innate immune response that protect healthy cells in the proximity of virally infected cells from possible infection.

In bacteria, a particular type of high MW ribonucleic acid named the ribosome commuter or 'riboswitch' ($_{rs}$RNA) was discovered that also regulates gene expression. The $_{rs}$RNA can be translated as part of a protein, depending on the presence in the cytoplasm of a specific regulatory molecule (Figure 1.3). The $_{rs}$RNA is self-folding and, soon after synthesis, folds itself into an inactive conformation. In the presence of an activator molecule – the concentration of which in the cytoplasm appears directly proportional to the amount of the protein within which the $_{rs}$RNA will be translated – the $_{rs}$RNA unfolds to expose a portion of the sequence that will be translated into protein (Blount and Breaker, 2006). Until now, 12 different classes of $_{rs}$RNA have been identified. According to Blount and Breaker (2006), $_{rs}$RNAs have been identified in several pathogenic bacteria such as *Brucella melitensis* (which encodes five classes of $_{rs}$RNA that regulate 21 genes), *Escherichia coli* (has four classes of $_{rs}$RNA which regulate 15 genes), *Haemophilus influenzae* (can translate five classes of $_{rs}$RNA which regulate 15 genes), *Helicobacter pylori* (has one class of $_{rs}$RNA which regulates 2 genes), *Mycobacterium tuberculosis* (has three classes of $_{rs}$RNA which regulate 13 genes), *Vibrio cholerae* (has five classes of $_{rs}$RNA which regulate 13 genes) and *Yersinia pestis* (has three classes of $_{rs}$RNA which regulate 11 genes). In the future, it may be possible that once activators of $_{rs}$RNA have become more widely recognized, then structural analogous of these activators can be synthesized which could act as antibiotics, insofar as they could bind to the $_{rs}$RNA preventing the unfolding.

Slowly, genes that encode specific RNA(s) are being discovered. The task is quite difficult because genes encoding only RNA(s) don't have clearly defined regions that identify the beginning and end of transcription in the same way protein-encoding genes do. Undoubtedly, genes encoding RNA(s) exert tight control on chromosomal activity, thus demonstrating an intrinsic relationship between heredity and the development and maintenance of healthy cells, tissues and whole organisms. Malfunction of one RNA gene could explain why some inherited genetic disorder can suddenly appear in a family pedigree without any previous evidence in preceding generations.

Actually, there is some evidence that the information which controls all metabolic events in eukaryotic organism is not totally dependent on the sequence of billions of pairs of nitrogenous bases contained within chromosomal DNA. A significant number of metabolic events appear to be governed by a complex mixture of special proteins (histones) and specific chemical markers (methyl, acetyl, phosphate, ubiquitin, for example), which are intrinsically embedded into the physical structure of chromosomal DNA. The assembly of all these components is named chromatin (a filamentous

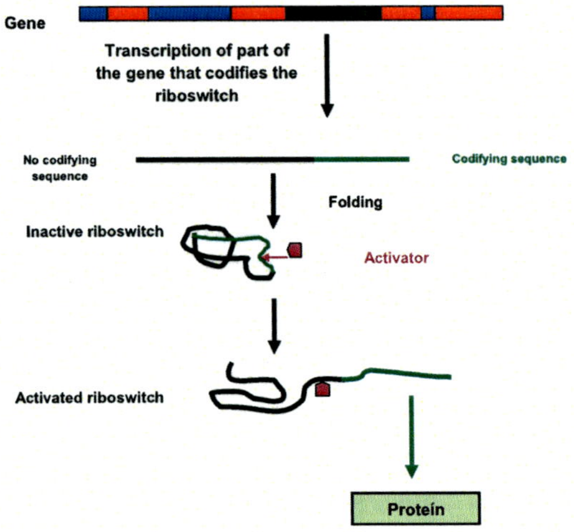

FIGURE 1.3 Sketch for riboswitch transcription from part of a gene. In the first step, an inactive cyclic riboswitch is formed. In the second step, an activator binds to the inactive riboswitch causing the RNA to unfold, revealing the sequence to be translated.

structure that controls access to DNA sequences). The parts of chromatin can condense or expand independently so that one portion of DNA can remain hidden whereas another is exposed for transcription. This is possible because DNA is wrapped around the histone proteins forming a dynamic structure. Such a configuration also explains how a high MW molecule like DNA can be inserted inside a small organelle such as the cell nucleus. Nowadays, there is the suspicion that chemical markers (methyl, acetyl, etc.) which are distributed not randomly along the length of the chromatin, but instead discreetly positioned to form a code named the **epigenetic code**, i.e., chemical information stored on DNA but independent of mechanisms that regulate DNA itself. There is evidence that the silencing of chromosome X in females involves the action of an epigenetic code. Here, the X_{ist} produces an $_{ant}$RNA that specifically binds to the X gene to be silenced. After binding, methyl markers are imprinted across the X chromosome. This methylation of DNA causes the histones to aggregate so that the overall chromatin structure condenses tightly around them, preventing the exposure of DNA sequence and transcription is blocked.

It has been estimated that possibly as much as 45% of human DNA is contaminated by viral genes or fragments of gene sequences, which became integrated into the genome early during human evolution. But, fortunately, these viral sequences remain silenced thanks to a high degree of methylation between cytosine-guanine pairs along the DNA strands (Felsenfeld and Groudinein, 2003). Transposons are also silenced by methylation. The methyl radicals are derived from folic acid and vitamin B_{12}, and methylation is considered the main mechanism of epigenetic regulation (Lodish et al., 1995). It has been well documented that the DNA of cancerous cells has a low degree of methylation, and it has been proposed that the activation of oncogenes may occur through the loss of methyl groups during cell division (mitosis and/or meiosis). It has also been suggested that toxic compounds ingested involuntarily (ambient pollution, for instance) or voluntarily (illicit drugs, smoking and self-medication) might also provoke an intense reduction in base methylation that induces a disease state.

Differently from the conventional genetic code in which the DNA architecture remains stable, in the epigenetic code the degree of methylation often varies. For example, when a section of chromatin condenses, an inhibitory effect carries along the chromosome up to a barrier, which separates portions highly methylated and silenced from those not methylated and active. The increased rate of cancer in older people results from the fragmentation of such barrier as a consequence of imperfect cell division.

It is speculated that the epigenetic code influences mechanisms related to growing, ageing and development of cancer. Moreover, mutations in the epigenetic code also contribute to other complex syndromes such as diabetes and schizophrenia. However, knowing the architecture of the epigenetic code, it should be possible for new approaches to treat these diseases. On the one hand, cells protect their DNA against mutations; but on the other hand, they add or erase epigenetic markers routinely. In principle, new drugs could interfere with the epigenetic code in order to inactivate or activate noxious genes, leading to the reversion of genetic damages connected to the ageing process, or those preceding the appearance of cancer. Besides, cell senescence – which induces ageing, inflammation and activators of cancer (Rodier and Campisi, 2011) – whose intensity is reduced by the presence of excess of 'p16' protein codified by gene $p16^{INK4a}$ (Blagosklonny and Hall, 2009). A possibility – still to be demonstrated in the future – would be controlling p16 protein biosynthesis through epigenetic markers in order to postpone ageing and the occurrence of illnesses. There is evidence that the distribution pattern of epigenetic markers (acetyl, methyl and ubiquitin, for instance) inside some chromosomes can contribute (or not) to the appearance of symptoms in humans – for long periods of time – related to depression, and social behaviour connected to the use of abusive drugs (cocaine, heroin and marihuana, amongst others). Following this idea, controlling the protein called the mammal target of rapamycin (mTOR) (a protein sensor for the amount of nutrients available inside the cells) would also allow for the manipulation of ageing/longevity mechanisms in humans. Thereby, nutrients are in high amount leading to increased cell division and growth (longevity enhanced), but the contrary would occur at nutrients in short supply (ageing enhanced). As a result, the inhibition of mTOR with drugs (rapamycin, for instance) should be a very interesting concept in treating elderly patients.

Undoubtedly, the epigenetic code will be elucidated. In the UK, the so-called Human Epigenome Project has begun under the direction of the Wellcome Trust Sanger Institute aiming to map all DNA methylation points.

Considering the three levels of the genomic machine – the expression of DNA into proteins, the control of gene expression by different types of RNA and the epigenetic information associated with chromatin architecture (in particular, the interaction amongst histones and chemical radicals) – all of which can now be manipulated, it is envisaged that there will be another biotechnological area emerging in the future called **genomic engineering.**

1.2.4 Virus

So far, we have briefly discussed the biomolecules that participate in hundreds of biochemical pathways that guarantee the survival of life on Earth.

Nevertheless, it must be remembered that in nature there are particular combinations of carbohydrates, lipids, proteins and nucleic acids (DNA or RNA) with no apparent vital-for-life characteristics (nutrition and generation of metabolic energy, amongst others). Such combinations are found in inanimate structures – when not interacting with cell – named generically as **virion**. In other words, a virion is a molecular aggregate unable to reproduce by itself.

After a virion penetrates the cell, it becomes active and is called a **virus.** The virus has the following characteristics: small dimensions (up to 250nm), one type of nucleic acid, a small amount of different types of proteins and no enzyme related to the energy generation as it is a strict intracellular parasite. However, there are exceptions such as *Cytomegalovirus* and MiMiVirus (**Mi**miking **Mi**crobe **Virus**) which have dimensions over 250nm – the MiMiVirus has 400nm, DNA and RNA, and more than 100 types of proteins (La Scola, 2003; Colson et al., 2017).

Hitherto, all viruses known – since the delta virus (the smallest) up to MiMiVirus (the biggest) – don't have ribosomes and the capability for producing ribosomal proteins. The latter characteristics separate the viral world from that of autonomous living organisms.

The discovery of MiMiVirus (an amoeba parasite, having a genome size of circa 1.2 million base pairs and 1,200 sequences that resemble genes. Some of these sequences can encode for DNA repair enzymes) would lead to the division of the living world into four domains, i.e., **eukaryotes** (*Arabidopsis thaliana, Homo sapiens* and yeasts, for example); **prokaryotes** including **bacteria** (*Bacillus subtilis, Mycobacterium tuberculosis, Escherichia coli*, amongst others) and **archaea** (*Aeropyrum pernix, Pyrococcus abyssi, Archeoglobus fulgidus*, for example); **virus**; and **MiMivirus** (Colson et al., 2017; Suhre et al., 2005).

The virion has its nucleic acid covered by a protein envelope (capsid) which protects the nucleic acid from hydrolytic enzymes and mechanical stress (caused by shear forces), and it is the machinery that injects the nucleic acid inside a host cell. The capsid is a regular array of one or a few types of protein. For example, the tobacco mosaic virus – the first virus to be identified by Martinus Beijerinck in 1898 – has RNA formed of 6,400 nucleotides and a capsid constituted by 2,130 identical protein subunits, each one composed of 158 amino acids (Stryer et al., 2019).

The replication cycle of a virus inside eukaryotic cells can be divided into seven phases as follows: first phase: the binding of the virus to a receptor outside the cytoplasmic membrane; second phase: the active penetration of the virus into the cell. The penetration can be by injection (e.g. bacteriophages), endocytosis (e.g. Epstein–Barr virus and human papilloma virus) or fusion (the capsid fuses with the cytoplasm membrane); third phase: the virus nucleic acid is freed in the cytoplasm; fourth phase: the nucleic acid links to cytoskeleton proteins (e.g. microfibrils and microtubules) and is translocated to the nucleus (eukaryotes) or to the replication zone (prokaryotes); fifth phase: the viral genome is replicated by the transcription machinery of the host cell, and the $_m$RNA molecules are translated into proteins using cell's ribosomal apparatus; sixth phase: nucleic acid and protein molecules self-assemble forming new virions; seventh phase: the escape of virions from the cell. The virions reach the internal surface of cytoplasmic membrane by anchoring themselves onto the cytoskeleton structure. If the viral capsid doesn't have an envelope, the virion particles agglutinate inside the cell until the cell bursts. If the capsid has an envelope (constituted by lipoproteins or glycoproteins), exit occurs by exocytose (the virion capsid fuses with the internal surface of cytoplasm membrane, followed by budding of the membrane. In this case, the capsid envelope is formed by constituents of the cell membrane).

Amongst the viruses studied to date, the nucleic acid can be single-stranded DNA (phage QX174, for instance), double-stranded DNA (e.g. phage T4 and Poxvirus of vaccinia), single-stranded RNA (e.g. influenza virus and tobacco mosaic virus) or double-stranded RNA (e.g. Reovirus). Moreover, the number of genes varies from 4 (phage Qβ) to 240 (Poxvirus of vaccinia) (Stryer et al., 2019).

Viruses that use DNA require several proteins belonging to the host cell for replication and gene expression. Viruses that use RNA have an additional problem because the non-infected host cell doesn't have RNA-dependent RNA polymerase to replicate the viral RNA.

Generally, two classes (I and II) of DNA viruses and four classes (III, IV, V and VI) of RNA viruses are recognized. **Class I:** contain a single molecule of double-stranded DNA, which inside the host cell diverts the nucleus enzymes for its replication and transcription into viral $_m$RNA. This class includes the adenoviruses (responsible for upper respiratory tract infections) and the herpesviruses (which cause various inflammatory skin diseases such as chickenpox). Some DNA viruses, the poxviruses (e.g. smallpox and vaccinia), can replicate in the host-cell cytoplasm because they carry their own enzymes for synthesizing viral $_m$RNA and DNA; **Class II:** refers to viruses (parvoviruses, as commonly named) containing one molecule of single-stranded DNA. Parvoviruses have either plus or minus strands of DNA, but never both. In both cases, the single-stranded DNA is copied inside the host cell into double-stranded DNA, which is then itself copied into $_m$RNA; **Class III:** refers to viruses (also known as reoviruses) having double-stranded RNA. The minus RNA strand acts as a template for the synthesis of plus strands of $_m$RNA, which occurs in the host-cell cytoplasm. The virions of this class have segmented genomes containing 10–12 double-stranded RNA segments, each of which encodes one or two polypeptides; **Class IV:** refers to viruses containing a single plus strand of RNA, which is identical to the viral $_m$RNA. Since the plus-strand RNA encodes proteins, it is infectious by itself. During replication of class IV viruses, the plus-strand RNA is copied into a minus strand, which then acts as a template for synthesis of more plus strands or $_m$RNA. Viruses such as the poliovirus synthesize proteins directly from the viral plus-strand RNA, whereas the so-called togaviruses (the virions are wrapped in a lipid envelope) synthesize an $_m$RNA first. Togaviruses cause illnesses such as yellow fever and viral encephalitis in humans. Moreover, the Sars-Cov-2 virus, which causes the Covid-19 disease, can be included in this class because it has a single strand plus RNA (Oliveira et al., 2020); **Class V:** refers to viruses having a single minus strand of RNA, and this sequence is complementary to that of the viral $_m$RNA. The genomic RNA in the virion acts as a template for the synthesis of $_m$RNA but doesn't encode proteins itself. In this class, two types of viruses can be distinguished. In viruses causing measles and mumps, the genome is formed by a single molecule of RNA. A virus-specific RNA polymerase present in the virion catalyses the synthesis of several $_m$RNAs, each encoding a single protein, from the genomic template strand. There are also viruses, typified by influenza virus (its RNA is fragmented into eight pieces), having segmented genomes; each segment acts as a template for the synthesis of a different $_m$RNA species. In most cases, each $_m$RNA produced encodes a single protein; however, some $_m$RNAs can be read in two different frames to yield two distinct proteins. For example, in the one single-stranded RNA viruses, the RNA fragmented viruses also have a virus-specific polymerase that catalyses the synthesis of the viral $_m$RNA. Thus the genomic RNA (a minus strand) in both types of class V viruses is not infectious in the absence of the virus-specific polymerase. The influenza RNA polymerase initiates the synthesis of each $_m$RNA by a unique mechanism. In the host-cell nucleus, the polymerase cuts off about 15 nucleotides from the 5′ end of a cellular $_{m}$RNA ($_{m}$RNA precursor); this oligonucleotide acts as a 'primer' that is elongated by the polymerase to form viral plus $_{m}$RNAs, using the minus-strand RNA as a template; **Class VI:** refers to the so-called retroviruses because the single plus RNA strand of their genome directs the formation of double-stranded DNA molecule, which, in turn, acts as the template for synthesis of viral $_{m}$RNA. The conversion of RNA to DNA is due to the reverse transcriptase (a specific viral enzyme) that copies the viral RNA genome into a single minus strand of DNA, which, in turn, serves as the template for the same enzyme to synthesize a complementary plus strand. The resulting DNA is integrated into the chromosomal DNA of the host cell. Finally, the viral DNA is transcribed by the cell's own machinery into RNA, which either acts as a viral $_{m}$RNA or becomes enclosed in a virion, thereby closing the retroviral growth cycle. The retroviruses are unique in that they can contain cancer-causing genes (named oncogenes). When these occur in an infected host cell, the cell is transformed into a tumour cell. Among the known human retroviruses are human T-cell lymphotropic virus, which causes a form of leukaemia, and human immunodeficiency virus (HIV), which causes acquired immune deficiency syndrome (AIDS). Both of these viruses can infect only specific cell types, primarily certain cells of the immune system and some brain and glial cells in the case of HIV.

When the virus entering the host cell is not properly neutralized by the immune system, it hijacks the protein synthesis apparatus of the cell for its own reproductive cycle. Sometimes, the components of viral capsid – a protein, lipoprotein or glycoprotein – exert toxic effects directly on the host cells and by extension to the whole organism. Insofar as the virus dominates the host-cell biochemical pathways, it dictates the power of life or death over the cell through the activation of necrosis and/or apoptosis controlling genes, which during the productive life of the cell must remain silenced.

Generally, an infected cell has a short life, except when the viral genome is incompletely expressed resulting in anomalous structures and metabolic functions in the host cell. As a consequence, the infected cells can acquire typical characteristics of cancerous cell, i.e., immortality, intense angiogenesis and uncontrolled proliferation. In some cases, the modifications suffered by the host cell are modest such as the reduction of capability to produce a certain hormone or enzyme.

In humans, viruses enter the body across mucosa (digestive, respiratory, ocular and/or genital) or through a skin wound. Sometimes clinical symptoms at the beginning of infection are not evident, being only evidenced when one or more organs are affected. For example, the poliomyelitis virus multiplies intensely inside the digestive tract, then enters into the bloodstream and reaches the central nervous system to destroy it resulting in paralysis (Güemes et al., 2016).

It must be remembered that the damage suffered by an organ results from the viral attack *per se* associated with the immune response of the body. The latter reflects self-defence of the organism against the invading virion.

The immune response of the body can be divided into two steps. The first step, named innate immune response, is due to the presence of specialized cells (macrophages and dendritic cells, among others) normally found in body fluids (lymph and blood) responsible for the first defence against the virion. The second step, called the adaptive response, requires the participation of cells – such as lymphocyte B (antibody producing), cytotoxic T-lymphocytes (destroy the infected cells) and memory cells (remain dormant inside the lymph nodes, becoming active only if the same virion infects the body again) – formed against a specific virion. Generating memory cells in the body without a real infection by a specific virus is the strategy used in preventive vaccination.

Generally speaking, a viral infection can be of a **persistent** type or **not**. In the case of **not persistent** type, the host dies (organs are intensely affected such as in encephalitis, pneumonia and hepatitis) or survives. Surviving implies that the body acquired temporary or perennial immunization.

In the case of a **persistent** type of infection, the organism doesn't have the capability for eliminating the virus completely, so equilibrium between the body and the virus is reached. Thereby, the infection can become latent or chronic. In the chronic form, the virus continues multiplying in the body beyond the acute phase, e.g., HIV virus and hepatitis C virus (observed in 80% of patients infected). In the latent form, when the propagation of the virus is blocked by the immune system, or the virus resides in any part of the body inaccessible to immune cells (central and/or peripheral nervous system), its pathogenic effect is quite diminished. However, it can be reactivated when the immune system is weakened (food deprivation, microbial infections, chemotherapy, radiotherapy, etc.) or when the body is exposed to noxious agents (environmental pollution, smoking, illicit drugs, stress, etc.).

The virus can suffer mutation enabling it to escape from the surveillance of the immune system or become resistant to antiviral medicines.

Some examples of viruses causing **persistent** infection: **(a) Herpes virus**: there are two types of *Herpes simplex*, i.e., type 1 (HSV-1) and type 2 (HSV-2) that manifest as lesions either on the facial lips or genitalia, respectively. Both multiply in epithelial cells of the body, thereafter migrating along nerve axons. HSV-1 normally causes only mild pain around vesicles on the lips. However, sometimes HSV-1 can cause quite serious problems such as ocular infection and encephalitis, requiring treatment with antiviral as acyclovir (inhibits replication of the viral genome). HSV-2, in turn, is more dangerous to newborns where mothers have been infected. This virus can be activated by sun exposure, mechanical stress (e.g. odontological intervention), fatigue and skin irritations; **(b) varicella virus and herpes zoster virus**: both infect sensory neurons; **(c) Epstein–Barr virus (EBV)**: it is transmitted by saliva and is found in 95% of adult humans. It lays dormant in memory B-lymphocytes and from time to time becomes activated and migrates into the saliva allowing transmission between people. However, in persons who are immune-compromised (e.g. transplanted patients), EBV can cause lymphoma or mouth cancer; **(d) Human Immunodeficiency Virus (HIV)**: this virus infects and multiplies preferentially inside cells expressing the CD4 cell surface marker (typically CD4+ and T4 lymphocytes, but also monocytes, macrophages and dendritic cells). The depletion of T4 cells reduces the capability of the body to orchestrate the immune response, leading to life-threatening opportunistic infections and certain types of cancers. There are now medicines available efficient in blocking all points of the HIV life cycle, such as penetration into the target cell, inhibition of viral proteases, DNA insertion enzymes and reverse transcriptase. The development of a vaccine against this virus is difficult due to its high mutation rate in the gene encoding the enzyme reverse transcriptase; **(e) Hepatitis B virus (HBV)**: infects hepatic cells and, when reactivated, can cause cirrhosis or hepatic carcinoma; **(f) Hepatitis C virus (HCV)**: is transmitted by blood transfusion. It produces proteins that inhibit interferons as well as shortening the half-life of CD4+ and CD8+ cells. Like HIV, HCV has a high rate of mutation, making development of an efficient vaccine challenging, and limiting the action of potential antiviral drugs. The best therapeutic protocol uses a combination of ribavirin and α-interferon; **(g) Measles virus and Rubella virus; (h) Human papilloma virus (HPV)**: is found in epithelial cells of genitalia, skin or larynx mucosa. The transmission is by sexual contact, with types HPV-16 and HPV-18 responsible for cancer of the cervix. This cancer can easily be diagnosed through the 'Papanicolau' test. If it is found early, then treatment is usually very effective; and **(i) SARS-CoV-2**: is a virus that infects pulmonary cells by linking its surface spike-protein to the ACE2 receptor (**A**ngiotensin-**C**onverting **E**nzyme 2) of the host cell. Then, a protease hydrolyses part of the spike-protein enabling the virus penetration into the cell. The viral single-stranded plus RNA molecule links to the host cell ribosome for producing all proteins needed for virus multiplication. This virus has a long RNA strand (near 30,000 bases) when compared to other RNA viruses, which could suffer random undesirable mutations during replication leading to an unviable or weakened virus progeny. However, the SARS-COV-2 presents two remarkable characteristics, i.e., the capability to correct any genome error during replication (similarly to human cells, but uncommon in RNA viruses) and the presence of genes that defend it against the attack of the body immune system.

Therapies using virus as drug vectors or as medicines *per se* have been developed.

Viral vectors are widely used in the so-called gene therapy, which consists of introducing specific genetic information into the cells of humans who have inherited mutations in a single gene (i.e. monohybrid traits such as haemophilia, Duchenne disease and thalassemia). There are currently three types of viral vectors, i.e., adenovirus, associate adenovirus and retrovirus (Lundstrom, 2018). The adenovirus is a double-stranded DNA virus in which structural genes are substituted by the desired healthy human gene to be transferred to the patient. The associate adenovirus is a single-stranded DNA virus in which the desired gene is inserted. This virus must be fused with the adenovirus in order to accomplish the delivery of the healthy gene to the patient. Retroviruses are single-stranded RNA viruses in which the protein-encoding genes are removed and substituted by the healthy human gene. All of the cited vectors are viruses deficient in their ability to replicate. The success of gene therapy depends on the capability of vectors in accommodating therapeutic high MW molecules, moving to the specific tissue in which the medicine must be delivered and avoiding stimulation of the immune response.

As therapeutic agents *per se* viruses could be used as 'antibiotics' – or the so-called phage-therapy – and as anticancer agents (oncologic virus-therapy).

Phage therapy is the use of phages to combat infections caused by bacteria. They are viruses that live exclusively inside bacteria and are called bacteriophages. This approach was envisaged by D'Hérelle at the beginning of 20th century (Martins et al., 2020). Bacteriophages can be divided into two groups, that is, lytic (destroy the host bacteria) and lysogenic (virus becomes integrated into the bacterial genome).

All bacteriophages have a similar mechanism to infect bacteria, i.e., recognition and adherence on the target bacteria followed by the injection of the genetic material into the bacterial cytoplasm. Once inside the bacterial cell, phage multiplies to a critical number whereby the cell lyses to release new phage particles. For the lysogenic phages, the cell lysis is not immediate because the phage genome integrates into the genome of the host cell; in this case, phage and bacteria can exchange genes by recombination.

When nucleic acid of the phages integrates into the bacterial genome, the virus is named a prophage.

Phage latency can be reverted when bacteria are exposed to extreme conditions such as high radiation or temperature.

Although phage therapy preceded antibiotic therapy by two decades, its use was not widespread beyond the Soviet Union and in particular Georgia because of the following reasons: (a) antibiotics are substances of well-determined chemical structure and can be produced in huge amounts at low cost; (b) uncertainty of its efficacy as medicine for curing diseases; (c) difficulty in producing reliable suspensions having one type of phage; and (d) lack of reproducible protocols for efficient industrial production at competitive cost. However, lysis bacteriophages present two not negligible advantages, that is, infecting only host bacteria not human cells and therefore only multiplying at the site of infection. Both aspects increase markedly their antibacterial action.

Although the use of phage therapy is still limited by factors such as the production of unwanted toxins and the activation of the immune system, two approaches have been attempted to bring phage therapy to the attention of the wider medical community. These approaches include the isolation and purification of the enzymes involved in bacterial lysis, as usable medicines as well as establishment of rigorous human clinical trials to determine efficacy. The future of phage therapy seems to be linked to combine its use with antibiotics to treat infections such as septicaemia.

American food companies have obtained FDA approval for the use of bacteriophages as indicators of *Listeria monocytogenes* present in foods for human consumption.

Oncologic virus-therapy – this was developed in the 1950s and gained favour in the 1990s with advances in immunology, biology and molecular genetics – is based on reprogramming viruses so that they target receptors only on the surface of tumour cells (Roberts et al., 2006). The virus chosen for this application can be either a wild species, which might have inherent specificity for tumour cells, or viruses genetically modified to alter target cell specificity. The use of wild species exploits the high protein diversity of tumour cell surface receptors and transcription factors in the cytoplasm. On the other hand, genetic modification of viruses can allow for specific targeting of receptors that are expressed only in tumour cells and not by normal cells. In this case, the strategies employed are inserting a sequence of nucleotides recognized by tumour cells as transcription factors into the virus genome; modifying at least one capsid protein; and inhibiting the expression of viral genes into proteins normally directed to neutralize the natural defences of healthy cells. According to Roberts et al. (2006), by introducing a specific gene (e.g. a transcript normally encodes a protein called apoptin) into the virus genome, it can induce apoptosis of the tumour cell. In spite of the great potential of virus therapy in medicine, its generalized use in humans is not appropriate because the viruses employed could cause adverse side effects, such as recombination with other viruses generating a variant highly transmissible among humans that could transform normally healthy cells into a disease state or stimulate uncontrollable immune responses.

Validation of oncologic virus therapy can be divided into three phases: phase 1 is to evaluate the body's tolerance to the medicine; phase 2 is to evaluate the efficacy, mode of administration and clearance of the medicine from the body; and phase 3 is a double-blinded trial of the medicine against current best therapy in order to further verify the efficacy of the medicine. Due to the complexity of the human body, it has been quite difficult to extrapolate available clinical data. Probably, oncologic virus therapy alone will not be sufficient to completely eradicate a tumour, although some promising results have already been demonstrated in combination with conventional anticancer protocols. The reason for combination therapy may prove more successful is the possibility of circumventing the resistance developed by the tumour against anticancer medicines. Progress in optimizing oncology viruses (Newcastle virus disease, measles, adenovirus, and parvovirus, amongst others) as well as in the formulation and delivery of these medicines (e.g. in liposomes) points to expanding the use of oncologic virus therapy in the near future (Roberts et al., 2006).

Based on the above facts, it can be concluded that viruses represent one of the most dangerous microorganisms humans and animals face. The simplicity of their molecular architecture – basically a nucleic acid wrapped up by a protein capsid – enable them to disperse and proliferate easily. Insofar as being inside host cells, their genes can be integrated into the host genome to hijack all the biochemical cell machinery, leading to the formation of thousands of virions. The virions can leave the cells either by cell lysis or by exocytose as particles. Nevertheless, as the knowledge of virology grows in areas such as the viral cycle, points that can be exploited for antiviral therapy become easily identifiable. These points are relatively few and consist of inhibition of key viral enzymes (e.g. reverse transcriptase and acid proteases of HIV), attenuation or modification of virulence (viruses used as drug vectors) and identification of immunogenic capsid molecules (development of vaccines).

1.3 BASIC TECHNOLOGIES OF THE BIOTECHNOLOGICAL PROCESSES

1.3.1 Monoclonal Antibody Technology

Monoclonal antibodies are produced using the cell fusion technique, which involves the combination of immunized animal organ cells (such as spleen tissue) and myeloma cells in a Petri dish. The fused cells are called hybridoma cells and can be identified and separated for further multiplication into a bioreactor.

Due to high specificity for an epitope, monoclonal antibodies are useful for detecting, quantifying and locating some substances – those having immunogenic capability – present in the body.

There are several applications for monoclonal antibodies (pregnancy tests and anticancer drugs, amongst others). At the moment, these proteins are widely used in the treatment of breast cancer. This is a malignance with a high potential for spread in women, which becomes less and less fatal in recent years. These positive results are due to improved diagnostics (digital mammograms, ultrasound and magnetic nuclear resonance imaging) and development of new medicines: **enzyme inhibitors** (such as anastrozole®, letrozole® and exemestane®, which are inhibitors of the enzyme aromatase; and lapatinib® , which is an inhibitor of tyrosine kinase) and **monoclonal antibodies** trastuzumab (Herceptin®) and pertuzumab that inhibit the surface receptor HER2 related to the Epithelial 2 human growth factor, and bevacizumab (Avastin®) that blocks VEGF angiogenesis factor. It must be remembered that the first medicine against breast cancer was **tamoxifen**, which is a competitor of the estrogen hormone by a cytoplasm protein and links to DNA and stimulates it to express proteins related to cell survival and growth (Moulder and Hortobagyi, 2008; Becker, 2015).

1.3.2 Bioprocessing Technology

Bioprocessing involves the use of living cells or components of their metabolic machinery (e.g. enzymes and organelles) to synthesize products, decompose substances and/or produce energy. Both prokaryotes and eukaryotes are used in this technology. Bioprocessing encompasses techniques such as microbial fermentation, mammal cell culture and biodegradation (microorganisms used to reduce environmental pollution).

From the 1990s, human proteins such as monoclonal antibodies and erythropoietin began to be produced on a large scale, becoming the mainstay products of the biotechnological pharmaceutical industry at that time.

Besides microbial fermentations and cell cultures, transgenic animals (cows, goats and pigs, among others) have been used for producing medicines. These livestock are called 'zooreactors'. The animals are genetically modified by introducing a piece of human DNA into the genome of an animal embryo, which is translated to a desired medicinal protein (Figure 1.4). The human α-1-antitrypsin and antithrombin are produced through zooreactors. Another way for animal manipulation is by cloning, which consists of transferring the nucleus of a somatic cell to an empty

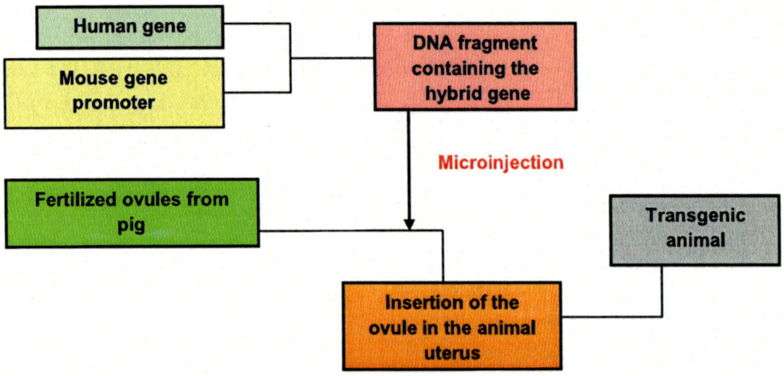

FIGURE 1.4 Sketch illustrating how a transgenic animal is made. The hybrid gene, for example, encoding human milk protein and the mouse promoter gene, is injected into the male pronucleus of a pig fertilized ovule. The ovule, in turn, is implanted into the uterus of the pig that gives birth to a transgenic animal, whose milk will be enriched in human milk protein.

ovule, which is then implanted into the uterus of a female of same species. Some years ago, the sheep 'Dolly' was generated by cloning. Human antithrombin, commercialized under the trade name ATRYN®, could be extracted from the milk produced by the livestock of female goats. Antithrombin is a protein found in the blood whose action is to inhibit coagulation and is absent in some people suffering from a genetic condition known as hereditary antithrombin deficiency. The absence of antithrombin leads to patients being highly predisposed to uncontrolled thrombosis (Zhu, 2012).

Besides ATRYN®, it is expected zooreactors will soon be generated for the production of the web spider protein (e.g. treads highly resistant to tension, useful in the manufacture of tennis racket), factor IX (a protein essential for blood clotting not produced by patients with the genetic disease haemophilia) and granulocytes colony-stimulating factor (G-CSF; a protein that stimulates the formation of marrow stem cells) used to bolster immunity in immune-compromised patients (Kumar et al., 2015).

The use of zooreactors reduces R&D costs for the biotechnological-pharmaceutical industries. The cost of laboratory manipulating hamsters' cells – normally employed in the production of monoclonal antibodies – is something between US $400 and US $500 million, while a female goat can produce a comparable amount of the same protein at a cost of just US $50 million. The market only for ATRYN® is about US $700 million/year.

Transgenic is a method for the direct transfer and expression of DNA across species barriers.

The use of animals (mice, rabbits, apes, etc.) in experiments aimed to evaluate products for human use (medicines, cosmetics, foods, chemicals, etc.) is now ethically questioned and banned in many countries. Through the transgenic method, it is possible to generate animals with desired characteristics related to the trait to be studied (e.g. illnesses and tissue modifications by physical and/or chemical agents), which implies that there can be a reduction in the numbers of animals required for product testing, as well as diminishing the suffering of the animals during the tests (Mukerjee, 1997). Moreover, transgenic animals would allow evaluating more precisely the physiopathology of several diseases and the development of new treatments and diagnostic tests, thereby reducing the development costs for new drugs.

Transgenic animals can be used as a source of organs for transplant into humans, overcoming the current shortages of human organs. The organs are recovered as soon as the animal is slaughtered so that tissue damage is reduced. In spite of the advantages xenotransplantation offers, there are some disadvantages which must be considered such as ethical questions, rejection of the transplanted organ and introduction of strange viruses and/or microorganisms (zoonotic infection) to the human recipient. The ethical issues would be circumvented if the reasons for the transplant were transparent in advance of the procedure. Rejection could be avoided if the organ from a transgenic animal expresses a human protein on its cell surface that will not stimulate the immune response, or if an animal gene that produces an immunogenic protein (e.g. 1,3-galactosyltransferase) is silenced.

1.3.3 Cell Culture Technology

Cell culture technology focuses on growing animal or plant cells in a specific culture medium. Only one plant cell can form a complete plant, so that its culture in an artificial medium constitutes an important tool for plant biotechnology, allowing many varieties of a certain plant species to be cultivated.

Regarding animal cell culture, three aspects must be considered.

The first refers to the use of insect cells for proliferating viruses, which can be used for controlling crop pathogens.

The second refers to the use of mammalian cells as genetic resources for improving traits of animals with economic value. Obtaining and storing bovine zygotes from mating bulls and genetically improved cows is a well-known example of this application of cell culture.

Finally, the culture of undifferentiated stem cells can divide indefinitely and can be a source of new stem cells and differentiated cells.

We can distinguish two types of stem cells: **tissue-specific stem cells** – immature cells found in all body tissues after division produce another differentiated cell – and **embryonic stem cells**.

The main difference between tissue-specific and embryonic stem cells is that embryonic cell can become any cell type of the body so it can be cultivated in a medium containing the essential factors required for differentiation, whereas the tissue-specific stem cell differentiates only into a cell of a particular tissue (e.g. liver stem cells differentiate only into liver cells).

Tissue-specific stem cells have been used for many years if one considers marrow transplantation for leukaemia treatment as an example. This technique is very successful – due to histological compatibility between donor and receptor – because the marrow has all stem cells needed to generate those normally present in the blood. Other examples include the transplantation of fetal neuron stem cells (for the treatment of brain diseases) and insulin producer β-cells (for the treatment of diabetes) (Zhou et al., 2018).

The first lineage of stem cells – an identical population of cells that maintain cell division indefinitely and with the capability for becoming any differentiated cell type – was achieved in 1998 from human embryos stored in fertilization clinics. There are now nearly 200 lineages of stem cells for research purposes. However, there are many ethical, legal and moral issues underlying the use of stem cells. The debate about embryonic stem cells, reproductive cloning and therapeutic cloning are all related (Levenberg, 2005). Other ways for obtaining cells similar to stem cells would be the reversal of adult cells into embryonic stages or by parthenogenesis (the activation of unfertilized human ovules to divide as normal embryos). The efficacy of both approaches still needs to be validated. Advances in regenerative therapies are highly linked to the understanding of how specific tissue stem cells could transform themselves into new functional tissues which don't belong to the layer of the same embryo. For example, how would conversion of a haematopoietic stem cell to liver tissue occur, if the liver stem cells come from the endoderm layer of embryos whereas the hematopoietic cells originate from the mesoderm layer of embryos. An important point to be mentioned is the direct introduction of stem cells into an organism would certainly result in cancer. Moreover, there is the suspicion that tumours have their own stem cells, which allow their proliferation and/or their recurrence after apparent remission following chemotherapy or radiotherapy (Nassar and Blanpain, 2016). In short, the therapeutic use of stem cells involves a precise understanding of the stimulation mechanisms (identification and coupling of signalling proteins and growth factors) that lead to the differentiation of embryonic cell into a particular type of adult cell. Despite the difficulties pointed out, some promising results have been obtained in the treatment of chronic ischaemic cardiomyopathy, lupus, diabetes type 1 and multiple sclerosis (Tang et al., 2018).

1.3.4 Tissue Engineering Technology

This technology results from the interaction between cell biology and material sciences leading to semi-synthetic tissues. Commonly, those tissues are formed by living cells (e.g. cartilage or skin cells) anchored onto biodegradable material. In this category, the so-called pharmacological stents used for opening blocked arteries can be considered.

Cultures of skin and cartilage tissues are promising procedures because they are bidimensional and low vascularized structures. However, producing tissues of multidimensional organs which are highly vascularized and composed of many cells types with low regenerating capability still demands intense studies, and is a formidable challenge to tissue engineering (Lavik and Langer, 2004). The products resulting from tissue engineering are already on the market and include Epicel® (used as a skin substitute in burn patients), Carticel® (injectable suspension of chondrocytes for cartilage repair) and Vascugel® (a reticulated matrix filled with endothelial cells of the patients undergoing vascular reconstruction. The healthy cells embedded in the device send signals to stimulate cells of the damaged vessel promoting growth, reduction of inflammation and cauterization) (Lysaghtet, 2008; Ramos and Moroni, 2020).

Finally, it must be remembered the possible use of silk – the same raw material used in textiles – as surgical sutures (low rejection by the human body), as implants in blood vessels (in this case, hardened silk is moulded as empty hollow fibres that can substitute sections of obstructed vessels) and as basement web sponges in the reconstruction of tissues and bones (Mehretra et al., 2019).

1.3.5 Biosensors Technology

This technology is a fusion of molecular biology with microelectronics. Basically, a biosensor is an electronic device containing a biological element (microorganism, antibody or enzyme) immobilized onto the surface of a transducer. When an enzyme is the biological element, the biosensor is called an enzyme electrode. The great advantage of a biosensor is the possibility to detect or measure substances present in quite low concentrations in a solution. When the device is immersed into a sample, an electronic signal directly proportional to the concentration of the substance is generated on the transducer surface. An electrical circuit amplifies the current intensity sufficient to be recorded. Biosensors will be discussed in detail elsewhere in this book.

Recently, sensors based on nanotechnology concepts have been developed. A monolayer of graphite having a width of 1 nm is coiled to form a carbon nanotube. Carbon nanotubes positioned between two microelectrodes become conductors (like any metal) when an electrical potential is applied. If a biomolecule such as an antibody is placed inside the nanotubes and the device is then put in contact with a sample containing the corresponding antigen, the resulting antigen–antibody interaction causes a change in the flux of current through the nanoweb generating a measurable variation in electrical potential. Such nanometer biosensors have been built that can measure the concentration of prostate-specific antigen (PSA) in blood samples. Other devices containing monoclonal antibodies specific for antigens of different diseases have also been built so that several antigens can be quantified simultaneously from a single blood sample (Gruner, 2006). New biosensors must pass the following research and development phases before reaching the market: concept, concept proof, prototype, product development and production. According to Gruner (2006), two companies stand out as leaders in developing such kind of devices: Motorola (Schaumburg, Illinois; www.motorola.com) that has a product at prototype phase; and Nanomix (Emeryville, California; www.nano.com) that has several products in product development phase.

The development of nanometer devices requires that the energy sources used to power the devices can function over long periods. An implantable nanometer device would require electrical power of the order of 10^{-6} or 10^{-9} W, and ideally, this power should be generated from the body in which the device was implanted. In theory, this energy could be harnessed from energy normally released by the body through muscles contracting and relaxing (mechanical energy), audible acoustic waves (vibration energy) and blood flux (hydraulic energy). The question is: how to convert all these forms of energy into electricity?

Wang (2008) has suggested that treads of hexagonal crystals made from zinc oxide can be manufactured in parallel to form a diameter of about 10 nm. Such a material would have piezoelectric and semiconductor properties. Small electrodes fixed at either ends of the compacted treads could then move from side-to-side in response to vibrations or hydraulic forces so that the assembly would work as a nanometer generator. According to the author, a device of 6 mm^2 can generate a current of 0.8 μA and a voltage of 10 mV for about 50 h.

Mehretra et al. (2019) have suggested that silk treads can be manufactured into thin pellicles, and when inserted below the skin, small silicon circuits or a diffraction net would change colour in response to changes in blood oxygen levels.

Undoubtedly, biosensor technology will become more and more of interest to the pharmaceutical and biotechnology industries, particularly when miniaturization processes are improved.

1.3.6 Genetic Engineering Technology

This technology is also known as recombinant DNA technology which allows genetic material from different origins to be covalently combined. In nature, genetic material is constantly recombining by crossing-over between chromosomes during meiosis, or by exchanging genetic material by conjunction, transformation or transduction. In each case, genetic variation occurs which allows for natural selection according to Darwin's Theory of Evolution.

Humans can exploit genetic variation either by selective mating or by using recombinant DNA techniques.

Selective mating takes individuals with desirable characteristics and uses these as sources of parental DNA for subsequent generations. Mating is carried out from one generation to another until the resulting offspring have the desired characteristics. In genetic engineering, DNA fragments (at least one fragment encodes the desired characteristic) from different species are joined together to produce recombined DNA. This DNA is amplified within a cell through a specific vector (plasmid or virus). These methods differ in that the process of recombination in mating is essentially random, whereas it is highly directional in genetic engineering.

Genetic engineering has been used to develop strains of *Saccharomyces cerevisiae*, e.g., a strain which floccules at low glucose concentration in fermentation mash. Yeast with this characteristic are quite important in the ethanol industry because at the ending of fermentation – when the glucose concentration in the mash is near zero – the flocculating cells sediment to the bottom of reactor allowing the ethanol to be removed by decanting. This procedure is far less expensive than separating the cells from the ethanol by centrifugation in terms of the energy required, time involved and costs for centrifuge repair and maintenance (Cavalheiro-Abreu and Monteiro, 2013).

The strategy to engineer this yeast strain is to construct a plasmid encoding the flocculating gene (FloI) – this gene remains silent in the wild-type yeast genome – the promoter from alcohol dehydrogenase (which is inhibited by glucose) and a selective marker gene for arginine permease (a membrane protein which controls arginine transport across the cytoplasmic membrane). When the engineered plasmid is introduced into the yeast cell by thermal shock, the plasmid integrates into the yeast chromosome by homologous recombination with the arginine permease gene. The genomes of the resulting recombinant yeast cells encode the FloI gene regulated by the alcohol-dehydrogenase promoter which, in turn, is dependent upon the concentration of glucose. Moreover, the natural gene for arginine permease remains silenced. Cells that do not form the channel of arginine permease are insensitive to canavanine (a molecule similar to arginine that blocks the cell growth) so that recombinant cells are capable of growing in a medium containing canavanine.

Undoubtedly, the advances promoted by genetic engineering have been tremendous, and it is now possible to create minimal life by genetic engineering as exemplified by the complete synthesis of the *Mycoplasma genitalium* genome, which when transferred to another bacterium of the same genus was expressed perfectly (Rohregger et al., 2020; Wang and Zhang, 2019). In spite of the popular media communication sensationalizing this advancement, in reality only the genome was synthesized and not the cell that received the DNA. However, undeniably the first step to manufacture 'synthetic life' has been made.

1.3.7 Protein Engineering Technology

This technology is a form of genetic engineering aimed at improving the physical and chemical properties of natural proteins (e.g. change the molecule pH_i, increase or decrease activity and change solubility in water or organic solvents, amongst others) or creating proteins with improved characteristics from those found in nature.

Protein engineering can be divided into two main branches: **enzyme engineering** and **abenzyme engineering**.

The aim of **enzyme engineering** is to modify the molecular structure of enzymes in order to improve, for instance, thermal stability, specificity to substrates or resistance to changes in solvent pH. The main objective is to increase the application of enzymes in industrial processes. Enzyme molecules can be modified directly through chemical reactions, or indirectly by genetic modification of the DNA that encodes the primary amino acid structure of the enzyme.

Abenzyme engineering aims to study and then apply antibodies as catalysts. Antibodies and enzymes have specificity as a common property but differ in the manner in which each protein binds to its respective substrate. Antibody–substrate interaction is by affinity (the substrate molecule – the antigen – has a three-dimensional structure that fits perfectly into the antibody binding site; the substrate isn't chemically modified as a result of the interaction), whereas enzyme–substrate interaction is by affinity with simultaneous modification of the substrate molecule. The creation of an **abenzyme** aims to produce a catalyst with improved properties such as highly specific protease (i.e. capable of recognizing and then hydrolysing only peptide bonds formed by arginine and glycine).

1.3.8 Antisense RNA Technology

The aim of this technology is to prevent the translation of proteins in cells. Basically, this technology uses a synthetic oligonucleotide complementary in base sequence to the target $_{m}$RNA to be silenced. Since the $_{m}$RNA is a single strand, the synthetic oligonucleotide will form a duplex by complementary base pairing rendering the $_{m}$RNA ineffective and preventing translation at the ribosome. This technology has opened the way for developing the so-called **metabolic engineering**, where the main aim is to produce high-value compounds, which are usually only produced as intermediate metabolites by cells. Antisense RNA technology has been developed as anticancer and antivirus therapies.

1.3.9 Chip DNA Technology

This technology has emerged from the interaction between molecular biology and semiconductor electronics.

DNA is a macromolecular biomolecule from which nanotechnology structures and devices can be manufactured.

A molecule of DNA consists of two polymers referred to as backbones aligned end-over-end, formed by a phosphate group which joins alternating deoxyribose sugar units. Each sugar molecule is covalently attached to a purine or pyrimidine nitrogenous base, and it is hydrogen bonding between these bases (A–T and C–G) on opposite backbones that moulds a molecule of DNA together. The polymer twists on itself forming a double-helix right (B-DNA) or left (Z-DNA) handed. Both DNA structures have a diameter of 2 nm and so DNA can be considered a natural nanotechnology structure. Nowadays, the tools offered by biotechnology – restriction enzymes (that cut the DNA molecule at definite points), ligases (enzymes that promote the formation of covalent bonds between DNA molecules) and PCR (a reaction that amplifies DNA into millions of copies) – allow native DNA molecules or newly synthesized DNA to be manipulated.

The three-dimensional structure of DNA can lead to several forms, one being a cube structure. If a protein is inserted within a nanometer cube, it can be immobilized in a particular structural conformation. Using such a matrix, the analysis of protein can be made at the X-ray crystallographic resolution. Additionally, if the protein were a cell receptor, low MW molecules can be titrated into the matrix and possible receptor ligand could be modelled at the atomic level (Shih et al., 2004).

It was suggested that the conversion between right-handed B-DNA to left-handed Z-DNA – the so-called B–Z shift (Figure 1.5) – could be used as an on–off switch, similar to 0–1 employed in electronic circuits designed on silicon chips. Theoretically, logic combinations of different types of DNA molecules – in this case, either natural DNA or synthetic DNA with well-defined length and distinct helix arrangement could be employed – should allow the design of 'nanometer machines' that could be used as intracellular sensors or 'biological computers'.

The computational concept involving DNA molecules must be treated cautiously. Certainly, a 'biological computer' would have a performance quite inferior to conventional electronic processors. For example, the speed of protein synthesis on a ribosome is about 100 operations per second, which is infinitely lower than a conventional computer that operates billions of operations per second. However, a 'nanometer machine' speaks the 'language' of living cells (Benenson et al., 2004).

The advantage of a biological computer would come from the possibility that it could work in a biochemical environment (or even inside an organism), interacting with the environment's natural biological molecules to enter and exit the computer. A computer such this could act as an autonomous 'doctor' inside a cell, for instance. It could identify signals related to a disease, which can be compared with preprogrammed 'medical knowledge' to generate an output signal, for example to release a drug (Figure 1.6) (Benenson et al., 2004).

The aspects discussed are still theoretical and may take decades to become reality.

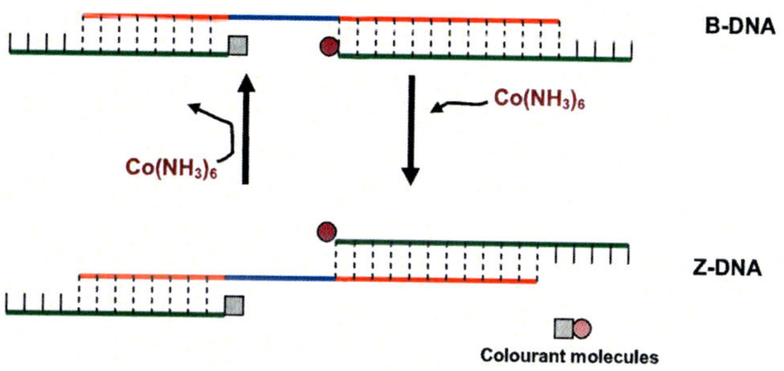

FIGURE 1.5 Conversion of B–Z forms of DNA catalysed by hexaammonium cobalt. The configurations B-DNA/Z-DNA would represent logic on–off key.

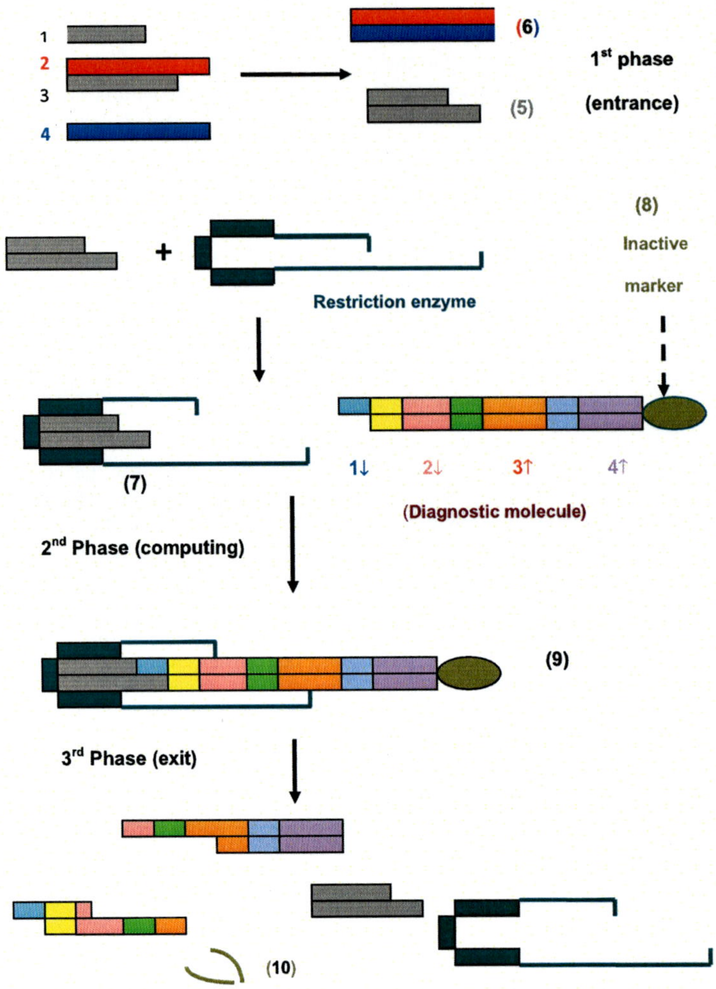

FIGURE 1.6 Sketch depicting the action of a **molecular computer** planned to generate an antisense DNA molecule (**10**) [avoiding the formation of $_{m}$RNA that would be translated into an abnormal protein]. The process occurs in three phases: first phase: oligonucleotides-sensors (**1** and **2–3**) detect anomalous $_{m}$RNA (**4**) which is neutralized by linking with a strand (**2**), resulting in the formation of an inactive complex (**6**). At the same time, strands **1** and **3** combine generating the complex (**5**); second phase: the complex (**5**) alerts the system about the defectiveness. A restriction enzyme binds to the complex (**5**) and to the diagnostic molecule (containing the inactive marker (**8**) and genes with activities below (**1↓ and 2↓**) or above (**3↑ and 4↑**) normal action). If all anomalous markers are present then complex (**9**) [complex (**5**) plus **diagnostic molecule**] is formed; third phase: the restriction enzyme digests the diagnostic molecule at specific positions, generating a positive diagnostic signal represented by the **antisense DNA (10)**, which, in turn, integrates into the cell genome to silence the faulty gene. Note that the **antisense DNA (10)** comes from the activation of the marker molecule (**8**).

Glass plates have been used as a substrate to attach long synthetic nucleotide molecules (a sequence of adenine, cytosine, guanine and thymine bases). DNA to be analysed is removed from the nucleus of a cell, marked with a fluorescent substance and put in contact with the glass plate. Sequences of the DNA molecule that were complementary to the nucleotides immobilized on the glass would bind by hydrogen bonding, essentially immobilizing the DNA of interest. It is thereby possible to analyse in only one step different sequences of nitrogenous bases (genes, in reality) belonging to DNA. A device like this constituted by a glass plate (support) and synthetic nucleotides ('circuit') was named 'DNA on a Chip' by analogy to electronic chips (a silicon plate in which a metallic circuit is impressed).

In fact, there is a variant of called 'gene chips' already on the market that are designed for sequencing a whole genome or for the identification of gene polymorphisms. The aim is to associate an error in the genome or an abnormal gene marker to the susceptibility of a person to develop a particular disease in the future. A product like this is marketed by Affymetrix® (human genome U95Av2; U.S. Pat. no 5,744,305; 5,445,934) (Rocha-Martins et al., 2020). However, it must be advertised that susceptibility to a disease is not synonymous that a person will actually develop the illness, simply because multiple genes interact in a biological system as well as the participation of several other elements (RNAs and biological molecules).

Although high-quality DNA sequencing is still expensive, the costs must be reduced. The cost of complete genome sequencing is about US $15,000, but must be reduced to US $1,000 if widespread sequencing is to be used as a diagnostic tool (Wu et al., 2018).

1.3.10 Biocomputing Technology

Bioinformatics is the organization and use of biological information and represents the interaction between mathematics, computing and molecular biology.

Nowadays a lot of data related to gene sequences and proteins, three dimensional structures of biological molecules and genome maps of many thousands of organisms are available. Using computational tools (algorithms, graphics, artificial intelligence, statistical programs, simulation and management of data banks), biocomputing allows mapping and comparing genomes, determining structures of proteins, simulating bond formation amongst molecules (e.g. designing new medicines based on cell receptor structures), identifying genes, evaluating the effect of mutations and determining phylogenetic relationships. Recently, a method for digitalizing microscopic images of human tissue allowed increased sensitivity in disease diagnostics by improving the sharpness of the images and by facilitating the exchange of information between pathologists. The implementation of this technology has revolutionized the screening of pathological specimens, where protocols were previously established over a century ago (Perkel, 2010).

A convergence between computing and biology universes can be seen by comparing biological and computer viruses. Biological viruses (BV) are composed of nucleic acids, the sequence of nitrogenous bases that make up the nucleic acids (A, T, C and G) to form a code, much like a computer program. Computer viruses (CV), on the other hand, are a program which is switched on after entering into a computational system. Both cause 'infections' to the host (cell or computer). CV deviate the performance of software for its own interest, causing malfunction on the overall computer hardware. Finally, BV and CV have in common the power of intense duplication, high mutation rate and can hide themselves indefinitely inside the cell or the computer (Bonfante et al., 2005). Nevertheless, the viral reproduction cycle inside a cell occurs according to a unique blueprint encoded by the viral genome. However, in computing science, the conception of program and data don't exist because a computer manipulates symbols. On other words, the data is a potential program and, conversely, the program is data.

1.4 BIOTECHNOLOGY AND APPLICATIONS

1.4.1 Medical

Biotechnology in medicine has applications in therapeutics and clinical diagnostics.

1.4.1.1 Diagnostics

Biotechnology has contributed to diagnostics by improving the speed and precision of laboratorial tests, e.g. using monoclonal antibodies, DNA probes, DNA chips and PCR (polymerase chain reaction).

1.4.1.2 Therapeutics

This area has been impacted by biotechnology through the development of animal and plant cell cultures and the use of genetically modified microorganism in the fermentation processes. Thanks to these developments, nowadays there are large quantities human-specific compounds available for therapeutic use such as interleukin-2 (T-cells activator), erythropoietin (erythrocyte activator) and tissue plasminogen activator (dissolves blood clots). The use of transgenic animals and plants in the production of drugs deserves special mention. Several therapies based on biotechnology that have been developed are now well studied: **(a) Substitution therapy:** substances produced or not produced in low quantity by the body (insulin and factor VIII, for instance) are now plentiful on the market due to the use of industrial processes derived from advances in genetic engineering and bioprocessing techniques; **(b) Gene therapy:** administers functional genes in patients deficient on these. However, this process requires further improvements because the functional genes must be introduced in the target cells at a correct position in the genome, as must respond adequately to specific physiological signals; **(c) Cell therapy:** defective cells are replaced by healthy cells. For example, autoimmune type 1 diabetes – a disease in which antibodies attack β-cells of the Islets of Langerhans – can now be cured by introducing the insulin-producing healthy cells entrapped in calcium alginate into the pancreas; **(d) Immune suppressive therapy:** use of monoclonal antibodies for blocking the action of T cells (one of the components of the human immune system) in order to avoid rejection of a transplanted organ; and **(e) Therapy against cancer:** oncogenes are silenced using monoclonal antibodies or antisense RNA.

Finally, let's remember the development of DNA type vaccines that are revolutionizing preventive medicine (Laddy, 2009). Another valuable approach in vaccinology is the use of adjuvant substances that enhance the immune response to a vaccine. Amongst the classic adjuvants are Freund's adjuvants, aluminium salts, vitamin E and monophosphoril lipid A, whereas new adjuvant such as plant saponins, CpG (non-methylated bacterial DNA) and interleukins have been introduced in vaccine formulations (Welder, 2009).

1.4.2 Environment

As well known that the Earth's environment has been polluted by human activities for centuries, and the rate of pollution has accelerated during the last 50 years. Biotechnology is an area which is envisaged to reduce these environmental problems as well as decrease the rate of pollution. The strategy would be based on the substitution of chemical processes that work with non-renewable raw materials and replace these with biological processes which handle renewable material. Moreover, biotechnological processes offer high specificity, precision and predictability.

Biotechnology can contribute to the environmental area through three ways, i.e., bioremediation, prevention of problems and monitoring.

Bioremediation is a natural, low-cost, high-yield and low-energy consuming process. Normally, genes belonging to microorganisms adapted to hostile environments (extreme cold, heat or salinity) are transferred to microorganisms intended to act on the ecosystem to be reclaimed.

Prevention of pollution is based on the development of **sustainable industry** which operates processes that are friendly to the environment – causing low or zero pollution and allowing maintenance of the environment.

Finally, environmental monitoring can be conducted using biosensors and/or biological markers. For example, the use of genetically modified phenol-metabolizing bacteria in the presence of phenols produces a specific protein, whose amount is proportional to the concentration of phenols present in the sample.

1.4.3 Livestock Raise and Agriculture

Biotechnology has been used to raise livestock and in agriculture to reduce production costs and improve quality and overall nutritional value of animals or crops.

Undoubtedly through genetic engineering and the use of cell cultures, biotechnology has a profound impact on agriculture. Some of the main characteristics introduced into crops are nutritional, delays in ripening to increase shelf life, resistance against parasites (bacteria, fungi, viruses, insects, etc.) and adversity to extremes in climate (prolonged frost and dry).

Nutritional improvement has focused on grains, fruits and vegetables to enrich these crops in proteins, minerals and vitamins. Such an approach has led to the so-called functional foods – having high levels of elements beneficial to health and/or enhancing the body physiological activity – such as green tea enriched with antioxidants (elimination of free radicals), broccolis and cabbage with high amount of glucosinolates (stimulation of anticancer enzymes). Moreover, tomatoes used in the food industry to make ketchup or concentrated extract (commonly used in many types of sauces) have been obtained from crops genetically improved either by selecting somatogenic variants – fruits having 30% less water are useful in the manufacture of concentrated tomato extract because these require less energy in the concentration unit operation; the American producer saves about US $35 million/year in energy consumption – or by applying genetic engineering (the metabolic processes in which fruit ripens and rots can now be manipulated using antisense RNA techniques). Just to mention, the tomato plant was the first genetically modified plant approved by FDA for human consumption and was protected by a patent filed by Calgene Co. Since 2012, a revolutionary technique for editing DNA with high precision – called **C**lustered **R**egularly **I**nterspaced **S**hort **P**alindronic **R**epeats, or in short CRISPR – has been widely impacted agriculture. For example, in the Paris mushroom (*Agaricus bisporus*) – the darkening process has been delayed; wheat (*Triticum aestivum*) – resistance against soil fungi has been increased; soya (*Glycine max*) – the content of monounsaturated fat in the edible oil has been reduced so that the content is now similar to that present in olive and canola oils; and potato (*Solanum tuberosum*) – reduction in acrylamide formation during frying (Hall, 2016; Andersen, 2015).

The pedigree of animals of economic importance has been improved over many centuries. Up to the 1970s, improvements in pedigrees were achieved by natural mating between males and females with the desired genetic characteristics (i.e. less fatty muscles). However, with the introduction of genetic engineering (after 1972), mating has now been largely substituted by embryo transplantation. In this technique, sperm and eggs are removed from breeding stocks and fertilized *in vitro* leading to embryos that are implanted to impregnate females, resulting in offspring which can then be selected for the desired trait. In this area, biotechnology has also contributed to improvements in overall animal husbandry by introducing precise diagnostic tests (kits of monoclonal antibodies) and increasing the numbers of vaccines and drugs for veterinary use.

The industrial production of foods must guarantee that the products are safe for humans and pets. This is achieved through rigorous quality control, either of raw material or the final products thanks to the tools offered by biotechnology based on monoclonal antibodies, biosensors and DNA probes. These devices allow quick, sensitive and precise analysis.

1.5 MATTERS OF BUSINESS

There are several definitions for biotechnology, most of them permeated with good intentions such as 'biology dedicated to the well-being of mankind', 'biological techniques that can give society products to generate wealth' or 'cures for illnesses that threaten human health'. However, a most realistic definition would be the 'art of making money from biological processes'. Without the pressure and discipline of the market, certainly biotechnology would be an elegant science but deprived of significant technological advancements.

Although biotechnology has been developing for many centuries, it was from new basic biochemical knowledge and innovations since the 1970s – elucidation of structure/activity mechanisms between macromolecules, invention of recombinant DNA and cell fusion techniques leading to the development of monoclonal antibodies – that impacted most profoundly the advancement of biotechnology. The effects of these scientific advancements spread from the laboratory to reach the industrial and financial sectors, which allowed entrepreneurs and bankers to see the potentially high profits that could be made by the burgeoning biotechnology business. Lawyers, parliaments, civil and governmental agencies also had to confront new regulatory problems, including those ethical and moral dilemmas, as well as issues of patent protection.

In the business world, biotechnology can be considered an atypical sector because commercialization is related to new discoveries based on basic science. In other words, biotechnology is attractive for venture capital. In sectors such as computation, telecommunication and electronics, technology is based on consolidated scientific theories. Conversely, consolidation in the biotechnology sector depends upon improvements in technology arising from the convergence of scientific investigation into the development of new products and/or processes, and financiers interested in marketing the potential bioproduct.

The *sine qua non* for consolidating any economic sector is the structuring of small, medium and larger corporations, which require investment capital. At the beginning of commercial development, only small enterprises appear because at this stage there is only discovery with the potential to be marketed, usually (but not always i.e. in the case of 'first to market' products) if the product or idea can be protected by patent. Generally, larger corporations generally don't invest at this stage because their aim is to derive the greatest profit from their investment in the shortest period. Such management behaviour usually precludes investment in early-stage biotechnological innovation.

The biotechnology sector is spread across the globe but its success is felt more so in the USA than any other country. The reasons for this maybe a combination of various factors, particularly business culture, such as entrepreneurial spirit for innovation and competition, availability of risk capital, flexibility of the economic system, extensive and sustained government support of basic research, a highly skilled and well qualified, motivated workforce and well-regulated capital markets. There are more than 300 successful enterprises involved in biotechnology, the pioneers being Collaborative Research (1961), Cetus (1971), Genentech (1976), Biogen (1977), Genex (1977) and Amgen (1979) (Rathmann et al., 1994).

To translate biotechnology research into a revenue-generating venture usually requires creation of a company (unless a patent is sold or licensed by the inventors directly to an existing company) with the following characteristics: (a) a skilled and diversified workforce; (b) an operating structure to handle renewable and recyclable material; (c) be of small or medium size; (d) uses processes requiring low-energy input; (e) experienced management well-tuned with the specific laws that regulate the business; (f) be able to satisfy consumer expectation versus bioproduct quality; (g) be compliant with health and safety measures to ensure safe operation; (h) be aware on the high financial risks involved (e.g. bioproducts tend to have a shorter market life than other products); (i) continuous P&D; and (j) realistic expectation to return the invested capital.

1.5.1 Structuring of a Biotechnology Enterprise

Generally speaking, there are four main aspects to consider when structuring a new biotechnology company: **(a) Preparation:** enthusiasm of people (the scientific inventors and the risk investor) for a marketable product and engagement of a renowned scientific advisory board headed by a CSO, a management team (CEO, CFO and a law office) experienced in the biotechnology sector. In this phase, it is important to define the scope and strategy to be adopted by company – e.g. the production of medicines for human or animal use or reagents for diagnostic tests. Moreover, the available capital (provided by risk investors) is usually small, generally sufficient for remuneration of the scientific and management teams. It is quite common for the members of the initial team to receive a small salary, with additional remuneration taking the form of stocks or future revenue shares in the company. Often, this stage in company development is called 'year zero' and is characterized by no liquidity and with a high probability that the company may go bankrupt (higher than 90% of probability). Obtaining the patent for the bioproduct and/or process in this phase becomes an important aspect for attracting more risk investors; **(b) Initial consolidation:** if the preparation phase is well done and the future bioproduct or service demonstrates commercial appeal, the embryonic enterprise will now enter the next operating year (named 'year I') with the objective of attracting new financial investors, which usually accept as guarantee on their investments a proportion of the company in the form of convertible preferred stocks (CPS). Additionally, the distribution of CPS amongst company employees also stimulates loyalty to the company. It is quite common in this phase of development for a leader investor (with highest invested capital) to perform due diligence – scrutinize the business plan, check that all the claims and analysis are correct, ensure there is freedom to operate in patent space and get to know the management team. In other words, the leader acts as an enticer for attracting more investors and, as a consequence, increases the company's capital and the prospect for success in the future. The administrative structuring is accelerated mainly on legal matters (organization of all documentation needed as well as a descriptive memorandum of the company). In spite of the positive expectation at this phase, the probability of bankruptcy is still about 65% because the company's liquidity is still quite low with the stock value typically valued at US $1 for tax purposes; **(c) Partial consolidation of the company:** this phase corresponds to years II and III of business and involves the company increasing the competition for buying its assets. The strategic plan is improved (inclusion of market estimates and return on investments), the organizational structure is confirmed (location of the manufacturing facility, the number of employers needed, the labour responsibilities inside the organization, the strategies of selling and selection of distributors is defined, amongst other considerations) and the bioproduct or service is considered near-to-market (normally, at this point the company is awaiting the approval from a regulatory body such as the FDA). The probability of bankruptcy is still high, near 40%, because liquidity in the value of the company stock is quite low, around US $5; and **(d) Full consolidation:** this corresponds to the years IV and V of operation when the bioproduct or service is on the market and is generating revenues for the company (and by extension, returning profits to the investors). At this point, the company is consolidated and primed for expansion by opening its capital to the public by floating ordinary stocks. The chance of bankruptcy is small (lower than 10%) and the stocks may have reached values over US $100.

1.5.2 Biopharmaceutical Industry

This kind of industry can be divided into two branches depending on whether the products are based on modern or traditional biotechnology manufacturing methods.

Modern biotechnology focuses on the production of mammal polypeptides as the main bioproducts. Manufacture of these bioproducts has disadvantages such as high production costs and typically difficult routes of administration requiring complex formulation and delivery development. Such bioproducts include anticancer medicines, blood coagulation factors, immune-stimulating factors, interferons and monoclonal antibodies (e.g. rituximab, infliximab and trastuzumab), amongst others. Probably, the most successful bioproducts are monoclonal antibodies if we consider the size of the market, which is around US $10 billion/year, at least. Moreover, the quality of monoclonal antibodies that can now be manufactured has greatly improved with the development of humanized antibodies which induce fewer adverse side effects than mouse- or rabbit-derived antibodies.

Traditional biotechnology is typically based on the production of bioproducts from microorganisms rather than chemical synthesis. Traditional biotechnology approaches have two advantages: first, fermentation using microorganisms is a well-established industrial process (in terms of equipment, good protocols for production and quality control of the bioproduct); secondly, the yield of the process can be increased by manipulating fermentation conditions or genetic characteristics of the producing strain.

Microorganisms are producers of a large variety of substances but usually in small amounts because regulatory mechanisms have evolved in order to avoid high expenditure of the energy costs involved in overproduction of primary or secondary metabolites, which in large amounts may also be toxic to cells. However, strains of industrial interest must have the capability of overproducing desired primary or secondary metabolites, which require specific metabolic pathways to be manipulated. Such manipulations can be achieved by changing culture conditions (temperature, pH, nutrients composition, presence or absence of inhibitors, etc.), by induced mutations (through chemical and/or physical agents) and/or by genome manipulation through genetic engineering.

Undoubtedly, the most spectacular advances in the overproduction of microbial metabolites have been achieved by genome modification. Classic mutation improvements involve exposing a microorganism to a chemical (azide, for instance) and/or physical (UV radiation) agent, followed by the separation of mutants that produce higher amounts of the desired metabolite. Following several iterations of mutation, finally, an overproducing strain can be isolated and be used at an industrial scale. In genetic engineering, exposure of the producing strain to a mutagenic agent is substituted by genome modification by introducing a gene (which encodes the proteins required for the biosynthesis of the product) into the genome of a heterologous host, typically a host well developed for industrial fermentation. In reality, both techniques are typically used to achieve strain improvement.

Microbial fermentations occupy an important position in the biotechnology industry because of the following reasons: (a) the high metabolic rate of microorganisms – in terms of substrate conversion into the desired product; (b) a large variety of simultaneous reactions inside a microbial cell; (c) easy adaptation of the cells to different culture media facilitating easy scale up from test tube to fermenter scale; (d) conversion of low-cost substrates (mainly carbon and nitrogen sources) into valuable bioproducts; and (e) capability of microorganism to biosynthesize specific enantiomers.

As referenced earlier, microbial fermentations can be used to produce primary metabolites – low MW molecules which are intermediates or final products of intracellular pathways – or secondary metabolites (higher MW molecules, generally formed after the exponential growth phase, having functions such as hormones, antibiotics, differentiation effectors and symbiosis stimulators, among others). The world market for primary and secondary metabolites is over US $20 billion and US $100 billion, respectively.

The main primary metabolites of industrial interest are: **(a) Amino acids:** 19 different compounds are produced at an industrial scale; L-aspartic acid, L-glutamic acid, L-lysine, DL-methionine and L-phenylalanine are produced in the greatest amounts. The main bacterial sources for these amino acids belong to the genera *Corynebacterium* and *Brevibacterium*; **(b) Nucleotides and nucleosides:** the most important are guanylic acid and inosinic acid; both are used as flavour enhancers and are produced mainly by *Bacillus subtilis*, *Corynebacterium glutamicum* and *Brevibacterium ammoniagenes*; **(c) Vitamins:** the main vitamins produced by fermentations of *Ashbya gossyppi*, *Propionibacterium shernanii* and *Gluconobacter oxydans* are riboflavin (vitamin B_2), cyanocobalamin (Vitamin B_{12}) and ascorbic acid (vitamin C), respectively; **(d) Organic acids:** the most important acids are citric acid (*Aspergillus niger*), acetic acid (*Acetobacter aceti*), itaconic acid (*Aspergillus terreus*), gluconic acid (*Gluconobacter suboxydans*) and lactic acid (*Lactobacillus delbrueckii*); **(e) Alcohols:** there are a large diversity of alcohols available on the market, some examples being: ethanol (*Saccharomyces cerevisiae*), butanol (*Clostridium acetobutylicum*), glycerol (*Candida glycerinogenes*), mannitol (*Candida magnoliae*) and xylitol (*Candida guilliermondii*); and **(f) Polymers:** the principal polymers produced from microbial fermentations are polysaccharides such as xantan gum (*Xanthomonas campestris*), dextran (*Leuconostoc mesenteroides*), pullulan/phytocolloids (*Spirulina platensis*) and polytrimethylene terephthalate (*Escherichia coli*).

There are a large variety of secondary metabolites produced commercially from microbial fermentation: **(a) Antibiotics:** undoubtedly the most famous and numerous types of secondary metabolites, for example, heterogeneous large and diverse group of molecules having MW ranging from 102 Da (cycloserine) to 2,500 Da (nisin and nystatin). There are over 6,000 different antibiotics described in the literature, of which nearly 160 are commercialized and include penicillin (*Penicillium chrysogenum* and *Aspergillus nidulans*), cephalosporin (*Cephalosporium acremonium*), clavulanic acid (*Streptomyces clavuligerus*) and monolactams (*Chromobacterium violaceum*); **(b) Antitumour agents:** drugs (mitomycin C, bleomycin, daunorubicin, doxorubicin, etc.) obtained from fungal strains; **(c) Inhibitors agents:** represented by drugs that inhibit enzymes. Probably statins are the most widely used drugs in this category and are used clinically to reduce cholesterol biosynthesis by inhibition of the liver enzyme 3-hydroxy-3-methyl-glutaryl-coenzyme A reductase (HMG-CoA reductase). The first statin on the market was lovastatin produced by *Aspergillus terreus*, which, in turn, was chemically converted to simvastatin (Zocor®). Based on the chemical structure of simvastatin, further analogues have been synthesized and include one of the most widely prescribed drugs in the world today, atorvastatin (Lipitor®); and **(d) Products used in livestock**

and agriculture: the main products can be classified as follows: **biopesticides** (kasugamycin from *S. kasugaensis* and polyoxin from *S. cacaoi*), **bioinsecticides** (nikkomycin and *Bacillus thuringiensis*), **bioherbicides and anti-helminths** (ivermectin and doramectin, both obtained from *Streptomyces* sp.), **coccidiostats** (salinomycin), **plant hormones** (gibberellins that regulate blooming, seed germination and stem elongation) and **anabolic agents for livestock of economic interest** (tylosin from *Streptomyces fradiae* and mikamycin from *S. mitakaensis*).

The biotechnology industry has experienced rapid growth as knowledge of cell biology improves as concepts and methodologies in genomics, proteomics and metabolomics become better understood. Many sensitive analytical equipment (e.g. microscopes and spectrometers) allow a more detailed study of internal cell structures (organelles, cytoskeleton, membranes, etc.) and the chemical nature of biomolecules, as well as the flux in their intracellular levels and mechanisms of action. Advancements in computing science and bioinformatics have amplified remarkably the capacity to store, compute and compare huge amounts of different types of biological data. The increased precision for studying three-dimensional structures of cell components and biomolecules now allows new biomolecules to be designed to target interactions with specific targets such as membrane receptors, ion channels and membrane carrier protein. As new genomes are sequenced, the potential to genetically modify increasing forms of life as valuable sources of bioproducts becomes possible.

1.6 SOCIAL ASPECTS OF BIOTECHNOLOGY

In Section 1.5, critical roles such as government funding and participation of venture capital play in technology transfer from basic science to industry, that is, from the university to the market, were emphasized. Biotechnology identifies through scientific discovery, develops, manufactures and commercializes products by manipulating biological materials. These, in turn, become high-value sources of bioproducts or resources for development platforms – e.g. evaluating toxicity and bioavailability, establishing therapeutic dose or standardizing the biological activity of bioproducts.

Eventually, the interaction between biotechnology and society comes down to two opposing ideas, i.e. profit versus safety and compassionate use of living beings, which might be balanced through the principles of ethics.

Ethics – understood as a philosophical reflection of morality – can be divided into three areas as follows: **(a) meta-ethics:** relates to questions about the nature of one's own ethics, for instance if a pursuit is, or is not, a science; **(b) normative ethics:** establishes a pattern over what is effectively good (correct) or bad (incorrect); and **(c) practical ethics:** applies the concepts of normative ethics to daily moral questions.

The ethics applied to biotechnology are called **bioethics** (ethics of life). So, bioethics can be defined as the systematic study of human behaviour in life and health sciences, evaluated using values and principles of morality.

The idea of bioethics is based upon the premise that life, science and technology are intrinsically connected. Thereby, the bioethics question must consider four basic principles, i.e., **autonomy** (right for free choices by a living being); **non-maleficence** (no damage or harm to an organism); **beneficent** (promotes well-being for everybody); and **justice**.

The justice principle is quite complex because politics and social questions are not mutually exclusive. The justice principle can be divided into **formal justice** and **material justice**. **Formal justice** considers treating all as equals using identical measures, and not equals under different measures. **Material justice** considers fair distribution amongst people according to necessity, merit, individual contribution (i.e. distribution of revenues amongst investors according to the initial proportion each investor made) and legal market statements.

To the picture presented above must also add concepts and behaviours that change over time. For example, abortion and in vitro fertilization were unthinkable 30 years ago but now quite acceptable, whereas human cloning (i.e. stem cell technology) – for egocentric matters (positive eugenics, lineage, gender choice, etc.) or to produce clones as sources of organs for transplantation – is still unacceptable. However, stem cell technologies should be the salvation of humanity when, e.g., the Earth's resources become exhausted or the solar system is on the brink of collapse. Such situations would push humans to long-distance space travel. As we know, humans are quite fragile in space conditions (zero gravity, intense radiation, freezing temperatures, oxygen availability, etc.). The survival of *Homo sapiens* might therefore depend on genetic modification and reproductive cloning.

Due to the complexity of biotechnology and society interactions, themes related to the privacy of genetic information and results of laboratory prognostic testing, mammalian stem cells cloning and using mammals (as test subjects for evaluating efficacy, therapeutic dose, toxicity and quality control of bioproducts) will be considered.

1.6.1 Use of Bioproducts

Vaccines are the most widely used bioproducts for prophylaxis against bacterial and viral diseases (human-to-human or animal-to-human diseases). Nowadays, many viruses are known to cause serious and potentially fatal human disease. For example, HIV, rotavirus and influenza viruses killed about 7 million, 700,000 and 1,200,000 humans respectively in 2003 (Azevedo et al., 2020). However, there are viruses such as Ebola, SARS (severe acute respiratory syndrome), fowl influenza and variola which were the cause of fewer than 1,000 fatalities in 2003 but have a devastating potential if more virulent mutants appear such as Sars-Cov-2 (causes the so-called Covid-19).

The vaccine against rotavirus – which causes acute infant diarrhoea (AID) – deserved special attention due to the high transmission rate of the virus between children. If not treated properly and promptly, AID can be fatal. The life cycle of rotavirus (composed of 11 double-strands of RNA) is as follows: the virus is transmitted to an infant through the mouth and attaches to the surface of intestinal epithelial cells. Once inside the epithelial cells, the virus multiplies and expresses a toxin which causes the cells to lyse, thereby releasing new viral particles into the intestine lumen for transmission by the anal–oral route. The subsequent diarrhoea can lead to dehydration, shock and death if not treated (Vasikari, 2006). In 1991, a vaccine under the trade name Rotashield® (Ayerst Co.) was launched into the market against rotavirus. Two years post-marketing surveillance of a worldwide immunization program found that 1 out of 11,000 children vaccinated developed a fold in the intestine, which, if not removed surgically, could be fatal. Considering the bioethics principle of no maleficence, the vaccine was removed from the market. Regulatory authorities based this decision on the following facts: **(a)** in a developed country, AID is rarely fatal due to excellent healthcare infrastructure, besides vaccination in these countries had only the purpose of avoiding the expense of hospitalization for 1 week; **(b)** even in undeveloped countries in which AID has a fatality rate of 1 out of 200 children, administering a vaccine that could cause 1 death out of 11,000 was still considered too great because the level of healthcare infrastructure would not be sufficient to treat the adverse side effect caused by the vaccine; and **(c)** it would be unacceptable to use a medicine banned from a developed country in an undeveloped one.

A solution to the Rotashield® problem came following the observation that children naturally infected by rotavirus did not present with intestine folding, which implicated the source of the vaccine might be the problem – the vaccine was made using attenuated monkey rotavirus (MR). The solution was to find another source of rotavirus. A couple of years later two new vaccines appeared on the market, i.e., Rotarix® (attenuated human rotavirus) produced by GlaxoSmithKline and Rotateq® (bovine rotavirus modified with part of the human rotavirus genome) manufactured by Merck.

As genomic techniques advance, it will be possible to personalize medicine in order to take into account how the genome of a patient might respond to drug treatments. This approach is called 'pharmacogenomics'.

Some drug producers have already shown interest in the production and commercialization of medicines focusing on specific human ethnic groups. The first product, a blood vessel relaxant (BiDil®), was approved by FDA in 2005 as an ethnic medicine. BiDil (a combination of hydralazine – an inhibitor of superoxide radicals, and isosorbide dinitrate – a stimulator of NO_2) was licensed for use to treat congestive cardiac failure which is common in African Americans (Duster, 2005). By labelling BiDil® as an 'ethnic-specific drug' allowed the pharmaceutical industry to reposition commercially obsolete drugs (hydralazine and isosorbide dinitrate) already surpassed by other medicines, e.g. enalapril® (an inhibitor of angiotensin enzyme converser). Besides, BiDil® cannot be considered a 'pharmacogenomic' drug because its mechanism of action has not been linked to the regulation of a specific gene. Therefore, it is incorrect to labelling BiDil® as an 'ethnic-specific drug'.

Based on the development of BiDil®, several other companies have also attempted to commercialize certain drugs by marketing improved efficacy in specific ethnic groups. For example, Iressa® (against lung cancer); Crestor® (cholesterol reducer); and AidVax® (vaccine against AIDS) would be more efficient in, respectively, Asians; African Caribbean, Asians and Hispanics; and African Caribbean and Asians.

Although BiDil® has therapeutic efficacy – it was clinically proved to reduce hospitalization and deaths in patients with congestive cardiac failure – labelling it as an 'ethnic-specific drug' is imprudent because of the affront to the bioethics principle of justice: 'it is discrimination establishing criterions for distributing estates, offering opportunities either educational or health according to genera, ethnics, physical conditions, nationality and social status'.

1.6.2 Genetic Privacy and Laboratory Prognostic Testing

After the human genome sequence was published, some questions immediately arose around individual genetic privacy as well as what could be considered 'invention' (patentable) or 'discovery' (not patentable). Both questions are related to the bioethics principles of autonomy and justice. A person must be autonomous to decide how their own genomic information and/or results from laboratory prognostic tests are handled. The principle of justice does not allow any kind of individual segregation based on genetic and laboratorial information. Moreover, this principle must be invoked every time a biological discovery – such as when a new gene, gene markers or mechanisms of gene expression (RNA transcription and translation) are discovered and a patent application is made. Only a clear distinction between discovery (comprehension and identification of any natural phenomenon or biological component) and invention (clear useful modification of a biological component made in a laboratory, including to an individual's own living cell) would avoid a breach of autonomy and justice principles.

Undoubtedly, any patent filing will evoke intense discussion, insofar as several aspects (religious, cultural tradition, degree of morality of certain human community, economical and/or political interests, legal backgrounds, etc.) must be addressed by legislators and supported by the goodwill of the scientific community. Patenting of natural elements (genes, chimeric animals, microorganisms and protein receptors, amongst others) raises the following questions: on what grounds can a gene be patented? How can a mouse or fish be patented since the genomes of these animals evolved and were not a creation of human invention? What happens to the scientific freedom to research when all oncogenes are patented? This might lead to researchers spending more time struggling in courts for the freedom to research rather than looking for cancer cures? What is the time limit before a patent expires?

In order to give an idea about the complexity involved in patenting biological material, let's take the following example: the University of Harvard in 1988 wanted to file a patent application with the USA Patent Agency to protect the so-called oncomouse, a rat genetically modified having a gene predisposing the rat to cancer. The new rodent became a valuable tool in cancer research. The argument for requiring the patent was based on the fact that the insertion of the oncogene into the rat genome was a human invention. The patent was granted taking into account the usefulness of the modified rat as a model for studying a severe human illness to discover new anticancer drugs. The patent application was granted because the patent considered the bioethics principle of beneficence. However, such an interpretation was not accepted by Canada and the European Community, who restricted the subject matter of the patent to future generations of the genetically modified rat only and not to future genetic modification of all rodent species. Such interpretation of the patent subject matter and any future claims enhances two aspects of patent law that are considered by patent agencies before a patent is granted. First, in biotechnology, the requisite 'evaluation of the utility of invention' must be considered prior to the 'innovation inherent to the invention', which is the opposite of what is considered in patent applications in other sciences. Second, a sequence of nucleic acid (i.e. an isolated and cloned gene) or any other genetic material is not considered a simple chemical substance. Moreover, biological information stored in databases, generally, is not patentable. For example, some years ago companies requested to patent sequences of nucleotides (expressed sequence targets (ESTs)) that could be used for sequence homology searching whole chromosomes to discover the positions of complete genes. However, the sequences of the ESTs had been deduced from data that was publicly available. In other words, only information would be patented in the case of ESTs. Thereby, the USA Patent Agency vetoed such patent requesting (Jensen and Murray 2005).

Genetic data obtained from genomics can be used with high probability to predict if a person is predisposed to develop a disease in the future – e.g. certain cardiovascular or pulmonary illness. Nevertheless, if the privacy of this information cannot be guaranteed and the information is in the public domain, then possible future illnesses could be considered pre-existent by an employer or health insurance companies. That information could be used as an excuse for an employer not to hire a person or an insurance company to increase the health insurance or life assurance premiums.

Nevertheless, the value of new prognostic tests to better inform patient choice and physician treatment options is widely considered to be invaluable. However, the huge amount of information related to a specific patient requires memory and computation powers of computers to be available. It is at this point that the question of how to guarantee patient privacy is a concern, if computing hardware stores data on data racks which have to be accessed simultaneously by multiple users.

In order to guarantee patient privacy, answers must be addressed to following questions: what is defined as a genetic disorder? What variants in a gene sequence should be considered normal or abnormal? What is defined as clinically normal and an abnormal physiological condition? Who is qualified to define normal and abnormal? A way to solve the privacy dilemma would be to partition information. For instance, an orthopaedic doctor does not need to know if the patient has a predisposition for developing cancer.

The partitioning information strategy requires software (the so-called contextual access criterions algorithms) capable of tracking electronic registration and matching this to security access. For instance, a life assurance agency would only have access to the type of work an individual is employed to do in order to predict the risk of death. This technology, although possible, is not available yet. Of course, the development of this technology will not be stimulated by life assurance agencies and other insurance companies. Perhaps, governments must encompass the ethical ramifications of this technology to establish specific privacy laws.

It might seem obvious that a patient has a vested interest in making available all of their genetic information to a doctor, who could then prescribe the best treatment possible. But if the doctor is acting on behalf of a health insurance agency, what are the guarantees that the treatment decision will only be made in the patient's health interest and not based on company operating decisions? If the patient would wish migrating from the actual assurance company to another, the company would persuade him not to do that menacing him to send the health file to the other, which, in turn, knowing past, present and even future diseases could either reject him or obligate him to afford an expensive premium. At last, it is essential and defying to establish what entities have the right to the information and decide the way how to use it.

It is expected that many of the above questions will be asked with increased intensity when the so-called Genome Project 1000[10] will be concluded in a near future (Rothstein, 2005; www.1000genomes.org).

Privacy is a subject so crucial that clinical tests for prognosis and/or diagnosis should also be considered in discussions involving medical ethics and morality.

Let's take laboratory tests designed to assess an individual's risk of developing an autoimmune disease – such as Addison's disease (malfunction of adrenal glands causing weakness, loss of weight and low arterial pressure), celiac disease (gluten allergy manifested in intestinal inflammation), multiple sclerosis (autoimmune disease affecting the myelin of the brain and CNS) and rheumatoid arthritis (chronic inflammation of joints) – by using monoclonal antibodies. Addison's syndrome, celiac disease, multiple sclerosis and rheumatoid arthritis can be predicted by detecting auto-antibodies in the blood against the enzyme 21-hydroxylase, the tissue transglutaminase, myelin and citrulline, respectively.

The advantage to an individual after having a positive result from a prognostic test would be that it will allow doctors to advise a patient to take preventive measures in order to avoid or minimize, at least, damage caused by the illness. However, before tests like these are used, ethical questions have arisen including: might doctors perform tests to identify illnesses that are still incurable? Will the patient realize that a positive test is only an indication that an illness might occur? How to minimize the risks of false-negative or false-positive test results that could lead to patients remaining either mistakenly reassured or unnecessarily worried? Will be the cost of routine screening tests justified by the number of persons that would be identified as having a high risk, and would a patient be beneficiated by available treatments? In the case of familiar autoimmune diseases, all members of a family might be tested if one of the family members had a positive prognostic test? What would be best for the other members of family: to live with the knowledge that the illness may occur in the future (if they also receive a positive test result) or with the anxiety of not knowing (for those that refuse to be tested)? Would a positive test lead to discrimination of people by employers and health insurance and life assurance companies? (Avouac et al., 2006). Unfortunately, answers to these questions are not simple and the advice of a genetic counsellor is crucial before such tests are undertaken.

1.6.3 Stem Cells and Cloning

Two types of stem cells have previously been discussed, i.e. embryo and tissue-specific. The fundamental question is how these stem cells can be ethically obtained. The tissue-specific cells can be simply extracted from body tissues without any problem if permission has been obtained from the donor. Embryo stem cells, however, by definition must be extracted from an embryo which could go on forming a fully developed human but is destroyed at the end of the extraction process. Such a situation has led to a fierce discussion on matters related to ethics, morals and philosophical principles. The polemic will be settled when the orthodoxy of religion and the dynamism of science find a reasonable compromise over a way to better manipulate embryonic cells.

The discussion started in 1998 when a lineage of embryo cells was cultured and differentiated into several human cell types. At this time, in vitro fertilization was already common practice, which had been demonstrated 20 years previously by the birth of Louise Brown. However, the cultivation of blastocytes for unlimited time rather than implanting the fertilized embryo into a uterus stimulated discussion over the use of these cells as a source of embryonic stem cells. The debate became fiercer when the question was proposed: what would be more disrespectful to life, destroy and discard embryos not to be implanted, or to destroy these cells in order to obtain human stem cells? If the conclusion was that both procedures are unethical, the next logical question would be that in vitro fertilization is unethical unless all fertilized embryos are implanted? The reasonable answer was using embryos that otherwise would not be implanted and therefore destroyed as a source of stem cells (in Brazil, such a decision was made in).

Modern biotechnology raises another topic for debate, related to cloning higher animals.

Cloning is defined as a process of clone propagation. A clone, therefore, is a collection of genetically identical individuals that originated from a single progenitor. There are three types of cloning: molecular cloning (represented by recombinant DNA techniques in which the 'clone' is the recombinant organism and the gene or piece of DNA is the recombinant molecule), cell cloning (represented by the production of monoclonal antibodies from hybridoma) and full cloning of an organism.

The term 'cloning' was introduced in 1903 by the botanist Herbert Webber in order to define the process of producing copies of identical plants from one progenitor by the grafting of plant parts. Conversely, animal cloning is the subject of fierce debate between scientists (rational thinkers) and philosophers (abstract thinkers) because in animal cloning an entire nucleus is removed from a differentiated cell (say, skin cells) and transferred to an ovule where the nucleus has been removed to activate cell multiplication but under the genetic control of the transplanted nucleus.

Historically, the first animal to be cloned – in 1958 by John Gurdon an Oxford biologist – was a frog (Mahowald, 2004). However, the first mammal to be cloned (February, 1997 by Ian Wilmut) was a sheep (named Dolly). Dolly was the result of inserting the nucleus from a mammary cell of an adult sheep into the ovule of another adult sheep from which the nucleus had been previously removed. The 'fertilized' ovule incubated for 6 days in a culture medium, followed by the implantation into a third sheep ('surrogate mother'). After 5 months, the normal gestation time for sheep, Dolly was born. Despite embryonic cells being used as a source of genetic material for cloning animals of economic importance (cows, oxen, goats and sheep) since the 1980s, Dolly was cloned directly from somatic cells of an adult animal. That is, it was demonstrated that undifferentiated cells could be obtained from differentiated tissues for the purpose of generating clones 'identical' to the donor nucleus. What was perplexing to the general public was the possibility of human cloning, and the possibility of selecting characteristics, the so-called positive eugenics. If one disregards the possibility of direct cloning of humans (reproductive cloning), whole cloning could be useful as an approach to circumvent disorders that are inherited from mitochondrial DNA, for example, infertility that cannot be overcome using in vitro fertilization, and obtaining lineages of embryonic cells cultivated in vitro for use in cell therapy and tissue engineering. In other words, the emphasis would be towards the so- called cloning therapeutics, which from an ethical point of view could be regarded as a beneficence principle.

It seems that the discussion on reproductive cloning (a matter of preoccupation by some philosophers) versus therapeutic cloning is now more focused on how to control misuse of the activity rather than the technique per se. That is, how to define the frontier between these aspects of cloning? Undoubtedly, the control must be based on ethics, because scientific and technological control can sometimes be tarnished by personal interest and/or by subjective judgment of scientists – and are not a guarantee of human dignity.

1.6.4 Use of Test Subjects

Over the last years, an ethical question has appeared in the biotechnology industry concerning the use of animals for testing bioproducts, either during production and quality control or commercialization. Each year millions of animals are sacrificed in order to evaluate the safety of pesticides, herbicides, cosmetics, drugs, food and raw material used in the manufacture of bioproducts. For example, before a pesticide is can be marketed, safety must be demonstrated e.g. measure of skin absorption, half-life in the environment, toxicity from inhalation and ingestion, in order to evaluate how long a person can be exposed to the pesticide, how much he can intake and how the substance distributes through the body. Such procedures require thousands of different animal specimens. Examples of traditional tests used to evaluate bioproducts before their use in humans include toxic kinetic tests, applied toxicology techniques and measures of acute systemic toxicity and LD_{50}.

The toxic kinetic test evaluates the absorption, distribution throughout the body, metabolism and excretion of chemical substances. After introducing the substance into a test animal, samples of blood, urine and excrement are taken over several days. Then, the animal is sacrificed in order to localize the substance and its metabolites throughout the body. An applied toxicology test evaluates the effects of a substance on skin, eyes and mucosal membranes by the appearance of redness, blisters and other signs of irritation. An example is Draize's test which is made by placing a solution of the substance to be tested into the eyes of a rabbit. This is considered to be one of the most painful tests that can be performed on an animal. The acute systemic toxicity test evaluates the effect of a bioproduct introduced into the test animal every 24h for 14 days. The LD_{50} test consist of introducing different doses of a bioproduct into groups of six animals in order to determine the minimal dose that kills half of the animal population.

It is clear that the ethical principles of no maleficence – taking into account the animals – and beneficence – considering the human aspect – are very opposed. The question is: how to be sure that a bioproduct is safe for human use – in terms of toxicity, iatrogenic properties, pharmacological efficacy, etc. – without testing on animals? The answer is now to find reliable alternatives to animal testing which are equally effective, are low or medium cost and can provide consistent and reproductive results.

Alternative tests could be based on: (a) the use of imaging analysis of organs through X-rays, magnetic nuclear resonance and/or positron emission tomography. Such approaches can reduce the number of animals needed to obtain conclusive data on the effects of a drug by 80%; (b) interruption of a painful test immediately after conclusive results have been obtained; (c) the use of more primitive organisms such as the zebra fish (*Brachydanio rerio*), the nematode (*Caenorhabditis elegans*) and the horseshoe crab (*Limulus polyphemus*); (d) the use of residual organs retrieved from slaughterhouses. Eyes of dead animals could substitute living rabbits in Draize's test, for instance; and (e) the use of cells harvest from specific human tissues (skin, lungs, muscles, membranes, etc.) and maintained in vitro cultures. An interesting variant of this procedure would be growing the tissue on a three-dimensional surface which reproduces, roughly, organs such as skin, lungs, kidney, eye and digestive system. This approach would reduce significantly the number of test subjects needed for conducting experimental tests in biotechnology.

It must be emphasized that the substitution of traditional animal testing protocols, which are well established and reliable, by new ones – in order to respect the principle of no maleficence regarding animals – must be made cautiously.

The validation of a new biological test must fulfil the rigor and complexity required by conventional clinical experiments. This was considered so important that an international meeting was organized in 1996 at Solna (Sweden) from which two regulatory agencies were created, one by the European Community (named European Center for Validation of Alternative Methods, or ECVAM) and the other by the USA (named Interagency Coordination Committee for Validation of Alternative Methods, or ICCVAM). Both agencies carry out 'pre-validation' studies to evaluate the practical potential of alternative methods to animal testing and establish well-defined protocols. In Europe, if the pre-validation phase is passed successfully then the ECVAM organizes laboratories amongst EU member countries to perform analysis of different blind marked compounds using the alternative test. The results are analysed by a committee comprised of scientists from different countries, members of industry and animal welfare protection groups. During the process of validation, ECVAM acts as an observer only. Thereby, if the alternative method evaluated was capable to determine precisely, consistently and reproducibly a relevant property of the compound, then the agency would formally endorse the use of the test. Since 2003, it has been illegal in the EU to test on animals any cosmetics or materials used to manufacture cosmetics, except where alternative procedures are not available.

In spite of the high costs in developing alternative protocols, the use of animals for testing is also expensive taking into account the cost involved in lineage selection, mating, maintaining and commercialization (storage, transportation, etc.).

1.6.5 Agriculture

As already discussed, recombinant DNA techniques – invented in 1973 by Stanley Cohen and Herbert Boyer, who successfully cloned a frog gene into *E. coli* – have become fundamental techniques in biotechnology. Since 1973, several genetically modified organisms (GMO) have been successfully commercialized for biotechnology processes. This methodology greatly impacted agriculture, when in 1996 Monsanto launched into the market a variety of glyphosate-resistant transgenic soya bean (still today the market-leading herbicide worldwide sold under the trade name Roundup).

The introduction of transgenic soya (sold under the trade name Roundup Ready) in the field enabled farmers to reduce production costs while increasing the productivity of the crop. The use of GMO in agriculture promoted an upgrade of the successful 'green revolution' – which began in 1950 and was based on the improvement of equipment, using more effective agriculture chemicals (herbicides and insecticides, amongst others) and selected seeds. Moreover, GMO is also more resistant to climate patterns variations of rain, snow, wind, heat, etc.

From an ethical point of view, GMO fulfils three basic principles, i.e., autonomy (freedom from deciding on the use of a modified plant or not), beneficence (higher productivity and lower environmental damage, for instance) and justice (more food for people in underdeveloped countries).

During the last two decades, a fierce debate has erupted between defenders and detractors of GMOs in agriculture. The quarrel is based on the ethical 'no maleficence' principle, in which some groups have been benefited at the detriment of other groups. In Brazil, for instance, one of the 27 states of federation did not allow the use of transgenic soya. It forbade the sowing and transportation of GMO soya across its territory. Nowadays, the state has recognized the advantages presented by transgenic soya and allows the farming of this crop.

The main objections against GMO plants are introduction into the nature of a non-natural organism, reduction of genetic diversity, involuntary transfer of new genes to native species, dispersion to other areas, no guarantee that disease-causing organisms don't become resistant to the biopesticide (toxin of *Bacillus thuringiensis*, for example) and safety for human consumption (in this case, opponents are worried about the potential allergenic effect on consumers).

Despite the scepticism spread amongst millions in the general public, the use of GMO plants is a reality. In the USA, Canada, Argentina and China, only genetically modified soya is cultivated. In the period 1996–2002, the area dedicated to GMO soya bean agronomy was about 60 million of hectares (Zhang et al., 2016).

It is worth mentioning that all plants consumed resulted from crossing between species. For example, *Beta vulgaris* (beet) versus *Beta procumbens* gave a hybrid resistant to nematodes; *Triticum aestivum* (wheat) versus *Agropyron elongatum* gave a hybrid resistant to desiccation. So, during centuries of human agricultural development, a wide array of plants that are insect resistant and herbicide tolerant have been developed. The introduction of GMO has only accelerated the selection of a desired characteristic into the crop.

A final consideration of the principles of ethics and morality in terms of consumer choices is that any food containing GMO ingredients must be labelled as GMO on the food packaging.

1.7 FINAL CONSIDERATIONS

Biotechnology can be defined as an assembly of techniques that use cells, cell organelles and biological molecules with the aim of developing and improving products of social (drugs for health care, for instance) and economic interest. Although biotechnology became a field of immense scientific and industrial interest, biotechnological processes have been used by humans for centuries in beer, bread, cheese, tannery and wine production.

As the genome machine has been understood at the levels of expression of DNA into proteins, control of gene expression by different types of RNA, and epigenetic information associated with chromatin architecture all of which can now be manipulated originating the so-called bioprocessing. Thereby, prokaryotic and eukaryotic organisms have been handled in order to obtain cells fitted for the production of a large variety of bioproducts (monoclonal antibodies, vitamins, hormones, statins, etc.).

Bioprocessing involves the use of living cells or components of their metabolic machinery (e.g. enzymes and organelles) to synthesize products, decompose substances and/or produce energy. Bioprocessing encompasses techniques such as microbial fermentation, mammalian cells culture and biodegradation (microorganisms used to reduce environmental pollution).

It must be borne out that bioprocessing is dependent on reliable analytical techniques (biosensors, enzyme kits, highly sensitive microscopes and spectrometers, amongst others) and bioinformatics (based on computing science for storing and handling a large amount of data). Moreover, the biotechnology industry has experienced rapid growth as knowledge on cell biology improves and methodologies in genomics, proteomics and metabolomics become better understood. These approaches lead to the detailed study of internal cell structures (organelles, cytoskeleton, membranes, etc.) and the chemical nature of biomolecules, as well as the flux in their intracellular levels and mechanisms of action.

Advancements in computing science and bioinformatics (represent the interaction between mathematics, computing and molecular biology) have amplified remarkably the capacity to store, compute and compare huge amounts of different types of biological data. The increased precision for studying three-dimensional structures of cell components and biomolecules now allows new biomolecules to be designed to target interactions with specific targets, such as membrane receptors, ion channels and membrane carrier protein. As new genomes are sequenced, the potential to genetically modify increasing forms of life as valuable sources of bioproducts becomes possible – besides, advances in genetic engineering led to create minimal life, being exemplified by the complete synthesis of the *Mycoplasma genitalium* genome, which when transferred to another bacterium of the same genus was expressed perfectly. Moreover, the miniaturization of electronic devices will enlarge sharply the bioprocessing precision.

Bioinformatics allowed obtaining a lot of data related to gene sequences and proteins, three-dimensional structures of biological molecules and genome maps of many thousands of organisms. Using computational tools (algorithms, graphics, artificial intelligence, statistical programs, simulation and management of data banks), biocomputing allows mapping and comparing genomes, determining structures of proteins, simulating bond formation amongst molecules, identifying genes, evaluating the effect of mutations and determining phylogenetic relationships. Recently, a method for digitalizing microscopic images of human tissue allowed increased sensitivity in disease diagnostics by improving the sharpness of the images and by facilitating the exchange of information between pathologists. The implementation of this technology has revolutionized the screening of pathological specimens, where protocols were previously established over a century ago.

Finally, the interaction between biotechnology and society comes down to two opposing ideas, i.e., profit versus safety and compassionate use of living beings, which might be balanced through the principles of ethics. For example, the use of animals (mice, rabbits, apes, etc.) in experiments aimed to evaluate products for human use (medicines, cosmetics, foods, chemicals, etc.) is ethically questionable. Since 2003, this approach has been banned in many countries as long as transgenic animals have been available in the market. Through transgenic it is possible to generate animals with desired characteristics related to the trait to be studied (illnesses, tissue modifications by physical and/or chemical agents, for example), which implies that there can be a reduction in the numbers of animals required for product testing, as well as diminishing the suffering of the animals during the tests. Moreover, transgenic animals would allow evaluating more precisely the physiopathology of several diseases and the development of new treatments and diagnostic tests, thereby reducing the development costs for new drugs.

REFERENCES

Andersen, M.M. Feasibility of new breeding techniques for organic farming. *Trends in Plant Science*, v.20, n.7, 426–434, 2015.

Ast, G. How did alternative splicing evolve? *Nature Review Genetics*, v.5, 773–782, 2004.

Avouac, J.; Gossec, L.; Dougados, M. Diagnostic and predictive value of anti-cyclic citrullinated protein antibodies in rheumatoid arthritis: a systematic literature review. *Annals of Rheumatic Diseases*, v.65, n.7, 845–851, 2006.

Azevedo, P.C.T.; Azevedo, P.P.C.; Ramos, S.W.F. Use of remdesivir for patients with Covid-19: a review article. *Revista da Associacão Médica Brasileira*, v.66, n.6, 838–841, 2020.

Becker, S. A historic and scientific review of breast cancer: the next global healthcare challenge. *International Journal of Gynecology and Obstetrics*, v.131, n.1, 536–539, 2015.

Benenson, Y.; Gil, B.; Bem-Dor, U.; Adar, R.; Shapiro, E. An autonomous molecular computer for logical control of gene expression. *Nature*, v.249, 423–429, 2004.

Blagosklonny, M.; Hall, M.N. Growth and aging: a common molecular mechanism. *Aging*, v.1, n.4, 357–362, 2009.

Blount, F.K.; Breaker, R.R. Riboswitches as antibacterial drug targets. *Nature Biotechnology*, v.24, n.12, 1558–1564, 2006.

Bonfante, G.; Kaczmarek, M.; Marion, J.Y. On abstract computer virology from a recursion theoretic perspective. *Journal in Computer Virology*, v.1, n.4, 25–37, 2005.

Cavalheiro-Abreu, A.; Monteiro, G. Solving ethanol production problems with genetically modified yeast strains. *Brazilian Journal of Microbiology*, v.44, n.3, 665–671, 2013.

Chalbatani, G.M.; Gharagouzloo, E.D.H.; Marmari, V. Small interfering RNAs (siRNAS) in cancer therapy: a nanobased approach. *International Journal of Nanomedicine*, v.14, 3111–3128, 2019.

Colson, P.; La Scola, B.; Levasseur, A.; Caetano-Anollés, G.; Raoult, D. Mimivirus: leading the way in the discovery of giant viruses of amoebae. *Nature Review Microbiology*, v.15, n.4, 243–254, 2017.

Coufal, N.G. L1 retrotransposition in human neural progenitor cells. *Nature*, v.460, 1127–1131, 2009.

Duster, T. Enhanced: race and reification in science. *Science*, v.307, 1050–1051, 2005.

FDA - https://www.fda.gov/drugs/drug-approvals-and-databases/drugsfda-glossary-terms#B

Felsenfeld, G.; Groudinein, M. Controlling the double helix. *Nature*, v.421, 448–453, 2003.

Fu, D.X. Non-coding RNA: a new frontier in regulatory biology. *National Science Review*, v.1, n.2, 190–204, 2014.

Garrett, H.R.; Grisham, C.M. *Biochemistry*. New York: Saunders College Publishing, 1995.

Gruner, G. Carbon nanotube transistors for biosensing applications. *Analytical and Bioanalytical Chemistry*, v.384, 322–335, 2006.

Güemes, C.G.A.; Youle, M.; Cantú, V.A.; Felts, B.; Nulton, J.; Rohwer, F. Viruses as winners in the game of life. *Annual Review of Virology*, v.3, 197–214, 2016.

Hall, S.S. Editing the mushroom. *Scientific American*, v.314, n.3, 56–63, 2016.

Jensen, K.; Murray, F. Intellectual property landscape of the human genome. *Science*, v.310, 239–240, 2005.

Kumar, R.; Yadav, A.K.; Singh, V.K.; Vandre, R.K.; Singh, K.P.; Vineeth, M.R. Transgenic animal technology: recent advances and applications. *Agriculture Review*, v.36, n.1, 46–53, 2015.

La Scola, B. A giant virus in amoebae. *Science*, v.299, n.5615, 2033–2035, 2003.

Laddy, D. J. Electroporation of synthetic DNA antigens offers protection in nonhuman primates challenged with highly pathogenic avian influenza virus. *Journal of Virology*, v.83, n.9, 4624–4630, 2009.

Lavik, E.; Langer, R. Tissue engineering: current state and perspective. *Applied Microbiology and Biotechnology*, v.65, n.1, 1–8, 2004.

Levenberg, S. Engineering blood vessels from stem-cells: recent advances and applications. *Current Opinion in Biotechnology*, v.16, n.5, 516–523, 2005.

Lodish, H.; Baltimore, D.; Berk, A.; Zipursky, S.L.; Matsudaria, P.; Darnell, J. *Molecular cell biology*. 3rd ed. New York: Scientific American Books, 1995. Chapter 11.

Lundstrom, K. Viral vectors in gene therapy. *Diseases*, v.6, n.2, 42–49, 2018.

Lysaghtet, M.J. Great expectations: private sector activity in tissue engineering, regenerative medicine, and stem cell therapeutics. *Tissue Engineering Part A*, v.14, n.2, 305–315, 2008.

MacManus, T.M.; Sharp, A.P. Gene silencing in mammals by small interfering RNAs. *Nature Reviews Genetics*, v.8, 737–747, 2002.

Mahowald, M.B. The human cloning. *Perspectives in Biology and Medicine*, v.48, n.2, 307–309, 2004.

Martins, W.M.B.S.; Toleman, M.A.; Gales, A.C. Clinical utilization of bacteriophages: a new perspective to combat the antimicrobial resistance in Brazil. *Brazilian Journal of Infectious Diseases*, v.24, n.3, 239–246, 2020.

Mehretra, S.; Chouhan, D.; Konwarh, R.; Kumar, M.; Mandal, B.B. Comprehensive review on silk at nanoscale for regenerative medicine and allied applications. *ACS Biomaterials Science & Engineering*, v.5, n.5, 2054–2078, 2019.

Moulder, S.; Hortobagyi, G.N. Advances in the treatment of breast cancer. *Clinical Pharmacology and Therapeutics*, v.83, n.1, 26–36, 2008.

Mukerjee, M. Trends in animal research. *Scientific American*, v.276, n.2, 86–93, 1997.

Nassar, D.; Blanpain, C. Cancer stem cells: basic concepts and therapeutic implications. *Annual Review of Pathology*, v.11, 47–76, 2016.

Oliveira, B.A.; Oliveira, L.C.; Sabino, E.C.; Okay, T.S. Sars-Cov-2 and the Covid-19 disease: a mini review on diagnostic methods. *Revista do Instituto de Medicina Tropical*, on line, https://doi.org/10.1590/s1678-9946202062044, 2020.

Park, E.; Pan, Z.; Zhang, Z.; Lin, L.; Xing, Y. The expanding landscape of alternative splicing variation in human populations. *The American Journal of Human Genetics*, v.102, n.1, 11–26, 2018.

Perkel, J. M. Digitizing pathology. *Bioscience Technology*, v.34, n.2, 8–12, 2010.

Rathmann, G. B. Biotechnology Startups. In: Moses, V.; Capre, R. E. (Eds.). *Biotechnology: the science and the business*. Chur: Hardwood Academic Publishers, 1994. 49–67.

Ramos, T.; Moroni, L. Tissue engineering and regenerative medicine. *Tissue Engineering Part C*, v.26, n.2, 91–104, 2020.

Roberts, M. S.; Lorence, R. M.; Groene, W. S.; Bamat, M. K. Naturally oncolytic viruses. *Current Opinion in Molecular Therapeutics*, v.8, 314–321, 2006.

Rocha-Martins, M.; Cavalheiro, G.R.; Matos-Rodrigues, G.E.; Martins, R.A. From gene targeting to genome editing: transgenic animals applications and beyond. *Annals of Brazilian Academy of Sciences*, v.87, n.2, 1323–1348, 2020.

Rodier, F.; Campisi, J. Four faces of cellular senescence. *Journal of Cell Biology*, v.192, n.4, 547–556, 2011.

Rohregger, R.; Sganzerla, A.; Simão-Silva, D.P. Synthetic biology and genetic manipulation: risks, promises and responsibilities. *Environment & Society*, v.23, on line version, 2020. https://doi.org/10.1590/1809-4422asoc20180196r3vu2020l4ao.

Rothstein, M.A. Genetic exceptionalism and legislative pragmatism. *Hastings Center Report*, v.35, n.4, 27–33, 2005.

Shih, M.W.; Quispe, J.D.; Joyce, G.F.A. 1.7-kilobase single-stranded DNA that folds into a nanoscale octahedron. *Nature*, v.427, 618–621, 2004.

Sorz, G. An expanding universe of noncoding RNAs. *Science*, v.296, 1260–1263, 2002.
Stryer, L.; Berg, J.M.; Tymoczko, J.; Gatto, G. *Biochemistry*. 9th ed. London: Palgrave MacMillan, 2019.
Suhre, K.; Audice, S.; Chaverie, J.M. Mimivirus gene promoters exhibit an unprecedented conservation among all eukaryotes. *Proceedings of New York Academy of Sciences*, v.102, n.41, 14689–14693, 2005.
Tang, J.N.; Cores, J.; Huang, K.; Cheng, K. Is cardiac cell therapy dead? Embarrassing trial outcomes and new directions for the future. *Stem Cells Translational Medicine*, v.7, n.4, 354–359, 2018.
Vasikari, T. Safety and efficacy of a pentavalent human-bovine (WC3) reassortant rotavirus vaccine. *New England Journal of Medicine*, v.354, 23–32, 2006.
Wang, J. Can man made nanomachines compete with nature biomotors? *ACS Nano*, v.3, n.1, 4–9, 2008.
Wang, F.; Zhang, W. Synthetic biology: recent progress, biosafety and biosecurity concerns, and possible solutions. *Journal of Biosafety and Biosecurity*, v.1, n.1, 22–30, 2019.
Welder, W.J.H. Vaccine adjuvants: current challenges and future approaches. *Journal of Pharmaceutical Sciences*, v.98, n.4, 1278–1316, 2009.
Wu, X.; Mu, F.; Wang, Y.; Zhao, H. Graphene and graphene-based nanomaterials for DNA detection: a review. *Molecules*, v.23, 2050–2073, 2018.
Zhang, C.; Wohlhueter, R.; Zhang, H. Genetically modified foods: a critical review of their promise and problems. *Food Science and Human Wellness*, v.5, n.3, 116–123, 2016.
Zhou, X.; Yuan, L.; Mao, Z. Recent review of the effect of nanomaterials on stem cells. *RSC Advances*, v.8, 17656–17676, 2018.
Zhu, J. Mammalian cell protein expression for biopharmaceutical production. *Biotechnology Advances*, v.30, 1158–1170, 2012.

2 Thermodynamics Applied to Biomolecules

Attilio Converti
Genoa University

Adalberto Pessoa Jr.
University of São Paulo

Jônatas de Carvalho Silva, Rodrigo Lira de Oliveira, and Tatiana Souza Porto
Rural Federal University of Pernambuco

CONTENTS

2.1 INTRODUCTION

Thermodynamics is one of the branches of physics that has wide application in biotechnology. It assesses the influence of temperature on different reactions involving biomolecules of pharmacological interest, either native or modified by different treatments, as well as their structure. Moreover, knowledge of kinetic and thermodynamic parameters is useful, not only for investigating the reaction mechanism and behaviour but also for successfully designing reactors and other equipment used in the pharmaceutical industry.

The use of thermodynamics in biochemistry has been emphasized since the 1950s, starting with the work of Krebs and Kornberg (1957), in which the authors developed an useful and extensive appendix containing Gibbs free energy values of reactions catalysed by different enzymes. Later, Jones (1979) published the first edition of the book 'Biochemical Thermodynamics'. In the 1990s and 2000s, Professor Robert Alberty made a considerable contribution to the topic in books (Alberty, 2003a, 2006a) and review articles (Alberty, 2003b, 2006b; Alberty et al., 2011).

Various biological processes can be understood through thermodynamic parameters such as activation energy (E_a), enthalpy (ΔH), entropy (ΔS) and Gibbs free energy (ΔG). Contrary to what one might imagine, such parameters can be estimated from the results of relatively simple experiments similar to those used for biochemical characterization. Directly associated with kinetics, thermodynamics is fundamental for understanding phenomena related to different biomolecules of pharmacological interest such as enzymes, natural pigments and antibiotics.

Enzymes are the most reported biomolecules in scientific studies on thermodynamic modelling. As is known, during enzyme-catalysed reactions, the biocatalyst undergoes simultaneous thermal denaturation, responsible for a change in its structure from a native and functional form to an inactivated one, which represents an important restriction in many biotechnological processes. Whereas the thermodynamic parameters of the reaction are related to its kinetics, the study of thermal denaturation is based on the determination of enzyme residual activity over time. Some relevant kinetic parameters of thermal inactivation, such as enzyme half-life ($t_{1/2}$), decimal reduction time (D) and thermal resistance constant (Z), can be determined using data already collected in the kinetic study of this phenomenon conducted at different temperatures. Therefore, a complete kinetic and thermodynamic analysis allows a better understanding of the enzyme reactivity with the substrate, the formation of the enzyme–substrate complex, the spontaneity of product formation and the protein denaturation process. The thermodynamic study allows researchers not only to optimize the yield of products but also to predict their stability during processing and/or storage.

Pigments, as natural constituents of certain foods or even added on purpose, are subjected to product preservation processes based on heat treatments, which make it necessary to study the phenomenon of their thermal degradation. Since the degradation of pigments can cause changes in the sensory and nutritional properties of food products, knowledge of the kinetic and thermodynamic parameters is essential for the development of new processes, in order to ensure obtaining safe products with maximum preservation of quality factors.

Antibiotics obtained through biotechnology suffer losses during their production, which is a serious problem for the pharmaceutical industry. Therefore, studying the degradation of these compounds can help to reduce losses and costs inherent to their production. However, a few kinetic and thermodynamic studies have been performed on this class of biomolecules. An example of the use of this approach was reported by Marques et al. (2009) for the formation and simultaneous degradation of clavulanic acid, an inhibitor of β-lactamases. Likewise, many other biomolecules of pharmaceutical interest can be affected by changes in temperature during their production.

This chapter will address aspects related to the thermodynamic properties of biomolecules, with emphasis on biotechnological processes, thus facilitating their understanding from the experimental procedures to the interpretation of estimated parameters.

2.2 THERMODYNAMIC CONCEPTS

Before starting the theory and equations, it is necessary to introduce some basic thermodynamic concepts so that their application to biomolecules is well interpreted.

The Zeroth Law of Thermodynamics addresses the concept of thermal equilibrium based on the direction of heat transfer. When two bodies at different temperatures come into contact, heat is spontaneously transferred from the hotter to the colder, until both achieve thermal equilibrium.

On the other hand, the First Law of Thermodynamics, which explains the reason for such a transfer, deals with the conservation of energy, which is neither created nor destroyed but is transformed. This concept of energy transfer is related to the variation of the so-called enthalpy (H), a thermal parameter that represents the general heat content of a system and reflects the number and type of chemical bonds in reactants and products. When a chemical reaction releases heat, it is called exothermic; since the heat content of products is lower than that of reactants, the enthalpy variation (ΔH) takes on a negative value. On the other hand, the reacting systems that receive heat from the medium are endothermic and are characterized by positive ΔH values.

The Second Law of Thermodynamics, in addition to dictating the limits to the conversion of heat into work, deals with the disorder of systems, thus indicating whether a process tends towards maximum disorder or not. These concepts are related to the so-called entropy (S), a parameter that represents the randomness or the disorder of the components of a chemical or biological system. Changes in the randomness of systems are expressed as entropy variation (ΔS), which has a positive value when there is an increase in randomness and vice versa. It is worth noting that thermodynamics does not determine the rate at which any reaction takes place, which, instead, is within the competencies of chemical kinetics. For example, a process, even spontaneous, can occur too slowly, at a negligible rate.

Gibbs free energy (G) represents the global energy content, which combines both enthalpy and entropy contributions at a given temperature. When a chemical reaction occurs at a constant temperature, the change in free energy, ΔG, is given by the difference between the change in enthalpy, ΔH, which reflects the type and number of chemical bonds as well as the formation and breakdown of non-covalent interactions, and the change in entropy multiplied by the absolute temperature ($T\Delta S$), which describes the variation in randomness, or level of disorder of the system. A process tends to occur spontaneously only if $\Delta G = \Delta H - T\Delta S$ is negative, that is, if it releases free energy. On the contrary, ΔG is positive when the process does not occur spontaneously. The former type of process is called exergonic, while the latter endergonic.

The application of these concepts to biological systems will depend on the joint analysis of the different energy contributions of reactions that occur simultaneously with. This is the case, for instance, of cells whose functioning basically depends on biomolecules such as proteins and nucleic acids. The free energy of formation of these biomolecules is positive, which means that they are less stable and more highly ordered than the mixture of their monomeric components (amino acids and nucleotides). For these endergonic and thermodynamically unfavourable reactions to occur, cells couple them to other exergonic reactions (for instance, ATP hydrolysis), so that the process as a whole is exergonic, that is, the overall sum of free energy variations is negative (Nelson and Cox, 2012).

As these concepts can be applied to many biomolecules, mainly enzymes, it was decided to consider them in a generic way. The aim is to allow the understanding of thermodynamic concepts and their interpretation in a simple way, providing pharmaceutical biotechnology researchers with complementary tools to characterize the biomolecules of their interest. In mathematical equations, the parameters are defined according to the process that is being studied. For example, in the case of enzymes, whose molar mass is often unknown before completing the characterization studies, it is preferred to refer to the enzyme activity (A), while in the case of chemical molecules of known composition, such as pigments and antibiotics, the concentration (C) is preferably used, but these differences do not affect the application of thermodynamic equations. For simplicity, in the following section, it was decided to indicate the biological activity of any biomolecules as A.

2.3 MATHEMATICAL MODELLING

In thermodynamic modelling studies on the behaviour of biomolecules, it is important to understand the phenomena that contribute to their final biological activity.

Many chemical and physical agents may influence their behaviour. Among them, temperature is the variable that has the greatest impact on the thermodynamic and kinetic features. Basically, every single molecule in the entire universe is affected by temperature. If on the one hand an increase in temperature can speed up chemical reactions, on the other it can degrade the reactants and/or products. Biocatalysts are not different, in that the reactions catalysed by them can be favoured, but they can be degraded more quickly.

This section will address the thermodynamics of biocatalysts as regards the influence of temperature on the rate of reactions they catalyse, as well as on their structure and stability. It should be remembered, however, that the equations describing degradation are applicable to any biomolecules, even those without catalytic action.

2.3.1 Reaction Thermodynamics

2.3.1.1 Activation Energy and Standard Enthalpy Variation of the Unfolding Equilibrium

Equations are used either to estimate parameters from laboratory data or to predict the behaviour of specific biomolecules under untested conditions. Generally, reactions catalysed by biocatalysts like enzymes (E) (see Chapter 15), which are responsible for converting a substrate (S) into a product (P), are characterized by the formation of an enzyme–substrate complex (E-S). However, according to the Arrhenius theory, such a reaction happens if the E–S complex has sufficient energy to overcome an energy barrier called activation energy (E_a).

Biocatalysts can accelerate reactions, even those less thermodynamically favoured by reducing their activation energy, i.e. lowering the energy barrier (Figure 2.1). It noteworthy, however, that at any temperature biocatalysts have no influence on the Gibbs free energy variation between reactants and products, that is, on the equilibrium constant. Therefore, if a reaction is thermodynamically unfavourable, it is not possible to reverse this condition using a biocatalyst.

Although the reaction rate is a function of the number of collisions between biocatalyst and substrate molecules, such a collision process has diffusion limitations, and only a fraction of all molecules will effectively collide. Arrhenius proposed Equation 2.1 to correlate the rate constant (k), i.e. the frequency of collisions resulting in a reaction, expressed as the reciprocal of time, to E_a and the absolute temperature (T):

$$k = ae^{-\frac{E_a}{RT}} \quad (2.1)$$

where a is the pre-exponential factor, a constant for each chemical reaction, while R=8.314 J/K/mol is the ideal gas constant.

In the case of enzymes, the actual rate constant is often difficult or even impossible to determine; therefore, the enzyme activity (A) is often used instead of k. So, Equation 2.1 can be written as Equation 2.2:

$$A = a'e^{-\frac{E_a}{RT}} \quad (2.2)$$

However, it should be remembered that, in this case, the pre-exponential factor (a') has the same unit as the enzyme activity and then a different meaning from that of the Arrhenius equation.

Applying the logarithm function, this equation can be linearized as (Equation 2.3):

$$\ln A = -\frac{E_a}{R} \cdot \frac{1}{T} + \ln a' \quad (2.3)$$

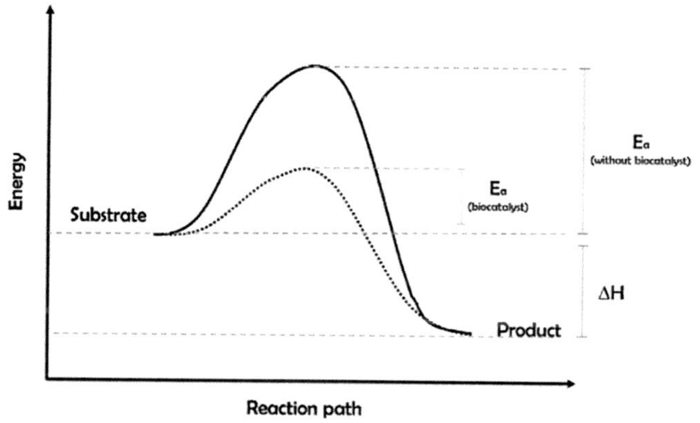

FIGURE 2.1 Energy profile of a reaction catalysed or not by a biocatalyst.

and the activation energy can be easily calculated from the slope of the straight line obtained by plotting the experimental results of ln A versus $1/T$ on a semi-log graph, as described in the following.

Analysing Equation 2.3, it can be noted that for each temperature there is a corresponding activity. In biochemical characterization of a biomolecule, often an enzyme, its performance is represented in a graph that shows the influence of temperature on its activity (Figure 2.2). This graph exhibits upward and downward curves and a central point of maximum activity at the optimum temperature, which allow us a better understanding of the underlying phenomena.

To understand the usefulness of Equation 2.3, it is important to make an analogy with the expression of a straight line (Equation 2.4):

$$\ln A = -\frac{E_a}{R} \cdot \frac{1}{T} + \ln a' \qquad (2.4)$$
$$y = a \cdot x + b$$

where the variables y and x correspond to ln A and $1/T$, while the slope (a) and the intercept (b) to $-E_a/R$ and ln a', respectively. Then, plotting in a semi-log plot the experimental data of ln A in the upward left curve of Figure 2.2 as a function of $1/T$, a straight line is obtained (Figure 2.3) from whose slope the E_a value (J/mol) can be easily calculated. The experimental procedure is described in Section 2.4.1.1 of this chapter.

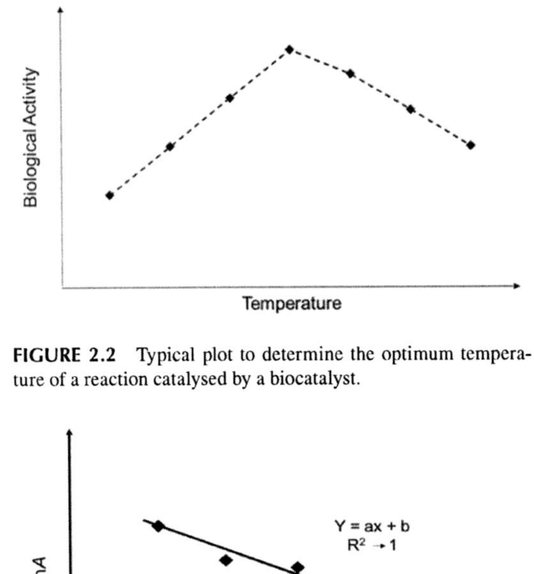

FIGURE 2.2 Typical plot to determine the optimum temperature of a reaction catalysed by a biocatalyst.

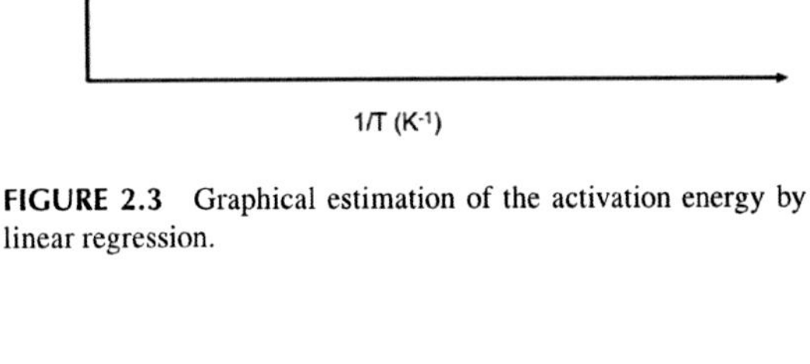

FIGURE 2.3 Graphical estimation of the activation energy by linear regression.

As explained above, temperature influences both the reaction rate and the molecular stability of the biocatalyst, which is then supposed to be simultaneously subject to an unfolding equilibrium. The energy involved in this equilibrium is known as standard enthalpy variation of the unfolding equilibrium (ΔH_u^o). Applying the Gibbs equation to it and taking into account the Michaelis–Menten theory, we obtain Equation 2.5:

$$A = \frac{a' \exp\left(-\frac{E_a}{RT}\right)}{1 + b\exp\left(-\frac{\Delta H_u^o}{RT}\right)} \qquad (2.5)$$

where b is an additional pre-exponential factor that depends on the standard entropy variation of the same equilibrium.

Since at temperatures below the optimum the unfolding equilibrium is shifted to the left, the exponential term in the denominator of Equation 2.5 becomes negligible with respect to unity; therefore, it simplifies to the Arrhenius-type Equation 2.2, or 2.3 in its linearized form. On the other hand, an increase in temperature shifts such an equilibrium to the right, i.e. towards the unfolded form of the biocatalyst, which becomes predominant at a temperature higher than the optimum, thus leading to the overall activity decrease described by the downward curve on the right in Figure 2.2. Under these conditions, Equation 2.5 simplifies to Equation 2.6:

$$A = \frac{a'}{b} \exp\frac{\Delta H_u^o - E_a}{RT} \qquad (2.6)$$

Since E_a was already calculated as described above, ΔH_u^o can be calculated by linear regression from the slope $\left(\frac{\Delta H_u^o - E_a}{R}\right)$ of the straight line obtained by plotting versus $1/T$ only the experimental values of ln A collected at temperatures higher than the optimum.

2.3.1.2 Enthalpy, Entropy and Gibbs Free Energy of the Reaction

Before proceeding with the thermodynamic aspects of biomolecules, it is important to see some kinetic concepts described in Chapter 15 that will help to understand further considerations.

The maximum rate (V_{max}) corresponds to the highest volumetric activity achieved by a biocatalyst. Based on the Michaelis–Menten equation, the catalytic constant (k_{cat}) is defined as the ratio of V_{max} to the enzyme concentration [E] according to Equation 2.7a:

$$k_{cat} = \frac{V_{max}}{[E]} \qquad (2.7a)$$

Such a first-order rate constant, also known as the turnover number, is equivalent to the number of substrate molecules converted into product per unit time (s^{-1}) by a single enzyme molecule operating under saturation conditions.

To better understand the meaning of this constant, it is important to remember, that the transition state of the E–S complex and the fundamental state of reagents are in thermodynamic equilibrium of association and disassociation, governed by the rate constants k_1 and k_2, respectively, and that product formation is the result of complex breakdown, governed by k_3, for a more deep understanding see the Chapter 15. The catalytic constant of the reaction can then be written as follows (Michel, 2013):

$$k_{cat} = \frac{k_1 \cdot k_3}{k_2 + k_3} \tag{2.7b}$$

When $k_3 \ll k_2$, this equation simplifies to Equation 2.7c:

$$k_{cat} = k_3 \cdot K_{eq} \tag{2.7c}$$

where $K_{eq} = k_1/k_2$ is the equilibrium constant. In other words, under these conditions, the catalytic constant depends on the association–disassociation equilibrium.

According to Eyring (1935), the frequency of E–S breakdown to product (k_3) corresponds to the vibrational frequency of the bond being broken; therefore, based on the principles of classical physics, it is equal to the frequency (s^{-1}) that appears in the quantum energy Equation 2.7d and 2.7e:

$$E^* = h \cdot v \tag{2.7d}$$

So:

$$E^* = h \cdot k_3 \tag{2.7e}$$

where E^* is the mean vibrational energy of the E–S bond and h the Planck constant (6.62×10^{-34} J s).

E^*, in turn, is given by the product of the Boltzmann constant ($k_B = 1.38 \times 10^{-23}$ J/K) and the absolute temperature (Equation 2.7f) (Michel, 2013):

$$E^* = k_B \cdot T \tag{2.7f}$$

Therefore, by equating Equations 2.7e and 2.7f, we have:

$$k_3 = \frac{k_B T}{h} \tag{2.7g}$$

Substituting this equation into Equation 2.7c results in:

$$K_{eq} = \frac{k_{cat} h}{k_B T} \tag{2.7h}$$

Under the equilibrium conditions described above, the Gibbs free energy can be described by Equations 2.8 and 2.9:

$$\Delta G = \Delta H - T\Delta S \tag{2.8}$$

$$\Delta G = -RT \ln K_{eq} \tag{2.9a}$$

Substituting Equation 2.7h into Equation 2.9a and equating the right member to that of Equation 2.8 gives:

$$-RT \ln\left(\frac{k_{cat} h}{k_B T}\right) = \Delta H - T\Delta S \tag{2.9b}$$

or in exponential form:

$$k_{cat} = \frac{k_B T}{h} e^{-\frac{\Delta H}{RT} + \frac{\Delta S}{R}} \tag{2.9c}$$

The next steps consist of applying the logarithm function again to separate the variable T from h and k_B and then the derivative function $f(T)$ to eliminate all constants:

$$\ln k_{cat} = \ln\left(\frac{k_B}{h}\right) + \ln T - \frac{\Delta H}{RT} + \frac{\Delta S}{R} \tag{2.9d}$$

$$\frac{d(\ln k_{cat})}{dT} = \frac{1}{T} + \frac{\Delta H}{RT^2} \tag{2.9e}$$

The same approach can be used in the linearized form of the Arrhenius-type equation (Equation 2.1) replacing k with k_{cat} (Equation 2.7a), and then applying the derivative form $f(T)$ to eliminate all constants:

$$\ln k_{cat} = -\frac{E_a}{RT} + \ln a' \tag{2.9f}$$

$$\frac{d(\ln k_{cat})}{dT} = \frac{E_a}{RT^2} \tag{2.9g}$$

Equalling the right terms of Equations 2.9e and 2.9g, the enthalpy of the biocatalyst-catalysed reaction can be calculated at a given temperature from the activation energy (Equation 2.10):

$$\Delta H = E_a - RT \tag{2.10}$$

It is important to note that, since k_{cat} is calculated from V_{max} (Equation 2.7a), which is determined under conditions of substrate saturation, also ΔG refers to them, and consequently also E_a (Equation 2.9f) and ΔH (Equation 2.10) must be determined under these conditions.

After the calculation of ΔG by Equation 2.9a, the entropy of the reaction can be calculated through the rearranged form of Equation 2.8, generating Equation 2.11:

$$\Delta S = \frac{\Delta H - \Delta G}{T} \tag{2.11}$$

2.3.2 Thermodynamics of Biomolecules Thermo-Inactivation

2.3.2.1 Enthalpy, Entropy and Gibbs Free Energy of Biocatalyst Thermo-Inactivation

Despite the acceleration of biocatalyst-catalysed reactions, an increase in temperature promotes not only the above biocatalyst unfolding but also the subsequent denaturation of its structure. In fact, at any temperature, all biocatalysts progressively lose their activity over time, and the higher the temperature the more pronounced this loss. It has been proposed that the rate of biocatalyst activity decay is proportional to its activity and then can be treated as a first-order reaction (Silva et al., 2018):

$$\frac{dA}{dt} = -k_d \cdot A \tag{2.12a}$$

where t is the time of exposure at a given temperature and k_d is the first-order denaturation constant.

Solving this differential equation gives Equation 2.12b:

$$\ln\left(\frac{A}{A_o}\right) = -k_d \cdot t \tag{2.12b}$$

The k_d value (s^{-1}) at a given temperature can be estimated as the slope of the straight line obtained by plotting in a semi-log plot the ratio of the residual activity to the initial activity (A/A_0) versus the exposition time. This experimental procedure is explained in more detail in Section 2.4.2 of this chapter by means of an example. As can be seen in Figure 2.4, straight lines are obtained, which all start from the origin of coordinates and have different slopes.

It interesting to note that, at $t=0$, $\ln(A/A_0)=0$ for all the straight lines, because at the beginning of biocatalyst exposure its denaturation has not yet started regardless of the temperature.

The previous hypothesis of kinetically considering inactivation as a first-order reaction does not preclude the possibility, to estimate thermodynamic parameters, of tacking it as an equilibrium strongly shifted towards the formation of the denatured form of the biocatalyst. Therefore, replacing k_{cat} with k_d in the K_{eq} expression in Equation 2.7h, a new equation is obtained (Equation 2.13):

$$\Delta G_d = -RT\ln\left(\frac{k_d h}{k_B T}\right) \tag{2.13}$$

which allows estimating the Gibbs free energy of thermo-inactivation (ΔG_d) at each temperature.

The activation energy of the biocatalyst thermo-inactivation (E_d) is another important parameter that can be estimated from k_d values at different temperatures by simply replacing k with k_d in the Arrhenius equation (Equation 2.1), generating Equation 2.14:

$$\ln k_d = -\frac{E_d}{R}\cdot\frac{1}{T} + \ln a'' \tag{2.14}$$

where a'' is the Arrhenius pre-exponential factor in this specific case.

It is important to highlight the conceptual difference between the standard enthalpy variation of the biocatalyst

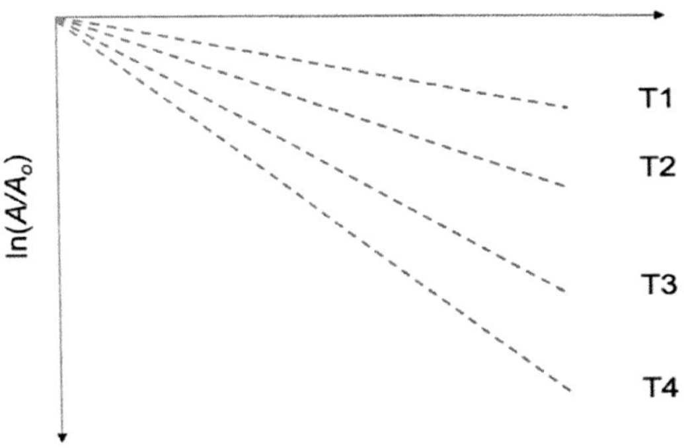

FIGURE 2.4 Straight lines of $\ln(A/A_0)$ versus time to be used for k_d estimation at different temperatures.

unfolding equilibrium and E_d. Although both energies are related to the biocatalyst inactivation, ΔH_u^o reflects the detrimental effect of a temperature increase on the biocatalyst conformation during catalysis, which makes the formation of the E–S complex difficult due to unfolding. Instead, E_d is the energy barrier that must be overcome for the biocatalyst denaturation to take place following unfolding and regardless of the presence of substrate. Nevertheless, both thermodynamic quantities lead to the same interpretation, that is, the higher their values, the greater the stability of the biocatalyst.

As for the reaction, the enthalpy and entropy of biocatalyst thermo-inactivation can be estimated using equations equivalent to Equations 2.10 and 2.11:

$$\Delta H_d = E_d - RT \tag{2.15}$$

$$\Delta S_d = \frac{\Delta H_d - \Delta G_d}{T} \tag{2.16}$$

2.3.2.2 Complementary Kinetic Parameters

Some other important kinetic parameters are usually estimated from the first-order rate constant of thermal inactivation, namely the half-life, the time of decimal reduction and the thermal resistance constant, which allow a better understanding of the biocatalyst denaturation process.

As explained earlier in the presentation of Equation 2.12a, a biocatalyst loses its activity over time at a rate that is greater the higher the temperature. Assuming that the activity loss is proportional to the value of k_d, it is possible to calculate the time required to achieve a selected decay percentage at a given temperature using Equation 2.12b. So, the time required to reduce the activity of a biocatalyst to half of its initial value ($A_0/2$), i.e. the half-life ($t_{1/2}$), is given by:

$$\ln\left[\frac{\left(\frac{A_0}{2}\right)}{A_0}\right] = -k_d \cdot t_{1/2} \tag{2.17a}$$

$$t_{1/2} = \frac{\ln 2}{k_d} \tag{2.17b}$$

Likewise, the time required to reduce it by 90% ($A_0/10$), i.e. the time of decimal reduction (D), is given by:

$$\ln\left[\frac{\left(\frac{A_0}{10}\right)}{A_0}\right] = -k_d \cdot D \tag{2.18a}$$

$$D = \frac{\ln 10}{k_d} \tag{2.18b}$$

Assuming that a biocatalyst is exposed to a temperature of 40°C, it will have a well-defined value of D, which can be reduced by 90% by increasing the temperature by a few degrees Celsius. So, the temperature increase necessary for the thermal denaturation curve to complete a logarithmic cycle and reduce the D value to $D/10$, i.e. the so-called thermal resistance constant (Z), can be calculated from the slope of the straight line obtained by plotting log D versus T according to the linearized form of the Bigelow model (Mafart et al., 2002):

$$D = D_{\text{ref}} 10^{\frac{(T_{\text{ref}}-T)}{Z}} \tag{2.19a}$$

$$\log D = -\frac{1}{Z}(T - T_{\text{ref}}) + \log D_{\text{ref}} \tag{2.19b}$$

where T_{ref} and D_{ref} are reference values of temperature and time of decimal reduction, respectively.

TABLE 2.1
Collagenolytic Activity (A) of a Fungal Collagenase at Different Temperatures (T)

T (K)	$1/T$ (K^{-1})	A (U/mL)	ln A
293.15	0.0034	24.60	3.16
303.15	0.0033	47.00	3.81
313.15	0.0032	116.30	4.77
323.15	0.0031	98.60	4.59
333.15	0.0030	80.90	4.43

U (μmol/min) indicates the units of enzymatic activity. 1 U is defined, in this case, as the amount of collagenase that catalyses the conversion of 1 μmol of substrate per minute.

2.4 EXPERIMENTAL PROCEDURES AND CASE STUDIES

Contrary to popular belief, the experimental procedure that allows the estimation of thermodynamic parameters of a biomolecule is relatively simple, since it is based on experiments that are commonly carried out for its biochemical characterization, in particular to evaluate the effect of temperature on its biological activity and stability. Therefore, general protocols for obtaining thermodynamic parameters related to biological reactions and degradation of biomolecules will be presented herein.

2.4.1 Thermodynamics of Enzyme-Catalysed Reactions

2.4.1.1 Experimental Procedure

1. The thermodynamic parameters of an enzyme-catalysed reaction can be determined using the enzyme activity results collected at different temperatures through experiments carried out to identify its optimal temperature, as already shown shown in Figure 2.2.
2. Initially, the substrate is subjected to proper conditions of pH and concentration according to the chosen methodology.
3. Tests of biocatalyst activity are conducted under the conditions described by the chosen methodology by incubating it in the substrate solution. The reaction must be carried out at different temperatures.
4. As a result, a graph like that of Figure 2.2 is obtained, which generally contains an upward and a downward curve.

2.4.1.2 Practical Example

Case Study 1

To get a solid idea of how to calculate the thermodynamic parameters of the reaction and biocatalyst inactivation, the following case study is provided.

Collagenases are enzymes that degrade collagen and have several industrial applications, both in the pharmaceutical industry, facilitating wound healing or debridement, etc., and in the food industry as a meat tenderizer (Alipour et al., 2016). Table 2.1 lists the results of collagen hydrolysis by a fungal collagenase expressed as starting collagenolytic activity at different temperatures. The choice to specifically work out only the values of the initial enzyme activity at each temperature instead of any activity is an attempt to make negligible any interference of the time-dependent thermal inactivation of the enzyme. It is noteworthy, however, that there is no rule for establishing a standard time after which such a starting activity should be determined, because it depends on the time the reaction takes to reach equilibrium. Therefore, its selection should be made on the specific kinetic characteristics of the reaction under consideration.

The data of Table 2.1 were used to build a graph of ln A versus $1/T$ composed of two straight lines (Figure 2.5), from whose slopes E_a and ΔH_u^o were calculated using Equation 2.3 and the logarithmic form of the Arrhenius-type Equation 2.6, respectively.

Note that the ascending straight line on the left of Figure 2.5 corresponds to the enzyme unfolding, prevalent at high temperatures, while the declining one on the right to the enzyme-catalysed reaction, which is practically free from interference from thermal inactivation at low temperatures. The two straight lines intersect at the optimum temperature for the reaction (40°C), and, in both cases, the closer the R^2 is to 1, the better the linear fit.

At temperatures lower than the optimum (20°C≤ T≤40°C), i.e. referring to the right portion of Figure 2.5, the unfolding equilibrium can be considered shifted to the left and then having a negligible influence on the enzyme activity. Under these conditions, the linear Arrhenius-type Equation 2.3 can be used:

$$\ln A = -\frac{E_a}{R}.\frac{1}{T} + \ln a'$$

$$y = -8{,}050 \cdot x + 30.478$$

$$\frac{E_a}{8.314\,\text{J/K/mol}} = 8{,}050\,\text{K}$$

$$\boldsymbol{E_a = 66{,}927.7\ J/mol = 66.93\,kJ/mol}$$

At temperatures higher than the optimum (40°C ≤T≤60°C), i.e. referring to the left portion of Figure 2.5, the unfolding

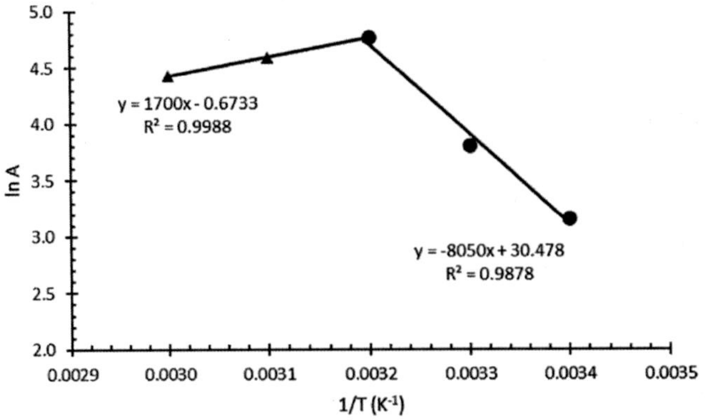

FIGURE 2.5 Arrhenius-type semi-log plot of the initial activity of a fungal collagenase versus the reciprocal absolute temperature.

equilibrium can be considered shifted to the right, thus affecting the enzyme activity. Under these conditions, the logarithmic form of Equation 2.6 can be used:

$$\ln A = \frac{\Delta H_u^o - E_a}{R} \cdot \frac{1}{T} + \ln \frac{a'}{b}$$

$$y = 1,700 \cdot x + 0.6733$$

$$\frac{\Delta H_u^o - E_a}{8.314\ \text{J/K/mol}} = 1,700\,\text{K}$$

$$\Delta H_u^o - E_a = 1,700\,\text{K} \cdot 8.314\ \text{J/K/mol} = 14,133.8\,\text{J/mol}$$

$$\Delta H_u^o = E_a + 14,133.8\,\text{J/mol} = (66,927.7 + 14,133.8)\,\text{J/mol}$$

$$\mathbf{\Delta H_u^o = 81,061.5\ J/mol = 81.06\ kJ/mol}$$

Small values of E_a are desired for a biocatalyst, because the energy required to trigger the reaction it catalyses is less. Contrariwise, higher values of ΔH_u^o are desirable because are indicative of a less favourable unfolding equilibrium of the biocatalyst and then of better performance of the system.

In a previous unpublished kinetic study (Silva and Porto, unpublished data), k_{cat} was found to be 2.9/s at 40°C (313.15 K). Then the other thermodynamic parameters can be estimated as follows:

a. Gibbs free energy (Equation 2.9a):

$$\Delta G = -RT \ln\left(\frac{k_{cat} h}{k_B T}\right)$$

$$\Delta G = -8.314\,\text{J/K/mol} \cdot 313.15\,\text{K} \cdot \ln$$

$$\left(\frac{2.9/\text{s} \cdot 6.626 \times 10^{-34}\,\text{Js}}{1.38 \times 10^{-23}\,\text{J/K} \cdot 313.15\,\text{K}}\right)$$

$$\mathbf{\Delta G = 74.05\,kJ/mol}$$

It should be noted that this parameter must be referred to a selected reference temperature, usually the optimum temperature, which in this specific case is 40°C, i.e. 313.15 K.

As is known, the Gibbs free energy reflects the spontaneity of a process, in the sense that the smaller the ΔG value, the more spontaneous the process.

b. Enthalpy (Equation 2.10):

$$\Delta H = E_a - RT$$

$$\Delta H = 66,927.7\,\text{J/mol} - (8.314\,\text{J/K/mol} \cdot 313.15\,\text{K})$$

$$\mathbf{\Delta H = 64.32\,kJ/mol}$$

Enthalpy is correlated with the efficiency of the E–S complex formation, in that, the lower its value, the more effective its formation.

c. Entropy (Equation 2.11):

$$\Delta S = \frac{\Delta H - \Delta G}{T}$$

$$\Delta S = \frac{64.32 - 74.05\,\text{kJ/mol}}{313.15\,\text{K}}$$

$$\mathbf{\Delta S = -31.06\,J/K/mol}$$

ΔS is correlated to the order (rigidity) degree of the E–S activated complex. The negative value of this parameter indicates that the structure of E–S complex in the transition state is more ordered than that of the reacting system composed of free enzyme and substrate.

2.4.2 Thermodynamics of Biocatalyst Thermal Denaturation

2.4.2.1 Experimental Procedure

1. Incubate the enzyme solution without substrate at a given temperature.
2. Withdraw samples at reasonable time intervals so that the number of samples is not less than four to provide a valid estimate. Usually, 60 min time intervals and a total duration of 240 min are selected, but for especially thermosensitive or thermostable enzymes the time intervals can be reduced or increased, respectively.

3. Determine the enzymatic activity in all the withdrawn samples.
4. Prepare a semi-log plot of the ratio between the residual enzyme activity after a given time (A) and the initial one (A_0) over time.
5. Draw the straight line that best fits the experimental data of $\ln(A/A_0)$ versus time, imposing $\ln(A/A_0)=0$ at the beginning ($t=0$) and consider $R^2>0.9$ as acceptable.
6. Estimate the first order thermal inactivation constant (k_d) from the slope of the above straight-line using Equation 2.12b.
7. Repeat the experiments in exactly the same way at different temperatures and estimate the k_d value at each temperature.
8. Plot a semi-log graph of $\ln k_d$ versus $1/T$ and determine the value of the activation energy of the biocatalyst thermo-inactivation (E_d) using Arrhenius-type Equation 2.14, considering $R^2>0.9$ as acceptable.
9. Calculate the other thermodynamic parameters ΔG_d, ΔH_d and ΔS_d using Equations 2.13, 2.15 and 2.16, respectively.
10. Calculate the complementary kinetic parameters $t_{1/2}$ and D using Equations 2.17b and 2.18b, respectively. Estimate the value of Z from the slope of the straight line obtained plotting log D versus temperature (°C) according to Equation 2.19b.

2.4.2.2 Practical Example

Case Study 2

Fibrinolytic proteases are enzymes that hydrolyse fibrin, a major component of thrombus. The formation of fibrin clots is usually regulated by the coagulation cascade; but when fibrin is not hydrolysed, it accumulates in blood vessels and usually causes thrombosis, leading to myocardial infarction and other cardiovascular disorders (Royston, 2013; Da Silva et al., 2020). These enzymes can be obtained by plants, animals, mushrooms, snake venom and microbial sources. High fibrinolytic activities have also been found in filamentous fungi, which have the peculiar advantage of producing them in large quantities mainly extracellularly, thereby making their extraction/purification easier (Nascimento et al., 2016; Shirasaka et al., 2012). Thus, the objective of this case study is to show the calculation of the thermodynamic (E_d, ΔH_d, ΔS_d and ΔG_d) and kinetic ($t_{1/2}$, D and Z) parameters related to the time-dependent thermal inactivation (denaturation) of a crude fungal fibrinolytic protease taken as a biocatalyst example. For this purpose, experiments were performed as described in Section 2.4.2.1 at temperatures of 37°C, 45°C, 55°C and 65°C by sampling every 30 min for an overall incubation time of 180 min.

A kinetic and thermodynamic study is usually based on a large number of experimental data, whose careful organization is essential to avoid errors. First, the experimental data of the residual protease activity determined at different temperatures and after different incubation times have been gathered in Table 2.2 either as such or expressed as $\ln(A/A_0)$. The latter values have then been plotted as a function of the incubation time, allowing to obtain, for each temperature, a straight line having as its origin the point of intersection of the Cartesian axes [$\ln(A/A_0)=0$, $t=0$], from whose slope the corresponding value of k_d was estimated (Figure 2.6).

TABLE 2.2
Results of Residual Activity (A) of a Fungal Fibrinolytic Protease at Different Temperatures (T) and Incubation Times (t)

T (°C)	t (min)	A (U/mL)	$\ln(A/A_0)$
37	0	14.80	0
	30	8.20	−0.590
	60	5.53	−0.984
	90	4.70	−1.147
	120	4.30	−1.236
	150	3.00	−1.596
	180	2.07	−1.968
45	0	14.80	0
	30	7.73	−0.649
	60	5.47	−0.996
	90	3.00	−1.596
	120	2.50	−1.778
	150	2.13	−1.936
	180	1.40	−2.358
55	0	14.80	0
	30	5.00	−1.085
	60	3.30	−1.500
	90	1.60	−2.224
	120	1.20	−2.512
	150	0.80	−2.918
	180	0.50	−3.387
65	0	14.80	0
	30	4.10	−1.283
	60	1.70	−2.164
	90	1.20	−2.512
	120	0.74	−2.995
	150	0.30	−3.898
	180	0.12	−4.814

U (μmol/min) indicates the units of enzymatic activity. 1 U is defined, in this case, as the amount of protease responsible for a 0.1 increase per hour in the absorbance.
A_0, starting value of A.

One can see in Figure 2.6 that the values of k_d, expressed in min^{-1} and listed in Table 2.3, were estimated with excellent correlation being R^2 always>0.98. Moreover, they progressively increased with temperature, which means that the biocatalyst time-dependent denaturation became increasingly rapid.

To estimate the activation energy of thermal denaturation of the biocatalyst (E_d), the above values of k_d determined at different temperatures were plotted in the Arrhenius type-plot of $\ln k_d$ versus $1/T$ illustrated in Figure 2.7.

The correlation between experimental data of k_d and Arrhenius-type Equation 2.14 was excellent, being the value of $R^2=0.9963$. The following mathematical steps show how

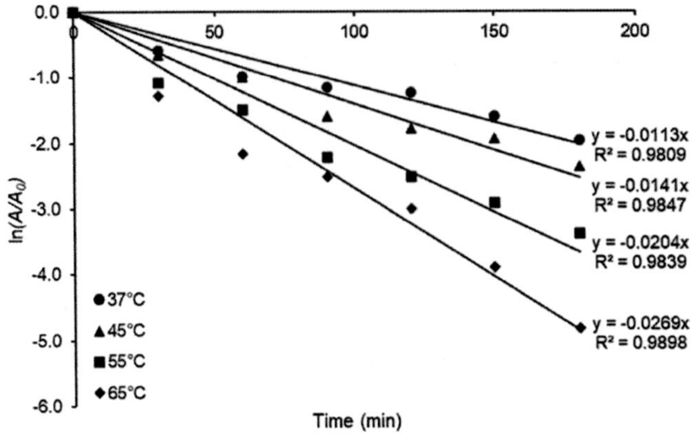

FIGURE 2.6 Semi-log plot of the ratio between the residual activity of a fungal fibrinolytic protease (A) and the initial one (A_0) versus time at different temperatures.

TABLE 2.3
First-Order Rate Constant of Thermal Inactivation (k_d) Determined Experimentally for a Fungal Fibrinolytic Protease at Different Temperatures

T (°C)	T (K)	$1/T$ (K^{-1})	k_d (min^{-1})	ln k_d
37	310.15	0.00322	0.0113	–4.4847
45	318.15	0.00314	0.0141	–4.2616
55	328.15	0.00305	0.0204	–3.8922
65	338.15	0.00296	0.0269	–3.6164

the activation energy of thermal inactivation of the biocatalyst (E_d) was calculated:

$$\ln k_d = -\frac{E_d}{R}\cdot\frac{1}{T} + \ln a$$

$$y = -3{,}319.4\cdot x + 6.2037$$

$$\frac{E_d}{8.314\,\text{J/K/mol}} = 3{,}319.4\ \text{K}$$

$$\boldsymbol{E_d = 27{,}597.49\,J/mol = 27.60\,kJ/mol}$$

Substituting the values of k_d and E_d in Equations 2.13, 2.15 and 2.16, ΔG_d, ΔH_d and ΔS_d were calculated at all tested temperatures, whose values are listed in Table 2.4.

For illustrative purposes, the procedure followed to calculate these thermodynamic quantities at a reference temperature of 37°C (310.15 K) is presented below:

a. Gibbs free energy (Equation 2.13)

$$\Delta G_d = -RT\ln\left(\frac{k_d h}{k_B T}\right)$$

$$\Delta G_d = -8.314\,\text{J/K/mol}\cdot 310.15\ \text{K}\cdot\ln$$

$$\left(\frac{1.88\times10^{-4}/\text{s}\cdot 6.626\times10^{-34}\ \text{J s}}{1.38\times10^{-23}\ \text{J/K}\cdot 310.15\ \text{K}}\right)$$

$$\boldsymbol{\Delta G_d = 98{,}176.81\ J/mol = 98.18\ kJ/mol}$$

Please note that the value of k_d (0.0113 min^{-1}) is here expressed in s^{-1} (1.88×10^{-4}s^{-1}) so that its unit is consistent with that of the Plank constant (J s). Obviously, the opposite could be done, that is to convert the unit of the Planck constant from J s to J min and use in this equation k_d expressed in min^{-1}.

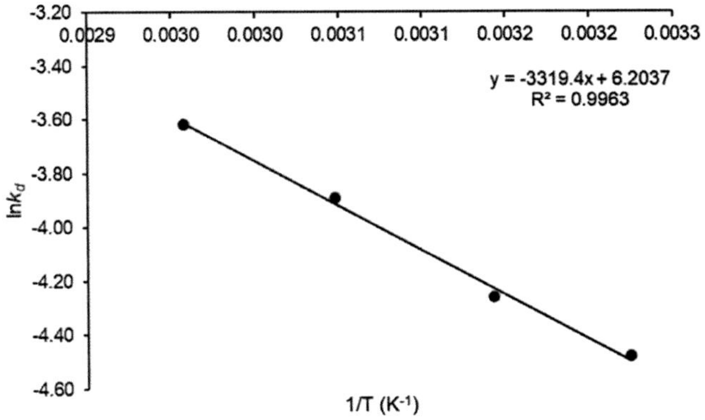

FIGURE 2.7 Arrhenius-type plot used to estimate the activation energy of thermal inactivation (E_d) of a fungal fibrinolytic protease.

TABLE 2.4
Thermodynamic Parameters for Fibrinolytic Protease Thermal Denaturation

Temperature (°C)	ΔH_d (kJ/mol)	ΔS_d (J/K/mol)	ΔG_d (kJ/mol)
37	25.02	−235.88	98.18
45	24.95	−236.49	100.19
55	24.87	−236.32	102.42
65	24.79	−236.76	104.85

TABLE 2.5
Complementary Kinetic Parameters ($t_{1/2}$ and D) of Thermal Denaturation of Fungal Fibrinolytic Protease

Temperature (°C)	$t_{1/2}$ (min)	D (min)	log D
37	61.34	203.77	2.31
45	49.16	163.30	2.21
55	33.98	112.87	2.05
65	25.77	85.60	1.93

b. Enthalpy (Equation 2.15)

$$\Delta H_d = E_d - RT$$

$$\Delta H_d = 27{,}597.49 \text{ J/mol} - \left(8.314 \text{ J/K/mol} \cdot 310.15 \text{ K}\right)$$

$$\boldsymbol{\Delta H_d = 25{,}018.90 \; J/mol = 25.02 \; kJ/mol}$$

c. Entropy (Equation 2.16)

$$\Delta S_d = \frac{\Delta H_d - \Delta G_d}{T}$$

$$\Delta S_d = \frac{25{,}018.90 - 98{,}176.81 \text{ J/mol}}{310.15 \text{ K}}$$

$$\boldsymbol{\Delta S_d = -235.88 \; J/K/mol}$$

E_d, which is the minimum energy amount required for the thermal denaturation process, in the specific case of the fibrinolytic protease employed in the current case study was 27.60 kJ/mol. This parameter is related to ΔH_d, which corresponds to the total amount of thermal energy required to denature the enzyme and is given by the number of non-covalent bonds broken during the denaturation process. According to Pace (1992), the energy required to remove a -CH_2 moiety from a hydrophobic or hydrogen bond is approximately 5.4 kJ/mol. Therefore, considering the low ΔH_d values obtained in this study ($24.79 \leq \Delta H_d \leq 25.02$ kJ/mol), it is possible to estimate that the formation of the transition state that led to the denaturation of the protease under examination, on average, involved the breaking of only about 4.6 of these bonds. The disruption of enzyme structure is often also accompanied by an increase in disorder or randomness, i.e. in entropy (ΔS_d). However, the negative values ($-236.76 \leq \Delta S_d \leq -235.88$ J/K/mol) obtained in this case study point to slightly less disorder of the transition state compared to the native enzyme, which suggests possible aggregation of enzyme molecules. Some authors explained such a compaction of enzyme molecule as the result of the formation of charged particles around them or of the ordering action of solvent molecules (Gummadi and Panda, 2003). Finally, ΔG_d, which indicates the spontaneity of a reaction, is the most accurate parameter to evaluate the phenomenon of thermal denaturation of a biocatalyst because it incorporates both the enthalpy and entropy contributions. High positive values of ΔG_d like those obtained in this study (98.18–104.85 kJ/mol) are typical of non-spontaneous reactions, which, in the case of thermal inactivation (denaturation), corresponds to high thermal stability of the biocatalyst.

Equations 2.17b and 2.18b were used to calculate the complementary kinetic parameters $t_{1/2}$ and D, respectively, whose values are summarized in Table 2.5. For illustrative purposes, the calculation of these kinetic parameters at a reference temperature of 37°C (310.15 K) follows.

d. Half life (Equation 2.17b)

$$t_{1/2} = \frac{\ln 2}{k_d}$$

$$t_{1/2} = \frac{\ln 2}{0.0113 \text{min}^{-1}} = \mathbf{61.34 \; min}$$

e. Time of decimal reduction (Equation 2.18b)

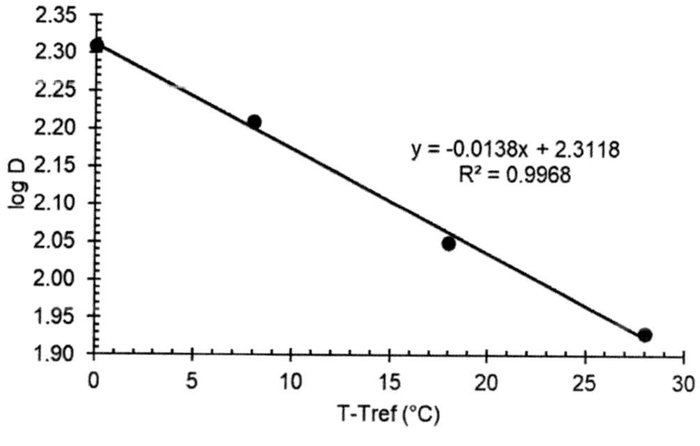

FIGURE 2.8 Thermal-death-time curve for the estimation of the Z value.

$$D = \frac{\ln 10}{k_d}$$

$$D = \frac{\ln 10}{0.0113\,\text{min}^{-1}} = \mathbf{203.77\ min}$$

A so-called thermal-death-time curve was obtained by plotting log D versus temperature (°C) (Figure 2.8). From the slope of the resulting straight line, a Z value of 73°C was calculated using Equation 2.19b. As can be seen, the value of the intercept on the ordinate axis (log D_{ref}=2.31) corresponds to the value of log D obtained at the reference temperature (37°C) (Table 2.5). In general, high Z values indicate high sensitivity to the duration of heat treatment, while low Z values high sensitivity to the temperature rise (Tayefi-Nasrabadi and Asadpour, 2008).

2.5 PHYSICAL AND CHEMICAL FACTORS INFLUENCING THERMODYNAMIC PARAMETERS

So far, the influence of temperature on enzyme-catalysed reactions and activity of other bioactive molecules has been discussed, as well as the thermal inactivation process and respective thermodynamic parameters. Before proceeding to analyse the influence of other physical and chemical variables, we must explain in greater detail what happens at molecular level to the biocatalyst as the temperature changes.

2.5.1 Temperature

At low temperatures, a biocatalyst exhibits high stability as its molecules are subject to low vibration levels. In such a low stage of excitation, its structure is relatively rigid and stable; therefore, the active site is less willing to react with the substrate. As the temperature rises, vibrations become more pronounced, and some non-covalent bonds, e.g. hydrogen bonds, Van der Waals forces, dipole-dipole and hydrophobic interactions, are broken. Hence, the enzyme structure becomes more flexible and undergoes conformational changes that make the active site more available to react. This trend is the one already shown in the ascending curve on the left of Figure 2.2 until the optimum temperature is reached, at which the biocatalyst activity achieves its maximum value. A further increase in temperature beyond the optimum value increases the vibration level and the number of broken non-covalent bonds to such an extent that the three-dimensional structure of the enzyme is compromised. Under these conditions, the enzyme activity is described by the declining curve on the right of Figure 2.2, which indicates that the higher the temperature, the more pronounced is enzyme unfolding.

Even though this cannot be considered a general rule, an increase in the rigidity of a biocatalyst induced by thermostability is often accompanied by a decrease in activity and vice versa (Porto et al., 2006). An optimal compromise should therefore be sought between opposing requirements of high activity and high thermostability to select the best conditions for biotechnological applications, especially in the pharmaceutical field.

The progressive biocatalyst thermo-inactivation over time can be described by the classical general scheme:

$$\mathrm{N} \underset{k_2'}{\overset{k_1'}{\rightleftharpoons}} \mathrm{U} \xrightarrow{k_3'} \mathrm{D} \qquad (2.20)$$

where N is the native biomolecule in its full-active state, U its reversibly unfolded state and D its final irreversible denatured state. In an intermediate range of temperature, some thermotolerant biocatalysts have a high k_2' value that shifts the equilibrium to the left. Therefore, in case of cooling, refolding occurs spontaneously.

At high temperatures, the above model can be reduced into a single step model (N → D), assuming that k_1', k_2', and k_3' are all first-order kinetic constants and $k_3' \gg k_2'$. In this case, practically no equilibrium between N and U is established during denaturation. Therefore, U molecules formed are preferentially converted to D instead of refolding to form N. In this case, the concentration of U becomes negligible, and denaturation, which depends only on k_1', becomes practically irreversible.

These behaviours are clearly revealed by ΔG_d, which indicates how much the original native conformation is still preserved or active. In most enzymes, the D form becomes predominant with increasing temperature, and ΔG_d exhibits decreasing values. Some other enzymes show increasing values of ΔG_d with increasing temperature, which indicates that they are resistant to a temperature increase, i.e. thermostable. Such a behaviour was observed for fibrinolytic protease thermal denaturation described in Case study 2 (Table 2.4).

2.5.2 pH

The pH, which expresses the $-\log_{10}$ of concentration of protons (H^+) in the reaction medium, is one of the chemical parameters that most influence the biocatalyst activity. Particularly, it influences the dielectric constant which is directly correlated with the polarization of biocatalyst molecules. In a solvent, biomolecules such as proteins are in fact subject to specific interactions that influence their dielectric relaxation mechanisms. In polar environments, there is a reorganization of solvent polar molecules along the electric field lines generated by dipoles present on the biomolecule surface. For example, proteins are made up of amino acids, whose side chains can carry functional amino or carboxyl groups that can undergo ionization. In solution, the biocatalyst surface becomes more negatively or positively charged depending on pH; hence, due to conformational changes, the active site may become more or less solvent exposed, which explains why a biocatalyst shows optimal performance at a specific pH value.

These phenomena directly affect all thermodynamic parameters. Gibbs free energy of the reaction can indicate in which pH range the biocatalyst is more prone to spontaneously catalyse the reaction. On the other hand, enthalpy can suggest the pH range in which the E–S complex is most efficiently formed. Finally, the negative or positive sign of entropy at a given pH can suggest if the E–S complex in the transition state is more or less ordered than the reacting system.

2.5.3 Substrate Concentration

As explained earlier, the rate of any biocatalyst-catalysed reaction depends on the number of effective collisions between biocatalyst and substrate molecules. Although at lower concentrations the molecules have more freedom of movement, the probability of effective collision decreases, and biocatalyst activity is reduced. Such a behaviour is explained in more detail in Chapter 15 devoted to enzyme kinetics. The higher the substrate concentration, the higher the reaction rate; however, under conditions of substrate saturation, the biological activity remains almost the same as the substrate concentration is significantly higher than that of the biocatalyst. Usually, an increase in substrate concentration leads to a higher ΔH value due to higher activation energy. This can also happen because the higher the concentration, the higher the viscosity of the medium and the less mobile the molecules. This being the case, more energy is needed to overcome the energy barrier of the reaction. An opposite trend is usually observed for ΔS, in that, the lower the concentration, the higher the mobility of molecules and then the higher the disorder degree. Since these opposite enthalpic and entropic effects offset each other, ΔG does not undergo any variation as function of substrate concentration, being a parameter depending only on the temperature.

2.5.4 Techniques of Biomolecule Improvement

The natural biological properties of a biomolecule can be modified to improve its performance in a bioprocess. In the specific case of enzymes, one of the most frequent problems is their poor long-term operational stability. To get around this problem, stabilizing additives or techniques such as immobilization and chemical modification can be used, which can significantly influence the thermodynamic parameters compared to the native form of the enzyme.

The immobilization technique is a viable tool for the recovery and reuse of biomolecules, mainly enzymes, in bioprocesses. Enzyme immobilization can be defined as a procedure capable of combining the kinetic, selectivity and stability properties of an enzyme with the physical and chemical properties of a carrier in a specialized formulation, with the primary objective of maximizing physical stability and activity (Basso and Serban, 2019). Since the multipoint covalent attachment of an enzyme to a support leads to greater rigidity of its structure, the improvement of structural enzyme stability depends on the immobilization method. This rigidification makes it possible to keep the relative distance among the amino acid residues unaltered during the conformational changes induced by any distorting agent (e.g. temperature), thus reducing the impact of denaturation on the structure. However, not all immobilization techniques are based on covalent bonding. For example, in the case of immobilization by entrapment within porous matrices, the increased biomolecule stability is due to its protection from interaction with other compounds or proteolytic enzymes present in the medium (as is the case of enzymatic cocktails or crude extracts), which would otherwise cause its inactivation or degradation (Mateo et al., 2007). The improvement of enzyme thermal stability can be kinetically confirmed by an increase in the so-called stabilization factor, which corresponds to the ratio between the half-life of the immobilized enzyme and that of the free enzyme at a reference temperature, or by an increase in the value of D induced by immobilization.

The structural stabilization of an enzyme induced by immobilization is accompanied not only by a reduction in the kinetic parameters related to its thermal inactivation, but also by an increase in the values of E_d and ΔH_d, which indicate that more energy is required for its denaturation. As is known, the breakdown of the enzyme structure due to thermal denaturation is often accompanied by an increase in the disorder or randomness degree of the system, i.e. by a large positive value of ΔS_d. As already observed for some enzymes (Dhiman et al., 2020; Karam et al., 2017; Oliveira et al., 2020; Silva et al., 2018), immobilization can also lead to a decrease in the ΔS_d value, which indicates that it gives greater order and integrity to the enzyme, as well as greater stability induced by inter and/or intramolecular forces (Pal and Khanum, 2011). The immobilized form of the enzyme is generally characterized by higher ΔG_d values than the free enzyme, which highlights a lower spontaneity of the thermal denaturation process, i.e. a greater thermostability induced by immobilization.

Another way to improve the thermostability of biomolecules, particularly proteins, is their chemical modification by conjugation with polymers, usually polysaccharides or polyols through electrostatic interactions (Wehaidy et al., 2018). For instance, Santos et al. (2019) reported the efficiency of the PEGylation process in improving the thermal stability of cytochrome c, a heme protein involved in the mitochondrial electron transport chain with bioelectrochemical applications involving hydrogen peroxide. The authors observed an increase in the values of E_d, ΔH_d and ΔG_d; as well as a decrease in those of ΔS_d, a behaviour that confirms improved thermostability. However, such an improvement does not always occur. For example, Abdel-Naby, Ibrahim and El-Refai (2016) observed that the conjugation of a bacterial keratinase with activated pectin led to a conformational shift of the enzyme towards a partially unfolded transition state.

2.6 FINAL CONSIDERATIONS

Knowledge of thermodynamic quantities is very important for a deeper understanding of the physicochemical or biochemical features of biomolecules with biological activity. In fact, they allow us to predict not only stability but also how a biomolecule will behave during its production or during storage following variations in temperature, pH or medium composition. Knowing these characteristics, it is possible to select the optimal conditions for the biomolecule production process or its use in biotechnological processes.

It is possible that, throughout this chapter, the presentation of thermodynamic equations may have initially seemed complex, but it is hoped that the case studies, thanks to the step-by-step detailing of the experimental protocols, have facilitated understanding of their application as well as interpretation of the data obtained.

Although the application of thermodynamic modelling of biomolecules with biological activity has been increasingly explored in recent years in the biotechnology area, most of the reported studies are restricted to a few categories of biomolecules such as antibiotics, pigments and enzymes. However, it is important to remember that a wide variety of other biomolecules of pharmaceutical interest can be subjected to an approach similar to that presented in this chapter. Thus, an extension of thermodynamic concepts to a broader spectrum is expected to contribute to a better understanding of the behaviour and application of biomolecules in general.

REFERENCES

Abdel-Naby, M. A.; Ibrahim, M. H. A; El-Refai, H. A. Catalytic, kinetic and thermodynamic properties of *Bacillus pumilus* FH9 keratinase conjugated with activated pectin. *International Journal of Biological Macromolecules*, v. 85, 238–245, 2016.

Alberty, R. A. Fundamental equation of thermodynamics for protein–ligand binding. *Biophysical Chemistry*, v. 104, n. 3, 543–559, 2003b.

Alberty, R. A. *Thermodynamics of Biochemical Reactions*. 1. ed. Hoboken, NJ: John Wiley & Sons, Inc., 2003a.

Alberty, R. A. *Biochemical Thermodynamics*. 1. ed. Hoboken, NJ: John Wiley & Sons, Inc., 2006a.

Alberty, R. A. Relations between biochemical thermodynamics and biochemical kinetics. *Biophysical Chemistry*, v. 124, n. 1, 11–17, 2006b.

Alberty, R. A. et al. Recommendations for terminology and databases for biochemical thermodynamics. *Biophysical Chemistry*, v. 155, n. 2–3, 89–103, 2011.

Alipour, H. et al. Therapeutic applications of collagenase (metalloproteases): A review. *Asian Pacific Journal of Tropical Biomedicine*, v. 6, n. 11, 975–981, 2016.

Basso, A.; Serban, S. Industrial applications of immobilized enzymes—A review. *Molecular Catalysis*, v. 479, 110607, 2019.

Da Silva, A. V. et al. Partial purification of fibrinolytic and fibrinogenolytic protease from *Gliricidia sepium* seeds by aqueous two-phase system. *Biocatalysis and Agricultural Biotechnology*, v. 27, 101669, 2020.

Dhiman, S. et al. Immobilization of mannanase on sodium alginate-grafted-β-cyclodextrin: An easy and cost-effective approach for the improvement of enzyme properties. *International Journal of Biological Macromolecules*, v. 156, 1347–1358, 2020.

Eyring, .H. The activated complex in chemical reactions. *The Journal of Chemical Physics*, v. 3, n. 2, 107–115, 1935.

Gummadi, S. N.; Panda, T. Purification and biochemical properties of microbial pectinases - A review. *Process Biochemistry*, v. 38, 987–996, 2003.

Jones, M. N. (Ed.) *Biochemical Thermodynamics*. 1. ed. Amsterdam, Netherlands: Elsevier, 1979.

Karam, E. A. et al. Production, immobilization and thermodynamic studies of free and immobilized *Aspergillus awamori* amylase. *International Journal of Biological Macromolecules*, v. 102, 694–703, 2017.

Krebs, H. A.; Kornberg, H. L. *Energy Transformations in Living Matter*. Berlin, Germany: Springer-Verlag, 1957.

Mafart, et al. On calculating sterility in thermal preservation methods: application of the Weibull frequency distribution model. *Acta Horticulturae*, v. 566, n. 566, 107–114, 2002.

Marques, D. A. V. et al. Kinetic and thermodynamic investigation on clavulanic acid formation and degradation during glycerol fermentation by *Streptomyces* DAUFPE 3060. *Enzyme and Microbial Technology*, v. 45, n. 2, 169–173, 2009.

Mateo, C. et al. Improvement of enzyme activity, stability and selectivity via immobilization techniques. *Enzyme and Microbial Technology*, v. 40, n. 6, 1451–1463, 2007.

Michel, D. Simply conceiving the Arrhenius law and absolute kinetic constants using the geometric distribution. *Physica A: Statistical Mechanics and its Applications*, v. 392, n. 19, 4258–4264, 2013.

Nascimento, T. P. et al. Purification of a fibrinolytic protease from *Mucor subtilissimus* UCP 1262 by aqueous two-phase systems (PEG/sulfate). *Journal of Chromatography B: Analytical Technologies in the Biomedical and Life Sciences*, v. 1025, n. 1, 16–24, 2016.

Nelson, D. L.; Cox, M. M. *Lehninger Principles of Biochemistry*. 6. ed. New York: W. H. Freeman, 2012.

Oliveira, R. L. et al. Immobilization of a commercial *Aspergillus aculeatus* enzyme preparation with fructosyltransferase activity in chitosan beads: A kinetic/thermodynamic study and fructo-oligosaccharides continuous production in enzymatic reactor. *Food and Bioproducts Processing*, v. 122, 169–182, 2020.

Pace, C. N. Contribution of the hydrophobic effect to globular protein stability. *Journal of Molecular Biology*, v. 226, n. 1, 29–35, 1992.

Pal, A.; Khanum, F. Covalent immobilization of xylanase on glutaraldehyde activated alginate beads using response surface methodology: Characterization of immobilized enzyme. *Process Biochemistry*, v. 46, n. 6, 1315–1322, June 2011.

Porto, T. S. et al. Kinetic and thermodynamic investigation on ascorbate oxidase activity and stability of a *Cucurbita maxima* extract. *Biotechnology Progress*, v. 22, n. 6, 1637–1642, 2006.

Royston, D. Anticoagulant and antiplatelet therapy. In: Hemmings Jr, H. C.; Egan, T. D. (Eds.). *Pharmacology and Physiology for Anesthesia*. 1. ed. Philadelphia: Elsevier, 643–667, 2013.

Santos, J. H. P. M. et al. PEGylation as an efficient tool to enhance cytochrome c thermostability: a kinetic and thermodynamic study. *Journal of Materials Chemistry B*, v. 7, n. 28, 4432–4439, 2019.

Shirasaka, N. et al. Purification and characterization of a fibrinolytic protease from *Aspergillus oryzae* KSK-3. *Mycoscience*, v. 53, n. 5, 354–364, 2012.

Silva, J. C.; Porto, T. S. Desenvolvimento de processo biotecnológico para produção e purificação de pectinases produzidas por fungos filamentosos para aplicação na clarificação do suco de frutas regionais. (unpublished data).

Silva, J. C. et al. Kinetic and thermodynamic characterization of a novel *Aspergillus aculeatus* URM4953 polygalacturonase. Comparison of free and calcium alginate-immobilized enzyme. *Process Biochemistry*, v. 74, 61–70, 2018.

Tayefi-Nasrabadi, H.; Asadpour, R. Effect of heat treatment on buffalo *(Bubalus bubalis)* lactoperoxidase activity in raw milk. *Journal of Biological Sciences*, v. 8, n. 8, 1310–1315, 2008.

Wehaidy, H. R. et al. Improving the catalytic, kinetic and thermodynamic properties of *Bacillus subtilis* KU710517 milk clotting enzyme via conjugation with polyethylene glycol. *International Journal of Biological Macromolecules*, v. 111, 296–301, 2018.

3 Expression Systems for the Production of Therapeutic Recombinant Proteins

Amanda Mizukami, Adalberto Pessoa Jr., and Kamilla Swiech
Universidade de São Paulo

CONTENTS

3.1 INTRODUCTION

The first proteins used for therapeutic purposes were extracted from human blood and from human or animal tissues. For example, they include clotting factors and human albumin derived from blood plasma, insulin derived from porcine pancreas and the enzyme glucocerebrosidase derived from human placenta. Developments of recombinant DNA technology in the 1970s, a topic discussed in Chapters 4 and 5, revolutionized medicine, as it allowed the production of these same proteins in genetically modified microbial cells and animals. Insulin was the first therapeutic recombinant protein produced by this technology. Recombinant DNA techniques have also allowed an increase in the safety profile of these products and the production of engineered proteins with characteristics superior to native proteins. Amongst these characteristics, proteins with a longer serum half-life and with less risk of triggering immunogenic reactions can be mentioned (Walsh, 2007).

In 1975, scientists developed hybridoma technology for the production of monoclonal antibodies (Köhler and Milstein, 1975), allowing the production of these products with high specificity, in larger scales and in a reproducible way. This technology also allowed the use of antibodies as a tool for characterizing and purifying proteins with high specificity, which in turn was essential in the development of production processes for recombinant proteins with greater complexity for therapeutic purposes. Therapeutic recombinant proteins and monoclonal antibodies belong to a category of drugs known as biopharmaceuticals or biologics, as well as products used in gene therapy based on nucleic acids (DNA and RNA) and products used in cell therapy based on genetically modified cells and tissues (Walsh, 2018).

Biopharmaceuticals are some of the highest profit grossing drugs developed by the pharmaceutical industry today. Amongst the products approved for therapeutic use from 2014 to 2018, there were 68 monoclonal antibodies, 23 hormones, 16 clotting factors, 9 enzymes and 7 vaccines (Walsh, 2018; Tripathi and Shrivastava, 2019). The biopharmaceutical market is in exponential growth and is forecast to increase by 8.7% in the period 2018–2025 reaching a value of US $446 billion. This trend is in part due to an increase in the elderly population, a better understanding of how to treat chronic diseases such as cancer and diabetes, and emerging diseases such as HIV/AIDS. Biopharmaceuticals offer interesting advantages when compared to drugs of synthetic origin: potent and effective action, minor side effects and potential to cure diseases instead of simply treating symptoms. The global demand for these products is led by monoclonal antibodies (mAb), which reached US $123 billion in sales in 2018 and are projected to grow by 10.8% (2018–2025) (Biopharmaceuticals, 2019).

3.2 DIFFERENCES BETWEEN SYNTHETIC DRUGS AND BIOPHARMACEUTICALS BASED ON RECOMBINANT PROTEINS

The vast majority of synthetic drugs have a molecular mass of approximately 500 Da (often less than 1,000 Da). Due to the small size, any chemical modification can dramatically alter the pharmacological activity and potentially lead to a new drug for a new application. Conversely, recombinant biopharmaceuticals usually share the same biological function as the natural protein or peptide, despite differences in amino acid primary structure amino acid residues. That is,

TABLE 3.1
Main Differences between Drugs of Synthetic Origin and Biopharmaceuticals Based on Recombinant Proteins

Synthetic Drugs	Biopharmaceuticals
Produced by chemical synthesis	Produced by biotechnological processes
Low molecular weight (less than 1,000 Da)	High molecular weight (up to 150 kDa)
Well-defined structure	Heterogeneous and complex structure
Most of the time it is process-independent	Heavily dependent on the process
Possible to be fully characterized	It is difficult to fully characterize the molecular structure and heterogeneity
Low susceptibility to contamination during production	High susceptibility to contamination during production
Stable	Unstable and sensitive to external conditions
Majority are non-immunogenic	May be immunogenic MW >10 kDa
They can be administered in several ways including oral	Mostly administered via parenteral route
Mechanism of action well elucidated	Complex mechanism of action, in many cases not fully understood
In the body, they may be give rise to pharmacologically and toxicologically active metabolites	In the body, they are degraded into non-active compounds
Often associated with side effects	Due to the specific action, it is less likely to cause side effects
Limited clinical study	Extensive clinical studies
Relatively low cost	High cost (approximately 22 times higher than synthetic drugs)

Source: Adapted from Sekhon, B.S., Saluja, V., *Biosimilars*, 1, 1–11, 2011; Kubo, T., Random peptide library for ligand and drug discovery, in: Gopalakrishnakone, P., Cruz, L., Luo, S. (Eds.), *Toxins and Drug Discovery*, Springer, Dordrecht, 207–230, 2017.

a small difference does not result in a new therapeutic product. Table 3.1 lists other main differences between drugs of synthetic origin and biopharmaceuticals based on recombinant proteins.

Protein-based biopharmaceuticals have a complex three-dimensional structure, formed by secondary, tertiary and sometimes quaternary structures. Different regions of the molecule have different functions and the biological activity depends on the correct conformation of the protein. Synthetic drugs can be completely characterized due to their well-defined molecular structure using simple analytical techniques. Defining the sequence of amino acids that forms a protein is simple; however, establishing the conformation of the protein, involving all structures (secondary, tertiary and quaternary) and its folding is complex and requires a large number of analytical techniques for correct characterization.

3.3 MAIN SYSTEMS FOR EXPRESSION OF THERAPEUTIC RECOMBINANT PROTEINS

The manufacturing process of a pharmaceutical product is one of the most regulated and strictly controlled production processes. To obtain a commercial production license, industry must demonstrate to the regulatory agencies that not only is the product safe and effective, but also that the production process is developed according to high standards of safety and quality. Choice of an expression system is the first and one of the main key aspects for efficient production of a functional and safe protein. This choice should be based on factors such as (Mizukami et al., 2018):

- Main characteristics of host cells
- Capacity to produce functional proteins (proteins with adequate post-translational modification profiles)
- Bioprocess costs
- Product effectiveness and safety
- Regulatory aspects

Recombinant proteins are traditionally produced in microbial cells (bacteria, fungi and yeasts), animal cells (mammals and insects) and plant cells. They can also be produced in transgenic animals and plants, but these systems will not be addressed here. Microbial systems are attractive due to low cost, high productivity, rapid development and lack of concern with viral contamination of both the producing cells and the final product. Despite these advantages, these systems are not suitable for the production of complex proteins that require extensive post-translational modifications, normally performed by mammalian cells. Plant cells can be considered as an alternative platform for production instead of microbial and animal cells because they are eukaryotic (but with a better protein processing capacity than mammalian cells), easily grown in high densities and rarely contaminate the product with endotoxins or other pathogens (prions and virus).

Animal cells have the ability to produce more complex proteins due to a better ability to carry out post-translational modifications, such as glycosylation. The glycosylation of a therapeutic recombinant protein plays a crucial role in its effectiveness, half-life and antigenicity (Tripathi and Shrivastava, 2019). More than half of the recombinant proteins currently approved are produced in animal cells. Despite this, they require complex cultivation conditions and have moderate yields of recombinant protein. Table 3.2 presents the main advantages and disadvantages of each expression system.

Between 2014 and 2019, 62 of the 71 biopharmaceutical products on the market were therapeutic recombinant proteins and of these 84% were produced in mammalian cells, 1 was produced in transgenic systems, 5 were produced in *E. coli* and 4 in *S. cerevisae* (Tripathi and Shrivastava, 2019).

TABLE 3.2
Main Advantages and Disadvantages of the Expression Systems Currently Used for the Production of Therapeutic Recombinant Proteins

	Advantages	Disadvantages
Bacteria	High levels of expression High level of growth Simple cultivation conditions Low susceptibility to shear stresses Low-cost culture medium Simple and well-established gene modification protocols Many parameters can be changed to optimize expression Easy scale up	Protein production in inclusion bodies (*E. coli*) Refolding protocols often inefficient Inefficient disulphide formation and bridging and protein folding Minimal post-translational modifications[a] Possibility of product contamination with endotoxins
Yeast	Good levels of expression Relatively high growth level Low susceptibility to shear stresses Simple cultivation conditions Low-cost culture medium Simple and well-established gene modification protocols Able to perform post-translational modifications Efficient protein folding Endotoxin free Easy scale up	Glycosylation profile not identical to mammals Possibility of forming immunogenic glycans Tendency to hyperglycosylate proteins
Insect cells	Good levels of expression (especially for intracellular proteins) Moderate growth level Moderately easy and well-established gene modification Able to perform post-translational modifications Efficient protein folding Endotoxin free Moderate scalability	More complex growing conditions High-cost culture medium High susceptibility to shear stresses High amounts of baculovirus required for large-scale production Viral infection leads to cell lysis and possible degradation of the protein of interest. Glycosylation profile not identical to mammals. Possible presence of insect-specific glycoforms. Many residues of mannose and paucimanose Do not perform sialylation
Mammalian cells	Good levels of expression Moderate growth level Moderately easy and well-established gene modification Proteins are secreted Able to perform post-translational modifications similar to human cells Efficient protein folding Endotoxin free Moderate scalability	Complex cultivation conditions High susceptibility to shear stresses High-cost culture medium Susceptibility of viral contamination of the process and the final product
Plant cells	Moderate growth level Simple and low-cost culture medium Low susceptibility to shear stresses Proteins are secreted Able to perform post-translational modifications Endotoxin free Easy scale up	Moderate levels of expression Glycosylation profile not identical to mammals. Possible presence of plant-specific glycoforms Do not perform sialylation

[a] Some strains of *E. coli* have already been genetically modified to perform N-glycosylation, but the production yield is still very low.

The main characteristics of the expression systems that are being used for the production of therapeutic recombinant proteins will next be addressed in more detail.

3.3.1 Bacteria

The first biopharmaceutical (recombinant protein) approved for use in humans was insulin produced in *E. coli* by the company Genentech in 1982. After this approval, many other proteins were produced in this system: interferons, growth hormones, glucagon, granulocyte colony-stimulating factor and interleukin-2.

E. coli is considered a model prokaryotic cell for genetic studies, and, therefore, its molecular biology is well characterized. This bacterium grows at high rates (generation time 1–3 h) in simple and low-cost culture media. As a result, the cultivation schedule is simple and well established, capable of expressing the protein of interest at high levels, in

TABLE 3.3
Examples of Therapeutic Recombinant Proteins Produced in Bacteria (Jozala et al., 2016; Park et al., 2011)

Biopharmaceutical	Clinical Use	Bacteria	Approval Date
Interferon-α2a and b (Roferon-A)	Chronic hepatitis C, chronic myelogenous leukaemia, hairy cell leukaemia, Kaposi's sarcoma	*E. coli*	1986
Aldesleukin (Proleukin)	Melanoma and renal cancer treatment	*E. coli*	1992
Oprelvekin (interleukin 11) (Neumega)	Prevention of severe thrombocytopenia (patients in chemotherapy)	*E. coli*	1997
Glucagon	Hypoglycaemia	*E. coli*	1998
Denileukin diftitox (interleukin-2 and diphtheria toxin fusioned) (Ontak)	T-cell lymphoma treatment	*E. coli*	1999
Tasonermin (cytokine) (Beromun)	Soft sarcoma treatment	*E. coli*	1999
Anakinra (Kineret)	Rheumatoid arthritis treatment	*E. coli*	2001
Calcitonin (salmon calcitonin) (Fortical)	Postmenopausal osteoporosis treatment	*E. coli*	2005
Insulin (Exubera)	Diabetes mellitus treatment	*E. coli*	2006
Ranibizumab (Mab fragment) (Lucentis)	Age-related macular degeneration	*E. coli*	2006
Pegloticase (Krystexxa)	Chronic refractory gout	*E. coli*	2010
Xeomin (Incobotulinumtoxin A)	Cervical dystonia, blepharospasm	*Clostridium botulinum*	2010
Glucarpidase (bacterial carboxypeptidase G2) (Voraxaze)	Control of methotrexate concentration in patients with deficient renal function	*E. coli*	2012

some cases reaching up to 30% of the total protein content of cells. These attractive characteristics made *E. coli* the main expression system for the production of biopharmaceuticals for many years (Berlec and Strukelj, 2013; Gupta and Shukla, 2016).

E. coli cells often produce the protein of interest in intracellular inclusion bodies, which are aggregates of partially folded insoluble proteins (due to the inability to form disulphide bridges properly and the presence of a high content of hydrophobic regions). *E. coli* rarely secretes the protein of interest into the culture medium. The high density of inclusion bodies makes them easily separated from the rest of the cellular components by centrifugation. Despite this, it is necessary to solubilize these inclusion bodies (with strong denaturing agents such as urea, solvents and detergents), subsequent removal of denaturing agents (using diafiltration or dialysis, for example) and refolding of the protein of interest. These processes can be laborious, expensive and are often inefficient. More downstream steps are needed to purify the protein of interest from native proteins. Through molecular biology, the production of the protein of interest can be directed to the periplasm (a cell compartment located between the inner and outer membrane), where the formation of disulphide bridges is facilitated. This technique, however, as well as others that have been used to prevent the formation of inclusion bodies and induce soluble protein expression is not widely used on an industrial scale, with most recombinant proteins produced in *E. coli* as inclusion bodies (Berlec and Strukelj, 2013; Tripathi and Shrivastava, 2019; Gupta et al., 2019; Malekian et al., 2019).

Despite the easy and efficient production of simple recombinant proteins, this expression system is not able to adequately produce recombinant proteins that require post-translational modifications, mainly glycosylation. Some efforts are being directed to genetically modify this bacterium to be able to perform N-glycosylation. Researchers have identified a new mechanism of N-glycosylation in the bacterium *Campylobacter jejuni* and have shown successful transfer to *E. coli* cells (expression of glycosyltransferases) (Schwarz et al., 2010). However, the yields obtained for glycosylated proteins are extremely low (Jaffé et al., 2014). Another important disadvantage of the production of biopharmaceuticals in *E. coli* is the presence of lipopolysaccharides (LPS) on the cell surface. The pyrogenic nature of LPS makes removal essential during downstream processing, and processes to achieve this are successful and have become widely used in industry (Magalhães et al., 2007).

Although the vast majority of human proteins are glycosylated, the lack of glycans in some of these proteins does not compromise their biological activity. For example, the non-glycosylated form of IL-2 produced in *E. coli* has identical biological activity to the native protein. This form of IL-2 (aldesleukin) was produced and marketed under the tradename Proleukin, for the treatment of patients with metastatic melanoma (Moulton, 2015). In cases like this, *E. coli* production may be economically viable.

In addition to the production of recombinant proteins in *E. coli*, other strains of bacteria have been studied in recent years, such as *Pseudomonas*, *Bacillus subtilis* and *Streptomyces* (Table 3.3). These strains already have industrial applications, whether in the food and veterinary industries or in the production of antibiotics.

3.3.2 Yeasts

Yeasts can be considered as the simplest eukaryotic cells. Their molecular and biochemical characteristics are well known, as are cultivation techniques. Among the strains available for the production of biopharmaceuticals, *Saccharomyces cerevisiae* and *Pichia pastoris* stand out (Table 3.4). The first approved biopharmaceutical produced in *S. cerevisiae* was human insulin in 1987 by Novo

TABLE 3.4
Examples of Therapeutic Recombinant Proteins Produced in Yeasts (Jozala et al., 2016; Vogl et al., 2013)

Biopharmaceutical	Clinical Use	Yeast	Approval Date
Insulin (Humulin)	Diabetes mellitus treatment	*S. cerevisiae*	1982
Hirudine (Refludan, Revasc)	Anticoagulant	*S. cerevisiae*	1997
Platelet derived growth factor-BB (Regranex)	Treatment of neuropathic, chronic and diabetic ulcer	*S. cerevisiae*	1997
Hepatitis B surface antigen (Engerix)	Hepatitis B vaccine	*S. cerevisiae*	1998
Recombinant interferon-alpha 2b (Shanferon)	Hepatitis C and cancer treatment	*Pichia pastoris*	2002
Recombinant human insulin (Insugen)	Diabetes therapy	*Pichia pastoris*	2004
HPV vaccine (Gardasil)	HPV vaccine	*S. cerevisiae*	2006
Rasburicase (Ranibizumab)	Treatment of leukaemia, lymphoma and tumour lysis syndrome	*S. cerevisiae*	2007
Somatropin (GH) (Valtropin)	Neutropenia treatment	*S. cerevisiae*	2007
Recombinant hepatitis B vaccine (Shanvac)	Hepatitis B prevention	*Pichia pastoris*	2009

Nordisk (Nielsen, 2013). Since then, this strain has been used to produce other insulin analogues, human serum albumin, hepatitis B vaccine and VLPs (virus-like particles) for vaccination against human papillomavirus (Mizukami et al., 2018). The first biopharmaceutical produced in *P. pastoris* was approved in 2009 – Kalbitor by Dyax Corp., the Kallikrein inhibitor (Meehl and Stadheim, 2014).

S. cerevisiae has some characteristics of prokaryotic cells desirable for biopharmaceutical production such as ease of growth in simple and low-cost culture medium and ease of genetic manipulation; as well as characteristics of eukaryotic cells such as proteolytic processing, folding capacity facilitated by the formation of disulphide bridges and some post-translational modifications. It has a robust cell wall which provides resistance to shear stresses. Conversely, it has limited secretion capacity and can produce proteins with excessive and/or irregular glycosylation profile, impairing therapeutic activity. Production levels achieved are around 5% of the total cellular protein content, values well below those obtained for *E. coli*. *E. coli* and has a long history of industrial application not only in the pharmaceutical industry but also for the manufacture of bread and beer (Berlec and Strukelj, 2013; Mizukami et al., 2018).

Pichia pastoris (*Komagataella phaffii*) is a methylotrophic yeast, that is, capable of using methanol as a sole source of carbon and energy. It also has well-established handling techniques. Gene expression can also be induced by methanol, making possible control over the timing of protein production possible. Recombinant proteins can be produced at high levels (more than 30% of the total cellular protein content or up to 14 g/L) and be efficiently secreted, facilitating downstream processing steps because of the low amount of host cell proteins. It can perform post-translational modifications common to eukaryotic cells, including glycosylation, proteolytic processing and disulphide bond formation. Unlike *S. cerevisiae*, *P. pastoris* reaches high cell densities (more than 130 g/L of dry matter) in simple and low-cost culture media. Some researchers consider the need to use large amounts of methanol in industrial fermentations to induce expression a disadvantage of this system (Berlec and Strukelj, 2013).

Glycoproteins expressed in yeast often contain high mannose content. This profile usually leads to rapid elimination from the bloodstream, and because of this, proteins produced in this system will have a shorter serum half-life. This limitation can, however, be overcome by knocking out the mannosyltransferase gene (Gupta and Shukla, 2017). In addition, some sugars added by yeasts can elicit immunogenic responses (Gemmill and Trimble, 1999; Ghaderi et al., 2012). However, engineered strains capable of promoting a glycosylation profile more similar to human proteins have already been developed (Hamilton and Gerngross, 2007). As an example, the GlycoSwitch® platform that was developed in *P. pastoris* for the production of glycoproteins can be mentioned, where the hypermannosylation gene (OCH1) was removed and glycosyltransferase and glycosidase genes were introduced (Laukens et al., 2015; Effer et al., 2019). Although efficient, the strategies still lead to the production of recombinant proteins at low levels, which limits commercial production (Tripathi and Shrivastava, 2019). An interesting application of yeasts involves the production of glycoproteins for vaccination, where the immunogenicity of mannose residues can be advantageous for efficient therapeutic action.

3.3.3 Insect Cells

The application of insect cells for the production of therapeutic recombinant proteins is more recent compared to microbial and mammalian cells. The first commercial product produced using a baculoviral expression system and Hi-Five lepidopteran cells (*Trichoplusia ni*) was approved in 2007 – GlaxoSmithKline's Cervarix – a vaccine against human papillomavirus that causes cervical cancer. The recombinant protein PAP-GM-CSF (fusion protein – prostatic acid phosphatase and granulocyte-macrophage colony-stimulating factor) used in the formulation of Provenge autologous vaccine (Dendreon, approved in 2010) is also produced using a baculovirus expression system in Sf21 cells (*Spodoptera frugiperda*). This is also produced in insect cells (a lineage derived from *Spodoptera frugiperda* Sf9- ExpressSF) using baculovirus in the adenoassociated vector (AAV), Glybera, is intended for the treatment of lipoprotein lipase deficiency

TABLE 3.5
Examples of Therapeutic Recombinant Proteins Produced in Insect Cells (Palomares et al., 2015)

Biopharmaceutical	Clinical Use	Insect Cells	Approval Date
Porcilis Pesti (Intervet-Schering-Plough)	Classic swine fever virus	Sf9	2004
Cervarix	Cervical cancer	Trichoplusia ni	2007
Provenge	Prostate cancer	Sf21	2010
Glybera	Lipoprotein lipase deficiency	Derived from Sf9 cells	2012
Flublok	Seasonal influenza (vaccine)	Derived from Sf9 cells	2013

and was approved in 2012. Insect cells and baculoviral vectors have also been considered for the production of human vaccines based on VLP and veterinary proteins (Mizukami et al., 2018).

In comparison to mammalian cells, insect cells have simpler cultivation conditions (cultivation at room temperature, 26°C–29°C, do not require the addition of CO_2 for example), have greater growth rates, increased tolerance to variations in osmolarity and reduced risk of toxic by-products being formed. Because they demand more complex culture conditions than bacteria and yeasts, they have moderate scalability. Until recently, insect cells were considered to have low susceptibility to viral pathogens, especially those common to mammalian cells. However, recently it has been reported that several insect cell lines were persistently infected with adventitious viruses (arboviruses and other viruses associated with insects). These findings have raised serious concerns about the safety of biological products produced in insect cells. As a result, new insect cell lines free of adventitious agents were isolated to be used in research and also in the commercial production of recombinant proteins and vaccines (Geisler and Jarvis, 2018).

Insect cells can produce a protein of interest at high intracellular levels (30%–50% of total cellular proteins) due to potent promoters used by baculoviruses. In this system, the expression of recombinant proteins occurs after infection of a culture at high cell densities, and episomal replication does not need selection of the inserted transgenes (Walsh, 2007). This expression system has the disadvantage, however, of the need to produce high amounts of viral vectors for the production of recombinant protein on a large scale, which demands time and a relatively high cost. In addition, viral infection leads to cell lysis, which releases a large amount of proteases to the supernatant that can degrade the protein of interest.

In addition to the production of recombinant proteins using baculoviral vectors, it is also possible to carry out this production in genetically modified insect cells in a stable manner. This approach has been carried out in S2 cells of *Drosophila melanogaster* S2 and Sf9 cells of *Spodoptera frugiperda*. These cells have already been used to express a large amount of recombinant proteins in the academic field. The levels achieved in these systems are lower than the levels achieved using baculovirus (0.1–50 mg/L) (Kollewe and Vilcinskas, 2013). *Drosophila melanogaster* S2 cells are being used for the production of a vaccine against dengue, which is now in clinical studies (Coller et al., 2011).

Insect cells have the ability to perform some post-translational modifications, although with a different profile from those performed by mammalian cells. There is the possibility of the presence of insect-specific glycoforms and many residues of mannose and paucimanose. They do not perform sialylation (Kollewe and Vilcinskas, 2013) and N-glycosylation, a fact that can be solved by the introduction of mammalian glycosyltransferase genes into insect cells, which can be co-expressed with the protein of interest (Le et al., 2018). Examples of therapeutic recombinant proteins produced in insect cells can be seen in Table 3.5.

3.3.4 Plant Cells

Similar to bacteria and yeasts, plant cells can be easily grown in simple culture media at high levels, being easily scalable. They are robust cells that can withstand shear stresses because of the cellulose cell wall. They are not susceptible to contamination and are consequently free from transmission of human pathogens and endotoxins. Plant cells can be considered for the expression of proteins that can be harmful or toxic to mammalian cells (Hidalgo et al., 2018). It has the ability to synthesize complex proteins with a glycosylation profile similar to mammalian cells (Paul and Ma, 2011). However, there is the possibility of the presence of plant-specific glycoforms. Plant cells do not perform sialylation of proteins. The recombinant enzyme glucocerebrosidase (Elelyso, Protalix BioTherapeutics) was approved in 2012 and was the first to be produced commercially using plant cells. This product is produced in genetically modified carrot cells after transformation with *Agrobacterium* and is the proprietary expression system of the company Protalix (Fox, 2012). These cells are easily grown in disposable bioreactors (plastic bags) requiring low costs.

Another interesting aspect of this carrot cell production platform is the possibility of oral administration of genetically modified cells. The company Protalix is conducting pre-clinical studies for this purpose, with the production of the enzyme β-glucocerebrosidase. This product is at a promising stage of development and takes the advantage that the plant cell wall is resistant to enzymatic degradation in the digestive tract. Once released and absorbed, the active form of the enzyme can reach the bloodstream and perform its function. In addition to the advantage associated with improving the patient's quality of life (frequent infusions are not necessary), this oral biopharmaceutical would have a lower cost as it does not require extensive protein

TABLE 3.6
Examples of Therapeutic Recombinant Proteins Produced in Plant Cells (Owczarek et al., 2019)

Biopharmaceutical	Clinical Use	Plant Cells	Approval Date
Newcastle disease virus protein	Poultry vaccine	Plant cell cultures	2006
Taliglucerase alfa	Gaucher disease	Carrot	2012
Human epidermal growth factor	Burn treatment	Barley seed (*Hordeum vulgare*)	2012
Human growth hormone	Deficiency treatments	Barley seed (*Hordeum vulgare*)	-

purification processes (Tekoah et al., 2015). Other recombinant proteins produced in the Protalix proprietary system are in a clinical study phase. Despite the facilities for cultivation and scaling up, the yields of recombinant protein obtained in plant cells are still lower than those obtained in mammalian cells (Paul and Ma, 2011). Examples of therapeutic recombinant proteins produced in plant cells can be seen in Table 3.6.

3.3.5 Mammalian Cells

As already mentioned, most therapeutic recombinant proteins approved for use in humans are now produced in mammalian cells. The main advantage of using these cells as an expression system resides in the fact that these cells are capable of carrying out post-translational modifications (glycosylation, carboxylation, hydroxylation, among others), as well as protein folding in an authentic way, generating a protein with similar characteristics to those of native human proteins. The vast majority of protein biopharmaceuticals have been produced in CHO (Chinese hamster ovary) cells, but cells from rodents have also been used, for example, BHK (Baby hamster kidney) and NSO (murine myeloma). The first approved biopharmaceutical produced in mammalian cells (CHO) was tPA (tissue plasminogen activator) in 1986. Mammalian cell culture is addressed in detail in Chapter 12.

Human cell lines have gained prominence as possible alternatives to other mammalian cell platforms, as they produce recombinant proteins with post-translational modifications even more similar to native proteins, thereby avoiding an immune response against non-human epitopes such as N-glycolylneuraminic acid and α-galactose type bonds (Swiech et al., 2012).

Glycosylation is a highly regulated process and varies during differentiation, development and under different physiological conditions, such as cell culture. Despite the safety and efficiency of recombinant proteins produced in murine cells, non-human cells lack specific glycosidases, glycosyltransferases, sugar donors and other enzymes capable of protein modification and promoting the correct glycosylation of the protein. Other chemical moieties that bind to protein glycans that introduce chemical groups or bonds between monosaccharides that are not found in human glycosylation include N-glycolylneuraminic acid and Gal (1 → 3) Galβ1-(3) 4GlcNAc terminations, which can be highly immunogenic and/or be rapidly eliminated from circulation, shortening a product's serum half-life (Ghaderi et al., 2012).

When biopharmaceuticals were first launched onto the pharmaceutical market in the 1980s, protein production levels were below 0.5 g/L. The levels of current industrial processes are now around 2.5 g/L, with a projected growth to 3.25 g/L (Rader and Langer, 2014/2015).

Most lineages used are originally anchorage-dependent cells which have been adapted for suspension culture using fetal bovine serum-free media. This has facilitated scaling up, improving the reproducibility of the process and the safety profile of the products. Despite the ability to generate a product with acceptable therapeutic properties, the use of mammalian cells to express recombinant proteins also has some limitations. While other expression systems, such as yeasts and bacteria, allow the production of the target protein at high levels in low-cost culture media, mammalian cells need complex and high-cost culture media for the maintenance of growth and subsequent protein expression. They are also very susceptible to shear stresses that exist in bioreactors. There is also a risk that the process can be contaminated with infectious agents (viruses, bacteria, fungi) capable of causing diseases in humans, and a theoretical risk that DNA from the host cell could integrate into the human genome. These issues impact the complexity and subsequent cost of the production process and final product. Examples of therapeutic recombinant proteins produced in mammalian cells can be seen in Table 3.7.

3.4 DEVELOPMENT OF BIOPROCESSES FOR THE PRODUCTION OF THERAPEUTIC RECOMBINANT PROTEINS

After choosing the expression system (a topic covered in Chapters 4 and 5), the development/choice of a culture medium begins, which must be capable of both promoting high cell growth and high production levels of the protein of interest. Normally, culture media are specific to each cell line because they take into account the metabolic and cell processes. With the culture medium defined, next the choice of the bioreactor, the operating conditions (temperature, pH, dissolved oxygen concentration) and modes of operation (batch, batch-fed, continuous or perfusion) begin. These steps are part of upstream development that has, as its final objective, the robust production of a functional protein with high yields and productivity. The main steps and variables that can be studied/optimized in upstream development are shown in Figure 3.1.

After upstream development, the choice and establishment of unit operations that will be part of the purification

TABLE 3.7
Examples of Therapeutic Recombinant Proteins Produced in Mammalian Cells (Zhu, 2012; Swiech et al., 2012)

Biopharmaceutical	Clinical Use	Mammalian Cells	Approval Date
Kovaltry (octocog alfa)	Haemophilia A	BHK	1993
Benepali (etanercept)	Moderate or severe rheumatoid arthritis, psoriatic arthritis	CHO	1998
Replagal	Fabry disease	HT-1080	2003
Elaprase	Hunter syndrome (Mucopolysaccharidosis II, MPS II)	HT-1080	2007
Arzerra (ofatumumab)	Chronic lymphocytic leukaemia	NS0	2009
Actemra (tocilizamab)	Systemic juvenile idiopathic arthritis	CHO	2010
Belatacept	Prevention of acute rejection in adult kidney transplant patients	CHO	2011
Yervoy (ipilimumab)	Metastatic melanoma	CHO	2011
Benlysta (belimumab)	Systemic lupus erythematosus	NSO	2011
Nuwiq (simoctocog alfa)	Haemophilia A	HEK	2015
Rekovelle (follitropin delta)	Ovarian stimulation	PER.C6	2016
Inflectra (infliximab-dyyb)	Rheumatoid arthritis, plaque psoriasis and IBD	CHO	2016
Lartruvo (olaratumab)	Soft tissue sarcoma	NSO	2016
Taltz (ixekizumab)	Plaque psoriasis and psoriatic arthritis	CHO	2016
Alprolix (eftrenonacog alfa)	Haemophilia B	HEK	2016
Refixia (non-acog beta pegol)	Haemophilia B	CHO	2017
Shingrix	Zoster vaccine	CHO	2017
Rebinyn (rh coagulation factor IX)	Haemophilia B	CHO	2017
Lifmior (etanercept)	Rheumatoid arthritis, Juvenile idiopathic arthritis and Psoriatic arthritis	CHO	2017
Truxima (rituximab)	Follicular lymphoma and diffuse large B-cell Non-Hodgkin's Lymphoma (NHL); chronic lymphocytic leukaemia (CLL); severe active rheumatoid arthritis and granulomatosis with polyangiitis and microscopic polyangiitis	CHO	2018
Aimovig (erenumab-aooe)	Migraine	CHO	2018
Tremfya (guselkumab)	Plaque psoriasis	CHO	2018
Retacrit (epoetin alfa-epbx)	Anaemia	CHO	2018
Trogarzo (ibalizumab-uiyk)	Human immunodeficiency virus type 1 (HIV-1) infection	NSO	2018
Herzuma (trastuzumab)	Breast and gastric cancers	CHO	2018
Zessly (infliximab)	Gastroenterological, rheumatological and dermatological diseases	CHO	2018

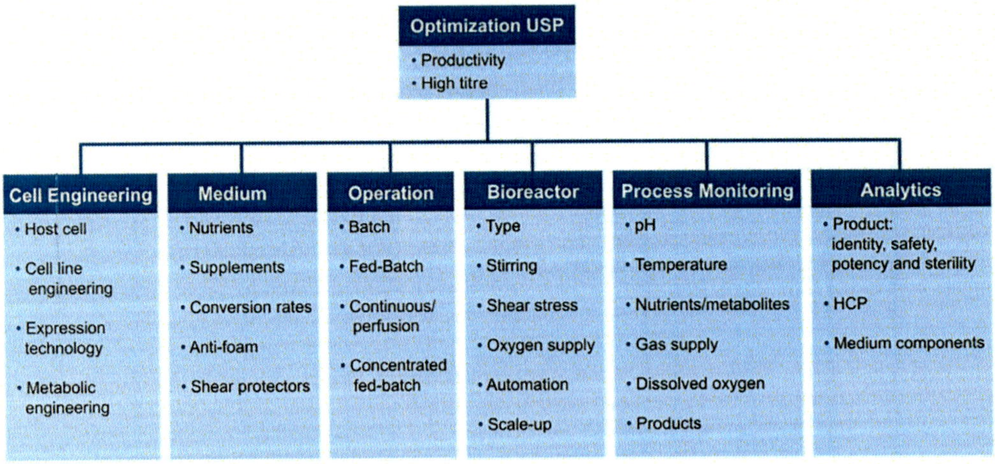

FIGURE 3.1 Optimization steps and parameters in upstream development. (Adapted from Gronemeyer, P., Ditz, R., Strube, J., *Bioengineering*, 1(4), 188–212, 2014.)

stage (downstream development) begin, which must ensure that the final product is free of contaminants and with quality and functionality compatible with regulatory requirements. Figure 3.2 lists examples of points that should be considered at this stage of development. This topic will be addressed in detail in Chapter 13. Recombinant cells that consistently express high levels of the protein of interest and processes that give rise to safe and reproducible purified products are

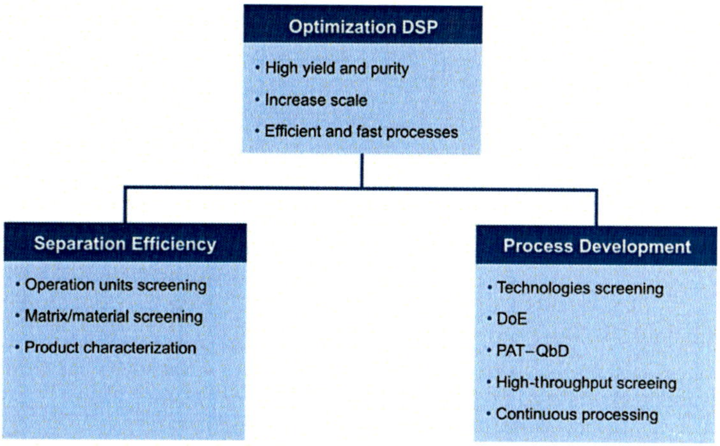

FIGURE 3.2 Optimization steps and parameters in downstream development. An integration of simulation and modelling unit operations along with the use of mini-plant facilities is also applied in this process. (Adapted from Gronemeyer, P., Ditz, R., Strube, J., *Bioengineering*, 1(4), 188–212, 2014.)

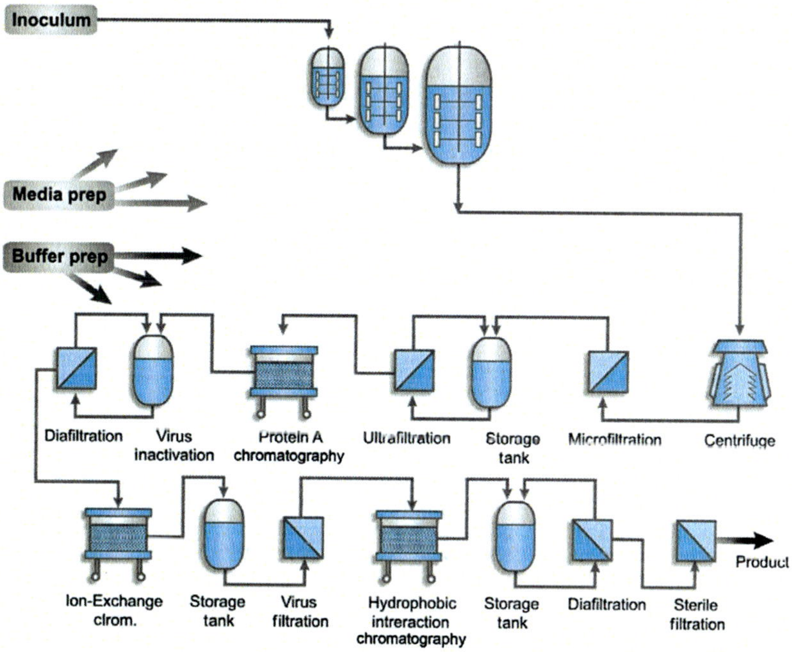

FIGURE 3.3 Schematic manufacturing process of monoclonal antibodies from mammalian cells. (Adapted from Gronemeyer, P., Ditz, R., Strube, J., *Bioengineering*, 1(4), 188–212, 2014.)

developed to support preclinical and clinical development stages of new therapeutic recombinant proteins.

This entire development stage occurs ideally based on the principles of experiment design (DoE) and quality by design (QbD) (see Chapter 21). In DoE, several process variables can be modified in a single set of experiments. A small number of experiments are necessary to evaluate the effects of different process variables, allowing the most relevant variables that give the most efficient levels of production to be selected. QbD is a principle where product quality is integrated into the production process by understanding the associated risks and includes strategies to mitigate those risks during manufacture.

Process Analytical Technology (PAT) is also an important tool in this development process, helping to ensure that product quality is in compliance with specifications through design, analysis and production control to maintain critical quality and performance attributes. Critical quality attributes (CQAs) are properties or characteristics that must be within an appropriate specification to ensure the desired product quality, safety and efficacy. In order to ensure that the process will result in the desired quality, critical parameters that can affect CQA must be controlled (Tripathi and Shrivastava, 2019). All of these principles and tools can lead to the development of more robust and efficient processes.

The final objective of this development stage is large-scale production of the recombinant protein in compliance with regulatory and quality aspects. Figure 3.3 shows, by way of example, the main unit operations that are part of a commercial production process of a monoclonal antibody using mammalian cells.

3.5 FINAL CONSIDERATIONS

The biopharmaceutical market has outgrown other sectors of the pharmaceutical market in recent years and is a trend expected to continue in the foreseeable future. This is because advances in scientific research have led to the discovery of molecular mechanisms of diseases, thus allowing targets and possible therapies to be identified. Consolidation and advances in the use of recombinant DNA technology have also made it possible to expand the range of products produced by this technology, as well as advances in the efficiency and quality of production. The therapeutic efficacy of recombinant proteins can now be improved, by precisely controlling glycosylation profiles of the resulting proteins, either by selecting host lines with better characteristics or by engineered cell lines for this purpose.

REFERENCES

Berlec, A.; Strukelj, B. Current state and recent advances in biopharmaceutical production in Escherichia coli, yeasts and mammalian cells. *Journal of Industrial Microbiology & Biotechnology*, v.40, n. 3–4, 257–274, 2013.

Biopharmaceuticals - A Global Market Overview, Report, 2019. https://www.researchandmarkets.com/reports/4769497/biopharmaceuticals-a-global-market-overview

Coller, B.A.; Clements, D.E.; Bett A.J.; Sagar, S.L.; Meulen, J.H.T. The development of recombinant subunit envelope-based vaccines to protect against dengue virus induced disease. *Vaccine*, v.29, n. 42, 7267–7275, 2011.

Effer, B.; Lima, G.M.; Cabarca, S.; Pessoa, A.; Farías, J.G.; Monteiro, G. L-Asparaginase from E. chrysanthemi expressed in Glycoswitch: effect of His-Tag fusion on the extracellular expression. *Preparative Biochemistry & Biotechnology*, v.4, 1–7, 2019.

Fox, J.L. First plant-made biologic approved. *Nature Biotechnology*, v.30 (6), 472–472, 2012.

Geisler, C.; Jarvis, D. L. Adventitious viruses in insect cell lines used for recombinant protein expression. *Protein Expression and Purification*, v.44, 25–32, 2018.

Gemmill, T.R.; Trimble, R.B. Overview of N- And O-linked oligosaccharide structures found in various yeast species. *Biochimica et Biophysica Acta*, v.1426, n. 2, 227–223, 1999.

Ghaderi, D.; Zhang, M.; Hurtado-Ziola, N.; Varki, A. Production platforms for biotherapeutic glycoproteins. Occurrence, impact, and challenges of non-human sialylation. *Biotechnology & Genetic Engineering Reviews*, v.28, 147–175, 2012.

Gronemeyer, P.; Ditz, R.; Strube, J. Trends in upstream and downstream process development for antibody manufacturing. *Bioengineering*, v.1, n. 4, 188–212, 2014.

Gupta, S. K. and Shukla, P. Advanced technologies for improved expression of recombinant proteins in bacteria: perspectives and applications. *Critical Reviews in Biotechnology*, v.36, n. 6, 1089–1098, 2016.

Gupta, S. K. and Shukla, P. Sophisticated cloning, fermentation, and purification technologies for an enhanced therapeutic protein production: a review. *Frontiers in Pharmacology*, v.8:419, 2017.

Gupta, S. K.; Dangi, A. K.; Smita, M.; Dwivedi, S. Effectual bioprocess development for protein production. *Applied Microbiology and Bioengineering*, 203–227, 2019.

Hamilton, S.R.; Gerngross, T.U. Glycosylation engineering in yeast: the advent of fully humanized yeast. *Current Opinion in Biotechnology*, v.18, n. 5, 387–392, 2007.

Hidalgo, D.; Sanchez, R.; Lalaleo, L.; Bonfill, M.; Corchete, P.; Palazon, J. Biotechnological production of pharmaceuticals and biopharmaceuticals in plant cell and organ cultures. Review. *Current Medicinal Chemistry*, v.25, n. 30, 3577–3596, 2018.

Jaffé, S.R.P.; Strutton, B.; Levarski, Z.; Pandhal, J.; Wright, P.C. Escherichia coli as a glycoprotein production host: recent developments and challenges. *Current Opinion in Biotechnology*, v.30, p.205–210, 2014.

Jozala, A.F.; Geraldes, D.C.; Tundisi, L.L.; Feitosa, V.A.; Breyer, C.A.; Cardoso, S.L.; Mazzola, P.G.; Oliveira-Nascimento, L.; Rangel-Yagui, C.O.; Magalhães, P.O.; de Oliveira, M.C.; Pessoa Jr., A. Biopharmaceuticals from microorganisms: from production to purification. *Brazilian Journal of Microbiology*, v.47, Suppl 1, 51–63, 2016.

Köhler, G.; Milstein, C. Continuous cultures of fused cells secreting antibody of predefined specificity. *Nature*, v.256, 495–497, 1975.

Kollewe, C.E.; Vilcinskas, A. Production of recombinant proteins in insect cells. *American Journal of Biochemistry and Biotechnology*, v.9, n. 3, 255–271, 2013.

Kubo, T. Random peptide library for ligand and drug discovery. In: Gopalakrishnakone, P.; Cruz, L.; Luo, S. (Eds.). *Toxins and Drug Discovery*. Dordrecht: Springer, 207–230, 2017.

Laukens, B.; De Wachter, C.; Callewaert, N. Engineering the Pichia Pastoris N-glycosylation pathway using the glycoswitch technology. *Methods in Molecular Biology (Clifton, N.J.)*, v.1321, 103–122, 2015.

Le, L.T.M.; Nyengaard, J.R.; Golas, M.M.; Sander, B. Vectors for expression of signal peptide-dependent proteins in baculovirus/insect cell systems and their application to expression and purification of the high-affinity immunoglobulin gamma Fc receptor I in complex with its gamma chain. *Molecular Biotechnology*, v.60, n. 1, 31–40, 2018.

Magalhães, P.O.; Lopes, A.M.; Mazzola, P.G.; Rangel-Yagui, C.Y.; Penna, T.C.; Pessoa, A., Jr. Methods of endotoxin removal from biological preparations: a review. *Journal of Pharmacy & Pharmaceutical Sciences*, v.10, n. 3, 388–404, 2007.

Malekian, R.; Sima, S.; Jahanian-Najafabadi, A.; Moazen, F.; Akbari, V. Improvement of soluble expression of GM-CSF in the cytoplasm of Escherichia coli using chemical and molecular chaperones. *Protein Expression and Purification*, v.160, 66–72, 2019.

Meehl, M.A.; Stadheim, T.A. Biopharmaceutical discovery and production in yeast. *Current Opinion in Biotechnology*, v.30, 120–127, 2014.

Mizukami, A.; Caron, A.L.; Picanco-Castro, V.; Swiech, K. Platforms for recombinant therapeutic glycoprotein production. *Methods in Molecular Biology (Clifton, N.J.)*, v.1674, 1–14, 2018.

Moulton, G.G. *Fed-Batch Fermentation: A Practical Guide to Scalable Recombinant Protein Production in Escherichia Coli*. Cambridge: Woodhead Publishing, 2015.

Nielsen, J. Production of biopharmaceutical proteins by yeast: advances through metabolic engineering. *Bioengineered*, v.4, n. 4, 207–211, 2013.

Owczarek, B.; Gerszberg, A.; Hnatsuszko-Konka, K.A. Brief reminder of systems of production and chromatography-based recovery of recombinant protein biopharmaceuticals. *BioMed Research International*, v.2019, 2019.

Palomares, L.A.; Realpe, M.; Ramírez, O.T. An overview of cell culture engineering for the insect cell-baculovirus expression vector system (BEVS). In: Al-Rubeai, M. (Eds.). *Animal Cell Culture. Cell Engineering*. v.9. Cham: Springer, 2015.

Park, J.; Lee, M.S.; Harrison, A.R. Profile of Xeomin® (incobotulinumtoxinA) for the treatment of blepharospasm. *Clinical ophthalmology (Auckland, N.Z.)*, v.5, 725–632, 2011.

Paul, M.; Ma, J.K. Plant-made pharmaceuticals: leading products and production platforms. *Biotechnology and Applied Biochemistry*, v.58 (1), 58–67, 2011.

Rader, R.A; Langer, E.S. Biopharmaceutical manufacturing: historical and future trends in titers, yields, and efficiency in commercial-scale bioprocessing. *BioProcessing Journal*, v.13, n. 4, 2014/2015.

Schwarz, F.; Huang, W.; Li, C.; Schulz, B.L.; Lizak, C.; Palumbo, A.; Numao, S.; Neri, D.; Aebi, M.; Wang, L-X. A Combined method for producing homogeneous glycoproteins with eukaryotic N-glycosylation. *Nature Chemical Biology*, v.6, n. 4, 264–266, 2010.

Sekhon, B.S.; Saluja, V. Biosimilars: an overview. *Biosimilars*, v.1, 1–11, 2011.

Swiech, K.; Picanco-Castro, V.; Covas, D.T. Human cells: new platform for recombinant therapeutic protein production. *Protein Expression and Purification*, v.84, n. 1, 147–153, 2012.

Tekoah, Y.; Shulman, A.; Kizhner, T.; Ruderfer, I.; Fux, L.; Nataf, Y.; Bartfeld, D.; Ariel, T.; Gingis-Velitski, S.; Hanania, U.; Shaaltiel, Y. Large-scale production of pharmaceutical proteins in plant cell culture-the protalix experience. *Plant Biotechnology Journal*, v.13, n. 8, 2015.

Tripathi, N.K.; Shrivastava, A. Recent developments in bioprocessing of recombinant proteins: expression hosts and process development. *Frontiers in Bioengineering and Biotechnology*, v.7, 420, 2019.

Vogl, T.; Hartner, F.S.; Glieder, A. New opportunities by synthetic biology for biopharmaceutical production in Pichia Pastoris. *Current Opinion in Biotechnology*, v.24, n. 6, 1094–1101, 2013.

Walsh, G. *Pharmaceutical Biotechnology: Concepts and Applications*. Chichester: Wiley. 2007.

Walsh, G. Biopharmaceutical benchmarks 2018. *Nature Biotechnology*, v.36, n. 12, 1136–1145, 2018.

Zhu, J. Mammalian cell protein expression for biopharmaceutical production. *Biotechnology Advances*, v.30, n. 5, 1158–1170, 2012.

4 Molecular Biology
Tools in Industrial Pharmaceutical Biotechnology

Marina Gabriel Fontes and Gisele Monteiro
Universidade de São Paulo

CONTENTS

4.1 INTRODUCTION

One of the best known and widely used drugs in the world is aspirin; the active ingredient of this drug is salicin that was isolated in 1828 from the bark of the willow tree, despite the use of willow as an anti-inflammatory or pain reliever having been recorded in medical texts for millennia. Yet, it took another 143 years to elucidate the mechanism of action of aspirin as an inhibitor of cyclooxygenases – step-limiting enzymes of the inflammatory response. This serendipitous approach of discovering new drugs is known as forward pharmacology, and its advantage lies in the fact that the effectiveness of the compound is already known, although the active principle and its mechanisms are unknown. However, this methodology is based on trial and error, which can be expensive and takes time to obtain a drug which can be produced industrially.

With the development of chemistry and pharmacology as scientific disciplines, it was possible to increase the complexity and number of drugs discovered. After joining this knowledge, several drugs capable of improving quality and life expectancy began to appear increasingly. These advances in pharmaceutical sciences also advanced another way of discovering new drugs – reverse pharmacology also known as target-directed drug discovery. This methodology is based on finding possible targets that are important for the disease in question and then designing drugs that act in a specific and effective way against that target. Thus, from the characterization of a biomolecule (the target) that is involved in the development, maintenance or susceptibility to a disease, the search for molecules that can inhibit or modulate the activity of the target begins. Only then will the effectiveness of the lead molecule be progressed through lengthy and expensive drug development pipelines.

With the advent of molecular biology in the 1970s, it became possible to characterize genetic material, culminating with the publication of the human genome sequence in 2003. So far, the number of whole genome sequences has exceeded 23 thousand. Many of these genomes are of organisms of medical, agricultural and industrial interest. With the impressive and exponentially increasing amount of information available regarding biological molecules, reverse pharmacology has been gaining major industry investments while investment in serendipitous forward pharmacology has been in decline since the mid-1990s (Lazo, 2008).

The use of recombinant DNA technology has been crucial in providing sufficient material (i.e. protein) for structural biology that has subsequently provided a fundamental understanding of structure–function relationships between targets and potential drug-like molecules. An important advance in structural biology has been that the molecular structure of drug receptors can be composed of many subtypes, each leading to possibly different biological outcomes in response to an interaction with a molecule. This can be exemplified in cases of anti-inflammatories that affect both cyclooxygenase 1 and 2, causing adverse effects.

Cyclooxygenase (COX) has two slightly different proteoforms, called COX-1 and COX-2. These enzymes transform arachidonic acid, a substance formed from lipids present in the cell membrane released by the action of phospholipase A2, into two types of compounds, prostaglandins and thromboxanes. The role of these mediators in inflammation and pain, as well as in several other physiological processes (such as blood coagulation), is widely accepted. In the early 1990s, the discovery of the two subtypes of COX created expectations for the establishment of new drugs that would retain the properties of existing COX inhibitors, but with reduced adverse side effects, the main one being gastrointestinal complications.

The different distribution of COX subtypes led to the hypothesis that COX-1 would be 'constitutive', that is, it would always be present in every cell throughout the body and was responsible for important physiological functions, while COX-2 would be 'induced' only during the inflammatory response. Thus, the side effects of non-steroidal anti-inflammatory drugs (NSAIDs) should only be caused by non-selective inhibition of COX-1. Structural analysis and interaction studies of NSAIDs with COX (Figure 4.1) allowed the design of compounds that specifically inhibited COX-2, called Coxibs (the first drug to market in this group was Celecoxib) (Drews, 2000). Thus, it was possible to exemplify a significant reduction in gastrointestinal side effects by rational design of molecules that inhibit specific receptor subtypes.

Examples of the use of recombinant DNA technology in drug design include the heterologous expression of G protein-coupled receptors in fungi. Other examples of products obtained by recombinant DNA technology and used for therapeutic purposes are TPA anticoagulant (tissue-type plasminogen activator), coagulation factor VIII and erythropoietin (a hormone that stimulates erythrocyte production).

Insulin was originally produced on a large scale for human use by purifying this protein from swine. Insulin is now produced by heterologous expression in genetically modified bacteria or yeasts. This latter method is more efficient and has the advantage of eliminating the possibility of contaminants or infectious agents that may be present in the purified proteins of animals. One of the greatest expectations in this field is the possibility of obtaining mutants with activity much superior to the native protein but without certain adverse effects (Watson 1992 and 2006).

Another example of the application of heterologous protein expression is gene therapy by the transfection of retroviruses, or other types of virus, into cells taken from the patient in order to replace a mutant gene, or to treat multifactorial diseases such as cancer and heart diseases. The first gene therapy performed to replace a mutant gene was adenosine deaminase (ADA) deficiency therapy, which involves implanting retroviral transfected T lymphocytes expressing the missing protein in the carriers of this disease. This has been considered the prototype of successful gene therapies.

This chapter aims to present important molecular biology techniques and concepts in the study, development and production of various compounds of industrial pharmaceutical interest.

4.2 THERAPEUTIC TARGETS

Diseases of interest to the pharmaceutical industry are mostly polygenic, that is, the disease phenotype results from the expression of several genes and consequently several biomolecules influence the clinical picture. In addition, there is redundancy of biological functions due to gene duplication events or through adaptive responses, i.e., increased expression of another gene encoding a protein with similar biological function to compensate for the lack of activity of the inhibited target. The analysis of therapeutic targets against which drugs have already been successfully developed shows that 95% of targets are proteins, of which 45% are membrane receptors (Yao and Rzhetsky, 2008). These data reinforce the relationship between the efficacy of a drug being associated with a broad spectrum of genes affected by it. For example, cells respond to external stimuli in the form of signal amplification (Figure 4.2).

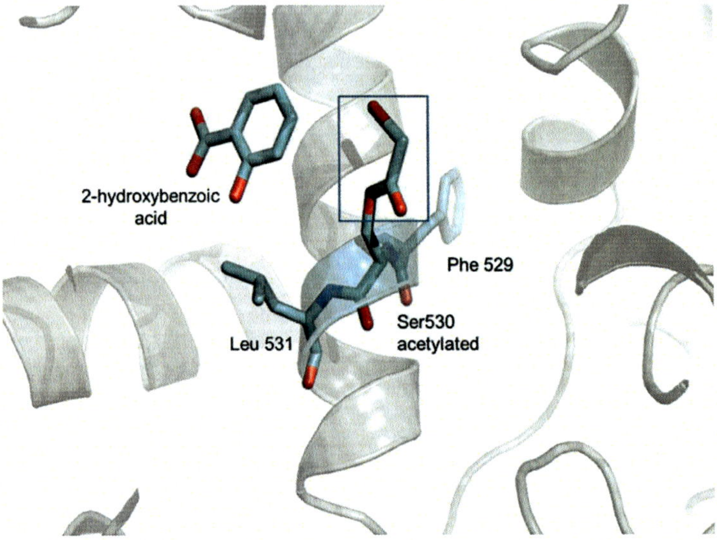

FIGURE 4.1 Tertiary structure of cyclooxygenase 1 (COX-1, PDB ID: 1PTH) inactivated by aspirin. Acetylation of serine from the active site leads to the inactivation of COXs. Also shown is the 2-hydroxybenzoic acid group in the active site region, released after the acetylation of Ser530.

Cell signal	Number of activated molecules
Signal reception E.g. Binding of epinephrine to G protein-bound receptor	1 molecule
Transduction Inactive protein G Active protein G	10^2 molecules
Inactive adenylate cyclase Active adenylate cyclase	10^2 molecules
ATP cAMP	10^4 molecules
Inactive protein kinase A (PKA) Active PKA	10^4 molecules
Inactive phosphorylase kinase Active phosphorylase kinase	10^5 molecules
Inactive glycogen phosphorylase Active glycogen phosphorylase	10^6 molecules
Response Glycogen Glucose-1-phosphate	10^8 molecules
Release of glucose from the liver into the blood.	

FIGURE 4.2 The signalling cascade of the external stimulus to the intracellular modification. Signalling is a process of signal amplification that culminates with proper metabolic response. Example of the action of an epinephrine molecule resulting in the activation of 1 million glycogen phosphorylase molecules, releasing approximately 100 million glucose 1-phosphate from the liver into the bloodstream.

Thus, when a receptor is activated, a cascade of responses is initiated within the cell. A receptor may interact with several signal amplifying molecules, such as a protein kinase or a second messenger converting enzyme (for example, adenylate cyclase, which generates cAMP). The second messenger also leads to the activation of kinases that, through a post-translational modification called phosphorylation, activates or represses the action of enzymes and/or transcriptional factors. Enzymes are responsible for rapid and transitory cell responses. Transcription factors, in turn, bind to regions upstream of target genes that have a consensus binding sequence in their promoters (called a promoter element) that, when activated, leads to the synthesis of more mRNA that ultimately translates into protein. The increased expression of proteins will lead to an adaptive metabolic response to the external signal. Thus, by inhibiting a single receptor responsible for a signalling pathway cascade, it can be possible to prevent more than 10^8 biomolecules (depending on the pathway – Figure 4.2) from being biosynthesized by metabolic pathways, which would dramatically change the function of a cell.

An important goal of post-genomic research is to understand the cellular machinery responsible for regulation and communication since most diseases influence the levels of these processes. Bioinformatics tools have been of great value in analysing drug targets. A study conducted with therapeutic targets of 919 drugs approved by the Food and Drug Administration (FDA) showed that the genes involved in responding to these drugs had characteristics in common with each other and relatively divergent from the rest of the genome. Among these characteristics, there is little non-synonymous polymorphism (these proteins are highly conserved in the human species), they are tissue specific (present only in certain types of tissues of the organism), have high connectivity (influence and/or interact with many proteins) and have high metabolic specificity, that is, they influence a single or few types of global cellular response.

Another study from 2009 of 148 human proteins targeting drugs has shown that there are other important features: targets are usually outside the cell compartments; have long half-life; are glycosylated and usually secreted; are commonly enzymes with oxidoreductase, lyase or transferase activity; and are involved in anchoring, signalling and cellular communication processes (Bakheet and Doig, 2009). To better understand this cascade of events that might influence the druggability of therapeutic targets, let's now look at how communication, signalling and cellular regulation work.

4.3 CELLULAR COMMUNICATION

Our multicellularity imposes an important function, intercellular communication. In complex organisms, there are specialized organs – the glands – responsible for producing activating or inhibitory stimuli of intracellular pathways to maintain the health of the organism. Conversely, cells also have stimulus-activated suicide mechanisms (apoptosis), so that the death of a small number of cells preserves the health of the whole organism. Another clear example of the importance of cellular communication is evidenced by cells that undergo malignant (tumorigenic) transformation and do not communicate with the organism (but communicate with each other), not responding to the stimuli that would control their proliferation, differentiation and location, which in many cases leads to the death of the organism (Alberts B 2008).

Even for unicellular organisms like yeast and bacteria, there is cellular communication. This phenomenon is known as quorum sensing and is a process of decision-making by a decentralized unicellular group that ultimately culminates in the coordination of gene expression in the population. Quorum sensing has been well described in *Escherichia coli* and *Pseudomonas aeruginosa* bacteria that coordinate gene expression in response to population density (in the former) or biofilm formation and expression of virulence factors (in the latter). Communication between cells is made possible by the biosynthesis and secretion into the growth environment of signalling molecules that upon reaching a given concentration influence cellular receptors and modulate the gene expression of the population.

Yeasts use signalling communication as a way to inhibit cell proliferation and prepare cells of the opposite mating types for sexual reproduction, giving rise to diploid cells. Under stress conditions, these diploid cells sporulate (enter into meiosis, generating haploids), now with the possibilities of forming new gene combinations by recombination processes, thereby increasing genetic variation in the population and, consequently, the possibility of survival in a hostile environment. This process occurs by molecules and mechanisms extremely similar to those of animal cells, but obviously less complex.

In multicellular organisms, communication can be performed at small or long distances. Cellular contact is a type of communication in which cells exchange information (ions, small signalling molecules such as cAMP or Ca^{+2}) through cell junctions. Adhesion proteins and interaction with the extracellular matrix are also vital for viability, proliferation control and differentiation. Other types of communication involve extracellular signals that can be sent by neighbouring cells or arrive at their destination through an electrical stimulus or through the bloodstream, travelling long distances.

The common feature between the different types of cellular communication is the fact that they all involve some kind of molecule responsible for connecting a messenger and transferring the information to other cells. These molecules are called cellular receptors, and their biological function is always related in keeping the intracellular environment in tune with the needs imposed by the external environment – homeostasis. There are intracellular and cell surface receptors.

Intracellular receptors require the external signal to be able to cross the plasma membrane. Two examples of signalling molecules with this ability are steroid hormones and nitric oxide. In the case of steroids, these are hydrophobic molecules that can pass through the lipid bilayer of the membrane and then interact with the receptor in the cytoplasm or nucleus. Receptors that are bound by steroid hormones become activated and themselves then bind to enhancer sequences on DNA leading to activation or repression of target genes. In the case of nitric oxide, because it is a gas dissolved in aqueous bodily fluids, it is able to cross biological membranes by diffusion and can reach soluble guanylate cyclase in the cytosol, its receptor inside the cell, increasing the generation of the second messenger cGMP. This messenger has as its main target protein kinase G, which phosphorylates serine and threonine residues of target proteins, leading to modulation of protein activity. This second messenger linked to guanylate cyclase has an important role in vasodilation and is the target for NO-releasing drugs such as nitrovasodilators which are used as a medicine for cardiac treatments.

Cell surface receptors are characterized into three main types: those connected to ion channels, those associated with G protein and those coupled to enzymes.

Receptors attached to ion channels are involved in rapid responses, such as during synaptic transmission between nerve cells and other electrically excitable cells (such as those present in nerves or muscles). This type of signalling is due to neurotransmitter binding to the receptor, which leads to the opening or closing of the associated ion channel by changing the permeability of the plasma membrane to ions and preparing the next cell to receive the signal. Examples of receptors attached to the ion channels are those which bind neurotransmitters: acetylcholine, glycine, GABA (γ-amino-butyric acid) and glutamate. Other receptors are linked to voltage-dependent channels, which respond to neuronal electrical stimuli. Anxiolytic medications (benzodiazepines and barbiturates) exert their clinical effect by modulating these receptors.

The most interesting feature of these receptors is that the ion channel is inside the receptor, that is, the molecules responsible for interacting with the ligand are around the channel, forming a protective ring. The transmembrane domains of these receptors have amphipathic characteristics to interact with the plasma membrane but also allow the passage of charged molecules. In addition, the nature of the ions to be transported through the channel is specific, due to the amino acid composition in the hydrophilic portion of the amphipathic helix. For example, glycine and GABA receptors have positively charged amino acids distributed along the channel wall, consistent with the transport of chlorine (Cl^-) ions they carry out.

G protein-coupled receptors are generally indirect regulators of the activity of ion channels or enzymes. Communication between the receptor and the target is performed by a protein bound to GTP called a G protein. The receptor is a trimer (three subunits) activated by GTP nucleotide binding, which can be hydrolysed by the G protein itself into an activity which is self-regulating since the GDP-linked G protein is inactive. Thus, when the receptor is activated by the ligand it changes its conformation and induces the change of conformation in the G protein which leads to the substitution of GDP by GTP in the α subunit. The α-subunit attached to GTP then dissociates from the $\beta\gamma$ dimer and can move through the membrane and thereby activate the target. The GTPase activity of the G protein itself, or of the helper proteins, terminates the activation process by hydrolysing GTP to GDP in the α subunit. One of the most well-known targets is adenylate cyclase which is responsible for the synthesis of the second messenger cAMP. Many drugs target these receptors such as opioids, adrenergic agonists (vasopressor amines), etc.

Enzyme-coupled receptors are direct activators, as are those coupled to ion channels. They function as enzymes in their cytoplasmic domain or associate directly with the target enzymes. They are mostly protein tyrosine kinases or associated with kinases (tyrosine or serine/threonine kinases) that carry signal transduction by the phosphorylation of other proteins that can be activated or repressed by this modification. An example is the insulin receptor. Examples of drugs that target these receptors are insulin, erythropoietin and interferon.

From the description of the types of receptors, we can conclude that there are many different ways of transmitting the signal. Rapid transmissions involve only modifications of enzymes that lead to activation or repression and do not require altered gene expression. Enzymes associated with receptors mainly bound to ion channels are good examples. In these cases, the responses are transient, and the cell is able to respond to the external stimulus in seconds, at most in minutes. When more long-lasting responses are required, the cell needs alternative proteins that respond to and adapt to the stimulus; in such cases, there is a need for protein synthesis and induction of gene expression, leading to a response to last hours. In these situations, as in the case of hormonal stimuli, G-associated and enzyme-associated receptors are primarily involved.

The action of the receptors occurs in order to amplify the signal to several auxiliary molecules through second messengers. Among these secondary messengers are Ca^{+2}, cAMP, cGMP, IP_3 (inositol 1,4,5-triphosphate) and DAG (1,2-diacylglycerol). The second messengers will modulate protein kinases and phosphatases. These phenomena result in a cascade of phosphorylations that enhance or inhibit enzyme activity, or induce or repress transcriptional factors that will act in the modulation of gene expression appropriate to the signal received by the receptor.

Interestingly, the action of the ligands will depend on the set of genes expressed by that cell and may have completely different actions in different tissues. Take the effects of epinephrine as an example. In heart cells, epinephrine causes the beat rate to rise; but in liver cells, epinephrine causes the mobilization and breakdown of glycogen to release glucose into the bloodstream. The receptor is the same, the linker is the same, but the physiological response is completely dependent on the differentiated protein pool in the specialized cells of the organism.

We will now see how different cell types but with the same genome can respond completely differently to the same external stimulus, by modulating and regulating the expression of the genome.

4.4 EXPRESSION GENE REGULATION

The definition of a gene is not trivial. According to the central dogma of biology, a gene is responsible for transmitting genetic information through a molecule of messenger RNA that will be translated into a protein. Thus, we can say that DNA stores information, RNA is the intermediary that carries information and proteins are the result, the effectors of information. However, several examples deviate from this rule. One of the main components of ribosomes is a ribozyme, an RNA with catalytic function, responsible for making the peptide bonds that form the proteins. Thus, the ribosomal ribozyme is by itself the effector and will not be translated into protein. Viruses such as HIV store genetic information in an RNA molecule (rather than DNA) and are therefore called retroviruses. A more comprehensive definition of a gene would be as a hereditary unit. This hereditary unit can be represented in a simplified way as in Figure 4.3.

A gene must have at least two functional regions: regulatory or promoter, and structural or coding. The coding region is the message that will be transcribed into RNA at the time and in the amount governed by the regulatory region. The main characteristic of a coding region is the presence of an open reading frame (ORF), that is, a sequence of three nucleotides (codons) not randomly interrupted by translation termination codons.

Like DNA, RNA synthesis is directed from the 5′-to 3′-portion. This is imposed by the activity of the enzyme responsible for the synthesis, the RNA polymerase, which makes phosphodiester bonds by adding a triphosphate nucleotide (specifically the 5′-phosphate) to the 3′-hydroxyl of the previous nucleotide. RNA polymerase uses template DNA strand ($3' \rightarrow 5'$) for annealing complementary bases and thus polymerizes RNA in the $5' \rightarrow 3'$ direction. Because of this property, the RNA molecule is identical to the DNA encoding strand, with thymine replaced by uracil. Another crucial difference is the nature of the nucleotides added by RNA polymerase, the ribonucleotides. By convention, any DNA or RNA sequence must be written from the 5′-end to the 3′-end.

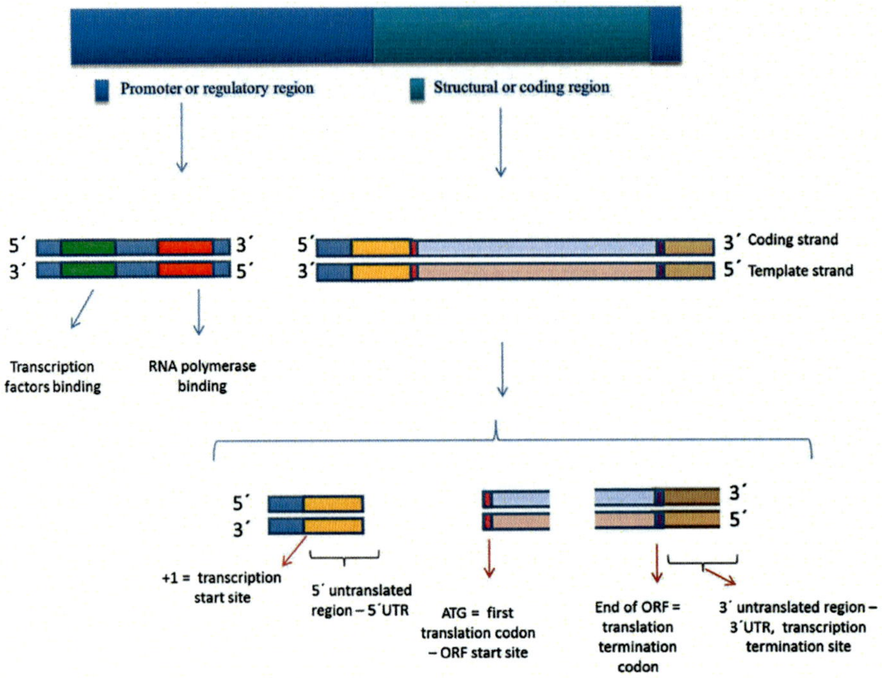

FIGURE 4.3 Gene – an inherited unit. To be considered a gene, two functional regions are required: the promoter region and the coding region. Within the promoter region, RNA polymerase binding sequences are required to the transcriptional factors action. The yellow region represents the 5′-untranslated sequence – showing that the transcript is larger than the protein – and the +1 position represents the beginning of transcription rather than translation (codon 1 = methionine-encoding ATG); furthermore, it has the 3′ untranslated portion, important for the termination of RNA transcription and processing. ORF = open reading frame.

For this polymerization, a template is needed but not a primer. RNA polymerase initiates the synthesis of RNA in the transcription initiation region and is capable of polymerizing a new strand without the need for an initiator. This region is conventionally called +1 and does not correspond to the translation initiation codon of the Open Reading Frame (ORF), the ATG. Sequences transcribed prior to ATG are called the 5′-untranslated region and are important in RNA processing and for ribosome binding and assembly. Likewise, RNA polymerase does not terminate transcription in the translation termination codon, but in a transcription termination sequence which is located further downstream. This region, after the termination codon, is termed the 3′ untranslated region and is important for RNA processing, efficient transcription termination and RNA stability; consequently, 3′ UTR is a regulatory region.

In order for RNA synthesis to occur, there is a need for the DNA strands to be denatured, since the RNA polymerase will use the template strand to anneal the ribonucleotides during the polymerization, generating a transient DNA–RNA hybrid. At the beginning of transcription, the DNA region covered by the RNA polymerase corresponds to approximately 75 base pairs (bps), in which it completely protects the DNA sequence that is either single stranded or annealed to RNA. During elongation, the RNA polymerase moves across the DNA strand in a seemingly discontinuous fashion. This has been associated with the stability of the RNA–DNA hybrid and with the movement of the elongation complex factors. This region in which the DNA is unpaired and the synthesis of RNA occurs is called the transcription bubble and comprises on average 20–40 bps (Nudler et al., 1997).

The transcriptional process can be divided into three main steps: initiation, elongation and termination. The main regulatory point in transcription occurs in the initiation process.

A crucial factor for RNA synthesis to occur is the various interactions between nucleic acids (both DNA and RNA) and the proteins required for transcription. These interactions are specific and transient, composed of many weak interactions, mainly formed by hydrogen bonds. An exception to this rule is deoxyribonucleotide thymine that has a methyl (non-polar) group exposed in the region of most frequent interaction between DNA and proteins, the main groove. In the region of the minor groove of the double helix, interactions between DNA and protein also occur, but in this case the differences between the bases are not so obvious.

Another fact that assists in this specificity is that the DNA-binding sequence for the protein is usually duplicated and palindromic. This is because transcriptional factors are typically dimeric, thus allowing increased selectivity since both subunits will have to interact with the sequences providing stability to the bonds.

Transcriptional factors have characteristically modular structures. They can be separated into regions of DNA binding and transcription activation. DNA binding domains are typically small (about 60–90 amino acids), as well as the sequences recognized by them, at about six nucleotides. Activation domains are usually involved in interactions with the transcription machinery, mediators or other proteins as co-activators.

The best-known DNA binding domains are helix-turn-helix, zinc finger, leucine zipper and helix-loop-helix.

The helix-turn-helix domain, as the name itself suggests, is composed of two small α-helices (7–9 amino acid residues) separated by a β-strand of about 20 residues. The homeotic proteins (involved in the regulation of the expression of genes important for the embryonic development) of higher eukaryotes and the Cro repressor of bacteriophage 434 provide examples of transcriptional factors containing that domain. In these proteins, an α-helix fits into the major double-helix DNA groove and contacts the exposed base pairs. Meanwhile, the second helix accommodates itself on the first and makes contact with other proteins of the transcription apparatus.

The zinc finger domain consists of approximately 30 amino acid residues, in which four residues (four cysteines, or two cysteines and two histidines) are responsible for the coordination of a zinc atom. This motif is very common, and some proteins have several of these repeated in their structure. Zinc atoms usually firmly approximate α-helices and β-strand of these proteins and probably lead to folding in favour of DNA recognition and binding. There are many examples of transcriptional factors with this domain such as Spl (steroid receptor).

Leucine zipper proteins are so-called due to motif repeats composed of four or five leucine residues precisely separated by seven amino acid residues. These domains confer hydrophobic faces through the interaction of the leucine zippers leading to the formation of the dimer. These dimers may be from the same protein (homodimers) or from different proteins (heterodimers). Immediately adjacent to the leucine zipper is an N-terminal domain rich in positively charged amino acids (arginine and lysine) that makes up the DNA binding region. Interestingly, leucine zipper mutations prevent both dimerization and DNA binding, suggesting that only dimers have the ability to recognize and bind promoter elements. Mutations in the positively charged N-terminal domain abolish DNA binding ability only, but not dimerization. The transcription factors c-fos and c-jun (proto-oncogenes) are examples of genes encoding leucine zipper proteins.

Finally, the family of helical-loop-helix transcriptional factors (HLH) is similar to leucine zipper proteins because they bind to DNA only as homodimers or heterodimers, and their DNA binding domain is also composed of positively charged amino acid residues. This family contains important members involved in the regulation of cell growth and differentiation such as the MyoD protein, which is important in the differentiation of muscle cells. Another member is the c-myc proto-oncogene.

The mechanisms of recognition of DNA sequences by transcriptional factors are still poorly understood, but what is known is that in these classes of domains, always an α-helix binds to the main DNA groove.

4.5 TRANSCRIPTIONAL SYSTEM IN PROKARYOTES

Despite having three types of RNA (the messenger, the ribosomal and the transfer RNA), prokaryotes have only one RNA polymerase. This enzyme is composed of two subunits (encoded by the *rpoA* gene), one α subunit (encoded by the *rpoB* gene) and one β′ (encoded by the *rpoC* gene). The holoenzyme also has a subunits called σ (encoded by the *rpoD* gene). Although it is part of the RNA polymerase, there are different types of subunit, each with a different composition (Figure 4.4).

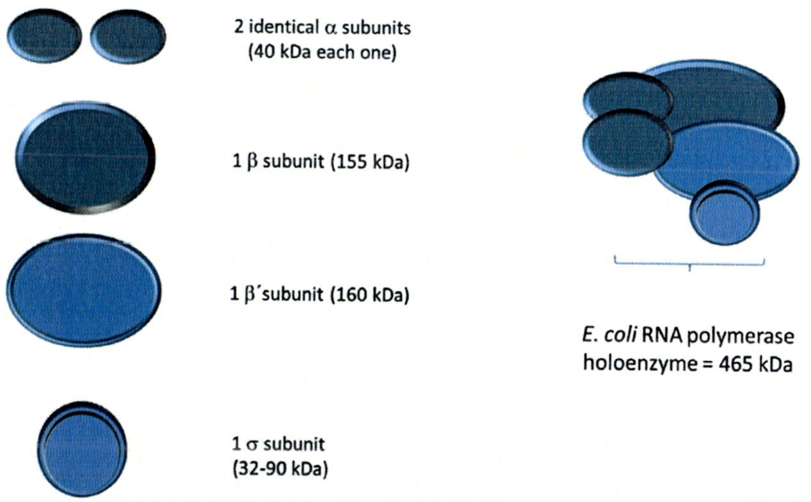

FIGURE 4.4 *Escherichia coli* RNA polymerase is formed by several subunits. It can be separated into two main regions: the catalytic core formed by the subunits α2ββ′ and σ factor. The catalytic core is always the same, but the σ factor, responsible for the holoenzyme binding in the promoter regions, may vary.

The holoenzyme (α2ββ′σ) can be divided into enzymatic core (α2ββ′) and σ factor. Only the holoenzyme is capable of initiating transcription. The α subunits are important in the formation of the holoenzyme complex, recognition of promoters and attachment of some transcriptional activators. The β and β′ subunits form the catalytic centre where phosphodiester bonds form between ribonucleotides of the growing RNA chain. The α subunit is responsible for the binding specificity and recruitment of RNA polymerase to the promoter regions. After reaching eight to nine bases of synthesized RNA, the σ factor is released and only the core enzyme continues the transcription. Thus, the core RNA polymerase can synthesize RNA from template DNA but is not capable of initiating transcription alone.

Thus, one of the most common mechanisms used by prokaryotes to regulate initiation of transcription is by substitution with different σ factors. The σ factors alone are not able to bind to DNA; however, when present in the holoenzyme, they change their conformation and bind to the promoter sequences, increasing the affinity of RNA polymerase for DNA in that region and promoting the initiation of RNA synthesis. Soon after, the σ factor is released and the core of the enzyme continues the synthesis, moving to the elongation phase until a transcription termination sequence is found.

4.5.1 Promoter Structure and Operon

Promoters are regions of DNA that regulate gene expression. In these sequences, the regulatory factors of transcription and the RNA polymerase holoenzyme will bind. Regulatory factors can act negatively, and so are called repressors, or positively, the activators. For each promoter, there are specific binding sequences for both RNA polymerase and transcriptional factors, called promoter elements. The 'strength' of a promoter will be governed by maintaining the consensus sequence of elements for binding of the transcription factors and σ factors.

E. coli RNA polymerase is the most well-known and studied. The most abundant factor in this polymerase is σ^{70}. This factor binds to two conserved promoter elements, both of six nucleotides, at positions –10 (TATAAT consensus sequence) and –35 (TTGACA consensus sequence); noting that the nucleotide that initiates the RNA strand is called +1. Variations in this sequence result in poor recruitment of the holoenzyme to the promoters since the σ^{70} factor binds less strongly.

As the σ factor is essential for the specific recognition and binding of RNA polymerase to the promoter sequence, a strategy amongst prokaryotes for transcriptional regulation is the use of alternative σ factors. In cases where there is a need for a major shift in gene expression, such as during a drastic change in growth conditions (e.g. during a heat shock), alternative σ factors (such as σ^{32}, names vary according to molecular weight) replace the σ^{70} factor in holoenzyme, resulting in a change in the set of promoters that will be actively transcribed. These classes of promoters that are regulated by the same transcriptional factors are called regulons. For example, the proteins responsible for protecting cell viability during heat shock are regulated by the σ^{32} regulon.

In addition to the holoenzyme binding sequences, there are other regulatory elements in the promoters called operators. In such sequences, the repressors (or activators) that interact with the RNA polymerase will bind in order to prevent (or enhance) their ability to bind to the promoter and initiate transcription.

Prokaryotes have a peculiar organization of their genes. Those involved in the same metabolic pathway are grouped in the same transcriptional unit and are regulated by the same promoter. Thus, the primary transcript (the RNA is unprocessed) contains information not for one protein, but for several proteins that participate in the same cellular process, for example, synthesis of an amino acid or metabolism of a carbon source. This organization of a promoter followed by several ORFs arranged in a cis (side-by-side) configuration, resulting in the transcription of a polycistronic RNA is called an operon (Figure 4.5).

Some promoters are actively transcribed under any cellular condition, as they encode for biomolecules essential

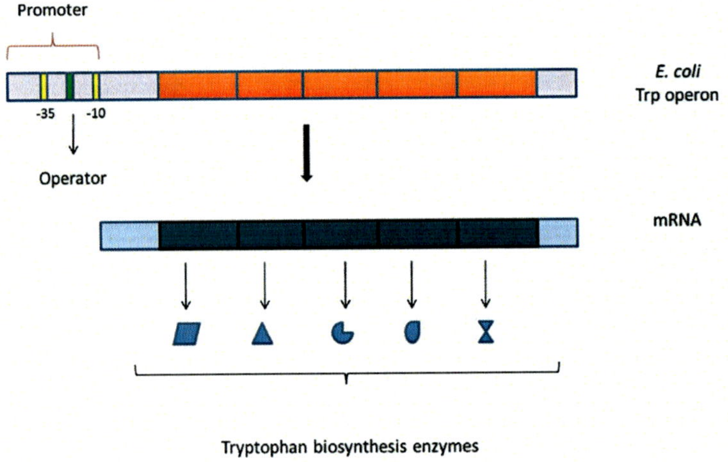

FIGURE 4.5 Schematic representation of the operon organization in prokaryotes. Representation of tryptophan biosynthesis; highlighted, the regions –10 and –35 (in yellow) and the operator region (in green).

for cell survival, and are called constitutive genes. These genes will have the –10 and –35 promoter elements highly similar to the consensus sequence. The promoters of these genes are not regulated by transcriptional factors that lead to repression or induction of transcription. However, the cell does need to regulate levels of expression so as not to make the energy expenditure unfeasible. Thus, several genes involved with specific functions in the cells will undergo strong regulation of their gene expression, having their promoters under positive or negative control.

As we saw earlier, cells communicate and adapt to the external environment through signals picked up and transduced by a cascade of intracellular events. For prokaryotes, these signals are commonly small molecules (ligands) that interact with transcriptional factors to induce expression, when they turn off a repressor or bind an activator; or they repress gene expression, when they bind a repressor or turn off an activator (Figure 4.6). To understand specifically how these processes occur, we will take some classic examples of how operons are regulated in bacteria.

Microorganisms (yeasts and bacteria) undergo a very important type of global gene regulation that allows for rapid growth and proliferation when conditions are favourable, but also allows for growth to slow when conditions are harsh (Lewin, 2007). This phenomenon is regulated by the presence of glucose in the environment and is called catabolic repression. Catabolic repression ensures that only genes responsible for rapid growth and metabolism of glucose are expressed when glucose is present in excess, while repressing genes for defence, adaptation and metabolism of alternative sources of carbon. The *lac* operon is the set of genes grouped as a single transcriptional unit responsible for producing the proteins involved in the metabolism of lactose, an alternative carbon source to glucose for bacteria. In this example, we will see how the actions of a repressor and an activator precisely regulate metabolism when there is no preferential source of carbon (glucose) but there is lactose in the medium that can be metabolized.

In bacteria, catabolic repression is regulated by a protein called CAP (catabolite activator protein), also known

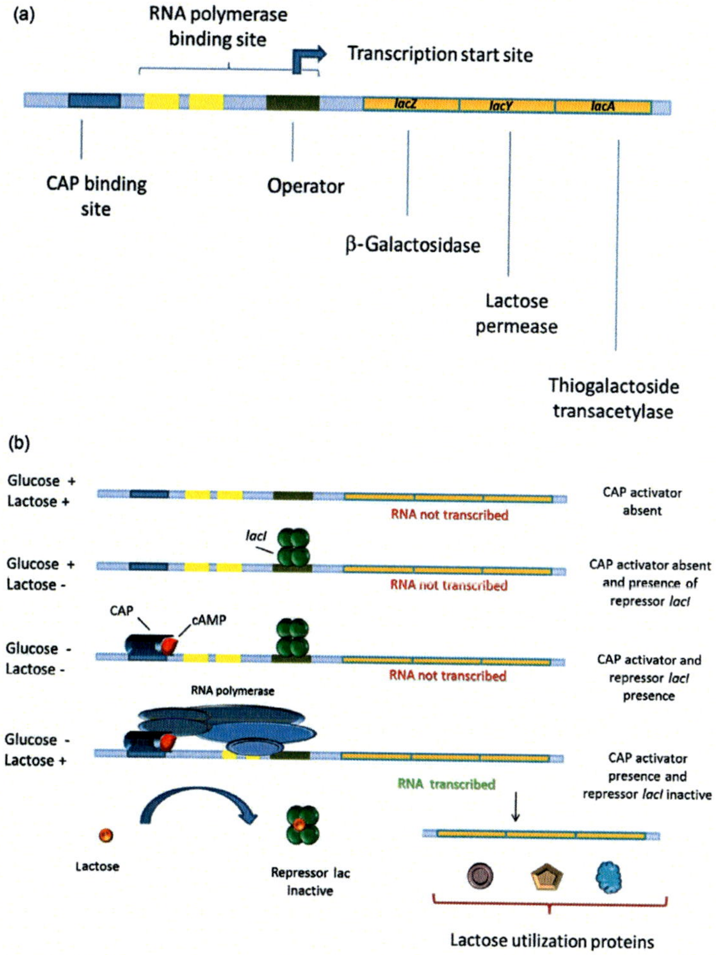

FIGURE 4.6 Regulation of the operon *lac*. (a) Structure of the operon, highlighting the binding regions of the holoenzyme RNA polymerase (in yellow), the CAP protein binding site (in dark blue) and the operator where the *lacI* repressor binds (in green). (b) Different metabolic conditions of *Escherichia coli*, showing the interaction between CAP activator and *lacI* repressor in the coordination of *lac* operon expression.

as CRP ('cAMP receptor protein'). When glucose is at low levels in the medium, the protein adenylate cyclase which is responsible for the synthesis of the second messenger cAMP is activated. Thus, in the absence of available glucose, cAMP levels increase. The cAMP binds to the CAP protein and is then able to bind to DNA and activate the transcription of the *lac* operon. However, lactose is not the only alternative source of carbon that can be used by the bacteria. Therefore, for the *lac* operon to be activated not only must the levels of glucose be low, but the levels of lactose must be high.

For activation of the *lac* operon when glucose levels are low and lactose levels are high, the *lac* operon undergoes a second level of regulation through a repressor called *Lac* (encoded by the *lac I* gene). This repressor is always bound to the promoter in the absence of lactose and has a binding

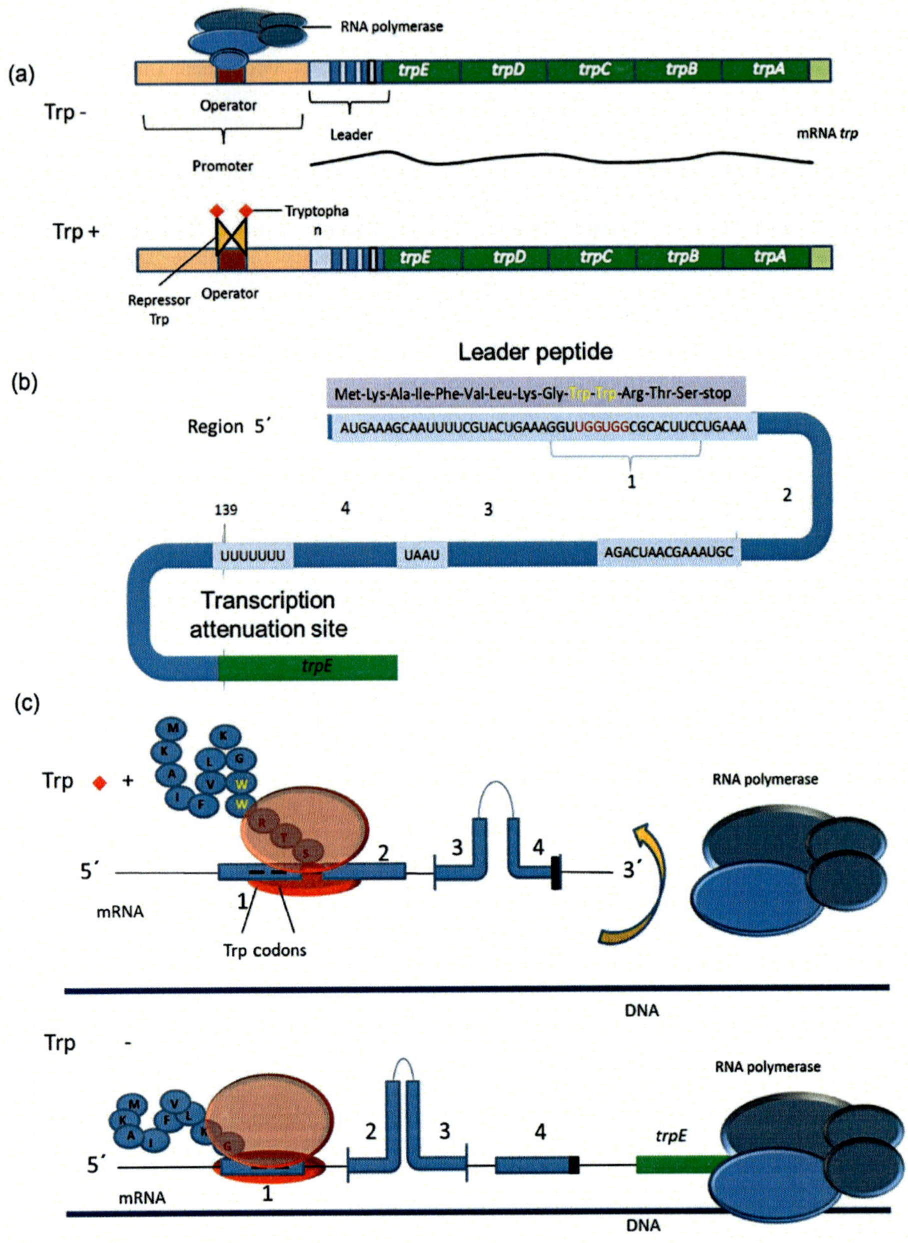

FIGURE 4.7 Regulation of the *trp* operon. (a) Structure of the *trp* operon showing the five coding genes of the biosynthetic pathway proteins (in green), the operator in the promoter region where the Trp repressor binds (in brown) and the leader sequence (in blue), also called trpL (highlighted with black outline the region called attenuator). (b) Structure of the trpL showing the leader peptide sequence and the regions forming secondary structure generating the attenuator or non-attenuator clamp. (c) Schematic representation of the attenuation.

site for that sugar in its protein structure. When lactose is present in high amounts, the lactose binds to the repressor which then dissociates from the operator. In addition to the CAP protein (an example of positive regulation), the *Lac* repressor regulates *lac* operon expression, causing the cell to express only enzymes required for lactose metabolism when glucose levels are low. This is an example of negative regulation (Figure 4.6).

Another interesting example of transcription regulation in prokaryotes is represented by tryptophan operon and is called transcriptional attenuation. *E. coli* has the ability to synthesize all 20 of the common amino acids required for protein synthesis. The genes responsible for encoding the enzymes required for amino acid biosynthesis are again usually grouped into operons, and the transcriptional unit is only activated when the cellular requirement exceeds the available amino acids supply. The tryptophan (*trp*) operon is also an example of negative regulation and provides an example of coordinated transcription and translation that is only possible in prokaryotes due to the absence of cellular compartmentalization (Kaberdin and Bläsi, 2006).

The *trp* operon consists of five genes encoding the enzymes required for tryptophan biosynthesis, an operator sequence that is the site for binding of the repressor (called the Trp repressor) and precedes the promoter and a sequence called the attenuator (Figure 4.7). The process of regulating the Trp repressor is opposite to the Lac repressor, that is, when the ligand is not present, the repressor is dissociated from the promoter. This is because the Trp repressor is a homodimer which has binding sites for amino acid tryptophan itself. Thus, when this amino acid is abundant, the Trp repressor has tryptophan bound in the protein structure which changes the conformation of the repressor, increasing its binding strength to DNA. With the Trp repressor bound in the operator, transcription of the *trp* operon is down-regulated (Figure 4.7a). A second level of regulation is imposed in conjunction with the Trp repressor, the transcription attenuation process. In this process, transcription is usually initiated but abruptly stopped before the RNA polymerase reaches the operon genes. The attenuation occurs because of the positioning of the leader sequence following the start of transcription and precedes the operon genes (Figure 4.7b). This leader sequence has two codons for the amino acid tryptophan, and DNA regions numbered 1–4 that have complementarity to each other. When there is abundant tryptophan, the ribosome will be able to quickly translate peptide 1 (which contains the two trp codons in a row) and will then pass into region 2 before the RNA sequence 3 is transcribed. Thus, region 3 will pair with the newly transcribed region 4, forming a hairpin clamp, much like a terminator (see below for transcription termination procedures), causing the RNA polymerase to dissociate from the operon before it transcribes the first gene in the biosynthetic pathway (trpE). When tryptophan levels are low, the ribosome cannot translate sequence 1 of the leader peptide because insufficient tryptophan is present in the intracellular pool to allow translation of the two trp codons in sequence 1. Sequence 2 is therefore not associated with the ribosome and can pair with sequence 3, forming a non-attenuating clamp. Thus, RNA polymerase continues the synthesis of the mRNA encoding the tryptophan biosynthetic proteins (Figure 4.7c).

4.5.2 Transcription Termination

The termination of transcription occurs when RNA polymerase encounters transcription termination sequences encoded by the DNA template. At this point, the RNA polymerase dissociates from the template strand and releases the newly synthesized RNA strand. Prokaryotes have two characteristic forms of transcription termination: dependent or independent of Rho (ρ) protein.

Rho-independent terminators do not require any additional factors and are therefore called intrinsic terminators. This form of termination involves two elements in the DNA sequence, an inverted repeat of approximately 20 nucleotides and a segment of about 8 bps composed by A:T. These elements only affect RNA polymerase after transcription, that is, they modify the growing RNA chain by the formation of a stem-loop, often called a termination clamp, due to the matching of the repeated and inverted bases (Figure 4.8). This clamp probably disrupts the interaction between RNA polymerase and the DNA template, leading to the release of the RNA polymerase.

Rho-dependent terminators have no characteristic elements in their DNA sequence. The rho-factor is a ring-shaped hexamer that binds to the newly synthesized single-stranded RNA. These sequences have only one characteristic: they are rich in C and poor in G and are always positioned before the terminator. Similar to the independent rho-factor terminators, there is clamp formation, but the RNA polymerase only pauses upon finding it. The most accepted rho-dependent termination model suggests that this factor binds to the free 5′-end of an RNA in the C-rich sequence; this factor walks along with RNA polymerase but is faster than the enzyme. Thus, when the RNA polymerase pauses in the termination clamp, the factor ρ reaches it and this interaction dissociates the RNA polymerase from the

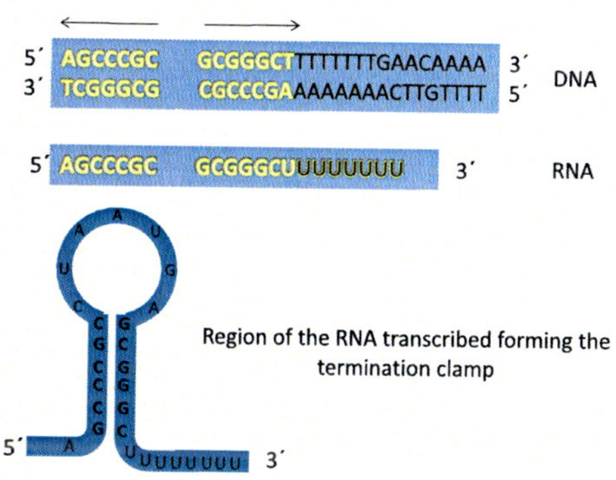

FIGURE 4.8 Structure of ρ-independent terminators. In yellow the repeated and inverted sequences that will pair in the transcribed RNA. A rod-shaped structure called a transcription termination clamp.

template and releases the synthesized RNA. Although *E. coli* has few Rho-dependent terminators, factor ρ is essential for bacterial viability, suggesting the importance of this factor in the transcriptional process.

4.6 TRANSCRIPTION REGULATION IN EUKARYOTES

In order to understand the difference between the transcriptional system in prokaryotes and eukaryotes, we must remember that in eukaryotes there is cellular compartmentalization. Eukaryotes have organelles, including mitochondria or chloroplasts which have their own genome. This implies an increase in complexity of eukaryotic gene regulation.

First, eukaryotes increase the possibilities of regulatory levels. As we saw earlier in cellular communication, depending on the signal and the type of response (both quantity and time), a cell will either change gene expression or not. Eukaryotic cells can perform localized changes through post-translational modifications such as glycosylation, phosphorylation, methylation and ubiquitination, amongst others. Cells can also redirect proteins from one compartment to another, from the cytoplasm to the nucleus or to mitochondria (etc.), changing the molecular partners and modulating the effector response to the external or internal stimulus.

However, despite so many possibilities, synthesizing proteins still means considerable energy expenditure for any cell. Thus, even for eukaryotes, the main level of cellular metabolic regulation is still at the transcriptional level. In addition, we have previously seen that most of the effective drugs on the market act on cellular receptors, the signalling reagents that culminate in transcriptional regulation. Moreover, most diseases of pharmacological interest are polygenic, probably because their effectors are under the same level of transcriptional regulation. Understanding gene expression in both prokaryotes and eukaryotes also helps us to genetically manipulate organisms for biotechnology purposes, as discussed below.

Unlike prokaryotes, eukaryotes have three types of RNA polymerases specific for different gene pools. RNA polymerase I transcribes genes encoding rRNAs (ribosomal RNA), RNA polymerase II transcribes mRNAs (messenger RNA) and RNA polymerase III transcribes tRNAs (transfer RNA) and other small nuclear and cytoplasmic RNAs. Our focus will be on RNA polymerase II, which covers most of the transcribed genes regulated in a eukaryotic genome.

In eukaryotes, transcription initiation is the most frequently regulated stage. RNA polymerase is first recruited to the promoter regions to activate transcription. A significant difference between prokaryotes and eukaryotes then begins to emerge: in prokaryotes the σ factor binds to the core of the RNA polymerase, forming the holoenzyme that will then bind to the DNA; in eukaryotes, several proteins are recruited to the promoter region, and this enzyme complex – called the pre-initiation complex – is responsible for signalling the recruitment of RNA polymerase II to the promoter. The holoenzyme forms on top of the transcription initiation site. To describe a 'minimal promoter' of eukaryotes, we need to include the basic elements so that this regulatory sequence is no longer recognized by the holoenzyme RNA polymerase, but by the basal transcription machinery that will form the pre-initiation complex in that region (Figure 4.9).

The pre-initiation complex is composed of TFIIx (general transcriptional factors of RNA polymerase II, where x denotes letters to differentiate one enzyme complex from another) and RNA polymerase II. The promoter recognition factor is the TBP (TATA-binding protein) protein that is in the TFIID complex. As the name itself denotes, TBP recognizes the TATA box sequence, the only more or less fixed position sequence in the eukaryotic promoters (about –25), responsible for positioning RNA polymerase II at the transcription initiation site (+1). Mutations in the TATA box region, contrary to expectations, do not completely abolish gene transcription but lead to a variation in transcription initiation site (in addition to a drastic reduction in the rate of initiation of the transcription process). The BRE sequence – which is the binding site for TFIIB transcription general factor, the primer (Inr) and the downstream

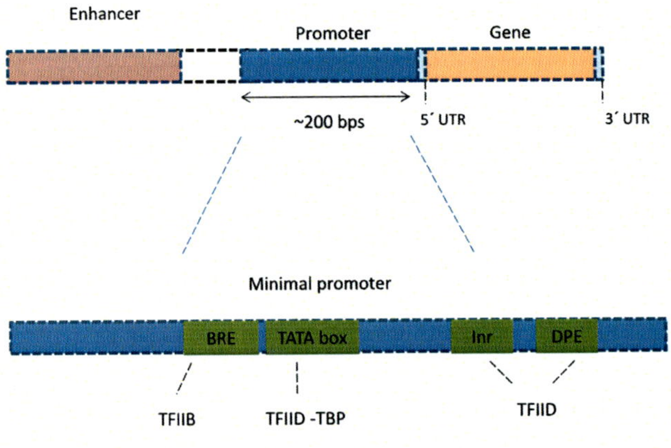

FIGURE 4.9 Structure of genes transcribed by RNA polymerase II from eukaryotes. In particular, the necessary promoter elements for the in vitro recognition by RNA polymerase II, constituting what is called minimum or essential promoter.

promoter element (DPE), required for the binding of TFIID – completes the minimal elements for in vitro transcription mediated by RNA polymerase II (Figure 4.9). Some promoters, such as those that regulate histone expression, do not have TATA Box and are therefore called 'TATA less'. In these promoters, TFIID is capable of interacting with the Inr sequence to correctly position RNA polymerase II at the transcription initiation site. This is done through the ability of some TAFs (TBP associated factors) to directly recognize Inr.

In addition, enhancers, upstream and response elements (regulated by transcription factors such as activators or repressors) mediate the efficiency with which that promoter is able to initiate transcription. Two upstream elements are found frequently in eukaryotic promoters: CAAT box and GC box. Mutations in CAAT box lead to a strong reduction in promoter efficiency under their control. CAAT box is normally at position –75, but may be at greater distances from the initiation site. GC Box is about –90 and is very common in constitutively expressed genes. The vast majority of the promoters will have at least one of these elements, CAAT and GC, to initiate transcription efficiently. The number of these elements is variable but the greater the number, the stronger the promoter.

In vivo basal transcription machinery needs many other components. Several protein complexes are required to form this machinery and initiate transcription. Initially, TBP, regulated by factors called TAFs, binds to DNA in the TATA box sequence. TBPs and TAFs are components of TFIID, the first general transcription factor to bind to the promoter. Next, TFIIA and TFIIB are ligated to create a protein complex capable of binding complexed RNA polymerase II to TFIF. Finally, TFIIE and TFIIH are ligated to terminate the assembly of the pre-initiation complex (Figure 4.10).

More important than the binding order of the TFIIx complexes (variable in many genes, according to experimental evidence) is an understanding that each plays a crucial role which is still poorly understood in the expression of most eukaryotic genes transcribed by RNA polymerase II. They assist in the recognition of promoters, correct positioning of RNA polymerase at the transcription initiation site, assist in opening the DNA strands to allow the initiation of transcription and release the RNA polymerase from the promoter to initiate elongation of the mRNA strand.

A limiting factor in gene transcription is the binding of TBP to the TATA box region. The binding of TFIID (containing TBP) leads to a distortion of the DNA strand that is probably recognized by RNA polymerase II and other TFIIx as an identifier of the promoter in active transcription, in addition to approaching upstream sites for subsequent assembly of the pre-initiation complex (Cairns, 2009) . Unlike most proteins that recognize and bind DNA, TBP binds to the minor groove of the double helix and this particularity is responsible for significant DNA twisting and approximation of upstream and enhancer sequences.

After the formation of the transcription pre-initiation complex in the promoter, RNA polymerase II needs to gain access to the template DNA at the transcription initiation site, an activity attributed to the TFIIH complex. This complex possesses a DNA helicase as one of its subunits can hydrolyse ATP by unfolding the DNA strand, allowing RNA polymerase II to initiate RNA transcription. Next, it is necessary for RNA polymerase II to be released from the pre-initiation complex and from the promoter region to initiate elongation of the RNA strand. A limiting step for this to occur is the C-terminal tail (known as CTD=C-terminal domain) phosphorylation of RNA polymerase II. This function has been attributed to the kinase subunit also present in TFIIH, in addition to other kinases (Glover-Cutter et al., 2009; Hong et al., 2009). Some studies have suggested the importance of CTD phosphorylation by TFIIH not only in releasing RNA polymerase II from the initiation phase but also in adding a cap 5′, transcription termination and 3′ end processing, as well as stabilizing the polymerase during transcript elongation. In the elongation step, topoisomerases reform the double helix following behind in the transcription bubble.

Incoherence arises in this mechanism if we consider how compacted DNA is in a eukaryotic nucleus: how does TBP reach the TATA box and distort the DNA for the recruitment of the rest of the basal transcription machinery? This is possible because more proteins are involved than just the mediators that are shown in Figure 4.1. Also involved are the so-called histone-modifying enzymes and the chromatin remodelling complexes. These enzymes have greater access to the nucleosome structure and make modifications such as acetylation and phosphorylation of histones, destabilizing the compact chromatin structure allowing access for TBP binding to the promoter.

Mediators are protein complexes involved in ensuring proper interaction between basal transcription machinery and activators or repressors, and with chromatin remodelling complexes. The first signal is the binding of the activator to enhancer regions (the enhancers), which then initiates the recruitment of the mediator complex, the chromatin remodelling complex and the basal transcription machinery, thereafter activating target gene expression. Only genes that are constitutively expressed do not require activators because these genes are constantly expressed in all tissues and cell types, as they are part of the minimal components for maintenance of cell viability and function. Typically, constitutively expressed genes are located in regions of easily accessible chromatin for TBP, in which the histone complex has a different conformation from the rest of the genome.

Contrary to popular belief, DNA compacted in the form of chromatin is not a passive, fixed structure in which proteins enter and leave to read and transmit information. Studies have shown that in fact, transcriptionally active genes in vivo are clustered in transcription islands, i.e. instead of each promoter recruiting RNA polymerase II and forming the transcription pre-initiation complex at that site, promoters are located in different regions of the genome that are grouped into nuclear compartments together with so-called 'transcription factories' to form pre-assembled transcription machinery. This is possible because DNA forms transient loops which allow genes that may be positioned in very distant loci (even on different chromosomes) to come into contact with the pre-assembled transcription machinery (Weintraub and Groudine, 1976; Osborne et al., 2004, Misteli, 2009).To exemplify the practical and technological

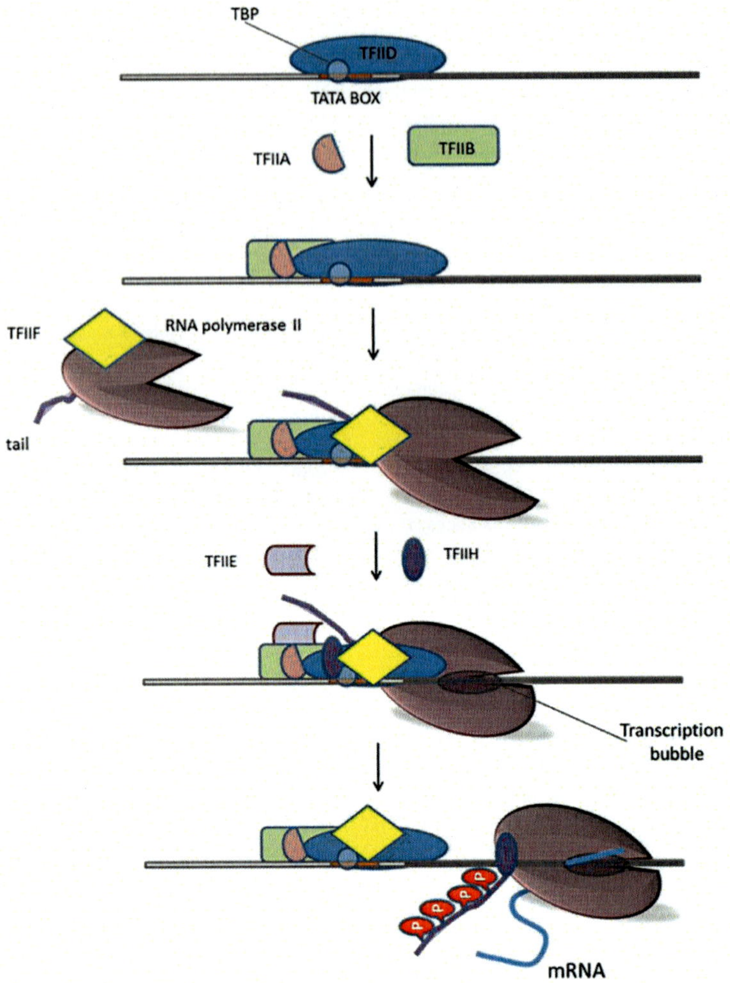

FIGURE 4.10 Sequence of events from the assembly of the transcription pre-initiation complex in a eukaryotic promoter. The order in which the factors are recruited may vary, but TBP contained in TFIID is the first protein to bind to DNA and signal that it is a promoter in active transcription, leading to recruitment of other transcriptional factors.

importance of understanding regulatory mechanisms of gene expression, let us see how the transcription basal machinery can be a powerful tool in the construction of genetically modified microorganisms for the production of industrially significant protein inputs. A technique called gTME (Global Transcription Machinery Engineering) generates random mutations in components of the basal transcription machinery, leading to a diversity of transcripts at the global level. Through this technique, selected mutants have been obtained for high tolerance to ethanol and high concentrations of glucose, a condition found in fermentation tanks for alcohol production. Mutants that exhibited these tolerances had modified isoforms of the *S. cerevisiae SPT15* gene encoding TBP (substitutions of amino acid residues: Phe177 for Ser, Tyr195 for His and Lys218 for Arg). Consequently, the rearrangement for selected gene expression was caused by changes in RNA polymerase preferences by promoter elements, leading to an adaptation of this yeast to the conditions imposed during growth. This engineered yeast was more efficient for industrial use in the production of bioethanol, with a 70% increase in productivity (Roeder, 1996; Alper et al., 2006).

In summary, unlike bacteria whose main artifice for regulating gene expression is the extensive use of repressors, eukaryotes activate the transcription of their genes. The promoters are normally silenced by the chromatin structure and require all transcription initiation machinery, mediators and activators to be transcribed.

4.7 RNA PROCESSING

The stability of RNA can be considered as a second level of regulation for gene expression. Factors that influence stability are mainly related to modifications at the 5′ and 3′ ends of the mRNA molecule. In eukaryotes, there are also other processing levels related to the exons (coding regions) and introns (intergenic regions) that generate greater variability of transcripts and consequently of effectors.

Unmodified transcripts after the end of their synthesis are called primary transcripts. Prokaryote transcripts undergo few changes after transcription and are generally still translated during their synthesis. Although to a lesser extent, some prokaryote mRNAs undergo processing. Recently, 2 poly-A polymerases have been characterized in *E. coli*, and the addition of small poly-A chains (15–60 adenylate residues) changes the stability of the RNA. In contrast to eukaryotes, in prokaryotes the larger the poly-A tail, the lower the mRNA stability. This change is attributed to an increase in affinity of RNAs with 3′ → 5′ exonuclease activities, such as PNPase (polynucleotide phosphorylase) and RNAse II.

RNAs that do not undergo polyadenylation, most of them in prokaryotes, protect their 3′ untranslated region (3′ UTR) from the action of exonucleases by forming secondary structures, such as clamps or hairpins. The untranslated 5′-portion (5′UTR), which does not undergo cap addition (see below), can be regulated by various other mechanisms; among the most important is the presence of regulatory ribosome-binding sequences, binding of non-coding RNAs and temperature. Inefficient binding of ribosomes or interaction with translation repression factors leads to a decrease in stability, making this RNA more susceptible to the action of exonucleases such as RNAse E, which is considered to be one of the main regulating enzymes for the decay of mRNAs in bacteria. Mutant bacteria for the gene encoding RNAse E have a two to threefold increase in stability of total mRNAs. This ribonuclease is also responsible for processing the 5′-portion of the primary rRNA transcript through an endonuclease activity (O'hara et al., 1995).

Temperature changes govern the formation of secondary structures in different 5′UTRs of prokaryotic mRNAs important for thermal shock, and these structures are being both inhibitory and activators of degradation, depending on cellular condition. A classic example is the mRNA encoding σ^{32} factor, the main component of the holoenzyme RNA polymerase of bacteria during heat shock. Such mRNA forms a secondary structure inhibiting translation by preventing recognition of the ribosome at the translation initiation site. When there is an increase in temperature, this structure is destabilized, leading to the translation of the mRNA and production of the alternative σ factor, necessary for the adaptation of *E. coli* to the new growth conditions.

In eukaryotes, as soon as the mRNA elongation phase begins, the factors associated with RNA polymerase II are replaced by the elongation and processing factors of the 5′-UTR region. The main modification is an addition of a 5′ cap (helmet) through three sequential reactions. The first is the removal of a phosphate from the 5′-end of the transcript by RNA triphosphatase. Next, the addition of a guanine by the enzyme guanylyl transferase through an unusual 5′-5′-linkage occurs. Finally, methylation of this GTP is catalysed by methyl transferase. The addition of a 5′-cap occurs when the mRNA strand is 20–40 nucleotides in length. After this modification, the enzymes responsible for the addition of a cap 5′ dissociate to leave a new enzyme complex that is responsible for rearranging exons and introns called the spliceosome.

A spliceosome is the cellular machinery responsible for removing introns and shuffling exons, allowing maturation of mRNA by forming different combinations of sequences that can increase the possible number of proteins that can be transcribed using only one encoding gene. Most eukaryotes have a gene arrangement that intersperses coding parts (the exons) and non-coding parts (the introns) (Figure 4.11). Thus, the primary transcript has intergenic regions that do not encode proteins or RNAs with function that need to be removed before the mRNA is exported to the cytoplasm and can be translated. Thus, after the addition of cap 5′, the second step in mRNA processing in eukaryotes is the removal of the introns and the rearrangement (splicing) between the exons, which is performed by the spliceosome ribonucleoprotein complex. In this way, the mRNAs can be processed by placing the exons in the order they are in the genome, or in different combinations, as shown in Figure 4.11.

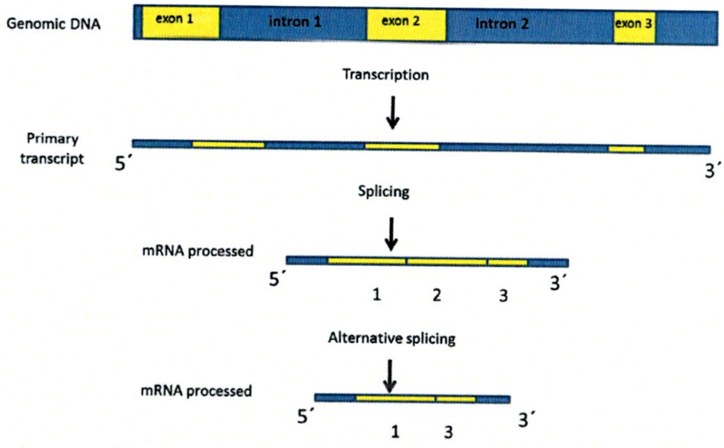

FIGURE 4.11 Structure of eukaryotic genes, showing the presence of exons (regions that will encode proteins or RNAs with function) intercalated by the introns (intergenic regions that are removed from the mRNA during processing). The machinery responsible for removing the introns and combining the exons is the spliceosome.

The last step of mRNA processing is closely related to the termination of transcription in eukaryotes, the addition of the poly-A tail. When RNA polymerase reaches the end of the gene it is transcribing, it encounters specific sequences that are recognized by the polyadenylation complexes. The binding of these complexes, called cleavage and polyadenylation specificity factor (CPSF) and CstF (cleavage stimulation factor), triggers the cleavage of the transcript, the addition of the poly A tail to the 3′ end and the subsequent termination of transcription. The enzyme responsible for poly-A tail synthesis is poly-A polymerase (PAP), which adds about 200 nucleotides of adenine to the 3′ end after cleavage. It is unclear how polyadenylation influences the termination of transcription by RNA polymerase, but it is well established that transcription of the polyadenylation signal is required for termination. Polyadenylation of the 3′ end of the mRNA has been involved with the stability of the mRNA, but in a directly proportional manner (unlike the examples in prokaryotes), i.e., the longer the poly-A tail, the more stable the mRNA.

Thus, after the addition of a 5′cap, removal of introns and polyadenylation of the 3′ end, the mRNA is considered mature and can be exported from the nucleus to the cytoplasm where it will be translated on the ribosomes. After this step, several levels of regulation will still be imposed to control the activity of newly synthesized polypeptide chain, but this is a subject that we will not address here.

The regulation of gene expression is what gives a unique identity to different cell types. Although they all have the same genome, each one will express a particular set of genes and this depends on which cellular contacts, communication systems and proteins already present – since they were defined during the cellular differentiation process, some as early as the embryonic stage. Knowing mechanisms of gene regulation and expression helps us to understand the biological function of genes and their products, their cellular interactions and when (often also how much) these proteins are necessary for the cells. We know what genes are, but we still have a lot to discover about what they actually do in the cell, this is the great challenge of the post-genomic era.

4.8 FINAL CONSIDERATIONS

Many diseases of interest to the pharmaceutical industry have a genetic basis and involve intricate networks of communication and cell signalling. Not coincidentally, most of the widely used drugs target proteins, half of them with cellular receptor function. Communication and cellular signalling are events of extreme importance for adaptation to the external environment and are crucial in multicellular organisms. Cell signalling involves receptors that interact with a ligand, this interaction changes the structural conformation of the receptor and leads to a cascade of intracellular events that ultimately affect gene expression and protein pool of cells in response to the stimulus. The main regulatory mechanisms of gene expression occur at the transcriptional level in the synthesis of RNA since it is very expensive energetically to synthesize proteins. Like DNA, RNA synthesis has a 5′ → 3′ single sense, the template strand being read by RNA polymerase and the encoding strand that will dictate the sequence of ribonucleotides. The regulation of gene expression in prokaryotes and eukaryotes differs in several aspects. The first is the absence of intracellular compartmentalization of prokaryotes, allowing transcription to be coupled to translation in many cases. Prokaryote RNA polymerase is only recruited to the promoters in the holoenzyme form, i.e., in the presence of the σ factor subunit recognition and promoter. Factor σ binds to conserved regions of bacterial promoters called –10 and –35. The more similar the consensus sequence is to these regions in the promoters, the greater will be the ability to recruit the holoenzyme and initiate transcription, a key process in the regulation of transcription. The process of transcription regulation most widely used by prokaryotes is the use of transcription repressors. Constitutive genes do not undergo regulation and have –10 and –35 regions very close to the consensus sequence. On the other hand, eukaryote promoters use mainly activators. This is because they require chromatin remodelling that normally represses the genes physically; this rearrangement is responsible for exposing important promoter sites such as the TATA Box, allowing the binding of the TBP and the formation of the pre-initiation complex of transcription and recruitment of RNA polymerase II. In addition, eukaryotes can regulate cell localization of repressors and activators, their activity by phosphorylation or other post-translational modifications, using kinases and second messengers that will modulate the strength, amplitude and duration of the response to the stimulus, in addition to control mRNA processing and stability. Prokaryotic mRNAs are translated into the primary transcript form undergoing little or no modification upon completion of their synthesis. Eukaryotic mRNAs undergo addition of a 5′ cap, removal of introns and exon organization, as well as 3′ end polyadenylation.

The expression regulation is essential to promote biotechnological advances, especially in pharmaceutical products. Several heterologous proteins are toxic for host cells and need to be tightly controlled. The knowledge about gene expression regulation is crucial to stabilize industrial platforms to perform biotechnological protein production.

REFERENCES

Alberts, B. (Coord.) *Molecular Biology of the Cell*, 5th edition, New York: Garland Science, 2008.

Alper, H., Moxley, J., Nevoigt, E., Fink, G.R., Stephanopoulos, G. Engineering yeast transcription machinery for improved ethanol tolerance and production. *Science*, v. 314, 1565–1568, 2006.

Bakheet, T.M. and Doig, A.J. Properties and identification of human protein drug targets. *Bioinformatics*, v. 25, 451–457, 2009.

Cairns, B.R. The logic of chromatin architecture and remodelling at promoters. *Nature*, v. 461, 193–198, 2009.

Drews, J. Drug discovery: a historical perspective. *Science*, v. 287, 1960–1964, 2000.

Glover-Cutter, K., Larochelle, S., Erickson, B., Zhang, C., Shokat, K., Fisher, R.P., Bentley, D.L. TFIIH-associated Cdk7 kinase functions in phosphorylation of C-terminal domain

Ser7 residues, promoter-proximal pausing, and termination by RNA polymerase II. *Mol. Cell Biol.*, v. 29, 5455–5464, 2009.

Hong, S.W., Hong, S.M., Yoo, J.W., Lee, Y.C., Kim, S., Lis, J.T., Lee, D.K. Phosphorylation of the RNA polymerase II C-terminal domain by TFIIH kinase is not essential for transcription of Saccharomyces cerevisiae genome. *Proc. Natl. Acad. Sci. U S A*, v. 106, 14276–14280, 2009.

Kaberdin, V.R. and Bläsi, U. Translation initiation and the fate of bacterial mRNAs. *FEMS Microbiol. Rev.*, v. 30, 967–979, 2006.

Lazo, J.S. Rear-view mirrors and crystal balls: a brief reflection on drug discovery. *Mol. Interv.*, v. 8, 60–63, 2008.

Lewin, B. *Genes IX*, 9th edition, Sudbury: Jones & Bartlett Publishers, 2007.

Misteli, T. Self-organization in the genome. *Proc. Natl. Acad. Sci. U S A*, v. 106, 6885–6886, 2009.

.Nudler, E., Mustaev, A., Lukhtanov, E., Goldfarb, A. The RNA-DNA hybrid maintains the register of transcription by preventing backtracking of RNA polymerase. *Cell*, v. 89, 33–41, 1997.

O'hara, E.B., Chekanova, J.A., Ingle, C.A., Kushner, Z.R., Peters, E., Kushner, S.R. Polyadenylylation helps regulate mRNA decay in Escherichia coli. *Proc. Natl. Acad. Sci. U S A*, v. 92, 1807–1811, 1995.

Osborne, C.S., Chakalova, L., Brown, K.E., Carter, D., Horton, A., Debrand, E., Goyenechea, B., Mitchell, J.A., Lopes, S., Reik, W., Fraser, P. Active genes dynamically colocalize to shared sites of ongoing transcription. *Nat. Genet.*, v. 36, 1065–1071, 2004.

Roeder, R.G. The role of general initiation factors in transcription by RNA polymerase II. *Trends Biochem. Sci.*, v. 21, 327–335, 1996.

Watson, J.D. (Coord) *Recombinant DNA*, 2nd edition, New York: W. H. Freeman, 1992.

Watson, J.D. (Coord.). *Tradução de Passaglia, Luciane. Biologia Molecular do Gene*, 5 edição, Porto Alegre: Artmed, 2006.

Weintraub, H. and Groudine, M. Chromosomal subunits in active genes have an altered conformation. *Science*, v. 193, 848–856, 1976.

Yao, L. and Rzhetsky, A. Quantitative systems-level determinants of human genes targeted by successful drugs. *Genome Res.*, v. 18, 206–213, 2008.

5 Molecular Biology Tools
Techniques and Enzymes

Marina Gabriel Fontes and Gisele Monteiro
Universidade de São Paulo

CONTENTS

5.1 INTRODUCTION

In this chapter, we will understand how the discovery and use of enzymes led to the genetic manipulation of different organisms, giving rise to the production of recombinant proteins which have had a wide range of purposes, from functional and structural studies, to commercialization of proteins for food, drugs and cosmetics, amongst other applications. We will consider recombinant DNA technology as an example of the use of enzymes in molecular biology for biotechnological purposes.

One of the most important technologies that have emerged with the advances of molecular biology has been the production of recombinant proteins. Recombinant proteins may be defined as an isolated and expressed gene product with or without modification, in the same organism of origin or in another. For the expression of proteins from a different origin of the host organism, the name heterologous expression is given.

Recombinant proteins and heterologous expression open up a myriad of possibilities and allow 'potentially' any biopharmaceutical to be produced using these technologies. This potential will depend on how complex the gene product is, i.e., whether it consists of many exons and introns, generating mRNAs that may undergo alternative splicing or result in proteins that require post-translational modification for their function, such as phosphorylation, glycosylation, oxidation (the formation of disulphide bonds) or binding to cofactors. These and other pre-requisites, if possible, should be known in advance for the successful production of the recombinant protein. The most classic example of successful heterologous production is that of human insulin in the bacterium *Escherichia coli*, commercially released in 1982.

From the definition of recombinant proteins, we arrive at the first necessary step for the genetic manipulation of any organism: to know and isolate the gene of interest. This implies a constraint to the number of possible genes that can be used for the production of recombinant enzymes. Only those genes where the location on a genome sequence is known can be isolated and modified, hence the immense importance of whole-genome sequencing (WGS) projects in this initial characterization (Alberts, 2008). We will understand this dependence by exemplifying the first technique necessary for gene isolation, the polymerase chain reaction (PCR).

5.2 ISOLATING A GENE SEQUENCE-DNA POLYMERASE

In order to isolate a gene sequence, it is necessary to know its sequence. This allows us to differentiate the gene of interest from the huge number of other genes encoded by the genome of an organism. This delimitation is possible through the use of nucleotide primers that can recognize the gene encoded on the genomic DNA template by complementary base pairing and act as a target for DNA polymerases, the enzymes responsible for duplicating the genetic material through the replication process. So let's remember how DNA replication 'in vivo' occurs to understand what happens in the 'in vitro' reaction.

DNA polymerases require: a substrate for the polymerization reaction (a template), primers, deoxyribonucleotides (dNTPs, i.e. dATP, dTTP, dCTP and dGTP) and divalent cations, usually magnesium (cofactor). The template is required for the DNA polymerase to replicate an identical copy of the genetic material, and for that, it requires a prior sequence to 'read', pair the dNTPs and then polymerize the nascent chain. This fact refers to an important property of the DNA replication process, being semiconservative, that is, a new DNA strand is synthesized from and paired with a pre-existing strand. Thus, when replication occurs, it is necessary to open the double helix of chromosomal DNA, forming the replication bubble, thereby providing the necessary template to the DNA polymerase. This opening of the strands is catalysed by the enzyme DNA helicase, and the stress caused by this opening is alleviated by the enzyme topoisomerase (Cooper and Hausman, 2009).

As in RNA synthesis, DNA is directed in the 5′ → 3′ direction. A crucial difference is that RNA polymerase is able to add ribonucleotides to a nascent RNA strand from the template without the need for a primer. DNA polymerases, however, require such a primer which will provide the free 3′OH end for the phosphodiester linkage to the 5′-phosphate end of the next dNTP to occur. The enzyme that synthesizes these primers in vivo is an enzyme called the primase and is responsible for the production of small random sequence RNAs that will serve as free 3′OH ends for DNA polymerases. In the laboratory, when we know the gene sequence of interest, we define the primer sequence which will be chemically synthesized by specialized companies.

Thus, upon the opening of the strands, annealing of the primers provides the substrate for the DNA polymerase to initiate the polymerization of the nascent chain, thereby adding the dNTPs. However, another no less important property of DNA polymerase is that the polymerase does not initiate DNA replication in vivo anywhere in the genome at any time. The synthesis is guided by a series of signals (for example, by cyclins) that recruit DNA polymerase to specific genomic sequences called the origin of replication. Each organism has a specific DNA replication origin for the recognition of its endogenous DNA polymerase. This is not necessary in vitro since there is no packaging of the template in the form of chromatin. In short, DNA replication occurs through three defined steps: (a) opening of the template strands, followed by (b) pairing of the primers and (c) polymerization reaction of the nascent DNA chain.

In the earliest in vitro DNA syntheses, researchers mimicked replication occurring in vivo using temperature cycling, i.e., high temperature to open the strands, decrease the temperature for annealing the primers and further change in the reaction temperature for polymerization of the nascent chains using *E. coli* DNA polymerase. However, this method was extremely labour intensive and costly since the DNA polymerase used was thermosensitive and denatured at each opening cycle of the DNA strands at 95°C, having to be added every time manually after the annealing cycle of the initiators.

This was changed when in 1986, K. Mullis and colleagues developed the polymerase chain reaction (PCR), which won him the Nobel Prize in 1993. In the mid-1965, Thomas D. Brock discovered an organism called *Thermus aquaticus*, which was able to colonize geysers in Yellowstone National Park from where it was isolated. This thermophilic bacterium is able to survive in environments with temperatures up to 80°C, and, consequently, its enzymes are heat resistant. Thus, Mullis et al. created the automation of the DNA synthesis reaction in vitro with the use of Taq polymerase and called this synthesis as PCR. With the manufacture of thermocyclers, this technique has become one of the most widely used as a molecular biology tool worldwide (Figure 5.1).

In addition to Taq polymerase and thermocyclers, another crucial tool to be developed was the automation of chemical synthesis of oligonucleotides. This allows specific sequence primers capable of delimiting the gene of interest to be rapidly and cheaply synthesized. To design sequence-specific primers, we have to know where the gene begins and ends. Thus, using a sequence database that contains all of the genome sequences published in the scientific literature, we can design specific primers by finding the translation start sequence (i.e. from the ATG forward) and the translation termination codons (i.e. from the TAA, TAG or TGA reverse) (Lewin, 2007) . It is important to note that any sequence deposited in the database is represented by a single strand represented always in the 5′ → 3′ direction. This is a convention, since the complementary strand can be deduced, because we know that A matches with T and G matches with C. This convention has a direct implication for primer design: The reverse primer must be represented in the direction of 5′ → 3′ as well as the forward primer. In general, the primers are represented indicating the direction of the synthesis. Thus, to draw a specific primer, as the database sequence is already in the sense of synthesis (5′ → 3′), simply copy the beginning of the strand DNA to about 14 bases, ending in C or G (see discussion follow). However, the reverse primer sequence will be the sequence complementary to the 3′ end in the reverse direction, hence its complementary reverse primer name (see example in Figure 5.1).

But, why do the primers define the length of the gene sequence that can be isolated? This is due to the exponential nature of the PCR reaction. In the first cycle, for each template DNA molecule, only two new strands will be partially delimited by the primers and synthesized. However, these will already be in equal number to the initial template DNA and will also serve as a template in the second PCR cycle. As the replication cycles continue, the original template will become insignificantly diluted in the midst of so many new copies. For a quantitative idea, after 30 cycles of PCR, 1 initial template genomic DNA molecule will yield approximately 60 'large' partially double-stranded copies, and about 1 billion (1,073,741,770) copies of the primers delimited double-stranded target sequence.

A generic PCR programme capable of amplifying any given template using most primers (since each, depending on the size and content of CG, will have a specific annealing temperature) can be described as follows:

- Initial template denaturation cycle – 1 initial cycle of 5 min at 95°C.
- Exponential polymerization cycle – from 25 to 30 cycles consisting of:
  - Denaturation of the DNA template: 1 min at 95°C.
  - Annealing (pairing) of the primers: 45 s at 55°C.
  - Extension of new DNA strands (target sequence): 1 min/Kb at 72°C.
- Final extension cycle – 1 cycle of 7 min at 72°C.

After this standard reaction, we can visualize the final product by agarose gel electrophoresis chromatography. Due to the negative charges of DNA conferred by the phosphate groups, the amplified products will migrate to the positive pole when subjected to an electric field. This migration will be as faster as the size of the molecule is smaller because it will be easier to cross the mesh of the agarose gel. Staining the agarose gel with a fluorescent dye that intercalates DNA (such as ethidium bromide) allows the amplified DNA to be

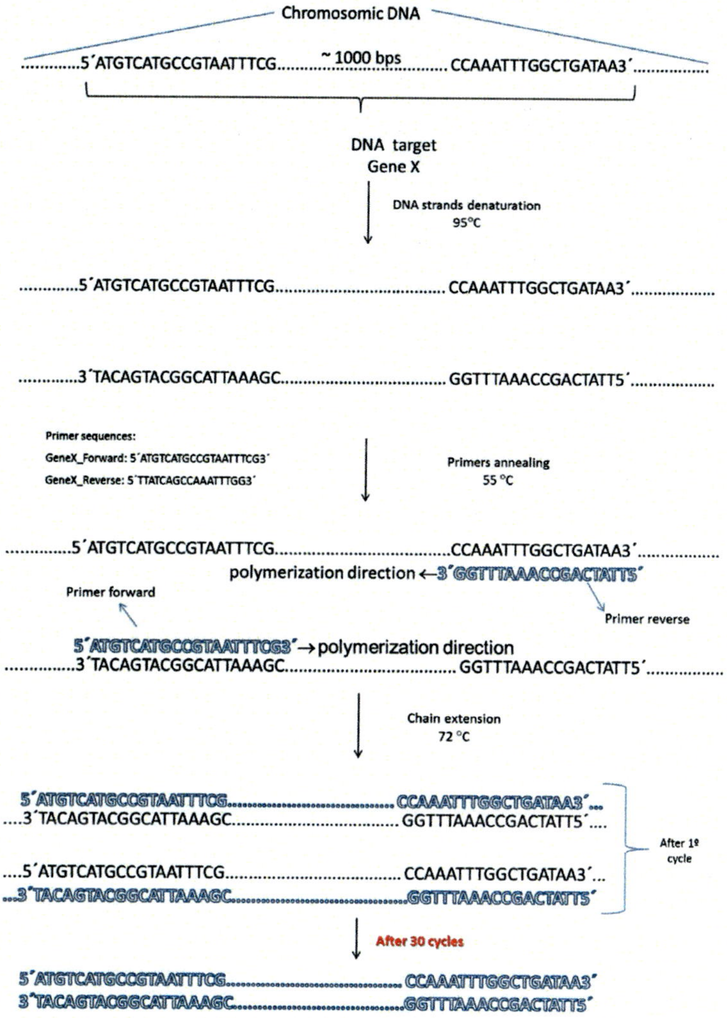

FIGURE 5.1 Isolation of genes by polymerase chain reaction (PCR). Thermocyclers vary temperatures by allowing the reaction to occur in three main steps: 95°C for template DNA denaturation, 55° primer annealing (this temperature is variable) and 72°C for the extension of the new DNA strand. Thanks to the thermostable properties of Taq polymerase, DNA synthesis in vitro revolutionized molecular biology.

viewed under a UV light in comparison to a DNA standard of known molecular weight (Figure 5.2).

Several parameters influence the efficiency of the PCR reaction. Let us begin by analysing the first step necessary to perform the cloning of a gene, the definition of the sequence of the primers. According to the size and composition of the sequence, such primers will have variable annealing temperatures (Tm = melting temperature). In order to establish the best Tm for both primers and complementary DNAs that will be ligated (annealing step of reaction), an average value can be calculated using the following general formula:

$$\text{Tm} = 4°\text{C} \times (\text{number of G's and C's in the sequence}) + 2°\text{C} \times (\text{number of A's and T's in the sequence}),$$

valid for DNAs <14 bps

The companies that chemically synthesize the oligonucleotide primers also provide free online programmes for these calculations. The melting temperature expresses the point at which half of the primer sequence molecules will be paired to the complementary DNA in the template and half will not. Thus, for PCR reactions this temperature is used during the annealing cycle of the reaction. If the temperature used during the annealing phase is greater than the Tm of the primers, less than half of them will anneal to the template during the extension phase of the DNA strand, decreasing the chances of amplifying the target gene but increasing the specificity (i.e. reducing the chances of non-specific binding to other DNA sequences). If the temperature is less than Tm, more than half will be paired and may generate non-specific (partial pairing) binding to other regions of the genome; remember that your DNA template is initially immense and diverse.

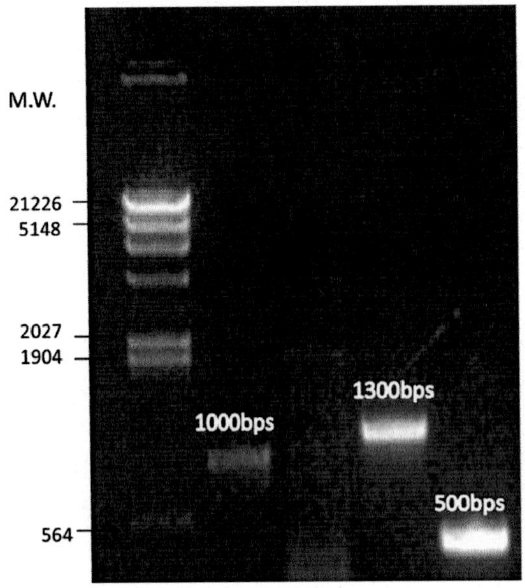

FIGURE 5.2 Agarose gel electrophoresis for the visualization of nucleic acids. 0.8% agarose gel in TAE 1X buffer (Tris-acetic acid-EDTA) stained with ethidium bromide. Visualization made by exposing the gel to ultraviolet light. M.W., molecular weight standard (in base pairs, bps).

This is one of the most important parameters for effective PCR. When a standard cycle is attempted, and there is no amplification of the target gene, the annealing temperature in the cycle must be changed. The first attempt is to obey the Tm of the primers, remembering that there are two, one forward and one reverse. Thus, since we normally have different Tm for each primer, the smaller of them is used. Since the sequence of the primers will be determined by those of us who are cloning the gene, it is a good idea when 'designing' primers is to keep the Tm of both primers nearly the same, varying at most 5°C between them.

Another crucial factor influencing the efficiency of the reaction, still with respect to the primers, is the 3′ end composition. Remember that this end paired with the template will be the substrate for the DNA polymerase. Thus, the primers should preferably terminate at C or G, because the bases of these nucleotides pair through three hydrogen bonds, while A and T pair by two hydrogen bonds. The higher the content of C and G at the 3′ end, the stronger the annealing and the more efficient the formation of the substrate for DNA polymerase.

This observation reflects a third point in the design of primers, the content of C and G relative to that of A and T. Some genes have regions of translation initiation extremely rich in CG which greatly increases Tm. The maximum Tm value that we can use when we are designing the primer is the temperature optimum of the DNA polymerase enzyme, that is, 72°C for the Taq polymerase. If the Tm of the primers is above the optimum for the polymerase, then the annealing temperature would have to be reduced to avoid chain extension during the annealing stage and to avoid likely premature denaturation of the polymerase. However, an annealing temperature below the Tm of the primers will generate a large number of non-specific amplification products that may be difficult to resolve on an agarose gel. To modulate Tm without altering the subsequent sequence of the amplified product, we can add bases at the 5′-portion of the primer. Thus, if Tm is too high, we add A and T in the 5′-portion; if Tm is too low, we add C and G to that end. This possibility of manipulating the 5′-end is extremely useful for cloning purposes, such as the addition of adapters with restriction enzyme recognition sequences (see below). Another way to manipulate Tm is to increase or decrease the size of the primer. If too low, we can include more bases of the target sequence, increasing specificity; if it is too high, we can decrease the size by placing fewer bases corresponding to the target sequence. An average size of primers with good amplification characteristics is around 18–30 bases, with Tm between 52°C and 65°C.

Another important parameter to initiate a cloning is to know the target DNA gene structure and establish the source of the template. This is because genes from higher eukaryotes have intronic regions. Thus, genomic DNA can only be used as a template if it is derived from organisms with little or no intron, such as prokaryotes or yeast *S. cerevisiae*, where rare genes are composed of exons and introns. For organisms with the genomic structure composed of both elements, we perform another cloning strategy using a special DNA polymerase enzyme, reverse transcriptase. This enzyme has all the characteristics of a conventional DNA polymerase but has the ability to use RNA as a template for the synthesis of a complementary strand of DNA, which is called cDNA.

In Chapter 4, we have seen that mRNA undergoes processing in order to remove the introns and to rearrange exons, adding a cap to the 5′-end and a poly-A tail to the 3′ end. This process is so fast that if we isolate total mRNA from a cell at any stage of growth, virtually all mRNA molecules will be in the processed form. We are able to use these mRNA molecules (called mature transcripts) as the initial template for cloning complex gene products by means of reverse transcriptase.

The strategy works the following way: we have to know if the gene of interest is expressed in a given tissue. Then, we extract RNA from these cells and use this as the template for the RT-PCR (reverse transcription-polymerase chain reaction). Taking advantage of the characteristic of poly-A tail in the processed mRNA, we use as primers oligo dT, that is, sequences of several deoxythymidine that will pair with the poly-A tail, serving as the initiator for all mRNAs. Reverse transcriptase will synthesize cDNAs from all mRNAs present in the template sample, generating mRNA-cDNA hybrids. This sample is then treated with a ribonuclease, the RNAse H, able to recognize these RNA–DNA hybrids, degrading only the RNA and leaving single-strand cDNA. This single-stranded cDNA can be used as the template for a common DNA polymerase, which also will use the small RNA fragments left by RNAse H as primers for the synthesis of double-stranded cDNA. We now have double-stranded cDNA, compatible to be used in the PCR reaction as a template to amplify the target gene using specific synthetic primers (Figure 5.3). With the target gene now isolated from the DNA template and in high

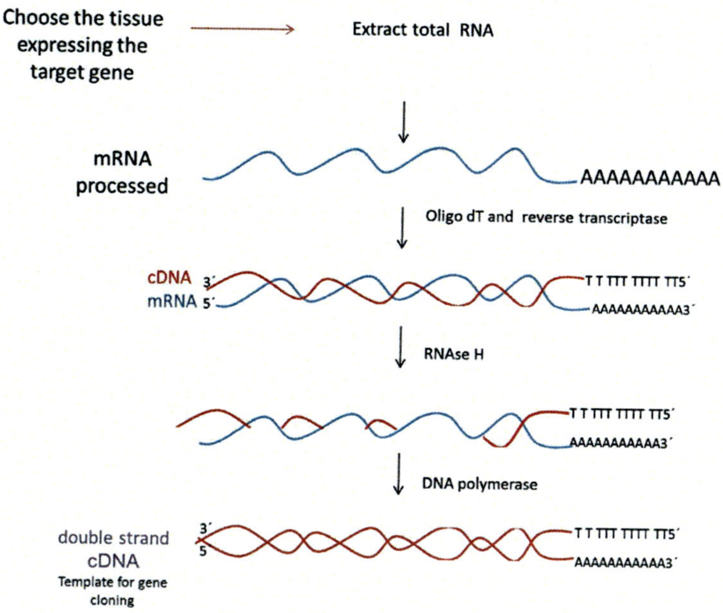

FIGURE 5.3 A strategy to obtain a template for the isolation of genes with complex gene structure. When a desired gene product is formed by exons and introns, DNA polymerase reverse transcriptase is used. This is because this enzyme recognizes RNA as template to synthesize a complementary strand of DNA, then called a cDNA. The cDNA can be used subsequently as a template for Taq polymerase for the isolation of the gene of interest.

quantity, we can start the next steps of cloning, i.e., insertion of the gene into cloning and expression vectors. We will first describe some techniques for obtaining DNA sequences – used in genome projects and in several other applications – as we refer to these throughout this chapter.

Initially, researchers began to understand the nucleotide sequence for only a region of a gene. This then expanded to determine the nucleotide sequence for entire genes, then to whole chromosomes and finally to the complete genomes. The first genome to be deciphered was that of the bacterium *Haemophilus influenzae* in 1995. The following year, the first complete genome of a eukaryote, the yeast *Saccharomyces cerevisiae*, was published, and on 14 April 2003, the sequencing of a human genome with coverage of 99.99% was announced.

In the 1970s, the researcher Frederick Sanger proposed a method for sequencing that is still used (albeit with modifications) today. This technique uses one of the most important properties of DNA polymerases, the fact that these enzymes recognize as a substrate only a free 3′OH end paired in the DNA template. From this premise, Sanger developed the method of sequencing based on the use of dideoxyribonucleotides (ddNTPs) (Figure 5.4a). Without the free 3′OH end to carry out the phosphodiester bond with the next 5′-triphosphate nucleotide, the polymerization of the nascent chain always stops at the position where the ddNTP was incorporated, allowing its identification (Figure 5.4b). Because there is a mixture between the four dNTPs and one ddNTP, in a certain proportion, when a ddNTP is randomly incorporated there will be the termination of chain synthesis, hence the name of the sequencing technique – the chain termination method or Sanger sequencing. The principle of the sequencing reaction is the same as PCR: there are cycles of temperature, template, Taq polymerase, Mg^{+2}, dNTPs and additionally ddNTPs to terminate chain synthesis. By performing this procedure in individual tubes each with a different known ddNTP (ddATP or ddTTP or ddCTP or ddGTP), it is possible to separate those fragments generated by the reaction on a high-resolution polyacrylamide gel capable of distinguishing strands with only one base difference in size. Visualization can be done with radiolabeled primers (Figure 5.4b).

Currently, we still use Sanger sequencing, but with some modifications. The most important of these is the fact that we can now perform a single sequencing reaction, by marking the ddNTPs with different fluorophores. This method is called dye terminator sequencing. Thus, in the same tube, we put the four dNTPs and the four labelled ddNTPs so that an optical reader can identify fluorescent fragments: blue when incorporating a ddATP, red when incorporating a ddTTP, green when incorporating a ddCTP and yellow when incorporating ddGTP (illustrative example only). The separation method continues to be electrophoretic by size with resolution of a base of difference, but not more in polyacrylamide gels, but by capillaries containing resins (for example, agarose). There is no longer any need to use radiolabeled primers since optical readers detect the fluorophores of ddNTPs and thus do not require this labelling.

An important observation to be made at this time is that the sequencing is by polymerization and reading of a single strand (although the template is double stranded, it is denatured in the sequencing reaction), and therefore only one primer is placed in the reaction mix. Normally, in order to confirm and have a duplicate of the reaction, we

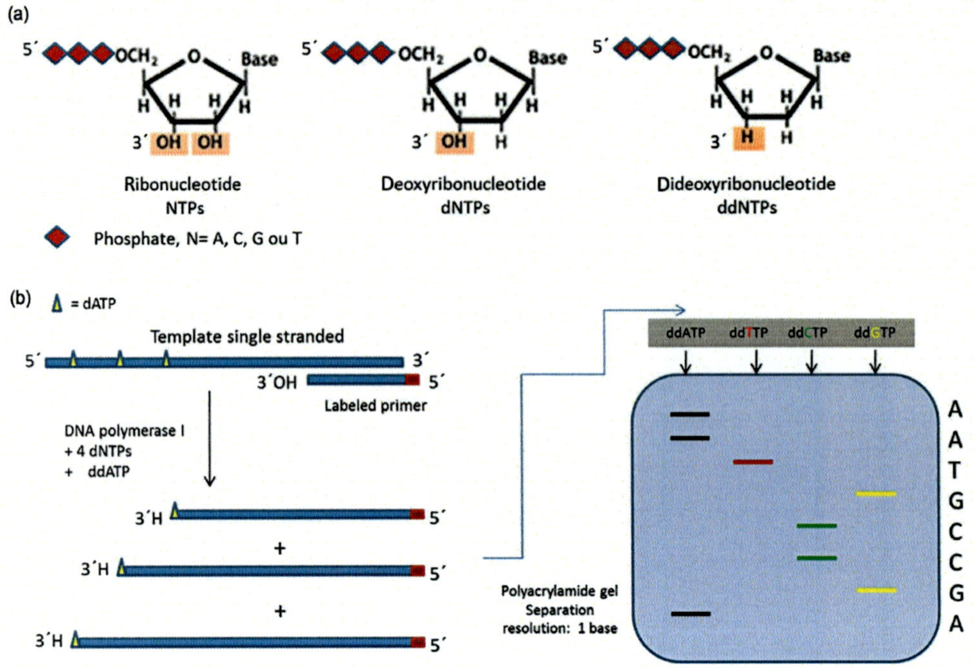

FIGURE 5.4 Sequencing based on chain termination. (a) Structure of nucleotides showing the differences between ribonucleotides, deoxyribonucleotides and dideoxyribonucleotides. (b) Sanger method: in four separate tubes, the single-stranded DNA polymerization is performed using dideoxyribonucleotides (ddNTPs, where N represents A, C, G or T). Due to the absence of the hydroxyl at the 3′ end, every time a ddNTP is incorporated there will be termination of the synthesis. By analysing the fragments produced by the four reactions using high-resolution gel electrophoresis, the sequence can be determined.

make sequencing of the direct strand, thus using the reverse primer, and in another tube, we do the sequencing of the reverse strand using the forward primer. Sample buffers for sequencing use denaturing agents, so that in capillaries only single strands of DNA are separated by size.

5.3 MANIPULATING THE GENE OBTAINED BY PCR

When we isolate a gene or a codifying sequence of gene (mRNA → cDNA), we take a risk because of errors caused by the insertion of aberrant bases by the Taq polymerase during synthesis. The error rate of this enzyme, which can generate a mutation in the PCR product, is 1 error for every 9,000 nucleotides incorporated (Tindall and Kunkel, 1988). The Taq polymerase enzyme lacks an exonuclease proofreading activity 3′ → 5′, so it does not correct errors incorporated after a certain distance from the synthesis site. This increases the enzyme's processivity, that is, how many nucleotides it incorporates per second (about 1,000 base pairs in less than 10 s), but leads to a considerable error rate across large sequences that subsequently be cloned. An alternative when large genes or gene products are to be cloned is to use another thermostable DNA polymerase, but one which retains high fidelity because it possesses 3′ → 5′ proofreading activity, like *Pfu* DNA polymerase. This enzyme was isolated from *Pyrococcus furiosus* and has an error rate of 1 in 1.3 million nucleotides incorporated. However, as expected, this enzyme has lower processivity, taking from 1 to 2 min to polymerize 1,000 base pairs.

A useful property of the Taq polymerase enzyme is the ability to add the nucleotide deoxyadenosine to the 3′ end of the newly synthesized strand in a template-independent manner, because of terminal transferase activity. Therefore, all PCR products generated using Taq polymerase have at their 3′ end an additional and unpaired deoxyadenosine. From this characteristic emerged a generation of cloning vectors for PCR products, without the need to use restriction enzymes (see below). This type of cloning is called TA cloning, due to the pairing between A of the 3′ end of the PCR product generated by the Taq polymerase, and T of the 5′-end encoded in the vectors, usually plasmids (circular extra-chromosomal DNAs common in microorganisms) (Figure 5.5). After annealing between the TA ends of the vector and the PCR product, a phosphodiester linkage has to be made to ligate these pieces of DNA together. For this reaction, two enzymes are currently used: DNA ligase or topoisomerase.

The DNA ligase commonly used in molecular biology is T4 DNA ligase isolated from bacteriophage T4. DNA ligases catalyse the covalent linkage between two DNA strands through the formation of a phosphodiester bond (Watson, 1992). For this, it is necessary that the two pieces of double-strand DNA are complementary and joined by hydrogen bonds, thus generating the substrate for the enzyme to bind. Two DNA strands can be ligated independent of pairing

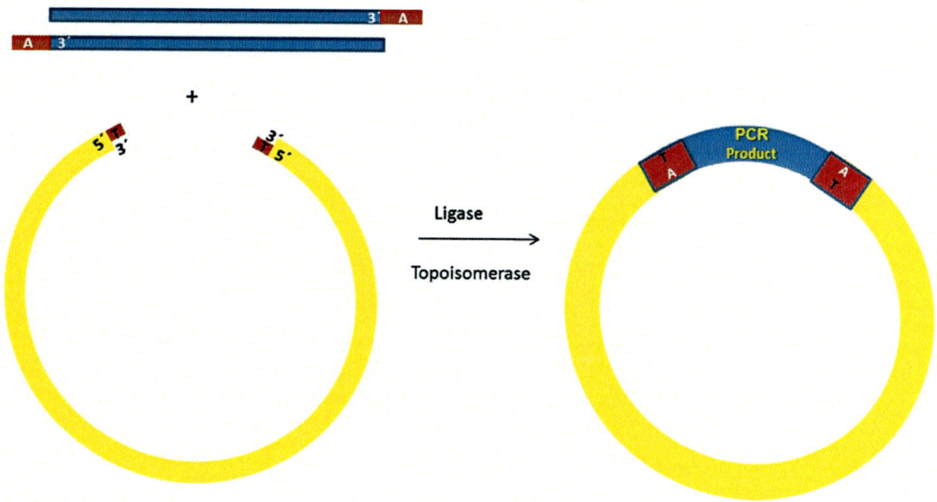

FIGURE 5.5 TA cloning. Taq polymerase has adenosine transferase activity that adds this nucleotide to the 3′ ends in DNA products resulting from the synthesis catalysed by this enzyme (PCR products). From this characteristic cloning vectors were developed which have at their 5′-end an unpaired thymidine nucleotide. Thus, by adding the PCR product to these vectors there will be TA pairing which can be sealed by the addition of DNA ligase or topoisomerase.

between complementary sequences. These reactions are called blunt or abrupt end ligations but are less efficient than 'sticky' end ligations between complementary DNA pieces.

Thus, in the example shown in Figure 5.5, four phosphodiester linkages were made to join all the strands. Such binding and positioning of the strands are the limiting steps in ligation efficiency catalysed by DNA ligases. These enzymes usually have an optimum temperature of activity around 25°C. However, the annealing temperature, also called the melting temperature (Tm, the term also used for the PCR reaction primers) between the complementary pieces, varies by size and is usually less than 25°C. Low temperatures slow the movement between the molecules facilitating the approximation between strands that will be ligated. Therefore, in order to have an efficient ligation reaction, there must be a balance between the optimal temperature of ligase activity and the Tm of the complementary strands. This can be improved by an intermediate temperature around 16°C and a long incubation time around 16 h (also referred to in laboratories as 'overnight'). Very long times of reaction favour the concatenation of plasmids, in contrast to the short times that decrease the possibility of encounter and stable pairing of the strands.

Another important feature of the DNA ligation reaction is the dependence of the ligase enzyme for Mg^{+2} and ATP. DNA ligases have a lysine side chain residue as the active site. The amine group of this side chain hydrolyses ATP and binds to the AMP, and only then, the enzyme is able to bind to the $5'PO_4$ terminus, transferring the AMP to that end of the DNA. In this way, the 3′ end hydroxyl is capable of attacking $5'PO_4$-AMP, releasing AMP and forming the phosphodiester bond (Figure 5.6).

An ATP-independent DNA ligation mechanism can be achieved with the use of topoisomerases. Such enzymes, as mentioned above, are responsible in vivo for alleviating the stress on DNA strands caused by the opening of the helices during the formation of a replication or transcription bubble. In vivo, topoisomerases bind to the tensioned strands and, through a tyrosine residue, cleave one of these forming a covalent intermediate with the DNA through a phosphotyrosine linkage. The opposite end of the cleaved DNA has a free hydroxyl and is strongly held within the enzyme structure. Relaxation is achieved by twisting one of the strands through the cleavage site, and the hydroxyl – at the opposite end – now attacks the phosphotyrosine linkage, releasing the enzyme. The energy of the phosphodiester bond released during cleavage is retained by phosphotyrosine binding but is released during the attack on the hydroxyl to create a phosphodiester bond that religates the DNA strands. Thus, topoisomerases are capable of making phosphodiester bonds without the need for ATP.

The TOPO-TA cloning vectors were developed using the properties of topoisomerase enzymes. These vectors have 5′-ends with a mismatched deoxythymidine (to pair with the A left at the 3′ end of the Taq PCR products) covalently linked to *Vaccinia* virus topoisomerase I via a phosphotyrosine linkage. Thus, when we mix the PCR-deoxyadenosine product and the deoxythymidine-vector, the hydroxyl group is able to attack the phosphotyrosine linkage, releasing energy so that the topoisomerase enzyme catalyses phosphodiester bond formation.

The strategy of using cloning vectors is necessary because the process of isolating a gene in sufficient quantity to be useful and then storing the DNA without inducing mutations is expensive and inefficient. The most common type of vector is plasmid, which is naturally occurring and self-replicating pieces of DNA found in bacteria. Hence, plasmid cloning vectors offer the opportunity to generate large quantities of a DNA sequence of interest for future use which can be stably maintained within a bacterial host. A recombinant plasmid vector that contained a cloned piece of DNA can be inserted by transformation into bacteria, which will then replicate the DNA of interest without modifying

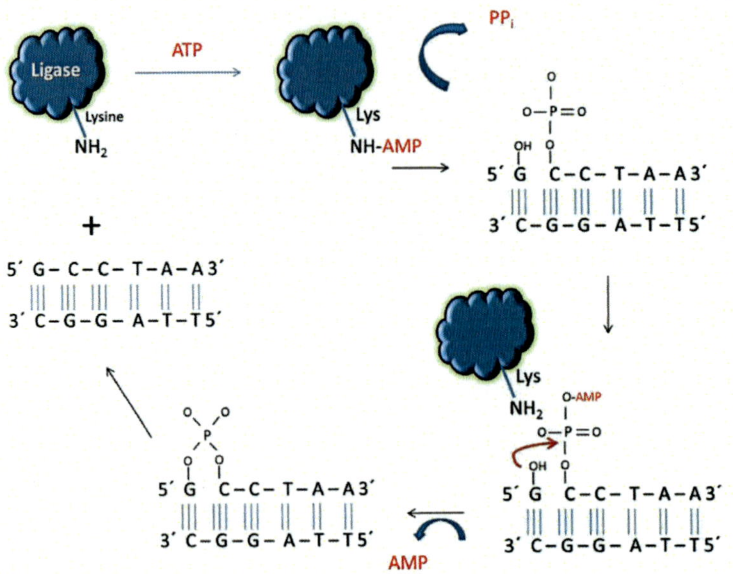

FIGURE 5.6 Mechanism of action of DNA ligase. The lysine side chain residue of the enzyme reacts with ATP releasing PPi. The bound AMP is then transferred to the 5′-end phosphate which can now be attacked by the 3′ hydroxyl releasing the AMP, the free enzyme and generating the phosphodiester bond.

it, because within the cells the DNA polymerases are highly accurate and there are mechanisms of DNA repair. For the plasmid vectors to be replicated by bacterial hosts, the vectors have to have some characteristics that we will see later, but the most important thing is to have an origin of bacterial replication. Only with this sequence is bacterial DNA polymerase able to recognize the plasmid as a 'self' DNA and thus replicate it.

The transformation of bacteria can be done through many different techniques, including electroporation or thermal shock which uses the same principle: both generate transient pores in the bacterial wall and plasma membrane, allowing the passage of exogenous DNA. The process is inefficient; however, we can easily obtain millions of cells for transformation, increasing the likelihood of this rare event occurring. Hence, when we obtain a bacterial colony on an agar plate of selective growth medium, it represents one bacterial cell that has been transformed with one plasmid, but this one cell has been able to survive the whole transformation process, in contrast to millions of cells that died. This single transformed cell then grows and divides, forming a colony unit of billions of clones within which the plasmid also replicates and is maintained in a high copy number. These genetically modified bacteria can be stored in a low temperature with cryoprotectant (e.g. 8% glycerol) and can remain viable for about 20 years.

To recover plasmid DNA from transformed bacterial cells, several techniques for plasmid purification can be used; one of the most well-known is alkaline lysis. In this process, the cells are lysed in buffer with detergent and sodium hydroxide, and then a concentrated solution of ammonium acetate at pH 4.0 is added, rapidly returning the pH to 7.0. Due to the large size and high viscosity of nucleic acids, these and other high molecular weight biomolecules including cell debris form a precipitate which can be collected by centrifugation. As the plasmid DNA is small by comparison, it remains in solution and can be precipitated by the addition of ethanol. After the ethanol has dried, the plasmid can be suspended in water or buffer and used for subsequent manipulations. There are several commercial plasmid DNA extraction kits on the market; all in essence are based on these principles.

The step of bacterial transformation is mandatory for all cloning procedures, regardless of the fate of the plasmid. For example, a yeast expression plasmid should have two origins of replication: one of yeast, but also one in bacteria. This is because the DNA ligation process is not 100% efficient. Not all plasmids will receive the insert (PCR or restriction/digestion product); some will bind the ends themselves and return to being circular, without incorporating the target gene. In addition, there would be no way to separate closed vectors without insert from those with insert, since the quantity of them is very low. Thus, after every binding reaction there is a need for cloning into bacteria to obtain sufficient quantities of isolated plasmid for later use. Another important procedure to be performed after obtaining the plasmid (and hence the target gene product) is to sequence the insert to be sure that no errors were incorporated during the cloning procedure, even using high-fidelity DNA polymerase enzymes.

5.4 RESTRICTION ENZYMES AND EXPRESSION VECTORS

The year 1970 was a milestone for recombinant DNA technology; this was the year a class of site-specific endonucleases (enzymes that promote DNA cleavage) called restriction enzymes were discovered. Hamilton Smith and Daniel Nathans made this discovery and received the Nobel Prize in Physiology or Medicine in 1986. Restriction enzymes

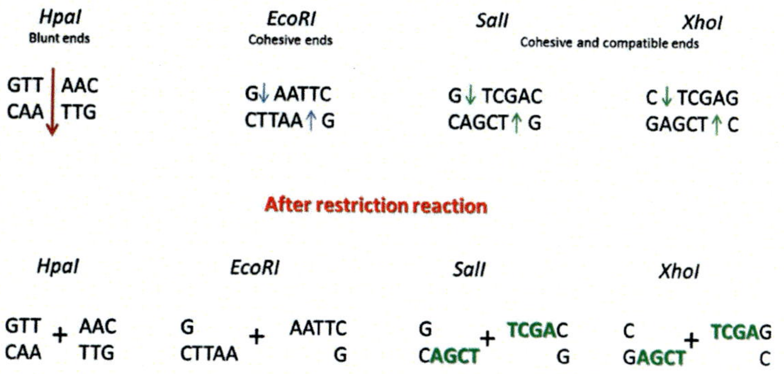

FIGURE 5.7 Restriction enzymes, types of cut ends produced.

are endonucleases that recognize very specific sequences of DNA, usually six base pairs, and are palindromic, i.e., reading the enzyme recognition site in the 5′ → 3′ direction of the two strands is identical. These sequences are targets of the dimeric form of the restriction enzymes, making recognition more specific and allowing cutting of the two DNA strands at the same time.

Restriction enzymes are present in different bacterial species and strains. Their biological function is related to defence against viral infections. Viral DNA entering a bacterium can be recognized as foreign by a restriction enzyme which will then digest the viral DNA, preventing its multiplication which will eventually kill the bacterial cell. Bacterial DNA is immune from the action of restriction enzymes by several mechanisms, either through protection by stabilizing proteins that surround the bacterial chromosome, or by methylation of DNA nitrogenous bases at the enzyme recognition sites (Nelson and Cox, 2000).

In the laboratory, these enzymes can produce different types of cuts in DNA strands generating blunt ends or cohesive ends (Figure 5.7). With blunt ends, there will be no complementary base pairing to facilitate ligation, which greatly diminishes the efficiency of this reaction. However, blunt ends may be useful when one only wants to linearize DNA, leaving no free nucleotides at the ends of each strand, which would then become preferential targets of degradation enzymes called the DNAses. This procedure can be used when you want to introduce sequences in an integrative way (see Host) in chromosomal DNA of a cell, for example. Other enzymes produce cohesive ('sticky') ends, single strands that greatly facilitate cloning by enabling the pairing of complementary bases between different DNA sequences.

Cloning and expression vectors have regions where gene sequences can be inserted. These regions encode various recognition sequences for many restriction enzymes and are called multiple cloning sites (MCS). This allows insertion of DNA sequences at very specific sites of the vector, for example, adjacent and downstream to a promoter region necessary for the expression of the inserted gene. Nowadays, with the ease of obtaining synthetic primer oligonucleotides, specific sites for restriction enzymes are added in the 5′-portion of the primer without altering in any way the reading frame of the amplified and cloned gene. Thus, in designing the primers, we can choose the restriction enzymes that will be subsequently used to cut the ends of the PCR product and produce cohesive ends with the vector of interest (Figure 5.8).

Using this strategy, we can opt for two types of cloning: directional and non-directional. In non-directional cloning, we used a single enzyme to cut the vector and the ends of the insert. This can also be done through the use of different enzymes that produce compatible cohesive ends, such as *SalI* and *XhoI* (Figure 5.7). The problem with this type of cloning in expression vectors is that the DNA of interest may insert in opposite orientation, with the reading frame arranged in the opposite direction to the transcription start site, leading to non-expression of the gene product. The same problem occurs in TA clones (Figure 5.9). In addition, non-directional cloning generates many false positives, that is, vectors self-ligate between their cohesive ends, becoming circular and without insert. One can determine whether the sequence of interest was introduced and in the right direction by restriction enzyme mapping, PCR and sequencing.

When it is possible, and this depends on the vector we have available in the laboratory, we can perform directional cloning by engineering different restriction enzyme sites at the 3′ and 5′ ends, with cohesive ends incompatible with each other (Figure 5.10). For this, the MCS of the vector must have more than one restriction enzyme option for cloning. We choose two restriction enzymes in the vector, one to receive the 5′-end of the insert, further upstream, just after the promoter and having the transcription initiation site; and another to receive the 3′ end, further downstream in the MCS and incompatible with restriction enzyme recognition site chosen for the 5′-end. (Figure 5.8).

After PCR, this sequence (the insert) can be introduced into cloning vectors and transformed into bacteria. The DNA is recovered by plasmid preparation and then the cloning vector+gene_X can be cut with the restriction enzymes. To obtain only the insert without contaminating vector DNA, the different DNA species can be separated by agarose gel electrophoresis. Using ethidium bromide dye, we can visualize DNA bands on the gel and using a molecular weight marker as a reference; the gel band of a size corresponding to gene X can cut out of the gel and the DNA purified from agarose.

To confirm the presence of an insert in the vector after ligation and cloning (directional or not), we can extract the vector+gene_X from the bacterium by plasmid preparation

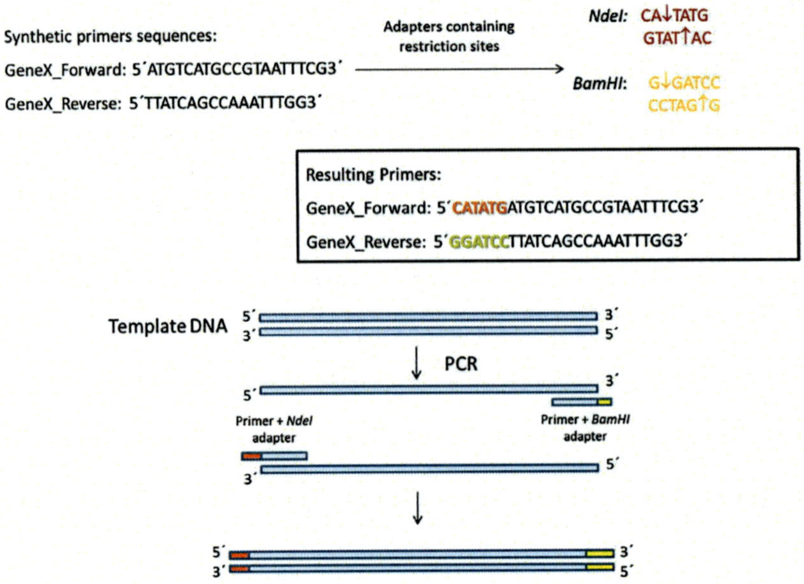

FIGURE 5.8 Insertion of adapters containing restriction enzyme sites compatible with cloning into vectors. The sequences are added to the 5′-ends of the synthetic primers without modifying the target gene sequence.

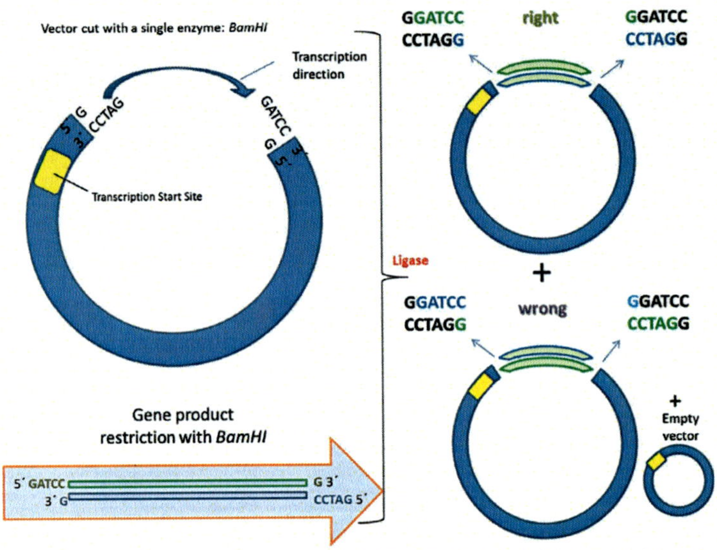

FIGURE 5.9 Non-directional cloning in vectors. If the vector is intended not only for the maintenance of the gene isolated but also for the expression of the gene product, the insert must be positioned correctly, with its reading frame open in the same direction of transcription, i.e., adjacent to the promoter contained in the vector. Since the cohesive ends are compatible with each other in the vector (by the use of a single restriction enzyme), the possibility of recirculating the empty vector is high.

(described previously), and using commercial primers that anneal to sequences flanking the gene product insertion site in the vector, we can perform a PCR reaction. These primers, shown for example in Figure 5.11, include the T7 promoter primer (so-called to be located in the T7 promoter region) and the T7 terminator primer (so-called to be located in the T7 transcriptional terminator region), and are usually separated in the vector by a distance of 100–200 base pairs. If gene_X is approximately 1,000 base pairs in size, a PCR product of approximately 1,100–1,200 base pairs will be amplified. For vectors that self-ligated during the ligation reaction (hence, gene_X was not inserted), the PCR product will have a size corresponding to the distance separating the two commercial primers, i.e. between 100

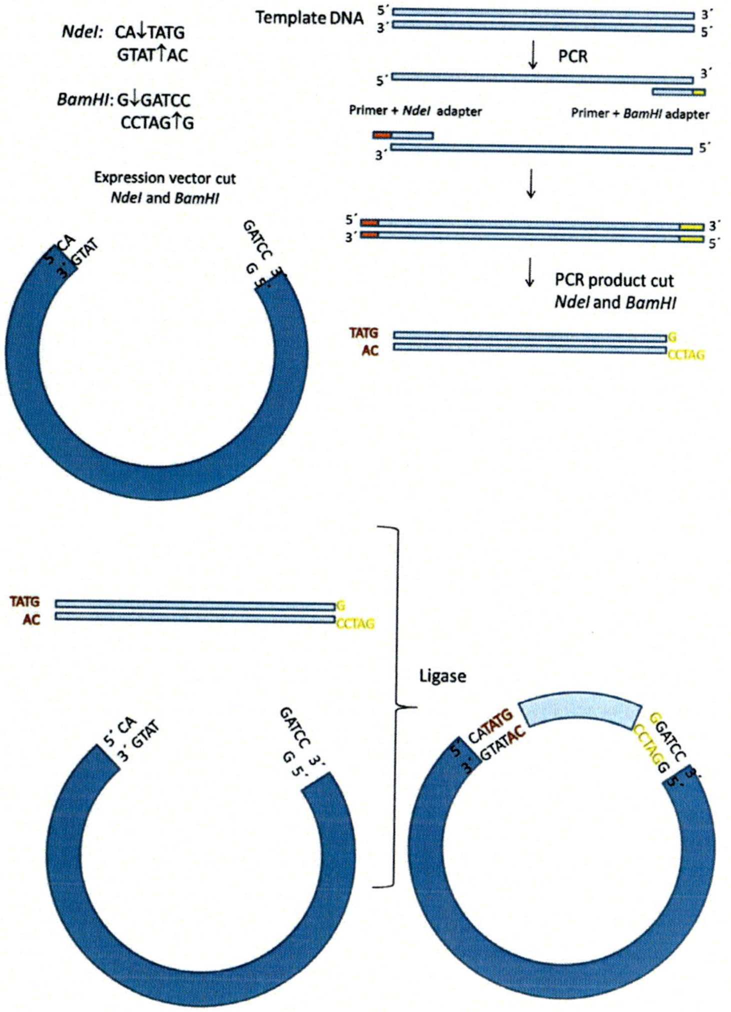

FIGURE 5.10 Directional cloning in vectors. Through the use of two restriction enzymes that produce cohesive ends incompatible with each other, the insert can only be inserted into the vector in the correct position.

and 200 base pairs. These primers are also used to sequence the construct before inserting it into the host.

As we have seen so far, all of the procedures described depend on the sequences of both the gene to be cloned and the vector to be used. To exemplify the production of recombinant protein, we will assume that the plasmid is an expression vector in bacteria. We will now describe the main characteristics of the cloning region of an expression vector (Figure 5.11).

The first region we will consider is the promoter that controls when and how much the protein of interest will be expressed by the host organism, in this case a bacterium. This region may vary from vector to vector, but some important features are common between them (Novagen, 2011). The main feature is that the promoter must be 'strong', that is, have the capacity to efficiently initiate transcription, producing high amounts of mRNA and, consequently, the protein of interest. The second – and no least important – is that the promoter encodes regulatory regions in order that the promoter be activated but only under certain conditions, for example in the presence of an inducer. This is because the high production of a protein, which in most cases is not endogenous (heterologous production, such as human insulin in bacteria), can lead to an energy imbalance within the bacterial cell by recruiting much of the cellular machinery towards the biosynthesis of a single product that typically is not useful to the bacterium (at least not in such a large quantity). Another factor to be considered in heterologous production is that the product, in this case the protein, is not a constituent of the natural metabolism of the host cell and therefore can be toxic. If this happens and there is no repression of this expression throughout the cell growth, these cells will die before reaching a cell mass (high population density) sufficient for the isolation of the protein of interest. If the toxicity is high, neither host bacteria containing the plasmid can be obtained without a strong repression

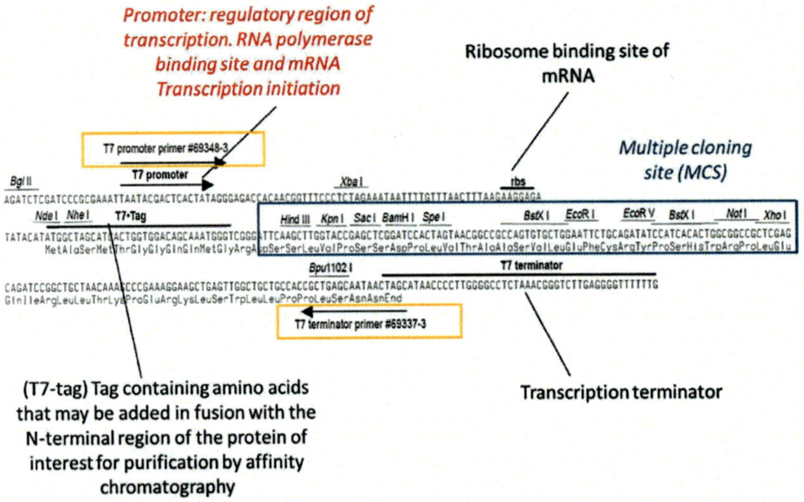

FIGURE 5.11 Region of cloning of expression vectors, Example I.

of gene expression, since these cells will also die soon after the transformation process.

In example I (Figure 5.11), the promoter region of the vector is a sequence called the T7 promoter; this name is given because this promoter region comes from the T7 phage. Being a viral promoter, its activity is very high and is able to redirect protein biosynthesis within a bacterial cell so that as much as 50% of the total protein produced by the cell is of target recombinant protein origin. However, how is a viral promoter recognized by a bacterial RNA polymerase, if the relationship is species-specific? In fact, the T7 promoter is not recognized by the bacterial RNA polymerase at all, but by the T7 RNA polymerase from the phage. These vectors can only be transformed for the expression of proteins in bacteria that have the bacteriophage DE3 (lysogenic) integrated into the bacterial genome, including the gene encoding the T7 RNA polymerase. In these bacteria (known as DE3), the T7 RNA polymerase gene is also under strong expression control by the *lacUV5* promoter. Note that now the promoter controlling the expression of T7 RNA polymerase is a bacterial promoter and not from a bacteriophage. The *lacUV5* promoter undergoes negative control by the Lac repressor so that when there is no lactose in the culture medium, there is no T7 RNA polymerase transcription; consequently, there is no transcription of the recombinant protein.

Because lactose can generate secondary metabolites or divert bacterial metabolic pathway, a synthetic analogue of the lactose breakdown product, IPTG (isopropyl-γ-thiogalactoside), can be used as a signalling molecule that regulates the lac repressor. When the cells reach a sufficient population density to obtain large amounts of recombinant protein, IPTG is added to the culture medium. This analogue will bind to the Lac repressor, and by shutting it down from the *lacUV5* promoter, the *lacUV5* promoter will begin to transcribe the T7 RNA polymerase. After translation of this enzyme, which is virtually simultaneous with prokaryote transcription (due to the lack of cell compartmentalization), the T7 RNA polymerase will bind to the T7 promoter of the expression vector. Thus, active transcription of gene_X mRNA is initiated which will result in the production of high concentrations of the target protein.

Let us now examine in our example vector (Figure 5.11) the multiple cloning site, MCS (also called 'polylinker'). This region will guide our primer design since the region encodes multiple restriction enzyme sites, which will be encoded as adapters in the primers. In this example, it is possible to do directional cloning, since the MCS has many restriction enzyme site options. So to choose the enzymes we need to make some decisions. First, whether the protein will be expressed in its native form or as a recombinant protein in fusion with the T7-tag.

The T7-tag is a sequence encoding 11 amino acids that will function as an epitope that can be recognized by an antibody immobilized onto a resin, thereby allowing the recombinant protein to be purified by relatively few and simple chromatographic steps. The choice of a T7-tag depends on some prior information known about the target protein, for example, whether the N-terminal region (where the T7-tag fusion will be located) is crucial for dimerization, activity or interaction with an enzyme substrate. There are vectors in which the tag used for purification purposes is attached to the C-terminal of the protein, in cases where the N-terminal region cannot be modified. There are other cases where neither the N- nor the C-terminal ends of the protein can be modified; therefore, we have to produce the protein without the tag-fusion, i.e. in its native form.

The next decision to be made when deciding upon the choice of enzymes available in MCS is the restriction map of the gene of interest. Take, for example, the yeast gene involved in iron metabolism, *YMR134W* (Figure 5.12). From the restriction map of this gene, we will find that two enzymes (*HindIII* and *EcoRI*) cannot be used during cloning in the example vector, because these enzymes will cut

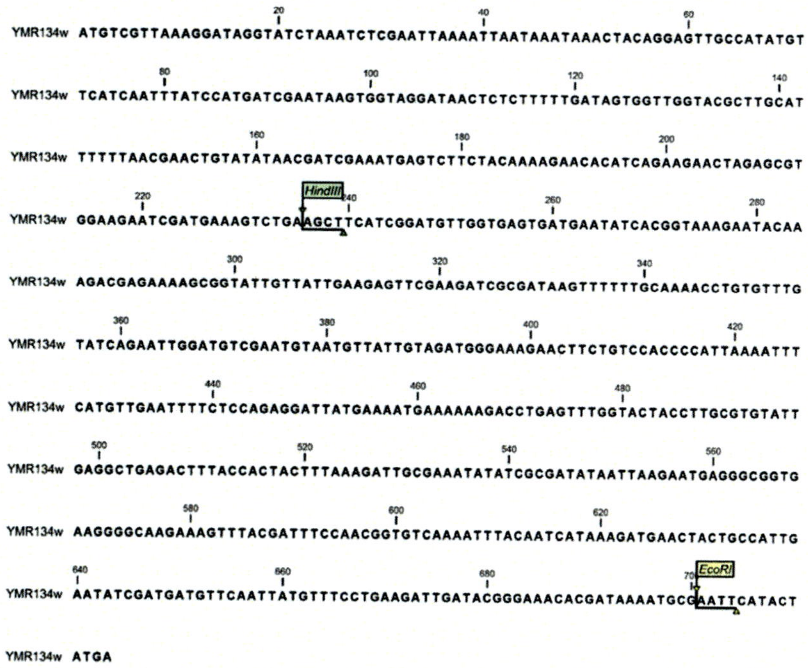

FIGURE 5.12 Restriction map of *YMR134W* from *Saccharomyces cerevisiae.*

into the gene, producing fragments that will not encode the entire protein.

We will now go over the steps we have been discussing by using, just as a simple example, the gene which encodes nisin, a bacteriocin of *Lactococcus lactis* (GI: 42521637) (Figure 5.13). Because of its small size, there are no enzyme restriction sites commonly used in molecular biology within the coding sequence, so we can now perform directional cloning in the example vector using any of the sites that are suitable for our purpose. We will first assume that both N- and C-terminal regions of nisin are important for activity, so we will produce the bacteriocin in its native form. Note that nisin may be a toxic product to *E. coli*, our host bacterium (although nisin as bacteriocin against Gram-positive bacteria and *E. coli* is Gram-negative), so the vector must be under strong transcriptional control so that only when we add IPTG, nisin will accumulate within the cell.

To produce native nisin, we have to remove the T7-tag of the vector, since any option to use the MCS will place the nisin gene in the same reading frame as the T7-tag, producing a fusion protein. On the other hand, we cannot eliminate the ribosome binding sequence (rbs), as it is an efficient binding site for ribosomal assembly in the transcribed mRNA. Remember that the start of transcription begins where the RNA polymerase binds, i.e., the T7 promoter (and not the ATG, which indicates the start of the translation). Thus, we have only two enzyme options remaining after the T7 promoter and the rbs sequence to insert into the direct primer: the *NdeI* and *NheI* enzymes. For the reverse primer, we can choose any restriction enzyme after the T7-tag, as long as the site is not after the terminator T7 (an essential sequence for efficient transcription termination).

Remember that your reverse primer sequence contains the translation termination codon (in the case of nisin, TAA), so no matter which MCS enzyme you will use, nisin will not have any fusion in the C-terminal region. Thus, for cloning of native nisin, we have the following options:

- *Direct primer:* 5′ATGCAT ATGAGTACAAAAGA TTTTAAC 3′ (with NdeI site) or 5′GCTAGC ATGAGTACAAAAGATTTTAAC 3′ (with NheI site).
- *Reverse primer:* 5′AAGCTT TTATTTGCTT ACGTGAATAC 3′ (with HindIII site) or KpnI, SacI, BamHI, SpeI, BstI, EcoRI, EcoRV, NotI, XhoI, Bpu1102I.

When one has so many options, it is worth remembering that it is possible to digest the DNA simultaneously with the two restriction enzymes chosen for directional cloning, provided the optimum temperature and the buffers for both enzymes are the same or at least compatible.

However, if we wish to produce nisin in fusion with the T7-tag to facilitate purification, we must use the restriction enzymes that succeed the tag-coding sequence, i.e., between *HindIII* and *Bpu1102I*. As there is no internal site for the restriction enzymes in the nisin gene, we can choose any two that have compatible temperatures and buffers, for example, *HindIII* and *XhoI*, which digest DNA at 37°C using pH 7.9 buffers (composed of 50 mM NaCl, 10 mM Tris-HCl, 10 mM $MgCl_2$, 1 mM dithiothreitol).

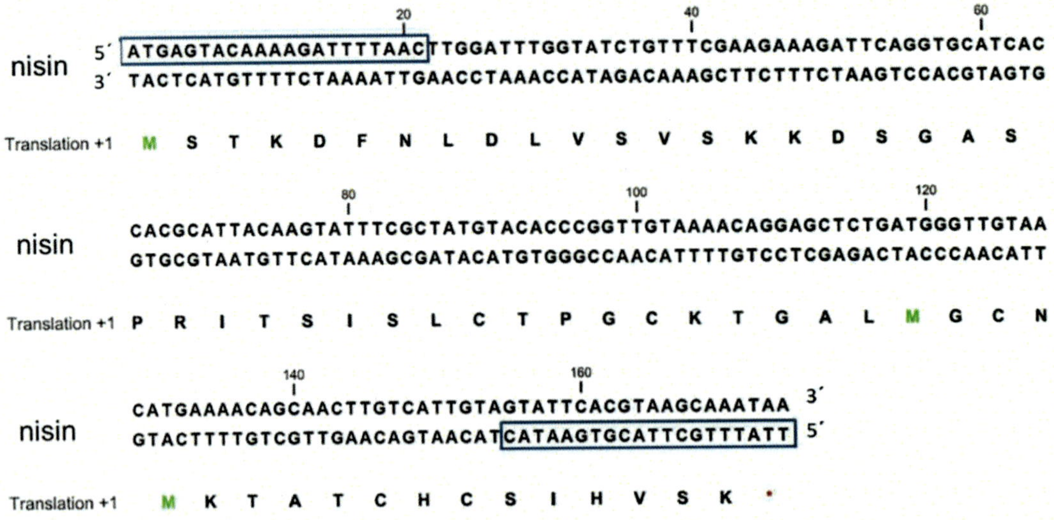

FIGURE 5.13 Coding region of the nisin peptide of *Lactococcus lactis*. The DNA is shown in an unconventional form (double strand) and just below the translation in the reading frame +1 (translation +1). Outlined in the blue boxes are the sequences that will be used to design the primers for gene amplification by PCR, always written in the 5′ → 3′ direction.

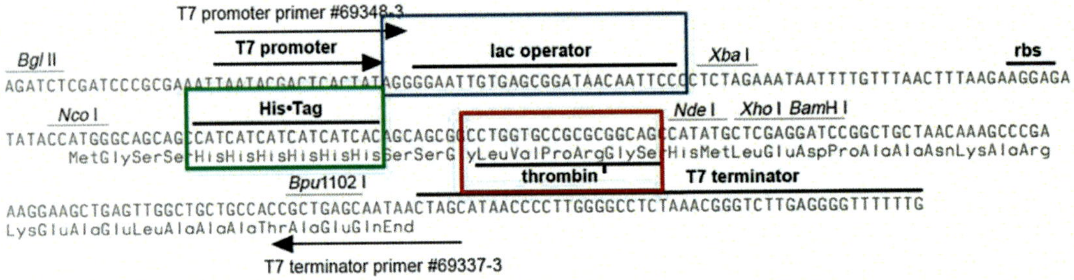

FIGURE 5.14 Region of cloning of expression vectors, Example II.

Each vector will have its advantages, disadvantages and options appropriate for each application. If we analyse the vector shown in Example I which we have used so far, we can discern the following advantages: a cloning option for both the native and a fusion protein with the T7-tag, and a MCS with many restriction enzyme sites. However, immunoaffinity purification of recombinant proteins is expensive, and the resins that contain immobilized antibodies are of little durability because the tertiary structure of the antibodies is readily susceptible to denaturation. There are, however, vectors that offer another option and a widely used fusion involves a histidine tag (Figure 5.14). This tag is a tail that consists of six histidine residues, which can (like the T7-tag) be fused to the N- or C-terminal region of the protein of interest (in the case of the example in Figure 5.14, to the N- terminal). Histidine, due to the imidazole ring present in the side chain, has a metal affinity. Thus, it is possible to purify histidine tail fusion proteins by metal affinity chromatography, the most commonly used of which is nickel.

Another difference between the vectors depicted in Figures 5.13 and 5.14 is the possibility to cleave the fusion tag after purification. In the second example, there is an amino acid sequence recognized by a very specific protease, thrombin. This site is absent in the vector of Figure 5.13, which restricts us to purify the pure protein in its native form by conventional methods which are more laborious and less efficient than affinity chromatography. In the case of the vector shown in Figure 5.14, we can express the fusion protein with the histidine tag, purify the protein using metal affinity chromatography and then incubate the pure protein with the thrombin to cleave the tag.

The vector shown in Figure 5.14 also encodes a *lac* operator within the T7 promoter, affording additional regulation using IPTG and glucose (see *Molecular Biology: Tools in Industrial Pharmaceutical Biotechnology*). Thus, when we add IPTG to the culture medium, both the *lacUV5* promoter responsible for regulating transcription of T7 RNA polymerase, and the T7 promoter which is responsible for

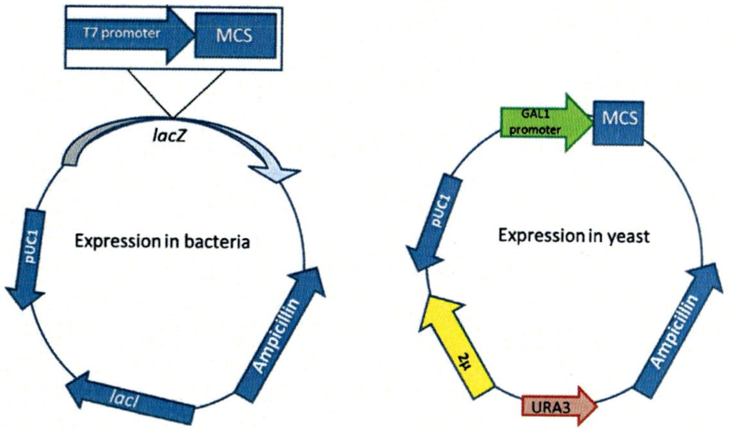

FIGURE 5.15 Similarities and differences between expression vectors used in different host organisms. Even vectors that will not be used for expression in bacteria need to have an origin of replication and a bacterial selective marker to enable cloning and multiplication before introduction into the final host chosen for protein expression.

regulating the expression of the target gene, will be activated because the *Lac* repressor will be inactivated. In cases where expression of a product is toxic to the host, this vector is more suitable because expression of the recombinant protein is additionally controlled. The disadvantage of the vector shown in Figure 5.14 is the poor options for restriction enzyme recognition sequences encoded in the MCS.

Now, let's look at the vectors in a more global way and relate essential characteristics for cloning and expression (Figure 5.15). One of the most crucial sequences encoded on plasmid vectors is an origin of replication for the host organism which will be used not only for protein expression but also for cloning. Accordingly, when expression is performed using bacteria as hosts, the origin of replication of a bacterial plasmid (in the example case, pUC1) is sufficient to efficiently clone and express recombinant protein. Conversely, in the case where yeast cells are used as expression hosts, there is a need for a yeast DNA replication origin (in the example of Figure 5.15, 2 μ) in addition to the bacterial origin of replication (pUC1 for the vector cloning enabling construction process).

Another essential feature of the vector is that it encodes for a selection marker, such as antibiotic resistance or auxotroph. These sequences will allow the selection of organisms that have undergone genetic modification to grow in selective culture medium. In the case of ampicillin resistance, a gene encoding a β-lactamase enzyme is induced by a constitutive promoter. Hence, a bacterial host that is successfully transformed by the plasmid encoding the ampicillin resistance gene will produce β-lactamase, and the cells will be resistant to ampicillin added to the culture medium.

Auxotrophic markers (dependence on the addition of an organic compound for cell survival) are regions of DNA containing a constitutive promoter regulating the transcription of a step-limiting gene that codes for the biosynthesis of some metabolite necessary for survival. In the example shown in Figure 5.15, the *URA3* gene encodes the enzyme orotidine-5′-phosphate (OMP) decarboxylase, which is responsible for the sixth step in the biosynthesis of pyrimidines (such as uracil). Lab yeast strains typically encode several auxotrophic markers, the most widely used being deficiencies in *URA3*, *TRP1* (tryptophan biosynthesis), *LEU2* (leucine biosynthesis), *HIS3* (histidine biosynthesis) and *LYS2* (lysine biosynthesis). Most yeast plasmids will, therefore, encode genes for the biosynthesis of any of these compounds. The selection is made by not adding the auxotrophic marker to the synthetic culture medium, as depicted in the example (Figure 5.15), transformed yeasts must be able to grow in synthetic medium without uracil.

Another crucial factor when producing recombinant proteins is the promoter region of the plasmid which will be responsible for regulating the expression of the gene of interest. This region must belong to the host (except for the T7 promoter, but it is only recognized by bacteria with lysogenic bacteriophage DE3, as explained previously). This is because the host RNA polymerase needs to recognize the promoter sequence to initiate transcription. Thus, if yeast is chosen as the host cell, the promoter encoded on the plasmid that will be responsible for regulating transcription of the recombinant gene must be of yeast origin. Note that the coding region, i.e. the sequence that will give rise to the recombinant protein, may come from any organism: humans, bacterial plants, protozoa, etc., but the regulatory region can only be from the host. Note also that the promoter may be any regulating promoter belonging to the host organism.

As shown in Figure 5.15, the promoter used is *GAL1*, a strong yeast promoter regulated by the presence of galactose in the growth medium. Using this vector to express a recombinant protein in yeast, it will produce high amounts of the target protein when grown in the presence of galactose, since this sugar (an analogue of IPTG) is the inducer responsible for the activation of this promoter.

An additional screening method may be present in vectors (bacterial, yeast and others), selection by reporter gene. The lacZ gene produces the β-galactosidase enzyme which recognizes X-gal (bromo-chloro-indolyl galactopyranoside) as the substrate and, upon cleaving, releases 5-bromo-4-chloro-3-hydroxyindole, a blue compound. A screening strategy inserted into some vectors is to use β-galactosidase as the reporter gene to indicate the success of cloning within the plasmid. The promoter (which will control the expression of the recombinant protein) and the MCS are encoded within the same reading frame of the gene encoding β-galactosidase. When the target gene is inserted into this region, the gene encoding the β-galactosidase will become out of frame and the enzyme will not be produced. In the absence of β-galactosidase, bacteria that have been transformed with self-ligated vectors will turn blue, whereas plasmids where the gene of interest has been cloned into the MCS will be white.

In all the examples discussed above, we have used plasmid vectors for protein expression. However, only bacteria and yeast are able to maintain and express plasmid vectors, since these vectors naturally occur in these cell types. In human cells, plants, insects and others, there is no replication of the plasmid when the cell divides. We can transform these cell types with extrachromosomal DNA, but the expression will be only transient. When the cells divide, there will be no replication of the plasmid (or episomal DNA), which will eventually be diluted in the cell population. Transient expression is very useful in studies of biological function of genes or vaccines based on DNA, but impracticable for the production of recombinant proteins on a large scale.

One way to solve this predicament is to transform the cells with integrative plasmids. These plasmids will have sequences capable of inserting into genomic DNA, just as lysogenic viruses integrate their DNA into a host genome. In mammalian cells, vectors based on retrovirus sequences are used, which will flank the gene sequence of interest, integrating the foreign DNA into the chromosomal DNA of the host. In plants, the Ti plasmid isolated from *Agrobacterium tumefaciens* was modified and is now used to integrate sequences of interest into the genomic DNA of plant cells.

5.5 FINAL CONSIDERATIONS

In order to express a recombinant protein, one has to know the gene sequence and the gene architecture. This includes knowing whether or not there are introns, whether there is an alternative splicing process and what proteoform we are interested in cloning. Choose the template to isolate the gene or the gene product (genomic DNA or cDNA). Choose the host organism, mainly according to the needs of post-translational modifications, ease of genetic manipulation and cultivation. Choose the vector that is appropriate to the host. Design primers, isolate the gene by PCR, insert the gene into the cloning vector and transform bacteria to obtain large quantities of faithful copies of the gene of interest. Cut the insert (choose directional or non-directional cloning), purify gel products, do a vector+insert ligation reaction and transform bacteria to clone the construct. Recover the recombinant vector from the bacteria, select the correct clones, sequence the region of the vector where the foreign sequence was inserted and insert the construct into the host organism. Standardize the conditions for expression and purification and, finally, test for the desired protein activity.

Molecular biology tools allow the manipulation of cells and microorganisms in a controlled way to potentially achieve any biopharmaceutical production. The production of recombinant proteins is one of the most important biotechnological advances gained by the use of recombinant DNA technology.

REFERENCES

Alberts, B. (Coord.) *Molecular Biology of the Cell*, 5th edition, New York: Garland Science, 2008.

Cooper, G.M. and Hausman, R.E. *The Cell: A Molecular Approach*, 5th edition. Washington, DC: ASM Press, 2009.

Lewin, B. *Genes IX*, 9th edition, Sudbury: Jones & Bartlett Publishers, 2007.

Nelson, D.L. and Cox, M.M. *Lehninger Principles of Biochemistry*, 3rd edition, New York: W.H. Freeman and Company, 2000.

Mullis, K., Faloona, F., Scharf, S., Saiki, R., Horn, G. and Erlich, H. Specific enzymatic amplification of DNA in vitro: the polymerase chain reaction, Cold Spring Harb, Symp. Quant. Biol., v 51, Pt 1, 263–273, doi: 10.1101/sqb.1986.051.01.032,1986.

Novagen. *pET System Manual*, 11th edition, Darmstadt, Germany: EMD Biosciences, 2011.

Tindall, K.R. and Kunkel, T.A. Fidelity of DNA synthesis by the Thermus aquaticus DNA polymerase. *Biochemistry*, v. 27, 6008–6013. doi:10.1021/bi00416a027, 1988.

Watson, J.D. (Coord) *Recombinant DNA*, 2nd edition, New York: W. H. Freeman, 1992.

6 Bioinformatics Applied to the Development of Biomolecules of Pharmaceutical Interest

Renata Bannitz Fernandes
Universidade de São Paulo

Carlos Alexandre Breyer
Biobreyer Pesquisa e Desenvolvimento Científico LTDA

Leonardo Schultz da Silva
Universidade Estadual Paulista (UNESP) – Campus do Litoral Paulista (CLP)

Pedro Henrique Camargo Penna
Universidade Federal do ABC

Antônio Sérgio Kimus Braz
Universidade Federal do ABC

Marcos Antonio de Oliveira
Universidade Estadual Paulista (UNESP) – Campus do Litoral Paulista (CLP)

CONTENTS

6.1 INTRODUCTION

The term biological drug, in general, refers to compounds of biological origin with therapeutic activity, but, unlike small molecule natural products, biological drugs are orders of magnitude larger in terms of size and complexity and include macromolecules such as proteins and nucleic acids. A classic example of a biological drug is insulin, which has been used clinically to treat diabetes since the 1920s when this hormone has been derived from bovine and porcine sources. However, since the 1980s, industrial fermentation of genetically modified microorganisms has become an economically viable approach for the scalable production of insulin and other therapeutic proteins to meet a worldwide demand. The biopharmaceutical market is now the most profitable sector of the pharmaceutical industry and is forecast to grow over the coming decades.

This growth has been made possible, at least in part, by the use of computer-aided approaches in the discovery and development of biomolecules, which this chapter will now focus.

Bioinformatics is the application of informatics to solve biological problems and involves the fusion of many disciplines including molecular biology, biochemistry, biophysics, databases, computer simulation, statistics and big data analysis. The field of bioinformatics has grown significantly, both in relation to the number of researchers but also the computing power and data storage capacity that is increasingly available. This accelerated growth can be attributed to three main points: the popularization of inexpensive new generation sequencing technologies (NGS); an increase in the number of 3D protein structures resolved and made available to the community; and increased and affordable access to computational power, generated by major advances related to processing in the graphics processing unit (GPU) and the advent of cloud computing.

Most of the scientific questions tackled by bioinformatics, in general, can be divided into problems related to nucleic acids (analysis of DNA or RNA data, in general big data derived from NGS), problems associated with understanding protein structure and function (structural analyses relating to the functioning and molecular interactions involving proteins/analysis of protein interaction networks) and problems for which an answer can be sought by linkage of data generated from the two previous queries.

Principals will be discussed in this chapter using some examples of biopharmaceuticals where bioinformatics has played an important part in the discovery or development of these medicines.

6.2 DATABASES

A biological database is a large and organized set of data, usually associated with software designed to update, query and retrieve data components stored in the system. In general, biological databases provide a wide variety of relevant biological information, ranging from genomic sequences to complex metabolic pathways, and are powerful bioinformatics tools used to conduct biomedical and biological research by researchers around the world. For example, the Kyoto Encyclopedia of Genes and Genomes (KEGG) database is a useful starting point for molecular biology and biochemistry studies since this database provides comprehensive information about metabolic pathways, including gene sequences, protein sequences and networks of metabolic interactions. In the following, several important databases containing biomolecules will be described.

6.2.1 Databases of Small Molecules

Small molecule databases generally comprise molecules that have a molecular mass of less than 900–1,000 u.m.a. Many of the molecules present in small molecule databases are not actually biomolecules, but these banks do provide 2D and 3D structures of small biomolecules. In addition, because this type of database contains a great diversity of chemical structures, they are extremely useful tools in the search for structures that can mimic specific regions of larger biomolecules. The most representative databases of small molecules available to the public are PubChem maintained by NCBI, and ZINC housed at University of California, San Francisco (UCSF).

PubChem contains information on more than 100 million chemical compounds available from more than 700 sources including, but not limited to individual academic researchers, companies and research centres. The data in PubChem comprises the two-dimensional structure of the compounds, and sometimes information related to bioactivity, toxicity, physical-chemical indicators, metabolic pathways in which the compound is involved and listings of the compounds in scientific articles and patents. The bank is continuously updated by the NCBI.

The ZINC databank contains information on more than 990 million chemical compounds, of which more than 750 million may be available for purchase. The ZINC data comprises the two-dimensional structure of the compounds, and sometimes information regarding pharmacological characteristics, toxicity, physicochemical indicators and commercial suppliers. Approximately 25% of compounds listed in the ZINC databank have information on the 3D structure.

6.2.2 Natural Products Databases

Natural products are a rich source of compounds for drug development, the most famous being penicillin, an antibiotic produced by the fungus *Penicillium notatum* identified in 1928 by Alexander Fleming. This family of compounds is extremely diverse in relation to biological activity and chemical structure, with a wide variety of uses such as human and veterinary medicines and in agriculture. Natural products are generally considered as secondary metabolites produced by marine and terrestrial bacteria, fungi, plants and invertebrates.

Considering all drugs approved between January 1981 and September 2019, 23.5% were natural products or derivatives thereof. Importantly, many synthetic compounds in clinical use were inspired by natural product chemical scaffolds. Databases of natural products have been the cornerstone of drug discovery programmes, and some widely used databases are shown in Table 6.1[1]

TABLE 6.1
Biomolecule Databases

Programme	Description	Source
	Small Molecules Databases	
PubChem	Database of chemical compounds with a great diversity of molecules, periodically updated with 2D structures of the molecules and information about them, including patents.	https://pubchem.ncbi.nlm.nih.gov/
ZINC	Database of chemical compounds with a great diversity of molecules, updated periodically with 2D and 3D structures of the molecules and information about them, including vendors.	https://zinc.docking.org/
	Natural Product Databases	
SciFinder	Database of chemical compounds. It has one of the largest curated collections of chemical compounds, including natural compounds	https://sso.cas.org/as/eGS8d/resume/as/authorization.ping
DNP	It is considered one of the most complete databases of natural products, originated from the famous *Dictionary of Organic Compounds*	http://dnp.chemnetbase.com/
GNPS	It is an open-sourced database focused on information on natural products from mass spectrometry experiments	https://gnps.ucsd.edu/ProteoSAFe/static/gnps-splash.jsp
SuperNatural II	Database with more than 300,000 natural products, but not updated for many years	http://bioinf applied.charite.de/supernatural_new/index.php
ZINC	Database of commercially available compounds for virtual screening that contains over 230 million purchasable compounds in ready-to-dock	http://zinc.docking.org/
3DMET	Curated database of 3D structures of natural products	https://www.genome.jp/dbget-bin/www_bfind?3dmet
KNApSaCK	Database of plant natural products	http://www.knapsackfamily.com/KNApSAcK/
NuBBEDB	Database of natural products from Brazilian biodiversity	https://nubbe.iq.unesp.br/portal/nubbe-search.html
MarinLit	Updated database of natural products of marine origin	http://pubs.rsc.org/marinlit/
COCONUT	Open-sourced database containing natural products from more than 50 databases and containing almost 430,000 unique molecules	https://coconut.naturalproducts.net/
	Nucleic Acid Databases	
INSDC	International collaboration between the main nucleotide sequence databases in the World	http://www.insdc.org
ENA	The first nucleotide sequence database created. Headquarters in Europe	https://www.ebi.ac.uk/ena/browser
GenBank	The largest nucleotide sequence database. Used by the Human Genome Project and based in the USA	https://www.ncbi.nlm.nih.gov/genbank
DDBJ	Nucleotide sequence database based in Japan	https://www.ddbj.nig.ac.jp
NDB	Nucleotide sequence database, with a structural emphasis	http://ndbserver.rutgers.edu
	Peptide Databases	
BIOPEP-UWM	It has two areas: a sequence database and tools to evaluate proteins as bioactive peptide precursors	http://www.uwm.edu.pl/biochemia/index.php/pl/biopep
PepBank	Database of bioactive peptides, in which it is possible to search and retrieve peptides with similar sequences	http://pepbank.mgh.harvard.edu/
StraPep	Structural database of bioactive peptides	http://isyslab.info/StraPep/index.php
MBPDB	Comprises bioactive peptides derived from mammalian milk proteins	http://mbpdb.nws.oregonstate.edu/
PeptideDB	Metazoan signalling peptides	http://www.peptides.be/
PlantPepDB	Curated database of plant peptides	http://223.31.159.8/PlantPepDB/index.php
DBAASP	Curated database of peptides with antimicrobial activity, including peptides acting on the SARS-CoV-2 virus	https://dbaasp.org/
$CAMP_{R3}$	In addition to patented sequences, structures and peptides, this database has signature sequences of peptides with antimicrobial activity	http://www.camp.bicnirrh.res.in
Brainpeps	Peptides that are able to cross the blood-brain barrier by several mechanisms, indicating its importance for therapy and diagnosis of diseases. This database contains data from the literature, both positive and negative for the transport of peptides across this barrier	http://brainpeps.ugent.be

(Continued)

TABLE 6.1 (Continued)
Biomolecule Databases

Programme	Description	Source
	Protein Databases	
UniProtKB	The most complete protein sequence database. It has the SwissProt category with manually curated entries and, the TrEMBL category containing computer-generated annotations	https://www.uniprot.org/
RCSB PDB	3D structure database of nucleic acids and proteins	https://www.rcsb.org/
EMDataResource	Exclusive database for structures obtained by electron microscopy	https://www.emdataresource.org/index.html
EMDB	Exclusive database for structures obtained by electron microscopy	https://www.ebi.ac.uk/pdbe/emdb/
CATH	Database focused on the classification of protein structures in four hierarchical levels: class (C), architecture (A), topology (T) and homologous superfamily (H)	https://www.cathdb.info/
SCOP	Database focused on the classification of protein structures, similar to CATH but with some important and interesting differences in the form of classification	http://scop.mrc-lmb.cam.ac.uk/
Pfam	Database of protein/domain families	https://pfam.xfam.org/
	Other Relevant Biological Databases	
LipidBank	Database of natural lipids, with structural, functional and nomenclature data.	http://lipidbank.jp
LMSD	Database of natural lipids, which also provide specific bioinformatics tools for lipid analysis.	https://www.lipidmaps.org/
CSDB	Database of carbohydrates from plants, fungi and bacteria. Contains functional, structural and nomenclature data.	http://csdb.glycoscience.ru/database/index.html
SMBP	Database of specific bioinformatics tools for analysis of secondary metabolites.	https://www.secondarymetabolites.org/
KEGG	Database of molecular interactions, metabolic pathways and network relationships	https://www.genome.jp/kegg/pathway.html
BioGRID	Metabolism, processing of genetic information, processing of environmental information, cellular processes, human diseases and drug development	https://thebiogrid.org/
IntAct	Experimental information bank of protein-protein interactions	https://www.ebi.ac.uk/intact
STRING	Experimental information bank of protein- protein interactions	https://string-db.org/
MINT	Experimental information bank of protein- protein interactions	https://mint.bio.uniroma2.it/

6.2.3 Nucleic Acid Databases

Nucleic acids are macromolecules responsible for storing genetic information and regulating protein synthesis. The sequence of nucleotides forming DNA or RNA molecules contains extremely relevant information from a pharmaceutical point of view; for example, these sequences provide information on proteins that are either druggable targets or future therapeutic proteins. With the advancement of DNA sequencing techniques, more and more nucleotide sequences are being determined and deposited in databases. This number doubles every 18 months, and in August 2020, the NCBI GenBank Sequence Database contained over 217 million gene sequences. The advancement of sequencing techniques can be exemplified by sequencing of the SARS-CoV-2 genome, which was performed by Brazilian researchers in just 48 h after the confirmation of the first case in Brazil, an essential factor for tracking infection routes, developing vaccines and effective treatments through the discovery of unique biological targets for this organism.

The International Nucleotide Sequence Database Collaborations (INSDC) is an international collaboration between the World's leading nucleotide sequence databases: European Nucleotide Archive (ENA) in the United Kingdom, GenBank in the United States of America and DNA Data Bank in Japan (DDBJ). This initiative aims to collect, store and exchange information on nucleotide sequences, which include DNA, RNA and cDNA, deposited in each of these databases and guarantees access to users in any country.

Established in 1980 by the European Bioinformatics Institute (EBI), ENA was the first nucleic acid database to be created. ENA provides data deposit and access services across the spectrum of nucleotide sequence data. It currently has more than 920 million sequences that are deposited by genome sequencing centres, individual researchers or patent offices.[3]

GenBank is a database of nucleotide sequences and peptide translations. Launched in 1982, initially with 606 sequences, GenBank is maintained by the National Institutes of Health (NIH) and the National Center for Biotechnology Information (NCBI) and is updated every 2 years. GenBank has already helped promote scientific discoveries around the world, including the complete sequencing of the human genome (Human Genome Project). DDBJ is located in Japan and contains approximately 85 million annotated sequences since its launch in 1987. DDBJ also collects nucleotide sequence data from individual researchers and sequencing centres. In total, the International Nucleotide Sequence Database Collaborations have approximately 2.5 billion sequences deposited.[4]

In summary, the data sent by the scientific community to the three organizations are synchronized daily. All sequence entries have a sequence description, scientific name of the organism and bibliographic references. This data can be accessed on the respective online platforms of the organizing institutions for free, by users from all over the World.

In order to organize and store information about the three-dimensional structures of nucleic acids, the Nucleic Acid Database (NDB) was founded in 1992 by researchers at Rutgers University and Wesleyan University in the USA, with funding from the National Institutes of Health, National Science Foundation and Department of Energy. NDB is an online portal that, in addition to primary data, provides access to information about three-dimensional structures of nucleic acids, such as geometric data, structural classifications, motifs, bioinformatics tools and information from complexes between nucleic acids and other biomolecules. Currently, the portal has approximately 11,000 nucleic acid structures and their complexes deposited.[5] The main nucleic acid databases are shown in Table 6.1.

6.2.4 Peptide Databases

Bioactive peptides are low molecular weight protein fragments of 2–20 amino acids that exhibit a physiological effect in vivo. Although there are cell-free peptides in nature, most known bioactive peptides result from the enzymatic cleavage of parental proteins. Bioactive peptides play important roles in most biological processes in all life forms and are also of great biological, medical and industrial importance. Bioactive peptides are considered the new generation of biologically active regulators and play important roles in human health, affecting the digestive, endocrine, cardiovascular, immune and nervous systems. Several peptides were discovered from their biological activities in vitro or based on in silico analyses and sequence similarity. With the growing importance of bioactive peptides, several databases are dedicated to this class of molecules. These databases provide information and analytical resources for the scientific community, increasing the efficiency of discovering and developing new biopharmaceuticals and favour the development of compounds with high therapeutic value. Some of these databases are described in Table 6.1.

6.2.5 Protein Databases

Proteins are macromolecules found in all organisms. With the advent of genome sequencing techniques and interpretation of genetic data, as well as techniques for predicting or obtaining the 3D structure of proteins, access to information about proteins is essential in modern drug discovery. The large volumes of protein data are distributed across several databases, and the following are the most important of these.

The Universal Protein Resource Knowledgebase (UniProtKB) is an open-access database that offers the scientific community more than 195 million protein sequences (release 2020/03) annotated with mandatory information such as the amino acid sequence, name or description of the protein, taxonomic information and references. Additional information can also be added to the annotation, such as the active site in the case of enzymes. This database is divided into two categories: SwissProt (more than 560,000 entries) contains manually annotated sequences and information extracted from the literature and TrEMBL (almost 185 million entries) which contains computer annotated entries in advance of manual annotation.

Each entry in UniProtKB has an alphanumeric access code of 6–10 characters. As an example, the access code P00805 refers to the *Escherichia coli* L-Asparaginase 2 protein (EcA2), used as a biopharmaceutical for the treatment of Acute Lymphoblastic Leukaemia. Through this code, it is possible to obtain data suitable for bioinformatics analysis such as information on the function (catalytic activity, parameters of biochemical activity, molecular function, biological processing), nomenclature and taxonomy, subcellular location, pathology and biotechnology, structure and sequences. This entry has been part of SwissProt since 1986, and by June 2020, it had undergone 184 updates which is an indication of the accuracy of this database.

Structural biology emerged in 1958 with the atomic structure of myoglobin, obtained by John Kendrew (Figure 6.1). Scientists realized the potential of this structural information and the field grew rapidly. However, the files containing the coordinates of the atomic structure were very large, which made it difficult for researchers to share this information with others around the world, since this was a time before the World Wide Web had been conceived. The Protein Data Bank (PDB) was created in 1971 to solve this problem. Researchers who obtained a structure deposited the coordinates in the PDB, which other researchers could then have access. In 1973, the PDB contained only nine deposited structures.

The Research Collaboratory for Structural Bioinformatics (RCSB) has been responsible for managing the PDB since 1998. This worldwide and freely available database is updated weekly, and in July 2020, there were more than 175,000 3D structures of large biological molecules such as proteins and nucleic acids deposited. These structures are obtained through different techniques: X-ray diffraction, nuclear magnetic resonance (NMR) and electronic cryomicroscopy (Cryo-EM).

Each deposited structure has a unique alphanumeric identification code (PDB ID) of four characters, which is used to identify the structure in the database itself but is also quoted in scientific articles and other databases. In addition to the PDB ID, the database provides various other information regarding the structure, such as depositors, work in which the structure was published, organism of origin, date of deposit, details of the molecule (number of chains, number of amino acids, presence of mutations, presence of ligands) and experimental data (resolution, technique used to obtain the structure, geometric quality and stereochemical parameters of the data obtained, amongst other information). The structure summary page also offers options to download files such as the sequence in FASTA format and the coordinates using PDB format, which can be used in 3D structure visualization and analysis programmes.

662 NATURE March 8, 1958 VOL. 181

A THREE-DIMENSIONAL MODEL OF THE MYOGLOBIN MOLECULE OBTAINED BY X-RAY ANALYSIS

By Drs. J. C. KENDREW, G. BODO, H. M. DINTZIS, R. G. PARRISH and H. WYCKOFF
Medical Research Council Unit for Molecular Biology, Cavendish Laboratory, Cambridge
AND
D. C. PHILLIPS
Davy Faraday Laboratory, The Royal Institution, London

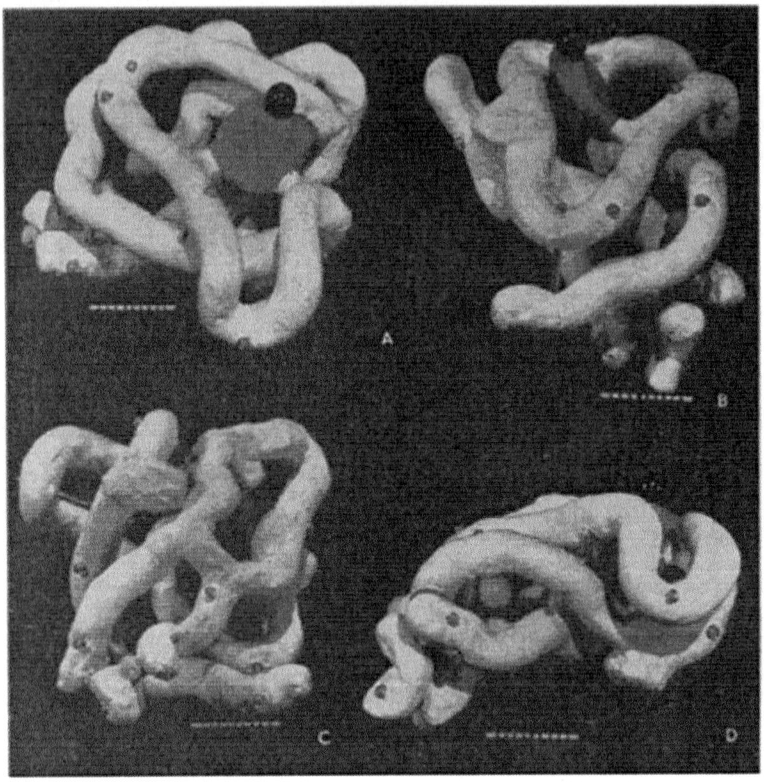

Fig. 2. Photographs of a model of the myoglobin molecule. Polypeptide chains are white; the grey disk is the hæm group. The three spheres show positions at which heavy atoms were attached to the molecule (black: Hg of p-chloro-mercuri-benzene-sulphonate; dark grey: Hg of mercury diammine; light grey: Au of auri-chloride). The marks on the scale are 1 Å. apart

FIGURE 6.1 First protein structure determined. In 1958, J.C. Kendrew and colleagues obtained the structure of myoglobin using the X-ray diffraction technique, marking the beginning of the structural biology of macromolecules. (Extracted from Kendrew, J.C., Bodo, G., Dintzis, H. et al. *Nature* 181, 662–666, 1958. License number 4960251182841.)

The Cryo-EM technique has gained much prominence in recent years due to considerable improvement in the resolution of structures obtained from relatively small amounts of pure protein (compared to more classical X-ray crystallographic methods). The highlight of this technique can be evidenced by the 2017 Nobel Prize in Chemistry awarded to researchers Jacques Dubochet, Joachim Frank and Richard Henderson for developing Cryo-EM for the determination of biomolecule structures in high resolution. EMDataResource is an exclusive database for structures obtained by electron microscopy and also functions as a source of news, events, software tools and method validation for the community. Every deposited structure has an identification number composed of five numeric digits.

For example, the SARS-CoV-2 coronavirus spike protein (S-protein) is a capsid glycoprotein that interacts with specific receptors on host cells (e.g. ACE2 in humans), thereby mediating membrane fusion and entry of the virus into the host cell. This trimeric glycoprotein has different conformations, which are related to the invasion capacity of the virus, with the open conformation most favourable for interaction with ACE2. The open and closed conformations of the S-protein have been obtained by Cryo-EM and can be accessed using the identification numbers 21457 (open conformation) and 21452 (closed conformation).[6]

The Electron Microscopy Data Bank (EMDB) is a public database of 3D structures of both subcellular structures and macromolecular complexes obtained by electron microscopy. This database has more than 11,000 maps and is part of EMDataBank.

Protein structures can reveal the complex organization of this class of macromolecules. The CATH database is focused on the classification of protein structures in four hierarchical levels: class (C), architecture (A), topology (T) and homologous superfamily (H) (https: //www.cathdb.info/). Class (C) is the simplest classification level, which describes the composition of the secondary structure of each protein domain. Architecture (A) describes the general arrangement of the secondary structure in space, topology (T) is related to the order of connection and arrangement of the secondary structures, and homologous superfamily (H) is based on the combination of sequence and structure similarities to provide evolutionary relatedness between proteins. From the analyses carried out using 3D structures deposited in this database, it is possible to obtain various information such as the position of conserved sites and how these are related to protein function and evolutionary aspects.

Pfam is a database of protein families, with 18,259 entries. Each family is represented by multiple sequence alignments analysed probabilistically using Hidden Markov Model (HMM). The model obtained is then used to search for homologous sequences in large protein sequence databases. These computational tools are important to predict functions of uncharacterized sequences. This is fundamental considering that the speed at which the scientific community obtains protein sequences is much faster than the speed at which the function of a protein can be determined by empirical experimentation.

6.2.6 Other Relevant Biological Databases

The LipidBank was created in 1989 by members of The Japanese Conference on the Biochemistry of Lipids (JCBL). The LIPID MAPS Structure Database (LMSD) was created in 2003 and is maintained by the National Institutes of Health (NIH) and National Institute of General Medical Sciences (NIGMS). Both databases are made freely available to the public and include information of all natural lipids including fatty acids, glycerolipids, sphingolipids, steroids and vitamins, in addition to bioinformatics tools, protocols, standards and specific publications for biologically relevant lipids.[7]

The Carbohydrate Structure DataBase, created by researchers from the Institute of Organic Chemistry of the Russian Academy of Sciences, is a database that aims to provide structural, bibliographic, taxonomic, NMR spectroscopic data and other information on glycan and glycoconjugate structures from bacteria, plants and fungi. It is a new online platform, the result of a merger in 2015 of databases for bacterial carbohydrate structures (Bacterial Carbohydrate Structure DataBase) which was created in 2005 and vegetable and fungal carbohydrate structures (Plant & Fungal Carbohydrate Structure DataBase) created in 2013. Its main source of information on carbohydrates is by review of approximately 8,500 literature references from 1941 to 2019, and approximately 22,400 compounds deposited from 11,000 different organisms.[8]

Finally, information on secondary metabolites, which are biomolecules of high pharmaceutical interest, can also be accessed through a database. The Secondary Metabolite Bioinformatics Portal (SMBP) is a database of specific bioinformatics tools for the analysis of secondary metabolites, including the main secondary metabolite databases. The SMBP is maintained by researchers from the New Bioactive Compounds Section of the Technical University of Denmark.[9]

6.3 STRUCTURAL PROTEIN ANALYSIS

Data on the three-dimensional structure of proteins provides us with a whole range of possible analyses like the identification of important domains and sites in the target protein and the visualization of certain properties and identification of types of interaction with a certain ligand. The 3D structure of a given protein can also be extrapolated to similar ones with the same evolutionary origin. Another possibility to work with 3D protein structures is to perform simulations and predict behaviour, conformational changes, movement, energy calculations, etc. But one of the possibilities with greatest impact from a biopharmaceutical perspective is the possibility to predict interactions between the 3D structure of a protein and small molecules or protein ligands.

6.3.1 Visualization Programmes and Analysis of Three-Dimensional Structures

The analysis of structural characteristics of biomolecules is very important to understand structure-function relationships. For example, understanding the interactions between enzyme and substrate is essential in the design of potential enzyme inhibitors and in the development of molecules with superior characteristics (biobetters). With the advent of DNA sequencing techniques and the development of molecular biology tools, the number of three-dimensional structures of biomolecules has increased exponentially. A quick analysis of the PDB database shows that the number of structures deposited jumped from 13,589 in 2000 to 158,965 in 2019 (https://www.rcsb.org/stats/growth/growth-released-structures). This increase in the number of available structures has led to the creation of several tools for structural analysis, ranging from simple online graphic representation programmes to more complex programmes capable of measuring distances and angles between atoms, predicting electrostatic and hydrophobic surfaces and other tools capable of performing molecular docking simulations. Table 6.2 shows the main programmes for visualization and analysis of three-dimensional structures of molecules.

The protein structure PDB data files contain all structural information (relative coordinates including distances and angles between the various atoms determined experimentally) of the macromolecule, including information

TABLE 6.2
3D Structure Visualization and Analysis Programmes

Programme	Functionality	Source
PyMOL	Visualization of molecules, editing and creation of images and videos with publication quality. It also measures distances and angles and overlaps structures	pymol.org/2
Discovery studio	Visualization of molecules, editing, measurement of distances, angles, overlapping structures and molecular docking	www.3ds.com/products-services/biovia
Blender	Visualization of molecules, lipophilic and electrostatic potentials	www.bioblender.org/
BRAGI	Visualization of molecules	bragi.helmholtz-hzi.de/
CCP4mg	Visualization of molecules, creation of images and films	www.ccp4.ac.uk/MG/
Chimera	Molecular visualization, molecular dynamics, hydrogen bond identification and ligand screening interface (DOCK)	www.cgl.ucsf.edu/chimera/
Cn3D	Visualization of molecules, simultaneously displays structure, sequence and alignment, with annotation and alignment editing features	https://www.ncbi.nlm.nih.gov/Structure/CN3D/cn3d.shtml
EzMol	Molecular web-based visualization tool, simple to use, designed especially for occasional users. The final preview template can be downloaded for publication or saved for later use	http://www.sbg.bio.ic.ac.uk/~ezmol/
iMol	Visualization of molecules, measure distances and angles, superimpose structures, calculate r.m.s.d. between coordinates of atoms, structurally align chains.	www.pirx.com/iMol/index.shtml
Jmol	Free open-source molecule viewer for students, educators and researchers in chemistry and biochemistry	jmol.sourceforge.net/
MOE	Generates high-quality images, contains a wide range of cutting-edge applications for modelling proteins and small molecules for drug discovery	www.chemcomp.com/
MVM	Visualization of molecules can be used to display proteins, nucleic acids, oligosaccharides, small and macromolecules. It has an intuitive interface	www.zmmsoft.com/
VMD	Visualization of molecules, creation of animations and analysis of biomolecular systems using 3D graphics. It is used to visualize and analyse the results of molecular dynamics simulations	http://www.ks.uiuc.edu/Research/vmd/
ICM-Pro	Visualization of molecules, execution of molecular modelling, coupling of ligands and receptors, virtual screening of ligands, amongst other functions	www.molsoft.com/icm_pro.html

on molecules associated with the macromolecules such as water, ions and ligands. In general, the programmes are able to represent molecules in different styles, for example: lines, with each bond between atoms displayed as a line (Figure 6.2a); balls and sticks, where spheres represent the position of each atom in the molecule, joined by lines (Figure 6.2b); cartoon, representing the secondary structure of the molecule (Figure 6.2c); surfaces, representing only the surface of the molecule. In addition, it is possible to generate images composed of several representations (Figure 6.2d).

In addition to the visualization of molecules, three-dimensional biomolecule analysis programmes also allow numerous other analyses. For example, comparison between structures is an important tool; Figure 6.3a shows an alignment between *E. coli* (EcA) and *Erwinia chrysanthemi* (ErA) L-asparaginases, and both enzymes are biopharmaceuticals used for the treatment of leukaemia. The r.m.s.d. (root-mean-square deviation of atomic positions)=0.89 Å, which represents the deviation of the average square root of the atomic positions of the carbons (backbone) between the overlapping proteins. This value can be calculated by a series of three-dimensional protein analysis programmes. This number indicates the structural similarity of the molecules, the lower the value, the more similar the structures are.

Another important feature of some structural analysis programmes is the evaluation of interactions between atoms. For example, the presence of hydrogen bonds can be determined by evaluating the distance between a possible receptor and a hydrogen donor (Figure 6.3b). Likewise, it can be possible to predict the presence of disulphide bridges between cysteines (Figure 6.3b) or other ionic interactions. In addition, analyses of molecules' surface can offer other important information, for example, understanding the electrostatic surface affords information on charge distribution which may indicate probable binding sites of molecules and protein-protein interactions. Generally, molecular graphics represent the electrostatic surface of a protein using different colours according to electrostatic potential, where red represents regions of negative potential, blue represents regions of positive potential and white shows regions that are neutral (Figure 6.3c). Hydrophobicity at the surface of a molecule plays a fundamental role in the stability of protein structures and is important to predict patterns of protein-protein and protein-inhibitor interactions (Figure 6.3d). These surface patterns can be calculated by various biomolecule analysis programmes, for example PyMOL, Chimera and Discovery Studio.

6.3.2 Protein Modelling. Why Model?

The 3D structure of a protein is the basis for working with structural bioinformatics, either to search for potential ligands or to predict structural aspects by protein simulations. However, 3D structures are often not available.

Despite the growing number of protein structures resolved experimentally, this number remains limited. If we consider the enormous growth in DNA and RNA sequence databases, this limitation is even more evident. Achieving experimental protein structures whether by NMR, X-ray crystallography or Cryo-EM is still ideal; however, all these

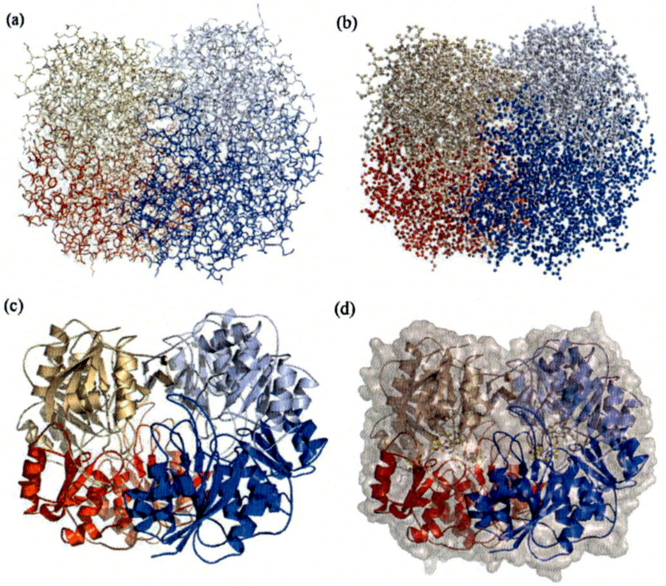

FIGURE 6.2 Examples of molecular structure visualization. (a) Representation of *Escherichia coli* L-Asparaginase protein (EcA2) as lines. (b) Representation of EcA2 protein using balls and sticks. (c) Representation of the EcA2 protein as a cartoon. (d) Representation of the EcA2 protein as a cartoon together with the molecular surface (60% transparency) and active site of the enzyme as balls and sticks. ECA PDB code: 3ECA. EcA2 is a tetrameric enzyme, in the images of each monomer is represented as a different colour. The images were generated using the PyMOL programme.

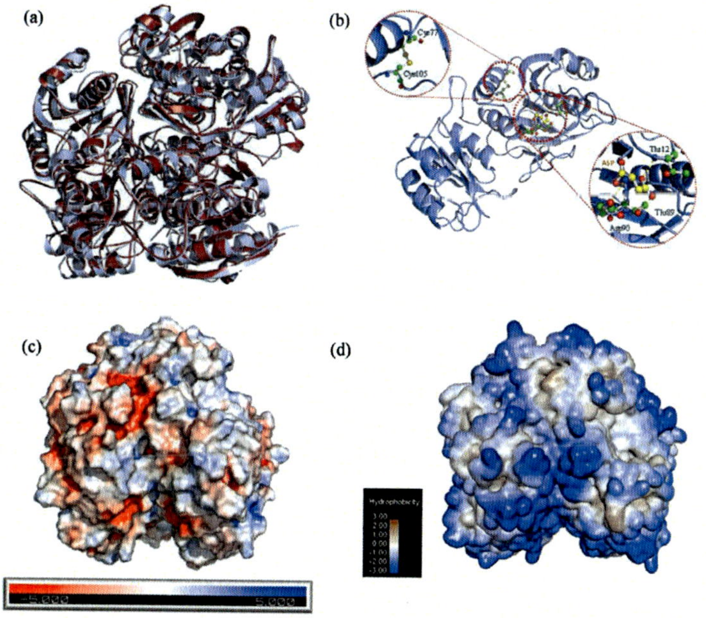

FIGURE 6.3 Structural analysis. (a) Overlapping of *E. coli* (PDB code 3ECA) and *E. chrysanthemi* (PDB code 1077) L-asparaginases. (b) Representation of hydrogen bonds between amino acids at the active site of EcA2 and its substrate asparagine. A disulphide bridge between cysteines 77 and 105 is also represented. The image shows an enzyme monomer in the cartoon representation, amino acids of the active site, cysteines (C: green, N: blue, O: red and S: orange) and asparagine (C: yellow, N: blue, O: red and S: orange) presented as balls and sticks. (c) EcA2 electrostatic surface. (d) EcA2 hydrophobic surface. Images (a–c) were generated by the PyMOL, while (d) was generated using the Discovery Studio programme.

techniques are expensive and have several experimental bottlenecks that are usually difficult to solve (amount of protein required, protein solubility, sensitivity to conditions used in the determination methodologies and structural considerations, amongst others).[10,11] Even proteins of special interest, for example, human proteins or proteins of human pathogens, are still very unlikely to have many structures resolved experimentally. Hence, there is a need to create theoretical models of a target protein,[10] and happily today, it is possible to create theoretical models using bioinformatics.

The issue of structural prediction is one of the major and most challenging themes in structural biology because of the intricate ways proteins can fold and interact with ligands (whether other proteins or small molecules). To illustrate how challenging structural prediction can be, there is a biannual contest to evaluate the most promising new structural protein modelling programmes: the CASP (Critical Assessment of Techniques for Protein Structure Prediction).

6.3.2.1 Basis of Methods

There are two basic methods for creating a theoretical model of proteins: ab initio methods and homology-based methods. The ab initio methodology uses only criteria physical simulations to predict the structure of the protein. This type of technique is still computationally very expensive and has been applied more to predict structures of small proteins and peptides. The precision of the ab initio method, although it has improved tremendously in recent years, still requires experimental support on a case-by-case basis.[12]

Homology is a relatively simple, powerful, but often misunderstood evolutionary concept. Homology means similarity by common origin. It is basically a qualitative concept, is never quantitative and does not imply any assumption regarding function.[13] In the context covered in this chapter, homology can be applied to both the primary structure (sequence of amino acids) and tertiary structure (three-dimensional) structure of a protein. There is a direct relationship between primary sequence identity and structural similarity amongst proteins based on the r.m.s.d of the backbone (root-mean-square deviation).[14] This means that the closer two proteins get in terms of sequence identity, the greater the structural similarity between them. This simple observation allowed the homology modelling method to be created.

6.3.2.1.1 Classical Homology Modelling

In this methodology, it is necessary that the 3D structure for a protein called the template protein is known. This 3D structure can be found by searching the PDB database using the primary amino acid sequence.[13] A flowchart describing the main steps of this methodology is shown in Figure 6.4. Template selection is usually based on the degree of sequence identity to the protein for which the 3D structure is not known (called the query sequence), the closer the better but avoiding values less than 40%. Another important factor may be the level of overlap in the alignment between the primary sequences. This can be determined using various open-access sequence alignment programmes (muscle, t-coffee, clustal) (Figure 6.5). Other important aspects that must be identified during the choice of the template protein and that may place special restrictions on the modelling step

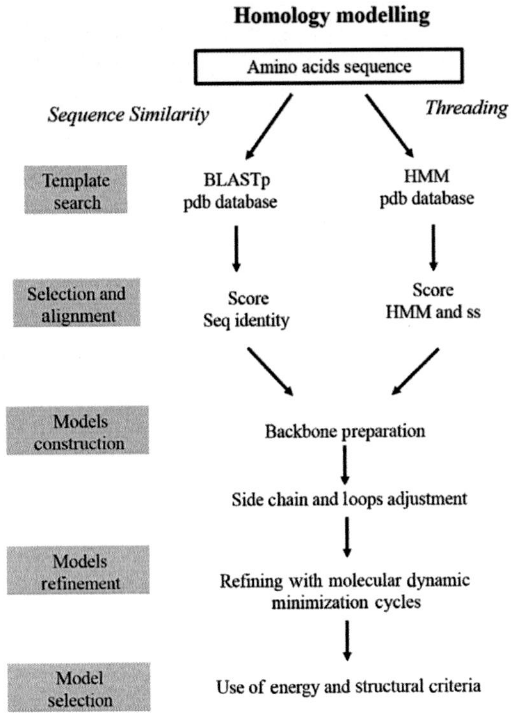

FIGURE 6.4 Flowchart of homology modelling methodologies indicating differences and similarities in key stages.

TABLE 6.3
Programmes for Creating Protein Models

Programmes	Functionality	Source
CASP	Protein structure prediction centre	https://predictioncenter.org/index.cgi
t-coffee	Sequence alignment programme	http://www.tcoffee.org/Projects/tcoffee/index.html
Muscle	Sequence alignment programme (standalone and/or server)	https://bioinformaticshome.com/tools/msa/descriptions/MUSCLE.html https://www.ebi.ac.uk/Tools/msa/muscle/
HHpred	Sequence alignment programme	https://toolkit.tuebingen.mpg.de/tools/hhpred
PHYRE 2	Protein folding predictions (standalone and/or server)	http://www.sbg.bio.ic.ac.uk/~phyre2/html/page.cgi?id=index
Modeller	Template search and alignment programme	https://salilab.org/modeller/

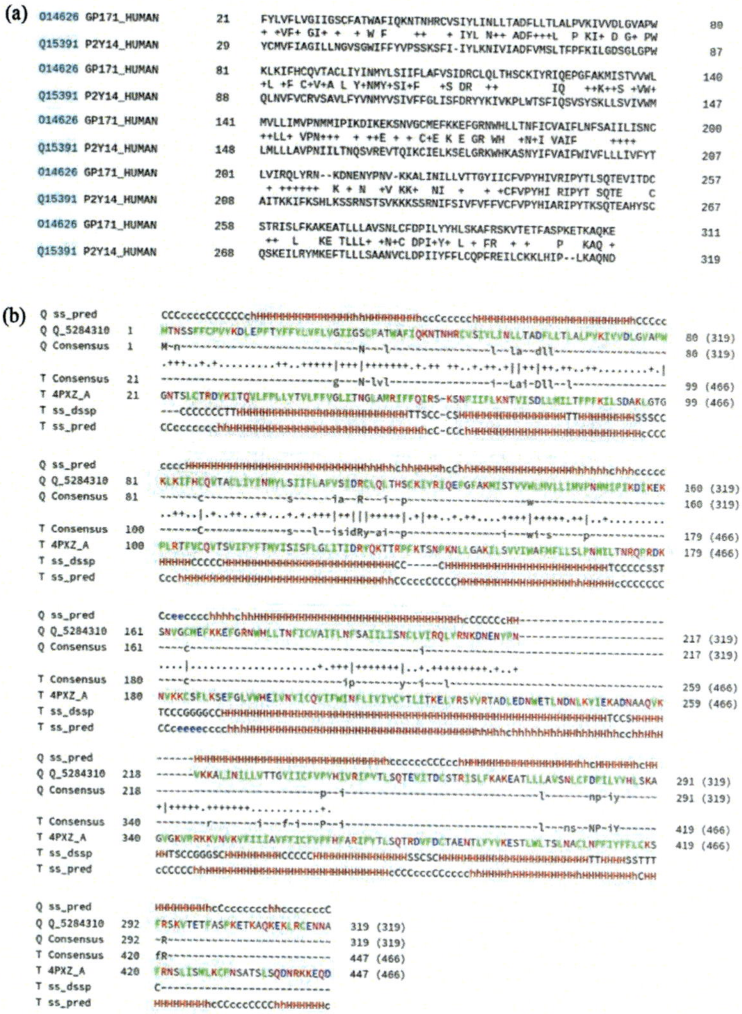

FIGURE 6.5 Sequence alignment comparison between human GPR171 query sequence and human P2Y14 selected as the template sequence. Modelling methodology: (a) Homology and similarity. (b) Threading with HMM and secondary structure.

are quaternary structure, presence of cofactors, ligands and important post-translational modifications such as disulfide bridges.[10]

The modelling process itself is achieved using programmes like Modeller that use the .pdb file of the template protein and the primary amino-acid sequence alignment between the query and template sequences. Any amino acid side chains in the template protein are removed, the residues are renamed and elements such as insertion and deletion of residues corresponding to the query protein are added to create an outline of the query protein. Then the programme adds the atoms of the amino-acid side chains corresponding to the query protein. Cycles of molecular dynamics steps are next performed to adjust the side chains and to correct angles to minimize destabilizing free energy in the predicted protein structure. Generally, hundreds of models are created and, several criteria can be used to select the best models such as DOPE (Figure 6.6).

6.3.2.1.2 *Threading Variation Homology Modelling*

We do not always find sequences with a sequence identity greater than 40%. Still, it is possible to create good models below this limit of sequence identity if we use a special type of homology modelling called threading. This type of modelling takes into account a type of deep homology. It is not based so much on elements of primary structure but rather on conserved elements of the tertiary structure.

To better explain this methodology, it is first necessary to remind ourselves of the structural classification of proteins. For some time, databases such as CATH, have been dedicated to classifying all available protein structures in the PDB. It has been found that there are no more than 1,400 different types of protein folding patterns, and it is unlikely that many more are left to be discovered based on artificial folding patterns that have already been created using ab initio[15] prediction programmes. The vast majority of proteins

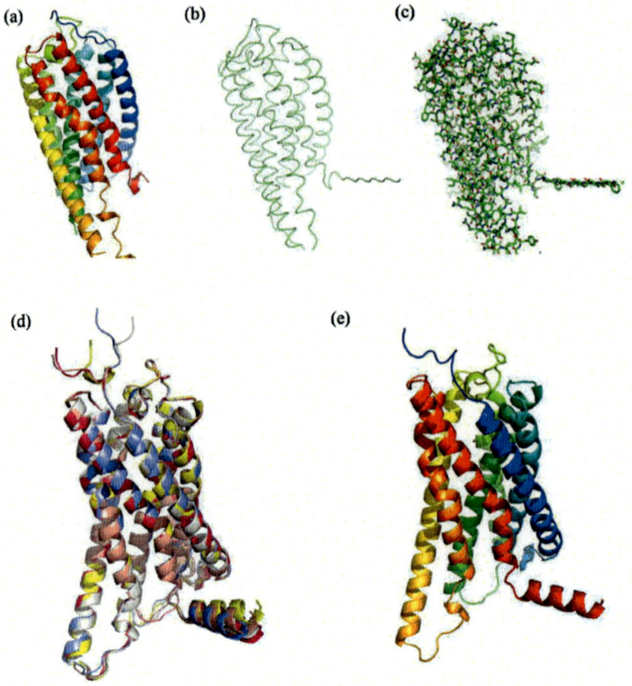

FIGURE 6.6 Construction of a model for query protein sequence GPR171. (a) Human P2Y14 template visualized by cartoon, coloured from the N-terminal region (blue) to C-terminal (red). (b) Initial backbone of the model viewed by ribbon. (c) Reconstruction of the side chain of the initial model visualized by sticks. (d) Composition of five refined models, visualized by cartoon, coloured by model. (e) Best model visualized by cartoon, coloured from N-terminal (blue) to C-terminal (red). The images were generated using the PyMOL programme.

are, therefore, the product of divergent evolution from a small number of structural types.

Even if the query protein has no significant identity with any known protein, it is very likely that it will fit into one of the folding patterns already known. This can be quickly assessed by using secondary structure and HMM prediction alignments. Threading modelling only differs from classical homology modelling when creating the alignment, but from that point on follows all of the same steps used in classical homology for constructing and then selecting models (Figures 6.4 and 6.5).

Due to more stringent alignment criteria and greater sophistication for model selection, it is recommended that *threading* modelling is used as a standard. For example, HHPRED or PHYRE 2 programmes.

6.3.2.1.3 *Other Methodologies*

Structural prediction methodologies are expected to change significantly in the coming years with advancements in artificial intelligence (AI) and quantum computing, which can tremendously accelerate ab initio predictions. AI-based methodologies such as ALPHAFOLD have emerged, which won the CASP13 protein structure prediction competition in July 2019.[16]

6.3.3 *Contextualization of Structural Properties*

Proteins are not fixed and immobile structures as crystallography data might suggest. Although techniques such as NMR and Cryo-EM present structures with a certain degree of movement, this is still not sufficient to explain certain behaviours and biological roles. There are movements inherent to proteins that generate structural variations which can be fundamental to explain biological functions.[10,11,17,18]

The intramolecular interactions within proteins, as well as the intermolecular bonds between proteins, ligands and solvents, are based on the physicochemical and structural characteristics of molecules such as hydrogen bonds, van der Waals interactions, ionic interactions, hydrophobic interactions, cation-type interactions π, interactions involving aromatic rings of the type π-π and coordination with metal ions. From these interactions and covalent bonds, phenomena arise that are of great interest to structural biology such as protein folding, stable structure, movements and ability to interact with other molecules.[10,19,20] There are several ways to represent and calculate these interactions. The ideal would be to use quantum physics calculations, but when it comes to systems with tens or hundreds of thousands of atoms, these types of calculations become impractical in terms of computational cost. The use of molecular mechanics (MM) using force fields to calculate potential energy makes it possible to represent molecular interactions described above, less computationally intensive and, therefore, more feasible.[19,21]

Simulations with molecular mechanics, whether molecular dynamics (MD) or normal mode analysis (NMA), to simulate the behaviour of proteins have many applications, for example: predicting steric hindrances caused by different ligands and structure and function effects caused by mutations. It should be noted that mutations can be both natural or induced, thus allowing mutations to be linked with a biological effect.[19,21,22]

6.3.3.1 MD

In molecular dynamics, the physical movement of atoms and molecules, or even macromolecules, is simulated where they will interact for a certain period of time. The trajectory of the atoms is solved numerically using equations of motion for the interacting particles. In order to better represent the biological conditions in MD simulations, in addition to the protein and molecules of interest, molecular representations of the solvent (water), ions and, if applicable, membranes (lipid bilayer and water) are used (Figure 6.7).[19,21]

6.3.3.2 NMA: Normal Modes Analysis

The normal modes analysis consists of an approximation of the free energy of the system dynamics to the Hooke Harmonic potential which, analytically, represents all of the movements of the various regions of the protein around an energetic minimum, including trajectories and frequencies. Many of the movements related to the biological function of proteins are collective movements, for example, the movements of domains that occur at high time scales. It is worth mentioning that an important difference between NMA and MM is that in NMA, the equations of motion are solved analytically rather than numerically and so we do not achieve a real-time picture of protein interactions.[23,24]

6.3.3.3 Comparison: MD vs NMA

MD provides a clear picture about the structural and energetic dynamics of the systems being studied, as well as the interactions between a protein and a ligand. Some effects we want to observe demand a relatively long computational simulation time which, even with availability of GPUs, is still difficult to implement on a large scale. NMA can be a good alternative to observe major conformational changes with a low computational cost. It should be noted that NMA is a methodology that requires more user experience to implement.[18,19,21,23]

6.3.3.4 Use

There are a large number of programmes for MD and NMA (Table 6.4) but amongst the open-access programmes, a few have a user-friendly graphic interface. Assembly of simulation systems is not usual, but there are tools that help, for example, the CHARMM-GUI platform (http://www.

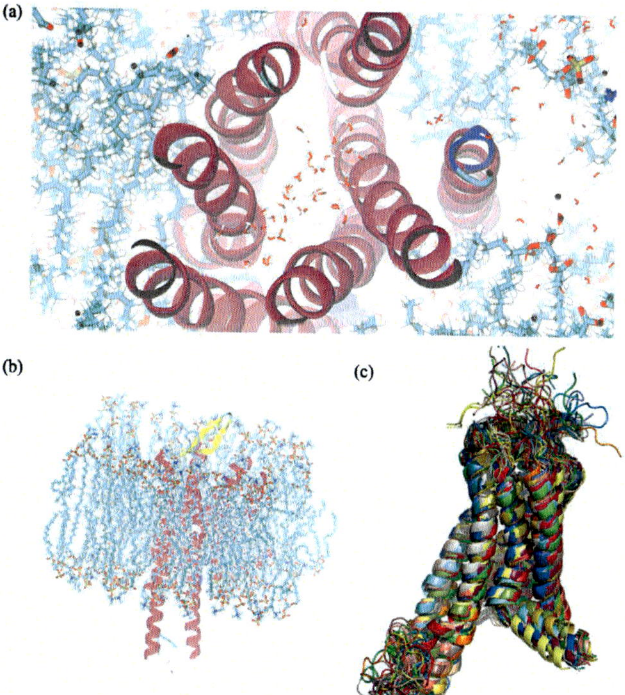

FIGURE 6.7 Molecular dynamics of the OXTR protein, lipid bilayer and water. (a) Water entering the protein cavity during the simulation, the protein is visualized using a cartoon and coloured according to secondary structure (purple helix), lipid POPC is represented by sticks and water is represented as lines, both are coloured by type of atom. (b) Side view of the system hiding water, protein is visualized by a cartoon and coloured according to secondary structure (alpha-helices in purple and beta leaves in yellow), lipid POPC is represented by sticks and coloured by type of atom. (c) Conform generated during MD simulation of OXTR and visualized using a cartoon with each conformer represented with a different colour. The images were created by PyMOL and VMD programmes.

TABLE 6.4
Programmes for Molecular Dynamics and Protein × Protein Docking

Programme	Functionality	Source
Charmm-gui	Web platform for assembling molecular mechanics systems. It allows the most varied types of systems to be assembled. Generates inputs and scripts for a range of simulation programmes such as CHARMM, GROMACS, AMBER and others	www.charmm-gui.org
GROMACS	GROningen MAchine for Chemical Simulations Open source molecular mechanics package, designed for simulation of systems containing proteins, nucleic acids and lipids, implemented for use in CPU and GPU	www.gromacs.org
AMBER	Assisted model building using energy refinement	ambermd.org
CHARMM	Chemistry at HARvard Macromolecular Mechanic) Another programme package for molecular simulations with broad support for a wide variety of methods	www.charmm.org
ICM-Pro	A suite of biomolecular simulation programmes.	http://www.molsoft.com/icm_pro.html
HDOCK	Protein vs Protein Docking server	http://huanglab.phys.hust.edu.cn/software/hdocklite/ http://hdock.phys.hust.edu.cn/
ZDOCK	Protein vs Protein Docking server	http://zdock.umassmed.edu/
HADDOCK	Protein vs Protein Docking server	https://bianca.science.uu.nl/haddock2.4/
HEX	Protein vs Protein Docking Software with GPU implementation and server	http://hex.loria.fr/

charmm-gui.org/), which builds the various molecular modelling systems step by step and also generates compatible scripts to run a range of different simulation programmes.

6.3.3.5 Interaction Models: Protein vs Ligand

Understanding the movement of proteins and their effects is essential to describe, in a more realistic way, the interactions between proteins and proteins with small ligands. The three basic models to describe these interactions are the classic key-lock, the induced fit and the conformational selection models. In the classic lock-and-key model, the receptor protein and ligand must be rigid and their binding interfaces must be perfectly compatible. In the induced fit model, binding in the protein is flexible and the interacting ligand induces a conformational change at the site of the interaction. In the conformational selection model (MWC), there is not a single rigid conformation for protein, but a vast array of conformational states that coexist in equilibrium with different distributions where the ligand can selectively bind to the most suitable conformational state, shifting the balance to this state. Currently, it is believed that what occurs physically is a mixture of conformational selection followed by induced adjustment.[25,26] Figure 6.8 shows an example of human AGTR1 docking with different ligands and possible protein/ligand interactions.

6.3.3.6 Docking: Protein vs Small Molecule

The molecular docking strategy consists of a technique used for structure-based drug design, in which the aim is to model the intermolecular interactions of molecules with a given target protein. The technique in this case is based on one or more three-dimensional models of the target protein to be studied and can be used both to understand how

(a)

(b)

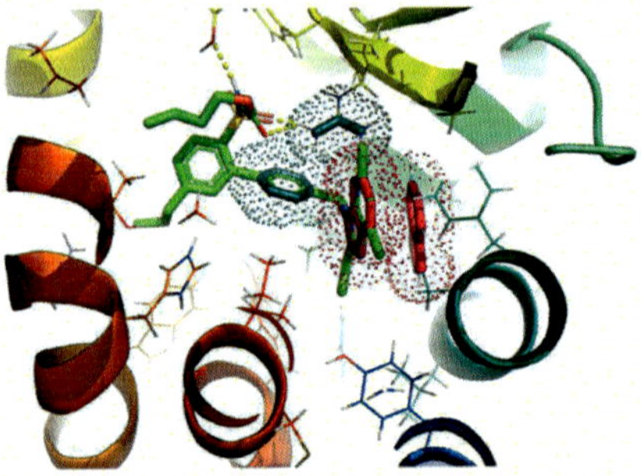

FIGURE 6.8 Docking of human AGTR1 protein visualized using a cartoon, coloured from the N-terminus (blue) to the C-terminus (red). (a) Different chemical ligands selected by docking, coloured surface visualization by charge. The overlap of several ligands is represented by sticks, with the spatial position and connection of each with the protein calculated by molecular docking. (b) AGTR1 binding site indicating the main interactions with a given molecule, hydrogen bonds are represented by dashed yellow lines, π-π stacking interactions are represented as pink dots and cation-π interactions with cyan dots (surface view). The images were generated using the PyMOL programme.

interactions between binding agents with a known effect and the target protein occur, and to prospectively analyse potential new molecules that bind to the target protein and present desired characteristics – such as specific activation or inhibition of the target protein, associated with pharmacologically desired physical-chemical characteristics.

In a concise and general way, the strategy consists of determining, either by a priori knowledge or by inference based on physical characteristics, which region present in the 3D model of the treated target protein will be treated as a 'pocket', i.e. the region in which the binders interact or should interact directly. Once the 'pocket' has been defined, for each molecule that is to be analysed, a routine follows in which a probable 3D conformation must be generated. In the sequence, a search algorithm must be applied to find possible conformations of the molecule within the 'pocket'. After this, an evaluation and selection of possible positions of the molecule within the 'pocket' must be made, with the aid of the analysis of intermolecular interactions between the molecule and the regions of protein which constitute the 'pocket'. Scoring functions are widely applied in molecular docking strategies, aiming at a quantitative value that reflects not only the interactions between the protein and the target molecule, but also takes into account entropic and entrapment factors involved in the interaction of molecules and proteins, such as electrostatic interactions, Van der Walls interactions, strain of the molecule's pose, solvation/desolvation energy of the molecule and the protein, amongst other values. All these factors are aimed at a correlated representation of the binding energy of the protein molecules in question. There are several free and commercial software focused on molecular docking, each one presenting its peculiarities in relation to the score function and in relation to the way they structure the search and evaluation of the poses of the molecules in the 'pocket'.

This type of technique is widely used in the searching for new molecules with potential pharmacological value, in projects in which billions of compounds are tested against a target protein. Regarding the development of biomolecules, it is worth mentioning that docking molecules have applications in the detection of targets for small molecules of biological origin with known general effects – for example: phenotypic tests; in the detection of small molecules of biological origin that can bind to protein targets with a desired effect on the target; in the search for biologically derived small molecules that mimic the action of molecules known to bind to a target protein; and finally, in the search for and optimization of molecules derived from biological molecules that bind to specific target proteins. The technique can also be coupled with other techniques such as molecular dynamics to find answers to complex questions such as prediction of the biophysical effect of substitutions on small molecules of biological origin and guided search for high-value substitutions in biological molecules in general, including macromolecules.

6.3.3.7 Docking: Protein vs Protein

Much of the regulation in a cell's biological activity involves interactions between proteins or with other molecules. Understanding and predicting the way in which proteins interact is essential in pharmacology.[10]

Experimental information on how proteins interact physically is more difficult to obtain compared to tertiary structure, due to difficulty in obtaining crystals of protein complexes. With the advancement of Cryo-EM, more and more valuable structural information about these interactions has been obtained.[10,11,28]

There are several in vitro and in vivo experimental techniques that can provide information on protein *vs* protein interactions, and these have been extensively been reviewed by Meyerkord and Fu (2015).[28] Databases which provide information on protein interactions include BioGRID, IntAct, STRING and MINT; however, information on interaction interfaces is sparse and so the use of in silico methods to predict these interfaces becomes an extremely interesting issue for bioinformatics with many applications. This importance becomes clear because like CASP, there is the CAPRI (Critical Assessment of PRediction of Interactions) initiative that selects the best programmes for predicting protein vs protein interaction interfaces based on experimental X-ray crystallography and Cryo-EM data.

In relation to protein vs protein docking, even though there are two rigid structures, the total number of interaction possibilities is large. In addition, there is the issue of interaction models of induced fit and conformational selection that make the problem orders of magnitude more complex.

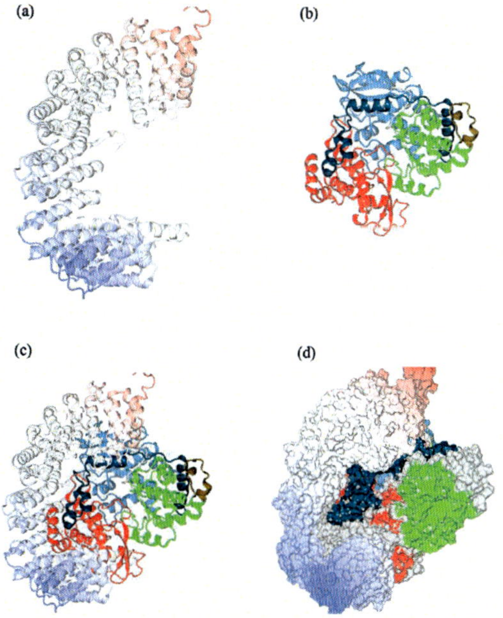

FIGURE 6.9 Docking protein vs protein. (a) Human KPNB1 protein, visualized using a cartoon, coloured from the N-terminus (blue) to the C-terminus (red). (b) Protein NS5-RdRp DENV, visualized using a cartoon with structural domains coloured. (c) Lower energy structure of the NS5-RdRp DENV NS protein complex with a lipid bilayer and water. (d) Lower energy structure of the NS5-RdRp DENV complex (surface view). The images were generated using the PyMOL programme.

Predicting the interface where proteins might interact can be done using ab initio approaches. The way in which programmes calculate interactions and how conformational space is searched varies widely between programmes, as does the way in which complex scores are calculated. An example of protein vs protein docking can be seen in Figure 6.9.

Some of these programmes have important differentials such as modelling the complex proteins and entering experimental information from SAXS (small angle X-ray scattering), which provides low-resolution data on the protein complex envelope in solution, allowing models closer to the real one to be more accurately predicted. Others use GPUs, which greatly speeds up the number of possible protein conformer tests.[29]

6.4 EXAMPLES

Here, examples will be discussed that highlight the importance of bioinformatics tools for the discovery and development of new biopharmaceuticals. The first example highlights the importance of mining genomic sequences and three-dimensional structures of proteins in the development of the meningococcal B vaccine. As the second example shows how the drugs that inhibit the ACE protein used to treat hypertension were developed from a natural molecule (peptide) obtained in the venom of the jararaca snake. Then, we show how modelling tools were fundamental in the development of nucleotides and nucleosides antiretroviral drugs which have revolutionized the treatment of HIV infection. Finally, we demonstrate how bioinformatics was used with synthetic biology in discovery and development of artemisinic acid, a biopharmaceutical now used in the treatment of malaria.

6.4.1 Bioinformatics in Vaccine Discovery

After clean water, the use of vaccines has been the most successful medical intervention strategy in history. It is estimated that every year vaccines prevent approximately 2.5 million deaths and contribute greatly to the quality of life. Vaccines aim to artificially stimulate a specific immune response, and as a result, long-term immunity against a pathogen is generated, which can prevent development and possibly transmission of a disease. A classic formulation of a vaccine is to use an inactive or attenuated pathogen, for example, vaccines against diseases such as measles, mumps and chickenpox. Attenuated vaccines do not always provide the desired protection because antigenicity can be lost and there can be safety concerns because possible reverse mutations may increase virulence.[30]

With the development of recombinant DNA technology in the 1970s, there was a major revolution in the field of vaccines. Through these techniques, heterologous large-scale production of pathogen proteins and their application in vaccine formulation has been achieved. The first examples of the use of recombinant DNA technology to produce vaccines were against hepatitis B and human papillomaviruses.

The advent of DNA sequencing techniques in the 1990s and 2000s allowed sequencing of entire genomes, providing quick access to mine for proteins that may be sufficiently antigenic to induce immunity, thus providing a bioinformatics-guided route to the development of new vaccines. To emphasize the speed of such advances, COVID-19 disease was first described in December 2019, and by mid-February 2020, the total genome of the SARS-CoV-2 that causes COVID-19 was released and possible vaccine candidates identified.[31] Over the course of the last few decades, advances in X-ray, NMR and especially Cryo-EM crystallography techniques have also increased the speed of determining three-dimensional structures of proteins, thus enabling the determination of the structure of antigens and their epitopes. Together, genomic and structural data have revolutionized vaccine development and created the so-called reverse vaccinology and structural vaccinology.

In reverse vaccinology, in silico searches of genomes are performed to identify genes that encode proteins with the potential to generate an immune response, such as surface proteins conserved between different strains. After this selection, the genes encoding for these proteins are expressed in a heterologous host and the proteins purified. The antigenicity of these purified proteins is next assessed by measuring antibody production following injection of the proteins into animal models. Structural vaccinology, on the other hand, exploits the three-dimensional structure of a possible antigen. For example, structural vaccinology can determine if the same protein from different strains of the same pathogen has structural variants of the protein. To induce a strong immune response, it is necessary to use all variants that have differential epitopes in the vaccine, which in most cases makes the manufacturing process more difficult and more expensive. Despite this, through structural analysis and molecular biology tools, it is possible to add all these epitopes in a single molecule. These strategies were used to develop a vaccine against Meningococcal B (MenB) caused by the pathogen *Neisseria meningitidis*.

The vaccines used for *N. meningitidis* serogroups A, C, Y and W135 are composed of bacterial capsule polysaccharide conjugated to a carrier protein. However, this strategy cannot be used to develop a vaccine for *N. meningitidis* serogroup B (MenB), as the capsule polysaccharide is identical to the polysialic acid [α (2–8) N-acetylneuramine] present in many human glycoproteins.[32] To develop a specific vaccine for MenB, reverse and structural vaccinology strategies were used. The *N. meningitidis* genome was first analysed in search of antigens with the potential to generate an immune response. In this in silico search, 600 potential vaccine candidates were identified, and of these, 350 were successfully expressed in *Escherichia coli* and used to immunize mice. Of the 350 potential vaccine candidate proteins, only 28 induced neutralizing antibodies,[33] from which 3 induced broad protection against many pathogenic strains. These antigens were named NHBA (Neisserial heparin-binding antigen), fHbp (factor H binding protein) and NadA (Neisseria adhesin A). In addition to these, antigens designated GNA (genome-derived Neisseria antigens) 2091 and GNA1030 were also selected, although they were not effective in all assays. The antigens were then used in the formulation of a new vaccine, called 4CMenB. It is worth mentioning that to facilitate production and reduce costs, NHBA-GNA1030 and GNA2091-fHbp antigens were produced as fused peptides.[34] The 4CMenB vaccine was first approved for use in the European Union in 2013 and then

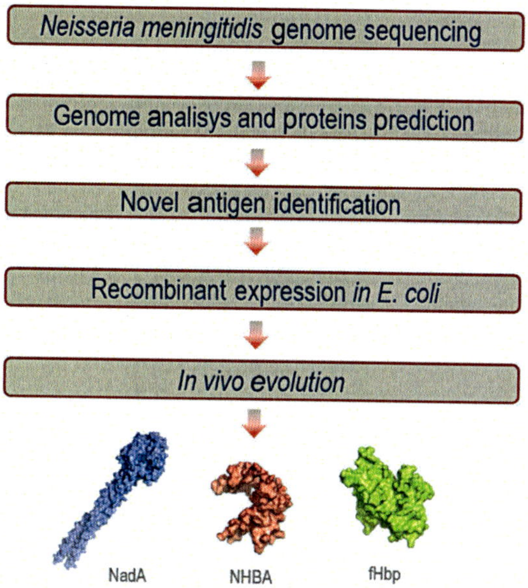

FIGURE 6.10 Search for specific antigens to produce the MenB vaccine. The genome of the virulent strain *N. meningitidis* MC58 was sequenced, and more than 2,000,000 proteins were predicted. Using bioinformatics algorithms, possible surface antigens were identified. The most promising proteins were expressed recombinantly in *E. coli*, purified and tested for ability to induce neutralizing antibodies. The NadA (PDB code: 6EUP), NHBA (PDB code: 2LFU) and fHpb (PDB code: 3KDV) antigens generated an immune response against the largest number of *N. meningitidis* strains. These proteins are part of the 4CMenB vaccine formulation. The images were generated by using the PyMOL programme.

in more than 35 countries, including the USA, Canada, Australia, Chile, Argentina and Brazil. The 4CMenB vaccine development flowchart is shown in Figure 6.10.

Despite the enhanced capacity of the fHbp protein to generate an immune response, more than 500 primary amino acid sequence variants have been reported in the literature, which are classified into three distinct classes (1, 2 and 3). Notably, these different classes do not induce cross-immunity with each other, which can reduce the neutralizing capacity antibodies raised against fHbp. To overcome poor antigenicity of some sequence variants, 4CMenB vaccine development has used proteins in class 1, which is the most abundant class found across *N. meningitidis* serotypes. To induce a strong protective immunity, specific antigens from all classes must be included in the vaccine, but this makes the manufacturing process complex and expensive.[32]

In this context, the use of structural vaccinology becomes important, with the fHbp antigen being the main example of using tools with a three-dimensional structure for the development of more efficient antigens. In this case, a chimeric fHbp antigen was created, where a single molecule contained the antigenic epitopes of all three fHbp variant groups.[32,35]

The development of this fHbp variant was designed based on its three-dimensional structure which was resolved using NMR and X-ray crystallography to reveal that the protein structure is conserved between the three classes and is composed of a central two β-barrels connected by a loop. For the development of chimeric fHbp, an evaluation of the epitopes recognized by specific monoclonal antibodies was initially carried out. Figure 6.11a shows the amino acids important for antibody recognition.[32,35]

Analysis of epitopes present in the three classes shows the presence of distinct regions which are recognized by antibodies. Figure 6.11a demonstrates that the epitopes of class 1 proteins, and class 2 and 3 are mapped in non-overlapping regions, which suggested that amino acids from class 2 or 3 proteins can be inserted into the structure of variant 1 using molecular biology techniques. This approach has been evaluated with single, double and triple mutations derived from the sequences of class 2 and 3 proteins incorporated into

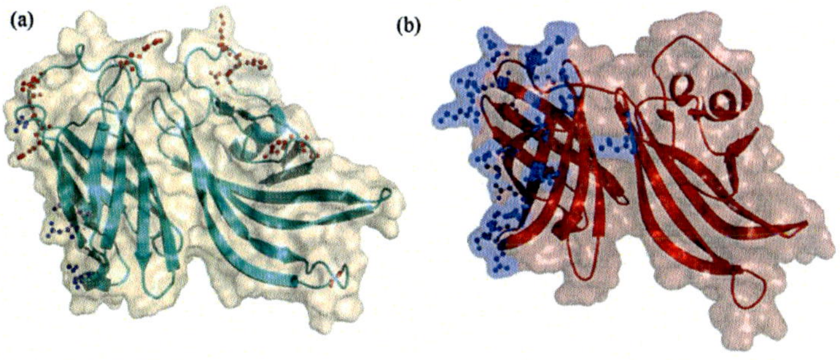

FIGURE 6.11 Structural analysis of fHBP and chimeric fHBP. (a) Analysis of the three-dimensional structure of *N. meningitidis* fHBP from strain MC58 (PDB code: 2KCO), presented in a light blue as a cartoon, with a 60% transparency straw surface. The immunogenic amino acids of class 1 are highlighted in red (ball and sticks) (D^{25}, H^{26}, K^{27}, T^{56}, Y^{57}, G^{121}, E^{146}, G^{147}, G^{148}, R^{149} e R^{204}) and in blue (ball and sticks); the immunogenic amino acids are modified in classes 2 and 3 (A^{174} – K in 2 and 3, K^{180} - R in 2 and 3; D^{192} – E in 2 and 3), and Q^{216} (S or G in 2 and S in 3). (b) Analysis of the three-dimensional structure of chimeric fHBP (PDB code: 2Y7S), represented by a cartoon in red, with a 60% transparency red surface. In blue, the amino acids substituted in the recombinant protein are shown by their corresponding in class 2 and 3 proteins ($I^{134} \rightarrow L$, $A^{135} \rightarrow G$, $S^{140} \rightarrow A$, $D^{142} \rightarrow N$, $K^{143} \rightarrow Q$, $E^{146} \rightarrow D$, $R^{148} \rightarrow K$, $T^{150} \rightarrow E$, $A^{172} \rightarrow T$, $A^{173} \rightarrow K$, $D^{191} \rightarrow E$, $A^{194} \rightarrow S$, $D^{196} \rightarrow E$, $P^{199} \rightarrow A$, $R^{203} \rightarrow S$, $S^{208} \rightarrow L$, $S^{210} \rightarrow D$, $L^{212} \rightarrow R$, $N^{214} \rightarrow G$, $Q^{215} \rightarrow S$, $A^{216} \rightarrow E$, e $K^{229} \rightarrow R$). The images were generated using the PyMOL programme.

class 1 sequences. However, this approach failed because none of the new hybrid antigens were able to induce a broad immune response, indicating that changing just a few amino acids was not enough to create immunodominant epitopes capable of raising protective antibodies class 2 and 3 proteins. The poor outcome of this approach demonstrated that it would be necessary not only to introduce the main amino acids of each epitope but also to recreate the entire epitope surface, mimicking the three-dimensional structure of the protein. This was a challenging prospect and so researchers chose to instead modify antibody recognition regions (epitopes) rather than just single amino acids, especially in the C-terminal region of class 2 and 3 proteins where most epitopes are located. The C-terminal region of class 1 proteins was divided into 11 sections of approximately 900–2,000 Å,[31] which corresponded to the average size of an epitope.[32,35] Within each distinct region, the amino acids with side chains exposed to the solvent were replaced by the corresponding sequences from 2 to 3 proteins regardless of the position in the primary sequence, in order to maintain the three-dimensional structure. These modifications generated a total of 54 variants which were successfully expressed in *E. coli*, purified and used to immunize mice; and the sera generated were tested for the presence of bactericidal antibodies against MenB strains carrying the fHbp 1, 2 and 3 variants. Several molecules elicited a wide immune response, and one recombinant protein raised neutralizing antibodies against MenB strains carrying all three fHbp class proteins. Mutations did not lead to significant changes in the structure of the molecule (Figure 6.11b). In fact, the work done with this fHbp protein demonstrates the need for in-depth structural studies for the successful development of a chimeric molecule since individual amino acids essential for an epitope were less important than preserving the total surface structure of the epitope itself.[32,35]

6.4.2 Use of Peptides in New Drug Development

The peptides exert several biological functions acting as chemical messengers, hormones, intra and extracellular mediators and inhibitors or stimulators with high specificity. They are also involved in various physiological processes and arouse a great interest in researchers and the pharmaceutical industry. Peptides have important characteristics for the development of drugs such as good efficacy and selectivity, good safety and tolerability, prior knowledge of metabolism and, in general, short time to market.[36]

Peptides represent a unique class of pharmaceutical compounds, positioned between small molecules and proteins, but with different biochemical and therapeutic characteristics. The first peptide to be used as a medicine was insulin in 1920 as replacement therapy for diabetic patients. Using peptides of natural origin as medicines has several disadvantages, such as low half-life (susceptibility to proteases), low absorption (high molecular mass), low transport (absence of specific transport systems) and induction of the immune response.[37]

Strategies for the development of peptides as therapeutic molecules are constantly evolving to overcome these impediments. By the 1950s, it was possible to synthesize peptides through structural knowledge and chemical synthesis, for example, oxytocin. The search for peptides from unconventional natural sources, such as arthropods or reptile poisons, has become a very popular strategy and resulted in the discovery of the peptide Teprotide, isolated from the venom of the *Bothrops jararaca* and was first in class of angiotensin-converting enzyme (ACE) inhibitors used to treat hypertension. The genomic revolution has gone beyond the characterization of proteins (sequences, structures, catalytic sites of enzymes) to now make the search for specific peptide ligands as potential drug candidates possible. The field of peptidomimetics has emerged in which the limitations of naturally occurring peptides can be overcome, resulting in peptides with higher stability, good bioavailability, greater affinity and selectivity being discovered. Lisinopril is a classic example of a peptidomimetic molecule, where the structure of the original peptide is modified in order to maintain the desired biological activity but minimize the disadvantages of the original molecule. The development of Lisinopril into a biopharmaceutical is next described.[37]

The angiotensin-converting enzyme (ACE) is a central component of the renin-angiotensin system, responsible for controlling blood pressure. ACE converts the hormone angiotensin I into angiotensin II, which causes vasoconstriction and increased blood pressure. In this way, ACE inhibitors act to control hypertension by lowering blood pressure. The development of the first orally available ACE-inhibiting drug is considered one of the first examples of rational drug design.[38]

In the 1960s and 1970s, venom of the snake *B. jararaca* and its potent ACE inhibiting effects were extensively studied. The molecules responsible for this effect were named bradykinin-potentiating factor (BPF) and were characterized as a mixture of peptides, but the most potent was Teprotide. However, the clinical use of this molecule was limited because activity was lost when administered orally.[37]

Teprotide was then studied using the peptide scanning technique in order to understand how each amino acid contributed to the biological activity of the peptide. Pharmacophoric groups responsible for the inhibitory activity of the peptide were two prolines at the C-terminal region of the peptide and a carbonyl group close to the second proline. At that time, the ACE structure had not been resolved, but the structure for carboxypeptidase A (CPA) was known and was used as a study model for Teprotide pharmacophoric group determination since both enzymes have metals in the active site. CPA inhibition studies showed that the carboxyl group of the benzylsuccinic acid-based inhibitor interacted with Zn^{+2} at the active site. By extrapolating this idea, it was subsequently demonstrated that the carbonyl group of the benzylsuccinic acid-based inhibitor also inhibited the ACE protein by binding with Zn^{+2}. Subsequently, the peptide was again modified in order to obtain a more potent ligand to Zn^{+2} from ACE by replacing the carbonyl group with a sulfhydryl group, and a new medicine was developed and launched in 1981 under the trade name Captopril, revolutionizing the treatment of cardiovascular diseases.[37]

Captopril is still used clinically but has several adverse side effects (skin rash, dry cough, headache and loss of taste). Replacing the carboxylate group with a thiol decreased drug affinity for Zn^{+2} to overcome many of the adverse side effects of Captopril, whilst pharmacological activity could be retained by the introduction of hydrophobic

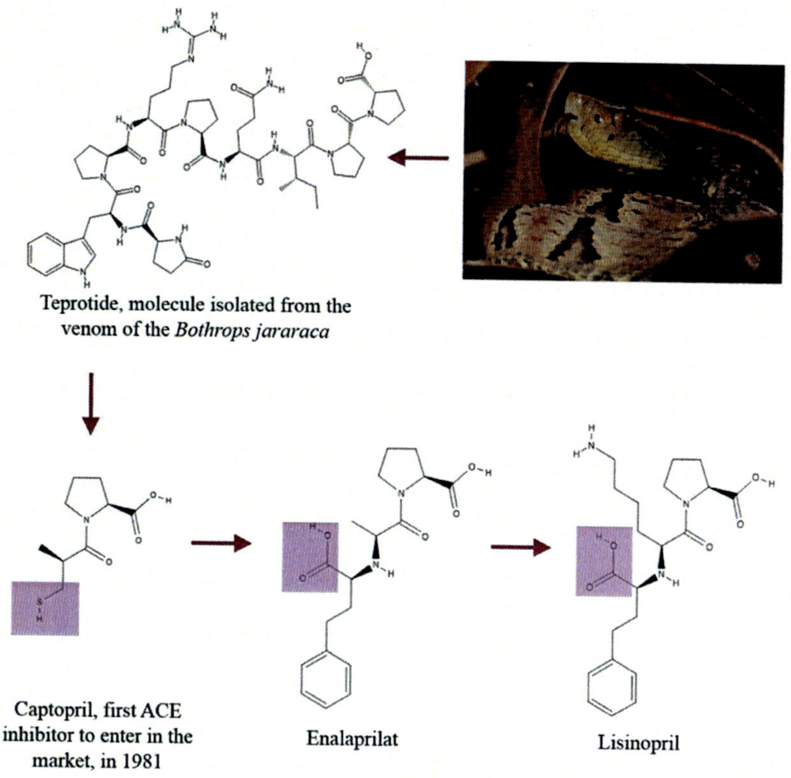

FIGURE 6.12 Drug development of ACE inhibitors. The Teprotide peptide was isolated from venom of the Brazilian snake *B. jararaca* and ACE inhibitory activity demonstrated. Low bioavailability via the oral route proved an obstacle to widespread clinical use. Captopril was developed by short peptide scanning and local structural modification of Teprotide. Additional chemical modifications generated Enalapril and Lisinopril. The regions highlighted in purple are the groups responsible for binding to the zinc atom at the ACE active site. The image of the *Bothrops jararaca* snake was obtained at: https://pt.wikipedia.org/wiki/Jararaca-da-mata. The molecules were generated using the MolView programme (https://molview.org/).

groups that interacted with the catalytic pocket. Enalapril and Lisinopril are second-generation orally available ACE inhibitors approved by the FDA in 2001 and 2003, respectively (Figures 6.12 and 6.13).[39]

6.4.3 Human Immunodeficiency Virus Reverse Transcriptase Inhibitors

Viruses can cause many mild to severe or life-threatening diseases in humans, for example, the common cold and different types of flu, measles, mumps, rubella, chickenpox, AIDS, severe acute respiratory syndrome (SARS) and haemorrhagic fevers (e.g. Ebola virus disease). Viruses are obligate intracellular parasites that use the metabolic machinery of a host cell to complete their life cycle. One of the greatest challenges in antimicrobial chemotherapy is the discovery of antiviral drugs with selective toxicity for the virus (that do not cause damage to the host cell) without the emergence of drug resistance.[40]

A protein unique to the virus or structurally very different from a host protein can be a potential target for selective drug toxicity. Human immunodeficiency virus (HIV) causes AIDS. It is a single-stranded RNA virus which must first convert the single-stranded RNA genome into complementary DNA (cDNA) in order to synthesize proteins. The enzyme reverse transcriptase (RT) has RNA-dependent DNA polymerase activity that copies the RNA template to cDNA which can then integrate into the host's chromosome.[41]

The RT enzyme is a heterodimer composed of p66 and p51 subunits. The p51 subunit has no catalytic activity and has a structural role only. The p66 subunit contains two domains, one with polymerase activity and the second with RNase H activity. The polymerase domain is subdivided into four subdomains, called: fingers (residues 1–85 and 118–155), palm (residues 86–117 and 156–236), thumb (residues 237–318) and connection (residues 319–426). The RNase H domain is a single domain consisting of residues 427–560 (Figure 6.14) (PDB: 3V4I).[42]

Most anti-HIV therapy protocols include at least two nucleosides or nucleotide analogue RT inhibitors, which include Combivir and Tenofovir, and one non-nucleoside or nucleotide analogue inhibitor, which include Nevirapine and Etravirin (Figure 6.14). RT nucleoside or nucleotide analogue inhibitors mimic and compete with natural substrates for polymerization into the newly synthesized DNA strand. These analogues do not contain a 3′–OH group and so act as chain terminators once incorporated into the growing DNA chain. Like natural nucleotides, analogues also need to be converted to 5′-triphosphate nucleotides by host cell kinases in advance of being incorporated into the DNA chain. In contrast, non-analogous nucleoside or nucleotide

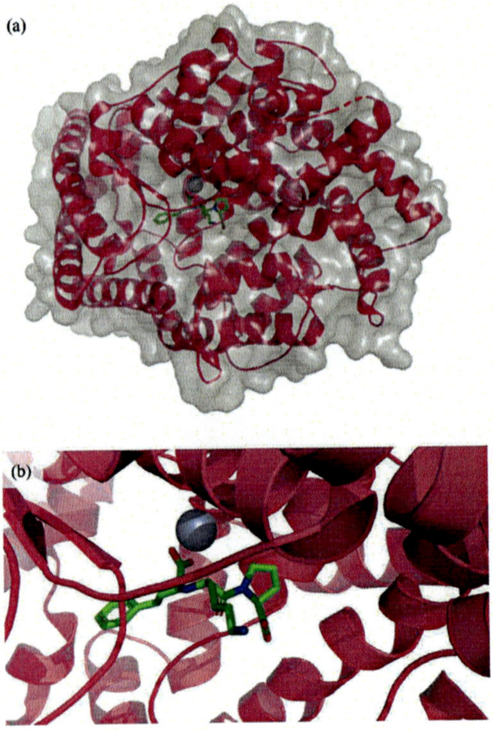

FIGURE 6.13 Structure of the angiotensin-converting enzyme protein containing Lisinopril. (a) Structure of the angiotensin-converting enzyme protein (angiotensin-converting enzyme – ACE) containing the peptide Lisinopril at the active site (code PDB 1O86). ACE is represented by cartoon (magenta) and surface (gray), the peptide Lisinopril is represented as a stick (C: green, N: blue and O: red) and the zinc atom is represented by a sphere (lilac). (b) Detail of the active site containing the Lisinopril peptide interacting with the ACE zinc atom, thereby functioning as a protein inhibitor. Note the carboxylate group interacting with the zinc atom. These images were generated using the PyMOL programme.

RT inhibitors are allosteric inhibitors, which bind to hydrophobic sites close to the active polymerase site, which causes structural changes that impair DNA synthesis.

HIV virions have emerged which are resistant to these inhibitors, and this resistance is associated with amino acid substitutions in the RT enzyme. The use of protein structure modelling has been used to explain the molecular mechanisms of enzyme resistance, and in the search for new compounds which inhibit these drug-resistant enzymes.

Derudas and collaborators built a database of 3,600 acyclic nucleoside analogues[43] which were screened using computational methods for docking with the co-crystal structure of HIV RT with the antiretroviral drug tenofovir diphosphate (PDB: 1Q05). A total of 984 compounds gave a hit against this screen as defined by a deviation of the r.m.s.d. <1 Å between the base of the coupled compounds and the base of the tenofovir diphosphate linker. Two subsequent virtual screens were performed, with the objective of finding molecules that remained phosphorylated under physiological conditions and with reduced affinity for host DNA-dependent RNA polymerase. At the end of the screening, five molecules were selected, including ganciclovir (ganciclovir is used clinically to treat human cytomegalovirus infections and bound tightly to RT in the initial screening with a r.m.s.d. <0.18 Å). Five of the 984 compounds met the criteria for these two subsequent screens but failed to show significant inhibition of RT in later empirical experimentation. Synthetic derivatives of these five molecules, however, did show higher inhibition of HIV RT, demonstrating how bioinformatics can provide leads for targeted medicinal chemistry (Figure 6.14).

6.4.4 Applied Bioinformatics Synthetic Biology: Production and Discovery of New Biopharmaceuticals

Synthetic biology is a field of biological sciences with the objective to produce new metabolic pathways in cells (potentially creating 'cell factories') to produce products of interest. Expansion of synthetic biology has only been possible by using comparative bioinformatics methods to interrogate large numbers of genome, transcriptome and proteome sequences to uncover the metabolic potential of cells.

The determination and analysis of the most diverse genomes revealed that the organisms that produce compounds able of behaving as biopharmaceuticals possess all the necessary genes involved the regulation, biosynthesis, export and even protection against compound harmful effects, co-located in genome-specific regions called biosynthetic gene clusters (BGCs).[44] Figure 6.15 illustrates two BCGs of *Streptomyces roseochromogenes* and *Streptomyces* sp. (SANK 60404), which are responsible for the synthesis of antibiotics of the amino coumarin classes and nucleoside derivative.

BGCs can be identified in the genomes based mainly on the presence of biosynthetic genes for the different classes of secondary metabolites. The clustering of genes involved in the production of a specific metabolite in BGCs greatly facilitates the characterization of biosynthetic pathways, allowing not only the elucidation of complete biosynthesis pathways for compounds with pharmacological activity but also the possibility of discovering new enzymes whose activity biochemistry is more efficient or able to generate new products with pharmacological properties.[44] Many of these BGCs appear cryptic because the predicted metabolites have never been isolated from the producing organism, suggesting regulatory systems for secondary metabolites remain to be uncovered. Several bioinformatics tools have been developed to identify BGCs such as antiSMASH (antibiotics and Secondary Metabolite Analysis SHell), MIBiG (Minimum Information about a Biosynthetic Gene Cluster) and ClustScan (Cluster Scanner).[45,46]

An important milestone in synthetic biology was the heterologous biosynthesis of artemisinin in the yeast *Saccharomyces cerevisiae*.[47] Artemisinin is a sesquiterpene lactone, originally extracted from the plant *Artemisia annua* and is one of the most effective treatments of malaria. Artemisinin is produced from the mevalonate biochemical pathway, an important metabolic pathway for the production of sterols in animals, plants and fungi.

Using synthetic biology methods, it has been possible to engineer the endogenous mevalonate pathway of *A. annua* into *S. cerevisiae* (being a yeast, *S. cerevisiae* produces the membrane lipid sterol ergosterol). This was

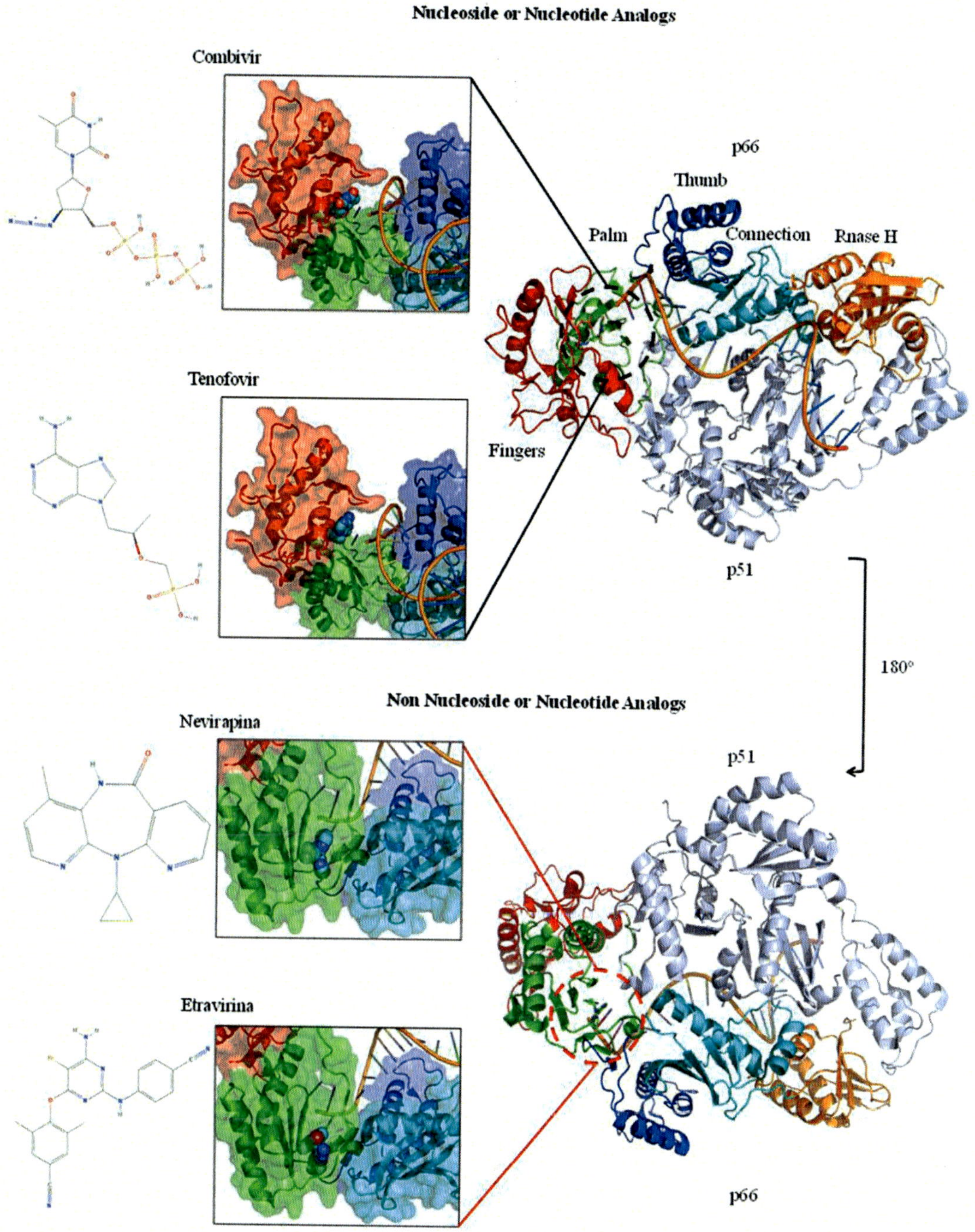

FIGURE 6.14 Crystal structure of the RT enzyme and chemical structures of some inhibitors. The crystal structure of the RT enzyme is represented as a cartoon, with the p51 subunit coloured in light blue and the p66 subunit coloured according to each domain: fingers in red, palm in green, thumb in dark blue, connections in cyan and RNase H in orange (PDB: 3V4I). The nucleoside analogues Combivir (PDB: 3V4I) and Tenofovir (PDB: 1Q05) are represented in spheres located at the catalytic site of RT, while the non-nucleoside analogue inhibitors Nevirapine (PDB: 1S1U) and Etravirine (PDB: 3MEC) are represented as spheres at the allosteric site of the RT enzyme. The images were generated using the PyMOL programme.

achieved by inserting two genes from the *A. annua* artemisinin synthesis pathway: amorphadiene synthase (ADS) and a multifunctional cytochrome P450 monooxygenase (CYP71AV1), responsible for three metabolic steps leading to the biosynthesis of artemisinic acid (AA), which is the immediate precursor of artemisinin. During fermentation of the engineered *S. cerevisiae* heterologous host, AA is secreted into the extracellular medium which greatly facilitates purification of this intermediate at low cost. The conversion of AA to artemisinin can be carried out in a simple

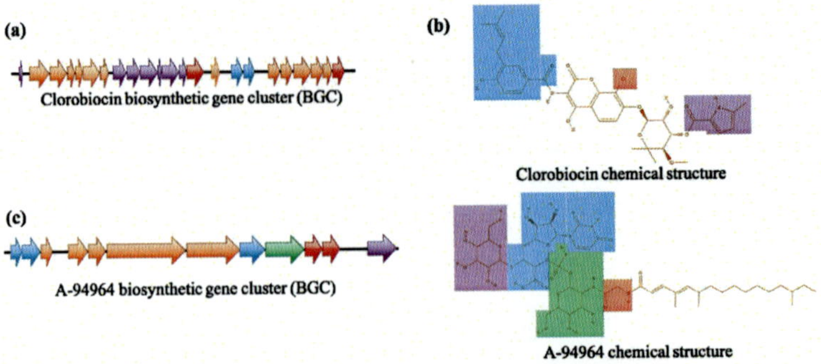

FIGURE 6.15 Example of BCGs from the Gram+ *S. roseochromogenes* and *Streptomyces* sp. involved in antibiotic biosynthesis. The figures on the right represent the chromosome segment (black line) containing the genes (represented by arrows) (a and c) involved in the synthesis of the antibiotics chlorobiocin (a) and A-94964 (c) and their respective chemical structures (b and d, respectively). The colours of the arrows in the chromosome diagram represent which segment of biopharmaceuticals the gene product is involved in.

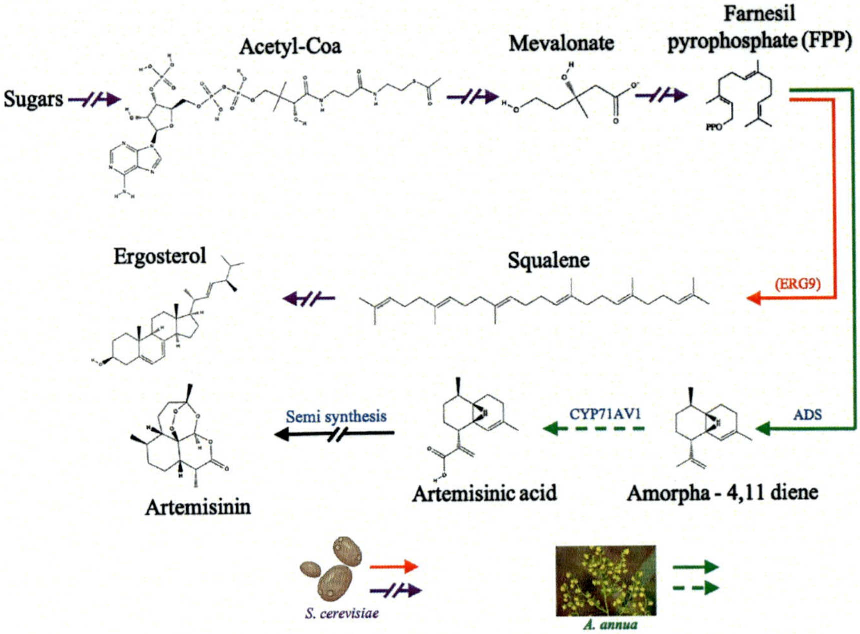

FIGURE 6.16 Simplified representation of the modification of the mevalonate pathway in *S. cerevisiae* for the biosynthesis of artemisinic acid. From the metabolism of sugars, acetyl-CoA is produced, which is converted into mavelonate for the synthesis of ergosterol by *S. cerevisiae*. In this metabolic pathway farnesyl pyrophosphate (FPP) is formed, which is converted into squalene and then to ergosterol. FPP is also a substrate for the amorphadiene synthase enzyme (ADS) of *A. annua* involved in the production of artemisic acid. Downregulation of ERG9, the endogenous enzyme of *S. cerevisiae* responsible for the conversion of FPP to squalene allows FPP to be shunted to increase biosynthesis of artemisic acid. In the figure, the interrupted purple arrows represent several biochemical reactions carried out by endogenous enzymes of *S. cerevisiae*. The red arrow indicates the ERG9 enzyme, which is downregulated. The solid green arrow represents the biochemical step performed by ADS, and the dashed green arrow represents reactions catalysed by CYP71AV1, both from *A. annua*. The interrupted black indicates the steps of chemical semi-synthesis to obtain artemisinin.

way by semi-synthetic methods, which again greatly facilitates downstream processing. By upregulating expression of at least 11 genes related to ergosterol biosynthesis and reducing the expression of ERG9 (the enzyme responsible for the conversion of farnesyl pyrophosphate to squalene) (Figure 6.16), higher amounts of farnesyl pyrophosphate could be made available for the biosynthesis of AA (ADS and CYP71AV1), significantly increasing amounts of AA produced. The pharmaceutical company Sanofi improved the production of AA, and in 2013, the World Health Organization (WHO) announced that the production of artemisinin using synthetic biology and semi-synthetic approaches was safe. Current production of artemisinin by Sanofi accounts for approximately 25% of the worldwide demand for the drug.

Another example of synthetic biology that has been used to shunt biosynthetic pathways in heterologous hosts for the industrial production of important biopharmaceuticals is in the manufacture of opiates, including codeine and morphine by *S. cerevisiae* (Figure 6.17).[48] This example demonstrates

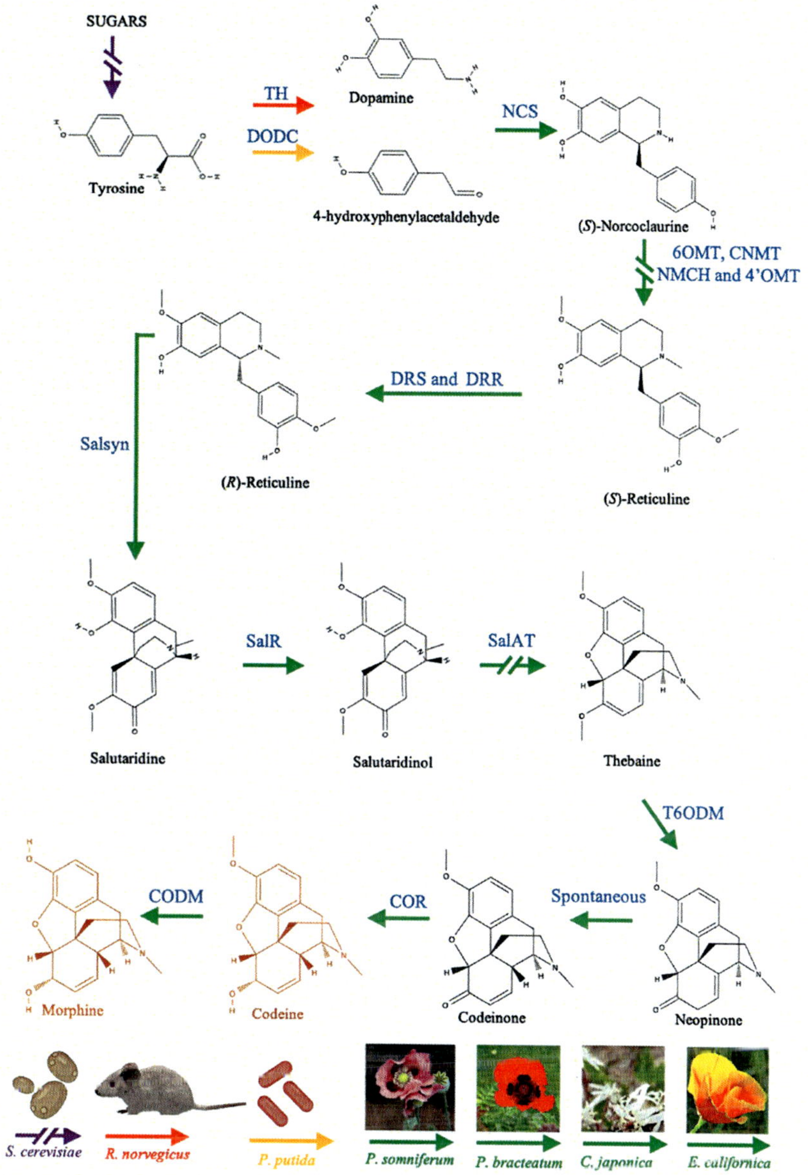

FIGURE 6.17 Biosynthetic pathway of morphine from the opium poppy *Papaver somniferum* engineered into the yeast *S. cerevisiae*. The synthesis begins with the condensation by norcoclaurine synthase (NCS) of two Tyr derivatives, dopamine and 4-hydroxyphenylacetaldehyde, producing the primary intermediate (S)-norcoclaurine. (S)-norcoclaurine is converted to (S)-reticulin through the successive action of norcoclaurine 6-O-methyltransferase (6OMT), CNMT (CNMT), N-methylcoclaurine 39-hydroxylase (NMCH) and 39-hydroxy N-methylcoclaurine 4′-O-methyltransferase (4′OMT). (S)-Reticulin is a branch point intermediate in the biosynthesis of several structural subgroups of BIA. Exclusively, codeine and morphine biosynthesis require epimerization of (S)-reticulin. Salutaridine, the first tetracyclic alkaloid promorfinan, is formed via intramolecular carbon-carbon-phenol coupling of (R-retriculin catalysed by cytochrome P450 monooxygenase salutaridine synthase (SalSyn). Salutaridine reductase (SalR) reduces the keto group at C7 of salutaridine to produce salutaridinol, which undergoes stoichiometric transfer from an acetyl group to the C7 hydroxyl by acetyl Salaridinol 7-O-acetyltransferase -CoA-dependent (SalAT) to form 7-O-acetate of salutaridinol. The spontaneous loss of the acetyl group results in a rearrangement to thebain, the first pentacyclic alkaloid in morphine. Thebain is O-demethylated by thebain-6-O-demethylase (T6ODM) in neopinone, which is spontaneously converted to codeinone. Codeinone reductase (COR) reduces codeinone to codeine and is demethylated by codeine-O-demethylase (CODM) producing morphine (structural formula highlighted in orange). In the figure, the continuous arrows represent a biochemical step dependent on a single enzyme, while the broken arrows show metabolic steps performed by more than one enzyme. The purple arrow represents metabolic steps performed by yeast enzymes, the red arrow represents a *R. novergicus* heterologous enzyme, the orange arrow represents *P. putida* and the green arrows represents heterologous enzymes from the plants *Papaver somniferum*, *Papaver bracteatum*, *Coptis japonica* and *Eschscholzia californica*.

the versatility of synthetic biology to exploit enzymes from many sources in a single 'cell factory'. Production levels have been significantly increased by deletion of genes encoding NADPH oxidases/reductases (*ari1, adh6, ypr1, ydr541c, aad3, gre2* e hfd1) whose products convert one of the initial intermediates of the pathway (4-hydroxyphenylacetaldehyde) into acid derivatives or alcohols that inhibit cell growth and, thereby, reduce titres of the opiate products.[49] An additional benefit of removing this bottleneck for enhanced biosynthesis of opiates was the accumulation of the shunt product benzylisoquinoline (BIA), which is a starter unit for the biosynthesis of more than 3,000 plant natural products which are now accessible for new lead discovery programmes.

The examples given above illustrate just a small portion of the great potential synthetic biology offers for the production of biopharmaceuticals using the large volume of data generated by sequencing projects. Creating new metabolic pathways or shunting known pathways in heterologous host organisms not only maximizes the production of pharmaceuticals but also offers benefits to the environment compared to chemical processes.

6.5 FINAL CONSIDERATIONS

Advances in bioinformatics methods have kept pace with the extraordinary rate at which sequencing platforms can generate gene data, whereas empirical experimentation to verify gene or protein function has not. Bioinformatics has compensated for this gap in knowledge by developing homology searching methods to predict function, whether this be at the gene, transcript or protein levels. There is a danger that predicted functions based on homology to sequences or structures in databases which themselves have not been experimentally validated, and it could overreach the true biological function of a sequence or protein. Even using the most sophisticated distance homology methods, no known function can be attributed to an increasing number of sequences – the so-called 'hypothetical' genes and proteins.

Artificial Intelligence (AI) may help close the gap between generating poor sequence annotation and experimental validation of function. An example of this has been the resounding success of the AlphaFold system, which uses AI to predict 3D protein structures. AI has also demonstrated potential in simplifying QM calculations for small molecules which make possible protein simulations involving protein-protein interactions and ligand docking allowing billions of molecules to be screened for new lead discovery by the pharmaceutical industry.

Thus, the development of bioinformatics, both in relation to databases and computational tools, is an increasingly clear and powerful path for the advancement of scientific development, which includes the discovery of biopharmaceuticals in a more accelerated manner, contributing to mitigate pandemics, and more specific bio-drugs, efficient for genetic/infectious diseases, and also counting on a more sustainable production.

REFERENCES

1. Newman, D. J. & Cragg, G. M. Natural products as sources of new drugs over the nearly four decades from 01/1981 to 09/2019. *J. Nat. Prod.* **83**, 770–803 (2020).
2. Cochrane, G., Karsch-Mizrachi, I., Takagi, T. & International Nucleotide Sequence Database Collaboration. The International Nucleotide Sequence Database Collaboration. *Nucleic Acids Res.* **44**, D48–D50 (2016).
3. Amid, C. et al. The European nucleotide archive in 2019. *Nucleic Acids Res.* **48**, D70–D76 (2019). doi:10.1093/nar/gkz1063
4. Ogasawara, O., Kodama, Y., Mashima, J., Kosuge, T. & Fujisawa, T. DDBJ database updates and computational infrastructure enhancement. *Nucleic Acids Res.* **48**, D45–D50 (2019). doi:10.1093/nar/gkz982
5. Coimbatore Narayanan, B. et al. The nucleic acid database: new features and capabilities. *Nucleic Acids Res.* **42**, D114–D122 (2014).
6. Walls, A. C. et al. Structure, function, and antigenicity of the SARS-CoV-2 spike glycoprotein. *Cell* **181**, 281–292.e6 (2020).
7. Watanabe, K., Yasugi, E. & Oshima, M. How to search the glycolipid data in 'LIPIDBANK for Web' the newly developed lipid database in Japan. *Trends Glycosci. Glycotechnol.* **12**, 175–184 (2000).
8. Egorova, K. S., Kondakova, A. N. & Toukach, P. V. Carbohydrate structure database: tools for statistical analysis of bacterial, plant and fungal glycomes. *Database* **2015**, bav073 (2015).
9. Weber, T. & Kim, H. U. The secondary metabolite bioinformatics portal: computational tools to facilitate synthetic biology of secondary metabolite production. *Synth. Syst. Biotechnol.* **1**, 69–79 (2016).
10. Petsko, G. A. & Ringe, D. *Protein Structure and Function Pib P (Primers in Biology)*. (Oxford University Press, 2008).
11. Safdari, H. A., Pandey, S., Shukla, A. K. & Dutta, S. Illuminating GPCR signaling by Cryo-EM. *Trends Cell Biol.* **28**, 591–594 (2018).
12. Wu, S., Skolnick, J. & Zhang, Y. Ab initio modeling of small proteins by iterative TASSER simulations. *BMC Biol.* **5**, 17 (2007).
13. Patthy, L. *Prote=in Evolution*. (Wiley-Blackwell, 2007).
14. Chothia, C. & Lesk, A. M. The relation between the divergence of sequence and structure in proteins. *EMBO J.* **5**, 823–6 (1986).
15. Kuhlman, B. et al. Design of a novel globular protein fold with atomic-level accuracy. *Science* **302**, 1364–1368 (2003).
16. AlQuraishi, M. AlphaFold at CASP13. *Bioinformatics* **35**, 4862–4865 (2019).
17. Geraldo, M. T., Takeda, A. A. S., Braz, A. S. K. & Lemke, N. Bending-twisting motions and main interactions in nucleoplasmin nuclear import. *PLoS One* **11**, e0157162 (2016).
18. Merz Jr, K. M., Ringe, D. & Reynolds, C. H. *Drug Design: Structure- and Ligand-Based Approaches*. (Cambridge University Press, 2010).
19. Leach, A. *Molecular Modelling: Principles and Applications*. (Pearson, 2001).
20. Verli, H. *Bioinformática: Da Biologia à Flexibilidade Molecular*. (Sociedade Brasileira de Bioquímica e Biologia Molecular, 2014).
21. Liu, X. et al. Molecular dynamics simulations and novel drug discovery. *Expert Opin. Drug Discov.* **13**, 23–37 (2018).
22. De Conto, V., Braz, A. S. K., Perahia, D. & Scott, L. P. B. Recovery of the wild type atomic flexibility in the HIV-1 protease double mutants. *J. Mol. Graph. Model.* **59**, 107–116 (2015).
23. Cui, Q. & Bahar, I. *Normal Mode Analysis: Theory and Applications to Biological and Chemical Systems*. (Chapman and Hall/CRC, 2005).
24. Floquet, N. et al. Collective motions in glucosamine-6-phosphate synthase: influence of ligand binding and role in ammonia channelling and opening of the fructose-6-phosphate binding site. *J. Mol. Biol.* **385**, 653–664 (2009).

25. Du, X. et al. Insights into protein–ligand interactions: mechanisms, models, and methods. *Int. J. Mol. Sci.* **17**, 144 (2016).
26. Csermely, P., Palotai, R. & Nussinov, R. Induced fit, conformational selection and independent dynamic segments: an extended view of binding events. *Trends Biochem. Sci.* **35**, 539–546 (2010).
28. Meyerkord, C. L. & Fu, H. *Protein-Protein Interactions: Methods and Applications (Methods in Molecular Biology)*. (Humana, 2015).
29. Yan, Y., Tao, H., He, J. & Huang, S.-Y. The HDOCK server for integrated protein–protein docking. *Nat. Protoc.* **15**, 1829–1852 (2020).
30. Liljeroos, L., Malito, E., Ferlenghi, I. & Bottomley, M. J. Structural and computational biology in the design of immunogenic vaccine antigens. *J. Immunol. Res.* **2015**, 1–17 (2015).
31. Khailany, R. A., Safdar, M. & Ozaslan, M. Genomic characterization of a novel SARS-CoV-2. *Gene Rep.* **19**, 100682 (2020).
32. Dormitzer, P. R., Grandi, G. & Rappuoli, R. Structural vaccinology starts to deliver. *Nat. Rev. Microbiol.* **10**, 807–813 (2012).
33. Seib, K. L., Zhao, X. & Rappuoli, R. Developing vaccines in the era of genomics: a decade of reverse vaccinology. *Clin. Microbiol. Infect.* **18**, 109–116 (2012).
34. Serruto, D., Bottomley, M. J., Ram, S., Giuliani, M. M. & Rappuoli, R. The new multicomponent vaccine against meningococcal serogroup B, 4CMenB: Immunological, functional and structural characterization of the antigens. *Vaccine* **30**, B87–B97 (2012).
35. Scarselli, M. et al. Rational design of a meningococcal antigen inducing broad protective immunity. *Sci. Transl. Med.* **3**, 91ra62–91ra62 (2011).
36. Fosgerau, K. & Hoffmann, T. Peptide therapeutics: current status and future directions. *Drug Discov. Today* **20**, 122–128 (2015).
37. Lenci, E. & Trabocchi, A. Peptidomimetic toolbox for drug discovery. *Chem. Soc. Rev.* **49**, 3262–3277 (2020).
38. Acharya, K. R., Sturrock, E. D., Riordan, J. F. & Ehlers, M. R. W. Ace revisited: a new target for structure-based drug design. *Nat. Rev. Drug Discov.* **2**, 891–902 (2003).
39. Natesh, R., Schwager, S. L. U., Sturrock, E. D. & Acharya, K. R. Crystal structure of the human angiotensin-converting enzyme–lisinopril complex. *Nature* **421**, 551–554 (2003).
40. Thibaut, U. Bioinformatics and rational drug design: tools for discovery and better understanding of biological targets and mode of action of drugs. *Scand. J. Gastroenterol.* **37**, 95–99 (2002).
41. Tarasova, O., Poroikov, V. & Veselovsky, A. Molecular docking studies of HIV-1 resistance to reverse transcriptase inhibitors: mini-review. *Molecules* **23**, 1233 (2018).
42. Sarafianos, S. G. et al. Structure and function of HIV-1 reverse transcriptase: molecular mechanisms of polymerization and inhibition. *J. Mol. Biol.* **385**, 693–713 (2009).
43. Derudas, M. et al. Virtual screening of acyclovir derivatives as potential antiviral agents: design, synthesis, and biological evaluation of new acyclic nucleoside protides. *J. Med. Chem.* **60**, 7876–7896 (2017).
44. Smanski, M. J. et al. Synthetic biology to access and expand nature's chemical diversity. *Nat. Rev. Microbiol.* **14**, 135–149 (2016).
45. Trosset, J.-Y. & Carbonell, P. Synthetic biology for pharmaceutical drug discovery. *Drug Des. Devel. Ther.* **9**, 6285 (2015). doi:10.2147/DDDT.S58049
46. Starcevic, A. et al. ClustScan : an integrated program package for the semi-automatic annotation of modular biosynthetic gene clusters and in silico prediction of novel chemical structures. *Nucleic Acids Res.* **36**, 6882–6892 (2008).
47. Ro, D.-K. et al. Production of the antimalarial drug precursor artemisinic acid in engineered yeast. *Nature* **440**, 940–943 (2006).
48. Galanie, S., Thodey, K., Trenchard, I. J., Filsinger Interrante, M. & Smolke, C. D. Complete biosynthesis of opioids in yeast. *Science* **349**, 1095–1100 (2015).
49. Pyne, M. E., Kevvai, K., Grewal, P. S., Narcross, L., Choi, B., Bourgeois, L., Dueber, J. E. & Martin, V. J. J. A yeast platform for high-level synthesis of tetrahydroisoquinoline alkaloids. *Nat Commun* **11**, 3337 (2020, July 3). doi: 10.1038/s41467-020-17172-x. PMID: 32620756; PMCID: PMC7335070.

7 Bioprocesses
Microorganisms and Culture Media

Ricardo Pinheiro de Souza Oliveira and Adalberto Pessoa Jr.
Universidade de São Paulo

CONTENTS

7.1 INTRODUCTION

The success of a bioprocess depends on the clarity of four basic aspects: the microorganism, the culture medium, the way that the bioprocess is conducted and product recovery steps.

The initial stage of bioprocess development refers to the choice of microorganism, either genetically modified or wild type. Then, the growth phase and the generation of bioproducts are measured as functions of the culture conditions. Generally, small-scale cultivations are performed using shake flasks from 250 to 500 mL in volume. The macro and microelements, inducers, growth factors, pH, temperature, inoculum concentration and pre-inoculum preparation must be determined to allow by maximum growth and productivity. Important parameters, such as cell growth, final cell concentration, and product yield, substrate consumption and bioproduction rate and productivity as a function of time, are commonly used to express the performance of the bioprocess (Doran, 2013). After obtaining such information from shaker-scale cultivations, the first step of bioprocess scale-up starts and involves studies in bioreactors, usually with volumes of 1–5 L. These cultivations will provide more accurate measurements, monitoring and/or controlling especially related to product and metabolite concentrations, variation and/or control of pH, temperature, antifoam addition, dissolved oxygen concentration and CO_2 concentration at the bioreactor exit (Doran, 2013). In addition, before starting the bioprocess operation, it is important to decide the best operating system, for example, batch, fed or continuous batch, as well as whether it will be necessary to recycle cells and control cell viability (Doran, 2013). Then, the bioprocess can be further scaled up to a pilot-scale bioreactor (usually ten times larger – 10–50 L), since the main objective is to evaluate how the whole bioprocess behaves in a bigger bioreactor. This step of the scale up is not simple, performance loss is common and it is necessary to pay particular attention to oxygen transfer parameters for aerobic microorganisms.

Often the geometry of small-scale bioreactors, that is, the diameter, agitation speed, aeration flow, bioreactor height and number of impellers (and distance between them), is proportional to those of a large-scale vessel, but even so the results of the cultivations (cell growth and production curves) can still be very different. When loss of productivity occurs after scaling up the process, it is necessary to rethink the criteria used for scaling up in such a way that productivity parameters remain similar. In this case, the geometry between the bench-scale and pilot bioreactors must be compared. If the geometries are equivalent, probably the oxygen supply is responsible for differences in performance. That is, the microorganism has to be supplied with oxygen in a similar way in the two systems. This comparison can be made based on comparing the dissolved oxygen concentration during cultivation at both scales, or according to the initial k_La (volumetric oxygen transfer constant) value (see more details in Chapter 11). If it is guaranteed that the cultivation conditions on a pilot scale are the same as the bench scale, that is, with a homogeneous supply of oxygen and nutrients, in general the cultivation scale-up usually provides satisfactory results (Pessoa et al., 1996). As soon as the scale-up conditions of the bioprocess are adjusted, the cultivation conditions become known and well established. Therefore, in the next step, the same parameters will be used for scaling up the process to an industrial scale. In this case, all facilities must be checked and adjusted, including steam and cooling water supplies, sterilization and rigor in the preparation of culture media and the process control network. Special attention must be paid when the bioprocess is conducted with pathogenic microorganisms (see Chapter 20). All stages of bioprocess must be carried out aseptically; therefore, the method used to achieve sterilization depends on the stage of inoculum propagation, on the type of microorganism involved, on the selectivity of the culture medium to the microorganism of interest, on the type of product to be produced, etc. Finally, an important part of the bioprocess is downstream processing, presented in more detail in Chapter 13. The success of downstream processing depends on the type of product and the medium composition (Doran, 2013) and so the choice of the most suitable culture medium is important, not only for the microorganism to grow and produce the biomolecule of interest, but also to allow efficient downstream processing, i.e. to provide high yield and purity of the desired product. After obtaining the bioproduct, it must be packaged and made available for sale.

In addition, depending on the bioproduct, such as biopharmaceuticals and diagnostic kits, an assessment of safety and effectiveness will also be necessary. In these cases, studies begin with tests on isolated cells (*in vitro* tests), and if the results are promising, the following tests are carried out on animals and then on humans before being released for health care application (Doran, 2013). The basic steps involved in a bioprocess are illustrated in Figure 7.1.

7.2 GENERAL CHARACTERISTICS OF MICROORGANISMS

The microorganisms used in bioprocesses can be obtained from natural resources, from mutations of these wild type strains (induced by classical or genetically modified methods (see Chapters 4–6) or acquired from recognized culture collections (Table 7.1). The microorganisms deposited in these institutions are listed in catalogs, many of them online, which includes information about the origin of the strain, preservation and cultivation conditions such as nutrient specifications or limitations, pH and temperature.

Culture collections can offer microorganisms with known characteristics, but they do not always contain all desired characteristics. While it is much cheaper to buy a culture than to isolate it from nature, an industrially important microorganism can also be found after exhaustive bioprospecting of different natural environments (air, soil, water, etc.). An important factor is the purchase of culture media, as these must strictly meet the needs of the process and guarantee its reproducibility (Stanbury et al., 2016).

Microorganisms of industrial interest, if possible, should have a high specific growth rate, do not produce undesirable metabolites, have a known oxygen demand, form little foam during cultivation and be able to metabolize low-cost substrates. In the case of peptide and protein therapeutics, it must also be taken into account that protein structure does not suffer any changes during cultivation at different scales.

The cultivation of microorganisms can start in different ways since the strain can be maintained on solid medium (inclined universal bottles or Petri dishes containing nutrient agar), in liquid medium or in lyophilized form. It is of fundamental importance that the microorganism, or stock culture, is kept in conditions that ensure that it does not change as a function of time. The stock culture is used for the preparation and development of the inoculum or starter culture, and then cell cultivation and biomolecule production.

Some characteristics are highly desirable in microorganisms for industrial use, such as: (1) the ability to accumulate products in the environment without being inhibited by them;

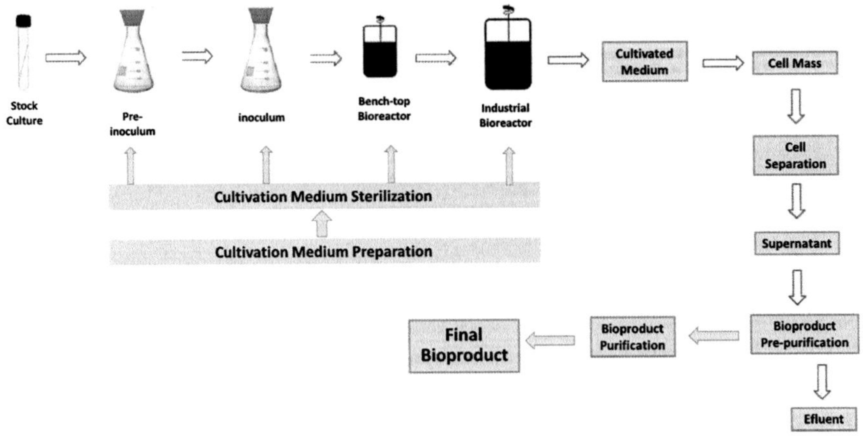

FIGURE 7.1 General flowchart for the development of a bioprocess.

TABLE 7.1
Major Culture Collections in Different Countries

Culture Collection	Website
American Type Culture Collection (ATCC)	http://www.lgcstandards-atcc.org/?geo_country=gb
The British Antarctic Survey Culture Collection	http://www.cabi.org
National Collection of Plant Pathogenic Bacteria	http://ncppb.fera.defra.gov.uk
National Collection of Type Cultures (NCTC)	http://www.phe-culturecollections.org.uk
National Collection of Industrial Food and Marine Bacteria (NCIMB Ltd)	http://www.ncimb.com
National Collection of Yeast Cultures (NCYC)	http://www.ncyc.co.uk
Deutsche Sammlung von Mikroorganismen und Zelkulturen (DSMZ)	https://www.dsmz.de
Centraalbureau voor Schimmelcultures (CBS)	http://www.cbs.knaw.nl
Japan Collection of Microorganisms (JCM)	http://jcm.brc.riken.jp/en/
Collection Nationale de Cultures de Microorganismes (CNCM)	https://www.pasteur.fr/en/cncm
Czech Collection of Microorganisms (CCM)	http://www.sci.muni.cz/ccm/index.html
Colecci6n Española de Cultivos Tipo (CECT)	https://www.uv.es/cect
Fundação Andre Tosello - Tropical culture collection	http://fat.org.br/

Source: Adapted from Stanbury, P.F., Whitaker, A., Hall, S.J., *Principles of Fermentation Technology*, 3rd edition, Elsevier, New York, 2016, ISBN 978-0-08-099953-1, https://doi.org/10.1016/C2013-0-00186-7.

(2) not to produce compounds that inhibit an organism's own metabolism which would affect the bioprocess; (3) remain physiologically stable throughout all the stages involved in a bioprocess from laboratory to industrial scale; (4) not to require too complex process conditions including growth over a range of optimal pH and temperature values and; (5) to produce extracellular products to simplify the downstream process.

7.2.1 Cell Concentration Measurement

The amount of cell mass (biomass) can be determined by direct methods, such as determining the number of cells visually using a light microscope and a Petroff-Hausser chamber or Neubauer chamber, number of colonies formed on a solid medium (colony forming units, CFU), direct electronic counting of cells (Coulter counter, flow cytometry), dry weight (this is the best way to determine the growth of filamentous microorganisms such as fungi or actinomycetes), turbidimetry of microorganisms grown in liquid broths, centrifuged volume (packed cell volume) and viscosity. Cell concentration can also be measured by indirect methods, as these may have a direct relationship with the number of cells, or with the dry mass such as cell components, e.g. total concentration of N or C, ATP, DNA, RNA, protein content or carbon dioxide production rate.

During cell culture, it is important that fast and reliable methodologies are used to determine the concentration of the components of interest, such as cells, substrates and products, because the ability in interpretation of the results will allow better control of the overall bioprocess. The most modern bioreactors, in addition to pH, temperature, dissolved oxygen and carbon dioxide, have automatic controls for various other parameters such as the concentration of some substrates – such as glucose and methanol – and some other products – such as ethanol and organic acids. Modern bioreactors also have resources for online and approximate monitoring of cell concentration, mainly based on reading optical density, which facilitates rapid identification of the growth phase of the microorganism. However, it is desirable to have a quick methodology that makes it possible to accurately calculate cell concentration, and for this, it is necessary to take a sample from the bioreactor and further analysis.

7.2.2 Standard Dry Mass Curve

A widely used methodology, both for bacteria and for yeasts, is reading the optical density (OD) of the sample at 600 nm with subsequent calculation of the cell concentration in g/L, using a previously constructed correlation curve (OD versus concentration cell). A new curve must be made for each microorganism, as well as for each time that the culture medium is changed. The following is an example of calculating the OD curve as a function of the concentration in g/L, using the yeast *Scheffersomyces (Pichia) stipitis* NRRL Y-7124 grown in two different culture media: hemicellulosic hydrolysate of rice straw and in a synthetic medium.

The yeast was kept on malt extract agar at 4°C. To make the standard dry mass curve, *S. stipitis* cells were grown in 50 mL of fresh rice straw hemicellulosic hydrolysate (pH 5.5) containing 30 g/L of xylose supplemented with 3.0 g/L of extract of yeast. After 24 h of cultivation in 250 mL Erlenmeyer flasks, at 30°C with aeration by shaking at 200 rpm, the cells were separated from the liquid medium by centrifugation (4000 × *g*, 20 min at 4°C) and washed twice with distilled water followed by centrifugation. The cell mass was then resuspended in distilled water to a final volume of 25 mL, and 5 mL aliquots (in triplicate) of this suspension was transferred to previously weighed porcelain containers. Drying was then carried out in an oven at 105°C for 24 h. The porcelain containers were then cooled in a desiccator and weighed. The dry mass of the cells was obtained by the difference in weight of the porcelain containers (with and without cells). The cell concentration of the original suspension (X_{SM}), in g/L, was calculated according to

Equation 7.1. In parallel, aliquots of the original suspension were serial diluted, and for each dilution, the absorbance reading was measured at 600 nm. The cell concentration of each dilution of the original suspension (XSD) was then calculated according to Equation 7.2.

$$X_{SM} = \left(\frac{P_c - P_v}{V_s} \right). \tag{7.1}$$

where X_{SM} = concentration of cells in the original suspension (g/L); P_c = porcelain container mass with cells (g); P_v = empty porcelain container mass; V_s = sample volume of the original suspension (L).

$$X_{SD} = \left(\frac{X_{SM}}{f} \right). \tag{7.2}$$

where X_{SD} = concentration of cells in the diluted samples from the original suspension (g/L);

f = dilution factor.

From these data, a graph was constructed plotting the cell concentration values on the abscissa [X] axis and the absorbance (600 nm) on the ordinate [Y] axis, thus obtaining by linear regression, an equation for the line.

The standard dry mass curve for the yeast *S. stipitis* grown in a synthetic medium was constructed in the same way as the standard dry mass curve for the yeast grown in hemicellulosic hydrolysate described above; however, the cells did not require washing. The composition of the synthetic medium was known and is shown in Table 7.2.

TABLE 7.2
Composition of the Synthetic Medium Used to Culture the Yeast *S. stipitis* for Subsequent Determination of a Dry Mass Growth Curve

Components	Concentration (g/L)
$(NH_4)_2HPO_4$	3.0
$MgSO_4.7H_2O$	1.0
KH_2PO_4	19.0
Yeast extract	3.0
Xylose	30.0

The salts and yeast extract were prepared separately in concentrated solutions (10 and 50 times, respectively, in relation to the concentration value indicated in Table 7.2). These concentrated solutions were sterilized separately in an autoclave at a temperature of 121°C for 20 min. The xylose solution was also prepared in a concentrated solution (eight times) and sterilized in an autoclave for 20 min, however, at a temperature of 111°C (to prevent xylose degradation). The appropriate volumes of the stock solutions were then combined aseptically to provide a broth composed of ingredients at the desired final concentrations.

7.2.2.1 Results

Figure 7.2 shows the standard curve of *S. stipitis* when grown in rice straw hemicellulosic hydrolysate and in a synthetic medium.

The curves showed good correlations (R^2 = 0.987 and 0.9988), indicating that in the absorbance range between 0.15 and approximately 0.7, the equations can be used to estimate the concentration of cells during cultivation in both hydrolysate and synthetic media.

7.2.3 Duplication and Generation Time

The generation time of a microorganism is defined as the time required for a generation to occur, that is, the formation of two cells from just one, under standard nutritionally favourable conditions. In the case of prokaryotic microorganisms (bacteria), duplication occurs by binary fission in geometric progression, also called logarithmic growth phase (exponential growth).

In the case of filamentous fungi and yeasts (eukaryotic microorganisms) and actinomycetes (prokaryotes), there is no multiplication by binary fission; therefore, this concept of generation time for the number of cells does not apply. However, the concept of 'doubling time' is used, which is when an increase in the mass of cells is used as a parameter of cell growth; and in this case, one can also observe exponential growth, that is, the mass of cells doubles in geometric progression. Additionally, the doubling time of a microorganism increases with increased cell size. The value ranges are found experimentally, which for yeasts are typically from 1.15 to 2.00 h for filamentous fungi, from 2.0 to 6.9 h, and for plant cells, from 20 to 40 h (Smith, 1985) (see Chapter 9).

As can be seen in Table 7.3, the generation time for *E. coli* in the laboratory is low (i.e. rapid growth) but some

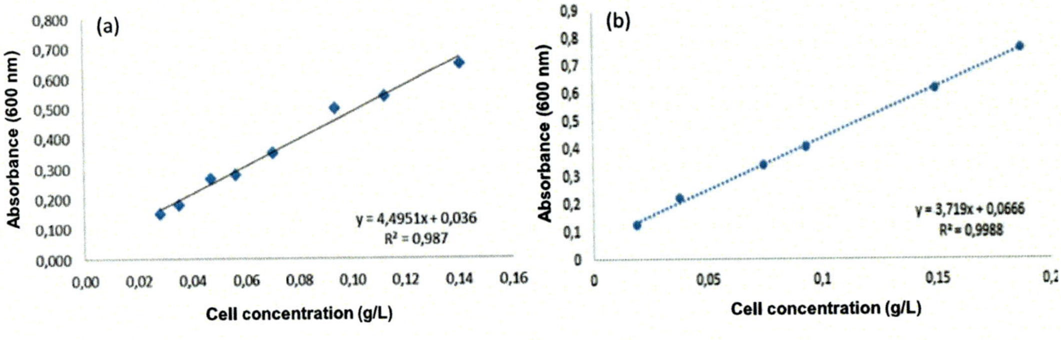

FIGURE 7.2 Relationship between absorbance and cell concentration of *S. stipitis* NRRL Y-7124 grown in fresh rice straw hemicellulosic hydrolysate (a) and in a synthetic medium (b).

TABLE 7.3
Generation Times for Some Bacteria

Bacterium	Medium	Generation Time (min)
Escherichia coli	Glucose–salts	17
Bacillus megaterium	Sucrose–salts	25
Streptococcus lactis	Milk	26
Streptococcus lactis	Lactose broth	48
Staphylococcus aureus	Heart infusion broth	27–30
Lactobacillus acidophilus	Milk	66–87
Rhizobium japonicum	Mannitol–salts–yeast extract	344–461
Mycobacterium tuberculosis	Synthetic	792–932
Treponema pallidum	Rabbit testes	1,980

pathogenic bacteria, for instance, *Mycobacterium tuberculosis* and *Treponema pallidum*, have long generation times; this is an important advantage related to virulence.

7.3 CULTIVATION MEDIA OF INDUSTRIAL INTEREST

Living cells are composed of different components; however, carbon and nitrogen atoms are essential, especially in the production of biomolecules as these are constituents of proteins and nucleic acids. The importance of carbon is of such consequence that the method in which a cell obtains reduced carbon for subsequent catabolic reactions can be used to classify different types of nutrition: autotrophic or heterotrophic. Autotrophic bacteria use carbon dioxide as the main source of carbon and are able to synthesize organic molecules. On the other hand, heterotrophic bacteria use only organic carbon sources. In addition, there are different nutritional patterns in nature, such that all organisms can be categorized into one of the following four groups: photoautotrophs, photoheterotrophs, chemoautotrophs and chemoheterotrophs. Organisms that use light as an energy source and carbon dioxide as the main oxidized carbon source are called photoautotrophic. In this case, carbon dioxide, in the presence of water, is reduced into carbohydrates and oxygen. Cyanobacteria, algae and green plants use hydrogen atoms from water to reduce carbon dioxide and form carbohydrates and oxygen (anoxygenic process). The anoxygenic process can also be performed by green and purple sulphur photosynthetic bacteria, using sulphur compounds or hydrogen to produce organic compounds. Photoheterotrophs, like green and purple non-sulphur bacteria, use organic compounds as a carbon source and light as an energy source, but cannot use carbon dioxide as a source of energy. Chemo-lithoautotrophs use carbon dioxide as their main carbon source and oxidation of inorganic compounds such as hydrogen sulphide, sulphur, ammonia, nitrites, hydrogen gas or iron as an energy source. However, it is important to note that most bacteria, fungi and animal cells are chemo-organoheterotrophs.

Culture media used in pharmaceutical biotechnology are chemical preparations that contain nutrients which provide the necessary organic and inorganic constituents for inoculated microorganisms to grow and generate primary or secondary metabolites. The way media are formulated, i.e. in liquid or solid forms, depends on the optimal growth characteristics for a particular type of microorganism, the desired product and the most profitable process. The culture media must contain all the nutrients – in balanced and adequate amounts – in order to allow optimum growth and production. Its composition will vary depending on the strain of the microorganism, as these may require organic or inorganic sources of nutrients. When the development of a new culture medium for an unknown microorganism starts, the best way to prepare the formulation is to evaluate the conditions in which the microorganism was isolated and to simulate a medium with nutrients similar to the environment. Additionally, the culturing of microorganisms requires careful control of various environmental factors (pH, temperature, aeration, agitation) which normally are maintained within narrow limits (https://www.biologydiscussion.com/microorganisms/culture-medium-for-microorganisms/54997).

Culture media are designed to meet the best nutritional demand of the microorganism for production of the required biomolecule. The selection of these media depends for most large-scale processes on the cost, availability, easy final treatment of the effluent and stability during long periods of storage time. When protein therapeutics are being produced, the medium also has the function to ensure the structure of the protein is not modified during the bioprocess thereby so that the function of the protein is not compromised.

With regard to the composition of the culture medium, microorganisms of industrial interest require macro and microelements for their growth and metabolism. In particular, the main elements, or macroelements, such as carbon, nitrogen, oxygen, hydrogen, sulphur, phosphorus, magnesium and potassium, are needed in concentrations greater than 10^{-4M} (Behera et al., 2019). On the other hand, microelements such as zinc, copper, manganese, calcium, sodium and growth factors are also needed, but in concentrations below 10^{-4M} (Behera et al., 2019).

With regard to the composition of the culture medium, microorganisms of industrial interest require macro and microelements for their growth and metabolism. In particular, the main macroelements such as carbon, nitrogen, oxygen, hydrogen, sulphur, phosphorus, magnesium and potassium are needed in concentrations greater than 10^{-4M} (Behera et al., 2019). On the other hand, microelements such as zinc, copper, manganese, calcium, sodium and growth factors are also needed, but in concentrations below 10^{-4M} (Behera et al., 2019).

In addition to the concentrations of the macro and microelements, the proportions between the elements must also be considered. For example, the C:N ratio, the ratio of the mass of the carbon atom to the mass of the nitrogen atom available in the culture medium must be carefully calculated mainly because cells, proteins and nucleic acids are basically made up of these two elements. The nitrogen atom is found in most molecules in a cell and has both structural and biochemical functions. The main examples are amino acids, nucleic acids, ATP, cell wall structures and organelles. However, as mentioned earlier, the nitrogen source can be organic or inorganic, depending on the microorganism.

In industrial cultivations, some carbon sources widely used are sugar cane molasses (sucrose, glucose and fructose), starch (glucose), barley, rye, wheat, oats, potatoes, sweet potatoes, manioc, sorghum), powder whey (lactose), etc. Other important sources of carbon which are also widely used include cellulose and hemicellulose hydrolysates from sugarcane bagasse (glucose, xylose and arabinose) and cheese whey (lactose). However, for the purpose of calculating the concentration of the carbon source in the case of heterotrophic microorganisms, it must be considered that there is also consumption due to cellular respiration for energy generation and elimination in the form of CO_2. In other words, the amount of carbon source added must basically include three main aspects: increase in cell mass, production of the target biomolecule and generation of energy and release of carbon dioxide.

Regarding nitrogen sources, microorganisms can use both inorganic sources such as ammonium salts (sulfate, nitrate and phosphate), as well as organic sources (urea). There are nutrients that are both sources of carbon and nitrogen, for example, peptone, yeast extract, amino acids and soy extract. Knowing the importance of adding the correct proportion of carbon and nitrogen to the culture medium, it is necessary to make the correct calculation of these components and to know the desired C:N ratio. This relationship will depend on the required product. For example, if the goal is to produce only cells to be used as a protein source in human or animal food, then we must know what is the ratio of C and N atoms in the structure of the cells of interest.

Although approximate values of the C:N ratio for different types of cells can be found in literature, a more accurate way to obtain this value is through analytical methods and calculations as shown below for *E. coli* BL21 (DE3). Initially, the bacteria must be cultivated in a known medium, such as LB medium (Luria Bertani) until cell mass is obtained, for example, at least 1.0 g/L. Then, the cells are collected from the fermentation broth by centrifugation, and the cell pellet resuspended and washed at least three times with deionized water and then dried to a constant mass. From this moment on, only *E. coli* cells should be sent for element composition analysis, especially with regard to the concentrations of carbon and nitrogen. Ideally, analysis should also be made of other constituents of the cell such as mineral salts and ions, to facilitate additional formulation of the culture medium. If the use of a complex medium is foreseen, it is also recommended that the element composition of the medium is obtained. Table 7.4 shows the elemental composition of *E. coli* BL21 cells (DE3).

If the goal is to produce a given biomolecule in high concentrations, for example, a protein, the C:N ratio of this protein must be calculated and the culture medium formulated taking this proportion into account. That is, it makes no sense to add random amounts of C and N when you know the target product to be generated. Table 7.5 shows the C:N ratios for some proteins that can assist in media formulation.

TABLE 7.4
Elementary Composition for Dry Cell Mass of *E. coli* BL21 (DE3)

Element	% (g/g)[a]
C	45.8
H	7.1
O	29.0
N	13.6
P	1.6
S	0.54
K	0.29
Na	0.52
Mg	0.16
Cl	0.41
Ca	5.9×10^{-2}
Fe	1.7×10^{-2}
Cu	1.8×10^{-3}
Mn	4.0×10^{-3}
Zn	2.1×10^{-2}
C:N ratio	3.36

[a]Analysis of *E. coli* BL21 (DE3) cells grown in LB medium, performed at the Analytical Center of the Chemistry Institute of the University of São Paulo, Brazil.

To calculate the C:N ratio, it is first necessary to find the exact chemical formula of the protein, since the structure and composition of a protein can vary between different organisms and under different growth conditions. For example, *E. coli* can produce two different molecules of L-asparaginase – asparaginase I and II. So, to calculate the C: N ratio it is necessary to specify which protein is of interest. To find the chemical formula, access the website https://www.ncbi.nlm.nih.gov/protein/, search for the protein of interest and find the chemical structure and the amino acid sequence. Then go to https://web.expasy.org/protparam/ and get a report with various information about the protein, such as the size of the molecule, the number of amino acids and the molecular formula from which it is then possible to find out the number of atoms carbon and nitrogen (GRAVY). With these data, divide the number of C-atoms by the number of N-atoms from the chemical formula of the protein to obtain the C:N ratio. To calculate the C:N ratio, the number of C and N atoms can be used, as well as the atomic mass, for example, multiplying the number of C-atoms by 14 and the number of N-atoms per 16. The amount of carbon and nitrogen needed to obtain the C:N ratio necessary to grow *E. coli* in order to produce L-asparaginase ($C_{1624}H_{2595}N_{439}O_{516}S_{12}$) can be determined (Table 7.5).

$C{:}N = 1{,}624 \div 439 = 3.69$ or $\rightarrow C{:}N = (1{,}624 \times 12) \div (439 \times 14) = 3.17$

Once the element composition of a cell to be cultivated and the biomolecule to be produced are known, the amount of the most suitable carbon and nitrogen source for the formulation of the culture medium must be calculated. An example of this calculation follows.

TABLE 7.5
C: N Ratio for Different Proteins

Protein and Formula	C:N Ratio Number C and N	C:N Ratio Atomic Mass C and N
L-asparaginase (native *E. coli*) $C_{1377}H_{2208}N_{382}O_{442}S_{17}$.	3.60	3.08
Glutaminase (*Streptomyces* sp.) $C_{2184}H_{3492}N_{618}O_{696}S_{18}Se_1$	3.53	3.02
Catalase (*Helicobacter pylori*) $C_{2652}H_{3989}N_{723}O_{758}S_{17}$	3.66	3.14
Superoxide dismutase (*Cyanobacterium aponinum*) $C_{1040}H_{1519}N_{265}O_{307}S_4$	3.92	3.36
Peroxidase (*Klebsiella pneumoniae*) $C_{779}H_{1248}N_{208}O_{243}S_4$	3.74	3.21
Invertase (*Thauera* sp.) $C_{868}H_{1441}N_{273}O_{265}S_2$	3.17	2.72
IgG (*Homo sapiens*) $C_{2325}H_{3598}N_{620}O_{707}S_{17}$	3.75	3.21
IgM (*Trichechus manatus*) $C_{791}H_{1194}N_{200}O_{241}S_5$	3.95	3.39
Hexoquinase A, isoforma C (*Drosophila melanogaster*): $C_{2206}H_{3498}N_{608}O_{657}S_{27}$	3.62	3.10
Glicerol-3-fosfato desidrogenase 1, isoforma A (*Drosophila melanogaster*) $C_{1712}H_{2742}N_{456}O_{496}S_{19}$	3.75	3.21

Calculation based on the number of carbon and nitrogen atoms and on the atomic mass.

It is intended to produce 10 g of L-asparaginase from a high density culture (60.0 g/L) of recombinant *E. coli* in a bioreactor with a capacity of 100 L. Let us consider that all microelements and cultivation conditions (temperature, pH, dissolved oxygen, agitation, aeration, microelements, growth factors) are adequate and defined in the bioprocess, and that we only need to calculate the concentrations of the sources of C and N so that the relationship C:N in the initial culture medium is between 3.13 (C: N of L-asparaginase) and 3.37 (C: N of *E. coli*). The main source of carbon to be used is glycerol ($C_3H_8O_3$) at an initial concentration of 40.0 g/L, and the initial sources of nitrogen are peptone (10.0 g/L) and yeast extract (5.0 g/L). However, to support growth in high cell density, the components of the medium need to be supplemented with new sources of C and N. To do this, calculate the additional amounts of glycerol ($C_3H_8O_3$) to adjust the carbon concentration and ammonium sulfate [$(NH_4)_2SO_4$] to adjust the nitrogen concentration. Data: L-asparaginase ($C_{1624}H_{2595}N_{439}O_{516}S_{12}$) consists of 52.6% carbon and has a C:N ratio equal to 3.13, whereas *E. coli* has 45.8% carbon and a C:N ratio equal to 3.37. Yeast extract has 40% carbon and the C:N ratio equals 4.0. Peptone is composed of 55% carbon, and the C:N ratio equals 3.93. The calculations are based on the atomic mass of the C and N atoms.

Calculation:

Output products: cells + L-asparaginase

Cells:

$\text{Carbon}_{\text{cell}} = (60\ g_{\text{cell}}/L \times 100\ L \times 0.458\ g_C/g_{\text{cell}}) = 2{,}748.00\ g$

$\text{Nitrogen}_{\text{cell}} = (C{:}N = 3.37;\quad N = 2748 \div 3.37) = 815.43\ g$

L-asparaginase:

$\text{Carbon}_{\text{asparaginase}} = (10\ g_{\text{asp}} \times 0.526) = 5.26\ g$

$\text{Nitrogen}_{\text{asparaginase}} = (C{:}N = 3.13;\ N = 5.26 \div 3.13) = 1.68\ g$

Total of carbon = 2,753.26 g

Total of nitrogen = 817.11 g

C:N ratio in output products = 3.37

Input products: glycerol + peptone + yeast extract + ammonium sulphate

$\text{Carbon}_{\text{Glycerol}} = 40\ g/L \times 100\ L = 4{,}000\ g \times 0.391 = 1{,}564.00\ g$

$\text{Carbon}_{\text{pepton}} = 10\ g/L \times 0.55 \times 100\ L = 550\ g$

$\text{Carbon}_{\text{Yeast extract}} = 5\ g/L \times 0.4 \times 100\ L = 200\ g$

Total of carbon = 2,314.00 g

$\text{Nitrogen}_{\text{peptone}} = 550 \div 3.93 = 139.94\ g$

$\text{Nitrogen}_{\text{Yeast extract}} = 200 \div 4.0 = 50.00\ g$

Total of nitrogen = 189.94 g

Initial carbon balance = 2,314.00 − 2753.26 = −439.26 g

Initial nitrogen balance = 189.94 −817.11 = −627.17 g

Supplement C with glycerol:

$1\ g_{\text{glicerol}} \rightarrow 0.391\ g_{\text{carbon}}$

$X\ g_{\text{glycerol}} \rightarrow 439.26\ g_{\text{carbon}}$

$X = 439.26 \div 0.391 = 1{,}123.43\ g_{\text{glycerol}}$

$\text{Total of glycerol} = 1{,}123.43 + 4{,}000 = 5{,}123.42\ g_{\text{glycerol}}$

Supplement N with ammonium sulphate:

$1\ g_{\text{Ammonium sulphate}} \rightarrow 0.212\ g_{\text{nitrogen}}$

$X\ g_{\text{Ammonium sulphate}} \rightarrow 627.17\ g_{\text{nitrogen}}$

$$X = 627.17 \div 0.212 = 2{,}958.35\ g_{\text{Ammonium sulphate}}$$

RESULT – COMPOSITION OF THE CULTURE MEDIUM FOR 100 L

Peptone = 1,000 g
Yeast Extract = 500 g
Glycerol = 5,123.42 g
Ammonium sulphate = 2,958.35 g
C:N = 3.37

In addition to macro and microelements, most microorganisms also need growth factors (vitamins, nucleotides and amino acids) not only for cell development but also for the synthesis of some metabolites. For example, for glutamic acid biosynthesis, some microorganisms require limited concentrations of thiamine and/or biotin in the culture medium. One way to determine, approximately, the amount of nutrients in a medium, is to analyse these components that make up a cell.

7.3.1 Trace Elements and Minerals

There are components of the culture medium that are important but are needed in only minimal quantities. These compounds generally function as cofactors (potassium, magnesium, calcium and iron) or as electron donors or acceptors (sodium, zinc, copper, molybdenum, manganese and cobalt) in enzymatic reactions.

In addition to carbon and nitrogen, amino acids may need sulphur (e.g. cysteine contains a sulphydryl [-SH] and methionine has a sulphur atom in a thio-ester bond [-S-CH_3] and, therefore, must be added to the medium. Another atom of great importance for cells metabolism is phosphorus, added in the form of phosphate ions, as it is part of the structures of phospholipids, ATP, DNA and RNA.

Microorganisms interact with metal elements since they can affect growth rate, cell activity and survival. They can be toxic when in high concentrations, since they can catalyse oxide-reduction reactions, bind to peptides and proteins and induce precipitation. But they are also essential in the composition of some proteins like metallothioneins and phytochelatins. Some metals are essential to life, such as Na, K, Cu, Zn, Co, Ca, Mg, Mn and Fe. However, the cellular functions need some other metals, such as Cs, Al, Cd, Hg, and Pb (Losa and Bindschedler, 2018).

Traces of metal elements are usually present in water used to dilute growth media and are also found in most raw material ingredients of media. It is recommended to determine the concentration of some ions present in complex culture media, especially when the media contains composite ingredients such as sugarcane juice, wheat bran, corn, rice, soy or corn steep liquor, as these may contain metals in concentrations above the value tolerated by microorganism. Magnesium ion is usually added to culture media in the form of $MgSO_4.7H_2O$; however, depending on the required concentration, it can cause precipitation of other ions present in the medium, and as a consequence, the cultured cells will be nutrient deplete. In addition, excess salt can complicate treatment of the final waste, unnecessarily increasing the cost of the process.

Silica, added to the culture medium in the form of silicate, is important for many organisms. since its low concentration can compromise cell growth and its high concentrations above 100 mM (e.g. in the yeast *S. cerevisiae*) can be toxic too.

Therefore, trace elements that are not present in sufficient quantities must be supplemented into the culture medium. Also, it is recommended that the concentrations of some metals are determined in the raw materials entering bioreactors to avoid possible toxic effects which would also enhance process costs.

7.3.2 Growth Factors

There are ingredients that must be part of the culture medium because cells do not have the capacity to synthesise these, as is the case with certain growth factors. These ingredients are composed mainly of amino acids, necessary for the synthesis of proteins, purines and pyrimidines used in the synthesis of nucleic acids and vitamins used as coenzymes or functional groups of some enzymes. For example, vitamin B_6 (pyridoxine) forms part of the transamination and deamination reactions of amino acids, while vitamin B_1 (thiamine) is involved in decarboxylation reactions. For some microorganisms, these components are clearly important, for example, thiamine is essential for the growth of *Bacillus anthracis*, while *Lactobacillus casei*, a lactic acid bacterium that requires cysteine hydrochloride (0.01%), riboflavin (0.01%), pantothenic acid (0.05%), nicotinic acid (0.02%) and tryptophan (0.01%). However, the cultivation of microorganisms such as *Pseudomonas*, *Mycobacterium*, *Aspergillus* and *Penicillium* do not need addition of growth factors, as these organisms have the capacity for synthesis. On the other hand, some microorganisms that apparently only grow in complex and nutrient-rich media (blood, serum and milk) can also grow in defined culture media if all the essential components such as minerals and growth factors are known and added. Folic acid is important for the synthesis of nitrogenous bases, amino acids and vitamins, and the precursor for biosynthesis is vitamin p-aminobenzoic acid (PABA).

7.3.3 Synthetic or Chemically Defined Media

Culture media consisting of defined components are called synthetic media. The chemical composition is always well-known and can be reproduced at any time since it is prepared from purified ingredients. For this reason, for cells that perform well in such media, a very stable and reproducible production system is expected along with no general problems for recovery and purification of the final product. These media are preferred, even though they are expensive, because of greater financial savings in the product recovery stages reduce the overall bioprocess costs.

Defined media are used to grow many different types of cells, for example, photoautotrophs such as cyanobacteria. These types of cells use carbon dioxide as a carbon source (formed after adding sodium carbonate or bicarbonate to the culture medium), light as an energy source and nitrate or ammonia as a nitrogen source and minerals. Table 7.6 shows compositions of different microbial culture media that can be used as starting points to define a new medium. Because microorganisms, both wild types and recombinant cells, have their own characteristics, and they will also have specific nutritional requirements. As a result, it is very likely that the media shown in Table 7.6 needs to be modified. Variations in composition of the culture media are very common as demonstrated by a simple search for scientific articles where, for example, there are numerous and different means for cultivating *E. coli* and *P. pastoris*, in addition to many other microorganisms.

Some precautions must be taken when selecting a culture medium, especially in relation to the concentration of salts. The BSM medium (Bifidus selective medium) developed by Sigma-Aldrich is widely used for the cultivation of Bifidobacterium spp.; however, precipitations may occur during its preparation due to high salt content. In addition, nitrogen supply depends on strict pH control of the medium, as there is no ammonium salt in its composition. The FM22 medium (fermentation medium) widely used for the cultivation of *P. pastoris* also contains a high salt content, but it has nitrogen in its composition. Therefore, in order to optimize the composition of the culture medium, these details must be evaluated to avoid problems with nutrient availability.

7.3.4 Complex Media

Examples of synthetic media for culture of cyanobacteria, *Escherichia coli* (two different media compositions) and *Pichia pastoris*.

TABLE 7.6
Examples of Synthetic Media for Culture of Cyanobacteria, *Escherichia coli* (Two Different Media Compositions) and *Pichia pastoris*

Medium for Cyanobacteria[f]	Amount (g/L)
$NaNO_3$	1.5
$K_2HPO_4.3H_2O$	0.04
$MgSO_4.7H_2O$	0.075
$CaCl_2.2H_2O$	0.036
Citric acid	0.006
Ferric ammonium citrate	0.006
EDTA (Na_2Mg salt)	0.001
Na_2CO_3	0.002
Trace metal solution[b]	1 mL/L
Final pH 7.4	

Medium for *Escherichia coli* (High-Density Cultivation)[a,c]	
Glucose	5.0
KH_2PO_4,	13.3
$MgSO_4.7H_2O$	1.2
$(NH_4)_2HPO_4$,	4.0
Citric acid	1.7
Fe(III)citrate	60
$MnCl_2$	15.0
$Zn(CH_3COO)_2.2H_2O$	8.0
H_3BO_3	3.0
EDTA	8.4
$CuCl_2.2H_2O$,	1.5
Thiamin-HCl	4.5
$CoCl_2.6H_2O$	2.5
$Na_2MoO_4.2H_2O$	2.5
Final pH 6.8	

Medium for *E. coli* BL 21 (DE3) Vector pET-15b[a]	Amount (g/L)
Glucose	5.0
$MgSO_4.7H_2O$	1.2
KH_2PO_4	13.3
$(NH_4)_2HPO_4$	4.0
Citric acid	1.7
Iron (III) citrate	60.0 mg
$MnCl_2.4H_2O$	15.0 mg
$Zn(CH_3COO)_2.H_2O$	8.0 mg
H_3BO_3	3.0 mg
$Na_2MoO_4.2H_2O$	2.5 mg
$CoCl_2.6H_2O$	2.5 mg
$CuCl_2.2H_2O$	1.5 mg
EDTA.Na_2	8.4 mg
Tiamine.HCl	4.5 mg

Medium for *Pichia pastoris*[d]		Amount (g/L)
Glycerol		95.2
KH_2PO_4		9.4
YTM[e]		4.56
$(NH_4)_2SO_4$		15.7
$MgSO_4.7H_2O$		4.6
$CaCl_2.2H_2O$		0.28
Biotin		0.4 mg
	KI	207.5 mg
	$MnSO_4$	760.6 mg
	Na_2MoO_4	484 mg
	H_3BO_3	46.3 mg
YTM[e]	$ZnSO_4.7H_2O$	5.032
	$FeCl_3.6H_2O$	12.0
	H_2SO_4	9.2

[a] There are several proposed synthetic media for the cultivation of *E. coli*. We present different options in this table. It is recommended that the biotechnologist evaluate more than one option to find the most suitable medium of interest.
[b] Trace elements solution recommended for the culture and maintenance of cyanobacteria. Composition (sterile solution, per L) H_3BO_3 2,860 mg; $MnCl_2.4H_2O$ 1,810 mg; $ZnSO_4.7H_2O$ 222 mg; $Na_2MoO_4.2H_2O$ 390 mg; $CuSO_4.5H_2O$ 79 mg; $Co(NO_3)_2.6H_2O$ 49 mg.
[c] Riesenberg et al. (1991).
[d] Gurramkonda et al. (2010).
[e] Yeast trace minerals.
[f] Kumar et al. (2015).

TABLE 7.7
Alternative Substrates Used in Biotechnological Processes

Substrate	Microorganism	Product	Reference
Waste paper	*Lactobacillus casei*	Lactic acid	Marques et al. (2008)
Corn cob waste	*Bacillus* sp. KK2S4	L-asparaginase	Makky et al. (2014)
Cheese whey	*Lactobacillus plantarum*	Bacteriocins, lactic acid and DL-3-Phenyllactic acid (PLA)	Sabo et al. (2017) Rodríguez-Pazo et al. (2016)
Mech-Degla juice	*Lactococcus lactis* F-mou	Exopolysaccharides	Nehal et al. (2019)
Vine-trimming shoots	*Bacillus tequilensis*	Biosurfactants	Cortés-Camargo et al. (2016)
Corncobs	*Debaryomyces hansenii*	Xylitol	Vázquez et al. (2017)
Brewer's spent grain	*Lactococcus lactis* Tw11, *Enterococcus mundtii* Tw492	Bacteriocins	Paz et al. (2018)
Rapeseed straw	*Debaryomyces hansenii, Candida guilliermondii*	Xylitol	López-Linares et al. (2018)
Grape vinasse	*Lactococcus lactis*	Biosurfactants	Vera et al. (2018)

Generally, complex culture media contain limited qualitative and quantitative standardization of the raw material which require pre-treatment steps prior to use such as hydrolysis (acidic or enzymatic) and clarification (removal of insoluble solids). Table 7.7 shows different types of alternative substrates used in biotechnological processes. In the formulation of this type of culture medium, some important aspects must be considered: the nutritional requirement of the microorganism during the bioprocess; standardization of the raw material; some technological inconveniences inherent in the use of these raw materials – such as aeration and agitation (which generates foaming); extraction and purification of the desired product; and cost. In the case of complex cultivation media, it is important to take into account alternative sources of carbon and nitrogen.

Sugar cane molasses, a by-product of the sugar and alcohol industry, is an important source of sucrose, fructose and glucose; and is widely used in the production of bioethanol and organic acids. European countries generally use beet molasses as a carbon source, which contains about 50% sucrose. Cheese whey, another agro-industrial waste, is generated in large volumes by the dairy industry but is of concern from an environmental point of view. This residue is widely used in animal feed (a rich source of proteins) and in the production of biomolecules of food interest (Antunes et al., 2015; Martinez et al., 2012; Sabo et al., 2017). A strong limitation to the use of this substrate is due to the costs of storage, transportation and quality.

Other complex carbon sources can be obtained from different raw materials such as malt extract, sulphite waste liquor (waste from the paper industry), cellulose associated with lignocellulose compounds, hydrocarbons (methane), alcohols (methanol), poly alcohols (glycerol, mannitol and sorbitol), alkanes (paraffin) and vegetable or animal oils.

Nitrogen sources from complex media, whether organic or inorganic, are widely used in bioprocesses. Complex sources of N offer several nutritional advantages, since they are also generally sources of carbon, sulphur, phosphorus, cofactors and traces metals too (Allikian et al., 2019; Stanbury et al., 2016). However, the use of ammonium salts can lead to pH variation during microbial cultivation, since microorganisms when using the ammonium ion release hydrogen ions, acidifying the culture medium (Allikian et al., 2019). On the other hand, organic sources of N are represented by amino acids, proteins and urea.

Generally, complex sources of nitrogen originate from vegetable flours (soy, peanuts and cotton) and animals (fish or meat), vinasse and corn steep liquor. In addition, yeast extract and casein are also considered excellent sources of high-quality peptides and cofactors; however, due to the high cost, use is generally restricted for the growth of less fastidious microorganisms.

7.4 ANTIFOAM AGENTS

Foam is a type of colloidal dispersion in which the dispersant is a gas. The main cause of foaming is, in general, microbial activity as a function of amino acids released from the disruption of cells during cultivation, and by the mechanical action caused by pumping culture medium, agitation and aeration. The carbon dioxide formed rises towards the surface, and when it reaches the interface with atmospheric air, it encounters resistance of an elastic layer formed mainly by proteins and peptides. This layer under pressure of carbon dioxide expands, forming bubbles and produces foam that can lead to overflow of bioreactors and interfere with industrial performance (Figure 7.3).

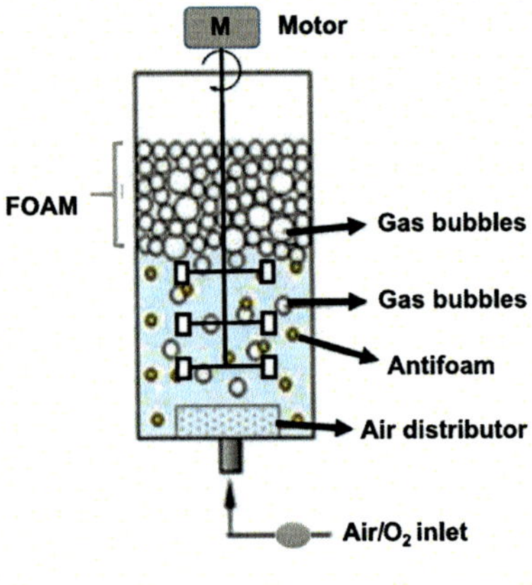

FIGURE 7.3 Bioreactor: agitation, aeration, foam formation and antifoam.

At the beginning of the 20th century, mechanical methods (paddle wheels, perforated spiral canals and centrifuges) were used to combat the foam formed in various industrial processes. These methods consumed excess energy and raised the cost of obtaining the final product. The emergence of chemical substitutes provided for a reduction in costs mainly because these chemicals performed well, even when added in small quantities.

Foams are structures that appear in the form of bubbles, and antifoams are emulsions made up of hydrophobic groups. When foams and antifoams come into contact, the bubbles are rapidly destroyed (Karakashev and Grozdanova, 2012).

In the period from 1940 to 1970, antifoams were basically 'oil' but later, the composition consisted of 'oil + hydrophobic groups'. Both compositions have the function of destabilizing foam through its hydrophobic regions (Figure 7.4) (Karakashev and Grozdanova, 2012).

The mechanism of action of many antifoams is based on surfactant effects on the surface elastic layer. Surfactants cause the elastic layer to stretch, making it thinner thereby reducing resistance to the pressure of carbon dioxide. In other words, antifoams control the surface tension of the liquid as they diffuse rapidly at the gas–liquid interface, favouring the collapse of the bubbles. As a result, the elastic layer breaks before the bubbles start to grow, minimizing foam formation. However, antifoams also act on bubbles that have already formed, releasing the gas contained inside the bubbles (Figure 7.4). Additionally, it is important to bear in mind that depending on the type of surfactant, the antifoam can act quickly in a few seconds, or be slow and take a few hours to destroy the foam.

Antifoams are developed to destroy different types of foam and, therefore, have different chemical compositions. They can consist of polyglycols, fatty acids, mineral and vegetable oils or amines. Silicone-based antifoams are widely used, such as dimethylpolysiloxane, the composition of which is a mixture of linear siloxane polymers fully methylated and belongs to a group of polymeric organosilicones commonly called silicones. Silicone-based antifoam is extremely effective, can be dosed in small quantities and has a long duration of action. Silicone-based antifoams are chemically stable, inert, non-toxic and easy to handle at room temperature. There are a large group of silicone-based antifoams that are usually effective in removing foams but can be detrimental to the bioprocesses by affecting oxygen transfer (upstream) or damage tangential filtration membranes (downstream). Substitute and less damaging antifoams can be composed of a balanced solution of copolymeric polyglycols (polyether glycol) dispersed in ethanol and water.

Antifoams can cause opposite effects, that is, in very low concentrations their effects can be negligible, while high amounts can provide foam stabilization (Karakashev and Grozdanova, 2012). This concentration depends on the geometry of the bioreactor and the culture conditions (type of microorganism, agitation, aeration, composition of the culture medium and type of final product). A good antifoam does not leave toxic residues in the bioreactors, has a prolonged action and has thermal stability under working conditions.

The use of antifoam on an industrial scale must consider some safety aspects, such as the use of personal protective equipment to avoid contact with eyes and skin (waterproof aprons, safety glasses and rubber gloves). The eyes should be washed immediately with abundant water, for 15 min, in case of contact with the antifoam, with the eyelids held away from the eyeballs, to ensure a thorough wash, as there is a risk of serious eye damage. If swallowed, vomiting should not be caused due to the risk of inhalation. In case of spillage, isolate the area; cover with earth, dry sand or other non-combustible material; remove with suitable utensils; and wash the area with plenty of water. When handling, use a dust mask, panoramic glasses and PVC gloves. It is important to note that the antifoam should be stored away from sources of heat and open flames.

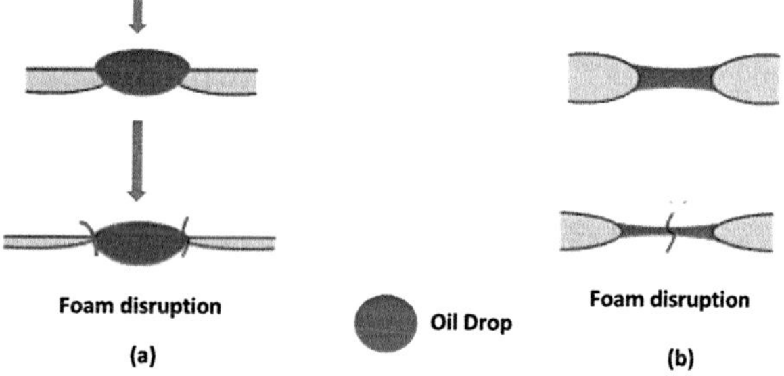

FIGURE 7.4 Antifoam mechanisms: (a) oil drop enters the surface of the foam film; (b) oil particle bridges the foam film surface forming an oil bridge. (Based on Routledge, S.J., *Comput. Struct. Biotec J*, 3(4), e201210014, 2012, https://doi.org/10.5936/csbj.201210014.)

The points of the production process where the antifoam must be added should preferably drain off before the place where the foams are formed in the industrial circuit, such as before feeding the culture medium into the bioreactor, or directly in the inoculum. The dosage will depend on the severity of the problem caused by the foam, the operating conditions of the system and specific tests at the production site.

Antifoams are widely used during the production of ethanol, fermented beverages, the food industry (e.g. in the processing of potatoes), water and effluent treatment, textile industries, breweries and, above all, in the process of production of recombinant proteins of pharmaceutical interest. They can affect a bioprocess in different ways, especially those conducted on a large scale, as they can alter the availability of dissolved oxygen in the medium, which is a problem for aerobic cultivation, can affect the structure of some proteins – especially those with highly surface hydrophobic – and can also clog continuous flow centrifuges, heat exchangers and filtration membranes (Routledge, 2012).

The exact composition of antifoaming agents is usually unknown, but some properties have been investigated, for example, the effects on foam height as a function of time and the influence on the k_La values (see Chapter 11). The influence of the antifoam on oxygen transfer in the culture medium can be measured as a function of k_La (volumetric oxygen transfer coefficient). At low concentrations there is a reduction in the value of k_La, while at high antifoam concentrations this value increases. This is because the oxygen is inside air bubbles that, when disrupted, allows the oxygen to dissolve in the medium. The smaller the size of the bubble, the larger is its specific area and the greater the amount of oxygen that can be transferred to the medium. Once the antifoam changes the permeability and size of the bubble, it also affects the transfer of oxygen (Routledge, 2012).

Some studies have assessed the influence of antifoams during protein production, specifically the effects of the antifoam on protein denaturation. Antifoams containing silicone oil and polypropylene glycol (PPG) have been found to affect the production of recombinant *E. coli* β-galactosidase fusion protein. The PPG/silicone oil mixture reduced the cell-specific growth rate, and the volumetric and specific activity of β-galactosidase increased as a function of PPG and PPG/silicone oil mixture concentrations. These observations highlighted possible effects of different antifoam compositions and concentration on cell and enzyme production (Routledge, 2012).

It was also observed that an increase in the concentration of an antifoam consisting of alkoxylated fatty acid esters in a vegetable base caused cells numbers of recombinant *Saccharomyces cerevisiae* and *P. pastoris* expressing a fusion protein to increase, whereas an increase in the concentration of an antifoam composed of polyalkylene glycol decreased the number of *S. cerevisiae* cells; on the other hand, the addition of silicone at a concentration of up to 8% v/v did not affect *S. cerevisiae* cells when cultivated in YPD medium.

Regarding regulatory issues, it is necessary to provide evidence that antifoam is absent in the final product, especially when the product is an injectable biopharmaceutical.

As we can see, antifoams can affect a biotechnological process in different ways. The antifoam can act directly on cells, as well as on the solubility of oxygen. Therefore, before choosing the antifoam to be added to a microbial culture, especially in large-scale cultures, it is recommended to carry out preliminary tests with different compositions and concentrations and to evaluate cell stability and possible damage the equipment.

7.5 pH CONTROL IN THE BIOPROCESS

Microorganisms can be classified according to their optimal pH. Those that grow best at pH ranging from 5 to 8 are classified as neutrophils, those that grow best at pH below 5.5 are classified as acidophiles and those that grow best above pH 8.5 are called alkaliphiles.

In biotechnological development, it has become standard practice to buffer solutions to regulate pH. However, attention has to be drawn to the different effects of various buffer systems on the consumption of sugars by microorganisms and, in particular, to the trend of the pH curve on cell growth since cell permeability can be affected. pH is a parameter that regulates several cellular functions, mainly because intra- and extracellular enzymes depend on pH for catalytic activity. The performance of a bioprocess can be greatly impaired due to fluctuations in pH. Buffering of pH during cultivation is, therefore, essential to ensure high yields and productivity. Changes in pH can be influenced by several factors such as temperature, cell growth, lactic and acetic acid production as well as CO_2 levels. Ammonium ion consumption and acetic acid production are, however, the two most important contributors to changes in pH during cell cultivation. Before starting, the bioprocess pH sensors must be calibrated (Figure 7.5).

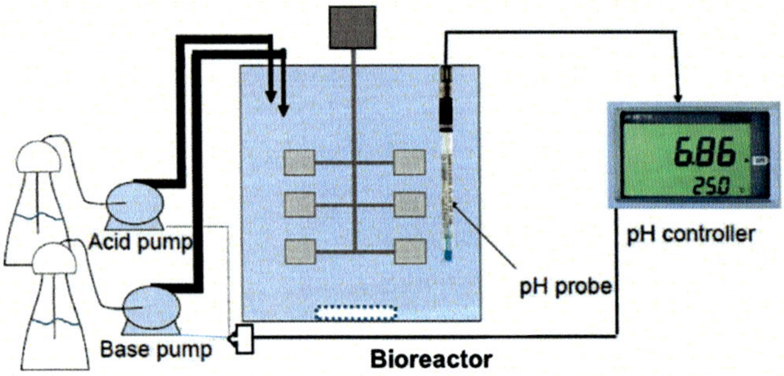

FIGURE 7.5 Bioreactor with pH control system.

Many buffers can be used on an industrial scale to ensure optimal yields; for example, acetate buffer can enter cells without affecting their viability. There are buffers that are able to control pH only in the extracellular medium. When pH control is achieved using acids or bases, such as HCl and NaOH, it may happen that only one of the ions (anion or cation) is able to penetrate a cell but still be able to exert a buffering effect.

There is also the possibility of controlling pH with some components of the culture medium such as proteins, peptides and amino acids. Commonly used buffers include K_2HPO_4, KH_2PO_4, $CaCO_3$, NH_4Cl, Borate-Na-hydroxide, soda-bicarbonate, citrate-NaOH, citrate-citric acid, acetic acid-Na-acetate, lactic acid-Na-lactate, succinic acid-Na-succinate and citric acid-Na-citrate. However, it is common to control pH with alkali (NaOH, KOH, NH_4OH, bicarbonate solution) solutions, thereby increasing pH during acidogenesis and with acids (H_2SO_4, HCl) to neutralize the mainly NH_3 formed bases. Both acid and base solutions can also be used to keep the pH at the same level at all times by using an automatic controller (Figure 7.5).

A drop in pH can be caused by overfeeding the substrate, and under such conditions, the cells can produce organic acids, such as acetate, as a byproduct. Conversely, a lack of carbohydrates can induce cells to consume proteins as carbon sources, generating NH_3 as a byproduct causing the pH to rise. However, when cells produce proteins (enzymes, antibodies), they consume ammonia as a nitrogen source and, consequently, the pH drops. It is important to highlight that monitoring and calculation of the mass of acid, base or buffer added to the bioreactor during cultivation to control pH will allow correlations between the consumption of the carbon and nitrogen source with cell growth.

An important example of pH control in the field of pharmaceutical biotechnology is in relation to the production of penicillin by *P. chrysogenum*. That is, the pH value must be closely monitored and controlled in both the growing phase and the production phase. In cultures carried out in bioreactors, the pH value must be strictly controlled between 4.5 and 5.5 especially at the beginning of the exponential growth phase, as there is an accelerated consumption of glucose and ammonia with a consequent drop in pH value which can inhibit cell growth and delay penicillin production. During the stationary growth phase and with the bioreactor maintaining the appropriate culture conditions, penicillin is produced. In this phase, ammonia formation generated from the metabolism of amino acids occurs and the pH increases to values close to 7.0 that must be maintained until the end of the process, keeping in mind that the best range for penicillin production ranges from 6.8 to 7.8. As there is a constant trend of increasing the pH in this growth phase, sulphonic acid is used to keep the pH within the desired range.

Yeast cells can tolerate a pH range from 4.0 to 8.5 but require an acidic environment for optimal performance between pH 4.0 and 6.0. Using bread production as an example, as dough begins to rise and fermentation takes place, the pH falls and the dough becomes more acidic. This fact can halt the fermentation process, and pH control is achieved by ingredients such as flour, milk and calcium carbonate which act as buffers (Table 7.8).

7.6 PRECURSORS

Precursors are substances that, when added to a culture medium, are used for the synthesis of a biotechnological product. Generally, precursors are metabolites that the producing micro-organism is not able to produce in sufficient quantities and are added either at the beginning (batch) or during (fed batch) the fermentation to increase the yield and quality of the target product.

A classic example of a precursor is phenylacetic acid, which is incorporated into the penicillin G molecule (Allikian et al., 2019). However, other substances can be considered precursors, such as chlorine and propionate used in the biosynthesis of chlortetracycline and riboflavin, respectively (Stanbury et al., 2017).

7.7 PRIMARY AND SECONDARY METABOLITES

There are two major categories of metabolites depending on the origin and function: primary and secondary.

Primary metabolism (or trophophase) is the metabolism of energy production and consumption for cell growth, reproduction and development. In glucose metabolism by aerobic organisms, there is immediate formation of pyruvic acid and energy, since the cells already have a pool of previously synthesized molecules, such as DNA and RNA.

Primary metabolites participate in respiration and photosynthesis, and are the decomposition and the polymeric products of metabolism. They are required for growth and cell maintenance. These metabolites are continuously produced during cell growth (equivalent to the log or exponential phase) (Figure 7.6). Primary metabolites are identical in most organisms and, generally, do not show any pharmacological actions or effects.

Primary metabolites can be classified as essential primary metabolites or as primary metabolic end products. Essential primary metabolites such as amino acids and nucleosides are produced in amounts that guarantee cell growth. Normally, microorganisms do not produce these metabolites in large quantities; however, when thinking about an industrial production process, it is necessary to understand regulation mechanisms and to adapt the process with a view to greater productivity. In contrast, primary metabolites are traditionally the end products of a bioprocess, even though they are not relevant for cell growth or maintenance. In fact, they can even be toxic, as is the case with ethanol. Another example is carbon dioxide, the final product from the cultivation of yeast such as *S. cerevisiae*, which is used in the manufacture of bread. Many other examples of primary metabolites can be found and include sugars, polysaccharides, nucleotides, proteins, nucleic acids, esters, lipids, amino acids, tricarboxylic acids, vitamins, nucleotides, enzyme, coenzyme, ethanol, acetone, lactic and citric acids, butanol, etc. Figure 7.6 show some structures of primary metabolites.

In 1910, the Nobel Prize winner Albrecht Kossel coined the term 'secondary metabolite', but it wasn't until 30 years later that the Czech botanist Friedrich Czapek described this concept. Secondary metabolism (also called idiophase) starts as soon as the exponential growth phase ends. Secondary metabolites are products of little relevance for

TABLE 7.8
Examples of Buffers, Acids and Bases Used to Control the pH of Cell Cultures

Cell	Product	Equipment	Buffer, Acid, Base	Concentration
Aspergillus and *Penicillium*	Tanase	Shaker Bioreactor	Tris-HCl	100–500 mM
Aspergillus and *Penicillium*	Protease Xylanase	Shaker	HCl	1–2 M
Haemophilus influenzae b (Hib), a	Vaccine against meningitis	Bioreactor	NaOH	5 mol/L
Pichia pastoris	Recombinant proteins from *Plasmodium vivax*	Shaker	K_2HPO_4/KH_2PO_4	1 mol/L
Talaromyces amestolkiae	Red colourant	Shaker	H_2SO_4 or NaOH 1 M.	1 mol/L
Pichia pastoris Glycoswitch SuperMan5 (his-)	Glicosylated L-asparaginase	Bioreactor	NH_4OH	28%
Pichia pastoris Glycoswitch SuperMan5 (his-)	Glicosylated L-asparaginase	Shaker	Potassium phosphate buffer	0.1 M
Escherichia coli BL21 (DE3)	L-asparaginase (double mutant)	Bioreactor	H_2SO_4 and NH_4OH	14%
CHO	TSH (thyroid-stimulating hormone)	Bioreactor	Injection CO2	-
Mucor subtilissimus	Protease with fibrinolytic activity	Bioreactor	NaOH/HCl	1 M
Escherichia coli	VHH antibody fragments for jararaca snake bite	Bioreactor	NaOH/HCl	2 M
Saccharomyces cerevisiae	Ethanol	Bioreactor	KOH	2 M

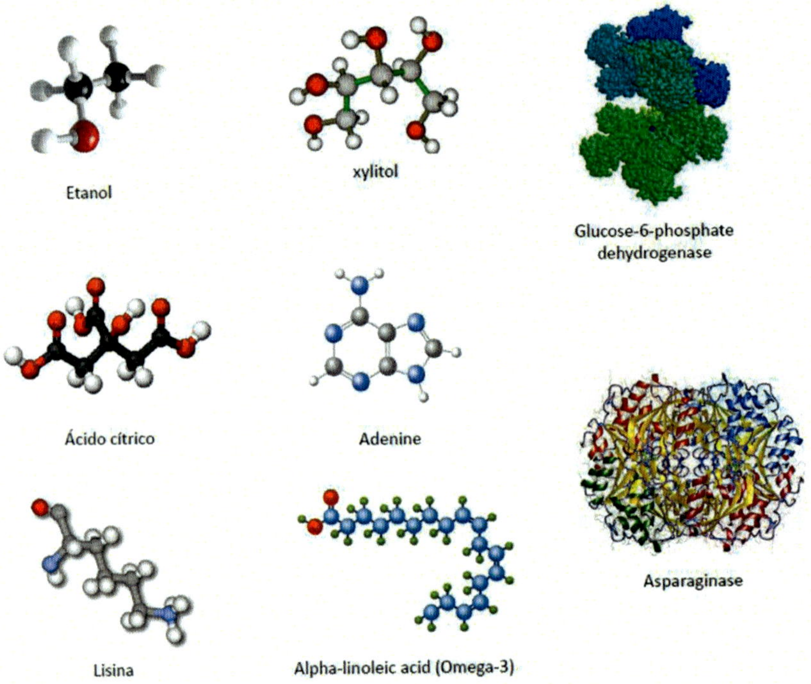

FIGURE 7.6 Primary metabolites.

the survival of living organisms, as these are not linked to energy production and are produced when there is accumulation of waste in the medium, as well as a lack of nutrients. Secondary metabolites are often produced in response to some stress generated in the culture medium and use the products of primary metabolism as precursors, for example, short-chain compounds such as carboxylic acids (acetyl-CoA) or amino acids (Figure 7.7).

Although secondary metabolites do not directly participate in normal growth, maintenance, development or reproduction of an organism, these metabolites are important for additional functions such as protection from predation, pathogenicity, competition, species interaction and communication (Fox and Howlett, 2008; Derntl et al., 2017). The discovery of penicillin, a broad-spectrum antibiotic, provided the first great exemplification that secondary metabolites

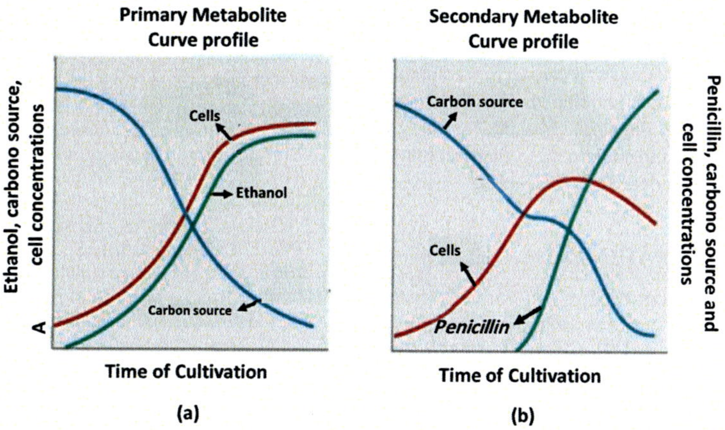

FIGURE 7.7 Primary and secondary metabolites as a function of cell growth phase. (a) Product (ethanol) associated with cell growth; (b) product (penicillin) not associated with cell growth.

were of biotechnological importance. Some secondary metabolisms have, for example, anticancer and immunosuppressive activities. Over the next eight decades since the discovery of penicillin, over 2 million secondary metabolites have been described. It is important to highlight that an individual microorganism can biosynthesize more than one secondary metabolite, and an example of this is the simultaneous production of 35 different anthracycline molecules by *Streptomyces*. Since secondary metabolites are produced in the stationary phase of cell growth, and under specific stress conditions, it is challenging to produce these in industrial quantities, which is the opposite for the industrial production of primary metabolites. In addition to antibiotics, there are secondary metabolites with relevant commercial value, such as gibberellins, alkaloids, toxins, fatty acids, ketones, alcohols, growth factors, steroids, carotenoids, anthocyanins, morphine, codeine, essential oils, monoterpenes, diterpenes, lemongrass oil, ricin, concanavalin A, and curcumin.

Importantly, the production of secondary metabolites by filamentous spore forming bacteria (the actinomycetes) and filamentous fungi is common, although many microorganisms have yet to be shown to produce secondary metabolites at all (Figure 7.8).

7.7.1 Overproduction of Secondary Metabolites

Although the production process for secondary metabolites is complex, studies on production optimization on bench (shaker and bioreactor) and industrial scales follow procedures similar to the optimization of processes for the production of primary metabolites. In other words, all important parameters for microbial cultivation must be evaluated in order to increase productivity and yield. By using genetic engineering, it is possible to isolate mutants less sensitive to end-product inhibition, thereby increasing the opportunity to overproduce secondary metabolites. Glucose is the most

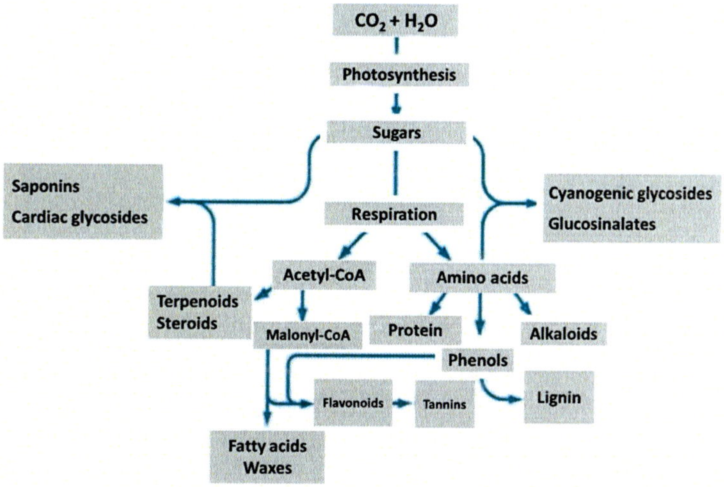

FIGURE 7.8 Secondary metabolites derived from primary metabolites.

commonly used carbon source in microbial fermentations, but glucose can inhibit production of several antibiotics like penicillin, streptomycin, bacitracin, chloramphenicol and puromycin. Likewise, ammonia is a commonly used nitrogen source in microbial fermentations and can act as a catabolite regulator to overproduce some antibiotics. The importance of secondary metabolites to the biopharmaceutical industry is highly relevant, and new and modern molecular techniques must be explored to overproduce such molecules.

7.8 FINAL CONSIDERATIONS

The success of a bioprocess depends on correctly defining parameters including the microorganism, the culture medium, the way that the bioprocess is conducted and the product recovery steps. The microorganisms can be obtained from natural resources, from mutations (induced by classical or genetically modified methods) or acquired from recognized culture collections. However, to monitor the production of biomolecules, it is necessary to monitor at least three important constraints such as cell concentration, substrate concentration and concentration of the final product. However, the overall performance of the bioprocess depends primarily on the quality and quantity of cells present in the process, in addition to cell viability. As a result, it is necessary that the amount of cell mass (biomass) can be determined by direct and fast methods. Another important aspect is that because the carbon (and often nitrogen) atom is the structural backbone of biomolecules, the proportion of carbon to other atoms such as nitrogen need to be precisely calculated. Also, as culture media are designed to meet the best nutritional demand of the microorganism for the production of a target biomolecule, the selection of these media depends for most large-scale processes on the cost and availability of ingredients. With regard to the composition of the culture medium, microorganisms of industrial interest also require microelements for growth and metabolism, in particular, trace elements, minerals and growth factors. Other parameters including antifoam agents, pH control and precursors can greatly influence the choice of culture medium to produce cells, thereby generating biomolecules with high productivity.

REFERENCES

Allikian, K., Edgar, R., Syed, R., Zhang, S. Fundamentals of Fermentation Media. In: Aydin Berenjian (ed.), *Essentials in Fermentation Technology* (41–84). Springer, Cham, 2019. ISBN 978-3-030-16229-0.

Antunes, S., Freitas, F., Alves, V.D., Grandfils, C., Reis, M.A.M. Conversion of cheese whey into a fucose- and glucuronic acid-rich extracellular polysaccharide by Enterobacter A47. *Journal of Biotechnology*, 210(20), 2015, 1–7.

Behera, S.S., Ray, R.C., Das, U., Panda, S.K., Saranraj, P. Microorganisms in Fermentation. In: A. Berenjian (ed.) *Essentials in Fermentation Technology. Learning Materials in Biosciences* (25–28). Springer, Cham, 2019. https://doi.org/10.1007/978-3-030-16230-6_1. ISBN 978-3-030-16230-6.

Cortés-Camargo, S., Pérez-Rodríguez, N., Oliveira, R.P.S., Huerta, B.E.B., Domínguez, J.M. Production of biosurfactants from vine-trimming shoots using the halotolerant strain Bacillus tequilensis ZSB10. *Industrial Crops and Products* (Print), 79, 2016, 258–266.

Derntl, C., Kluger, B., Bueschl, C., Schuhmacher, R., Mach, R. L., and Astrid, R. Mach-Aigner transcription factor Xpp1 is a switch between primary and secondary fungal metabolism. *PNAS*, 114(4), 24 January 2017, E560–E569. https://doi.org/10.1073/pnas.1609348114.

Doran, P.M. *Bioprocess Engineering Principles*. 2nd edition. Academic Press, Cambridge, MA, 2013.

Fox, E.M. and Howlett, B.J. Secondary metabolism: regulation and role in fungal biology. *Current Opinion in Microbiology*, 11(6), 2008, 481–487. https://doi.org/10.1016/j.mib.2008.10.007.

Gurramkonda, C., Polez, S., Skoko, N., Adnan, A., Gabel, T., Chugh, D., Swaminathan, S., Khanna, N., Tisminetzky, S., Rinas, U. Application of simple fed-batch technique to high-level secretory production of insulin precursor using Pichia pastoris with subsequent purification and conversion to human insulin. *Microbial Cell Factories*, 9, 2010, 31.

Karakashev, S.I. and Grozdanova, M.V. Foams and antifoams. *Advances in Colloid and Interface Science*, 176–177, 2012, 1–17. https://doi.org/10.1016/j.cis.2012.04.001.

Kumar, R.R., Ramesh, D., Mutanda, T., Rawat, I., Bux, F. Thermal behavior and pyrolytic characteristics of freshwater Scenedesmus sp. *Biomass*, 37, 2015, 1383–1391. https://doi.org/10.1080/15567036.2011.605428.

López-Linares, J.C., Romero, I., Cara, C., Castro, E., Mussatto, S.I. Xylitol production by Debaryomyces hansenii and Candida guilliermondii from rapeseed straw hemicellulosic hydrolysate. *Bioresource Technology*, 247, 2018, 736–743.

Losa, G. and Bindschedler, S. Enhanced tolerance to cadmium in bacterial-fungal co-cultures as a strategy for metal biorecovery from e-waste. *Minerals*, 8(4), 2018, 121. https://doi.org/10.3390/min8040121.

Makky, E.A., Loh, Y.C., Karim, M.R. Purification and partial characterization of a low molecular weight L-asparaginase produced from corn cob waste. *Biocatalysis and Agricultural Biotechnology*, 3, 2014, 265–270.

Marques, S., Santos, J. A. L., Gírio, F. M., Roseiro, J. C. Lactic acid production from recycled paper sludge by simultaneous saccharification and fermentation. *Biochemical Engineering Journal*, 41, 2008, 210–216.

Martinez, F.A.C., Balciunas, E.M., Salgado, J.M., Domínguez González, J.M., Converti, A., Oliveira, R.P.S. Lactic acid properties, applications and production: a review. *Trends in Food Science & Technology*, 30, 2012, 70–83.

Nehal, F., Sahnoun, M., Smaoui, S., Jaouadi, B., Bejar, S., Mohammed, S. Characterization, high production and antimicrobial activity of exopolysaccharides from Lactococcus lactis F-mou. *Microbial Pathogenesis*, 132, 2019, 10–19.

Pessoa Jr, A., Vitolo, M., Hustedt, H. Use of K_La as a criterion for scaling up the inulinase fermentation process. *Applied Biochemistry and Biotechnology*, 57–58, 1996, 699–709.

Riesenberg, D., Schulz, V., Knorre, W. A., Pohl, H. D., Korz, D., Sanders, E. A., et al. High cell density cultivation of Escherichia coli at controlled specific growth rate. *Journal of Biotechnology*, 20, 1991, 17–28. https://doi.org/10.1016/0168-1656(91)90032-Q.

Rodríguez-Pazo, N., Da Silva Sabo, S., Salgado-Seara, J.M., Arni, S.A., Oliveira, R.P.S., Domínguez, J.M. Optimisation of cheese whey enzymatic hydrolysis and further continuous production of antimicrobial extracts by Lactobacillus plantarum CECT-221. *Journal of Dairy Research*, 83, 2016, 402–411.

Routledge, S.J. Beyond de-foaming: the effects of antifoams on bioprocess productivity. *Computational and Structural Biotechnology Journal*, 3(4), 2012, e201210014. https://doi.org/10.5936/csbj.201210014.

Sabo, S.S., Pérez-Rodríguez, N., Domínguez, J.M., Oliveira, R.P.S. Inhibitory substances production by Lactobacillus plantarum ST16Pa cultured in hydrolyzed cheese whey supplemented with soybean flour and their antimicrobial efficiency as biopreservatives on fresh chicken meat. *Food Research International*, 99, 2017, 762–769.

Smith, J. E. *Biotechnology Principles: Aspects of Microbiology*. American Society for Microbiology, Washington, DC, 1985. 119 p.

Stanbury, P.F., Whitaker, A., Hall, S.J. *Principles of Fermentation Technology*. 3rd edition. Elsevier, New York, 2016. https://doi.org/10.1016/C2013-0-00186-7. ISBN: 978-0-08-099953-1.

Vázquez, G.B., Pérez-Rodríguez, N., Salgado, J. M., Oliveira, R.P.S., Domínguez, J.M. Optimization of salts supplementation on xylitol production by using a synthetic medium or corncob hemicellulosic hydrolysates and further scaled up. *Industrial & Engineering Chemistry Research*, 56(23), 2017, 6579–6589.

Vera, E.C.S., Azevedo, P.O.S., Domínguez, J.M., Oliveira, R.P.S. Optimization of biosurfactant and bacteriocin-like inhibitory substance (BLIS) production by Lactococcus lactis CECT-4434 from agroindustrial waste. *Biochemical Engineering Journal*, 133, 15 May 2018, 168–178.

8 Sterilization in Pharmaceutical Biotechnology

Valéria de Carvalho Santos Ebinuma, Fernanda de Oliveira, and Nathalia V. Veríssimo
São Paulo State University

Jorge Fernando Brandão Pereira
University of Coimbra

Adalberto Pessoa Jr.
University of São Paulo

CONTENTS

8.1 INTRODUCTION

Sterilization is a fundamental unit operation in any aseptic bioprocess but it is especially relevant in processes used in biopharmaceutical industries. Sterilization is used at different stages of drug production, regardless of whether the final product is intended for sterile use or not. The control of contaminants, particularly in drugs of biological origin, guarantees the safety and quality necessary for pharmaceutical products to be marketed in compliance with the requirements of regulatory agencies. There are different methods to achieve sterilization that ensure biomedicines meet regulatory requirements, each with its specific uses and limitations. This chapter will address these issues, focusing on restrictions and adaptations required when applying sterilization or aseptic procedures to the manufacture of pharmaceutical biotechnology products.

Biopharmaceuticals can be used either as sterile or non-sterile formulations. However, the biopharmaceuticals currently available on the market are mostly injectable and have an absolute requirement for sterility and absence of toxic or immunogenic substances in the formulation (Muheem et al., 2016). The maintenance of injectable sterility has always been a complex process for the pharmaceutical industry, since there is a regulatory requirement to provide a high standard of hygiene, safe and control of possible contaminants and particles present in industrial environments. The manufacture of biopharmaceuticals is complex since many peptides and proteins, used as therapeutics, are produced by the fermentation of genetically modified microorganisms, and the sterilization process is not as simple as the production operations usually carried out in manufacturing processes of synthetic and semi-synthetic drugs. During the biosynthesis of the target biomolecule, genetically modified microorganisms can generate undesirable by-products that must be removed or inactivated. Considering that biomolecules are less resistant than organic and inorganic molecules (synthetic or semi-synthetic), sterilization techniques commonly used in traditional pharmaceutical industries (such as heat sterilization) cannot be applied in most bioprocess operation units. It is thus necessary to adapt current pharmaceutical practices to suit the production of biopharmaceuticals, guaranteeing quality and safety are equivalent to the standard for conventional drugs, without compromising the integrity and therapeutic action of the biomolecules.

This chapter provides theoretical and scientific concepts for sterilization focusing on pharmaceutical bioprocesses. It also provides in-depth knowledge of sterilization techniques available and their impact on the properties of biopharmaceuticals. An introductory overview of some terms and definitions related to this field-of-study is provided, as well as a section focused on fundamental concepts, applications and limitations of different types of sterilization. Two sections of this chapter are dedicated to sterilization processes with a focus on upstream and downstream stages. The decision to individually analyse sterilization according to respective stages was based on the distinct characteristics of these processing stages. Upstream steps focus mainly on sterilizing equipment and culture media to prevent contamination of the biological process itself. Sterilization of the separation, purification and formulation stages (i.e., downstream) have the main objective of obtaining a product that retains therapeutic activity but with safety and quality standards required by international regulatory agencies [e.g., US Food and Drug Administration (FDA, United States of America), European Medicines Agency (EMA, Europe), World Health Organization (WHO, worldwide) and National Health Surveillance Agency (ANVISA, Brazil)]. A final subsection presents the validation requirements and methods for both upstream and downstream stages, essential to assure the quality and safety of biopharmaceuticals.

8.2 TERMS AND DEFINITIONS

Many terms like sterility, disinfectant and antiseptic are used interchangeably to indicate a decrease/removal of (micro)organisms and other contaminants. However, it is important to distinguish them, as each term refers to the application and degree of safety of a specific hygienic procedure:

- Sterilization is the destruction of all viable organisms and microorganisms (including spores). In the case of drugs, these are considered sterile with a 10^{-6} sterility guarantee. This value means that for every 10^6 sterile items in a final batch, there is a probability that there is only one viable microorganism (ANVISA, 2010);
- Antisepsis and disinfection are processes that inhibit growth or reduce the number of microorganisms. Antisepsis relates to procedures performed on living tissues (such as applying alcohol to a patient's skin before an injection), while disinfection is associated with methods performed on inanimate objects (such as, the use of disinfecting agents on surfaces and equipment). Asepsis, on the other hand, is a term applied to designing hygiene methods and procedures to prevent contamination by microorganisms (McDonnell, 2020).

For a complete understanding of the sterilization processes in the biopharmaceutical industry, it is important to be familiar with the terms related to the stability and resistance of the samples, namely:

I. *Thermolabile:* indicates that a substance is sensitive to heat (i.e. high temperatures can degrade it).
II. *Photolabile:* indicates that the substance is sensitive to light/radiation and can lose its integrity if irradiated under some wavelengths.
III. *Others:* include compounds sensitive to moisture, air, or specific substances, which also require special care in their handling.

In addition to understanding the definitions related to processes and samples, considering that this chapter focuses on bioprocesses, there are important terms that describe the contaminants present in the biopharmaceutical formulations, namely:

a. *Microorganisms:* although this term is being widely used in this book, we reinforce this concept as a key to understand a sterilization process. Microorganisms are a class of microscopic life which include bacteria, protozoa, fungi and viruses (although there are controversies about the classification of viruses as 'living being'), which vary in size from 1 nm to 100 μm. Microorganisms are present in virtually all environments, and there is no generic procedure for their removal or inhibition (Tortora, Funke, Case, 2013).
b. *Spores:* this term is used to classify the small reproductive structures of organisms such as bacteria, protozoa, fungi, algae and certain plants. Certain spores, particularly those of microorganisms, can be dormant for long periods of time (even years), as they can also survive under very unfavourable environmental conditions. Microbial spores are quite resistant to sterilization processes and have an average diameter between 1 and 50 μm (Wolken, Tramper, Van Der Werf, 2003).
c. *Toxins:* this term is used to classify peptides or proteins that can impair normal physiological functions (Claessens, 2015).
d. *Antigens, immunogens, allergens, pyrogens:* these terms are widely used in the biopharmaceutical industry (Blumenthal and Rosenberg, 2004):
   I. *Antigens:* substances capable of reacting with an antibody;
   II. *Immunogens:* antigens that generate an immune response when inserted into a host. Immunogens can also be subdivided according to the type of immune response they generate, such as allergens and pyrogens.
   III. *Allergens:* immunogens that induce specific immune responses called hypersensitivity (allergy).
   IV. *Pyrogens:* immunogens that induce specific immune responses of hyperthermia (fever).

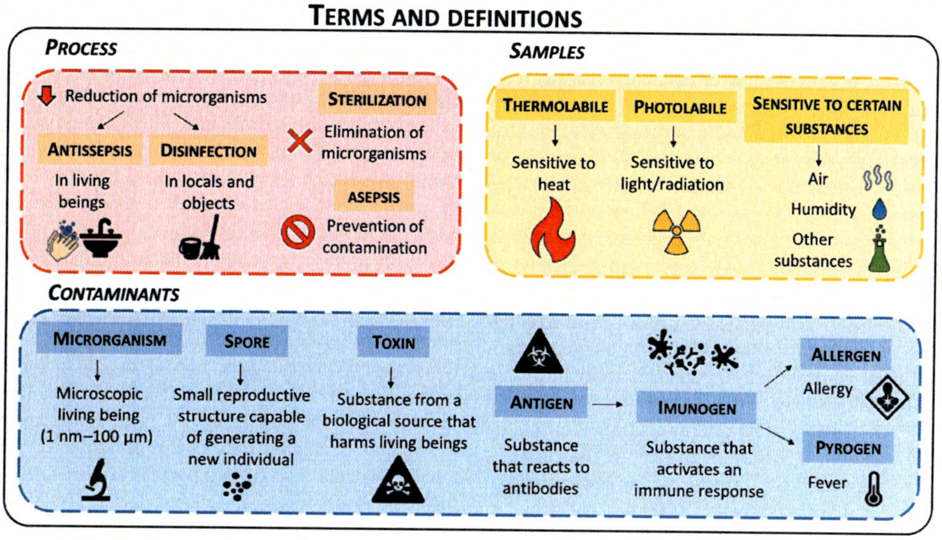

FIGURE 8.1 Summary of terms and definitions associated with contamination control processes in bioprocesses.

Figure 8.1 illustrates a summary clarifying each term and definition.

8.3 TYPES OF STERILIZATION

8.3.1 Physical Sterilization Methods

8.3.1.1 Heat

Heat is the most widely used method for sterilization, as it is safe, simple and inexpensive. This method is divided into moist and dry heat sterilization processes. Dry heat destroys microorganism by the oxidation of chemical constituents, whilst moist heat kills microorganisms mainly by increasing hydration inside cells which causes the irreversible denaturation of molecules essential for survival – e.g. enzymes and cell membranes (Doran, 2013). Because of these differences in the mechanism, moist heat requires shorter times and lower temperatures for sterilization. For example, sterilization at 121°C for 30 min in moist heat is equivalent to 6 h at the same temperature in a dry heat process.

It is important to note that dry heat is not recommended for sterilizing liquid formulations, because there is a loss of material through evaporation. Wet heat is not recommended for 'powder' (solids) formulations, since the moisture swells the material (ANVISA, 2010). Although heat sterilization is not applied directly to biopharmaceuticals, these procedures are widely used to sterilize heat-resistant materials and excipients that can be added aseptically to the formulation, as well as to eliminate microorganisms and spores from materials, reagents, equipment and facilities.

8.3.1.2 Ionizing Radiation

Sterilization by ionizing radiation consists of exposing contaminated materials to high energy emissions to ionize atoms and thereby generate free radicals that will cause cell damage. Ionizing radiation is a very advantageous method to sterilize thermolabile samples, as it does not require high temperatures, in addition to being fast and very effective. However, depending on the energy used, this method can degrade certain biomolecules particularly if these are photolabile. Before use, it is imperative to know the properties of the biological sample to set out what type of radiation, dose and exposure time are most suitable for sterilization and preserving the biopharmaceutical. The absorbed reference dose is 25 kGy, although variations are permitted as long as sterility of 10^{-6} is guaranteed and the sample does not deteriorate (ANVISA, 2010).

UV radiation is absorbed by many cellular compounds, especially nucleic acids, leading to significant cellular damage; however, the lethal effect of UV radiation will depend on the radiation dose being administered. The region of the UV spectrum with sterilizing action – also called the 'abiotic' region – is within the wavelength range from 220 to 300 nm. Electromagnetic ionizing radiations are mainly alpha (α), beta (β), gamma (γ), X-rays, cathode rays, as well as high energy protons, neutrons, and electrons (E-beam). The latter type of radiation has a wide variety of physical and biochemical effects on microorganisms, and the main targets leading to loss of cell viability are DNA and RNA molecules. Depending on the time of exposure, the damage caused by ionizing radiation can lead to efficient cell death (Doran, 2013).

Gamma radiation and the E-beam have added advantages of allowing products to be sterilized directly in their final packaging, as these radiation types have high penetrating power. However, it remains necessary to evaluate the stability of a biomedicine (biopharmaceuticals and other excipients in the formulation) as some molecules are sensitive to high-energy radiation. Furthermore, to work with these types of radiation, appropriate facilities and highly specialized personnel are required owing to the high risk to the operator. All these constraints increase the cost and complexity of ionizing radiation sterilization methods (Yaman, 2001).

8.3.1.3 Filtration

Filtration is an unit operation that involves the passage of a flow of fluid – liquid or gaseous – containing a mixture of insoluble particles or soluble molecules through a membrane that is permeable to some components (permeate or filtrate) whilst retaining other components (retained or concentrated) (see Chapter 13). Filtration operations, regardless of their application (e.g. sterilization, separation, purification), are generally classified according to the exclusion size of the species to be separated and/or the chemical composition of the main constituents of the membranes.

Sterilization by filtration is based on the mechanical removal of microorganisms and contaminants through membranes with different pore sizes. However, for proper sterilization of biopharmaceuticals by filtration, good product control must be achieved, as some microorganisms or pyrogenic substances may permeate and/or accumulate in the membranes causing unexpected contamination of the product. In addition, there are also some technical risks associated with incorrectly fitting the membrane or even the rupture of the membranes (Jornitz, 2019).

In order to sterilize a solution, the filtration membrane must have a pore diameter between 0.2 and 0.45 μm. When the medium passes through the membrane pores, microbial cells and other undesirable particles larger than the pore size are retained and can be collected. However, the pores are often partially blocked at the start of filtration, causing a sharp reduction in the flow of the permeate (Doran, 2013). To minimize this issue, the feed flow can be filtered in a tangential mode or by crossflow filtration (see Chapter 13).

In a traditional pharmaceutical filtration – compounds of very small organic and inorganic molecules – the separation of the active compounds from contaminants, such as microorganisms and pyrogenic macromolecules, is facilitated by the large difference in size between the compounds. For biopharmaceuticals, however, the selection of the most-adequate membranes and filtration conditions is critical, which requires an in-depth knowledge of the macromolecule structure of the therapeutic in comparison to protein-based contaminants – namely size and charge, as the biopharmaceutical and unwanted compounds can be very similar – e.g. protein isoforms. In addition to a careful selection of membrane pore size, the choice of filter material is also critical. Filtration systems are not supposed to release – by leaching – substances that compose their structure, such as fibres, especially in the case of possible contamination of injectable biopharmaceuticals. To overcome this potential problem, the sterilization operation uses material in the membrane whose compositions are based on glass, metal or polymers (ANVISA, 2010; Jornitz, 2019). In terms of membrane characteristics, almost all commercially available membranes for sterilization processes are those made of polymeric materials. The preferred choice of polymeric materials is essentially a result of their high thermal, mechanical and chemical stabilities; hydrophilic/hydrophobic properties; good resistance to washing and disinfection processes; and the large variety of pore dimensions, cartridges and configurations (Baker, 2012; Seader, Henley, and Roper, 2011). The main polymeric materials used in the manufacture of membranes are polypropylene, acrylonitrile, polysulfone, polytetrafluoroethylene (PTFE), nylon, aromatic polyamide and cellulose acetate (Seader, Henley, and Roper, 2011). Commonly, PTFE membranes are used for aqueous solutions filtration, whilst modified PTFE and nylon membranes are used for both aqueous and organic solutions. It is noteworthy that some membranes are unsuitable for filtering aqueous media, as is the case for hydrophobic PTFE membranes which should only be used to filter organic compounds which can be added during a bioprocess. In the specific case of sterilization by filtration of aqueous culture media, cellulose acetate membranes are widely used due to their low cost.

Further parameters should also be validated when using filtration for sterilization of biopharmaceutical on an industrial scale. In particular, the retention capacity of the membrane is influenced by the microbial load of the sample, the pressure and the flow rate of the filtration pumping process. It is necessary to establish parameters that ensure the sterility of the sample as well as flow rates and pressure conditions suitable for industrial processing. Very low flow rates or high operating pressures may render sterilization by filtration unfeasible on an industrial scale (ANVISA, 2010; Jornitz, 2019).

8.3.2 Chemical Sterilization Methods

Chemical agents are widely used to sterilize heat-sensitive equipment. Chemical sterilization methods use liquid or gaseous solutions – for example, hypochlorite, phenols, formaldehyde, ethylene oxide, ozone and sulphur dioxide – to destroy microorganisms (Doran, 2013). The most common example of a chemical sterilization process is the use of a 3% (w/v) sodium hypochlorite solution to decontaminate equipment and industrial facilities. However, when bleaches are used as chemical sterilizing agents, as is the case with sodium hypochlorite, these need to be carefully removed before adding the culture medium. Therefore, after the chemical sterilization procedures, adequate washing protocols of the equipment and/or installations must be carried out, in which sterile water and ethyl or isopropyl alcohol aqueous solutions (70%, v/v) are used to assure the removal of the toxic agents. Additionally, for some metallic materials, sterilization with bleach is not recommended due to corrosion issues.

As an alternative, ethylene oxide gas can be used to achieve chemical sterilization of industrial fermenters. Although ethylene oxide gas can be used as a chemical agent to sterilize thermolabile and photolabile pharmaceutical products, ethylene oxide is highly mutagenic and special care is required to ensure the removal of any residues from the final product. Ethylene oxide is flammable and requires caution when being used, for example, when sterilizing chambers fitted with a degasification system. Bearing in mind the low diffusion capacity of ethylene oxide, controlling temperature, humidity, gas concentration and pressure is imperative to ensure proper gas diffusion rates in the internal parts of the product (ANVISA, 2010).

Other chemical agents, such as oxidizers, bases and acids, can also be used in chemical sterilizing protocols. The most commonly used compounds are peracetic acid, hydrogen peroxide, formaldehyde and glutaraldehyde (Rutala and Weber, 2008). None of these chemical compounds are used directly to sterilize biopharmaceutical formulations, as most of these will affect the structure and biological activity of formulations.

8.4 STERILIZATION IN UPSTREAM OPERATIONS

Sterilization of upstream operations aims to ensure the culture medium is free from contaminants so that the medium remains sterile and only available as a nutrient source for the microorganism intended for cultivation. The choice of the most appropriate sterilization procedures is highly complex because characteristics of the culture medium, microorganisms and equipment must be considered. Heat sterilization in upstream processing is commonly used both at laboratory and industrial scales, as well as for sterilizing batch or continuous apparatus.

For some bioprocesses, mainly at an industrial level, partial elimination of microorganisms from equipment is sufficient to carry out the cultivation, e.g., bioprocesses where growth inhibitors are produced – namely, alcohols, acetic acid or lactic acid fermentation; and in processes for the production of antibiotics. In these cases, disinfection protocols that inactivate microorganisms are sufficient to guarantee adequate aseptic conditions. Although partial inactivation of microorganisms is a very common situation in the food industry, most biopharmaceutical processes are carried out under conditions free of any contaminating cells (Stanbury, Whitaker, and Hall, 2017). Therefore, bioreactors must be able to operate aseptically from days up to months.

8.4.1 Culture Media and Its Sterilization

From a general point of view and without considering the size of scale, most culture media are sterilized by moist heat. Although this process is widely used, it may result in the culture media overheating and, as a consequence, cause nutrient degradation and a change in pH values (Najafpour, 2007). This undesired nutrient degradation occurs when the energy of activation for thermal destruction of microorganisms (65–85 kcal/mol) is higher than the thermal destruction of the nutrients. Examples of culture media compounds that can be degraded by heat sterilization are vitamins A, B_1 and C – which have the energy of activation of 14.6, 26.0 and 23.1 kcal/mol, respectively (Villota and Hawkes, 2019) – and amino acids such as serine, threonine, methionine, tyrosine and histidine. Also, overheating of sugar-rich media may result in carbohydrate caramelization (leading to 'burnt' medium) and consequently reduction in availability as a carbon source for microbial growth. At the same time, the Maillard reaction can also take place in which the amino group of an amino can react with the carbonyl group of a reducing sugar causing a loss of carbon and nitrogen sources as well as darkening of the medium (Van Boekel, 1998). A further issue that can arise from wet heat sterilization of culture media is the formation of metal complexes and precipitation of metal salts leading to a decrease in the concentration of soluble salts essential for microbial growth.

Solid and/or semi-solid culture media are equally sensitive to overheating caused by sterilization. In such cases, the heating causes a sequence of catalytic reactions, particularly under acidic conditions, leading to the hydrolysis of some compounds in the medium – e.g., acid hydrolysis of cellulose producing glucose – and to the formation of cell growth inhibitors (Najafpour, 2007).

Sterilization of culture media on a laboratory scale is commonly carried out using an autoclave where temperature is one of the main factors responsible for altering the nutritional composition. Thermolabile nutrients must, therefore, be sterilized by other methods such as filtration or radiation (Liu, 2013).

8.4.2 Batch Heat Sterilization

The main advantage of batch sterilization is to guarantee simultaneous sterilization of the culture medium and the bioreactor, reducing the risks of contamination when the medium is transferred to the vessel.

Bioreactors – empty or loaded with the culture medium – as well as valves, filters and piping from upstream unit operations are generally sterilized by moist heat. On a laboratory scale, liquid culture media are sterilized in batches in containers in which they will be handled. In this case, autoclaves are used, generally, in cycles of 20–30 min for a closed batch sterilization system using steam under pressure. For batch sterilization, in a bioreactor or an autoclave, three phases are clearly distinguished: (a) *heating:* the increase of temperature; (b) *sterilization:* the temperature remains constant for a certain time; (c) *cooling:* decrease in temperature by adding cold water or using a refrigerated coil or jacket (Najafpour, 2007).

The steam takes up all of the internal space in a closed system – replacing all of the air contained inside the autoclave – and escapes through a discharge valve when in excess (Dion and Parker, 2013). After all the air is expelled – usually, from 10 to 20 min – the discharge valve is closed and the internal pressure and temperature are increased up to the sterilization operational conditions – in general, 121°C and 1 atm. At this point, heat exchange occurs between the steam and the material to be sterilized. For successful energy transfer, the vapour must condense on the surface of the material. The sterilization time is usually between 15 and 30 min, depending on the type, volume and characteristics of the material to be sterilized. For example, the sterilization of 1 L of culture medium at 121°C can be achieved in 20 min, while larger volumes may require longer sterilization times. Note that different waiting times are required for different temperatures of sterilization, namely, 20, 10 and 3 min for 121°C, 126°C and 134°C, respectively. Furthermore, it is estimated that a loss of liquid from the culture medium can occur by evaporation, of about 10%–15% (Dion and Parker, 2013). The final step in autoclaving is cooling, which starts by turning off the heating or by closing the steam inlet. The autoclave can only be opened when the internal temperature is close to that of the environment to protect against damage to glass materials and sensors placed inside the bioreactor – e.g., pH and dissolved oxygen probes – or even the loss of the culture medium inside the bioreactor (Stanbury, Whitaker, and Hall, 2017).

In order to ensure the safety of the operation, additional actions should be taken together with the standard sterilization protocol, namely: keeping an air inlet in the reactor containing a suitable filter and holding a device to disrupt the vacuum to prevent excessive pressurization (Doran, 2013). In fact, according to Dion and Parker (2013), the main mistakes associated with thermal sterilization are linked to the basic principles of sterilization in autoclaves, namely:

I. Tightly capped bottle tops and fully foil-wrapped container inlets are mistakes that impede the steam penetrating through materials;
II. Incorrect position of the items inside the autoclave chamber. Heavy loads placed in the upper basket of the autoclave lead to excessive steam condensation. The heavier is the load, the higher the rate of vapour condensation on the surface of the items which subsequently drains into the lower basket. Cotton plugs that seal Erlenmeyer® type bottles should not be soaked with condensed water at the end of the process. The heavier loads should always be placed in the lower basket of the autoclave. Materials arranged horizontally with no minimum distance or stacked on top of each other impede the contact of steam with the autoclave load.
III. *Other:* Inadequate air removal, deficiencies with autoclave sealing and low-quality steam – excess of non-condensable gases due to impurities in the water used produce it – are other common causes of failures in steam sterilization.

The effectiveness of sterilization by moist heat in an autoclave is checked and validated using biological and/or chemical indicators of sterility (Dion and Parker, 2013), as detailed below in Section 8.6.

On an industrial scale, when the culture medium is sterilized in batch, the medium is almost always inserted in the bioreactor and the heating of the system can be carried out by moist heat by direct bubbling of steam in the medium or indirectly by dry heat. In these cases, the culture medium inside the bioreactor must be mechanically stirred to ensure, as far as possible, a similar range of temperatures at all points of the system (Doran, 2013).

As previously addressed, dry heat sterilization is slower and less effective than wet heat sterilization; therefore, on an industrial scale, moist heating sterilizing protocols are preferred (Liu, 2013). Industrial sterilization apparatus generate moist heat as steam in boilers. The steam is distributed from boilers to equipment through thermally insulated galvanized steel or stainless steel pipes. Although the steam leaving the boilers is considered sterile because they are at high temperature and pressure, in some critical cases, filtration in cartridges with sterilizing membranes is used immediately before the steam is injected into the equipment, e.g., a bioreactor (Liu, 2013). Once sufficient steam has been introduced into the bioreactor, sterile air also needs to be introduced to ensure a slight positive pressure in the vessel which prevents the entry of contaminants. Note that filters are usually installed in the exhaust gas lines to retain the culture inside the bioreactor and to ensure that no contamination enters the vessel if a drop in operating pressure occurs (Doran, 2013).

Despite the common use of batch sterilization systems, it does have disadvantages, namely, overheating of the culture medium causes a loss of nutrients and, reduced energy transference and conservation can cause the sterilization process to fail. Therefore, depending on the characteristics of the bioprocess, batch heat sterilization process may not be viable for use on a large-scale (Najafpour, 2007).

8.4.3 Heat Sterilization in a Continuous System

Industrial bioreactors are generally designed for *in situ* steam sterilization under pressure, in which all the culture medium and the air entering the bioreactor must be sterile. There are two main methods of sterilizing the culture medium in a continuous system: (a) indirect heat using a heating coil or mantle; and (b) direct steam injection into the medium.

In principle, thermal sterilization in a continuous process comprises: (a) the heating time to achieve the desired sterilization temperature; (b) maintaining this temperature at a constant value for the required time – time in which the culture medium remains in circulation until complete sterilization; and (c) cooling of the medium (Stanbury, Whitaker, and Hall, 2017).

Figure 8.2 demonstrates this process. The wet heat exchange sterilization process starts with the inlet culture medium passing through the injection pump (B) and runs through the sterilization tube towards the first heat exchanger (TC1). This sterilized culture medium will raise the temperature of the new input medium and as a result will save energy. Then, the preheated inlet medium is sent to the steam direct injector (I), where the temperature rapidly increases up to the sterilization temperature. Thus, the medium is moved through the retention tube (RT) to the sterilization temperature. The temperature remains constant along the holding tube for as long as required to ensure reliable sterilization. The medium, already sterilized, but still at a high temperature, passes through the pressure reduction valve (V) and then goes to the heat exchangers to be cooled down to the desired temperature, passing first through TC1 and forwarded to TC2, where its temperature is reduced to the desired value. The cooling fluid, usually used in the TC2 exchanger, is cold water. The sterilized culture medium flows into the empty bioreactor, which should be sterilized beforehand. In this case – of direct steam injection into the culture medium – the medium is diluted in the order of 10%–20% (by volume) (Doran, 2013; Stanbury, Whitaker, and Hall, 2017).

A heat exchanger utilizing indirect steam can be used as a substitute for the direct steam injector (I) (Figure 8.2). The most suitable indirect heat exchangers are of a double spiral type and made of stainless steel, through which the medium passes. In this case, high temperatures are attained by steam passing through coils or heat exchanger plates, at the same time as the medium passes in the opposite direction – or countercurrent. As no direct contact between steam and the culture medium takes place, the risk of contamination is minimized and no dilution occurs (Stanbury, Whitaker, and Hall, 2017).

Compared to batch sterilization, the main advantages of the continuous sterilization process are savings in steam and

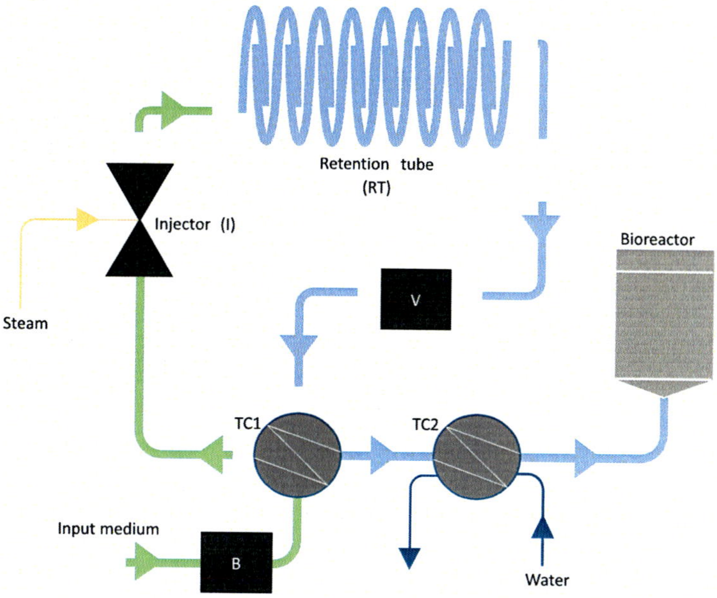

FIGURE 8.2 Schematic representation of a continuous heat sterilization process. B: injection pump. TC1 and TC2: heat exchangers. I: steam injector. TR: retention (or holding) tube. V: pressure reduction valve. (Adapted from Stanbury, P.F., Whitaker, A., Hall, S.J., *Principles of Fermentation Technology*, Butterworth-Heinemann, Elsevier, Oxford, UK, 2017.)

water, as well as shortening of heating and cooling times. Once retention times are reduced, degradation of nutrients is minimized or even prevented (Stanbury, Whitaker, and Hall, 2017).

8.4.4 Sterilization Kinetics of Microorganisms

For an efficient operation of heat sterilization – batch or continuous – the time necessary to achieve the desired reduction in the numbers of contaminating microorganisms must be estimated. As organisms do not die all at the same time, estimation of the retention time is critical (Dion and Parker, 2013). However, as mentioned above, depending on the time and temperature, heat sterilization can destroy/degrade/alter the nutritional composition of the culture medium. To minimize these negative effects, the retention time at the highest temperature should be as short as possible and is generally established according to the sterilization kinetics of each specific bioprocess (Doran, 2013).

Owing to the presence of a variety of contaminating microorganisms in the culture medium prior to sterilisation (N_0), the sterilization kinetics describes the rate of cell destruction by heat as a first-order chemical reaction. As the population of microorganisms (N) decreases over time, the rate is defined by Equation 8.1.

$$\ln \frac{N(t)}{N_0} = -k_d t \tag{8.1}$$

where N is the number of viable cells present in the culture medium after the sterilization, t is the sterilization time and k_d is the constant reaction rate constant or specific rate of cell death.

Since the sterilization kinetics is a first-order reaction, the constant for the mortality rate (k_d) follows Arrhenius' law (Deindoerfer and Humphrey, 1959). A graphical representation of the linear model for Equation 8.1 corresponds to a straight line with a negative slope towards k_d, as shown in Figure 8.3.

Figure 8.3 shows that the viable number of cells decreases exponentially over time as a function of the retention temperature. For example, it is possible to observe that as the temperature increases, for example from T_1 = 105°C to T_4 = 121°C, the slope of the line $(-k_d)$ increases. This means that the number of viable cells in a given sterilization time will decrease dramatically, even considering an

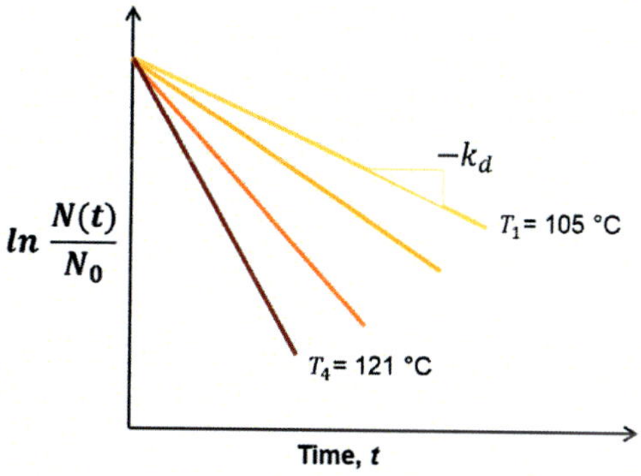

FIGURE 8.3 Decrease in cell viability as a function of time and sterilization temperature. (Adapted from Stanbury, P.F., Whitaker, A., Hall, S.J., *Principles of Fermentation Technology*, Butterworth-Heinemann, Elsevier, Oxford, UK, 2017.)

increase of just 16°C (Najafpour, 2007; Stanbury, Whitaker, and Hall, 2017).

In addition to the retention time, another critical factor in sterilization is temperature. As k_d depends on the temperature, its value is about 0.02/min at 100°C, the mortality rate increases 10 times at 110°C and 100 times at 120°C, according to Equation 8.2.

$$\ln k_d = \ln k_0 - \frac{E}{RT} \tag{8.2}$$

where k_0 is the mortality rate constant at a reference temperature also known as the Arrhenius constant (A), R is the gas constant, T is the absolute temperature and E is the activation energy of cell death. E values are between 60 and 70 kcal/mol of microbial cells, 100–150 kcal/mol of spores and 30–40 kcal/mol of culture medium containing vitamins and proteins (Deindoerfer and Humphrey, 1959; Stanbury, Whitaker, and Hall, 2017).

8.4.5 Air Sterilization

Air sterilization is of paramount concern and involves both the environment in which organisms are manipulated and the air entering the bioreactor as a source of dissolved oxygen to the cells. Most of the sterilization processes are addressed to ensure the asepsis of the culture media and equipment. Although the air can be sterilized through different methods, the most common is the filtration of the air stream (Doran, 2013).

HEPA filters (high-efficiency particulate air) are specially used in aseptic cabinets (e.g. laminar-flow hoods) or clean rooms and are recommended when manipulating cells to prepare inoculum. These filters are made up of cellulose acetate plates, which are 99.97% efficient in removing 0.3 μm medium diameter particles. The filters can be made from fibreglass, with over 99.97% effectiveness in removing particles larger than 0.5 μm (Perkowski, 1983)

The dimensions of microorganisms suspended in air are in the range of 0.5–1.0 μm, whilst dust particles which often transport microorganisms have diameters bigger than 4 μm. Considering that microbial spores are not normally associated with dust particles, then pore membranes with a diameter of 0.20, 0.22 or 0.45 μm are generally used to sterilize air. The filtration systems consist of cartridges with folded membranes within stainless steel holders (Stanbury, Whitaker, and Hall, 2017). The most commonly used material in the manufacture of membranes for air sterilization is PTFE (a hydrophobic material that can be sterilized using moist heat). Filters available on the market are quite resistant to sterilization, with the possibility of making 50–150 sterilization runs at 145°C for 30 min (Perkowski, 1983). Prior to supplying air to the bioreactor, the filter must be sterile to prevent microorganisms that have adhered to membranes from becoming detached and contaminating the growth medium.

In addition to sterilization of inlet gases entering the bioreactor, the outlet gases from the vessel must also pass through filtering processes. The control of exhaust air is increasingly required and is mandatory in the case of bioprocesses conducted with recombinant organisms, in order to maintain biosafety (i.e. accidental release of the recombinant organism into the environment) and to reduce the emission of potentially allergenic compounds (Doran, 2013).

8.5 STERILIZATION IN DOWNSTREAM OPERATIONS

Downstream processing includes all of the unit operations performed after obtaining the target biomolecule in the bioreactor and leads to the manufacture of the final commercial product (see Chapter 13).

Unlike sterilization procedures in upstream processes which aim to ensure that no cross-contamination could harm the production of the target molecule, sterilization in downstream operations must ensure and preserve the quality, purity and safety of the final product. In the case of biopharmaceuticals, in addition to destroying microorganisms, additional steps are also needed to inactivate or eliminate immunogenic antigens that have the potential to harm the patient. Biopharmaceuticals are more labile than small molecule medicines and so sterilization processes are more complex than those currently used by chemical–pharmaceutical industries. The low tolerance of most biomolecules to physical and chemical stresses, especially heat, limits the number of viable sterilization techniques that can be used and increases the number of operations and technologies to prevent contamination throughout unit operations.

A clear distinction exists between the most frequently practised methods of sterilization in upstream operations (usually batch or continuous heat sterilization) and those more complex methods applied in downstream steps, which are described in the next subsections.

8.5.1 Terminal Sterilization and Aseptic Processing

There are two possible approaches for sterilizing pharmaceutical formulations: (a) terminal sterilization – most common for traditional drugs; and (b) aseptic sterilization – commonly used in the case of biopharmaceuticals. While in terminal sterilization, the complete elimination of microorganisms occurs only at the end of drug manufacture and packaging, usually achieved using traditional thermal methods – dry heat and/or wet heat – in the case of biopharmaceuticals; the most common protocol is to apply aseptic processing across all or part of the production system to prevent contamination (Lim and Suh, 2015).

Aseptic processing is more complex and costly than terminal sterilization, since this processing requires asepsis of all equipment and rooms during all downstream stages, in addition to requiring validation of each stage of manufacture and qualified personnel involved in all of these operations (Lim and Suh, 2015). The need for more in-depth control combined with the complexity of biopharmaceutical manufacturing operations is clearly reflected in the final price of the bioproduct, which can be thousands of times more expensive than more traditional small molecule-based medicines (Puetz and Wurm, 2019).

Although aseptic processing is still the most popular strategy in the production of biopharmaceuticals, terminal sterilization methods are already being developed to reduce the complexity and high cost of manufacturing biopharmaceuticals. As previously mentioned, promising methods can be used for heat-sensitive molecules such as gamma radiation, natural light, microwaves and electron beam irradiation (Yaman, 2001).

8.5.2 Sterilization of Biopharmaceuticals in Downstream Processes

In the case of biopharmaceuticals, it is necessary to achieve high levels of purity, quality and sterility to meet the rigorous standards of health regulatory agencies (Dos Santos et al., 2018). However, as previously mentioned, it is not possible to perform the upstream steps in a completely sterile way, since biopharmaceuticals are actually produced by biological systems. Therefore, viable cells will always be present during the production of biopharmaceuticals; but, at some stage, after the cultivation, these contaminants must be inactivated and removed.

When a sterilization step is added before a manufactured batch of a particular biopharmaceutical is released onto the market, this is called terminal sterilization. Commonly used alternative involves sterilizing components of the biopharmaceutical formulation separately in the first stages of downstream processing (for example, after cell disruption/removal of cells), and, from that point on, an aseptic processing has to be maintained in subsequent unit operations. This strategy is very laborious as it requires asepsis to be maintained in all workplaces, equipment and supplies. However, this procedure is still the most widely used today by the biopharmaceutical industry.

Figure 8.4 represents a general sequence of how sterilization can be included in the downstream processing of biopharmaceuticals. This schematic representation explains how different sterilization methods should be framed for each unit operation. In the case of a system with aseptic processing, sterilization methods must be included to ensure that viable organisms are eliminated, while aseptic production is maintained until complete manufacture of the biopharmaceutical is achieved.

In the formulation stage, concentration and additions are made so that the final product has all of the required properties necessary for commercialization and use as a medicine. This might include adding preservatives, adjuvants, additives or other excipients. In the case of medicines, sterilization can be achieved before filling containers with the product or directly in the final packaging. In the case of aseptic processing, all excipients added to the formulation must be previously sterilized, usually by membrane filtration, to guarantee the product will not be contaminated.

If the biopharmaceutical is commercialized as a solid dosage form, a lyophilization step – performed under totally aseptic conditions – can be included to improve the preservation of the product. At this stage, the product is also filled and sealed under aseptic conditions, packaged and then labelled. These last two steps do not require aseptic conditions, as the product is already filled in a suitably sterile container.

Regardless of the sterilization approach selected – whether terminal sterilization or aseptic processing – the endgoal is to produce a medicine that complies with technical specifications and health legislation.

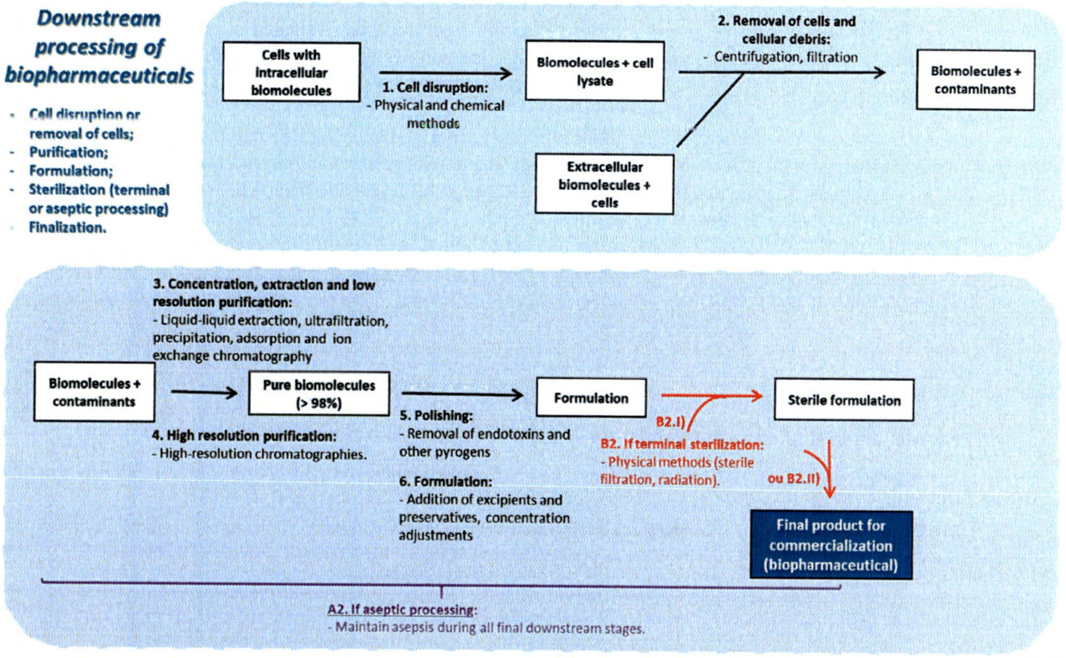

FIGURE 8.4 Scheme on different sterilization approaches (terminal and aseptic processing) in the downstream processing of biopharmaceuticals.

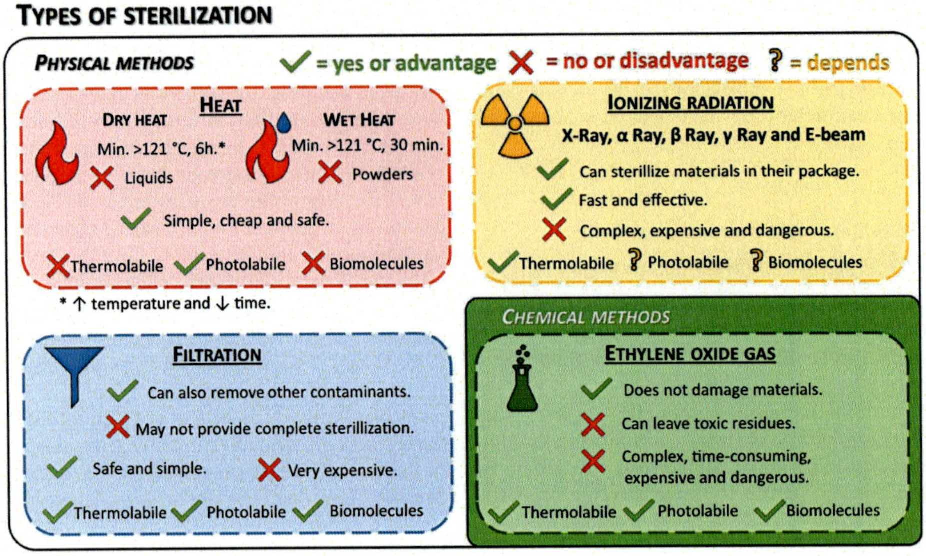

FIGURE 8.5 Applications, advantages and disadvantages of different types of sterilization procedures.

8.5.3 Sterilization of Biopharmaceuticals

In the final stages of the manufacturing process, there are several restrictions on the use of certain procedures and chemical agents since, in addition to ensure the elimination of microorganisms and spores, it is essential to maintain the integrity, activity and safety of the biopharmaceutical. Figure 8.5 summarizes the main applications, advantages and disadvantages of the different types of sterilization procedures that can be used in both upstream and downstream stages to obtain a biopharmaceutical.

8.6 VALIDATION OF STERILIZATION PROCESSES

In the production of any biopharmaceutical, it is essential to ensure consistency and safety of all processes and products. To achieve this, all stages must be validated, monitored and processes documented frequently to guarantee approval of the final product by regulatory agencies. For this, there are different elements of the system that must be validated, including facilities, operations and process performance.

Qualification of the installation is concerned with ensuring conformity of the equipment, auxiliary items and locations related to production and must be carried out periodically, or whenever changes occur that might impact the system, for example, replacement or addition of equipment. This qualification ensures that the equipment is functional and calibrated, and that production sites are clean and safe (ANVISA, 2010).

The qualification must demonstrate that the sterilization equipment is effective and can perform the required function under the defined conditions and intervals. Different parameters must be evaluated according to each type of sterilization. Table 8.1 indicates the factors analysed for each sterilization process.

In heat sterilization, the main aspect that must be verified is maintenance and distribution of the temperature in the sterilization chamber, to ensure that the minimum temperature is reached anywhere in the chamber. For methods that use radiation, it is necessary to check the distribution and penetration of ionizing rays in the samples. In the process of sterilization by filtration, a stricter environmental control is required, as in some cases the

TABLE 8.1
Parameters Evaluated in the Validation of Each Type of Sterilization

Type of Sterilization	Evaluated Parameters
Heat	Temperature and heat distribution
Radiation	Penetration and distribution of ionizing radiation
Filtration	Control of the aseptic environment, filter integrity, differential pressure and flow speed
Ethylene oxide gas	Relative humidity, temperature and gas concentration

Source: ANVISA - Agência Nacional de Vigilância Sanitária, CP n° 39, Propostas de revisão e atualização dos Métodos Gerais da Farmacopeia Brasileira, 30 Abr. 2010, Sec. 1, 97.

filtrate will be released to the external environment after processing. Additionally, the integrity of the filter membranes and their proper installation have to be checked, in addition to validating the differential pressure and the filtration flow velocity. Finally, for sterilization achieved using ethylene oxide gas, the relative humidity, temperature and gas concentration must be controlled (ANVISA, 2010; Shintani, 2011).

Qualification of sterilization performance comprises physical, chemical and microbiological monitoring of the final product so that the required level of sterility is achieved whilst maintaining the biological properties of the product. Process validation will confirm that the sterilization methods employed are consistent and effective. This process must be done for at least three consecutive cycles to demonstrate the reproducibility of the process (ANVISA, 2010; Shintani, 2011). Three approaches can be followed to qualify the performance of a sterilization process, listed in Table 8.2.

Validation of the sterilization performance using a bioburden method requires a broad knowledge and predictability of possible contaminants and their microbial load, as this approach is based on inactivation of the expected microbial load for a specific product and process. In a bioburden method, the inactivation of a reference microorganism is called the biological indicator – with an expected microbial load is monitored. On the other hand, in an overkill method, both a reference microorganism and a higher than expected microbial load are used. In this case, the objective is to challenge the sterilization system, thereby allowing a higher level of sterility to guarantee product safety and quality (ANVISA, 2010; Shintani, 2011).

Different biological indicators are available for different types of sterilization procedures. A biological indicator is, in fact, a preparation of specific organisms that have a stable and well-defined resistance when subjected to a sterilization process. These indicators are used to validate and monitor sterilization procedures under certain conditions. Highly resistant and persistent organisms are selected to ensure increased product confidence and safety. For example, biological indicators for autoclave sterilization may consist of test strips of spores of heat-resistant microorganisms which, after autoclaving, are aseptically transferred to the growth medium to confirm that they have been destroyed.

Three types of biological indicators can be used both to confirm sterility of the final products and in the process validation, and quality control is achieved at different stages of manufacture (ANVISA, 2010; Shintani, 2011) namely:

i. *Spores in support or carrier:* spores are incorporated into support or carriers – such as discs of filter paper, glass, plastic and others – and packed to maintain the viability of the inoculum.
ii. *Suspension of inoculated spores in representative units of the product:* in this type of biomarker, the spores are inoculated directly into a sample of the product or a product analogous to the real one, which has similar properties and behaviour.
iii. *Self-contained:* in addition to spores, this biological indicator contains culture medium and nutrients necessary for the growth of the microorganism, normally packaged in ampoules. With this, it is also possible to assess whether there is a recovery of contaminants after the sterilization process. It is also possible to add colorants or markers that indicate the absence/presence of the microorganism, facilitating the monitoring of the effectiveness of the sterilization process.

Selection of the most appropriate biological indicator will depend on the properties and application of the product, as well as the selected sterilization process. Thus, to guarantee a high degree of safety, biological indicators must use the most resistant/persistent microorganisms to the sterilization process being monitored. Table 8.3 shows the microorganisms usually selected as biological indicators for each type of sterilization process.

In heat sterilization processes, thermotolerant bacterial spores are selected, generally from *Geobacillus stearothermophilus* for moist heat and *Bacillus atrophaeus* for dry heat. *Bacillus atrophaeus* spores are also used as biological indicators to monitor radiation and ethylene oxide gas sterilization processes owing to their high resistance to several different types of stress. However, to monitor sterilization by filtration, indicator selection does not consider the resistance of the microorganism, but its size. In this case, the microorganism that is normally selected to assess sterile filtration is the smallest known bacterium called *Brevundimonas diminuta* (0.4 μm diameter and 1.6 μm length) (ATCC 19146) (Griffiths et al., 2000)

Besides biological indicators, chemical indicators can also be helpful in monitoring sterilization processes, such as autoclave tape which is impregnated with chemical compounds that change colour as a result of exposure to heat for a determined time/temperature. A change in colour of the tape itself can confirm that the necessary temperatures have been obtained during sterilization by moist heat,

TABLE 8.2
Different Methods to Perform the Qualification of Sterilization Performance of a Biopharmaceutical

Method	Principle
Bioburden	Inactivation of a natural microbial load
Bioburden and biological indicator	Combination of inactivation of reference microorganism associated with knowledge of microbial load
Overkill	Inactivation of reference microorganisms in above-natural loads

Source: ANVISA - Agência Nacional de Vigilância Sanitária, CP n° 39, Propostas de revisão e atualização dos Métodos Gerais da Farmacopeia Brasileira, 30 Abr. 2010, Sec. 1, 97.

TABLE 8.3
Most Common Biological Indicators for Different Types of Sterilization

Sterilization	Biological Indicator
Moist heat	Spores of *Geobacillus stearothermophilus*, *Clostridium sporogenes*, *Bacillus atrophaeus* and *Bacillus coagulans*
Dry heat	*Bacillus atrophaeus* spores
Radiation	*Bacillus atrophaeus* spores (niger variant)
Filtration	*Brevundimonas diminuta* ATCC 19146
Ethylene oxide gas	*Bacillus atrophaeus* spores

Source: ANVISA - Agência Nacional de Vigilância Sanitária, CP n° 39, Propostas de revisão e atualização dos Métodos Gerais da Farmacopeia Brasileira, 30 Abr. 2010, Sec. 1, 97.

guaranteeing a more reliable operation. Ideally, both types of indicators should be used for more complete monitoring of the sterilization process.

In order to guarantee sterility and safety assessment of injectable biopharmaceuticals, it is necessary to ensure the absence of pyrogens, i.e. endotoxins such as lipopolysaccharides (ANVISA, 2010). In this book, you can find a chapter (Chapter 14) that discusses these types of contaminants in more detail.

8.7 FINAL CONSIDERATIONS

Sterilization is an essential process in the manufacture of biopharmaceutical products, ensuring that they comply with health legislation. Sterilization can be achieved using different methods (e.g. dry heat, wet heat, filtration, radiation and ethylene oxide gas), which must be selected according to the stage of the process, specificity of the unit operation, the properties of the target biomolecule and consideration of most likely contaminating agents. In the case of obtaining biopharmaceuticals, sterilization can be terminal or by aseptic processing, which will essentially depend on the stability of the product of interest, and viability and effectiveness of sterilization processes available.

As a final consideration, we emphasize the importance of understanding the nature of the bioprocess to define the most appropriate method to guarantee an effective sterilization. In the upstream stages, the control of microorganisms has the essential objective of preventing cross-contamination that may harm the production of the biopharmaceutical, whilst in downstream stages, the aim is to eliminate any substance that may harm the quality of the medicine or health of the patient. Since we are discussing biopharmaceutical products, it is not enough to choose the most compatible sterilization method for the product and/or operation, but to maintain strict control of all sterilization processes and finished products, using properly validated protocols and biological indicators. Thus complex sterilization, asepsis and control systems are necessary to ensure quality and safety of biopharmaceuticals, which are required for approval by health regulatory agencies for use in humans.

REFERENCES

ANVISA - Agência Nacional de Vigilância Sanitária. CP n° 39, Propostas de revisão e atualização dos Métodos Gerais da Farmacopeia Brasileira. 30 Abr. 2010, Sec. 1, 97.

Baker, R. W. *Membrane Technology and Applications*. 3. ed. Chichester, UK: John Wiley & Sons, 2012.

Blumenthal, M. N.; Rosenberg, A. Definition of an allergen (immunobiology). In: *Allergens and Allergen Immunotherapy*, 3. ed. Lockey, R. F.; Bukantz, S. C.; Bousquete, J. (Eds.), New York: Marcel Dekker, Inc., 37, 2004.

Claessens, C. The human pathogens and toxins act and its regulations. *Canadian Journal of Medical Laboratory Science*, v. 77, n. 3, 7, 2015.

Deindoerfer, F. H.; Humphrey, A. E. Analytical method for calculating heat sterilization times. *Applied Microbiology*, v. 7, n. 4, 256–264, 1959.

Dion, M.; Parker, W. Steam sterilization principles. *Pharmaceutical Engineering*, v. 33, n. 6, 60–69, 2013.

Doran, Pauline M. *Bioprocess Engineering Principles*. 2. ed. Oxford, UK: Academic Press, 2013.

Dos Santos, N. V. et al. Liquid-liquid extraction of biopharmaceuticals from fermented broth: Trends and future prospects. *Journal of Chemical Technology & Biotechnology*, v. 93, n. 7, 1845–1863, 2018.

Griffiths, M. H. et al. Rapid methods for testing the efficacy of sterilization-grade filter membranes. *Applied and Environmental Microbiology*, v. 66, n. 8, 3432–3437, 2000.

Jornitz, M. W. *Filtration and Purification in the Biopharmaceutical Industry*. 1. edn. Boca Raton, FL: CRC Press, 2019.

Lim, S. Y.; Suh, M. Product and process innovation in the development cycle of biopharmaceuticals. *Journal of Pharmaceutical Innovation*, v. 10, n. 2, 156–165, 2015.

Liu, S. *Bioprocess Engineering: Kinetics, Biosystems, Sustainability and Reactor Design*. 1. ed. Oxford, UK: Elsevier, 2013.

Mcdonnell, G. E. *Antisepsis, Disinfection, and Sterilization: Types, Action, and Resistance*. 1. ed. Washington, DC: ASM Press, 2020.

Muheem, A. et al. A review on the strategies for oral delivery of proteins and peptides and their clinical perspectives. *Saudi Pharmaceutical Journal*, v. 24, n. 4, 413–428, 2016.

Najafpour, G. D. Sterilization. In: *Biochemical Engineering and Biotechnology*. Najafpour, G. D. (Ed.), Amsterdam, The Netherlands: Elsevier, 342–350, 2007.

Perkowski, C. A. Fermentation process air filtration via cartridge filters. *Biotechnology and Bioengineering*, v. 25, n. 5, 1215–1221, 1983.

Puetz, J.; Wurm, F. M. Recombinant proteins for industrial versus pharmaceutical purposes: a review of process and pricing. *Processes*, v. 7, n. 8, 476, 2019.

Rutala, W. A.; Weber, D. J. *Guideline for Disinfection and Sterilization in Healthcare Facilities (2008)*. Atlanta, GA: CDC-U.S. Center for Disease Control and Prevention, 2008.

Seader, J. D.; Henley, E. J.; Roper, D. K. *Separation Process Principles: Chemical and Biochemical Operations*. 3. ed. New York: John Wiley & Sons, 2011.

Shintani, H. Validation of sterilization procedures and usage of biological indicators in the manufacture of healthcare products. *Biocontrol Science*, v. 16, n. 3, 85–94, 2011.

Stanbury, P. F.; Whitaker, A.; Hall, S. J. Chapter 5 – Sterilization. In: *Principles of Fermentation Technology*. Stanbury, P. F.; Whitaker, A.; Hall, S. J. (Eds.), 3. ed. Oxford, UK:Butterworth-Heinemann, Elsevier, 2017.

Tortora, G. J.; Funke, B. R.; Case, C. L. *Microbiology: An Introduction*. 10. ed. Boston, MA: Pearson, 2013.

Van Boekel, M. A. J. S. Effect of heating on Maillard reactions in milk. *Food Chemistry*, v. 62, n. 4, 403–414, 1998.

Villota, R.; Hawkes, J. G. Reaction kinetics in food systems. In: *Handbook of Food Engineering*. Boca Raton, FL: CRC Press, 2019. Accessed 28 Nov 2020, Routledge Handbooks Online.

Wolken, W. A.; Tramper, J.; Van Der Werf, M. J. What can spores do for us? *Trends in Biotechnology*, v. 21, n. 8, 338–345, 2003.

Yaman, A. Alternative methods of terminal sterilization for biologically active macromolecules. *Current Opinion in Drug Discovery & Development*, v. 4, n. 6, 760–763, 2001.

9 Kinetics of Cell Cultivation

Aldo Tonso, Adalberto Pessoa Jr., and Michele Vitolo
Universidade de São Paulo

CONTENTS

9.1 INTRODUCTION

Kinetics deals with the rate changing of some parameters over time – such as consumption of reagents, formation of products and heat generation – in a process. In the case of a bioprocess, it is essential to monitor how these parameters vary, as cells are responsible for transforming elements of the culture medium into products useful for humanity. Usually, different rates are defined to adequately represent this variation. Thus, the purpose of this chapter is to provide elements that allow a better understanding of how a process producing a biopharmaceutical evolves and how this evolution can be mathematically estimated, using basic models. In bioprocesses, the main parameters to be quantified are the concentration of biomass (as it is the agent of all biotransformation), the limiting substrate (which is normally responsible for the end of cell growth) and of the product (which in the final analysis is what is of interest in any production process). In addition to these parameters, other variables are common to be monitored in the production of biopharmaceuticals. Among them, we find those related to the environment in which the cells grow, many of them being kept controlled at a constant level during cultivation, such as: temperature; pH; and dissolved oxygen (see Chapter 11).

Cell mass can be quantified in several ways, each of which has advantages and disadvantages, i.e. depending on the growth medium and cell type, the most appropriate methodology should be established. Biopharmaceuticals are frequently produced by animal cells (mainly due to the way they make post-translational modifications – see Chapter 12) and by yeasts (such as *Pichia pastoris*) or bacteria (like *Escherichia coli*).

It is frequent to express cell density as dry weight mass concentration, i.e. the concentration of cells after total water removal. This dry mass expression is relevant because of the high moisture content inside the cells. In general, the methodology for determining dry mass is carried out after filtering a cultivation sample on a suitable filter membrane (with a pore diameter of 0.45 or 0.22 μm). The cake cell membrane is washed with distilled water and then dried to constant mass using an oven from about 80°C to 100°C (until a constant mass is obtained), or a microwave oven for about 15 min at low power (Olsson and Nielsen, 1997). After drying, the sample is weighed so the biomass concentration is calculated by dividing the mass of cells (mass of membrane with cells minus the mass of the clean dry membrane) by the volume of filtered sample (g/L). As an alternative, in cases of difficult filtration, the cultivation media can be centrifuged and the cells are resuspended in distilled water before filtering (more information on cell concentration determination can be found in Chapter 7).

In the case of animal cell cultivation the method of cell counting is used in a Neubauer chamber (or haemocytometer) in which the number of cells occupying the whole volume of the chamber is counted by an optical microscope. Typically, the chamber has streaks marking 1 mm wide squares (with 1 mm^2 area) and 0.1 mm thick, so the count refers to 0.1 mm^3 volume. The conversion in cells per mL is made by multiplying the number of cells counted by 10,000 (1 mL = 10,000 mm^3). Such a method allows, in addition to the determination of cell concentration, the visualization of the cells (observing its morphology throughout the culture) and its viability, through the use of trypan blue dyes (Freshney, 2010).

An indirect way of quantifying the biomass concentration is by turbidity (or optical density) (Madigan et al., 2017), which is the absorbance of light passing through a sample (conveniently diluted) with appropriate dimensions (in a spectrophotometer cuvette). Usually, it is necessary to make a correlation between turbidity and dry mass concentration in different dilutions beforehand. There are also several automated methods for the determination of cell concentration, such as flow cytometers (Ormerod, 2008), based on the Coulter counter (Graham, 2003) or FACS (fluorescence-activated cell sorter) and image analysers, such as Countess II FL (Thermo). More information about cell concentration measurements can be seen in Chapter 7.

Monitoring the substrate concentration is essential to assess cell metabolism. Normally, the growth rate is reduced when there is a limitation in the concentration of a certain substrate (called limiting substrate), and the optimization of the bioprocess often involves controlling the supply of that substrate to the cell (see below and in Chapter 10).

The limiting substrate is, in most processes, the carbon and energy source, such as glucose, but it can also be another sugar or some amino acid (glutamine in animal cell cultures). From the same sample taken to quantify the biomass, it is possible to measure the concentration of the substrate, by traditional analytical methods (e.g., if the substrate is glucose, the concentration can be determined using the Fehling titration method, di-nitro salicylic acid method or Somogy method), by enzymatic kits (such as GOD-POD, based on the combined action of the enzymes glucose oxidase and peroxidase) or by liquid chromatography (HPLC). Ideally, it is desired to have this measurement in real time to allow actuation in the bioprocess, such as the control of a feed pump.

The monitoring of product concentration depends on its nature, and advanced analytical purification techniques may be necessary using liquid chromatography or electrophoresis (SDS-PAGE) (Kilikian and Pessoa, 2020). If the product is an enzyme, its concentration is usually expressed by volumetric enzymatic activity (IU/mL).

9.2 TYPICAL BATCH CULTIVATION PHASES

When cells come into contact with fresh culture medium, in flasks or bioreactors, cell growth begins. If no new nutrients are added during the cultivation, then this process is called batch (see Chapter 10) and typically goes through several growth phases (Butler et al., 2007), as can be seen in Figure 9.1:

a. *Lag phase*: a period in which no expressive variation in the concentration of biomass occurs, because the cells are preparing their machinery for reproduction and synthesizing enzymes to metabolize the nutrients in the culture medium;
b. *Exponential phase (or logarithmic phase)*: in which cells are growing without limitation and duplicating the concentration at regular intervals of time (doubling time T_d or generation time T_g) while maintaining their physiological state constant. This phase is very important in the study of kinetics and will be analysed later;
c. *Deceleration phase*: in which cells reduce their growth rate due to limitation (of a substrate) or inhibition (of a toxic by-product), which takes them out of the previous physiological state;
d. *stationary phase*: in which a variation in cell concentration is not observed but can result in a balance between growth and death;
e. *death phase*: in which a reduction in cell concentration is observed, depending on how the biomass is measured (total cells or viable cells).

The exponential phase deserves attention, since cells grow quickly and without limitations. Thus, the main goal of optimizing the bioprocess is to maintain the cultivation for the longest time in the exponential phase, in which the cells have the shortest doubling time. But, how can we determine when is the exponential phase? What is the numerical parameter that would indicate this phase? In the sequence is Table 9.1 which shows nomenclatures, expressions, meanings and units of all the parameters and variables used in this chapter and also in Chapters 10 and 11, and Table 9.2 shows a numerical example.

Let's analyse Table 9.2. Column A shows the time when the samples were collected (t). Note that the samples were taken at regular intervals. Column B shows the values of measured biomass concentration, represented by the letter X. Only by analysing this column, it can be seen that the cultivation had its biomass concentration doubling every 2h, in the period from $t = 0$ to 6h. That is, the doubling time (t_d) is 2h. However, after that, it can be seen that the cell population did not double in the same 2h interval and, therefore, left the exponential phase. So, what would be the numerical parameter that would show this?

In many chemical processes, it is usual to calculate the variation rates of a quantity as a function of time, which in our case would be $\Delta X/\Delta t$. Here, ΔX is the difference between the current value of X and the previous instant, and Δt is the time difference in this interval.

Now, observe the values of $\Delta X/\Delta t$ in column C. They increase until $t = 8$h and, then, fall. From this analysis, it is not possible to determine that the end of the exponential phase is at $t = 6$h. Therefore, the $\Delta X/\Delta t$ variation does not

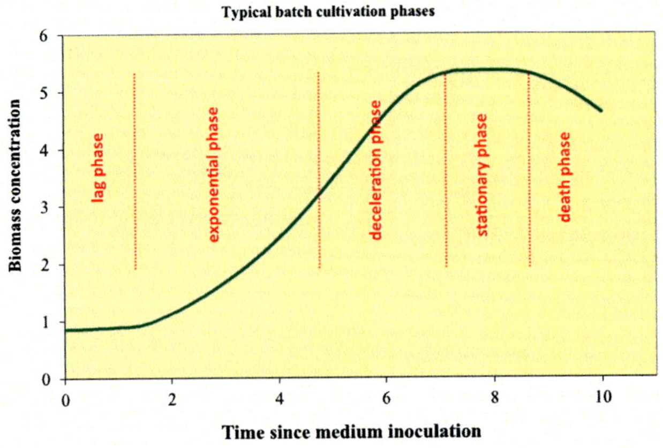

FIGURE 9.1 Typical curve of batch cultivation and the phases.

TABLE 9.1
Nomenclatures, Expressions, Meanings and Units of All the Parameters and Variables Used in this Chapter and Also in Chapters 10 and 11 of This Book

Nomenclatures/expressions	Meanings	Unit
X	Concentration of cell in bioreactor	g/L
x	Cell mass in bioreactor	g
X_0	Initial cell concentration	g/L
X_f	Final cell concentration	g/L
X_{in}	Cell concentration in inlet feed	g/L
X_R	Biomass concentration in the recycle stream	g/L
S	Concentration of substrate in bioreactor	g/L
s	Substrate mass in bioreactor	g
S_0	Initial substrate concentration	g/L
S_f	Final substrate concentration	g/L
S_{in}	Substrate concentration in inlet feed	g/L
P	Concentration of product in bioreactor	g/L
p	Product mass in bioreactor	g
P_0	Initial product concentration	g/L
P_f	Final product concentration	g/L
P_{in}	Product concentration in inlet feed	g/L
V	Volume of medium	L
F	Feed volumetric flow rate	L/h
F_{in}	Volumetric flow rate of inlet	L/h
F_{out}	Volumetric flow rate of outlet	L/h
F_R	Volumetric flow rate of recycle	L/h
Q_X	Volumetric productivity of cells	g/L/h
Q_P	Volumetric productivity of product	g/L/h
A	Recycled liquid fraction (recycle ratio)	Adimensional
B	Biomass concentration factor (in cell recycling)	Adimensional
$\varepsilon = (1 + A - AB)$	Recycling fraction	Adimensional
t	Cultivation time	h
t_d	Doubling time	h
t_B	Batch time	h
K_S	Saturation constant (Monod Model)	g/L
K_I	Inhibition constant (Andrews Model)	g/L
α	production constant associated with growth (Luedeking and Piret model)	Adimensional
β	production constant not associated with growth (Luedeking and Piret model)	h^{-1}
$\mu = \frac{1}{X} \cdot \left(\frac{dX}{dt}\right)_G$	Specific growth rate	h^{-1}
$r_X = \left(\frac{dX}{dt}\right)_G = \mu.X$	Variation of cell concentration due to growth	g/L/h
$k_d = \frac{1}{X} \cdot \left(\frac{dX}{dt}\right)_d$	Specific death rate	h^{-1}
$r_d = \left(\frac{dX}{dt}\right)_d = k_d.X$	Variation of cell concentration due to death	g/L/h
μ_{max}	Maximum specific growth rate (exponential phase)	h^{-1}
$\mu_S = \frac{1}{X} \cdot \left(\frac{dS}{dt}\right)_c$	Specific substrate consumption rate	g/g/h
$r_S = \left(\frac{dS}{dt}\right)_c = \frac{\mu.X}{Y_{X/S}}$	Total variation of substrate concentration due to consumption	g/L/h
$r_{SG} = \left(\frac{dS}{dt}\right)_G = \frac{\mu.X}{Y_G}$	Variation of substrate concentration due to growth	g/L/h

(*Continued*)

TABLE 9.1 (*Continued*)
Nomenclatures, Expressions, Meanings and Units of All the Parameters and Variables Used in this Chapter and Also in Chapters 10 and 11 of This Book

Nomenclatures/expressions	Meanings	Unit
$m_S = \frac{1}{X} \cdot \left(\frac{dS}{dt}\right)_m$	Specific substrate consumption rate due to maintenance (maintenance coefficient)	g/g/h
$r_{Sm} = \left(\frac{dS}{dt}\right)_m = m_S.X$	Variation of substrate concentration due to maintenance	g/L/h
$Y_{X/S} = \frac{r_X}{r_S}$	Observed biomass yield on substrate	g/g
$Y_{P/S} = \frac{r_P}{r_S}$	Observed product yield on substrate	g/g
$Y_{P/X} = \frac{Y_{P/S}}{Y_{P/X}}$	Observed product yield on biomass	g/g
$Y_G = \frac{r_X}{r_{SG}}$	True biomass yield on substrate	g/g
$\mu_P = \frac{1}{X} \cdot \left(\frac{dP}{dt}\right)_P$	Specific product formation rate	g/g/h
$r_P = \left(\frac{dP}{dt}\right)_P = \mu_P.X$	Total variation of product concentration due to production	g/L/h
STR	Stirred tank bioreactor	–
CV	Control volume	–

TABLE 9.2
Time of Cultivation, Cell Concentration and Some Calculations in a Cell Growth

Column A	Column B	Column C	Column D
Time of Cultivation (t)	Cell Concentration (X)	$\Delta X/\Delta t$	$(\Delta X/\Delta t)/X$
(h)	(g/L)	(g/L/h)	(1/h)
0	0.6		
2	1.2	0.3 [(1.2–0.6)/(2–0)]	0.25 [0.3/1.2]
4	2.4	0.6 [(2.4–1.2)/(4–2)]	0.25 [0.6/2.4]
6	4.8	1.2 [(4.8–2.4)/(6–4)]	0.25 [1.2/4.8]
8	7.6	1.4 [(7.6–4.8)/(8–6)]	0.18 [1.4/7.6]
10	8.0	0.2 [(8.0–7.6)/(10–8)]	0.03 [0.2/8.0]

represent properly the physiological state of the cells (quick growth and without limitations). This is because the variation of the population ($\Delta X/\Delta t$) is caused by the size of the population at each instant. As the population grows, the $\Delta X/\Delta t$ variation would be expected to become increasingly larger.

From this analysis, it can be concluded that the way to analyse the physiological state of the cells should be by dividing the velocity of appearance of new cells ($\Delta X/\Delta t$) by the number of cells that provided this growth. We then come to column D: $(\Delta X/\Delta t)/X$. We can observe that in the range of $t = 0–6$ h the growth phase is exponential and the value of $0.25\,h^{-1}$ is constant. Thus, with this new variable, we are able to determine how long the exponential phase lasts. Then, in the 6–8 h interval, the values shown in column D drop to $0.18\,h^{-1}$ and then to $0.03\,h^{-1}$.

As this calculation can be undertaken in any time interval, this rate of change is usually expressed at $\Delta X/\Delta t$ in a differential way, i.e. dX/dt, which is the rate of change of X (dX) in a tiny time interval dt. The derivative dX/dt is the value of $\Delta X/\Delta t$ calculated in a time interval that tends to zero, which was represented in a simplified way in column C of Table 9.2 by $\Delta X/\Delta t$.

For this reason, the variable μ, specific growth rate, is defined in bioprocess kinetics and represents the rate at which each cell is growing (hence the use of the specific term) and is represented in a simplified way in column D (Table 9.2). This variable is calculated according to Equation 9.1:

$$\mu = \frac{1}{X} \cdot \frac{dX}{dt} \tag{9.1}$$

where:

μ = specific growth rate (1/h)
t = cultivation time (h)
X = cell concentration (g/L)
dX/dt = derivative of X in relation to time (g/L/h)

As a way of homogenizing the expression, r_X is also usually defined as the growth rate, which is the variation in cell concentration due to growth, dX/dt (Equation 9.2).

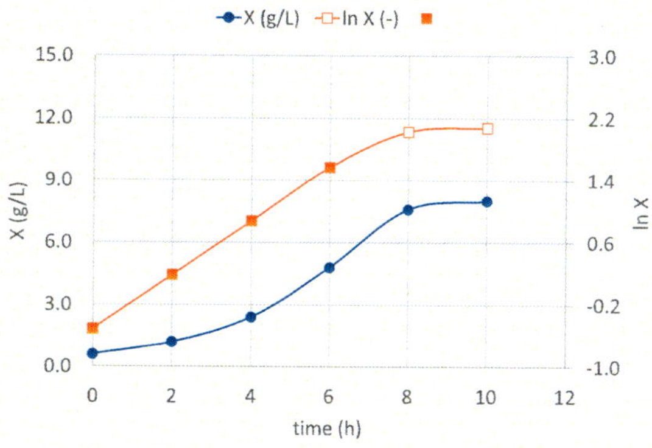

FIGURE 9.2 Cell concentration (X) (closed blue circles) and ln(X) (closed orange squares during exponential phase and open orange squares after that) as a function of time of cultivation (X data from Table 9.1). The ln X curve is a straight line and the tangent of its slope is μ_{max}. After the end of the exponential phase, the slope of ln X curve is reduced (open orange squares).

$$\mu = \frac{1}{X} \cdot r_X \tag{9.2}$$

where:

μ = specific growth rate (1/h)
t = cultivation time (h)
X = cell concentration (g/L)
r_X = growth rate (g/L/h)

Note that r_X shows the total variation in cell concentration, not specific, i.e. without taking into account how many cells are responsible for the variation.

Mathematically, $\frac{1}{X} \cdot \frac{dX}{dt} = \frac{d\ln(X)}{dt}$

then:

$$\frac{d\ln(X)}{dt} = \mu \tag{9.3}$$

This means that if you plot the value of ln (X) as a function of t, for each sample in Table 9.2, the slope of this graph will directly represent the value of μ. Furthermore, as long as the slope is constant, the cells are exponentially growing and μ is constant. We call this value as *maximum specific growth rate* (μ_{max}) referring to the exponential phase.

As can be seen in Figure 9.2, until t = 6 h, the ln X curve as a function of time is a straight line and the tangent of its slope is μ_{max}.

During the exponential growth period, μ has a constant value. This means that the derivative of ln X with time is constant and we can integrate Equation 9.3 in Equation 9.4:

In the exponential phase:

$$X = X_0 \cdot e^{\mu \cdot t} \tag{9.4}$$

where:

t = cultivation time since the start of the exponential phase (h)
X = cell concentration (g/L)
X_0 = cell concentration at the beginning of the exponential phase (g/L)
μ = specific growth rate (1/h)

From Equation 9.4, the doubling time (t_d, also known as generation/doubling time) can be mathematically correlated with a specific growth rate (μ). Doubling time, t_d, is the time interval for the cell population, initially in X_0, to reach double the concentration ($2X_0$). Thus:

$$2 \cdot X_0 = X_0 \cdot e^{\mu \cdot t_d} \tag{9.5}$$

Dividing Equation 9.5 by the value X_0 and applying the natural logarithm on both sides, we have Equations 9.6–9.8.

$$2 = e^{\mu \cdot t_d} \tag{9.6}$$

$$\ln 2 = \mu \cdot t_d \tag{9.7}$$

$$\mu = \frac{\ln 2}{t_d} \tag{9.8}$$

Thus, if in a certain situation, the population of bacteria is doubled in 1 h, the μ value is 0.69 h^{-1} (remembering that ln 2 = 0.69).

9.3 WITH WHAT DOES μ VARY? (MODEL FOR CELL GROWTH)

We have seen that μ, the specific growth rate, is an important parameter to indicate the metabolic state of the cell, and that while it is constant, the population is growing at an exponential rate. But, with what does μ vary? On what depends the value of μ?

It is known that organisms have an optimal temperature to grow, and it is common to find in the literature graphs of cell growth as a function of temperature that have a bell shape indicating an optimal temperature, or where growth is the highest. In these graphs, the ordinate is often called activity, but can also be the value of μ, as a function of temperature. Likewise, μ is a function of pH and other cultivation conditions, such as oxygen supply, shear stress (from agitation and aeration, see Chapter 11), nutrients and growth-inhibiting substances. Usually, in the production of biopharmaceuticals, the optimal pH and temperature are previously determined and these variables are kept

TABLE 9.3
Typical Values of Maximum Specific Growth Rate and Minimum Doubling Times for Different Organisms

Organism	Maximum Specific Growth Rate (h^{-1})	Minimum Doubling Time (h)
Escherichia coli (bacteria)	2	0.35
Aspergillus niger (filamentous fungi)	0.35	2
Saccharomyces cerevisiae (yeast)	0.17–0.35	2–4
HeLa cell (mammalian cell)	0.015–0.023	30–46

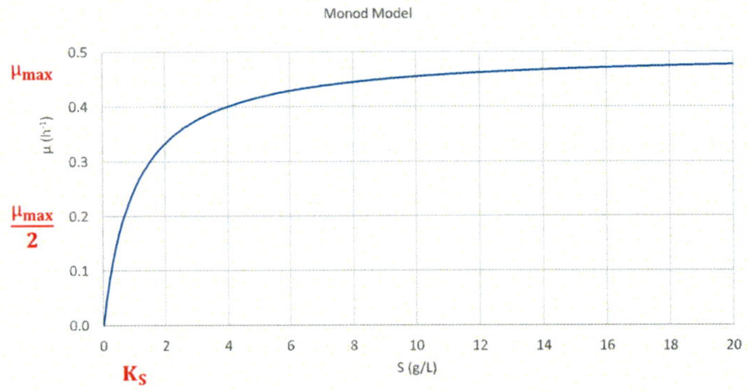

FIGURE 9.3 Representation of Monod model, with parameters values: $\mu_{max} = 0.5\,h^{-1}$ and $K_S = 1$ g/L.

constant during cultivation. Sometimes, the optimum value for growth does not coincide with the production value of the biomolecule, and the former is maintained until the concentration of cells is high and then the value is adjusted for the latter in the production phase. But keeping the optimum conditions for cultivation, the value of the specific growth rate (μ) depends mainly on two factors:

a. Organism

Smaller organisms, such as bacteria, allow the exchange of nutrients with the medium very quickly, and therefore, under ideal conditions, they grow in general with higher μ values. In addition, because they are prokaryotes, without a nuclear membrane, they allow duplication of genetic material more quickly. Among eukaryotes, yeasts and filamentous fungi generally grow more slowly than bacteria, and both present values of μ much higher than animal cells such as mammalian and insect cells. Table 9.3 shows typical values for maximum specific growth rate and minimum doubling times for different organisms.

The results shown in Table 9.3 show that to produce biopharmaceuticals it is more convenient to employ cells that grow more quickly in order to achieve high productivity. However, production is often carried out in animal cells, as they are capable of appropriate post-translational modifications (Butler et al., 2007). However, due to the small specific maximum growth rate of animal cells, it is common to use cell retention systems to maintain higher concentration within the reactor, as shown in continuous cultivations (see Chapter 10).

b. Limiting Substrate Concentration

It has been observed that the specific growth rate depends on the concentration of nutrients and usually one of them, called the limiting substrate (S), is responsible for the control of μ. Generally, at high concentrations of this substrate, μ is maximum and will not increase if the S concentration is increased. However, for lower values S, μ is reduced.

Jacques Monod (1910–1976, French biochemist), comparing this behaviour with the Michaelis–Menten model for enzymatic kinetics (see Chapter 15), proposed a model for cell growth (Monod, 1949), which represents a very significant part of the cultivations and is widely accepted (Equation 9.9 and Figure 9.3):

$$\mu = \mu_{max} \cdot \frac{S}{K_S + S} \tag{9.9}$$

where:

μ = specific growth rate (1/h)
μ_{max} = maximum specific growth rate (1/h)
S = limiting substrate concentration (g/L)
K_S = saturation constant and corresponds to the value of S for $\mu = \mu_{max}/2$ (g/L)

In this way, using the Monod model, with only two parameters to be adjusted, the behaviour of the cultivation can be estimated as the main substrate is being consumed. It is clear that this model represents a considerable simplification of the whole cellular metabolism, since there are thousands of intracellular reactions catalysed by enzymes which often depend on the concentration of this substrate. It would be naive, therefore, to assume that this simple model can represent all that happens within a cell. Even though, it is very common that it adequately represents the growth.

An example of how the specific growth rate can be perceived as the concentration of substrate in the culture medium decreases is shown in Figure 9.4. This data comes from an experiment conducted with *Kluyveromyces marxianus* yeast growing in a defined medium and having glucose as the only carbon source. Optical density (OD) is used to monitor the concentration of biomass. As can be seen from the slope of the curve of ln(OD) in the period from 1.0 to 6.5 h, the specific growth rate is constant and shows a value of μ_{max} of $0.58\,h^{-1}$. From 0.0 to 1.0 h, there is a lag phase (cell adaptation); however, after 6.5 h of cultivation, the specific growth rate decreases as glucose concentration reaches very low values and limits growth, which is in accordance with the Monod model.

On the other hand, when the Monod model does not properly represent cell growth, attempts are made to adjust the experimental data to alternative models. For example, in cultivation where high substrate concentrations inhibit cell growth, the Andrews model (Equation 9.10 and Figure 9.5) can be used:

$$\mu = \mu_{max} \cdot \frac{S}{K_S + S + \frac{S^2}{K_I}} \tag{9.10}$$

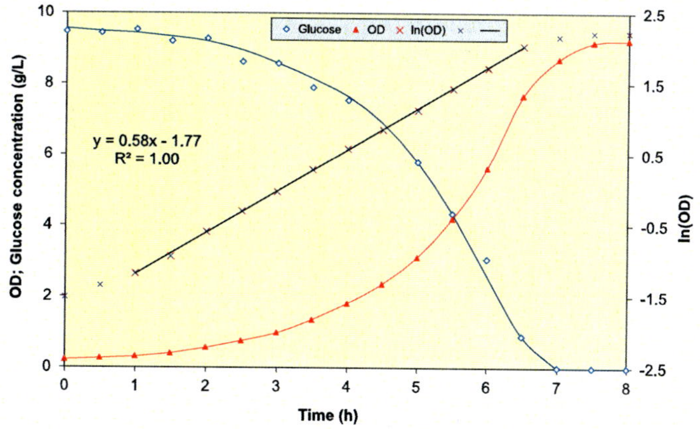

FIGURE 9.4 Experimental results of a yeast *Kluyveromyces marxianus* cultivation in a defined medium with glucose as an only carbon source: S, OD, ln(OD) = $f(t)$.

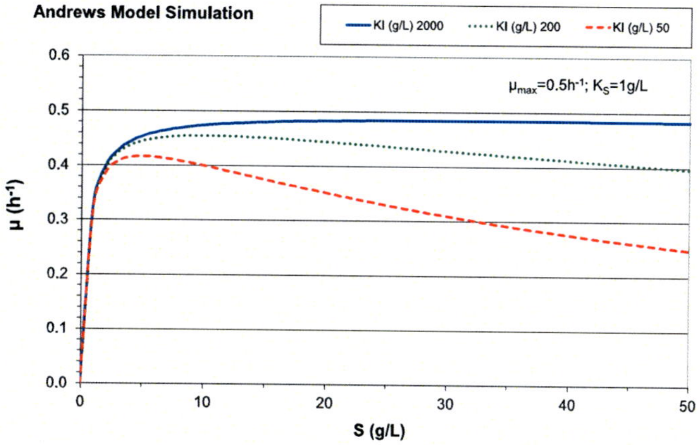

FIGURE 9.5 Representation of Andrews model, with parameters values: $\mu_{max} = 0.5\,h^{-1}$, $K_S = 1$ g/L and different values for $K_I = 50$, 200 and 2,000 g/L.

where:

μ = specific growth rate (1/h)
μ_{max} = maximum specific growth rate (1/h)
S = limiting substrate concentration (g/L)
K_S = saturation constant (g/L)
K_I = Andrews inhibition constant (g/L)

A possible modification to the Monod model is the inclusion of an inhibition factor due to a by-product generated by cell metabolism, such as the production of lactic acid by animal cells during glucose consumption. In this case, a new parameter is established and the equation is multiplied by a factor that is equal to 1 when no lactic acid is produced, but this value diminishes as the lactic acid concentration increases (Equation 9.11).

$$\mu = \mu_{max} \cdot \frac{GLC}{K_{GLC} + GLC} \cdot \frac{K_{LAC}}{K_{LAC} + LAC} \quad (9.11)$$

where:

μ = specific growth rate (1/h)
μ_{max} = maximum specific growth rate (1/h)
GLC = glucose concentration (g/L)
K_{GLC} = glucose saturation constant (g/L)
LAC = lactic acid concentration (g/L)
K_{LAC} = lactic acid inhibition constant (g/L)

Similarly, there are several other alternative models, which can be found in the literature (Butler et al., 2007).

9.4 WHAT DOES *S* VARY WITH? (MODEL FOR SUBSTRATE CONSUMPTION)

We have seen how growth is influenced by the concentration of the limiting substrate, and it is important to understand how substrate consumption occurs over time. Therefore, we will define the specific substrate consumption rate, μ_S, in a similar way to what was done with the growth.

$$\mu_S = \frac{1}{X} \cdot r_S \quad (9.12)$$

where:

μ_S = specific substrate consumption rate (1/h)
X = biomass concentration (g/L)
r_S = substrate consumption rate (g/L/h)

Note that, by convention, r_S is a positive value (how much substrate is being consumed). In the case of batch cultivation (where there is no feed), the value of S tends to fall, so that the variation as a function of time (the derivative of S) is negative.

Thus, it is defined:

$$r_S = -\frac{dS}{dt} \tag{9.13}$$

where:

r_S = substrate consumption rate (g/L/h)
dS/dt = derivative of substrate concentration in relation to time (g/L/h)

As we can see in Figure 9.6 which was generated with the data obtained from Figure 9.4 when substrate is consumed there is cell growth as well. The ratio between growth and consumption is maintained quite constant, a phenomenon that can be seen in the tendency line.

As this constant ratio is very common, a parameter known as $Y_{X/S}$ (cell mass yield coefficient) is usually defined:

$$\left(Y_{X/S}\right)_g = \frac{X_f - X_0}{-\left(S_f - S_0\right)} = \frac{\text{produced biomass}}{\text{consumed substrate}} \tag{9.14}$$

where:

$(Y_{X/S})_g$ = global cell mass yield coefficient
S_f = final substrate concentration (g/L)
S_0 = initial substrate concentration (g/L)
X_f = final biomass concentration (g/L)
X_0 = initial biomass concentration (g/L)

Note that the negative (−) sign in the denominator of the fraction is used to derive positive values in the calculation of $(Y_{X/S})$, because in batch cultivations the substrate concentration in the medium decreases over time and the difference between the final and initial concentrations is negative.

It is also useful to calculate the instantaneous conversion coefficient value from substrate to cell, when taking into account a tiny time interval. In this case, the slope at each pair of points, as in Figure 9.6, can be used, or the ratio between total or specific rates, according to Equation 9.15.

$$Y_{X/S} = \frac{r_X}{r_S} = \frac{\mu}{\mu_S} \tag{9.15}$$

where:

$Y_{X/S}$ = instantaneous cell mass yield coefficient
r_X = growth rate (g/L/h)
r_S = substrate consumption rate (g/L/h)
μ = specific growth rate (1/h)
μ_S = specific substrate consumption rate (1/h)

As already mentioned, the value of $Y_{X/S}$ usually remains approximately constant during the cultivation, so that a mathematical model, representing the substrate consumption, can be obtained, such as:

$$\mu_S = \frac{1}{Y_{X/S}} \cdot \mu \tag{9.16}$$

The model (Equation 9.16) can be interpreted as: all substrate consumption generates, proportionally, new cells. However, Marr et al. (1963) and Schulze and Lipe (1964) observed that in some cases there was substrate consumption without cell growth, or $\mu = 0$, and it was called as specific substrate consumption rate for maintenance (m_S). Later, this finding was immortalized as a Pirt model (1965). Such consumption refers to the generation of energy for the cell to remain viable, unrelated to cell growth. Thus, the Pirt Model is expressed as:

$$\mu_S = m_S + \frac{1}{Y_G} \cdot \mu \tag{9.17}$$

where:

μ_S = specific substrate consumption rate (1/h)
m_S = specific substrate consumption rate for maintenance (1/h)
Y_G = true biomass yield from substrate (which may be slightly different from $Y_{X/S}$
μ = specific growth rate (1/h)

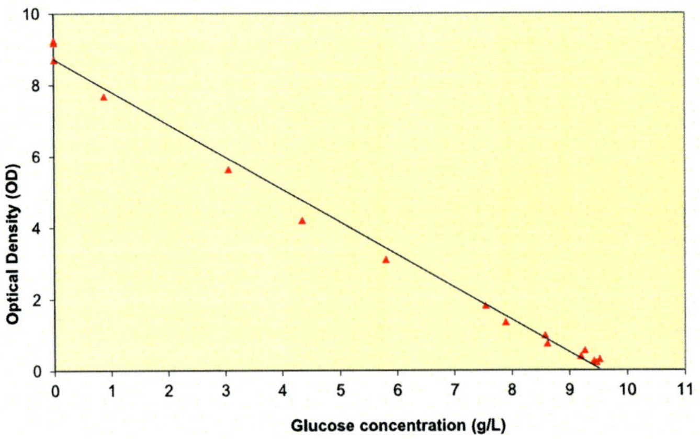

FIGURE 9.6 Experimental results of a yeast *Kluyveromyces marxianus* cultivation in a defined medium with glucose as an only carbon source: Optical Density (OD) = $f(S)$. (Data from Figure 9.4.)

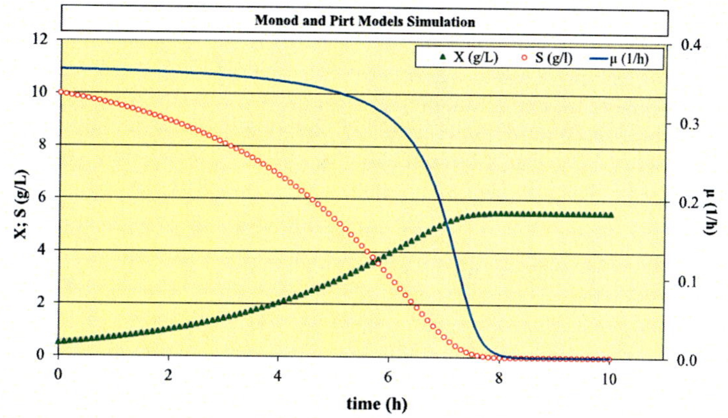

FIGURE 9.7 Mathematical simulation of a cultivation according to Monod (Equation 9.9) and Pirt (Equation 9.17) models, using initial $X = 0.5$ g/L; initial $S = 10$ g/L; $\mu_{max} = 0.4\,h^{-1}$; $K_S = 1$ g/L; $m_S = 0\,h^{-1}$; $(Y_{X/S})_t = 0.5$.

Therefore, the Pirt model (Equation 9.14) represents the substrate consumption by using two parameters, m_S and $(Y_{X/S})_t$, both of which can be calculated by correlating μ and μ_S from different cultivations at distinct times.

With the use of Monod (Equation 9.9) and Pirt (Equation 9.17) models, it is now possible to assign values to the parameters μ_{max}, K_S, m_S e Y_G and set up a simulation, according to Figure 9.7, that would represent a batch cultivation.

9.5 HOW TO CALCULATE THE PRODUCTIVITY OF A PROCESS?

Cell productivity (P_X) in a process is defined as the mass of cells produced per unit of bioreactor volume per unit of time, according to Equation 9.18.

$$Q_X = \frac{X_f - X_0}{t_f} \qquad (9.18)$$

where:

Q_X = cell productivity (g/L/h)
X_f = concentration of cells at the end of the considered time (g/L)
X_0 = initial cell concentration (g/L)
t_f = considered time of the process (h)

It should be noted that the highest productivity is not always achieved by reaching the maximum cell concentration and that the determination of the appropriate time to stop the process depends on maximizing the final cell concentration and/or productivity.

Similarly, the productivity in product (Q_P) of a process is calculated as the mass of product that is produced per unit volume of bioreactor per unit time, as shown in Equation 9.19.

$$Q_P = \frac{P_f - P_0}{t_f} \qquad (9.19)$$

where:

Q_P = productivity in product (g/L/h)
P_f = concentration of product at the end of the considered time (g/L)
P_0 = initial product concentration (g/L)
t = considered time of the process (h)

9.6 HOW DOES THE PRODUCT CONCENTRATION VARY (MODEL FOR PRODUCTION)?

In the production of biopharmaceuticals, there are several types of products, including enzymes, antibiotics and recombinant proteins, which have different units of measurement, namely: enzymatic activity, potency, title and mass concentration. In addition, each process of synthesis is particular to a specific process, which is sometimes coupled with cell growth; however, this is often controlled by an inducer (such as recombinant proteins expressed as *Escherichia coli* or *Pichia pastoris*). In other circumstances, as in the case of antibiotics, the synthesis is performed after growth (during the stationary phase) as they are secondary metabolites (not directly related to the primary metabolism of the cell, which produces energetic compounds for the cell to grow) (for more details, see Chapter 7).

So, there are few models that adequately represent the production processes. We will define the specific product formation rate, μ_P, in a similar way to what has been carried out with growth.

$$\mu_P = \frac{1}{X} \cdot r_P \qquad (9.20)$$

where:

μ_P = specific product formation rate (1/h)
X = biomass concentration (g/L)
r_P = product formation rate (g/L/h), which is the derivative of product concentration as a function of time (dP/dt)

A mathematical model to be used when producing biomolecules would hence be an equation, which defines μ_P as a function of the other variables in the process. Gaden (1959) analysed several bioprocesses and observed that they could be classified, essentially, into three categories: (a) processes whose product synthesis is associated with cell growth, as there is a simple chemical route for the synthesis; (b) processes in which this route is more complex; and (c) processes in which the synthesis is not associated with growth. Intriguingly, in the same year and in the same edition of the scientific *Journal of Biochemical and Microbiological Technology and Engineering*, two researchers (Luedeking and Piret, 1959), who studied the production of lactic acid by *Lactobacillus bulgaricus* at different pH values, came to the conclusion that it is possible to equate the instantaneous production of a product (r_P) by the addition of two parts: one associated with the rate of growth and the other not associated (associated only with cell concentration).

$$r_P = \alpha \cdot r_X + \beta \cdot X \tag{9.21}$$

where:

r_P = product formation rate (g/L/h)
r_X = growth rate (g/L/h)
X = cell concentration (g/L)
α = production constant associated with growth
β = production constant not associated with growth (1/h)

Dividing Equation 9.21 by cell concentration X, the Luedeking and Piret production model (Equation 9.22) is obtained:

$$\mu_P = \alpha \cdot \mu + \beta \tag{9.22}$$

According to the Luedeking and Piret model, if the production kinetics of a product presents a value of α greater than zero and β equal to zero, it can be said that the production is associated with growth. On the other hand, if the value of α is zero and β higher than zero, it can be said that the production is not associated with growth. In an intermediate way, when both α and β are higher than zero, it is said that the production is partially associated with growth (Figure 9.8).

It is clear that this simple model with only two parameters will not be enough to represent the complexity of the synthesis of a bioproduct. However, it is used to classify the production kinetics of the product and determine the bioreactor's operating strategies to maximize production (see Chapter 10).

9.7 FINAL CONSIDERATIONS

Kinetics is an important study of a bioprocess, which allows a more complete understanding of what occurs in a bioreactor. In this chapter, the basic elements for studying the kinetics of cultivation were introduced, such as specific growth, substrate consumption and product formation rates. The importance of analysing specific rates in relation to the simple derivative of the variables was demonstrated. Based on them, simple mathematical models were provided that represent the progress of the biotechnological process and make it feasible to optimize the biopharmaceutical production, by means of simulations of different operational conditions, as well as provide a deeper understanding of what is happening during cell growth.

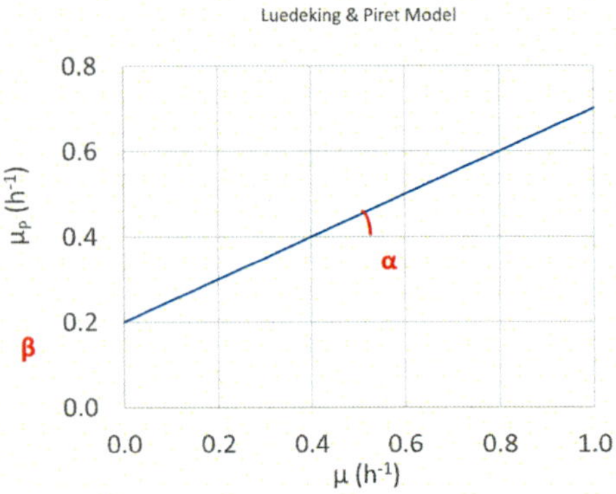

FIGURE 9.8 Representation of Luedeking and Piret model, with parameters values: $\alpha = 0.5$ and $\beta = 0.2\,h^{-1}$.

REFERENCES

Butler, M.; Castilho, L.; Moraes, A.; Augusto, E. *Animal Cell Technology – From Biopharmaceuticals and Gene Therapy*. New York: Taylor & Francis, 2007. ISBN 9780415423045.

Freshney, R.I. *Culture of Animal Cells: A Manual of Basic Technique and Specialized Applications*. 6th edition. New York: Wiley-Blackwell, 2010. ISBN 9780470528129.

Gaden, E.L. Fermentation process kinetics. *Journal of Biochemical and Microbiological Technology and Engineering*, v.1, n.4, 413–429, 1959.

Graham, M.D. The coulter principle: Foundation of an industry. *Journal of the Association for Laboratory Automation*, v.8, n.6, 72–81, 2003.

Kilikian, V.B.; Pessoa Jr, A. *Purification of Biotechnological Products: Operations and Processes with Industrial Application*. 2nd edition. São Paulo: Manole, 2020. ISBN 9788521219460.

Luedeking, R.; Piret, E.L. A kinetic study of the lactic acid fermentation: Batch process at controlled pH. *Journal of Biochemical and Microbiological Technology and Engineering*, v.1, n.4, 393–412, 1959.

Madigan, M.M.; Bender, K.S.; Buckley, D.H.; Sattley, W.M., Stahl, D.A. *Brock Biology of Microorganisms*. 15th edition. Pearson, 2017. ISBN 9780134268668. ISBN-13 9780134261928.

Marr, A.G.; Nilson, E.H..; Clark, D.J. The maintenance requirement of Escherichia coli. *Annals of New York Academy of Sciences*, v.102, p.536–548, 1963.

Monod, J. The growth of bacterial cultures. *Annual Review of Microbiology*, v.3, n.1, 371–394, 1949.

Olsson, L.; Nielsen, J.B. On-line and in situ monitoring of biomass in submerged cultivations. *Trends Biotechnology*, v.15, 517–522, 1997.

Ormerod, M.G. *Flow Cytometry — A Basic Introduction*. 2008. ISBN 978-0-9559812-0-3, disponível em: https://flowbook.denovosoftware.com/. Accessed on 02/18/2021.

Pirt, S.J. The maintenance energy of bacteria in growing cultures. *Proceedings of the Royal Society of London. Series B, Biological Sciences*, v.163, n.991, 224–231, 1965.

Schulze, K.L.; Lipe, R.S. Relationship between substrate concentration, growth rate and respiration rate of Escherichia coli in continuous culture. *Archives Mikrobiology*, v.48, 1–20, 1964.

10 Bioreactors
Modes of Operation

Inês C. Roberto, Adalberto Pessoa Jr., and Aldo Tonso
Universidade de São Paulo

CONTENTS

10.1 INTRODUCTION

Bioreactors can be defined as 'chemical reactors' where a series of reactions, catalysed by biological agents, are processed. Biological agents, also known as biocatalysts, can be basically of two types: (a) enzymes and/or cell organelles; and (b) living cells, which can be of vegetable, animal or microbial origin. However, in the present text, only those used with living organisms will be considered as bioreactors.

The main function of a bioreactor is to provide a controlled environmental condition whereby the biological agent is capable of maximum catalytic activity, by means of both cellular metabolism or a simple enzymatic reaction. To maintain a controlled environment to achieve maximum catalysis, bioreactors are equipped with tools for measuring and monitoring operational variables that affect the bioprocess, which may include temperature, pH, dissolved oxygen, agitation speed, input and output flows of nutrients, inducers, etc.

Multiple stages are required for the completion of a bioprocess, from the selection of the biocatalyst to product recovery (downstream processing). Once the biocatalyst has been selected, for example, microbial and mammalian cells, it is necessary to evaluate cell growth and metabolite production characteristics as a function of the environmental conditions (e.g. composition of the medium, pH and temperature – more details in Chapter 7). In general, at this stage, the cultivation is conducted on a small scale (250–1,000 mL flasks), and the performance of the strain is evaluated according to the knowledge of the process kinetics (volumetric and specific rates of the cell growth, substrate consumption and product formation and bioprocess yields). The next step is to conduct the process in a bench bioreactor (1–5 L), which is considered the first stage of scale up.

In addition to better control of the bioprocess, cultivations in bioreactors provide extra information such as: (a) degree of reactive mixture, which is of paramount importance to achieve uniform nutrient distribution; (b) cell requirements for dissolved oxygen to regulate cell metabolism; and (c) sensitivity of the biocalyser to shear (friction), which can affect cell viability.

In general, bioreactors are designed according to the mode of cultivation (for instance: in depth or on surface) and the configuration of the biocatalysts (free or immobilized). The in-depth cultivation mode, also known as submerged cultivation, is characterized by the presence of free water in the reaction system and often operates with soluble substrates. The surface cultivation mode, also called 'solid state cultivation/fermentation', is characterized by low free water content in the reaction medium and operates with predominantly insoluble substrates, including different types of cereals and agroindustrial residues.

Both modes of cultivation are of technological interest; however, submerged cultivations using the biocatalysts in a free form are the most commonly used on a large industrial scale, especially for the production of chemicals, fuels and biopharmaceuticals. On the other hand, the use of immobilized biocatalysts in the bioreactor provides a high biocatalyst load, which increases the reaction rate and simplifies product recovery. Immobilization involves the retention of cells or enzymes in a specific material (carrier) in order to prevent the biocatalysts from moving freely in the aqueous medium (see Chapter 17 for further details on immobilization).

For submerged cultivation, multiple types of bioreactors are available on the market and at a variety of scales. Some typical configurations are shown in Figure 10.1, and a concise overview of these bioreactors is shown in Table 10.1.

The mechanically stirred tank bioreactor (STR), also known as a mixture bioreactor, is the most commonly used industrial bioreactor due to its flexibility, simplicity of operation, good mixture efficiency and high dispersion capacity of gases (mass transfer). The performance of many bioprocesses depends on the efficiency of agitation/mixing of the reaction medium, and the STR-type bioreactor configuration is by far the most effective to satisfy these requirements. In such bioreactors, in addition to the gas disperser, which is used to distribute the air in the system (for aerobic cultivation), circulation of the fluid is provided by rotating devices (impellers) located along the height of the central axis of the motor. The impellers are designed to disperse gas into the liquid, promote mass and heat transfer and homogenize the suspension. In addition, these bioreactors are equipped with flat plates (baffles) arranged symmetrically in relation to the axis, which re-route the mixture flow to avoid vortex formation during agitation. Vortices generate circular stream lines around the agitation axis and, consequently, reduce the level of mixing and degree of turbulence.

The bioreactors (B), (C) and (D), illustrated in Figure 10.1, are of a pneumatic stirring type, i.e., the power required to achieve the desired degree of mixing, heat and mass transfer is supplied by the kinetic energy of the gas into liquid. For this purpose, pneumatic bioreactors are provided with gas spargers, usually in the form of a 'perforated tube', whose function is not only to supply air to the system (aeration) but also to allow movement of the liquid (mixture). Owing to a lack of mechanical agitation in these reactors, the shear stress is much lower than that in mechanically agitated reactors, which makes them more attractive for the cultivation of shear-sensitive cells (friction), such as mammalian cells.

The air-lift bioreactors, also called 'loop bioreactors', can be designed with internal devices (draft tube – Figure 10.1c) which allow the cyclical movement of the liquid and, as a consequence, generate a more efficient mixture compared to a bubble column (Figure 10.1b). In air-lift bioreactors, the upward flow zone, where the air bubbles are released, is known as the riser, while the downward flow zone is referred to as the downcomer. Important advantages of air lift over STR include lower capital cost and lower risk of contamination. However, this type of bioreactor can be limited in situations that require increased O_2 transfer, since the rate of gas flow is the only significant adjustable parameter during operation.

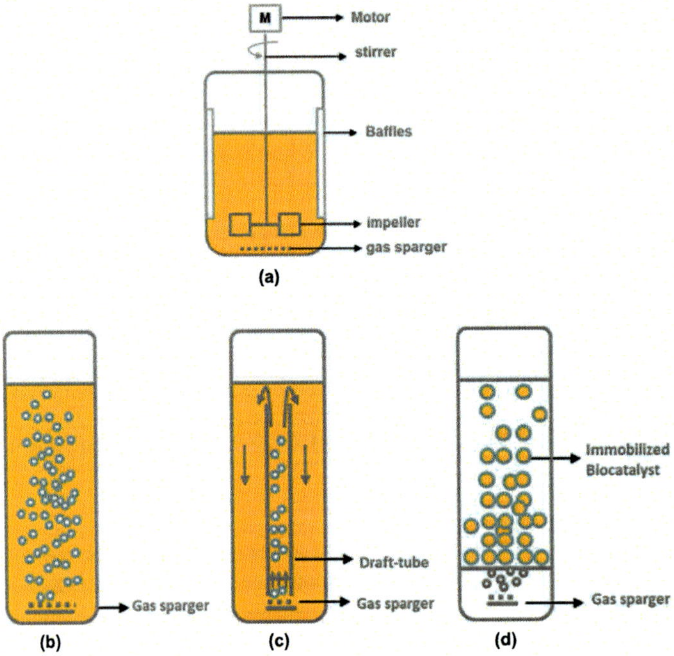

FIGURE 10.1 Typical bioreactor designs for submerged bioprocesses: (a) stirred-tank bioreactor, (b) bubble column bioreactor, (c) air-lift bioreactor, (d) fluidized-bed bioreactor with immobilized biocatalyst.

TABLE 10.1
Summary of the Specifications of Different Bioreactors Used in Submerged Cultivation

Type of Bioreactor	Key Components	Typical Geometry (H_T:D_T)	Hydrodynamic Characteristics
Mechanically stirred tank	Impellers Baffles Sparger	2:1	Good mixing efficiency High gas dispersion capacity High friction
Bubble column	Sparger	3:1–6:1	Low mixing efficiency Low gas dispersion capacity Low friction
Air lift	Sparger Draft tube	3:1–10:1	Good mixing efficiency Good gas dispersion capacity Low friction
Fluidized bed	Sparger	3:1–6:1	Good mixing efficiency Good gas dispersion capacity Low friction

For the fluidized-bed bioreactor, the key feature is to provide intense movement of the particles (immobilized biocatalyst) in the reaction environment. Fluidization is a function of the ratio between the biocatalyst volume and the reactor volume, and can be achieved by (a) injection of atmospheric air or inert gas (N_2 or CO_2) or; (b) recirculation flow of the liquid inside the bioreactor. The advantages of operating this type of bioreactor are the achievement of high productivity in the process, the ease of product separation and, the reuse of the biocatalyst. However, the most significant challenges to be faced are related to mass transfer within the particle and loss of catalytic activity during the immobilization process.

Although there is a huge variety of possible bioreactor settings, it is worth noting that the choice of a bioreactor must consider the morphological characteristics of the cells, the composition of the biocatalyst, the physical properties of the reaction medium and the biochemical characteristics of the process, as illustrated in Figure 10.2.

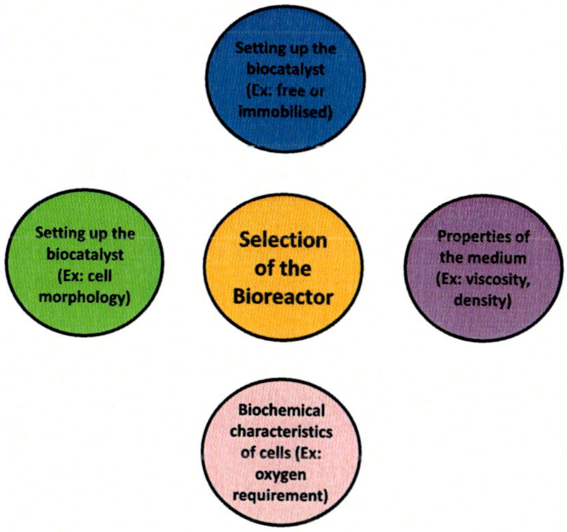

FIGURE 10.2 Main issues to be considered for the selection of a bioreactor.

As well as the best choice of bioreactor, the performance of any bioprocess is dependent on the mode of operation: batch, fed batch or continuous. The general characteristics of each mode of operation, as well as the material balances for each component of the system (substrate, cells and metabolic product) are described below.

10.2 GENERAL CHARACTERISTICS

As previously stated, the performance of a specific bioprocess depends not only on the appropriate choice of the cell the growth medium, the environmental conditions and the type of bioreactor but also on the mode of operation. There are three typical modes of operation of a bioprocess: (a) batch, (b) fed batch and (c) continuous. The difference between these is based on the addition and/or withdrawal of growth medium into the bioreactor. For instance, in a batch process all the substrate is fed at the beginning of the cultivation, whereas in the other processes the substrate is added throughout the cultivation. In the case of the continuous mode, the medium containing product, cell and eventually substrate is also withdrawn. Figure 10.3 illustrates the main operating modes of a bioprocess showing the mass flow (g/h) of substrate input ($F_{in}S_{in}$) and substrate withdrawal ($F_{out}S$) throughout the cultivation.

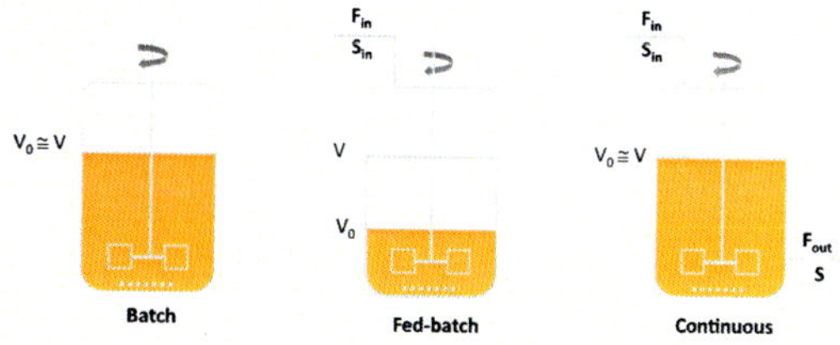

FIGURE 10.3 Illustration of different modes of operation: batch, fed-batch and continuous. F_{in}, volumetric flow rate of inlet feed (L/h); S_{in}, limiting substrate concentration in the inlet feed (g/L); F_{out}, volumetric flow rate in the outlet (L/h); S, limiting substrate concentration in the bioreactor (g/L); V_0, initial working volume (L); V, working volume (L).

10.2.1 Batch Mode

The batch mode process operates as a closed system, in which all nutrients are supplied at once at the beginning of the process, and no product/cell/substrate is removed during the cultivation. The process starts after inoculation of the culture medium with a calculated amount of cells and is terminated after depletion of the limiting substrate in the medium. Throughout the process, the cells consume the limiting substrate progressively to yield new cells (growth) and/or metabolic products. Therefore, the batch process operates in a transition-like mode, which is characterized by variation in cell concentration (X), limiting substrate (S) and metabolic products (P) as a function of time. The standard profile of a batch operation is illustrated in Figure 10.4.

The batch process is seen as quite safe for cultivation that demands a high degree of sterility because at the end of each batch, the bioreactor is washed and sterilized and fresh culture medium is added to receive a new inoculum (in some instances, the fresh medium is sterilized within the bioreactor). As it operates in a closed system, the risk of contamination during the bioprocess is quite low. Compared to the other operating modes, the batch mode is the most simplistic operating system, requiring no input or output media pumps or volume control.

An important aspect to highlight about this mode of operation is constraints imposed on cultivations susceptible to substrate concentration or any other components of the medium. These limitations are usually associated with some biochemical events, such as inhibition by the substrate or any metabolic product and the Crabtree effect, i.e., repression of respiratory flow induced by glucose, even in the presence of oxygen.

Substrate inhibition is associated with both physical (osmotic effects) and physiological (metabolic effects) phenomena. For instance, high substrate concentrations may cause osmotic changes to the cultivation leading to cell lysis, while catabolic repression can inhibit enzyme synthesis relevant for cell growth and/or inhibition of nutrient transport into the cells.

Inhibition of cell metabolism by the product occurs when the product is toxic to the microorganism, whereas the Crabtree effect represses respiratory metabolism of the microorganism owing to excess glucose in the medium, thereby reducing cell yield due to the production of undesirable metabolites. As a typical example, the Crabtree effect can be detected during the baker's yeast production process, when yeasts are grown in batch mode in a medium containing a high concentration of sugar. Excessive sugar in the culture medium, irrespective of sufficient dissolved oxygen for the oxidative metabolism of the cells, leads to the production of ethanol. This phenomenon is observed in most *Saccharomyces* species. A further disadvantage of the batch mode of operation is the long unproductive times between batches due to cleaning of the equipment, loading of the medium, sterilization of the system and unloading of the product, which impact overall productivity and, as a consequence, increase the costs of the process.

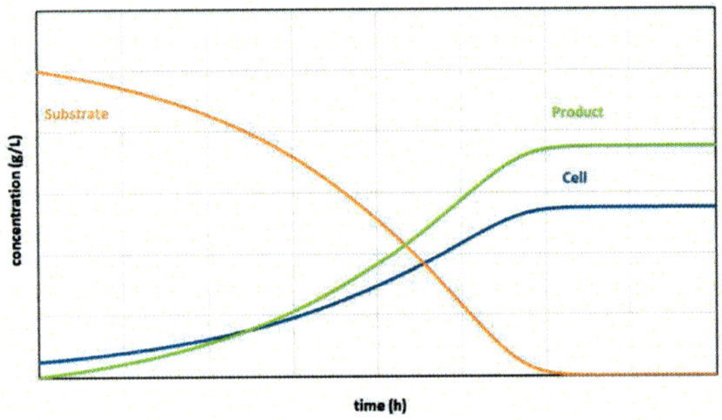

FIGURE 10.4 Typical cell growth curve profile, product formation and limiting substrate consumption of a batch mode operation.

Disadvantages of the batch process can be overcome by operating the bioprocess in other operation modes (fed batch or continuous), which are described below. It is worth mentioning that both the fed-batch and continuous modes start from batch cultivation. The initial batch cultivation will provide useful information about the kinetic behaviour of the microorganism (specific cell growth, limiting substrate consumption and production rates) and its fermentative performance, which are represented by the process parameters like yield factors and volumetric productivities. These parameters are of paramount importance for setting the initial operating conditions in fed-batch and continuous mode. In addition, the batch is the baseline for the comparison between the performances achieved in the different modes of operation. These comparisons enable the evaluation of the bioprocess costs and support the selection of operation mode. Examples of batch mode conducted bioprocesses are shown in Table 10.2.

10.2.2 Fed-Batch Mode

The fed-batch process is characterized by feeding one or more nutrients during cultivation while the products remain in the bioreactor until the process is completed. A typical fed-batch process operates with variable volume, and for this purpose, the process is started with a low initial volume of medium. In this first phase, the process is carried out in batch mode. At a specific moment in the batch, feeding of the bioreactor starts with a fresh culture medium, which can be performed continuously or intermittently. During feeding of the bioreactor, no medium is removed and when the desired volume of the bioreactor is achieved, the feeding is stopped. The fermented medium is withdrawn in order to recover the product and then the process is terminated. The behaviour of the variables (substrate, cells and product) as a function of time will depend on the feeding rate. For example, if the rate of substrate addition during feeding is higher than the rate of substrate consumption by the cells (demand for the substrate), substrate accumulation will occur in the medium and a second batch will be required to reduce the accumulated substrate concentration to a desired level. On the other hand, if the addition rate of the substrate during feeding is the same as the rate of substrate consumption by the cells, no accumulation of substrate will be detected during feeding and the process can be completed without a second batch. Figure 10.5 illustrates typical profiles of a fed-batch operation mode for different medium feeding strategies.

The benefit of operating in fed-batch mode is the possibility of controlling nutrients supplied to the cells, including

TABLE 10.2
Examples of Batch Mode Conducted Bioprocesses

Microorganism	Product	Substrate	Ref
Candida guilliermondii	Xylitol	Rice straw hydrolysate	Silva et al. (2006)
Pichia stipitis	Ethanol	Xylose	Silva et al. (2011)
Pichia pastoris	L-asparaginase	Glycerol	Pillaca-Pullo et al. (2020)
Lactobacillus rhamnosus	Lactic acid	Glucose	Ling et al. (2006)
Streptomyces clavuligerus	Clavulanic acid	Glucose	Marques et al. (2014)

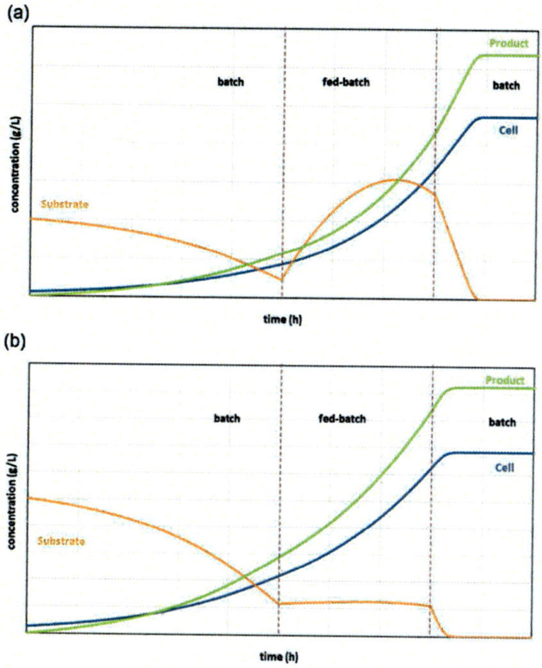

FIGURE 10.5 Typical profiles of a fed-batch fermentation process: (a) substrate feed rate above the rate of substrate consumption; (b) substrate feed rate equal to the rate of substrate consumption.

the sources of carbon and nitrogen, precursors or inducers, amongst others, through changes in the feed rates of the medium. This control enables the cultivation to be carried out with a concentration of substrate, or any other nutrient, that is favourable to the process. In this respect, restrictions imposed by the batch on cultivations sensitive to substrate concentration or any other component of the medium can be overcome by the fed-batch mode (Minihane & Brown, 1986; Lim & Shin, 2013).

A strong advantage of the fed-batch mode over batch mode is the potential for using high cell densities to achieve high concentrations of metabolic products, which makes this mode of operation economically quite attractive for industrial processing of various biomolecules, especially high-value biopharmaceuticals (Subramaniam et al., 2018). Also noteworthy is that the fed-batch mode uses batch kinetic models to set the appropriate feed rate for a specific nutrient in order to achieve the predicted outcomes for the bioprocess. In fact, there are two types of feed strategies for fed-batch mode operations: (a) open-loop system and (b) closed-loop system or 'feed-back control'. In the open-loop system, a pre-set feed flow rate is applied, which may be either constant or variable (linearly increasing, linearly decreasing, exponentially increasing or decreasing, etc.). In this system, the feed flow is calculated according to previous knowledge of the bioprocess. For this purpose, it is required to first of all carry out the bioprocess in a batch mode to estimate the kinetic properties of cell growth and product formation under experimental conditions being studied, and from the data obtained, the desired feed flow can then be set. In the feedback control strategy, changes in the feed flow are carried out throughout the process, thus requiring monitoring and continuous adjustment of one or more variables such as concentration of carbon and nitrogen sources, pH, dissolved oxygen concentration, aeration, agitation and inductor. In general, a feedback control strategy provides the most successful results in comparison to a pre-set feed flow; however, this strategy requires more complex automation systems which may demand more investment in equipment and manpower (Mears et al., 2017). Given the numerous feeding alternatives, each distinct choice will result in a different performance with respect to biomass accumulation and product synthesis.

Some bioproducts derived by fed-batch cultivations include amino acids, antibiotics, enzymes, microbial cells, organic chemicals, polysaccharides, proteins, tissue culture products and various recombinant DNA proteins (Lim & Shin, 2013). Table 10.3 shows some examples of bioprocesses and products from fed-batch cultures applying different feeding strategies.

A process being conducted in fed-batch mode may still operate in production cycles, which is then referred to as 'repeated fed-batch'. In this case, on completion of the process, only part of the fermented medium is withdrawn and a fresh feed stream is then restarted into the bioreactor. This procedure allows the operating time of the bioreactor to be extended, which is economically attractive. The number of cycles is dependent on the production capacity of the cells, which is quantifiable on the basis of bioprocess parameters (yield factors and volumetric productivities).

10.2.3 Continuous Mode

The continuous process runs as if it were an open system in which nutrients are added throughout the cultivation, and at the same time, broth with products are removed from the bioreactor. The key element of the continuous process is permanent operation (steady state), i.e., the concentration

TABLE 10.3
Bioproducts Derived from Microorganisms Cultivated in Fed-Batch Mode with Differentiated Feeding Strategies and in Varying Bioreactor Capacities

Microorganism	Product	Feeding Strategy	Goal	Bioreactor Capacity	Ref
Pediococcus acidilactici	Lactic acid	Constant feeding of glucose	To overcome product inhibition	2 L	Othman et al. (2017)
Streptomyces natalensis	Natamycin	Constant feeding of glucose	To overcome catabolite repression	7.5 L	Elsayed et al. (2019)
Recombinant *Escherichia coli*	Glutathione	Exponential feeding of glucose	To achieve high cell density	2 L	Li et al. (1998)
Bacillus licheniformis	Alkaline protease	Closed-loop control of ammonium, oxygen and glycerol	To increase protease production	4 L	Giesecke et al. (1991)
Corynebacterium glutamicum	L-lysine	Exponential feeding combined with pH-stat	To achieve high cell density with reducing the formation of acetate	2.5 L	Kim et al. (2004)
Escherichia coli	Plasmid DNA	Optimized feeding of glycerol	To achieve high volumetric plasmid titres	2,000 L	Listner et al. (2006)

of cultivation components (cell, limiting substrate and metabolic product) is kept constant over the continuous operation time. The prerequisite for this system to achieve steady state (or permanent regime) is to hold the cultivation volume constant. For this to happen, the inlet (feeding of the nutrient medium) and outlet (removal of the fermented broth) flow rates must be identical.

There are two possible strategies for cultivating microorganisms in continuous steady-state processes: (a) turbidiostat and (b) chemostat. In both cases, a constant volume and uniformed suspension of the cultivation in the bioreactor must be maintained.

In chemostats, the specific cell growth rate is controlled by the substrate addition rate, which is pre-set (external control). In the turbidiostat, the rate of substrate addition is automatically adjusted to a pre-set cell concentration, with culture turbidity continuously being measured (internal control), which requires more complex control systems.

A process in continuous mode also starts with the cultivation of the cells in batch mode. After a certain period of batching time, feeding of the bioreactor with fresh culture medium at a given flow rate begins, and with the same feeding rate, the fermented broth is withdrawn. In practice, a steady state often requires some time to be reached. The time taken to achieve a steady state depends on when the feeding starts and on the feeding flow used. Figure 10.6 illustrates the typical profile of a continuous mode operation.

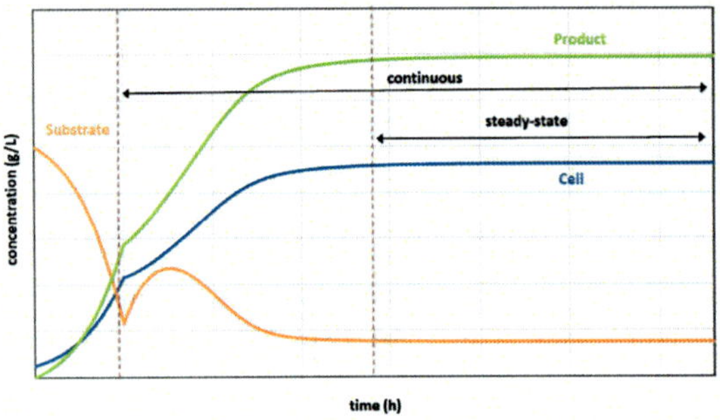

FIGURE 10.6 A typical profile of a continuous mode operation.

The possibility of keeping cultivation in a steady-state for prolonged periods of time provides some advantages to the continuous process, such as: (a) increasing volumetric productivity of the bioprocess, due to reduction in unproductive times characteristic of the batch mode, and because the bioreactor remains for most of the time with a high concentration of cells producing the biomolecule of interest in higher concentration; (b) producing a uniform fermented broth, which simplifies the design of product recovery operations; and (c) maintaining cells in a constant physiological state, which makes the continuous process as an excellent tool to study metabolic regulation mechanisms for optimizing nutritional and environmental conditions of the bioprocess and for the production of biopharmaceuticals that undergo post-translational modifications that require physiological state of the cells to remain constant.

Long-operating periods of time in continuous mode process are the cause of many problems, including increased potential for contamination of the bioprocess and possibility of spontaneous mutations that may result in the selection of less productive cells. Within this context, continuous cultivation is not widely used in the biopharmaceutical industry, although it is widely used for less aseptic processes like alcoholic fermentation, single-cell protein production and biological treatment of industrial effluents.

The continuous process can further operate with cell recycling, consisting of returning a fraction of the fermented medium containing the biomass to the bioreactor. This is especially useful when cells have a low specific maximum growth rate, limiting the maximum feed flow (as will be seen below). Maintaining a high biomass concentration in the bioreactor allows increased productivity of the bioprocess. Different cell retention mechanisms (internal and external) have been developed to achieve this biomass recycling.

Some examples of continuously conducted bioprocesses are shown in Table 10.4.

In view of the different alternatives for operating a bioprocess that are available, each mode of operation leads to distinct performances. For a better understanding of the bioprocess in each mode of operation, material balances must be described. Mass balance equations for cells, limiting

TABLE 10.4
Examples of Continuously Conducted Bioprocesses

Microrganism	Product	Substrate	Ref
Erwinia aroideae	L-asparaginase	Yeast extract	Liu and Zajic (1973)
Kluyveromyces fragilis	Single cell protein	Cheese whey	Ghaly et al. (2005)
Saccharomyces cerevisiae	Ethanol	Molasses	Perego et al. (1985)
Lactobacillus helveticus	Lactic acid	Sweet whey	Gätje and Gottschalk (1991)
Streptomyces clavuligerus	Clavulanic acid	Glucose	Marques et al. (2014)

substrate and metabolic product in the three modes of operation are outlined in the following topics. Mathematical simulations of the process can predict the behaviour of complex systems. Mass balances help to gather information about the rate at which a specific component accumulates, which concentration a particular product can be achieved and how long a conversion takes. Basic mathematical models relevant to understand balances are described in Chapter 9.

10.3 MASS BALANCES FOR DIFFERENT MODES OF OPERATION

Material balances are fundamental tools for evaluating any bioprocess, enabling monitoring of mass transformations that occur in bioreactors as a result of input and output flows of matter. In this sense, the operating modes should be treated as a system in which the basic element of the process is the equipment (bioreactor) where biosynthetic reactions occur, and where process connections are input and/or output mass flows of a specific component. To describe mass balance in a given system, the first step is to define the system boundary, also referred to as control volume (CV), as illustrated in Figure 10.7.

All bioreaction systems are governed by the same basic principles of nature – the conservation of mass and energy – which are mathematically translated by the balance between mass and energy, respectively. In this section, we will cover mass balances and their respective operating regimes:

a. *Closed system*: no mass entry and exit across the boundary of the system. The closed system is transient, i.e., variable values in the process vary over time (batch processes).
b. *Open system*: mass input and output across the system boundary. In the open system, a steady state in permanent regime can be achieved in which all process variables (temperature, concentration, flow, etc.) do not vary with time (continuous processes).
c. *Semi-open system*: entry, but no mass exit across the system boundary. The semi-open system, as well as the open system, runs on a transient regime (fed-batch processes).

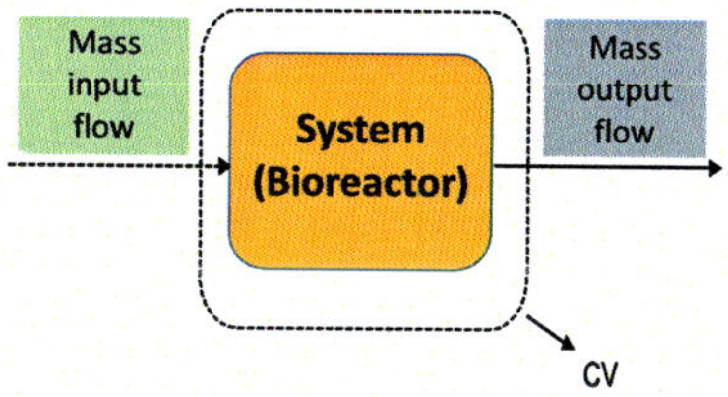

FIGURE 10.7 A typical system bounded by control volume with the input and output streams for mass balance.

The mass balance of the system, surrounded by a control volume, must take into account the input and output flow of each individual component, as well as the generation and/or consumption of these components as a result of the reactions inside the bioreactor. A general mass balance can be expressed as follows (Equation 10.1).

$$\begin{bmatrix}\text{Accumulation}\\ \text{rate}\\ \text{of component c}\end{bmatrix} = \underbrace{\begin{bmatrix}\text{Input rate}\\ \text{of component}\\ \text{c}\end{bmatrix} - \begin{bmatrix}\text{Output rate}\\ \text{of component}\\ \text{c}\end{bmatrix}}_{\text{Flows}} + \underbrace{\begin{bmatrix}\text{Generation}\\ \text{rate}\\ \text{of c}\end{bmatrix} - \begin{bmatrix}\text{Consumption}\\ \text{rate}\\ \text{of c}\end{bmatrix}}_{\text{Reaction}} \quad (10.1)$$

The nomenclatures of the parameters and the mathematical expressions used in the mass balances are presented in Table 9.1.

10.4 BATCH OPERATION

As previously described, a batch process operates as a closed system in which all nutrients are fed at once at the beginning of the process, and no product is withdrawn during the cultivation. Consequently, there is no mass input or output of any component across the boundary of the system (Figure 10.8).

To apply the general mass balance equation (Equation 10.1) in the system, illustrated in Figure 10.8, the following hypothesis will be considered: (a) the bioreactor operates as a perfect mixture, i.e., samples taken from any parts of the bioreactor share the identical composition; (b) the reaction volume is held constant – this assumes no significant addition of base or acid for pH control, no significant losses of liquid by evaporation and/or sample collection; and (c) cells are homogeneous, i.e., cells have identical size and chemical composition.

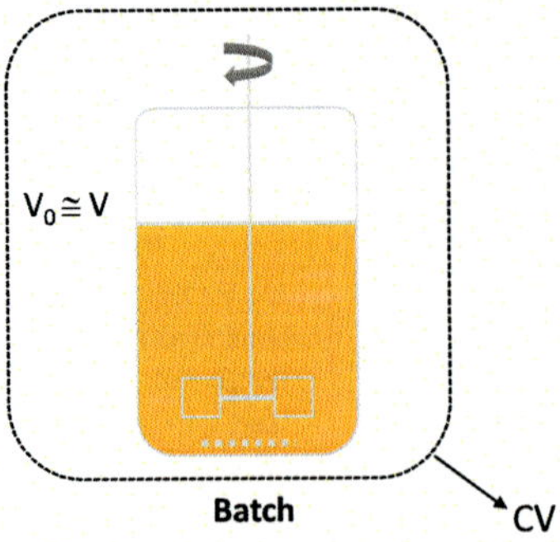

FIGURE 10.8 A batch process bounded by control volume.

10.4.1 Mass Balance for the Cells in the Batch Mode

From Equation 10.1, the cell mass balance can be written as presented in Equations 10.2 and 10.3. Bearing in mind that all terms in Equation 10.3 represent mass variations, the units are expressed in g/h.

$$\begin{bmatrix}\text{Accumulation}\\ \text{rate of}\\ \text{cells in bioreactor}\end{bmatrix} = \begin{bmatrix}\text{Input rate}\\ \text{of cells}\end{bmatrix} - \begin{bmatrix}\text{Output rate}\\ \text{of cells}\end{bmatrix} + \begin{bmatrix}\text{Generation}\\ \text{rate of cells}\\ (\text{growth})\end{bmatrix} - \begin{bmatrix}\text{Consumption}\\ \text{rate of cells}\\ (\text{death})\end{bmatrix} \quad (10.2)$$

$$\frac{d(XV)}{dt} = F_{in}X_{in} - F_{out}X + Vr_x - Vr_d \quad (10.3)$$

Since in batch operation mode there is no input or output of cells, it can be assumed that $F_{in}X_{in} = F_{out}X = 0$. Thus, using the rate concept shown in Table 9.1, the Equation 10.4 is obtained.

$$\frac{d(XV)}{dt} = V\mu X - Vk_dX \quad (10.4)$$

Splitting the terms of the derivative $(d(XV)/dt)$ and put the variables (VX) in evidence, one has Equation 10.5.

$$V\frac{dX}{dt} + X\frac{dV}{dt} = V(\mu - k_d)X \quad (10.5)$$

Assuming that the volume in the bioreactor is constant, the term $(dV/dt) = 0$; then Equation 10.6 is derived.

$$V\frac{dX}{dt} = V(\mu - k_d)X \quad (10.6)$$

Then, dividing the two members of Equation 10.6 by the volume of the medium (V), Equation 10.7 is obtained.

$$\frac{dX}{dt} = (\mu - k_d)X \quad (10.7)$$

Assuming that the rate of cell death k_d is negligible, Equation 10.8 – cell balance in the batch mode – is attained. Therefore, Equation 10.8 describes the overall cell balance in a batch mode cultivation. Based on this, multiple analyses can be made assuming simplifying hypotheses.

$$\frac{dX}{dt} = \mu \cdot X \quad (10.8)$$

Hypothesis 10.1

Maintain μ constant until completion of cultivation.

Assuming that (a) the process is started with cells in the exponential phase, i.e., $\mu = \mu_{max}$, and (b) this cell growth specific rate (μ_{max}) is maintained constant for most of the batch time, then Equation 10.8 can be integrated. The outcome is the relationship between batch time (t_B) and cell concentration (X). Hence, for the initial condition of $X = X_0$ at $t = 0$, and for the final condition of $X = X_f$ at $t = t_B$, Equation 10.9 is obtained.

$$\int_{X_0}^{X_f} \frac{dX}{X} = \mu_{max}\int_0^{t_B} dt \quad (10.9)$$

The integration of Equation 10.9 results in Equation 10.10, and consequently, the batch mode time is expressed by Equation 10.11.

$$\ln\frac{X_f}{X_0} = \mu_{max}t_B \quad (10.10)$$

$$t_B = \frac{1}{\mu_{max}}\ln\frac{X_f}{X_0} \quad (10.11)$$

The correlation between the final concentration of cells (X_f) as a function of batch time (t_B) can even be obtained (Equations 10.12–10.14).

$$\ln\frac{X_f}{X_0} = \mu_{max}t_B \quad (10.12)$$

$$\frac{X_f}{X_0} = e^{\mu_{max}t_B} \quad (10.13)$$

$$X_f = X_0e^{\mu_{max}t_B} \quad (10.14)$$

Since the operating costs of a batch process are impacted by the time of processing, the shorter the time it takes to achieve the target concentration of cells and/or product with high substrate conversion, the cheaper the bioprocess will be.

10.4.2 Mass Balance for the Limiting Substrate in the Batch

In order to assess this type of mass balance, the following remarks have to be addressed: (a) no substrate production during the batch; and (b) no substrate input or output flow into the bioreactor while the process is in progress. Then, based on the general Equation 10.1, the mass balance for the limiting substrate (S) can be written as shown in Equations 10.15 and 10.16.

$$\begin{bmatrix}\text{Accumulation}\\ \text{rate of substrate in the}\\ \text{bioreactor}\end{bmatrix} = -\begin{bmatrix}\text{Consumption}\\ \text{rate of substrate}\end{bmatrix} \quad (10.15)$$

$$\frac{d(SV)}{dt} = -\ Vr_S \quad (10.16)$$

Splitting the terms from the derivative, Equation 10.17 is obtained.

$$V\frac{dS}{dt} + S\frac{dV}{dt} = -\ Vr_S \quad (10.17)$$

However, assuming no variation in volume in the bioreactor, the term (dV/dt) = 0, then Equation 10.18 is obtained.

$$V\frac{dS}{dt} = -Vr_S \quad (10.18)$$

Dividing the terms of Equation 10.18 by the volume of the medium (V) and replacing r_S by $\mu_S X$, one gets the substrate balance in batch mode, as represented by Equation 10.19.

$$\frac{dS}{dt} = -\ \mu_S \cdot X \quad (10.19)$$

Hypothesis 10.2

Assume that the Pirt model is valid (see Chapter 9).

The substrate can be consumed for cell growth (generation of new cells), cell maintenance (replacement of cell components; osmoregulation, etc.) and formation of metabolic products (primary or secondary). However, growth is the cellular function that consumes the most amount of energy (ATP). Below the mass balance for the substrate is presented, which considers the Pirt model as valid. In this context, the rate of substrate consumption is a function of the growth (r_{SG}) and the maintenance rates r_{Sm}.

Based on the Pirt model, the Equation 10.19 can be written as Equation 10.20.

$$\frac{dS}{dt} = -\left(\frac{\mu}{Y_G} + m_S\right)X \quad (10.20)$$

Hypothesis 10.3

Considering that μ is constant until the cultivation is complete.

Based on the same hypothesis adopted for batch mode cell mass balance, the process is initiated with the cells in the exponential phase ($\mu = \mu_{max}$) and the maximum cell growth rate is maintained for most of the batch cultivation time. In such a case, the variable (X) can be replaced by Equation 10.14, described in the mass balance for the cells, to get Equation 10.21 further integrated at the initial condition of $S = S_0$ and $t = t_0$, and at the final condition in which $S = S_f$ and $t = t_B$, in order to obtain Equation 10.22. As long as the parameters remain constant, the time of batch mode cultivation can be expressed by Equation 10.23.

$$dS = -\left(\frac{\mu_{max}}{Y_G} + m_S\right)X_0 e^{\mu_{max} t_B} dt \quad (10.21)$$

$$\int_{S_0}^{S_f} dS = -\left(\frac{\mu_{max}}{Y_G} + m_S\right)X_0 \int_{t_0}^{t_B} e^{\mu_{max} t_B} dt \quad (10.22)$$

$$t_B = \frac{1}{\mu_{max}} \ln\left[1 + \frac{S_0 - S_f}{\left(\frac{1}{Y_G} + \frac{m_S}{\mu_{max}}\right)X_0}\right] \quad (10.23)$$

If the amount of energy consumed for the cell maintenance is negligible, the real mass growth yield (Y_G) is equal to the biomass yield ($Y_{x/s}$); therefore, under such conditions, Equation 10.23 can be written as Equation 10.24.

$$t_B = \frac{1}{\mu_{max}} \ln\left[1 + \frac{Y_{x/s}}{X_0}\left(S_0 - S_f\right)\right] \quad (10.24)$$

10.4.3 Mass Balance for the Product in Batch Mode

To evaluate the mass balance of a product, the following considerations shall be assumed: (a) no inflow or outflow of product during the process ($F_{in}P_{in} = F_{out}P_{out} = 0$); and (b) no product degradation during the batch. Based on Equation 10.1, the mass balance for the product (P) can be written as Equation 10.25.

$$\begin{bmatrix}\text{Accumulation}\\ \text{product rate}\\ \text{in the bioreactor}\end{bmatrix} = \begin{bmatrix}\text{Generation}\\ \text{rate}\\ \text{of product}\end{bmatrix} \quad (10.25)$$

Like the mass balance for the substrate (see Equation 10.16), the mass balance for the product can be expressed as in Equation 10.26.

$$\frac{d(PV)}{dt} = Vr_p \quad (10.26)$$

Splitting the terms of the derivative and using the expressions outlined in Table 9.1, the mass balance equation for the product can be described in Equation 10.27.

$$V\frac{dP}{dt} + P\frac{dV}{dt} = V\left(\mu_P X\right) \quad (10.27)$$

Making the assumption that the volume is constant and the term (d*V*/d*t*) = 0, then Equation 10.28 is obtained.

$$V\frac{dP}{dt} = V(\mu_P X) \tag{10.28}$$

Dividing the terms of Equation 10.28 by the medium volume (*V*) and putting (*X*) in evidence, Equation 10.29, which refers to the product balance in the batch mode, is obtained.

$$\frac{dP}{dt} = \mu_P \cdot X \tag{10.29}$$

Hypothesis 10.4

Considering that μ is maintained constant until the end of the cultivation.

Taking into account the same hypothesis adopted in the mass balance for cells and substrate, it can be assumed that the process was initiated with cells in the exponential phase ($\mu = \mu_{max}$) and that this rate is maintained during most of the time of batch cultivation mode. In this situation, the variable (*X*) can be replaced by Equation 10.14, described in the mass balance for the cells. Thus, Equation 10.30 describes the mass balance for the product.

$$\frac{dP}{dt} = \mu_P \left(X_0 e^{\mu_{max} t}\right) \tag{10.30}$$

In the case that the specific rate of product formation remains constant between the batches, Equation 10.30 can be integrated considering the initial condition of $P = P_0$ and $t = t_0$, and thus the batch time (t_B) can be described as a function of the final concentration of product, P_f, according to Equation 10.31. As a conclusion to this topic, Table 10.5 summarizes the mass balance equations for the batch mode operation

$$t_B = \frac{1}{\mu_{max}} \ln\left[1 + \frac{\mu_{max}}{X_0 \cdot \mu_P}\left(P_f - P_0\right)\right] \tag{10.31}$$

10.5 FED-BATCH OPERATION

As explained in Section 10.2.2, a fed-batch operation mode is characterized by feeding of one or more nutrients during the process, without withdrawal of products during the cultivation, and by numerous feeding possibilities. In this section, a general balance and some simplifying hypotheses are presented to allow specific evaluations, such as fed-batch process with variable volume and constant feed flow. A schematic display of a fed-batch operation in which the bioreactor is supplied with a limiting substrate S_{in} (g/L) at a specific volume flow rate F_{in} (L/h) is shown in Figure 10.9.

TABLE 10.5
Mass Balance Equations for Cells, Limiting Substrate and Metabolic Product during Batch Cultivation

Components	Mass Balance Equations	Equation
Cells	$\frac{dX}{dt} = \mu \cdot X$	(10.8)
Limiting substrate	$\frac{dS}{dt} = -\mu_S \cdot X$	(10.19)
Metabolic product	$\frac{dP}{dt} = \mu_P \cdot X$	(10.29)

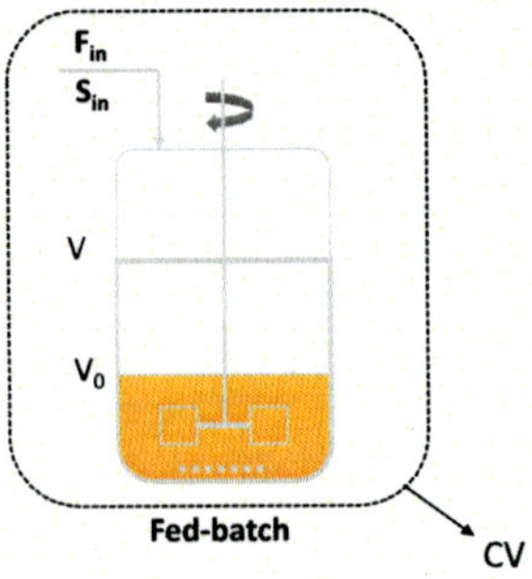

FIGURE 10.9 A fed-batch process bordered by control volume.

A fed-batch cultivation begins in batch mode where the volume is less than half of the working capacity of the bioreactor. Feeding with a fresh culture medium marks the beginning of the fed-batch mode. As a result, the concentrations of the variables (cells, limiting substrate and product) achieved in the batch will be taken as the starting point of the fed-batch.

10.5.1 Mass Balance for Cells in Fed-Batch Mode

Taking into account that there is no addition or withdrawal of cells during the feeding, the mass balance for the cells can be written as in Equation 10.32.

$$\frac{d(XV)}{dt} = Vr_x - Vr_d \tag{10.32}$$

Equation 10.32 shows that cell mass variation during feeding of the bioreactor is the difference between cells growing (Vr_x) and cells dying (Vr_d). Then, by splitting the variables of the derivative and substituting the terms related to growth and cell death, Equation 10.33 is obtained.

$$V\frac{dX}{dt} + X\frac{dV}{dt} = V(\mu - k_d)X \tag{10.33}$$

The division of all terms presented in Equation 10.33 by the volume of medium (*V*) results in Equation 10.34.

$$\frac{dX}{dt} + \frac{X}{V}\frac{dV}{dt} = (\mu - k_d)X \tag{10.34}$$

As the volume of medium in the bioreactor varies over time, the term (d*V*/d*t*) ≠ 0; and according to Figure 10.9, the term (d*V*/d*t*) = F_{in}, then Equation 10.35 is obtained.

$$\frac{dX}{dt} + X\frac{F_{in}}{V} = (\mu - k_d)X \tag{10.35}$$

Rearranging Equation 10.35, Equation 10.36 is obtained.

$$\frac{dX}{dt} = \left(\mu - k_d - \frac{F_{in}}{V}\right)X \qquad (10.36)$$

When dividing the flow rate F_{in} (L/h) by the volume of medium V (L), the dilution rate is obtained, whose unit is given in h^{-1}. This parameter is also referred to as the specific feed rate, and by convention, it is represented by the letter D. Note that when operating a fed-batch with constant feed flow and with varying volume, parameter D decreases during the bioreactor feed because of an increase in volume of the medium.

By replacing the F_{in}/V ratio by D and assuming that the death rate k_d is negligible, Equation 10.36 can be simplified to obtain Equation 10.37.

$$\frac{dX}{dt} = \mu X - DX \qquad (10.37)$$

Therefore, the cell balance in fed-batch mode is represented by Equation 10.38.

$$\frac{dX}{dt} = (\mu - D)\cdot X \qquad (10.38)$$

It is worth noting that in a fed-batch culture with no cell feeding (the most common case), the substrate feeding increases the cultivation volume. It is also important to state that Equation 10.38 differs from Equation 10.7 (batch balance) as the cell concentration in the fed-batch decreases with the addition of medium, causing dilution of the culture medium.

Hypothesis 10.5

Consider a constant inlet flow (F_{in}).

By examining Equation 10.38, it is possible to verify that for an increase in cell concentration to occur over the time of medium feeding (i.e., $dX/dt > 0$), the cell growth, represented by the term (μX), must be higher than the dilution rate of the medium, which is represented by the term (DX). On the other hand, if the dilution rate of the medium (DX) is higher than the growth rate (μX), the cell concentration can decrease over time, i.e., ($dX/dt < 0$). However, when there is no variation in cell concentration as a function of time, i.e., for $dX/dt = 0$, an equilibrium situation – also called a quasi-steady state – is encountered. In this condition, the specific cell growth rate (μ) is numerically equal to the dilution rate (D). As the process operates with variable volume, the value of μ decreases with the feeding time, as shown below.

Supposing that for a given constant flow rate F_{in}, the volume of the medium during feeding varies from V_0 to V. Then, for a constant flow rate, the variation in volume with time in the fed-batch is F_{in}, i.e., the term (dV/dt) = F_{in}. Thus, Equation 10.39 can be obtained.

$$F_{in} = \frac{dV}{dt} \qquad (10.39)$$

By separating the variables, one gets Equation 10.40.

$$dV = F_{in}dt \qquad (10.40)$$

As the feed flow, F_{in}, is constant, Equation 10.40 can be integrated with the initial condition of $V = V_0$ at $t = t_0$. Thus, during feeding Equation 10.41 can be written and after its integration Equation 10.42 is obtained.

$$\int_{V_0}^{V} dV = F_{in}\int_{t_0}^{t} dt \qquad (10.41)$$

$$V - V_0 = F_{in}(t - t_0) \qquad (10.42)$$

For the initial fed-batch time, one can adopt $t_0 = 0$, so Equation 10.42 can be written as Equation 10.43 or 10.44. Therefore, from Equation 10.44, the bioreactor filling time is calculated for a constant feed flow rate, F_{in}.

$$V = V_0 + F_{in}t \qquad (10.43)$$

$$t_f = \frac{V_f - V_0}{F_{in}} \qquad (10.44)$$

Hypothesis 10.6

Considering that the fed-batch mode is in a quasi-steady state for X.

For a quasi-steady state $dX/dt \approx 0$ and $\mu \approx D$, Equation 10.45 can be written to show that μ is inversely proportional to the time at which the bioreactor is filled and decreases as a function of the fed-batch operating time.

$$\mu = \frac{F_{in}}{V_0 + F_{in}t} \qquad (10.45)$$

It is also important to note that, even in the quasi-steady state condition, the total cell mass (x) increases, since it is a result of the multiplication between cell concentration (X) and volume of medium (V), i.e., $x = XV$.

10.5.2 Mass Balance for the Limiting Substrate in Fed-Batch Mode

Since the fed-batch mode is characterized by the feeding of a limiting substrate during cultivation, the mass balance for the limiting substrate can be written as presented in Equation 10.46, which shows that the mass variation of substrate is given by the difference between the substrate added during the feeding and its consumption by the cells.

$$\frac{d(SV)}{dt} = F_{in}S_{in} - Vr_s \qquad (10.46)$$

Separating the variables from the derivative and substituting the term related to the consumption of the limiting substrate, Equation 10.47 is obtained.

$$V\frac{dS}{dt} + S\frac{dV}{dt} = F_{in}S_{in} - V\mu_S X \qquad (10.47)$$

As the variation in volume is equal to the feed flow of the medium, i.e., $dV/dt = F_{in}$, Equation 10.47 can be written as Equation 10.48, and dividing all the terms by the volume (V), Equation 10.49 is obtained.

$$V\frac{dS}{dt} = F_{in}S_{in} - F_{in}S - V\mu_S X \quad (10.48)$$

$$\frac{dS}{dt} = \frac{F_{in}}{V}(S_{in} - S) - \mu_S X \quad (10.49)$$

By substituting the F_{in}/V ratio by D, the substrate balance is derived for the fed-batch mode (Equation 10.50).

$$\frac{dS}{dt} = D(S_{in} - S) - \mu_S X \quad (10.50)$$

Equation 10.50 differs from Equation 10.19 (batch balance), because the substrate concentration in the fed-batch is enriched by the difference in substrate concentration between the feed stream and that present in the tank. In case the feed stream concentration is equal to the tank ($S_{in} = S$), the equation of dS/dt is equal to the batch.

Hypothesis 10.7

Taking into account that m_S is negligible

As previously discussed in Chapter 9, Equation 9.17 (the Pirt model), in addition to cell growth, the substrate can be consumed for cell maintenance (m_S). However, if the limiting substrate is negligible for cell maintenance, Equation 10.50 can be reduced to Equation 10.51.

$$\frac{dS}{dt} = D(S_{in} - S) - \frac{\mu X}{Y_{x/s}} \quad (10.51)$$

Hypothesis 10.8

Consider that m_S, S, dS/dt, dX/dt are negligible.

As can be seen, Equation 10.51 illustrates how the substrate-limiting concentration (dS) varies as a function of the cultivation feeding time (dt). However, if the cultivation volume varies over time, then the D parameter is also variable and the integration of Equation 10.51 is more complex. In order to simplify this description, the same hypothesis used in the mass balance for cells is considered, i.e., an equilibrium situation (quasi-steady state). In this scenario, the substrate variation as a function of time should be null and void ($dS/dt \approx 0$). In addition, it is understood that at the start of the feeding, the cell concentration is so high that any substrate fed into the reactor is immediately consumed, allowing to state that $S \ll S_{in}$. Furthermore, it is possible to assume that the cell concentration remains constant during cultivation, i.e., $dX/dt \approx 0$, which leads to $\mu \approx D$. When applying these hypotheses, Equation 10.51 can be written as Equation 10.52. Hence, in a quasi-stationary state and operating with high cell densities – as in the case of recombinant *E. coli* cultivation for the production of biopharmaceuticals – Equation 10.53 can be written and cell concentration can be represented by Equation 10.54.

$$0 = DS_{in} - \frac{\mu X}{Y_{x/s}} \quad (10.52)$$

$$DS_{in} = \frac{\mu X}{Y_{x/s}} \quad (10.53)$$

$$X = \frac{DS_{in}Y_{x/s}}{\mu} \quad (10.54)$$

Despite having $dX/dt \approx 0$, the bioreactor volume increases over time and therefore the cell mass also increases. To express Equation 10.54 in terms of mass, the parameter D is replaced by F_{in}/V, and the two terms of the equation are multiplied by the volume (V). Thus, Equation 10.55 is obtained.

$$\mu XV = F_{in}S_{in}Y_{x/s} \quad (10.55)$$

As $(XV) = x$, Equation 10.56 can be applied.

$$x = \frac{F_{in}S_{in}Y_{x/s}}{\mu} \quad (10.56)$$

Equation 10.56 clearly shows that in the quasi-steady state, where $dX/dt \approx 0$; the specific growth rate (μ) decreases with the increase in cell mass (x).

With the understanding that $\mu = \frac{1}{x}\frac{dx}{dt}$, Equation 10.57 is obtained. Consequently, the variation in cell mass over time is given by Equation 10.58.

$$\frac{1}{x}\frac{dx}{dt}x = F_{in}S_{in}Y_{x/s} \quad (10.57)$$

$$\frac{dx}{dt} = F_{in}S_{in}Y_{x/s} \quad (10.58)$$

Equation 10.58 can be integrated within the initial condition of $x = x_0$ at $t = 0$ (start of feeding), and thereby Equation 10.59 is obtained.

$$\int_{x_0}^{x} dx = S_{in}Y_{x/s}F_{in}\int_{0}^{t} dt \quad (10.59)$$

By integrating Equation 10.59, the performance of cell mass as a function of the bioreactor feeding time (Equation 10.60) can be obtained.

$$x = x_0 + S_{in}Y_{x/s}F_{in}t \quad (10.60)$$

Equation 10.60 shows that, for the previously assumed hypotheses and at constant $Y_{x/s}$ and F_{in}, the total cell mass increases linearly with the feeding time of the bioreactor, as illustrated in Figure 10.10. This pattern will be upheld as long as the conditions initially imposed are maintained during the feeding of the bioreactor.

It is further possible to determine the cultivation time in the fed-batch mode (t_{FB}) by integrating Equation 10.60 under the initial condition of $x = x_0$ at $t = 0$ (start feeding) and the

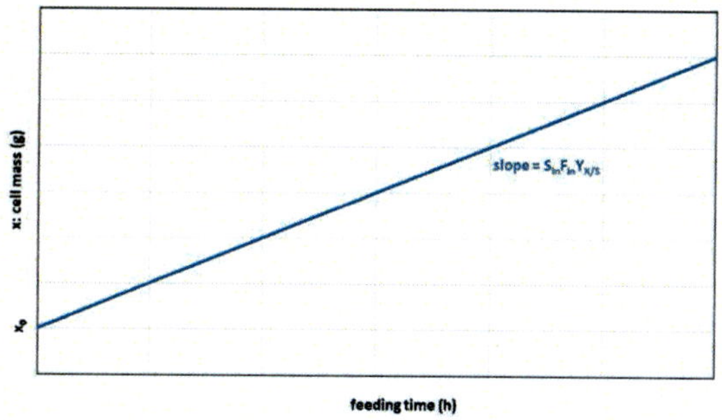

FIGURE 10.10 Cell mass profile during fed-batch cultivation with constant feed flow, according to hypotheses assumed for obtaining the quasi-steady state ($dS/dt \approx 0$ and $\mu \approx D$).

final condition with $x = x_f$ at $t = t_{FB}$. Hence, Equation 10.61 is obtained.

$$\int_{x_0}^{x_f} dx = S_{in} Y_{x/s} F_{in} \int_0^{t_{FB}} dt \tag{10.61}$$

After integration, Equation 10.61 can be written as presented in Equation 10.62, and the fed-batch mode cultivation time is given by Equation 10.63.

$$x_f = x_0 + \left(S_{in} F_{in} Y_{x/s}\right) t_{FB} \tag{10.62}$$

$$t_{FB} = \frac{x_f}{x_0 + \left(S_{in} F_{in} Y_{x/s}\right)} \tag{10.63}$$

Hypothesis 10.9

Assuming that $\mu = \mu_{max}$.

Since the fed-batch mode operates with a higher feed rate of substrate than the rate at which the substrate is consumed by the cells, the substrate accumulates during cultivation. If Monod's kinetics is valid (see Chapter 9, Equation 9.9 and Figure 9.3), the accumulated substrate concentration can be assumed to be much higher than the saturation constant, i.e., $S \gg K_S$. Under this condition, the specific growth rate of the microorganism is maximum, i.e., $\mu = \mu_{max}$. Thus, from Equation 10.37, described in the mass balance for the cells, Equation 10.64 is obtained.

$$\frac{dX}{dt} = \mu X - \frac{X}{V}\frac{dV}{dt} \tag{10.64}$$

Rearranging Equation 10.64, one obtains Equation 10.65.

$$\frac{dX}{dt} = \left(\mu - \frac{1}{V}\frac{dV}{dt}\right) X \tag{10.65}$$

Splitting the terms of Equation 10.65, one has Equation 10.66.

$$\frac{dX}{X} = \left(\mu - \frac{1}{V}\frac{dV}{dt}\right) dt \tag{10.66}$$

By integrating Equation 10.66 for the initial condition of $x = x_0$ at $t = t_0$ and $v = v_0$, and taking $\mu = \mu_{max}$, Equation 10.67 can be set.

$$\int_{X_0}^{X} \frac{dX}{X} = \mu_{max} \int_{t_0}^{t} dt - \int_{v_0}^{v} \frac{dV}{V} \tag{10.67}$$

Equation 10.67 in its integrated form can be written as Equation 10.68.

$$\ln\frac{X}{X_0} = \mu_{max}\left(t - t_0\right) - \ln\frac{V}{V_0} \tag{10.68}$$

Rearranging the terms of Equation 10.68, one finds Equation 10.69.

$$\ln\frac{X}{X_0} + \ln\frac{V}{V_0} = \mu_{max}\left(t - t_0\right) \tag{10.69}$$

By eliminating the Neperian logarithm of Equation 10.69, Equations 10.70 and 10.71 are obtained.

$$\frac{XV}{X_0 V_0} = e^{\mu_{max}(t - t_0)} \tag{10.70}$$

$$XV = X_0 V_0 e^{\mu_{max}(t - t_0)} \tag{10.71}$$

For a constant flow of feeding, the variation of volume is given by $V = V_0 + F_{in} t$, and the performance of the cell concentration during the bioreactor feeding can be described by Equation 10.72. This equation shows that cell concentration varies exponentially over time and might present a more moderate trend as the feed flow increases. Since the mass of cells (x) is given by the product between cell concentration and media volume ($X \cdot V$), Equation 10.71 can be written as Equation 10.73.

$$X = \frac{X_0 V_0 e^{\mu_{max}(t - t_0)}}{V_0 + F_{in} t} \tag{10.72}$$

$$x = x_0 e^{\mu_{max}(t - t_0)} \tag{10.73}$$

Equation 10.73 shows that under the hypothesis of $dS/dt > 0$ and $\mu = \mu_{max}$, for fed-batch cultivation with constant feed rate, the total cell mass increases exponentially as a function of feeding time, as illustrated in Figure 10.11. This pattern will be upheld as long as the set conditions are maintained during feeding of the bioreactor.

10.5.3 Mass Balance for the Product in Fed-Batch Mode

Taking into account that no product is added or withdrawn during feeding of the bioreactor and no product is degraded during cultivation, the mass balance for the product can be written as Equation 10.74.

$$\frac{d(PV)}{dt} = V r_p \tag{10.74}$$

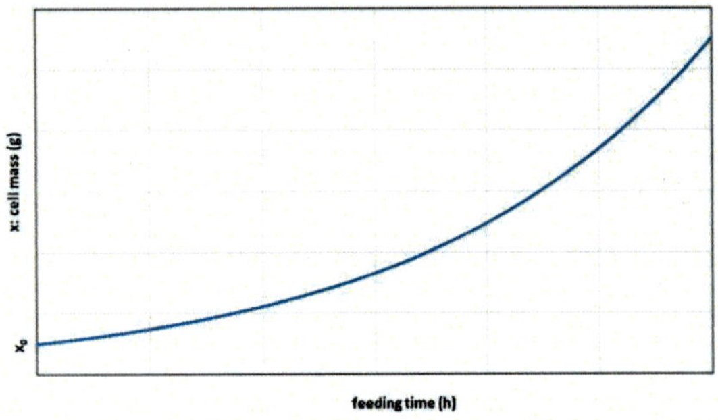

FIGURE 10.11 Cell mass formation during fed-batch cultivation mode with constant feed flow, assuming that $dS/dt > 0$ and $\mu = \mu_{max}$.

By separating the variables from the derivative and replacing the terms referring to growth and cell death, Equation 10.74 can be written as Equation 10.75.

$$V\frac{dP}{dt} + P\frac{dV}{dt} = Vr_p \tag{10.75}$$

Dividing the terms of Equation 10.75 by volume (V) and replacing $dV/dt = F_{in}$, Equation 10.76 is obtained.

$$\frac{dP}{dt} + P\frac{F_{in}}{V} = r_p \tag{10.76}$$

Knowing that $r_p = \mu_p X$, the variation in product concentration over time can be expressed by Equation 10.77 and by replacing $\frac{F_{in}}{V}$ by D, the product balance equation is thereby obtained in the fed-batch system (Equation 10.78).

$$\frac{dP}{dt} = \mu_p X - \frac{F_{in}}{V}P \tag{10.77}$$

$$\frac{dP}{dt} = \mu_p X - DP \tag{10.78}$$

Equation 10.78 demonstrates that product variation over time is given by the difference between the amount of product formed and the dilution rate.

Hypothesis 10.10

Taking into account that dP/dt is negligible.

Assuming a quasi-steady state where $dP/dt \approx 0$, Equation 10.78 might be written as presented in Equation 10.79 and the product concentration is given by Equation 10.80.

$$\mu_p X = DP \tag{10.79}$$

$$P = \frac{\mu_p X}{D} \tag{10.80}$$

Substituting cell concentration by Equation 10.54 in the quasi-steady state, Equation 10.81 is obtained.

$$P = \frac{\mu_p D S_{in} Y_{x/s}}{D\mu} \tag{10.81}$$

Rearranging Equation 10.81, one gets Equation 10.82.

$$P = \frac{\mu_p}{\mu} S_{in} Y_{x/s} \tag{10.82}$$

According to the kinetic description of metabolic products presented in Chapter 9, in a case where the product associated with growth $\mu_p = \alpha\mu$, Equation 10.82 can be written as Equation 10.83.

$$P = S_i Y_{x/s} \tag{10.83}$$

Knowing that $\alpha = Y_{p/x}$ and that $Y_{p/x} = \frac{Y_{p/s}}{Y_{x/s}}$, Equation 10.83 can be written as Equation 10.84.

$$P = S_{in} Y_{p/s} \tag{10.84}$$

Equation 10.84 shows that in the quasi-steady state, the product concentration is dependent on both the substrate concentration in the feed medium and the conversion factor from substrate to product. It is worth remembering that the quasi-steady state is a condition of the cultivation equilibrium, in terms of concentration of the variables ($dX/dt \approx 0$; $dS/dt \approx 0$ e $dP/dt \approx 0$), this situation can be achieved when the biomass concentration is high and when almost all of the substrate is completely depleted.

As previously demonstrated, the overall cell mass profile as a function of the bioreactor feeding time depends on the ratio between the substrate feeding rate and the substrate consumption rate by the cells. Thus, when assuming the hypotheses of $dS/dt \approx 0$ or $dS/dt > 0$, the total cell mass will exhibit, respectively, a linear or exponential behaviour over time, as shown by Equations 10.60 and 10.73. Then the behaviour of the metabolic product formation during a fed-batch process with constant feed flow will be dependent on the cell mass profile.

To describe product formation as a function of time, Equation 10.74 can be used and the term (PV) replaced by (p) to express the results in terms of mass to obtain Equation 10.85. Considering $r_p = \mu_p X$ and replacing the term (VX) by (x), Equation 10.86 and 10.87 can be obtained. Equation 10.87 shows that variation in product mass over time depends on how the cell mass (x) varies during cultivation. Thus, for a linear behaviour in total cell mass, given by Equation 10.60, we can write Equation 10.88.

$$\frac{dp}{dt} = Vr_p \tag{10.85}$$

$$\frac{dp}{dt} = V\mu_p X \tag{10.86}$$

$$\frac{dp}{dt} = \mu_p x \tag{10.87}$$

$$\frac{dp}{dt} = \mu_p\left(x_0 + S_{in}Y_{x/s}F_{in}t\right) \quad (10.88)$$

According to the kinetic description of metabolic products (see Chapter 9), in the case of a product associated with growth $\mu_p = \alpha\mu$, Equation 10.88 can be written as follows:

$$dp = \alpha\mu\ \left(x_0 + S_{in}Y_{x/s}F_{in}t\right)dt \quad (10.89)$$

Bearing in mind that for conditions describing a linear pattern of cell mass over time, the specific growth rate (μ) varies as described by Equation 10.56 and merging with Equation 10.89, one can write Equation 10.90. Therefore, Equation 10.90 can be shortened to Equation 10.91.

$$dp = \alpha\frac{F_{in}S_{in}Y_{x/s}}{x_0 + S_{in}Y_{x/s}F_{in}t}\left(x_0 + S_{in}Y_{x/s}F_{in}t\right)dt \quad (10.90)$$

$$dp = \alpha S_{in}Y_{x/s}F_{in}dt \quad (10.91)$$

By integrating Equation 10.91 under the initial condition of $p = p_0$ at $t = 0$ (start of feeding), one has Equation 10.92. This equation shows that, under a quasi-steady state condition and with $Y_{x/s}$ and F_{in} as constants, the total mass of product, whose synthesis is associated with growth, increases linearly over time when a fed-batch cultivation uses a constant feed flow.

$$p = p_0 + \alpha S_{in}Y_{x/s}F_{in}t \quad (10.92)$$

The following presents a description of the product mass as a function of time under the conditions that describe the exponential growth of cells ($dS/dt > 0$ and $\mu = \mu_{max}$). For this, Equation 10.73 must be merged with Equation 10.87 of the product mass balance. In this way, Equation 10.93 is obtained.

$$\frac{dp}{dt} = x_0 e^{\mu_{max}t}\mu_p \quad (10.93)$$

According to the kinetic description of metabolic products (see Chapter 9), in the case where a product is associated with growth, $\mu_p = \alpha\mu$, Equation 10.93 can be written as Equation 10.94. Rearranging this equation and assuming $\mu = \mu_{max}$, one gets Equation 10.95.

$$\frac{dp}{dt} = x_0 e^{\mu_{max}t}\alpha\mu \quad (10.94)$$

$$dp = \alpha\mu_{max}x_0 e^{\mu_{max}t}dt \quad (10.95)$$

Integrating Equation 10.95 for the initial condition of $p = p_0$ and $t = 0$ (start of feeding), one has Equation 10.96.

$$p = p_0 + \left(\alpha x_0 e^{\mu_{max}t} - 1\right) \quad (10.96)$$

Equation 10.96 shows that when assuming substrate accumulation ($dS/dt > 0$) and operating with $\mu = \mu_{max}$, the total mass of product, whose synthesis is associated with growth, increases exponentially over time when a constant feed flow is used.

Table 10.6 summarizes the mass balance equations for cells (X), metabolic product (P) and limiting substrate (S) during fed-batch cultivation with variable volume and constant feed flow.

The fed-batch mode has been the most widely used mode of operation on an industrial scale, especially for obtaining biopharmaceuticals. This is because in fed-batch mode the concentration of compounds (limiting substrate and/or inducers) can be controlled. This allows cell growth rate and product inhibition to be reduced by regulating the metabolism of overflow. Control of product inhibition has significantly increased the productivity of these systems, allowing cultivations with high cell densities as well as higher levels of metabolic products with high yields and productivity. The fed-batch mode has been used more and more frequently, as it is easier to operate than the continuous process (this is a particular advantage for validating biopharmaceuticals in FDA and EMA type regulatory agencies).

10.6 CONTINUOUS MODE OPERATION

As previously described, the continuous process operates as an open system in which nutrients are supplied throughout the cultivation, and at the same time, products are withdrawn from the bioreactor. A key feature of the continuous process is the possibility of operating in a steady state, in which the values of all process variables (temperature, concentration of components, flow, etc.) do not vary over time. A precondition for the system to achieve a steady state is that the volume of the cultivation is kept constant. For this, the volumetric flow rate of fresh medium and withdrawal of the reaction medium must be the same.

A continuous process is extremely versatile in terms of handling possibilities and can be operated as a single stage (only one bioreactor) or in multiple stages (several bioreactors in series). Both options can operate re-using cells (cell recycling), i.e., the cells being withdrawn in the output stream can be separated and brought back into the bioreactor (the option of internal recycling, e.g. spin filter, will not be discussed in this book). It is important to state that the

TABLE 10.6
Mass Balance Equations for Cells, Limiting Substrate and Metabolic Product during Fed-Batch Cultivation

Components	Equations	Number
Global	$F_{in} = \frac{dV}{dt}$	(10.39)
Cells	$\frac{dX}{dt} = (\mu - D)\cdot X$	(10.38)
Limiting-substrate	$\frac{dS}{dt} = D(S_{in} - S) - \mu_S X$	(10.50)
Metabolic product	$\frac{dP}{dt} = \mu_p X - DP$	(10.78)

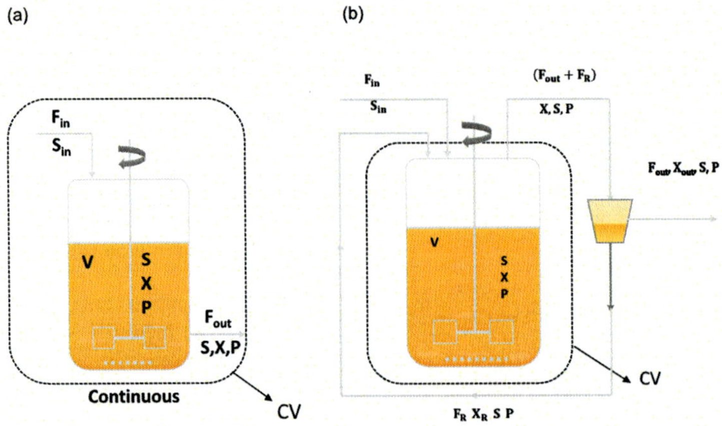

FIGURE 10.12 A continuous process bordered by control volume: (a) without cell recycle and (b) with external cell recycle.

most appropriate use of continuous operation depends on the nature of the process and its purpose.

A schematic depiction of a bioreactor operation in continuous mode without cell recycling (a) and with cell recycling (b) is shown in Figure 10.12.

There are two parameters which determine a continuous mode operation: (a) the specific feed flow rate (D), which is the ratio between feed volume flow rate (F_{in}) and the volume of the reaction medium (V), that is, $D = F_{in}/V$; and (b) mean residence time of the liquid in the bioreactor (t_R), determined by the ratio between the reaction medium volume (V) and the feed volume rate (F_{in}), that is, $t_R = V/F_{in}$. The mass balances presented below are described for the continuous process in a single stage (with and without cell recycling).

As shown in Figure 10.12a, in continuous single-stage cultivation without cell recycling, only the fresh culture medium, containing a limiting substrate (S_{in}) at a volumetric flow rate (F_{in}), is fed into the bioreactor. At the same time, the reaction medium containing S, X and P is withdrawn at a volumetric flow rate, F_{out}. It is worth pointing out that in order to keep the volume of the medium constant during the process, the volumetric flow rates in the inlet and outlet currents have to be equal ($F_{in} = F_{out} = F$).

In a continuous single-stage cultivation with cell recycling (Figure 10.12b), in addition to feeding the limiting substrate (S_{in}) at a volumetric flow rate F_{in}, the bioreactor is also fed with a fraction of the media concentrated with cells at a volumetric flow rate F_R. This is known as the recycle stream. For this procedure, a cell separator (filter, centrifuge or sedimentation tank) is attached to the bioreactor in order to concentrate the biomass that will be returned to the bioreactor. Cell recycling aims to increase the concentration of cells in the bioreactor, which in turn improves the rate of substrate uptake. In addition, the recycling of cells allows the system to operate at a higher dilution rate (D) when compared to single-stage cultivations without recycling, which is advantageous for the performance of the process in terms of volumetric productivity. The volumetric productivity parameter (Q_X) in the continuous is discussed below.

To evaluate a continuous process, a bioreactor operating with perfect mixing will be considered. Here, the concentration of all components in the (cells, limiting substrate and metabolic product) output stream of the bioreactor is identical to the concentration of these components in the reaction medium (within the bioreactor).

10.6.1 Mass Balance for Cells in Continuous Mode without Cell Recycling

In continuous mode cultivation without cell recycling, no cells are added to the input stream of the bioreactor, but cells are withdrawn from the output stream. Therefore, based on the general equation (Equation 10.1), the mass balance for the cells during a continuous process can be written as shown in Equation 10.97.

$$\frac{d(XV)}{dt} = -F_{out}X + Vr_x - Vr_d \qquad (10.97)$$

By splitting the variables from the derivative and substituting the terms related to cell growth and cell death, Equation 10.97 can be written as Equation 10.98.

$$V\frac{dX}{dt} + X\frac{dV}{dt} = -F_{out}X + V(\mu - k_d)X \qquad (10.98)$$

By dividing all the terms of Equation 10.98 by the volume of medium (V), and taking into account that the volume of the reaction remains constant over time ($dV/dt = 0$), Equation 10.99 is obtained.

$$\frac{dX}{dt} = \left(\mu - k_d - \frac{F_{out}}{V}\right)X \qquad (10.99)$$

To keep the volume constant during cultivation, the volumetric flow rate at the inlet (F_{in}) must be equal to the volumetric flow rate at the outlet (F_{out}). To simplify the rationale, the volumetric flow rate is designated as F. Moreover, it is known that the dilution rate (D) is given by the ratio between F/V, so Equation 10.100 can be written.

$$\frac{dX}{dt} = (\mu - k_d - D)X \qquad (10.100)$$

Taking into consideration that the cell death rate is negligible, i.e., $k_d \lll D$, Equation 10.100 can be simplified to

Equation 10.101, which represents the cell balance in continuous mode. This equation defines the overall balance of cells in continuous culture without cell recycling.

$$\frac{dX}{dt} = (\mu - D) \cdot X \tag{10.101}$$

Taking into account that the cultivation has achieved a steady state on a permanent regime, which means that the concentration of cells in the bioreactor does not vary over time, i.e., $dX/dt = 0$, Equation 10.102 is obtained.

$$0 = (\mu - D)X \tag{10.102}$$

By dividing the terms of Equation 10.102 by (X), rearranged Equation can be expressed as Equation 10.103 which shows that at steady state, the specific cell growth rate (μ) is controlled by the dilution rate (D).

$$\mu = D \tag{10.103}$$

Note that the dilution rate (D) is a hydraulic (physical), while (μ) is a biological parameter. In principle, the equality $\mu = D$ shows that the higher the value of D, the higher the value of μ. However, if the dilution rate (D) is increased to values which exceed the maximum cell growth capacity, i.e., above the μ_{max} value of the cells, the dilution rate of the bioreactor (D) will be higher than the maximum growth rate (μ_{max}); and as a consequence, the rate of cells withdrawal from the bioreactor will be higher than the growth rate. In this circumstance, the concentration of cells in the bioreactor will decrease over time, resulting in a total wash out. To prevent this issue, the maximum dilution rate should be lower than μ_{max}. From a theoretical point of view, it is possible to reach a steady state with $D = \mu_{max}$, but in this situation we would be working at the maximum limit of the growth rate. From a practical point of view, this condition is considered critical. In situations of operational instability, where there is an increase in the feed flow (F), a wash-out event is unavoidable. Thus, the value corresponding to the maximum specific cell growth rate (μ_{max}) can be defined as a critical value D (D_{crit}). The wash-out condition in the bioreactor means that the operating range of a continuous cultivation mode can be established, and for cultivation without cell recycling, the operating range is between zero and μ_{max}, thereby obeying the condition of $D < D_{crit}$.

10.6.2 Mass Balance for a Limiting Substrate in Continuous Mode without Cell Recycling

As shown in Figure 10.3, during continuous operation without cell recycling, there is inlet and outlet flow of limiting substrate, so based on the general equation (Equation 10.1), the mass balance for the limiting substrate (S) can be written as shown in Equation 10.104.

$$\frac{d(SV)}{dt} = F_{in}S_{in} - F_{out}S - Vr_s \tag{10.104}$$

By separating the variables from the derivative and replacing the substrate consumption rate by the expression defined in Table 9.1 (see Chapter 9), Equation 10.105 can be written.

$$V\frac{dS}{dt} + S\frac{dV}{dt} = F_{in}S_{in} - F_{out}S - V\mu_s X \tag{10.105}$$

Dividing all the terms of Equation 10.105 by the medium volume (V) and taking into account that the volume of the reaction is constant over time ($dV/dt = 0$), Equation 10.106 is obtained.

$$\frac{dS}{dt} = \frac{F_{in}}{V}S_{in} - \frac{F_{out}}{V}S - \mu_s X \tag{10.106}$$

whereas $F_{in} = F_{out} = F$ and $F/V = D$, Equation 10.106 can be set up as Equation 10.107 representing the overall mass balance for a limiting substrate in continuous cultivation without cell recycling.

$$\frac{dS}{dt} = D(S_{in} - S) - \mu_s X \tag{10.107}$$

Hypothesis 10.11

Taking into account that μ_S follows the Pirt model (see Chapter 9).

Assuming that the Pirt model is valid, the term for the substrate consumption rate ($\mu_S X = r_S$) in Equation 10.107 can be divided into two parts: one for the substrate consumption rate used for cell growth (r_{SG}), and the other for the substrate consumption rate used for cell maintenance (r_{Sm}). This gives Equation 10.108. Bearing in mind that in the condition of steady-state ($dS/dt = 0$) and $\mu = D$, this equation can be written as Equation 10.109.

$$\frac{dS}{dt} = D(S_{in} - S) - \frac{\mu X}{Y_G} - m_s X \tag{10.108}$$

$$D(S_{in} - S) = \left(\frac{D}{Y_G} + m_s\right)X \tag{10.109}$$

Dividing Equation 10.109 by D, one obtains Equation 10.110.

$$\frac{(S_{in} - S)}{X} = \left(\frac{1}{Y_G} + \frac{1}{D}m_s\right) \tag{10.110}$$

If the Y_G and m_S are constant, Equation 10.110 can estimate these parameters from (X) and (S) obtained in continuous cultivation mode without cell recycling, at different dilution rates (D), as illustrated in Figure 10.13.

Hypothesis 10.12

Considering that μ_S follows the Pirt model and m_S is negligible.

Equation 10.110 can also be further simplified, considering the substrate consumption rate for cell maintenance ($m_s = 0$) is negligible. In the absence of maintenance, $Y_G = Y_{x/s}$, Equations 10.111 and 10.112 are obtained.

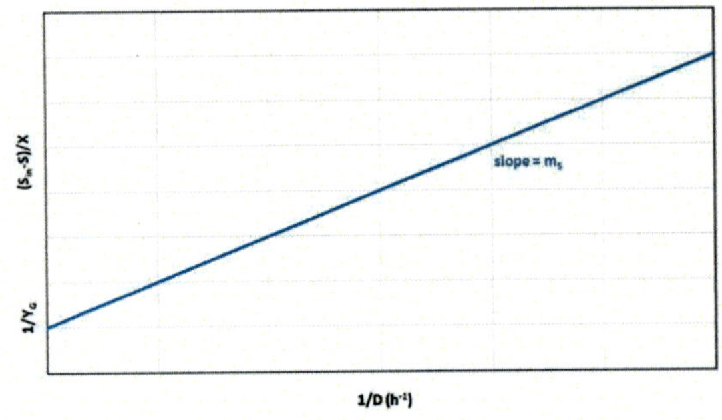

FIGURE 10.13 Graphical calculation of the maintenance coefficient (m_S) and true biomass yield (Y_G) using data from continuous process, at steady state, without cell recycle.

$$S = S_{in} - \left(\frac{X}{Y_{X/S}}\right) \tag{10.111}$$

$$X = (S_{in} - S) \cdot Y_{X/S} \tag{10.112}$$

Hypothesis 10.13

Considering that μ follows the model of Monod.

Figure 9.3 in this book presents the kinetic growth model proposed by Monod for microbial cultivations. If the Monod Model is valid, then the concentrations of (S) and (X) in a continuous process, without cell recycling and operating at a steady-state regime, can be expressed according to the kinetic parameters of the cultivation. According to Monod's equation, we can write Equation 10.113.

$$\mu = \mu_{max} \frac{S}{K_S + S} \tag{10.113}$$

Isolating the substrate concentration from Equation 10.113, one has Equation 10.114.

$$S = \frac{K_S \mu}{\mu_{max} - \mu} \tag{10.114}$$

In a steady-state condition, disregarding the cell death rate, $\mu = D$, then Equation 10.114 can be written as Equation 10.115.

$$S = \frac{K_S D}{\mu_{max} - D} \tag{10.115}$$

Equation 10.115 shows that the substrate concentration in a steady-state continuous system is a function only of the dilution rate, since K_S and μ_{max} are constant. Then, knowing the values of (S) at different dilution rates (D), the ratio between S and D can graphically be represented and an estimate of the kinetic constants K_S and μ_{max} using the method proposed by Lineweaver-Burk can be predicted (see Chapter 15).

By substituting Equation 10.115 in Equation 10.112, X can still be written as a function of D as presented in Equation 10.116.

$$X = Y_{X/S}\left(S_{in} - \frac{K_S D}{\mu_{max} - D}\right) \tag{10.116}$$

From Equations 10.115 and 10.116, assigning values to μ_{max}, K_S and $Y_{X/S}$ parameters, the behaviour of limiting substrate (S) and cell concentration (X) can be predicted as a function of dilution rate (D) during continuous, single-stage and non-recycling cell cultivation.

10.6.3 Mass Balance for the Product in Continuous Mode without Cell Recycling

During continuous operation without cell recycling, no product inlet flow takes place, but the product is withdrawn in the output stream; therefore, based on Equation 10.1, the mass balance for product (P) can be written as shown in Equation 10.117.

$$\frac{d(PV)}{dt} = -F_{out}P + Vr_p \tag{10.117}$$

By dissociating the variables from the derivative and replacing the term related to the volumetric rate of product formation with the expression defined in Table 9.1, Equation 10.118 can be derived.

$$V\frac{dP}{dt} + P\frac{dV}{dt} = -F_{out}P + V\mu_p X \tag{10.118}$$

Dividing all the terms of Equation 10.118 by volume (V) and taking into account that the volume of the reaction is constant over time ($dV/dt = 0$), Equation 10.119 is obtained.

$$\frac{dP}{dt} = -F_{out}P + \mu_p X \tag{10.119}$$

Bearing in mind that $F_{in} = F_{out} = F$, and $F/V = D$, Equation 10.119 can be written as Equation 10.120 to express the mass balance of the product in a continuous cultivation mode.

$$\frac{dP}{dt} = \mu_p X - DP \tag{10.120}$$

Equation 10.120 defines the mass balance for the metabolic product during continuous cultivation without cell recycling. Based on this equation, a series of assessments of the product's behaviour over time can be drawn up based on the kinetic models of product formation described in Chapter 9.

When a cultivation has achieved the steady-state status, it can be stated that the concentration of product in the bioreactor is constant over time ($dP/dt = 0$). Furthermore, it is known that the dilution rate (D) is calculated on the basis of F/V, so Equation 10.121 can be written, and after rearranging, one gets Equation 10.122.

$$0 = \mu_p X - DP \tag{10.121}$$

TABLE 10.7
Mass Balance Equations for Cells, Limiting Substrate and Metabolic Product for Continuous Cultivation Mode without Cell Recycling

Components	Equations	Equation Numbers
Cells	$\frac{dX}{dt} = (\mu - D)X$	(10.101)
Limiting substrate	$\frac{dS}{dt} = D(S_{in} - S) - \mu_s X$	(10.107)
Metabolic product	$\frac{dP}{dt} = \mu_p X - DP$	(10.120)

$$DP = \mu_p X \quad (10.122)$$

By dissociating the terms of Equation 10.122, the product concentration (P) can be expressed as a function of the dilution rate (D), as follows (Equation 10.123):

$$P = \frac{\mu_p X}{D} \quad (10.123)$$

Equation 10.123 can be applied to predict the behaviour of the metabolic product concentration during a continuous process, without cell recycling and with variation in the dilution rate of the cultivation. Some examples of representative kinetic models were discussed in Chapter 9.

Table 10.7 summarizes the mass balance equations for cells (X), metabolic product (P) and limiting substrate (S) for continuous mode cultivation and non-recycling cell cultures.

10.6.4 Mass Balance for Cells in Continuous Mode with External Cell Recycling

As shown in Figure 10.12b, in continuous culture with external recycling, cells from the outlet stream of the separator are redirected back into the bioreactor. This recycling stream, containing a concentrated biomass (X_R), will be recycled with a volumetric flow rate referred to as F_R. It is worth noting that the concentration factor (B) is defined by the ratio X_R/X and depends on the efficiency of the separator. The fraction of liquid to be recycled (A) is the ratio between the volume rate of the recycle (F_R) and the volume rate in the output current (F_{out}), i.e. ($A = F_R/F_{out}$). As $F_{in} = F_{out} = F$, you can write that $A = F_R/F$. The recycle parameters are summarized in Table 10.8.

Based on the general Equation 10.1, the mass balance for cells in the continuous process with cell recycling can be written as follows (Equation 10.124).

$$\frac{d(XV)}{dt} = F_R X_R - (F_{out} + F_R)X + Vr_x - Vr_d \quad (10.124)$$

By applying the recycle parameters listed in Table 10.8, it is possible to write Equation 10.125.

$$\frac{d(XV)}{dt} = AFBX - (F + AF)X + Vr_x - Vr_d \quad (10.125)$$

TABLE 10.8
Description of Recycle Parameters

Parameter	Description	Unity of Measurement
A	Recycled liquid fraction (recycle ratio)	Adimensional
B	Biomass concentration factor	Adimensional
$F_R = AF$	Volumetric flow rate of recycle	L/h
$X_R = BX$	Biomass concentration in the recycle stream	g/L
$\varepsilon = (1+A-AB)$	Recycling fraction	Adimensional

By separating the variables from the derivative and replacing the terms related to cell growth and death with previously defined expressions (see Table 9.1), Equation 10.125 can be written as Equation 10.126.

$$V\frac{dX}{dt} + X\frac{dV}{dt} = AFBX - (F + AF)X + V(\mu - k_d)X \quad (10.126)$$

Dividing all the terms of Equation 10.126 by the volume of medium (V), and taking into account that the volume of the reaction remains constant over time ($dV/dt = 0$), Equation 10.127 will be derived.

$$\frac{dX}{dt} = A\frac{F}{V}BX - \left(\frac{F}{V} + A\frac{F}{V}\right)X + (\mu - k_d)X \quad (10.127)$$

With the knowledge that $F/V = D$, we have Equation 10.128. This equation, therefore, defines the overall balance of cells in continuous culture with external cell recycling. From this balance, several predictions can be made using simple hypotheses

$$\frac{dX}{dt} = ADBX - (D + AD)X + (\mu - kd)X \quad (10.128)$$

Whereas in the steady-state $dX/dt = 0$, Equation 10.128 can be written as follows (Equation 10.129) and dividing the two members of this equation by DX, one has Equation 10.130.

$$-ADBX + DX + ADX = \mu X - k_d X \quad (10.129)$$

$$-AB + 1 + A = \frac{1}{D}(\mu - k_d) \quad (10.130)$$

Assuming a negligible occurrence of cell death, Equation 10.130 can be reduced to Equation 10.131.

$$D = \frac{\mu}{(1 + A - AB)} \quad (10.131)$$

Knowing that $A \leq 1$ and $B \geq 1$, one can infer that $D \geq \mu$. This shows that the continuous process with cell recycling can be operated with a dilution rate (D) higher than the specific cell growth rate (μ).

By naming ε the recycling fraction $(1 + A - AB)$, Equation 10.132 is obtained and one can estimate the critical D-value for continuous mode with cell recycling as in

Equation 10.133. Note that the D_{crit} value is increased by the inverse of the recycling fraction (ε), i.e., by $(1 + A - AB)^{-1}$.

$$D = \frac{\mu}{\varepsilon} \tag{10.132}$$

$$D_{crit} = \frac{\mu_{max}}{\varepsilon} \tag{10.133}$$

10.6.5 Mass Balance for the Limiting Substrate in Continuous Mode with External Cell Recycling

As shown in Figure 10.12b, during continuous cell recycle operation, there is limiting-substrate input into the feed stream of the bioreactor. Based on the general Equation 10.1, the mass balance for the limiting substrate (S) can be written as follows (Equation 10.134).

$$\frac{d(SV)}{dt} = F_{in}S_{in} + F_R S - (F_{out} + F_R)S - Vr_s \tag{10.134}$$

By applying the recycle parameters and replacing F_{in} by F, Equation 10.135 is obtained.

$$\frac{d(SV)}{dt} = FS_{in} + AFS - (F + AF)S - Vr_s \tag{10.135}$$

By separating the variables from the derivative and substituting the term related to substrate consumption for the expression defined in Table 9.1, Equation 10.135 can be written as Equation 10.136.

$$V\frac{dS}{dt} + S\frac{dV}{dt} = FS_{in} + AFS - (F + AF)S - V\mu_s X \tag{10.136}$$

By dividing all the terms of Equation 10.136 by the volume of medium (V), and taking into account that the volume of the reaction remains constant over time ($dV/dt = 0$), Equation 10.137 is obtained. Given that $F/V = D$, Equation 10.138 or 10.139 is obtained.

$$\frac{dS}{dt} = \frac{F}{V}S_{in} + \frac{F}{V}AS - \frac{F}{V}S - \frac{F}{V}AS - \mu_s X \tag{10.137}$$

$$\frac{dS}{dt} = DS_{in} - DS - \mu_s X \tag{10.138}$$

$$\frac{dS}{dt} = D(S_{in} - S) - \mu_s X \tag{10.139}$$

Equation 10.139 defines the overall balance of a limiting substrate in continuous cell recycle cultivation. Note that the mass balance equation for the limiting substrate remains unchanged in the recycle, being that Equation 10.139 is equal to Equation 10.107, defined from continuous cultivation without cell recycling.

Hypothesis 10.14

Taking into account that μ_S follows the Pirt model and m_S is negligible.

Similarly, to the assessment carried out for the continuous mode and without cell recycling, the Pirt model is assumed to be valid (see Chapter 9) and the rate of substrate consumption for cell maintenance is neglected. Then, Equation 10.140 can be described.

$$\frac{dS}{dt} = D(S_{in} - S) - \frac{\mu X}{Y_{X/S}} \tag{10.140}$$

When the continuous system is at steady state, $dS/dt = 0$ and $\mu = D\varepsilon$, Equation 10.141 can be written.

$$D(S_{in} - S) = \frac{\varepsilon XD}{Y_{X/S}} \tag{10.141}$$

By dividing Equation 10.141 by D and rearranging the terms, Equation 10.142 can be written.

$$X = \frac{Y_{X/S}\,(S_{in} - S)}{\varepsilon} \tag{10.142}$$

Comparing Equations 10.112 and 10.142, the cell concentration in continuous mode with cell recycling increases by the inverse of the recycling fraction, that is, $(\varepsilon)^{-1}$.

Hypothesis 10.15

Considering that μ follows Monod's model.

If the kinetics of Monod is valid, one can merge Equation 10.132 with Equation 10.114 and estimate the substrate concentration, as presented in Equation 10.143; thus this equation can be written as Equation 10.144.

$$S = \frac{K_S D\varepsilon}{\mu_{max} - D\varepsilon} \tag{10.143}$$

$$X = \frac{Y_{X/S}}{\varepsilon}\left[S_{in} - \frac{K_S D\varepsilon}{(\mu_{max} - D\varepsilon)}\right] \tag{10.144}$$

From Equations 10.115, 10.116, 10.143 and 10.144, values can be assigned to parameters μ_{max}, K_S, $Y_{X/S}$, A e B to predict the performance of the limiting substrate (S) and cell concentration (X) as a function of the dilution rate (D) in continuous single-stage cultivations with or without cell recycling. Figure 10.14 compares the behaviour of cell (X) and substrate (S) concentrations as a function of the dilution rate (D), obtained in continuous cultivation mode at steady state, with and without cell recycling.

Figure 10.14 shows that as the dilution rate (D) increases the cell concentration (X) tends to decrease, while the substrate concentration tends to increase. Note that when D approaches the D_{crit}, which corresponds to the value of 0.4 and 0.8 h^{-1}, for cultivation with and without cell recycling, respectively, a sharp drop in cell concentration occurs, concomitant with an increase in substrate concentration. When

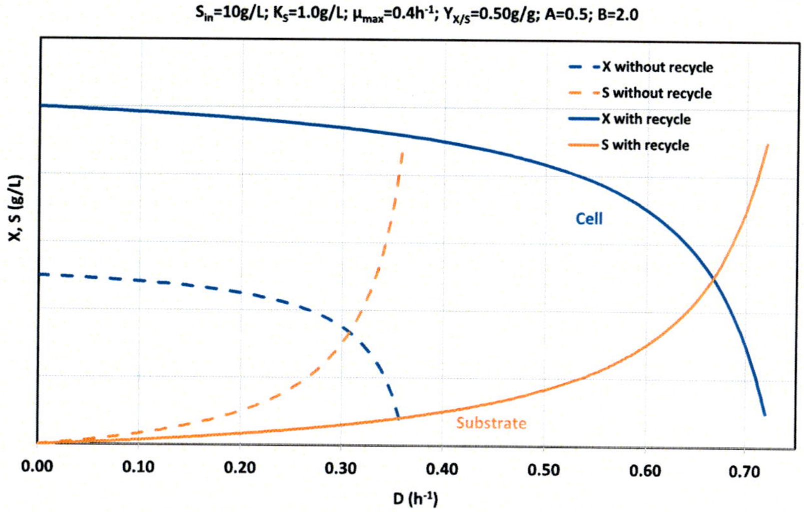

FIGURE 10.14 Profile of predicted cell and substrate concentrations in a continuous process at steady state with (lines blue) and without (lines orange) cell recycling.

$D = D_{crit}$, the cell concentration is zero and the substrate concentration (S) equals the value of the substrate concentration in the feed (S_{in}). This phenomenon indicates that the cells have been washed from the bioreactor (wash out). The main difference between the two systems is that in the continuous mode with cell recycling the operating range of the cultivation is extended, i.e., the system can operate with D higher than μ_{max}, which means obtaining higher yields in the cultivation.

10.6.6 Mass Balance for the Product in Continuous Mode with External Cell Recycling

As well as for the limiting substrate, the mass balance for the metabolic product remains constant as compared to the continuous mode without cell recycling. Thus, Equations 10.120 and 10.123 describe the mass balance for the product and its behaviour in the steady-state.

Table 10.9 summarizes the mass balance equations for cells (X), metabolic product (P) and limiting substrate (S) for continuous mode cultivation with external cell recycling.

TABLE 10.9
Mass Balance Equations for Cells, Limiting Substrate and Metabolic Product for Continuous Mode Cultivation with External Cell Recycling

Components	Mass Balance Equations	Numbers
Cells	$\frac{dX}{dt} = ADBX - (1+A)DX + \mu X$	(10.128)
Limiting substrate	$\frac{dS}{dt} = D(S_{in} - S) - \mu_s X$	(10.139)
Metabolic product	$\frac{dP}{dt} = \mu_p X - DP$	(10.120)

10.6.7 How to Calculate the Volumetric Productivity of a Continuous Process?

Volumetric productivity (Q) is a parameter of paramount importance for assessing the performance of any bioprocess, regardless of the mode of operation. This parameter expresses the mass of product formed (cell or metabolic product) per unit of bioreactor volume per unit of time and is considered one of the most important determining factors of bioprocess costs.

In a continuous process conducted in a bioreactor with perfect mixing, the reaction time is expressed by the mean residence time of the liquid in the bioreactor (t_R). This parameter is determined by the ratio between the volume of the reaction medium (V) and the feed volume rate (F) (Equation 10.145).

$$t_R = \frac{V}{F} \tag{10.145}$$

As $F/V = D$, the residence time is given by the inverse of the dilution rate (Equation 10.146).

$$t_R = \frac{1}{D} \tag{10.146}$$

Therefore, Equation 10.147 or 10.148 describes the cell mass productivity in the continuous mode process without cell recycling.

$$Q_X = \frac{X}{t_R} \tag{10.147}$$

$$Q_X = DX \tag{10.148}$$

Likewise, the productivity expressed as a metabolic product in the continuous process without cell recycling is given by Equations 10.149 or 10.150.

$$Q_P = \frac{P}{t_R} \quad (10.149)$$

$$Q_P = DP \quad (10.150)$$

Equations 10.116 and 10.144 are used to compare the volumetric productivities of the continuous process, in a single stage and in a steady-state regime, with and without cell recycling. These equations express the concentration of cells (X) as a function of the dilution rate (D), while Equation 10.151 represents the volumetric productivity in the system without cell recycling.

$$(Q_X)_{\text{without recycle}} = DY_{X/S}\left(S_{\text{in}} - \frac{K_S D}{\mu_{\max} - D}\right) \quad (10.151)$$

In the case of volumetric productivity in the system with cell recycling, Equations 10.144 and 10.148 are substituted and Equation 10.152 is obtained.

$$(Q_X)_{\text{with recycle}} = D\frac{Y_{X/S}}{\varepsilon}\left[S_{\text{in}} - \frac{K_S D\varepsilon}{(\mu_{\max} - D\varepsilon)}\right] \quad (10.152)$$

Figure 10.15 shows the cell volumetric productivity profile in a continuous process, in steady state, with and without cell recycling.

From Figure 10.15, it is clear that a continuous process with cell recycling is quite advantageous when one wants to produce biomass because of the high volumetric yields. In general, continuous processes have found a wide practical applications, for example in ethanol production, microbial biomass production and biological treatment of industrial effluents.

For biopharmaceuticals, the requirement of aseptic conditions for continuous mode cultivations is well recognized, and the high operating costs are offset by the high-added value of the product. A continuous mode operated aseptically has an important advantage of providing a product with uniform quality since the cells are in a constant physiological state and continuously produce the biopharmaceutical. This advantage stands out especially when dealing with post-translational modifications, which must be consistent because these modifications often directly influence the biological activity of the product. The expression 'the process is the product' is common, meaning that the features of the production process directly influence the characteristics of the final product. Furthermore, for biopharmaceuticals produced by animal cells, the low maximum specific growth rate ($\mu_{\max}$) of the cells heavily interferes with the productivity of the process, severely limiting the maximum feed flow rate in continuous processes without recycling (as seen in the definition of critical D). For this reason, it is very common in processes using animal cells to operate in continuous mode with recycling, commonly called 'perfusion'. Through recycling (external or internal), the maximum dilution rate can be raised by the inverse of the recycling fraction (ε), i.e., by the factor $(1 + A - AB)^{-1}$. With this significant increase in the flow rate of perfusion cultivation, the productivity of the process is also increased.

10.7 FINAL CONSIDERATIONS

In this chapter, different configurations of bioreactors have been described when considering cultivation to obtain products of interest. Special attention has been paid to the mode of operation of bioreactors to provide the best possible results, as specific bioprocess performance depends not only on the appropriate definition of the organism, the growth medium, the environmental conditions and the type of bioreactor, but also on the mode of operation. The three most commonly used operating modes in biotechnological industrial processes have been presented: batch, fed batch and continuous. The particular properties of each as well as the material balances for each component of the systems

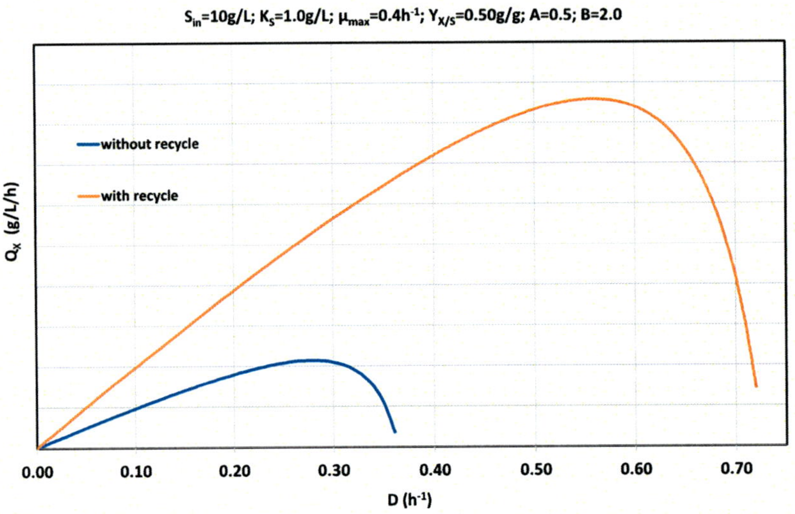

FIGURE 10.15 Steady-state volumetric productivity of cells as a function of dilution rate.

such as limiting substrate, cells and metabolic product have also been described. It is worth noting that differences between these modes of operation are related to addition and/or the withdrawal of growth medium during cultivation. One can conclude that there is no ideal mode of operation – i.e., one that meets all the requirements – because each has advantages and disadvantages, and for the best choice, each situation must be assessed carefully.

REFERENCES

Elsayed, E. A., Farid, M. A., & El-Enshasy, H. A. (2019). Enhanced natamycin production by *Streptomyces natalensis* in shake-flasks and stirred tank bioreactor under batch and fed-batch conditions. *BMC Biotechnology*, 19(1). doi:10.1186/s12896-019-0546-2.

Gätje, G., & Gottschalk, G. (1991). Limitation of growth and lactic acid production in batch and continuous cultures of *Lactobacillus helveticus*. *Applied Microbiology and Biotechnology*, 34(4), 446–449. doi:10.1007/BF00180568.

Ghaly, A. E., Kamal, M., & Correia, L. R. (2005). Kinetic modelling of continuous submerged fermentation of cheese whey for single cell protein production. *Bioresource Technology*, 96(10), 1143–1152. doi:10.1016/j.biortech.2004.09.027.

Giesecke, U. E., Bierbaum, G., Rudde, H., Spohn, U., & Wandrey, C. (1991). Production of alkaline protease with *Bacillus licheniformis* in a controlled fed-batch process. *Applied Microbiology and Biotechnology*, 35(6), 720–724. doi:10.1007/BF00169884.

Kim, B. S., Lee, S. C., Lee, S. Y., Chang, Y. K., & Chang, H. N. (2004). High cell density fed-batch cultivation of *Escherichia coli* using exponential feeding combined with pH-stat. *Bioprocess and Biosystems Engineering*, 26(3), 147–150. doi:10.1007/s00449-003-0347-8.

Li, Y., Chen, J., Mao, Y., Lun, S., & Koo, Y. (1998). Effect of additives and fed-batch culture strategies on the production of glutathione by recombinant *Escherichia coli*. *Process Biochemistry*, 33(7), 709–714. doi:10.1016/S0032-9592(98)00038-7.

Lim, H., & Shin, H. S. (2011). *Fed-batch cultures: Principles and applications of semi-batch bioreactors* (1–457) doi:10.1017/CBO9781139018777.

Ling, L. S., Mohamad, R., Abdul Rahim, R., Wan, H. Y., & Ariff, A. (2006). Improved production of live cells of Lactobacillus rhamnosus by continuous cultivation using glucose—yeast extract medium. *The Journal of Microbiology*, 44(4), 439–446.

Listner, K., Bentley, L., Okonkowski, J., Kistler, C., Wnek, R., Caparoni, A., & Chartrain, M. (2006). Development of a highly productive and scalable plasmid DNA production platform. *Biotechnology Progress*, 22(5), 1335–1345. doi:10.1021/bp060046h.

Liu, F. S., & Zajic, J. E. (1973). Fermentation kinetics and continuous process of L-asparaginase production. *Applied Microbiology*, 25(1), 92–96.

Marques, D A. V., de Carvalho Santos-Ebinuma, V., de Oliveira, P. M. S., de Souza Lima, G. M., Araújo, J. M., Lima-Filho, J. L., Converti, A., Pessoa Jr, A., Porto, A. L.F. (2014). Screening of wild type Streptomyces isolates able to overproduce clavulanic acid. *Brazilian Journal of Microbiology*, 45(3), 919–928.

Mears, L., Stocks, S. M., Sin, G., & Gernaey, K. V. (2017). A review of control strategies for manipulating the feed rate in fed-batch fermentation processes. *Journal of Biotechnology*, 245, 34–46. doi:10.1016/j.jbiotec.2017.01.008.

Minihane, B. J., & Brown, D. E. (1986). Fed-batch culture technology. *Biotechnology Advances*, 4(2), 207–218. doi:10.1016/0734-9750(86)90309-5.

Othman, M., Ariff, A. B., Wasoh, H., Kapri, M. R., & Halim, M. (2017). Strategies for improving production performance of probiotic *Pediococcus acidilactici* viable cell by overcoming lactic acid inhibition. *AMB Express*, 7(1), 215. doi:10.1186/s13568-017-0519-6.

Perego Jr., L., Cabral de S. Dias, J. M., Koshimizu, L. H., de Melo Cruz, M. R., Borzani, W., & Vairo, M. L. R. (1985). Influence of temperature, dilution rate and sugar concentration on the establishment of steady-state in continuous ethanol fermentation of molasses. *Biomass*, 6(3), 247–256. doi:10.1016/0144-4565(85)90044-7.

Pillaca-Pullo, O., Rodrigues, D., Sánchez-Moguel, I., Lopes, A., Pimenta, M., Basi, T., Feitosa, V., Zavaleta, A., Monteiro, G., Pessoa, A., & Vitolo, M. (2020). Recombinant L-asparaginase production using *Pichia pastoris* (MUT strain): establishment of conditions for growth and induction phases. *Journal of Chemical Technology and Biotechnology*. doi:10.1002/jctb.6540

Pirt, S.J. (1965). The maintance energy of bacteria in growing cultures. *Proceedings of the Royal Society of London. Series B, Biological Sciences*, 163(991), 224–231.

Silva, C. J. S. M., Mussatto, S. I., & Roberto, I. C. (2006). Study of xylitol production by *Candida guilliermondii* on a bench bioreactor. *Journal of Food Engineering*, 75(1), 115–119. doi:10.1016/j.jfoodeng.2005.04.001.

Silva, J. P. A., Mussatto, S. I., Roberto, I. C., & Teixeira, J. A. (2011). Ethanol production from xylose by *Pichia stipitis* NRRL Y-7124 in a stirred tank bioreactor. *Brazilian Journal of Chemical Engineering*, 28(1), 151–156. doi:10.1590/S0104-66322011000100016.

Subramaniam, R., Thirumal, V., Chistoserdov, A., Bajpai, R., Bader, J., & Popovic, M. (2018). High-density cultivation in the production of microbial products. *Chemical and Biochemical Engineering Quarterly*, 32(4), 451–464. doi:10.15255/CABEQ.2018.1394.

11 Agitation and Aeration
Oxygen Transfer and Cell Respiration

Aldo Tonso, Thiago Olitta Basso, Adalberto Pessoa Jr., and Ricardo Pinheiro de Souza Oliveira
Universidade de São Paulo

CONTENTS

11.1 INTRODUCTION

The importance of studying aeration in industrial bioprocesses is because most of these processes are aerobic, for example, for the production of different biomolecules such as amino acids, vitamins, organic acids, antibiotics, bacteriocins, enzymes, microbial and animal cells and recombinant proteins. In these processes, the oxygen transfer from the gas phase to the liquid phase is the key to its success. The efficiency of aeration depends on oxygen solubilization, diffusion rate into medium and bioreactor capacity to satisfy the oxygen demand of the cell population (Bandaiphet & Prasertsan, 2006; Buckland & Lilly, 1993).

Mixing is the process of achieving uniformity, i.e., it is a process by which substrates (from the liquid medium or a gaseous phase) are intimately dispersed throughout the reactor to afford cells that move through the reactor volume with, as far as possible, the same environment at every point. Also, mixing can be achieved in many ways, and the most common mixing technique in bioprocessing is mechanical agitation. Mixing is of paramount importance to promote the diffusion of oxygen into the culture medium. Even in anaerobic bioprocesses, such as for the production of lactic acid and solvents, where oxygen is not required, mixing is crucial to promote material dispersion, guaranteeing a homogeneous environment and increasing sustainable mass transfer rates.

Although oxygen can be considered a nutrient (Crueger & Crueger, 1984), it needs to be dealt differently compared to other nutrients. In fact, this occurs because it is not possible to store enough oxygen in the growth medium due to its low solubility, despite having a very important role in cell metabolism. In the respiratory chain of aerobic microorganisms and mammalian cells, oxygen is the final electron acceptor, where oxygen gas is reduced to water allowing reoxidation of the coenzymes that participate in the dehydrogenation reactions (along with glycolysis and the TCA cycle) and in the free-energy conservation reactions to generate ATP. ATP is thus necessary for various biosynthetic reactions, in cell maintenance and ultimately in cell growth, all of which require energy input (Tortora et al., 2005).

During an aerobic growth with a high cell growth rate, a high consumption rate of the carbon source is also expected, which results in a corresponding increased flux of electrons being transported in the electron transport chain (ECT), that ultimately results in the synthesis of ATP. For that reason, the availability of oxygen dissolved in the culture medium is extremely important, so that electrons can be transferred to molecular oxygen to produce water (Tortora et al., 2005). It is essential, therefore, to maintain appropriate availability of oxygen in the culture medium, either to avoid unnecessary use of energy to incorporate oxygen into the medium or to avoid oxygen limitation for cell respiration.

The main purposes of an agitation and aeration system in a bioprocess, especially for the production of biopharmaceuticals, are to:

- transfer oxygen from the gas phase to the liquid phase (where its solubility is low) as well as remove carbon dioxide from the liquid phase to the gas phase;
- keep cells in good physiological condition by maintaining dissolved oxygen in adequate level to ensure product homogeneity, such as correct glycosylation pattern of a glycoprotein;
- suspend solid particles (including cells and eventual insoluble solid substrates);
- promote the dispersion of immiscible liquids by forming emulsions or suspensions of extremely small droplets diameter (for example, lipid content in medium formulation);
- improve heat transport rates.

Agitation and aeration are closely associated and the importance of variables in each process can be illustrated by Figure 11.1.

Cells in aerobic culture take up oxygen from the liquid phase. In respiring cells, oxidative reactions involve the participation of molecular oxygen that is present in the liquid phase of the culture media, so that the dissolved oxygen (DO) concentration in this media exerts a great influence on the metabolic activity of the cells and, therefore, in their growth rates.

The response of the specific growth rate to dissolved oxygen concentration is analogous to other limiting substrates.

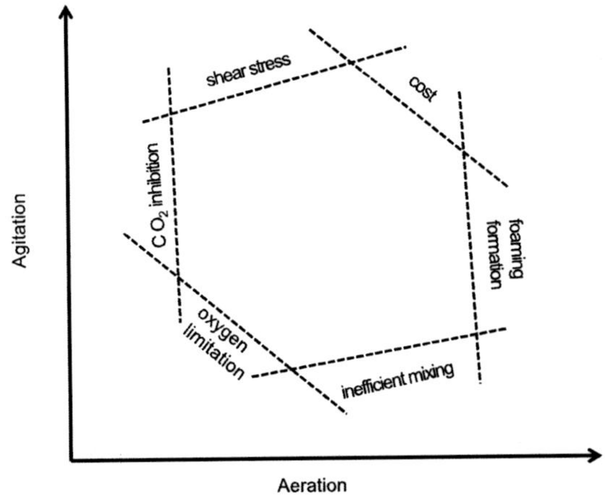

FIGURE 11.1 Operating limits for scale-up of cultivations in bioreactors.

Almost all of the biopharmaceuticals processes are conducted by chemoheterotrophic organisms, which obtain energy by the oxidation of electron donors in their environments. When a cell consumes 1 mole of glucose (180 g) in plain respiration, it requires 6 moles of oxygen (192 g). So, typically, the amount of carbohydrates requires a similar amount of oxygen to be consumed. Dissolved oxygen can be regarded as a nutrient similar to other dissolved nutrients in the liquid phase, such as the carbon and nitrogen sources, and so on. However, there is a striking difference between them: the solubility of oxygen is extremely low as compared to these other nutrients. For example, in a culture medium, we can easily maintain the concentration of a nutrient, such as glucose, at levels above 10,000 mg/L. On the other hand, we hardly can maintain a dissolved oxygen concentration above 10 mg/L. Therefore, there is a need to continuously transfer oxygen from the gas phase to the liquid phase of the culture medium in order to meet the 'cellular oxygen demand', i.e., the rate at which oxygen is consumed by cells in bioreactors. Table 11.1 displays data on the solubility of oxygen in liquid at saturation (i.e. the liquid-phase concentration of oxygen in equilibrium with the concentration of oxygen in the bulk gas phase) for different temperatures, presence of dissolved salts and varying values of partial pressure of oxygen in the gas phase.

Table 11.1 shows that the solubility of oxygen, i.e., the concentration of dissolved oxygen at saturation, decreases with increasing temperature or with increasing concentrations of dissolved salt, such as NaCl. Conversely, the concentration of dissolved oxygen increases with an increase of the partial pressure of oxygen in the gas phase, which could be provided by sparging the system with oxygen-enriched air or pure oxygen (partial pressure of oxygen equal to 1 atm) instead of air to improve oxygen transfer from the gas phase to the liquid phase, reaching 38.8 mg/L at equilibrium at 25°C in water, for example.

Even considering the advantage to have high values of oxygen solubility, we know that lowering the temperature would not be the most adequate strategy, because bioprocesses must operate at optimal growth and production temperature (typically under mesophilic conditions, in the range of 35°C), not at optimal oxygen transfer temperature. As an alternative, it would be interesting to work with

TABLE 11.1
Saturation of O_2

Temperature (°C)	NaCl Conc. (M)	p_{O_2}: Partial Pressure O_2 (atm)	C_S: Conc. O_2 Saturation (mg/L)	*H*: Henry Constant (mg/L/atm)
25	-	0.209	8.10	38.8
35	-	0.209	6.99	33.4
25	-	1.0	38.8	38.8
25	0.5	1.0	34.2	34.2
25	1.0	1.0	28.5	28.5
25	2.0	1.0	22.7	22.7

Values of dissolved oxygen concentration at saturation under different conditions.

higher oxygen partial pressures in the gas, enriching atmospheric air with pure oxygen. However, this must be done with caution, since high concentrations of dissolved oxygen can inhibit aerobic cells.

Table 11.1 also shows that other dissolved chemical species in the medium can reduce the concentration of dissolved oxygen, which can be troublesome since bioprocesses always involve culture media with various nutrients that can impact the concentration of oxygen. This can also happen with the metabolites released by the cells during the process. Alternatively, the influence of chemical species dissolved in the medium can be estimated. Schumpe (1993) describes a methodology for the prediction of gas solubility in electrolyte solutions, including common cultivation media components.

11.2 GAS PRESSURE AND OXYGEN PARTIAL PRESSURE

As previously presented, the solubility of oxygen in a liquid is in the order of a few mg per L, using air as the gas to aerate bioreactors, at 1 atm of pressure. More specifically, the solubility of oxygen in distilled water at 25°C, bubbling air, that contains a partial pressure of oxygen in the gas phase of 0.209 atm, is around 8.1 mg/L. A measurement of the relative dissolved oxygen tension can be done by using dissolved oxygen electrodes, either galvanic or polarographic types.

Oxygen partial pressure of the gas used to aerate bioreactors affects the value of the concentration of dissolved oxygen at saturation, i.e., its solubility. The equilibrium relationship between these parameters for dilute liquid solutions is given by Henry's law (Equation 11.1).

$$C_S = H \cdot pO_2 \tag{11.1}$$

where C_S=concentration of dissolved oxygen in the liquid at saturation (gO_2/m^3); H=Henry's constant ($gO_2/m^3/atm$); pO_2=partial pressure of oxygen in the gas phase (atm), which is defined in Equation 11.2.

$$pO_2 = P \cdot y \tag{11.2}$$

where y=mole fraction of oxygen in the gas phase; P=total gas pressure (atm)

So, the concentration of dissolved oxygen in the liquid at saturation, or, oxygen solubility, is given by Equation 11.3:

$$C_S = H \cdot P \cdot y \tag{11.3}$$

As can be seen in Table 11.1, varying the composition of a liquid media causes consequences on the solubility of oxygen. Moreover, by sparging pure oxygen instead of atmospheric air at the same total pressure and temperature, the solubility of oxygen increases by a factor of 4.8. Alternatively, the solubility can be increased by sparging compressed air at higher pressure. Both these strategies increase the operating cost of the bioreactor, and also it is possible that, in some cases, the cell culture suffers inhibitory effects from exposure to very high oxygen concentrations. Finally, one can observe that Henry's constant (H) is a function of the temperature and of the dissolved components in the liquid phase, such as salts and nutrients.

11.3 AERATION SYSTEMS

Several systems can transfer oxygen from the gas phase to the liquid phase of a culture medium in bioreactors (Badino Junior & Schmidell, 2020). Each one has its own particularities regarding costs, oxygen transfer rate and shear stress conditions. In fact, the oxygen transfer rate is the most important parameter influencing the design and operation of bioreactors in different scales (Thiry & Cingolani, 2002). The oxygen transfer rate is affected by several factors, such as geometry and characteristics of the vessel and its parts, like impellers and baffles, stirrer speed, liquid properties (density, viscosity, superficial tension, etc.) and gas pattern (including flow rate and sparger design); biomass characteristics, like species, concentration, aggregation level and morphology (Bandaiphet & Prasertsan, 2006; Eickenbusch et al., 1995).

According to the bioprocess and bioproduct, the main oxygen transfer systems in bioreactors should be considered (Figure 11.2):

- Superficial aeration (trays)
- Submerged aeration
  - Gas sparging with mechanical agitation (stirred tank reactor)
  - Gas sparging without mechanical agitation (bubble column, airlift)

11.3.1 Superficial Aeration

In this aeration system, atmospheric oxygen is transferred from the gas phase to the liquid phase via contact with the surface of the culture media, without gas sparging. This system is not used in industrial bioreactors, but it is important to be taken into consideration because it is currently used in small scale lab T-flasks, Roller Bottles, and Petri dishes. The oxygen transfer rate is quite small, but the system is simple to assemble because it is static and there is no shear stress to the cells. Shaker flasks in an incubator also use superficial aeration as a means of oxygen transference.

11.3.2 Submerged Aeration

In these systems, compressed air or a mix with pure oxygen gas is used and sparged in the bottom of the liquid column. They can be subdivided into aeration systems with or without mechanical agitation.

11.3.2.1 Aeration by Sparging with Mechanical Agitation (Stirred Tank Reactor)

By far, stirred aerated tank provides the most common oxygen transfer system in the biotechnology industry and in laboratory bioreactors, because they can achieve higher mass transfer rates than other sparged reactors due to the extra power input from the agitator. These reactors have the following parts: impellers – such as Rushton turbine or marine impeller – that are driven by motors and can move the liquid medium with cells; baffles that are used to improve the

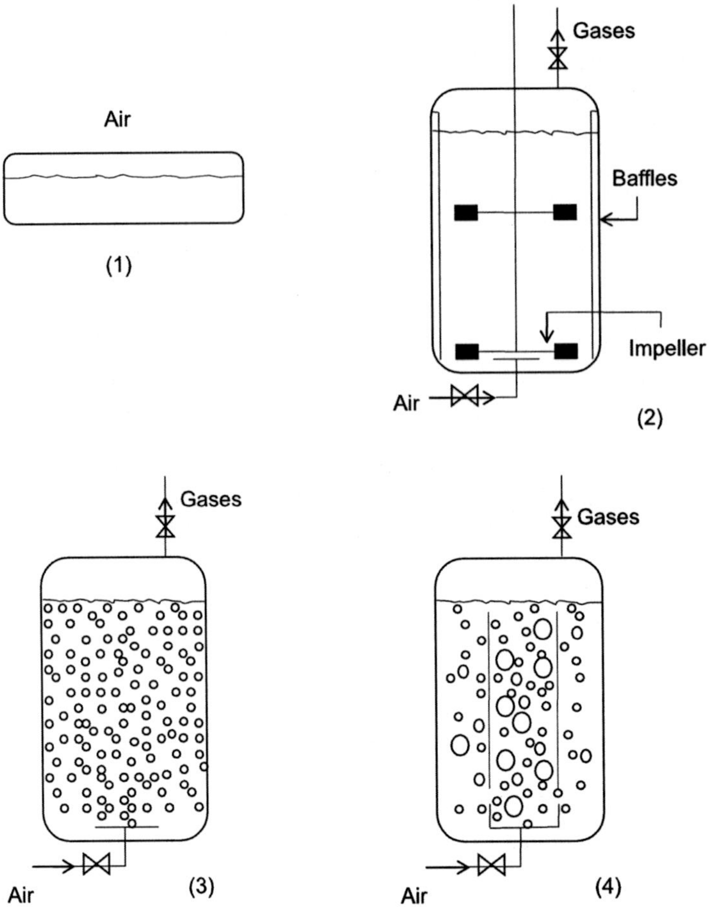

FIGURE 11.2 Main oxygen transfer systems in bioreactors. (a) tray (surface); (b) stirred aerated tank; (c) bubble column; (d) air-lift. (Adapted from Badino Junior, A. C., Schmidell, W. (2020). Agitação e aeração em bioprocessos. In: Schmidell, W. (Ed.), *Biotecnologia Industrial*. 2nd Edition, V.2: Engenharia Bioquímica. Blucher, São Paulo, Cap. 9, 241–310.)

mixing system and create turbulence in the fluid; and spargers that are open pipes with orifices and porous diffusers, known as micro spargers, that release bubbles in the liquid medium and promote the transfer of oxygen from the bubble to the liquid. The gas bubbles are broken and dispersed by the impellers, increasing the efficiency of oxygen dissolution and improving the transfer rate.

11.3.2.2 Aeration by Bubbling with No Mechanical Agitation

In order to avoid shear stress in the cells, some systems without mechanical agitation were developed. Bubble columns systems utilize spargers in different configurations that release small diameter bubbles of gas into the liquid, from the bottom to the top of the bioreactor, promoting the mixture of the culture media and the dissolution of oxygen. Ideally, such bioreactor should be high and narrow so that the bubbles can travel through the entire diameter and keep a long residence time to be in contact with the liquid.

Airlift systems are similar to bubble columns, but they improve the transfer rate and liquid homogenization because they they can define the patterns of liquid flow by creating two zones inside the tank, through a physical separation of upflowing and downflowing streams. Liquid circulation in airlift reactors results from the apparent density difference of the upflow and the downflow. Airlift reactors have been applied for the production of single cell protein and for plant and animal cell culture.

11.4 MECHANICAL AGITATION SYSTEMS IN BIOREACTORS

Mechanical agitation is a mixing operation to homogenize a suspension, maintain suspended solids, ensure heat transfer and improve the mass transfer rate of oxygen.

Baffles consist of vertical strips of metal mounted against the wall of the tank and have the function to reduce the gross vortexing and swirling of the liquid. Generally, four equally spaced baffles are sufficient to prevent liquid swirling and vortex formation. The impellers are mounted on a centrally located stirrer shaft, to promote mixing of the vessel liquid and gas phases. In stirred tank reactors, to achieve a suitable mixture of gases, such as oxygen, as well as to remove carbon dioxide produced from cellular metabolism, a wide range of impellers can be used (Buckland & Lilly, 1993; Doran, 2013). Figure 11.3 shows the main impellers used in industrial bioprocesses.

Impellers with blades aligned parallel to the stirrer shaft, such as the Rushton turbine, provide radial or horizontal flow (from the central axis out to the sides of the tank and back again) (Figure 11.4a), while impellers with inclined or pitched blades that make an angle of less than 90° with the plane of rotation, such as a propeller, have a characteristic of axial flow (up and down the height of the vessel) (Figure 11.4b). Typically, Rushton turbines are used in microbial cultivations, because they promote higher transfer rate and mixing (required for more oxygen demanding growth). On the other hand, propellers are used in animal cell cultures, because they impose smaller shear stress for more sensitive cells.

11.5 OXYGEN TRANSFER AND RESPIRATION: FROM GAS BUBBLES TO CELLS

When we think of a system of agitation and aeration, we are concerned with optimizing the supply of oxygen required for the respiratory activity of the cell, i.e., ensuring the transfer of oxygen from the gas phase to the liquid and its homogeneous dissolution in the culture medium, so that each cell can assimilate molecular oxygen for respiration and biosynthetic reactions (Doran, 2013; Vitolo et al., 2015).

In a well-mixed reactor, we can consider two processes occurring:

a. Oxygen transfer from the gas to the liquid
b. Oxygen uptake by cell

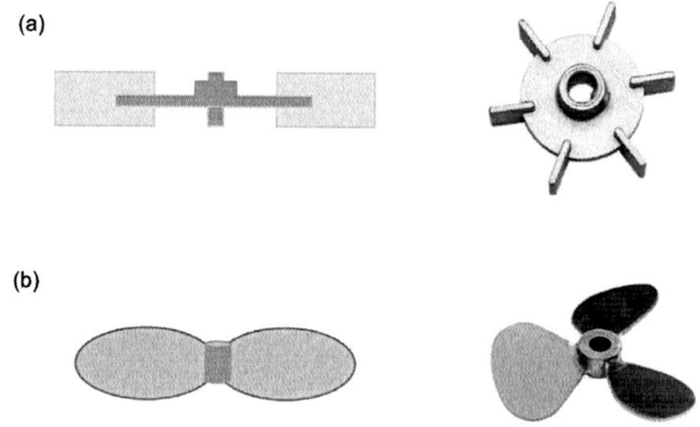

FIGURE 11.3 Types of impellers. (a) Rushton disc type turbine with six flat blades; (b) marine propeller turbine.

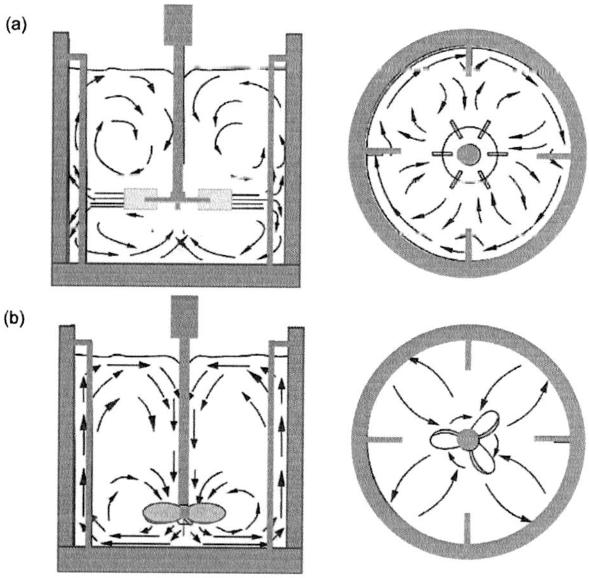

FIGURE 11.4 (a) Radial flow with disc-type impellers and flat blades, and (b) axial flow with impeller-type propellers in tanks with baffles.

Analysing the mass balance of oxygen on the liquid, we have:

$$\begin{matrix}\text{Accumulation in} & & \text{Rate of} & & \text{Rate of} \\ \text{the tank of} & = & \text{input of} & - & \text{output of} \\ \text{oxygen} & & \text{oxygen} & & \text{oxygen}\end{matrix}$$

$$\begin{matrix} & \text{Rate of} & & \text{Rate of} \\ + & \text{production of} & - & \text{consumption} \\ & \text{oxygen} & & \text{of oxygen}\end{matrix}$$

Considering that (input minus output) is the transfer rate to the system, and normally there is no production of oxygen, the variation over time of the concentration of dissolved oxygen in the medium (C) depends essentially on the difference between the rate of mass transfer of oxygen from the gas phase to the liquid phase (supply) and the consumption rate of oxygen by the cells (demand) (Figure 11.5), according to Equation 11.4.

$$\frac{dC}{dt} = \text{OTR} - \text{OUR} \tag{11.4}$$

where OTR is the rate of oxygen that is transferred from the gas phase to the liquid phase (oxygen transfer rate) and OUR is the rate of cell oxygen consumption for respiration (oxygen uptake rate).

The next sections will analyse each of these processes.

11.6 GAS-LIQUID MASS TRANSFER (OXYGEN SUPPLY): OTR

In aerobic cultures, oxygen molecules must overcome a series of transportation resistances before being utilized by the cells. The mechanism of oxygen transfer from the gas bubbles to the bulk of the liquid in which cells are suspended is hampered by various mass transfer resistances, such as the gas film resistance, resistance at the gas-liquid interface and resistance in the liquid film. However, considering a well-mixed reactor, we can consider that the limiting step of the oxygen transfer is the diffusion through the relatively stagnant liquid film surrounding the bubble.

When gas is sparged through a liquid, the interfacial area will depend on the size and number of bubbles present, which in turn depend on many other factors such as medium composition, stirrer speed and gas flow rate. These will affect the mass transfer coefficient (the inverse of respective resistances).

Mass transfers are regulated by mass transfer coefficients and driving forces for the transfer process. Specifically, Equation 11.5 represents the transfer of oxygen around a bubble (Badino Junior & Schmidell, 2020; Vitolo et al., 2015).

$$n_{O_2} = k_L \cdot (C_i - C) \tag{11.5}$$

where n_{O_2} is the mass flux (mass per area per time) of oxygen through the bubble interface (g/m²/h); k_L is the mass transfer coefficient in the liquid film around the bubble (m/h); C_i is the dissolved oxygen concentration at interface (mg/L or g/m³); and C is the dissolved oxygen concentration in the liquid medium (mg/L or g/m³).

Considering the difficulty to measure the mass flow, expressed by interface area, another variable has to be defined to assist this task: a, which is the interfacial area per liquid volume (m²/m³) (Equation 11.6).

$$a = \frac{\text{total interfacial area in bubbles}}{\text{liquid volume}} \tag{11.6}$$

$$n_{O_2} \cdot a = k_L \cdot a \cdot (C_i - C)$$

So, considering just the oxygen transfer rate (OTR or supply), the change in dissolved oxygen in the reactor (C) over time results in Equation 11.7.

$$\frac{dC}{dt} = \text{OTR} \tag{11.7}$$

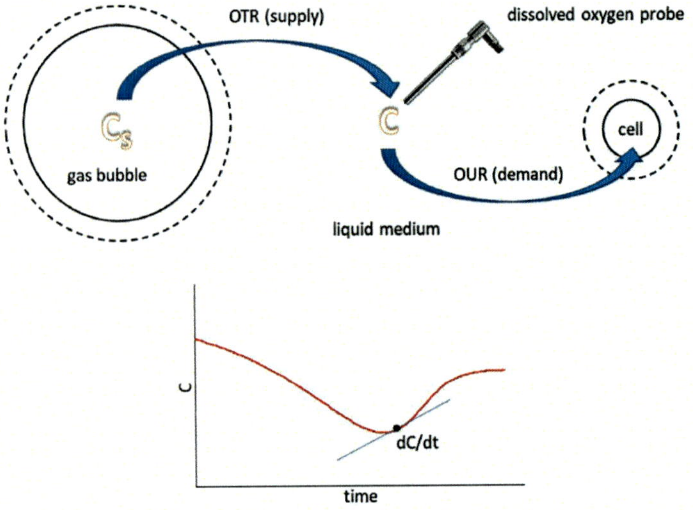

FIGURE 11.5 Oxygen supply and demand scheme (concentric dot-lined circles represent stagnant liquid film around bubbles or cells).

OTR is exactly $n_{O_2} \cdot a$, or mass transfer per volume of reactor per time (Equations 11.8 and 11.9).

$$\frac{dC}{dt} = n_{O_2} \cdot a \tag{11.8}$$

$$\frac{dC}{dt} = k_L \cdot a \cdot (C_i - C) \tag{11.9}$$

Now, k_La can be defined as the volumetric oxygen transfer coefficient, and we will see next how it can be useful to represent the intensity of oxygen transfer in different bioreactor systems.

On the other hand, C_i is difficult to be measured. However, taking into account that the most important resistance in the oxygen transfer from the bubble to the liquid is the stagnant liquid film surrounding the bubble, the value of C_i is approximately equal to C_S and it may be stated according to Equation 11.10:

$$(C_S - C) \cong (C_i - C) \tag{11.10}.$$

where C_s can be easily calculated by Henry's law (Equation 11.3).

So, the oxygen transfer rate (OTR) is given by the product of the volumetric oxygen transfer coefficient (k_La) by the driving force of this transfer ($C_S - C$) (Equation 11.11).

$$\frac{dC}{dt} = k_La \cdot (C_S - C) \tag{11.11}$$

If the oxygen rate from the gaseous to the liquid phase is to be optimized, the OTR must be increased. This can be done by making changes to the design of the bioreactor or its operation.

Regarding design, as seen above, there are different bioreactor types with different oxygen transfer systems. In each one, it is possible to change the geometry. Specifically, in the most common bioreactor (sparged, mechanically stirred-tank reactor), it is possible to change the type, the size and the number of impellers and the type of the sparger. In terms of operation, it is possible to change the transfer of oxygen during cultivation in different forms, which can be used to control the value of dissolved oxygen over time. The most important are:

- *increase agitation*: faster impeller will decrease the resistance in the liquid film due to the decrease of its thickness, increasing 'k_L'. Also, it breaks the bubbles, decreasing their size and increasing the relation surface/volume of each bubble, increasing 'a'.
- *increase air flow*: there will be a proportional increase in 'a'. The total gas flow rate is normally expressed as the liquid volume (in order to correlate with tank size) as 'gas volumes per bioreactor volume per minute', for instance 0.5 VVM.
- *enrich the gas (normally air) with pure O_2*: a higher fraction of oxygen in the gas (y) will increase C_s. This can be done by injecting pure oxygen in air in a controlled amount.

11.7 OXYGEN UPTAKE IN CELL CULTURES (OXYGEN DEMAND): OUR

Cells in aerobic culture take up oxygen from the liquid phase. The rate at which oxygen is consumed by cells (OUR) in bioreactors determines the rate at which oxygen must be transferred from the gas phase to the liquid phase (OTR).

The oxygen uptake rate (OUR) can be expressed as the product of the rate each cell consumes oxygen (Q_{O_2}) multiplied by the cell concentration (X) (Equation 11.12).

$$\text{OUR} = Q_{O_2} \cdot X \tag{11.12}$$

where OUR is the total consumption rate of oxygen by the cells per volume of bioreactor (mg O_2/L/h); Q_{O_2} is the specific oxygen consumption rate (mg O_2/g_{CDW}/h); X is the cell concentration (g_{CDW} /L) (where g_{CDW} is cell dry weight expressed in grams).

Many factors influence the specific oxygen consumption rate (Q_{O_2}), notably the cell species, the culture growth phase and the nature of the carbon source provided in the medium (McNeil & Harvey, 1990; Wang et al., 1979). For a certain organism, Q_{O_2} expresses the physiological state of the cell, and all these factors influence the growth, the substrate consumption and the respiration, considering that oxygen is consumed in parallel with the carbon/energy source consumption.

As seen in Chapter 9, the specific substrate consumption rate can be related to a specific growth rate, according to the Pirt model (Pirt, 1965) (see Equation 9.17), and Equation 11.13 can be obtained.

$$\mu_S = m_S + \frac{1}{Y_G} \cdot \mu \tag{11.13}$$

By analogy, considering that all the factors that affect the specific respiration rate also influence the specific growth rate; a relationship between cell respiration and cell growth can be written in terms of oxygen consumption, according to Equation 11.14.

$$Q_{O_2} = m_O + \frac{1}{Y_{X/O}} \cdot \mu \tag{11.14}$$

where m_O is the maintenance coefficient for oxygen (mg O_2/g_{CDW}/h); $Y_{X/O}$ is the cell biomass yield on oxygen (g_{CDW}/mg O_2); μ is the specific growth rate (h^{-1}).

With this simple relation of Q_{O_2} and μ, it can be assumed that the respiratory rate follows the growth rate and that there will be a maximum Q_{O_2} ($Q_{O_2 max}$) during maximum specific growth rate (at μ_{max}). On the other hand, to a fully respiratory metabolism, it is necessary that the concentration of dissolved oxygen (C, outside the cell) is sufficient to allow oxygen transfer to the mitochondria inside the cell. Thus, one can imagine a relationship between Q_{O_2} and C, in such a way that Q_{O_2} does not depend on C, as long as its value is higher than a certain value, called critical C (C_{crit}), as shown in Figure 11.6.

Typical values of C_{crit} are in the range between 0.15 and 0.60 mg/L, i.e., less than 10% of oxygen solubility (C_S) (Badino Junior & Schmidell, 2020).

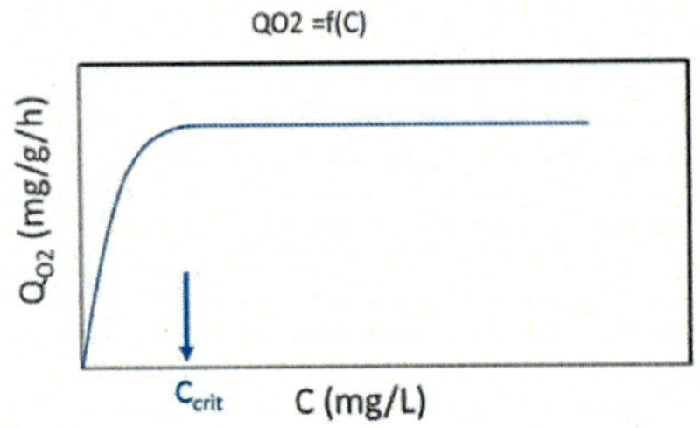

FIGURE 11.6 Correlation between the specific respiration rate of an organism and the concentration of dissolved oxygen.

11.8 INTEGRATING OXYGEN SUPPLY AND DEMAND

After the analysis of the oxygen supply (OTR) and demand (OUR) in the bioreactor, we can rewrite Equation 11.11 in its more complete form using Equation 11.15.

$$\frac{dC}{dt} = k_L a \cdot (C_s - C) - Q_{O_2} \cdot X \tag{11.15}$$

During a typical cultivation in bioreactor, the values of dissolved oxygen (C) vary very slowly. It means that dC/dt has a value much smaller, compared to the values of OTR and OUR.

In fact, the values of OTR and OUR are very similar because the system self-adjusts its value of C. If, for instance, OTR is greater than OUR, the variation dC/dt becomes positive, and C increases, decreasing automatically the transfer driving force ($C_S - C$) and consequently OTR (as in Equation 11.11). On the other hand, if OUR is greater than OTR, dC/dt is negative and C reduces its value, increasing the driving force and OTR.

The rational mentioned above means that throughout cultivation, at each instant, there is a value of C that makes the system in a 'pseudo steady-state' with respect to oxygen transfer.

During a small period of time, OUR≈OTR, $dC/dt \approx 0$ and $k_L a \cdot (C_S - C) \approx Q_{O_2} \cdot X$, so, Equation 11.16 is obtained.

$$C \approx C_S - \frac{Q_{O_2} \cdot X}{k_L a} \tag{11.16}$$

This can be observed by monitoring the dissolved oxygen probe value (typically expressed as a percentage of gas saturation) of a bioreactor. Looking at it for a short time (less than a minute), it seems that the value does not change. But throughout cultivation (hours or days), due to cell growth and change in respiration rate, changing slowly OUR, this value of C changes slowly.

As a numeric example, in a certain moment of the cultivation of *Pichia pastoris* (with cell concentration at 4 g/L), the oxygen probe showed circa 5 mg/L changing very slowly, as 0.3 mg/L/h (dC/dt). At that moment, OUR was calculated as 120 mg/L/h, similar to OTR. So OUR ≈ OTR ≫ dC/dt. This is called a pseudo steady-state, where we can consider dC/dt around zero during a small period of time.

11.9 DISSOLVED OXYGEN CONCENTRATION (C) PROFILE IN A CULTURE WITH CONSTANT OXYGEN TRANSFER CONDITIONS

We will now see what happens in a typical batch culture, in which the oxygen transfer conditions are kept constant, and throughout the cell growth, the C value varies slowly. We are therefore in a situation where the following parameters are kept constant throughout the entire cultivation: the frequency of turbine agitation; the total gas flow and the gas inserted into the system (usually air). Thus, $k_L a$ and C_S are kept constant. Under these conditions, what would happen to C as there is cell growth?

Based on what was seen of growth kinetics in Chapter 9, we can assume the following facts:

- While there is a high concentration of limiting substrate (higher than K_S), the specific growth rate (μ) is approximately constant and maximum. Therefore, according to Equation 11.14, Q_{O2} is also constant and maximum.
- As a result, X grows exponentially
- Therefore, OUR, or $Q_{O_2} \cdot X$, grows approximately exponentially.
- If $Q_{O_2} \cdot X$ grows exponentially, at each instant, there will be a value of C that satisfies Equation 11.15, so that C decreases slowly over the exponential phase (Figure 11.7).
- As the limiting substrate decreases, X continues to grow, but μ reduces (for example following the Monod model) so that the product $Q_{O_2} \cdot X$ reaches a maximum and decreases in sequence.
- As seen above, for each value of $Q_{O_2} \cdot X$, there will be a value of C, which will reach a minimum at maximum $Q_{O_2} \cdot X$ and then will increase (Figure 11.7).

From the above analysis, it is observed that in a typical cultivation, there will be a minimum value of C. Considering that, below the value of critical C, the value of Q_{O_2} decreases and the cells have limited respiration, it is essential to design the oxygen transfer system in such a way that growth is not impaired at any time. For this, $k_L a$ estimates are essential and the system must be prepared to guarantee a $k_L a$ value that ensures that C does not fall below C_{crit}. The following is a numerical example for the calculation of the value of C at the end of the exponential phase.

Consider an *Escherichia coli* culture that starts the exponential growth phase at 0.50 g/L with μ=0.30 h^{-1}. During this 5-h phase, it maintains a Q_{O_2} value of 200 mg $O_2/g_{cel}/h$. Calculate the value of dissolved oxygen at the end of the exponential phase, knowing that $k_L a$ was maintained at 100 h^{-1} and C_S=6.00 mg/L.

$$X \text{ at } t = 5\text{ h}: X = X_0 * \exp(\mu * t) = 0.50 * \exp(0.30 * 5)$$

$$= 0.50 * 4.48; X = 2.24 \text{ g/L}$$

$$Q_{O_2} * X = 200 * 2.24 = 448 \text{ mg}O_2/\text{L}/\text{h}$$

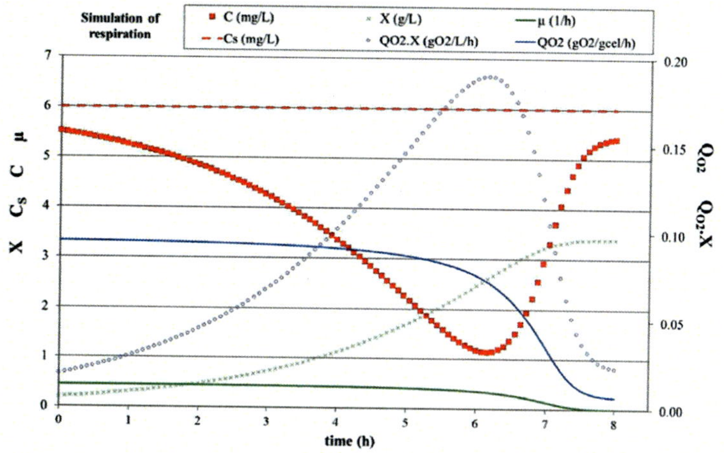

FIGURE 11.7 Simulation of a typical batch culture without variation in k_La and C_S, showing the profile of C and other variables over time.

At pseudo steady-state (Equation 11.16): $C = C_S - Q_{O_2} \times X/k_La$;

$$C = 6.00 - 448/100 = 6.00 - 4.48$$

$$= 1.52\,\text{mg/L}\,(\text{or } 25\% \text{ of air saturation})$$

11.10 CONTROL OF C

Despite the fact that most bioprocesses still use a more inexpensive transfer system – due to the low added value of their products – in the production of biopharmaceuticals, it is common to use more elaborate strategies. As we have seen, in a system with fixed k_La and C_S, the value of C varies throughout cultivation. This variation can affect the quality of the biopharmaceutical, such as the pattern of glycosylation in the case of glycoproteins (Restelli et al., 2006). In this direction, we looked for transfer systems that could keep the C value constant throughout the cultivation. There are two common strategies for this: changing k_La and changing C_S.

Changing k_La means that the dissolved oxygen control system receives the current value of C, compares it with a desired value (*setpoint*) and acts on the variation of the agitation frequency and/or the gas flow, so that k_La is changed and C remains at the desired value. The drawback of this strategy is that variations in the frequency of agitation (and to a lesser extent in the gas flow) can interfere with growth and production, mainly due to the shear stress imposed on the cells. Such interference is particularly detectable in animal cell cultures (host systems for several recombinant proteins that function as biopharmaceuticals) (see Chapter 12).

The other way to change the oxygen transfer to control the dissolved oxygen value C is through systems that can vary the composition of the gas that is inserted in the bioreactor, for example with a mixture of air and pure oxygen. Thus, if it is necessary to increase the transfer to reach the desired value of C, the system increases the pure oxygen flow rate and decreases proportionally the air flow rate, maintaining the total flow rate, but changing the value of y (mole fraction of oxygen in the gas phase) and consequently C_S, according to Equation 11.3. This strategy does not alter the shear stress, but it is more expensive to implement (requires, for example, solenoid valve systems or expensive Mass Flow Controllers MFC).

11.11 MEASUREMENT OF K_La

As seen in Section 11.6, volumetric oxygen transfer coefficient (k_La) is very important to represent the intensity of oxygen transfer in different bioreactor systems. Therefore, it is possible to estimate its value through mathematical correlations (Doran, 2013), but they require several parameters to be used. On the other hand, k_La can be estimated with simple experiments, such as:

- Dynamic method without the presence of cells;
- Dynamic method with the presence of cells (this method will be seen at the end of the Section 11.12 for Q_{O_2} estimation).

11.11.1 Dynamic Method without Cells

As seen in Equation 11.11, the value of the dissolved oxygen concentration in a tank without cells will tend to reach the saturation value (C_S) over time. The rate with which this value is reached will depend on the transfer conditions, represented by k_La. Thus, if from a situation of saturation with air, the gas sparged in the tank is changed to nitrogen, the C_S value drops to zero, causing a negative driving force and dC/dt will be negative. Following the drop in the value of C over time, it is possible to calculate the value of k_La for those transfer conditions.

This procedure must be carried out under the same conditions that will be used in culturing the cells, including temperature and tank pressure. By varying the transfer conditions, for example, agitation frequency or gas flow, information about k_La can be accumulated for each condition. Ideally, the sterile culture medium should be used, but this

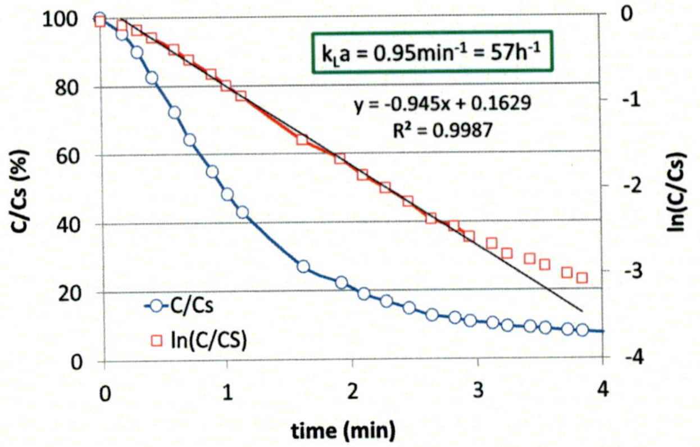

FIGURE 11.8 Dissolved oxygen concentration (C) and ln (C/C_S) as a function of the time of aeration. Equipment: Eppendorf bioreactor (New Brunswick BioFlo III). Experimental conditions: water; 30°C; 0.92 atm; volume 2L; 700rpm; 1.0 vvm aeration. Calculated: k_La=57 h^{-1}.

procedure is laborious (keeping the system sterile during the various measurements) and often only deionized water is used and the approximate value is obtained.

During the descent in the value of C due to the passage of nitrogen (whose oxygen fraction is zero), as C_S is zero, Equation 11.11 is represented by Equation 11.17.

$$\frac{dC}{dt} = k_La \cdot (C_S - C) = - k_La \cdot C \qquad (11.17)$$

Integrating, Equation 11.17 for the fall of C with nitrogen, Equation 11.18 is obtained.

$$\ln\left(\frac{C}{C_S}\right) = - k_La \cdot t \qquad (11.18)$$

Thus, k_La is calculated as the opposite of the slope of curve $\ln(C/C_S)=f(t)$, as shown in Figure 11.8.

Similarly, it is possible to calculate k_La in the rise of the C value, when changing the gas sparged from nitrogen to air. In this situation, C_S value rises to 100% (of saturation), and the integration of Equation 11.11 generates Equation 11.19.

$$\ln\left(1 - \frac{C}{C_S}\right) = -k_La \cdot t \qquad (11.19)$$

To facilitate the understanding of this method, we present the following example in which determinations of the k_La value are made, in water and at a defined temperature, for an ascension of C with air.

Assemble the bioreactor, install the dissolved oxygen and temperature sensors, add the adequate volume of water and adjust the temperature to the desired value. Then, the addition of gaseous nitrogen (N_2) is initiated by bubbling to remove all dissolved oxygen until the dissolved O_2 concentration reaches 0%. After this procedure, the bioreactor must be quickly adjusted to the desired conditions of agitation and aeration and, from this moment on, open the valve to add air and monitor the gradual increase in the concentration of dissolved oxygen (C) as a function of time until its concentration reaches almost saturation, that is, 100% of the value indicated by the sensor. This same procedure must be repeated for different agitation and aeration values so that different k_La values can be calculated. With this, for each aeration/agitation combination, a dissolved oxygen concentration curve (%) will be obtained as a function of the time (this step usually takes few minutes) which will allow calculations of several k_La values in 1 day.

According to integrated Equation 11.19, plotting the ln $(1-C/C_S)$ values as a function of the aeration time, the slope of the curve is determined and will correspond to the opposite of the value of the volumetric oxygen transfer coefficient (k_La). Table 11.2 presents some k_La values obtained under different conditions of agitation and aeration in a specific bioreactor.

11.12 MEASUREMENT OF $Q_{O_2} \cdot X$

In addition to estimating the efficiency of the oxygen transfer system, being able to measure the Q_{O_2} value is essential to

TABLE 11.2
Values of k_La Used in Microbial Cultures, Depending on the Conditions of Agitation and Aeration Obtained in Water in the Benchtop Bioreactor

Aeration (vvm)	Agitation (rpm)	k_La (h^{-1})
0.7	200	15
1.3	400	58
1.3	500	86
1.7	400	60
1.7	600	123
2.0	500	99
2.0	600	128
2.0	700	193
2.3	600	135
2.3	700	200
2.3	800	230

Equipment: Eppendorf bioreactor (New Brunswick mod. NBS-MF 105). Experimental conditions: water; 35°C; 0.92 atm; liquid volume 2 L.

understand the physiological state of the cells, since aerobic metabolism occurs simultaneously with the consumption of carbon sources, such as carbohydrates, and oxygen, and as we have seen, this occurs in parallel with cell growth. There are different ways of estimating Q_{O_2}, and we will see below the main three.

11.12.1 Dynamic Method

Considering that during a normal cultivation, the oxygen is, at the same time, transferred and consumed (Equation 11.15) and that these values are balanced, then dC/dt is almost null, and OTR is almost equal to OUR. That means that at each instant of cultivation, there is a value of C that causes this pseudo steady-state. Let C_0 be the value of C at a given time of cultivation, t_0 (see Equation 11.15). Then, at this time of cultivation, Equations 11.20 and 11.21 are obtained.

$$C_o = C_S - \frac{Q_{O_2} \cdot X}{k_L a} \quad (11.20)$$

$$k_L a \cdot (C_S - C_o) = Q_{O_2} \cdot X \quad (11.21)$$

If at time t_0, for a short period of time, the transfer OTR$=k_La\cdot(C_S-C)$ is cancelled; from this moment on, there will only be oxygen consumption, then OUR $= Q_{O_2} \cdot X$, and Equation 11.22 is obtained.

$$\frac{dC}{dt} = -\, Q_{O_2} \cdot X \quad (11.22)$$

Using a dissolved oxygen probe, which records the value of C over time, during this disturbance in which the transfer is cancelled, the value of $Q_{O_2} \cdot X$ can be estimated by the slope of the curve $C=f(t)$ in the section $t_0=>t_1$ of Figure 11.9.

The following is a numerical example for the calculation of $Q_{O_2} \cdot X$ value through the dynamic method.

Consider a cultivation that at a certain moment, dissolved oxygen probe shows 60% (pseudo steady-state). Suddenly, the oxygen transfer is interrupted and we observe a decrease in C as follows:

t (s)	0	5	10	15	20	25	30
C (%)	60	58	52	46	40	34	28

After 30 s of disturbance, oxygen transfer is restarted. To calculate the $Q_{O_2} \cdot X$ value, knowing that the oxygen solubility is 7 mg/L, we consider that after stabilizing the fall in the value of C, we have $\Delta C/\Delta t=-6\%/5$ s$=$ -1.2%/s. Considering that 100% of air saturation is 7 mg/L, $\Delta C/\Delta t=-1.2\%/\text{s}*7$ mg/L/100%$=-0.084$ mg/L/s. Converting the unit of measurement to hours, we have: $dC/dt=-0.084$ mg/L/s$\times$3,600 s/h$=-302$ mg/L/h. According to Equation 11.22, $dC/dt = -Q_{O_2} \cdot X$; then $Q_{O_2} \cdot X = -dc/dt = -(-302$ mg/L/h$)=302$ mg/L/h.

In addition to calculating $Q_{O_2} \cdot X$, it is possible through this method to calculate the k_La value. After this disturbance, the oxygen transfer is restarted at t_1, and the concentration of dissolved oxygen begins to rise, towards the previous value of C_0, relative to the consumption condition of that moment of the cultivation. During this ascent, the phenomenon can be used to estimate k_La.

Substituting the value of $Q_{O_2} \cdot X$ by k_La. (C_S-C_0) in Equation 11.15, valid for this moment of cultivation as seen above, Equations 11.23 and 11.24 are obtained.

$$\frac{dC}{dt} = k_L a \cdot (C_S - C) - k_L a \cdot (C_S - C_0) \quad (11.23)$$

$$\frac{dC}{dt} = k_L a \cdot (C_0 - C) \quad (11.24)$$

Integrating Equation 11.24 from t_1, Equation 11.25 is obtained.

$$\ln\left(\frac{C_0 - C}{C_0 - C_1}\right) = -k_L a \cdot (t - t_1) \quad (11.25)$$

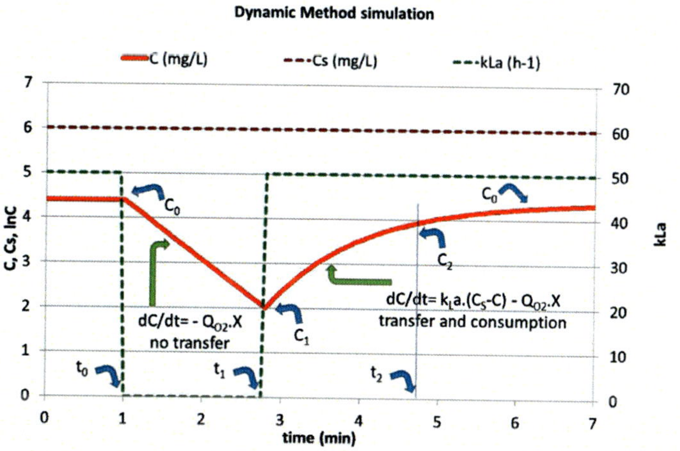

FIGURE 11.9 Simulation of dissolved oxygen concentration (C) as a function of the time in a dynamic method to measure $Q_{O_2} \cdot X$ (from t_0 to t_1) and k_La (from t_1 to t_2).

Thus, k_La is calculated as the opposite of the slope of curve $\ln[(C_0-C)/(C_0-C_1)]=f(t)$.

k_La can also be calculated with just two points of the ascent in the curve of C (C_2 and C_1) in time t_2 and t_1, as shown in Figure 11.9 and represented by Equation 11.26.

$$k_La = -ln\left(\frac{C_0 - C_2}{C_0 - C_1}\right)\Big/(t_2 - t_1) \qquad (11.26)$$

11.12.2 Gas Balance

Another way to calculate the value of $Q_{O_2} \cdot X$, or OUR, considers that there is no accumulation of oxygen in the liquid phase (culture medium) in the reactor and that all oxygen that enters the system is either consumed by the cells or leaves the tank. Thus, if it is possible to measure the mass flow of oxygen in and out of the system, the difference between them will be the total consumed by the cells in respiration. To measure the mass (or molar) flow of oxygen in the gas stream, it is necessary to measure the total flow rate and the molar fraction of oxygen in the gas (Figure 11.10). Particularly, if the input gas is air, we know the molar fraction of oxygen in it ($y_{air}=0.2095$). However, the measurement in the outlet gas requires expensive equipment, called a gas analyser, which typically measures the fraction of oxygen and carbon dioxide in a gas.

With this data, Equation 11.27 is obtained.

$$\text{OUR} = Q_{O_2} \cdot X = \left(\Phi_i \cdot y_{Oi} - \Phi_o \cdot y_{Oo}\right)/V \qquad (11.27)$$

where: $\text{OUR}=Q_{O_2} \cdot X$ = oxygen uptake rate (mmol O_2/L/h); Q_{O_2} = cell specific respiration rate (mmol O_2/g/h); X = cell concentration (g/L); V = volume (L); Φ = gas flow rate (mmol/h); y = molar fraction of O_2 (or CO_2) in inlet or outlet gas; subscripts: i = inlet gas; o = outlet gas; O = oxygen; C = carbon dioxide.

As it is also common to have the carbon dioxide molar fraction of the gas, it is also possible to calculate CER value ((Equation 11.28) and Respiratory Quotient (RQ), that is the ratio of the CO_2 produced to the O_2 consumed in respiration (Equation 11.29)).

$$\text{CER} = Q_{CO_2} \cdot X = \left(\Phi_o \cdot y_{Co} - \Phi_i \cdot y_{Ci}\right)/V \qquad (11.28)$$

$$\text{RQ} = \frac{\text{CER}}{\text{OUR}} = \frac{Q_{CO_2}}{Q_{O_2}} \qquad (11.29)$$

where: $\text{CER}=Q_{CO_2} \cdot X$ = carbon dioxide evolution rate (mmol CO_2/L/h); RQ = respiratory quotient; Q_{CO_2} = cell-specific carbon dioxide production rate (mmol O_2/g/h).

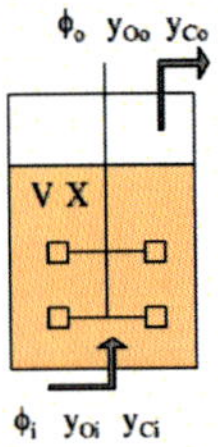

FIGURE 11.10 Scheme of variables in bioreactor for OUR measurement via gas balance method.

It should be noted that the RQ value can be very useful for monitoring the physiological state of cells, indicating whether they are respiring completely all the consumed oxygen, or producing fermentation products, such as lactic or acetic acid (Goudar et al., 2011). In addition, all of this information is acquired, automatically and all the time, without the need to manipulate the cultivation, as in the dynamic method.

The following is a numerical example for the calculation of $Q_{O_2} \cdot X$ value through Gas Balance.

Consider a 5-L cultivation of *Pichia pastoris* that at a certain moment, oxygen fraction of exhaust gas is 19% (0.19). Knowing that the total gas flow rate is 3 L/min (inlet and outlet), and it is air in the inlet, calculate $Q_{O_2} \cdot X$ at that moment (temperature = 25°C and pressure = 1 atm). First, we have to calculate the total gas flow rate in moles per minute. For this, we will use the ideal gas law ($P \times V = n \times R \times T$). Considering 1 min, we have: $n = P \times V/R/T = 1$ atm×3 L/0.082 (L×atm/mol/K)/(25+273) K = 0.123 mol. The total gas flow rate (Ø) is: Ø = 0.123 mol/min (=Ø$_i$=Ø$_o$). Using Equation 11.27, we have $Q_{O_2} \cdot X = \left(\Phi_i \cdot y_{Oi} - \Phi_o \cdot y_{Oo}\right)/V =$ (0.123×0.21 − 0.123×0.19)/5, (where $y_{Oi}=0.21$ is oxygen fraction of inlet air and V=5 L). Then: $Q_{O_2} \cdot X = 0.00049$ $\text{mol}_{O_2}/\text{L}/\text{min} = 0.030\,\text{mol}_{O_2}/\text{L}/h = 0.94\,\text{g}_{O_2}/\text{L}/h$.

11.12.3 Liquid Phase Mass Balance

If the value of k_La has already been calculated, it is also possible to estimate the value of $Q_{O_2} \cdot X$ based on Equation 11.21. Thus, for each moment of cultivation, using the C_0 value of a pseudo steady-state, we have Equation 11.30.

$$Q_{O_2} \cdot X = k_La \cdot (C_S - C_o) \qquad (11.30)$$

11.13 CRITERIA FOR SCALING UP THE BIOPROCESS

It should be noted that when calculating the cost of an industrial bioprocess, the oxygen transfer system can have a major impact, and it is necessary that the sizing is adequate, so that there is neither limitations in the transfer (avoiding C to reach below C critical and cells to lack oxygen) nor over-dimensioning, causing unnecessary costs. Based on experiments carried out on a laboratory scale, it is usual to design the dimensions and operating conditions of the industrial bioreactor using different scale-up expansion criteria, that is, designing the system on a larger scale while maintaining equal the value of the criterion parameter. Those parameters are as follows: volumetric oxygen transfer coefficient (k_La), power in the non-aerated system per volume, linear speed at the impeller tip or dissolved oxygen concentration (C).

11.14 FINAL CONSIDERATIONS

The majority of biopharmaceutical production processes are aerobic and carried out under aeration and agitation since they are important factors for adequate oxygen mass transfer, as well as mixing of nutrients, cells suspension and other heat and mass transfers in the system. Aeration is performed by bubbling air or a mixture of gases through the liquid, and the agitation makes the components of the cultivation homogeneous, such as the cells and the nutrients. Therefore, cells must be supplied with oxygen during growth, considering the very low solubility of oxygen in liquids. However, the productivity of many cultivations is limited by oxygen availability, and it is important to consider the factors which affect the bioreactor's efficiency in supplying oxygen to the cells. Usually the most important factor is the agitation responsible for reducing the size of the bubbles, increasing the specific surface, improving the transfer coefficient in the liquid phase and reducing the thickness of the liquid film around the bubble. The efficiency of the dissolved oxygen transfer process in the cultivation medium depends on both agitation and aeration, and the concept of k_La, volumetric oxygen transfer coefficient, was established. This parameter varies as a function of the intensity of agitation and aeration and can be adjusted according to the specific needs of each cultivation. The use of k_La as a parameter that aggregates the agitation and aeration variables in only one parameter has the advantage of being a quick method and can be measured (Pessoa et al., 1996). On the other hand, through the oxygen transfer system, it is possible to monitor the metabolism of the cells, specifically their respiration rate, which is deeply connected to substrate consumption rate and growth rate. Several methods to estimate oxygen transfer and respiration rate were reviewed here.

REFERENCES

Bandaiphet, C., & Prasertsan, P. (2006). Effect of aeration and agitation rates and scale-up on oxygen transfer coefficient, k_La in exopolysaccharide production from *Enterobacter cloacae* WD7. *Carbohydrate Polymers* 66, 216–228.

Buckland, B.C., & Lilly, M.D. (1993). Fermentation: An Over View. In: Rhem, H.-J. & Reed, G., (Eds.), *Biotechnology*. Vol. 3: Bioprocessing. Verlag Chemie, Basel.

Crueger, W., & Crueger, A. (1984). *Biotechnology: A Textbook of Industrial Microbiology*. Science Tech, Madison.

Doran, P. M. (2013). *Bioprocess Engineering Principles*. 2nd Edition. Academic Press, Cambridge.

Eickenbusch, H., Brunn, P.-O., & Schumpe, A. (1995). Mass transfer into viscous pseudoplastic liquid in large-diameter bubble columns. *Chemical Engineering Process*, 34, 479–485.

Goudar, C. T., Piret, J. M. and Konstantinov, K. B. (2011), Estimating cell specific oxygen uptake and carbon dioxide production rates for mammalian cells in perfusion culture. *Biotechnology Progress*, 27, 1347–1357. doi:10.1002/btpr.646.

McNeil, B., & Harvey L. M. (1990). *Fermentation: A Practical Approach*. IRL Press, New York.

Pessoa, A., Vitolo, M., & Hustedt, H. (1996). Use of Kla as a criterion for scaling up inulinase fermentation process. *Applied Biochemistry and Biotechnology*, 57, 699–709.

Pirt, S. J. (1965). The maintenance energy of bacteria in growing cultures. *Proceedings of the Royal Society of London. Series B, Biological Sciences*, 163(991), 224–231.

Restelli, V., Wang, M. D., Huzel, N., Ethier, M., Perreault, H., & Butler, M. (2006). The effect of dissolved oxygen on the production and the glycosylation profile of recombinant human erythropoietin produced from CHO cells. *Biotechnology and Bioengineering*, 94, 481–494. doi:10.1002/bit.20875.

Badino Junior, A. C., Schmidell, W. (2020). Agitação e aeração em bioprocessos. In: Schmidell, W. (Ed.), *Biotecnologia Industrial*. 2nd Edition, V.2: Engenharia Bioquímica. Blucher, São Paulo, Cap. 9, 241–310.

Schumpe, A. (1993). The estimation of gas solubilities in salt solutions. *Chemical Engineering Science*, 48(1), 153–158. doi:10.1016/0009-2509(93)80291-w.

Thiry, M., & Cingolani, D. (2002). Optimizing scale-up fermentation processes. *Trends in Biotechnology*, 20, 103–105.

Tortora, G. J., Funke, B. R., & Case, C. L. (2005). *Microbiologia*. Artmed, São Paulo.

Vitolo, M., Pessoa, Jr., A., Souza, G. M., Carvalho, J. C. M., Stephano, M. A., Sato, S. (2015). Biotecnologia Farmacêutica. In: Carvalho, J.C.M., Matsudo, M.C., Bezerra, R.P., Sato, S. *Tecnologia de Fermentações*. 1st Edition. Blucher, São Paulo, Cap. 3, 103–156.

Wang, D. I. C., Cooney, C. L., Demain, A. L., Dunnill, P., Humphrey, A. E., Lilly, M. D. (1979). *Fermentation and Enzyme Technology (Techniques in Pure and Applied Microbiology)*. 1st Edition. Wiley, New York.

12 Mammalian Cell Culture Technology

Patrícia Barros dos Santos
Empresa Brasileira de Hemoderivados e Biotecnologia

Patrícia Léo
Instituto de Pesquisas Tecnológicas do Estado de São Paulo

Ricardo Pinheiro de Souza Oliveira and Marco Antonio Stephano
Universidade de São Paulo

CONTENTS

12.1 INTRODUCTION

Mammalian cell culture technology has developed enormously over the past 30 years and has become an integral part of biotechnology, as it has emerged from the coalescence of several areas of knowledge such as biochemistry, cell biology, genetic engineering, protein chemistry, genomics and chemical engineering. This technology is based on the use of mammalian cell culture techniques in strictly controlled systems. It has been used for decades in the research and development of several biological products such as viral vaccines, recombinant proteins, monoclonal antibodies and, more recently, in cell and gene therapies. This *in vitro* culture system is also widely used in pharmacology and toxicology for the search for new drugs.

In 1907, an experiment in which nerve cells from an amphibian embryo were grown in a medium containing lymphatic fluid observed the growth of axons with long, thin fibres at nerve endings. This demonstrated, for the first time, the possibility of growing cells and tissues and is recognized as one of the greatest contributions to medical research.

The technique developed by Ross Harrison, and published in more detail in 1910, was reproduced by several other researchers and can be considered as a precursor to organ transplants and vascular surgeries (Witkowski, 1979). Although several researchers at that time postulated that tissues grown *in vitro* would not survive for long periods of time, Carrel and Burrows (1911) kept cells viable for more than 11 years simply by continually renewing culture medium. These authors were also the first to describe tissue culture methods (Carrel and Burrows, 1911) and detailed aseptic techniques necessary to avoid microbial contamination. Curiously, these aseptic techniques predated the discovery of antibiotics (Dutkowski, De Rougemont, and Clavien, 2008).

Cell cultures began to be used to obtain therapeutic products that required production in living cells, such as viral vaccines. In 1949, studies demonstrated that cell cultures from different origins could be used to replicate and then produce the polio virus on a large scale, using tubular 'roller'-type bottles. The application of this cultivation technique allowed the replacement of monkeys by *in vitro* culture, in addition to enabling the large-scale production of the polio virus (by Enders, Weller, and Robbins, 1949). In 1955, the virologist Jonas Salk announced the results of a large-scale national study that established safety and efficacy of a poliomyelitis vaccine (Tan and Ponstein, 2019). This killed poliovirus vaccine manufactured using a primary cell culture from monkey kidney was the first product with commercial interest generated from the cell culture technology (Weller et al., 2004).

An important step in the evolution of cell culture technology was the approval of the use of continuous cell lines by the FDA (Food Drug Administration) in 1979 (Petricciani, 1995). These continuous cell lines can grow indefinitely and some of these can be grown in liquid suspension, a fact that made it possible to scale up the process. The use of recombinant DNA technology paved the way for the development of new products by cell culture, with the possibility of improving stability, effectiveness and biological activity of therapeutic peptides by manipulating the genes that encoded their primary amino acid structure.

The simultaneous development of technologies for recombinant organisms and technologies for culturing mammalian cells allowed bioprocesses to be conducted in a submerged medium and to become the main platforms for obtaining biological products and/or biopharmaceuticals. Growing cells of higher eukaryotes (e.g. mammalian cells), due to their complex protein biosynthesis systems, enabled the expression of complex biomolecules whose biological activities and bioavailability are dependent on the tertiary structure or post-translational modifications (PTMs), e.g. presence of non-immunogenic glycosylation, which provides a prolonged therapeutic treatment in humans.

Important classes of biological products are biopharmaceuticals. The most commercially successful products are recombinant proteins, monoclonal antibodies and drugs containing nucleic acids that can be produced by prokaryotic (e.g. *Escherichia coli*) or eukaryotic cells (e.g. *Pichia pastoris*, *Saccharomyces cerevisiae* or mammalian cells: CHO, BHK-21, Vero).

The interest in the large-scale production of recombinant proteins was driven by the need to obtain products that offer greater safety and of sufficient quantity to meet the increasing demands for drugs that were not possible to be synthesized chemically. It is important to note that mammalian cells, in addition to producing proteins for therapeutic use, can also be used as products themselves, as in the case of tissue engineering and cell therapy.

In this chapter, aspects related to the technology of culturing cells of mammalian origin will be presented, as well as the production of biopharmaceuticals used in the treatment and prevention of autoimmune diseases, cancer, rare or neglected diseases, cell therapy, tissue engineering, diagnosis and replacement of the used animals in preclinical and cosmetic studies.

12.2 BRIEF HISTORY

The advent of recombinant DNA and hybridoma technology marked the beginning of a new age in pharmaceutical sciences. These innovations overcame difficulties related to the production of therapeutic proteins because they had a limited or impractical medical application due to the low amounts available in biological fluids or in the tissues of animal origin (Walsh, 2013). The following are examples of therapeutic and technological advantages from the use of recombinant proteins in medicine:

- *Availability*: Many therapeutic proteins are found in low quantities such as interferons and coagulation factor VIII at body fluids (e.g. plasma). It decreases the availability in quantity, and it may be economically or technically unfeasible to extract and purify some important proteins, which would limit the access of patients to the indicated treatment.
- *Product safety*: The direct extraction from biological sources has, in the past, resulted in the transmission of diseases. Some examples are the transmission of Creutzfeldt–Jakob disease by human growth hormone (GH) preparations derived from pituitary glands, harvested from human cadaver in the 1970s, and the introduction of blood-borne viruses such as HIV and hepatitis C to a high percentage of haemophiliacs, by plasmatic concentrates of coagulation factors (White, 2010). Those and many other examples of product contamination and subsequent infection of patients drove the development and application (now required by regulatory agencies) of viral inactivation stages during the production of biomedicines (Darling, 2002).
- *Process safety*: Some raw materials used to produce therapeutic proteins are extracted from biological sources such as urine and human plasma. For example, follicle-stimulating hormone, also called 'urofollitropin' (u-FSH) used in fertility treatment, is extracted from the urine of postmenopausal women. The use of purification techniques, such as immunoaffinity chromatography, provides a homogeneous and standardized u-FSH with a reduced batch-to-batch variability. However, there are difficulties related to supply and quality of the urine used as the raw material (Dembowsky and Stadler, 2001).
- *More efficiency*: Some techniques causes small changes as is the case of the insertion, exclusion or alteration of a single amino acid residue, but they can also be more significant and include alteration and/or elimination of an entire peptide sequence (equivalent to a domain of a protein, for instance).

An example of the positive impact of recombinant DNA technology is the sustainable production of insulin, used in the treatment of diabetes mellitus. Original production started in 1923 by extracting the hormone from the pancreas of oxen and pigs; however, many cases of allergic reactions due to the animal origin of the protein were observed. In 1978, scientists from one of the first biotechnology companies to be created, Genentech, used recombinant DNA technology to clone the gene encoding human insulin and expressed this gene in *Escherichia coli*, as it grows faster than any mammalian cell. They licensed this discovery to Eli Lilly and Company, which in 1982 became the first recombinant protein for human use approved by the FDA, the health regulatory agency in the United States, under the trade name Humulin™. With the therapeutic need to express increasingly complex proteins that require post-translational processing only available in eukaryotic expression systems, studies have been intensified in order to manufacture these therapeutics on a large scale using mammalian cells.

The first recombinant protein, registered in the USA in 1987, produced in a mammalian cell (a Chinese hamster cell line called 'CHO') was tissue plasminogen activator (tPA), a so-called clot-buster drug used to restore blood flow by dissolving blood clots that can block arteries in patients with acute myocardial infarction. Then, in 1989, the FDA approved human recombinant erythropoietin (Epogen®, Amgen Corporation), again produced in CHO cells. Human erythropoietin (EPOh) is a glycoprotein produced mainly by the kidneys and stimulates progenitor cells to proliferate and differentiate into erythrocytes, in addition to controlling haemoglobin synthesis. EPOh was first isolated and subsequently purified from urine in the 1970s. Studies on the application of this protein as a possible treatment for patients with anaemia, caused by chronic renal failure, stimulated the search for genes responsible for its biosynthesis and subsequent production using recombinant DNA methods to provide a sufficient material to proceed with clinical studies in the 1980s.

The EPOh therapeutic protein was one of the first generation of biopharmaceuticals in which the natural protein was replaced using a recombinant source. This biopharmaceutical differs in amino acid sequence from the native protein that allows two new N-glycosylation sites so that when produced in CHO cells, the recombinant protein contains five glycosylation sites instead of the three which are found in the native protein. The presence of two new carbohydrate chains gives the molecule an extended serum half-life (up to 21 h, compared to 4–6 h for the native molecule).

In the mid-1990s, the first monoclonal antibodies and their fragments were approved for human use. Within this category is the group that targets tumour necrosis factor alpha (TNF-α), a major cytokine in the inflammatory process implicated in autoimmune diseases such as rheumatoid arthritis and psoriasis. Another example is etanercept, an anti-TNF-α medication, one of the most widely prescribed biopharmaceuticals, approved in 1998 for the treatment of rheumatoid arthritis and again, produced in a CHO cell line. Etanercept is a fusion protein (150 kDa) of the human tumour necrosis factor (TNFR2/p75) p75 receptor and the human IgG1 Fc fragment. This protein inhibits the binding of TNF-α and TNF-β to TNF receptors on the surface of cells, rendering TNF biologically inactive and thereby preventing cell responses mediated by it. Before the release of these antibodies therapeutics, the only anti-rheumatoid agents available were nonsteroidal anti-inflammatory drugs (NSAIDs) and corticosteroids, which have many undesirable side effects from the prolonged use necessary to treat this condition.

Up to 1989, only 33% of approved biopharmaceuticals were manufactured in mammalian-based production cell lines. Between 2015 and 2018, this scenario changed completely, and 79% of registered products were 'mammalian cell derivatives', demonstrating the importance of this expression system. Much of this increase is due to monoclonal antibodies and the expressive approval of biosimilars (WALSH, 2018).

Even so, alternative more efficient or economic eukaryotic systems are continuously being sought for protein expression, for example plant and insect cell cultures.

In insect cell-based systems uses baculoviruses (a viral family that naturally infects insects) as cloning vectors to transfect insect cells. Insect cells are capable of PTMs, such as N-glycosylation, and express high intracellular levels of recombinant protein. However, the glycosylation profiles are still not the same as proteins expressed using mammalian cell systems. In 2007, the first biopharmaceutical for human use produced using an insect cell expression system was approved in the USA. The biopharmaceutical was a bivalent vaccine marketed under the trade name Gardasil (Merck Sharp & Dohme) containing the main proteins of the human papilloma virus capsid types 16 and 18, which account for about 70% of cervical cancer cases.

Plant cells are also alternatives to the use of animal cells because of their ability to offer a new production system and reduced risk of transmitting pathogens capable of infecting mammals. However, due to the fact that they have different glycosylation profiles to the native human proteins, the proteins produced in these systems have shown hyperallergenic reactions in toxicology studies when administered parenterally into laboratory animals.

The production of biopharmaceuticals using mammalian cells on laboratory, pilot and large scales is next discussed in this chapter; matters including cell origin, controlling authenticity, sterility, isolating producer clones and using bioreactors, as well as good practices (laboratory and manufacturing), will also be addressed.

12.3 MAIN PRODUCT CATEGORIES

12.3.1 Vaccines

Viral vaccines against polio, measles, rubella and mumps for human use, in addition to rabies, and foot-and-mouth disease for veterinary use, are important products produced efficiently by cell culture technology. For these purposes, primary cells, diploid cells, continuous cell lines and recombinant cell lines can be used. An important advance in the large-scale production of viral vaccines using anchorage-dependent cells occurred with the development of microcarriers in the late 1960s, as it enabled cultivation in agitated tanks, called 'Stirred Tank Reactors'.

12.3.2 Monoclonal Antibodies

Monoclonal antibodies are biomolecules with important applications for diagnostic and therapeutic purposes. Their importance as therapeutics agents have evolved recently, since murine antibodies were immunogenic and are now being replaced by human, humanized or chimeric antibodies. The main areas of application are in the treatment of acute rejection after organ transplantation, diagnosis and treatment of cancer, rheumatoid arthritis, leukaemia, asthma and multiple sclerosis.

The first commercialized antibody molecule was Orthoclone (J&J), a potent immunosuppressive agent effective in the therapy of acute renal allograft rejection, which recognized the CD3 receptor of T lymphocytes (SCHWAB et al., 1985). This and other antibodies with exclusively murine characteristics, when used as therapies for humans, induced immune responses in patients, a phenomenon called 'HAMA' (human antibody murine antibody).

Avastin® (bevacizumab) is a recombinant humanized monoclonal antibody produced by recombinant DNA technology. It was first approved for therapeutic use in the United States in 2004 and later in the European Union, in 2005. It is indicated for the treatment of patients with metastatic colorectal cancer because it causes the inhibition of angiogenesis (formation of new arteries), which is responsible for maintaining tumour viability. Inhibition occurs due to binding of the antibody to the vascular endothelial growth factor (VEGF), preventing binding to its cell surface receptors, an essential process to trigger the formation of new blood vessels in the tissues. Avastin is produced using CHO cells in 12,000-L bioreactors, and purification is performed by immunoaffinity chromatography, followed by ion-exchange chromatography.

12.3.3 Glycoproteins

Glycoproteins, found outside the plasma membrane, are molecules that contain a portion of protein and at least a portion of carbohydrate. They belong to another important group of products obtained from mammalian cells. The main examples are cytokines (for example, interferons and interleukins), haematopoietic growth factors (for example, erythropoietin), GHs, thrombolytics (for example, tissue plasminogen activator), coagulation factors (factor VII, factor VIII, factor IX, etc.) and recombinant enzymes (DNAse) (Orturk, 2006).

12.3.4 Cells and Tissues

This group of products includes successful examples of the growth of liver, kidney, skin, cartilage and bone tissues, in addition to the differentiation of haematopoietic cells for bone marrow transplantation and gene therapy.

Replacing tissues (such as bone or cartilage) or graft materials poses the risk of virus infections (such as HIV, hepatitis C) or rejection. Artificial implants have limitations due to allergic reactions caused by the material. Therefore, cell culture in order to generate new tissues, that is, cell-based substitutes could restore, maintain or improve tissue function.

12.3.5 Gene Therapy

Gene therapy is based on the transfer of genetic material through (a) a single gene, the most common current form of gene therapy; (b) fragments of coding sequences, as in RNA-based therapy; and (c) entire genome, as occurs in therapies based on stem cells or embryonic cells.

The expected result for all types of gene therapy is to modify a defective phenotype. The simplest strategy is to correct or compensate for abnormal gene expression caused by the faulty genotype. In addition, gene therapy can be designed to regenerate a diseased organ, either by reengineering tissues, by expressing genes that induce organ development or, in the case of stem cell therapy, by using reprogramming pluripotent cells to generate normal tissues. The successful application of this therapy depends on addressing several challenges, including (a) genetic information; (b) the host's immune responses; (c) the hereditary transmission of the disease and the organs that display the abnormal phenotype; and (d) the various mutations that result in the disease (O'connor and Crystal, 2006).

12.4 MAMMALIAN CELLS

12.4.1 Basic Characteristics

Eukaryotic cells have a nucleus and numerous organelles in their cytoplasm that enable a complex division of intracellular metabolic tasks. The following is a brief description of the eukaryotic cell.

- *Nucleus*: It contains chromosomes and is where DNA replication and RNA synthesis take place. A porous nuclear membrane isolates and protects the genome from structural damage and interference with transcription. This segregation allows for two crucial steps in the expression of genetic information: (a) Transcription of DNA sequences into RNA. When the cell is not dividing, the DNA is in the form of chromatin, which allows transcription into RNA, and (b) the messenger RNA (mRNA) leaving the nucleus for the cytoplasm is translated resulting in the synthesis of a protein (translation).
- *Cytoplasm*: The organelles present in the cytoplasm are highly adapted and specialized. As an example, there are the ribosomes (produced in the nucleolus), which float freely in the cytoplasm or bind to another organelle called the 'endoplasmic reticulum' (ER), composed of interconnected membranes forming a network that enables the transport of molecules for post-translational modification or trafficking to specific destinations such as the cell membrane. The rough ER (because it is associated with ribosomes) periodically binds to the nuclear membrane to receive the mRNA to be translated. In addition, there is a smooth ER that has no ribosomes attached and receives the synthesized proteins and traffics these to the Golgi apparatus, where post-translational processing (e.g. glycosylation) will take place. Other organelles associated with protein biosynthesis and therefore of importance for biopharmaceutical production using eukaryotic production systems are vacuoles, which are used for storage, and vesicles that are responsible for transporting the molecules through the cell, for example, carrying the ER proteins to the Golgi apparatus.

12.4.2 Post-Translational Processes

The type and extent of PTMs will depend upon the type of protein, and the effects PTMs often play in the biological function of the protein. Although glycosylation represents the most common modification, other PTMs are important in the development of biopharmaceuticals such as phosphorylation, acetylation, sulphation and amidation, hydroxylation and acetylation.

Although glycosylation represents the most common modification, other PTMs are important in the development of biopharmaceuticals such as carboxylation, hydroxylation, sulphation and amidation.

Glycosylation is the most common form of PTM associated with eukaryotic extracellular cell surface proteins. One example is the hormone erythropoietin (EPO), which stimulates bone marrow to produce red blood cells. The half life of non-glycosylated EPO in serum can be measured in minutes; however, when the protein is glycosylated, this time increases to nearly 2 h.

Side chains of carbohydrates linked to proteins are synthesized by glycosyltransferase enzymes, located mainly in the ER, and two types of glycosylation can occur: N-glycans and O-glycans. In N-glycosylation, the oligosaccharide is linked to the protein through the nitrogen atom of a resi due of the amino acid asparagine; however, in the case of O-glycosylation, the carbohydrate chains are linked through an oxygen atom belonging to hydroxyl groups present in serine or threonine residues. The monosaccharides most commonly found in the side chain are mannose, galactose, glucose, xylose and fucose, in addition to N-acetylgalactosamine, N-acetylglucosamine and sialic acid.

N-Glycosylation involves the transfer of a presynthesized oligosaccharide chain to an Asn residue found in a characteristic Asn-X-Ser or Asn-X-Thr or Asn-X-Cis sequence, where X represents any amino acid residue, with the exception of proline. However, it should be noted that not all N-binding sites are glycosylated on certain proteins. The determinants of O-glycosylation are less well known, and, in this case, the characteristic recognition sequences are not apparent in most cases. In general, it can be said that glycosylated proteins can be characterized by one or more N-glycans, others by one or more O-glycans, or have both types of glycosylation. Human EPO, for example, has three N-glycans and one O-glycan on the side chain.

The composition and structure of the carbohydrate side chain present in glycoproteins can vary from one protein to another, resulting in microheterogeneity of the mixture, which can be visualized through isoelectric focusing. However, when there is a variation in the glycosylation site, the result is a macroheterogeneous mixture.

Altered profiles of glycosylation/microheterogeneity are determined during clinical research. If the product presents an altered glycosylation profile, but is considered safe and efficient, the quality control of the final product will be based on the analysis of the microheterogeneity of carbohydrates to determine the consistency of the batch-to-batch process.

The quality and extent of protein glycosylation depend mainly on the producing cell and play a significant role in therapeutic activity. Cultivation conditions also influence glycosylation, such as agitation, aeration, pH, concentration of macronutrients (sugars, amino acids, vitamins and growth factors) and micronutrients (coenzymes and mineral salts), as well as any other PTMs.

12.4.3 Cell Types

Mammalian cells, relevant to biopharmaceutical production processes, can be divided into several groups, such as primary cells that are isolated from a tissue and then maintained in culture (primary culture), diploid cell lines, continuous cell lines (which originate from a primary culture, but can proliferate indefinitely and are kept in a cell bank and are often used as host cells for the expression of recombinant proteins) and hybridomas (which are cells obtained by fusing lymphocytes and tumour cells and are capable of secreting monoclonal antibodies).

For primary culture, the process starts by obtaining tissue samples directly from an animal and human, followed by mechanical dissociation and/or enzymatic digestion in order to obtain a suspension of single cells. The cells are transferred to sterile flasks and/or plates for cultivation as an adherent monolayer or as a suspension in the culture medium. After reaching a certain cell density, the cells can be dispersed by enzymatic treatment or simply diluted (in the case of suspension) and transferred again to subsequent flasks for culture in fresh culture medium, in a process called 'subculture or replating'.

This subculture is the origin of the so-called secondary cultures, and part of these cells can be stored by freezing in liquid nitrogen, after the addition of cryopreservatives (e.g. DMSO – dimethylsulphoxide). The cryopreserved culture remains as a back-up stock from where, when necessary, it is possible to obtain a sufficient number of cells to start a new series of subcultures for the production of cell mass. The primary culture can be subcultured several times. However, they have a maximum number of duplications (mitoses) (between 50 and 100 times), when the cells stop growing and start to die. This finite growth capacity is genetically determined event involving several different genes called 'senescence'. It is common to cells derived from normal mammalian tissues, and this have a set of characteristics, such as (a) diploid number of chromosomes (that is, 46 chromosomes for human cells), in which no deleterious damage has occurred at the chromosome level; (b) adherence: the cells need a surface to support and grow (termed 'dependent anchoring'); the growth phase extends until the cells reach a confluence phase, which is usually accompanied by contact with neighbouring cells which inhibits further growth; (c) finite number of divisions and survival in culture; and (d) non-malignant: normal cells are not cancerous, that is, they do not cause tumours when injected into mice (Freshney, 2005).

Not all cell types produce exclusively primary cell cultures, and they die after a limited number of 'passages'. Some cells acquire the characteristics of infinite growth, such a population is generally called a 'continuous or transformed lineage'. These cells have undergone some type of transformation and have lost control of growth by contact. Transformed cells can also lose their growth characteristics when they are attached to a surface and only start to grow when they are in suspension. These transformations are sometimes reflected in the chromosomes, changing the genotype of the cells.

There are techniques that cause cell transformation and 'immortalization', such as treatment with mutagenic substances or viruses. *In vitro* transformation does not always generate cells that divide permanently, but all cells that are isolated from tumours (e.g. HeLa, NAMALWA) can be maintained in a culture permanently (Figure 12.1).

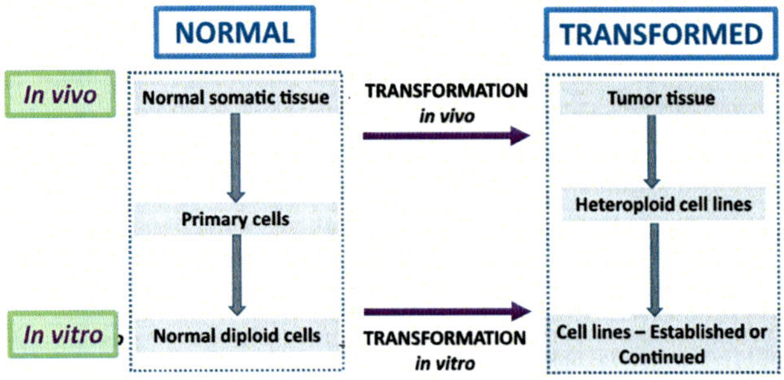

FIGURE 12.1 Normal and transformed cells.

TABLE 12.1
Examples of Cell Lines Important for Production and Research

Cell Line	Origin	Application
BHK (baby hamster kidney)	Isolated in 1961 from the newborn hamster kidney	Adherent cells, adaptable for growth in suspension. Production of foot-and-mouth virus
CHO (Chinese hamster ovary)	Isolated in 1957 from Chinese hamster ovary.	Adherent cell, adaptable for suspension. Used as a system for the expression of several recombinant proteins.
HeLa	It was isolated in 1951 from cervical carcinoma taken from the patient Henrietta Lacks.	First human lineage to proliferate in culture
NAMALWA	Human lymph tissue	Interferon alpha
MDCK	Canine kidney	Adhesive line used in the production of vaccines for animals
MDBK	Bovine kidney	Adherent strain used in the production of vaccines for humans and animals.
MRC-5	Human lung embryonic cells	Diploid cell used to produce vaccines
NS0 e SP2/0	Murine myeloma obtained from B lymphocytes	Production of monoclonal antibodies
Vero	African green monkey kidney, obtained in 1962	Lineage established with some normal diploid cell characteristics. Applied in vaccine production.
PERC.6	Human retinal embryonic cells	Immortalized lineage that produces high levels of recombinant proteins and viruses.

Examples of cell lines, important for research and protein expression, are shown in Table 12.1.

As mentioned earlier in this chapter, cell lines can be grown in the form of adherent monolayers or suspensions. Haematopoietic cells, transformed or tumour cell lines, are examples of cultures that can survive and proliferate without adhering to supports.

12.4.4 Hybridoma Cells

Hybridomas are conjugated cells produced by the fusion of lymphocytes, which have the capacity to secrete specific antibodies, and an immortal cell line. Lymphocytes cannot be maintained in culture for a long time as they did not survive. This problem was solved in 1975 by Köhler and Milstein, who fused murine myeloma cells (NSO or SP2/0) with lymphocytes isolated from the spleen of a mouse which had been immunized with a particular antigen. The cells generated in this way were hybrid cells (hybridomas) and had the characteristics of lymphocytes because they produced antibodies specific for the antigen, but could also be maintained in culture because the cells, like the myeloma cells, were now immortal.

Before the introduction of hybridoma technology, antibodies were isolated directly from blood serum, which contained a wide variety of different antibodies that reacted with different epitopes of the antigen and, as a result, these preparations were called 'polyclonal antibodies'. The application of these polyclonal antibodies is limited because of low purity and cross-reactions with similar antigens.

The basic technology for the production of monoclonal antibodies using hybridoma cells, developed by Köhler and Milstein, follows the steps described below:

1. *Fusion*: B lymphocytes, isolated from the spleen of mice previously immunized with a certain antigen, are fused with mouse or rat myeloma cells.
2. *Selection*: The hybrid cells are separated from the other cells with the help of a selective medium (HAT medium: mixture of hypoxanthine, aminopterin and thymidine).
3. *Isolation and clone selection*: After the isolation of a clone resistant to the selective medium, the hybridoma secreting the desired antibody must be selected.
4. *Maintenance of the clones*: The cells that produce the desired antibody can be propagated, cryopreserved and thawed whenever necessary.

It is possible to grow these cells in bioreactors or by producing ascitic fluid, in which the cells are injected directly into the peritoneum of mice, and after a few weeks, ascites fluid can be withdrawn from the cyst. The culture supernatant, or ascites fluid, then undergoes a purification process until pure fractions of the monoclonal antibody of interest are obtained.

The therapeutic application of monoclonal antibodies, mainly derived from mice, presents problems because of a series of side effects caused by the protein nature of the antibodies. This can be overcome with the use of new techniques that produce partially or fully humanized antibodies, or even just fragments of antibodies. In addition, these antibodies can be produced in other cell lines (CHO, for example) instead of NS0 cells or hybridomas.

12.4.5 Cell Culture Derivatives

12.4.5.1 Immunogenicity of Recombinant Proteins

Most traditional drugs are substances of low molecular weight and generally do not immunogenic. However, proteins are macromolecules and can trigger an immune response. The mechanisms of tolerance towards proteins normally found in the human body (called 'self-proteins') develop along with the immune system and are normally maintained throughout life by two forms of regulation: (a) preventing B and T lymphocytes from becoming sensitive to self-proteins and (b) effector cells that are also not activated by these self-proteins. Therefore, it can be assumed that a recombinant therapeutic protein obtained via the expression of a human gene/DNA sequence (for example, recombinant human hormones or cytokines, i.e. self-proteins) would be non-immunogenic in humans compared to therapeutic proteins produced from the genetic material of other animals (for example, murine monoclonal antibodies, i.e. non-self-proteins) which would be able to stimulate an immune response. Cases of humoral responses to therapeutic proteins obtained from human genetic sequences have been increasingly observed, which clearly demonstrates that factors other than amino acid sequences contribute to immunogenicity (Schernthaner, 1993; Jacquemin and Saint-Remy, 1998; Peces et al., 1996).

The incidence and clinical consequences of immunogenicity among biopharmaceuticals is one of the major concerns of regulatory agencies, as they vary according to the product (EMEA, 2006). Some factors can contribute to the immunogenicity of biopharmaceuticals (Mukovozov et al., 2008), such as:

- *Patient characteristics*: Patients' exposure to therapeutic proteins (i.e. non-self-proteins) often leads to the generation of neutralizing antibodies after a single or repeated exposure. Individuals with a compromised immune system (e.g. undergoing anticancer chemotherapy) may not develop an immune response.
- *Product characteristics*: Proteins of animal origin are often immunogenic in humans because the amino acid sequences in relation to human proteins are, in many cases, different (i.e. non-self). For example, recombinant bovine and porcine insulins (i.e. non-self) are more immunogenic in insulin-dependent diabetic patients than recombinant human insulin (i.e. self). Another class of therapeutic proteins that have varying degrees of immunogenicity are monoclonal antibodies. There are studies demonstrating that murine monoclonal antibodies, used in the treatment of various diseases, can cause negligible, tolerable or marked humoral response. However, to reduce immunogenicity, chimeric monoclonal antibodies or humanized antibodies can be developed (Hwang and Foote, 2005).
- *Expression systems*: Depending on the location, in the ER or in the Golgi complex, the activities of glycosyltransferases are different and explain the different glycosylation profiles of the same glycoprotein in different cell lines. The CHO strain is widely used for expression of recombinant proteins because its repertoire of glycosylating enzymes is similar to the human system (Monaco et al., 1996).
- *Post-translational modifications*: Enzymes that add glycans to proteins can generate different glycoform profiles. For example, dozens of glycoforms of human recombinant erythropoietin (RhEPO) have already been found (Neusüß, Demelbauer, and Pelzing, 2005). RhEPO and human endogenous EPO may show differences in their carbohydrate structures and in their isoelectric points, as they have different levels of sialic acids. Producing biopharmaceuticals in CHO cells and BHK cells with identical glycosylation profiles to human ones is very difficult.
- *Formulation*: Biopharmaceuticals are usually stored at 4°C in solution or lyophilized. Stabilizers can be added to the formulation to prevent protein degradation. The formulation, however, can increase the immunogenic potential and/or modify the biological activity of therapeutic proteins.
- *Storage and handling*: Even in closed containers, proteins remain susceptible to changes caused by external factors that can impact their immunogenicity, such as agitation, shear, changes in temperature, light, presence of gases and water vapour.
- *Containers and packaging*: The composition of the container (such as glasses, seals and lubricants) designed to protect the product from the external environment can interact with proteins, thereby effecting their biological activity and immunogenicity.
- *Chemical degradation*: Protein denaturation can occur through oxidation, deamination, extreme temperatures, surface interactions, freeze drying or storage. Denaturation can favour the formation of protein aggregates in polymeric structures by hydrophobic or covalent interactions.

In addition to all these factors, the frequency, duration and routes of administration of biopharmaceuticals also influence immunogenicity.

12.4.5.2 Basic Structure of Antibodies or Immunoglobulins

Antibodies are heterodimeric proteins composed of four polypeptide chains: two identical heavy chains (H) and two identical light chains (L) linked together by disulphide bonds to form a stable molecule. The amino acid sequences at the amino-terminal ends of both chains (H and L) vary between different antibodies and form the antigenic binding sites.

During the maturation of a B lymphocyte, rearrangements of the genes responsible for encoding the H and L chains occur to generate a DNA sequence that encodes an antigen binding site highly specific for a given epitope. When an antibody is hydrolysed by proteases, there is cleavage of the disulphide bonds to generate fragments of antigen-binding types (Fab) and F (ab′)2. The H chain region of antibodies is called a 'crystallisable fragment (Fc)' and determines the effector mechanism. Based on structural properties, each antibody chain is divided into variable domains and constant domains (Figure 12.2). In the Fc region, the heavy chain CH_2 and CH_3 constant domains contain binding sites for the components of the complement system and for the Fc receptors (FcR) found in cells of the immune system.

12.4.5.3 Types of Monoclonal Antibodies

The advent of murine hybridoma technology and the discovery of surface antigens expressed on cells have promoted an interest in the development of monoclonal antibodies (mAbs) for therapeutic use. The development of mAbs has enabled the production of four types of antibodies: murine, chimeric, humanized and fully human mAbs (Figure 12.3).

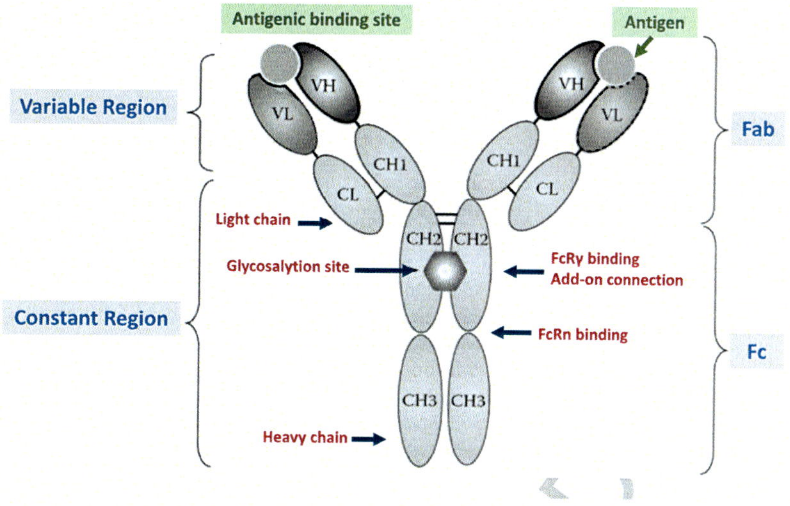

FIGURE 12.2 Typical structure of the IgG antibody.

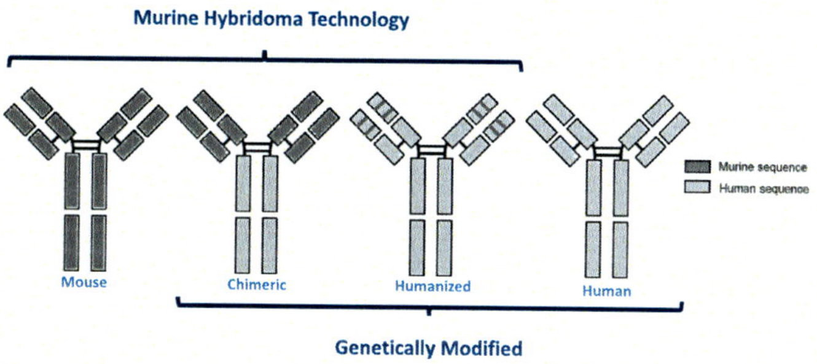

FIGURE 12.3 Different forms of therapeutic monoclonal antibodies (mAbs). (Adapted from Brekke, O., Sandlie, I. *Nat. Rev. Drug Discov.*, 2, 52–62, 2003.)

The types of antibodies are determined by the percentage of mouse and human fractions that make up the antibody molecules. The amino acid sequence of murine antibodies contains sequences from 100% of the mouse; chimeric antibodies are composed of about 35% of the murine sequence; humanized antibodies contain approximately 5% of the mouse amino acids; and in fully human mAbs, the amino acid sequences are 100% of human origin.

Genetic engineering techniques (phage display and PCR) are used to develop chimeric and humanized mAbs. Chimeric antibodies are constructed by cloning sequences encoding variable regions from mouse DNA and DNA sequences encoding constant regions derived from human DNA.

Humanized mAbs are predominantly derived from human DNA sequences, with insertions of murine DNA encoding variable and constant regions of the antibody. Fully human mAbs are produced by selection from antibody libraries *in vitro*, or by transgenic animals through the generation of human hybridomas.

12.4.5.3.1 Murine Monoclonal Antibodies

Initially, mAbs for human use were first generated from rats and mice using hybridoma technology. Although the use of murine hybridomas has offered a reliable source of large amounts of antibodies for clinical immunoassays, their use as therapeutics has now largely been ruled out because of high immunogenicity, which sometimes causes allergic and life-threatening anaphylactic reactions. After advances in molecular biology techniques and with the discovery of phages and the PCR technique, chimeric and humanized antibodies became possible. Some murine antibodies are approved by the FDA and used in cancer radiotherapy, such as ibritumomab marked with yttrium 90 and tositumomab marked with I-131.

12.4.5.3.2 Chimeric Monoclonal Antibody

Chimeric mAbs are genetically modified antibody molecules that contain mouse variable domains linked to the human domains of the two H and L chains, produced using transfected cell lines. Removal of constant regions of murine origin reduces the immunogenicity of murine antibodies and increases the serum half-life, while maintaining the same specificity as the natural murine antibody. In addition, chimeric mAbs can also interact more efficiently with the FcRγ receptor of human immune cells, which results in increased potency. Examples of FDA-approved chimeric antibodies are infliximab, rituximab, basiliximab and cetuximab.

12.4.5.3.3 Humanized Monoclonal Antibodies

These are antibodies produced by murine cells but the DNA encoding the protein sequences are modified to increase their similarity to antibodies produced naturally by humans. The most widely used method to humanize murine antibodies uses human CDRs (complementarity determining region) to replace those of murine origin. The protein sequences of antibodies produced in this way are only partially different from those antibodies that humans would naturally produce and, therefore, are potentially less immunogenic when administered to human patients. However, as the differences in fully human and humanized antibodies are small, they have a similar efficacy and safety.

The international common denomination for humanized antibodies ends in -zumab, for example omalizumab.

12.4.5.3.4 Human Monoclonal Antibodies (mAb)

These are antibodies produced by a single clone of a B lymphocyte. The antibodies produced by this B-cell clone have, therefore, the structure, physicochemical and biological properties, and antigen specificity and affinity. They are typically produced using transgenic mice, which are animals genetically engineered to contain human genes that express antibodies. After genetic modification, mice are immunized with target antigens to stimulate a humoral immune response *in vivo*. Immunization establishes a secondary repertoire of B cells that can be isolated and then fused with myeloma cells to produce hybridomas.

Non-immune synthetic libraries are used for screening antigen sequences, specific for the formation of fully human antibodies. Antibodies produced in this way have variable regions that are very similar to humans. For example, panitumumab and adalimumab have been approved by the FDA for the treatment of cancer and autoimmune diseases. Hybridoma technology is also used to produce fully human monoclonal antibodies.

12.4.5.3.5 Antibody Fragments

One of the most important factors for the clinical effectiveness of monoclonal antibodies is the ability to penetrate target tissues. Whole antibody molecules have been shown to diffuse only a little through the vascular space in solid tissues, such as tumours. To improve the tissue diffusion, a wide variety of antibody fragments can be developed; for example, antibody fragments without the Fc portion can be used to block soluble cytokines by binding to their receptors, while at the same time they can be rapidly eliminated from the systemic circulation. Antibody fragments can be, for example, Fab, scFv, 'minibodies' with molecular mass ranging from approximately 15 to 150 kDa.

The smallest antibody fragment that retains antigenic specificity of an entire antibody molecule is Fv, which is composed of VH and VL domains. Due to instability, when present in low concentrations, the VH and VL domains are covalently linked by peptide bonds and are called 'single-chain Fv' (scFv). ScFv homodimers can be joined by peptide bonds to form a molecular complex called a 'diabody', that is, a dimer of a scFv fragment. Minibodies consist of scFv homodimers linked by CH3; these are antibody fragments that have a biological activity.

12.4.5.4 Phage Display Technology

This laboratory technique is used to study protein–protein, protein–peptide, and protein–DNA interactions, and uses bacteriophages to connect proteins with the genes that encode them. Then the genes are cloned and expressed in a host, e.g. *E. coli* or yeast, to allow phage replication to occur. These displaying phages can then be screened against other proteins, peptides or DNA sequences, in order to detect an interaction between the displayed protein and other molecules. This technology generates a library of protein ligands, which can be tested for the ability to bind to a specific antigen. Using affinity chromatography, whose matrix contains the immobilized target molecules, only the phage that expresses the complementary protein is adsorbed. The phage is then eluted, for example, by reducing the pH of the elution buffer or by including a competitive ligand in the column buffer.

One of the most important applications of this technology is the production and screening of libraries in order to isolate/identify an antibody capable of binding to a desired target epitope. Two types of libraries can be generated: (a) 'immunological' libraries that are obtained by cloning the coding sequences of the antibody or fragment derived from B lymphocytes (usually isolated from the spleen) from donors previously immunized with the target antigen. A large number of positive clones are obtained from such libraries; and (b) Non-immunological libraries are produced in a similar way, but using non-immunized donor animals or human B lymphocytes as a source of antibody-encoding genes. This approach is necessary if immunization with the antigen of interest is not possible (for example, due to ethical considerations). Figure 12.4 represents the different technologies used to obtain human monoclonal antibodies.

12.4.5.5 Culture Media Used in Mammalian Cell Technology

The development of culture media for mammalian cells has been ongoing for several decades. The first attempts at culturing mammalian cells *in vitro* used biological fluids, such as serum and blood. Subsequently, the basic components of biological fluids were identified and used to produce defined media (with known components and concentrations) for cell growth. In parallel, another approach was developed that consisted of finding the minimum and essential ingredients for cell growth, which led to the development of the EMEM culture medium (Eagle's minimal essential medium). The composition of this culture medium is based on 13 amino acids, 8 vitamins, 6 ionic species and dialysed serum to provide the necessary, but undefined, components required for cell growth *in vitro*.

As new cell lines became available to the scientific community, other formulations of culture media were developed such as Dulbecco's modification of Eagle's medium (DMEM), Ham F12 and Roswell Park Memorial Institute (RPMI). Progress in understanding cell metabolism and

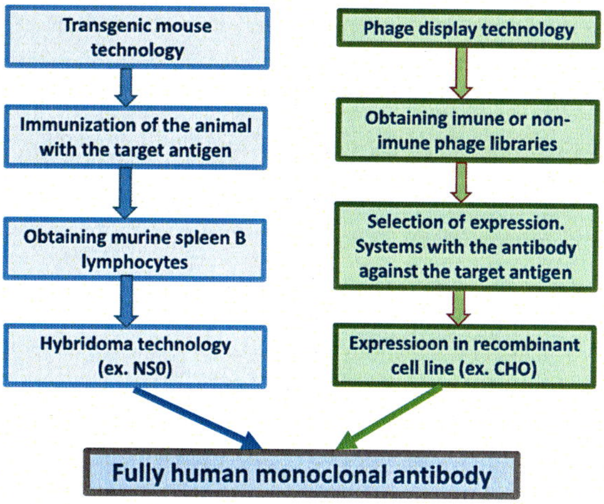

FIGURE 12.4 Technologies used to obtain human monoclonal antibodies.

growth factors has enabled the development of several serum-free formulations.

Basically, a culture medium for mammalian cells must provide all the nutrients necessary for growth and product formation (Table 12.2). In addition, it must have the buffering capacity in order to stabilize the pH (optimum pH 7.0–7.3) and provide adequate osmolarity (approx. 350 mOsm/L) avoiding damage to cell membranes. The presence of salts is essential to supply ions for cellular metabolism and to maintain the desired osmotic pressure.

There are one or more buffer systems (sodium phosphate buffer, HEPES and/or bicarbonate) commonly used to regulate pH. In some cases, the presence of a pH indicator (phenol red) in the medium helps in maintaining and monitoring the culture. The phenol red turns orange at pH 7.0, yellow at pH below 6.5, pink at pH 7.6 and purple at pH 7.8. An example of the usefulness of this indicator in media is to detect contamination with microorganisms (typically yeast derived from human manipulation of the cell culture). In cell cultures intended for cell and tissue therapy, the use of phenol red should be avoided.

TABLE 12.2
Description and Purpose of Cell Media Nutrients

Supplement	Description	Purpose
Serum	Fraction obtained from animal blood (bovine and horse)	Natural source of growth factors, lipids, vitamins, adhesion factors, metals. E.g. SFB.
Tissue extracts	Mixture of selected animal and plant components	Natural source of growth factors
Hydrolysates/peptones	Tissues and cells partially hydrolysed in acid or enzymatic form	Natural sources of peptides, amino acids, vitamins, carbohydrates. E.g. yeast extract
Hormones/growth factors	Cell proliferation. Active polypeptides, proteins, steroids among others.	Activates proliferation control pathways. E.g. IGF-1
Cell cycle regulators	Lectin-based mitogen and synthetic caspase inhibitors	Promote division and inhibit apoptosis in, e.g., primary crops
Adhesion and extracellular matrix factors	Cell adhesion/association components: gelatin, collagen, laminins, fibronectin and others	Promote adhesion or association with microcarriers, flasks
Antibiotics/antimycotics	Inhibitors of bacterial and fungal metabolism	Prevent, control and eliminate adventitious viruses and other contaminating microorganisms. E.g. penicillin and streptomycin
Vitamins and nutrients concentrates	Definitions of amino acids, vitamins and metal ions	Enriching the environment by increasing production capacity. 100 × non-essential amino acid concentrate
Simple salts and nutrients	Glutamine, glucose, pyruvate, Na_2HCO_3, $CaCl_2$	Supplementation of special media
Lipid concentrates	Stabilizes dispersions of sterols, fatty acids, lecithins and fat-soluble vitamins	Provides lipids to auxotrophic cells or reduces the need for biosynthesis, leading to more efficient metabolism in industrial processes
Selective agents	Toxins, analogues, inhibitors	Selection, amplification and maintenance of exogenous genes. E.g. MSX
pH buffers	HEPES, NaH_2PO_4, Na_2HCO_3	pH control allowing cultivation or production using cells.
Iron carriers	Natural or synthetic chelating agents associated with iron	Maintenance of intracellular iron concentration. Ex. human transferrin
High molecular weight synthetic polymers	Methylcellulose, polyvinylpyrrolidone, surfactants	It promotes viscosity, fluid rheology and shear protection. E.g. Pluronic® F68

A buffer system using carbon dioxide and bicarbonate acts similarly found *in vivo* in the blood. The carbon dioxide (produced by the cells or the external environment) in the gas phase dissolves in the culture medium and establishes an equilibrium with the HCO_3^- ions which acidify the medium (Equation 12.1). HCO_3^- has a low dissociation constant and tends to reassociate with the free H+cation, producing H_2CO_3.

$$H_2O{+}CO_2 \Leftrightarrow H_2CO_3 \Leftrightarrow H^+ + HCO_3^- \quad (12.1)$$

The partial pressure of atmospheric CO_2 directly regulates the concentration of dissolved CO_2, depending on the temperature. Therefore, the effect of increasing CO_2 pressure is neutralized by supplementing sodium bicarbonate in the medium (Equation 12.2). The increase in the concentration of bicarbonate ions allows a balance to be reached up to around pH 7.4.

HEPES, a stronger buffer, can control pH within physiological levels, in the absence of atmospheric CO_2. However, it is an expensive reagent and has toxicity depending on the cell and the concentration used.

$$NaHCO_3 \Leftrightarrow Na^+ + HCO_3^- \quad (12.2)$$

In most formulations, glucose is used as the main source of carbohydrate to obtain energy for biosynthesis. Alternative carbohydrates like fructose can also be added. Amino acids are also added for use in protein synthesis. Most amino acids are added at a concentration of approximately 0.1–0.2 mmol/L; however, glutamine is normally included at higher concentrations (2–4 mmol/L), as this is also a precursor that can feed directly to the TCA (tricarboxylic acid). In addition, synthetic media also contain vitamins, minerals and trace elements.

Animal serum contains proteins, growth factors, hormones, nutrients, lipids and minerals, and may even contain cell proliferation inhibitors. Currently, the use of serum in cell cultures for biopharmaceutical manufacturing processes is being discouraged or even banned by regulatory health agencies worldwide, due to the possible transmission of pathogens (viruses and prions, for example) that may not be detected by standardized tests and have been implicated as the cause of diseases in humans or other animals.

Many suppliers of fetal bovine serum (SFB) GMP (Good Manufacturing Practice), for example, document everything about the place of origin (free of foot-and-mouth disease or spongiform encephalopathy), sterilization (filtration and irradiation), and tests for the detection of bacteria, viruses, prions, mycoplasma and endotoxin limits (0.25 EU/mL), in addition to physical–chemical properties such as osmolarity and pH. This allows the serum to still be used in industrial pharmaceutical processes and for manufacturing media to grow mammalian and insect cells. The types of media are media with serum, serum-free media and medium without proteinaceous material. Serum media are the most complex. Although some cells only grow in a medium with serum, its use has some disadvantages (such as physiological variability, short validity, poor quality control, high cost, possible presence of contaminants, absence of standardization) and, finally, it provides many complications for downstream purification process, since it increases the protein content of the medium, usually being much larger than the final product produced by the cells. Serum-free media, on the contrary, have a more defined composition, simplifying downstream processes. Media without serum or any animal or protein constituents are less complex; however, the disadvantage is that the media can only be used for the cultivation of a limited number of cell types. Although serum-free media is the most viable alternative, it also has a disadvantage: each cell type will require a different medium, cell growth may be slower, and limited availability and safety cannot be completely ensured because the media may contain cell wall fragments and endotoxins. An alternative is to supplement the medium with recombinant albumin produced by yeast; however, this product may contain impurities, including constituents of the yeast cell wall and mycotoxins.

Traditionally, culture media are supplemented with approximately 5%–10% serum (for example, SFB or horse serum), in order to provide specific growth factors and to protect cells from the stress caused by shear.

The disadvantages of serum-containing media are (a) undefined composition, (b) high cost, (c) difficulty in downstream purification processes, (d) variations between batches, (e) the risk of contamination by viruses and (f) risk of prion transmission, such as bovine spongiform encephalopathy (BSE). Table 12.3 shows some characteristics of culture media with or without serum.

TABLE 12.3
Comparison of the Culture Medium with Serum and without Serum

	Medium with Serum	Medium without Serum
Composition	Not defined	Chemically defined
Variation between batches	Variation from batch to batch	Identical composition for all batches
Contamination	Potential source of contaminants (viruses, mycoplasmas and prions)	Low risk of contamination
Downstream processing	Complex	Simple
Biological activity of the final product	Decreases activity	Does not change
Validation and registration	More complicated	Easy
Viscosity of the medium	The albumin present, which increases viscosity and protects cells against shear stress	There are other protective substances that can be added in order to increase the viscosity of the medium

The serum can be partially replaced by the addition of transferrin, insulin, albumin or eventually, fibronectin as an adherence factor. The next step in developing a defined medium is to replace proteins of animal origin with salts or iron complexes, IGF-1 (insulin-like growth factor I), chemically defined concentrates of lipids, precursors or other agents such as fatty acids, biotin, choline, glycerin, ethanolamine, hormones and vitamins.

Hydrolysates and peptones are produced after an acidic or enzymatic hydrolysis of tissues (e.g. meat), cells (e.g. yeast) and seeds (e.g. soy). It is the oldest and most well-established way to cheaply add peptides, amino acids, vitamins, metals and carbohydrates. However, these additives have disadvantages, as they are complex mixtures and are not well characterized and, therefore, difficult to reproduce batches of the same quality.

Antibiotics (penicillin, streptomycin) are often added on a laboratory scale to prevent contamination but on an industrial scale, antibiotics are avoided because of the possibility of increasing the selection and spread of antibiotic-resistant microorganisms.

The growth rate and productivity of cells grown in serum-free culture media can decrease, while the sensitivity to shear stress can increase. To reduce these effects, high molecular weight polymers such as Pluronic F68 can be added to the media.

For the cultivation of primary cells, the use of serum media is common. For industrial production, it is essential to establish cell lines optimized in a serum-free medium, containing products derived from animals (e.g. bovine collagen, porcine trypsin) and chemically defined. Culture media with formulations developed on a case-by-case basis can yield cell concentrations from 5×10^6 to more than 10^7 cells/mL and 3–5 g/L in product titre (Wurm, 2004).

Sterilizing media used in cell culture ensure that the reagents are not a source of contaminants. Heat sterilization (using an autoclave operated at 121°C) is not a recommended option for many culture media, as it can cause degradation of components and biomolecules necessary for cell growth. The method of choice for sterilizing cell culture media and additives is mainly filtration. This is normally achieved by passage through filters containing 0.2-μm-diameter pore membranes.

12.4.5.6 Culture Medium Optimization

The complexity of culture media allows optimization of the composition. Classic culture media have been developed for small-scale, low-cell density cultures, and often with mandatory addition of serum as an important source of nutrients. With the rise and growth of the biotechnology industry, continuous efforts have been made to improve cell culture medium in order to maximize product titre and reduce costs.

Currently, most of the culture media for industrial biotechnological processes are free of serum and have higher concentrations of nutrients than classic media. This is due to the need to maintain high cell densities and to increase productivity. Even though there are high-quality and specific cell culture media on the market, it is necessary that this medium be improved *in loco* for better productivity of each culture of transfected cells.

An example of the need for optimization of culture media is the production of viruses for vaccines, which requires high cell densities for an efficient viral replication. However, after inoculation, nutrients must be in sufficient quantities to maintain the replication of the virus and the high levels of metabolic activity that often follow a viral infection.

To obtain higher densities of viable cells, the speed of cell growth and viability at different times are essential parameters for optimization, and the choice of the optimization method depends on the final product, the cell line and the process. The existing methods are as follows:

- *Component titration*: It is the classic approach for the development of culture media that involves carrying out a series of experiments to determine the 'dose response' of a cell line.
- *Mixing of media*: It quickly generates new media simply by mixing existing formulations. After evaluating the new formulations, additional tests are carried out with those that have obtained good performance. It is a quick method, but it does not provide a good understanding of the cell's metabolic behaviour during the process.
- *Analysis of cellular metabolism*: It provides important information by analysing changes in the medium during cultivation. When comparing the initial concentrations of the components with the final concentrations, the average consumption is determined. With the aid of mathematical models, the process can be described regarding the depletion of nutrients and the accumulation of metabolites.

Other approaches for optimization of culture medium monitor the cellular metabolic state by the analysis of metabolic flow and/or gene expression profiles. In the analysis of metabolic flow, the main biochemical reactions are translated into mass balance equations involving carbon and nitrogen atoms. Usually, only reactions of carbohydrate and amino acid catabolism are considered, as well as the production of biomass and product.

There are several factors that contribute to the definition of the 'ideal' formulation of nutrients, and each of these factors can alter the quantity of each component of the medium. For example, the composition of a culture medium suitable for batch production may inhibit cell growth or production in a fed batch or continuous culture system.

12.4.6 Basic Cell Culture Laboratory

Contamination by microorganisms is one of the biggest problems for the culture of mammalian cells. Bacteria, mycoplasma, yeasts and filamentous fungi can be introduced in different ways, such as operator, atmosphere, work surfaces, reagents etc. The use of the correct aseptic technique can provide an efficient barrier between microorganisms present in the environment and cell culture. Therefore, true asepsis is a combination of procedures that reduce the likelihood of contamination, in which all materials that come into direct contact with the culture need to be sterilized and handling must be carried out where there is no contact between culture and non-sterile external environment.

Some equipment is essential for performing aseptic techniques and maintaining cell cultures, such as microscope, pipettes for exclusive use, autoclaves, incubators and devices for water purification. Two other pieces of equipment are essential:

1. *Biological safety cabin*:

   The biological safety cabinets (laminar flow or biosafety cabinets) make it possible to obtain a high level of environmental sterility. To varying degrees, it can be designed to provide three basic types of protection: personal protection from harmful agents inside the cabinet, product protection to avoid contamination of the material, experiment or process and environmental protection from contaminants handled inside the cabinet.

   Biosafety cabinets are divided into three classifications according to the National Sanitation Foundation (NSF):

   - *Class I*: The cabinet is designed to provide personal and environmental protection; it does not protect the product from any contamination, as unfiltered air constantly enters the flow cabinet.
   - *Class II*: The design of the cabinet must meet the requirements for product, personal and environmental protection. This type of cabinet is widely used in clinics, hospitals, research and pharmaceutical laboratories.
   - *Class III*: These are gas-proof, designed for handling high-risk biological agents and are also called 'insulators'. The cabinets provide the highest level of protection for personnel, product and the environment.

   Most cell culture laboratories work with Class II biological safety cabinets. A subclassification for Class II cabinets is made according to the method by which the air volume passes through recirculation or exhaust. The subclasses are designated A1, A2, B1 and B2 and have different applications, according to the biosafety level required.

   The degree of cleanliness of biological safety cabinets is provided by a large motor, which pumps air from the environment through a high-efficiency filter (HEPA) before entering the work area of the cabinet. HEPA filters are designed to remove 99.97% for particles of larger than 0.3 µm.

   The location where the cabin is installed, ideally, should contain an independent air filtration system to ensure the efficiency and longer life of the HEPA filter installed in the cabin.

2. *CO_2 incubator*:

   The use of incubators that receive a supply of carbon dioxide gas aims to control three essential variables related to the environment necessary for the growth of mammalian cells: CO_2 level, temperature and relative humidity. The result is a balanced pH (7.2–7.4), stable temperature (37°C or 25°C–28°C for insect cell), high relative humidity (95%) and controlled CO_2 concentration (5%). The relevance of CO_2 is related to its widespread use in cell culture technology for pH regulation, in balance with the culture medium's bicarbonate.

   Some models of this equipment have a HEPA filtration system that, together with the use of materials (stainless steel), allows an efficient cleaning of the internal working area, thereby further reducing the risk of microbial contamination of the mammalian cell cultures.

12.4.7 Bioprocess and Mammalian Cells

One of the principles for achieving consistency and safety in the production of medicines for human use is the characterization and analysis of the raw material. In the case of biopharmaceuticals, the main raw material is the cell itself. More specifically, within the subject of this chapter are cells of mammalian origin.

A large number of cell lines already characterized are offered by commercial collections for various applications. The largest and best-known collections of cells lines are the American Type Culture Collection (ATCC), European Collection of Animal Cell Culture (ECACC) and the German Resource Center for Biological Material (DSMZ).

The characterization and analysis of cell banks in relation to identity, presence of microbial contaminants, viruses and the genetic stability of the gene that encodes the product are subject to regulation by health inspection bodies. In 1987, WHO published the first guidelines on the use of diploid or continuous cells for the production of biologicals (WHO, 1987). This document has recommendations for the creation of a master cell bank (MCB) and the need to perform several tests. In 2010, the World Health Organization published a new recommendation for the use of animal cell cultures as substrates for the production of biological drugs and, with this, complemented the standards of Good Laboratory Practice (GLP), mainly in quality control of cell substrates (WHO, Technical Report Series, 978, Annex 3, 2010).

Rapid advances in biotechnology, specifically products derived from recombinant DNA, have led regulatory agencies to develop necessary guidelines to be considered in the quality and safety assurance documents for these products. The FDA (CBER/FDA, 1993, 1997) and the International Harmonization Conference (ICH Q5D, 1998) have provided guidance on appropriate standards for the preparation and characterization of cell banks to be used for the production of biological materials. More recently, Geraghty and collaborators (2014) published a guideline of Good Practices in cell culture, which highlights the control of identity and contamination with mycoplasma, the biggest problems related to the use of primary, diploid and lineage cells. In the same way, Wrigley et al. (2014) describe how an MCB should be established in order to meet the legal requirements of regulatory agencies.

12.4.8 Main Guidelines for Obtaining Cell Banks

12.4.8.1 Obtaining and Controlling the 'Original' Cell

The 'original' source of cell line, which has not yet undergone genetic modification, or cells and tissues obtained from animals and humans, needs to provide an appropriate and traceable documentation in order to guarantee the fundamental characteristics for cell culture, such as identity of the cells for the intended purpose; purity, that is, a cell free of microbiological contamination, and phenotypic and genotypic stability after *in vitro* growth.

The prevention of microbial contamination of cells in a large-scale cultivation is carried out by complying with GMPs. Before preparing the master bank and working stocks, all cells received from recognized sources must be quarantined and tested for sterility (absence of bacteria and fungi) and absence of mycoplasma.

Several species of the bacterial genus *Mycoplasma* are common contaminants of mammalian cell cultures. Infection of cell cultures with mycoplasmas can generate different effects on cytogenetics, nutrient depletion in the culture medium, alteration of cell growth characteristics, inhibition of metabolism, interruption of nucleic acid synthesis, production of chromosomal aberrations, alterations in cell membrane antigenicity and interference with virus replication (Cheng, Shen, and Wang, 2007).

Tests for bacteria and fungi should be performed as described in the relevant pharmacopoeias. The tests recommended by the FDA to detect mycoplasma can be performed using a direct growth method using solid or liquid media to culture sensitive cells (for example, Vero and NIH-3T3), followed by DNA staining and PCR (CBER/FDA, 2010). This direct growth method, standard for industrial applications, is done by inoculating cell samples directly into an enriched broth, in order that possible mycoplasmas present in small quantities grow, with a subsequent detection in agar plates.

PCR-based assays can be used to detect mycoplasmas, as long as the result is compared to agar/broth growth and indicator cell culture tests. If in any case the indicator cell line culture procedure cannot be used, PCR-based assays are necessary.

12.4.8.2 Cell Bank Preparation

Once the quality controls of the 'original' cell have been done, the MCB can be produced. MCB is defined as an aliquot of a single set of cells (uniform composition), which was made from a selected cell clone under defined conditions, and stored at temperatures below –100°C.

The MCB is subject to characterization and quality control. After qualification, the Work Cell Bank (WCB) is generated from an inventory and the number of tubes used is dependent on production planning, that is, the number of batches required over a certain time period. Quality control parameters are repeated before each new tube from the WCB is used.

WCB is used for bioproduction. Whenever a batch of a certain product is started, a tube is thawed and the cells are replicated. Normally, 100 tubes containing 10^7 cells/vial of MCB and 100–500 tubes of WCB should be produced, depending on the production process. Cell banks are stored separately in cryogenic refrigeration systems (–156°C) or in liquid nitrogen (–196°C).

Although a costly process, organizing a cell bank for the production of biological products under GMP conditions, in addition to being a requirement of regulatory agencies, provides a safe and consistent source of cells sufficient to supply the length of the production run.

Like cells, raw materials of animal origin (for example, serum and trypsin) used in the production of biopharmaceuticals must be tested for contaminating bacteria, fungi, mycoplasmas and viruses. If the raw material is of bovine origin, for example, SFB, the raw material must have a certificate from the supplier confirming the animals used as a source of the raw material were raised in an area free of BSE (there is currently no test to demonstrate the presence of any prions).

12.4.8.3 Qualification Tests of the Cell Bank

MCB and WCB are generally tested following the recommendations in the ICH (Q5A; Q5D), FDA and WHO. Tests for identification of the cell line, detection of microbial and viral contaminants are performed when the MCB is established.

A summary of quality control tests applied to cell banks is as follows:

- Master cell bank:
    - Viability (0 and 24 h)
    - Sterility test – absence of bacteria and fungi
    - Karyotyping
    - DNA profile analysis (fingerprinting)
    - Isoenzyme analysis
    - Mycoplasma test
    - Absence of adventitious agents
    - Stability

  Work cell bank:
    - Viability (0 and 24 h)
    - Sterility test – absence of bacteria and fungi
    - DNA profile analysis (fingerprinting)
    - Mycoplasma test
    - Absence of adventitious agents
    - Stability

12.4.8.4 Karyotyping (Cytogenetic Analysis)

Cell line authentication is essential for all cell culture laboratories, including those for scientific and industrial research. It needs to be carried out at regular intervals after the MCB is established, so that the possibility of cross contamination between strains is eliminated. Karyotyping uses the chromosome content of cell lines. It is a direct method that confirms the species of origin of the cell and allows the detection of aberrations in the number of chromosomes and/or morphology. This technique is no longer recommended for continuous cell lines, especially those of rodent origin (for example, Sp2/0, NS0, CHO), for which only isoenzyme analysis is necessary. However, chromosomal characterization still applies to diploid cell lines.

12.4.8.5 Analysis of Isoenzymes

This method uses the property of isoenzymes to have specificities similar to a substrate, but with different molecular structures that effect electrophoretic mobility. Therefore, enzymes from different species will have a characteristic isoenzyme mobility profile. Identification of the species of origin of a cell can generally be determined by two isoenzyme tests, lactate dehydrogenase (LDH) and glucose-6-phosphate dehydrogenase (G6PD). However, other isoenzymes can be also used, including nucleoside phosphorylase (NP), malate dehydrogenase (MD), peptidase B (PepB), aspartate aminotransferase (AST) and mannose 6-phosphate isomerase (MPI).

The analysis of isoenzymes has the potential to detect the initial phases of cross-contamination between cell lines, that is, the period of time between the initial contamination event and eventual replacement of the original culture with the contaminating cell line. This can occur in cases where differences in growth parameters favour one of the cell types over the other. Evaluation of mixed cultures of rapidly growing Chinese hamster cells (CHO-K1) and a slower growing human cell line (MRC-5) revealed that extracts of the two cell types produced distinct LDH bands for each species, when each cell type was represented in at least 11% of the total number of cells (Figure 12.5).

12.4.8.6 DNA Fingerprinting

DNA contains regions known as satellite DNA that are apparently not transcribed. These regions are highly repetitive and vary in size. The functions of these regions are not completely known; however, hypervariability occurs because they have not conserved their sequences. When DNA is cleaved by endonucleases (restriction enzymes), specific sequences can bind to labelled probes of complementary DNA or can be amplified by PCR. Agarose gel electrophoresis can show variations in the sizes of satellite DNA fragments, called 'repetition site polymorphism' (RFLP), specific to the individual from which the cell is derived. With the aid of electrophoresis, each individual DNA sample provides a specific hybridization profile, revealed by autoradiography with radioactive or fluorescent probes. An alternative approach to multilocus analysis, previously presented, is the analysis of simple polymorphic loci called 'mini-satellites' (1–30 kb) or variable number of tandem repeats (VNTR). Microsatellite DNA sequences (two to four bases) are extremely small and, therefore, allow the identification and quantification of short tandem repeat (STR) sequences at a specific locus.

12.4.8.7 Genetic Stability of the Recombinant Mammalian Cell

The test for genetic stability of a recombinant cell has the main objective to verify a consistent production of the cloned protein and retention of production capacity during storage (ICH Q5B, 1996). For this, a comparative analysis of the genetic stability of producer cells (MCB) and cells maintained *in vitro* must be performed. The coding sequence of the recombinant protein is amplified by PCR. Amplified DNA or cDNA prepared from isolated mRNA is used as a template for DNA sequencing analysis. The integrity of the gene expression plasmid is also analysed by restriction mapping and Southern blotting to determine the number of copies of the plasmid and to detect any insertion or deletion sequences.

The preparation of cell banks is an important point of reference for any project that uses cell culture as a critical component. However, it is also important to know that cells are subject to variation when thawed. For this reason, the specification of the culture medium, cultivation conditions and subculture are important to standardize results from different laboratories, over time. In addition, for critical applications such as virus production for vaccines, it is recommended to run several cell passages up to the expected limit for use, establishing 'extended' cell banks at intervals of several passages, and then to compare the characteristics of these banks in parallel in order to determine any change in culture over time (Figure 12.6).

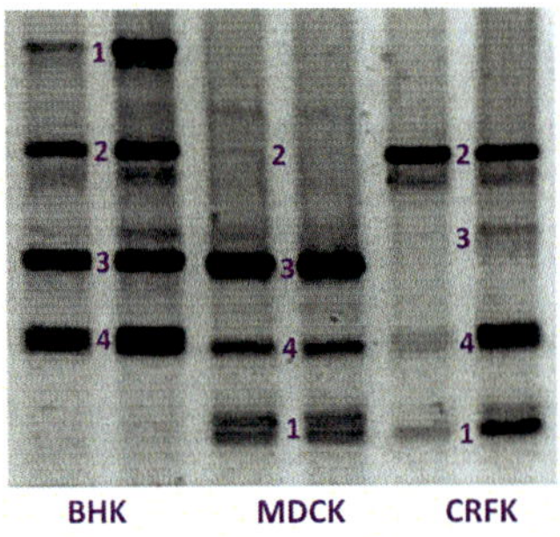

FIGURE 12.5 Sensitivity of isoenzyme analysis for the detection of interspecies cell line cross-contamination. 1: Lactate dehydrogenase, 2: glucose-6-phosphate dehydrogenase, 3: malate dehydrogenase, 4: phosphoglucoisomerase. (Photo courtesy of Dr. Celso Pereira Caricati from the Cell Culture Section of the Butantan Institute, 2003. BHK baby hamsters kidney, MDCK Mardi-Darby Canine Kidney, CRFK Crandell Reesr Feline Kidney.)

12.4.9 Best Practices in Cell Culture

The maintenance of a quality standard is fundamental to all good scientific practice and essential to ensure the reproducibility, safety, credibility, acceptance and proper application of any product. Cell culture is inherently prone to variability, creating difficulties for proper control, making standardization a critical activity to ensure confidence in the data and products obtained by cell-based systems. Proper handling of reference materials and the development of qualified cell banks are vital parts of the standardization process.

The application of Good Practices in Cell Culture aims to reduce uncertainty in the development and application of *in vitro* procedures, through the establishment of principles and techniques and are based on principles of GLP. A number of areas using cell culture technology, such as basic research, new product development, quality control, diagnosis, biopharmaceuticals, vaccines and therapeutics (gene therapy, cell therapy and tissue engineering), can develop a guide of good practices in cell culture.

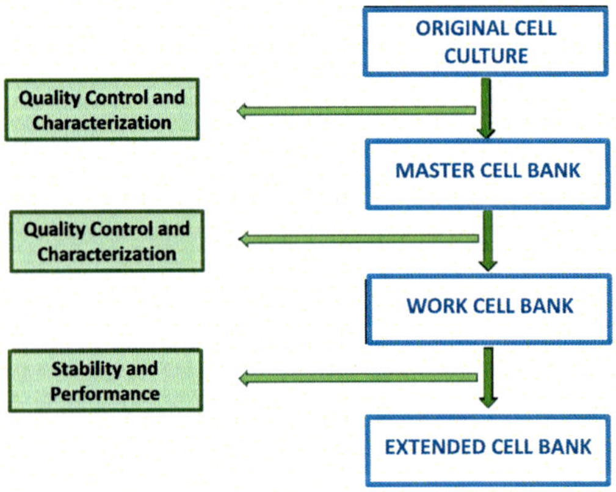

FIGURE 12.6 Bank of cells.

TABLE 12.4
Typical Karyotype Variation but Not Modal Number with In Vitro Passages

Lineage/Cell Type	Karyotype	Variation of the Number of Chromosomes by Cells, Determined in Different Culture Times
MCR-5/human diploid fibroblast	46	43–50
HeLa/human cervical carcinoma	60	53–70
RK13/rabbit cells similar to the epithelium	66	57–73
L20B/recombinant mouse cell expressing human polio receptor	45	37–51
Vero/green monkey kidney fibroblast	58	50–65

The variabilities found during cell cultivation, as a consequence of different forms of handling, are genetic instability and lack of productivity. Instability can be observed even in cells considered highly stable genetically and phenotypically after several passages of diploid strains (MRC-5 and WI-38), associated with a decreased telomere size. Continuous strains can present different karyotypes according to the number of passages, establishing a frequency around an average of chromosomes (Table 12.4).

Lack of reproducibility of results may be the result of variations in the techniques used for the manipulation of cell cultures. Situations that should be avoided are plating different amounts of cells, carrying out subcultures at different stages of culture or after reaching confluence (e.g. MDCK and Caco-2) and changes in cultivation conditions (e.g. culture medium, CO_2 concentration in the chamber). In addition, some critical parameters must be monitored such as cell viability and number of passages/duplications.

Cell viability at a given point of cultivation can be determined using several different methods, such as

- Integrity of the cell membrane with trypan blue dye and subsequent counting of non-stained and living cells using a Neubauer chamber;
- Enzymatic activity using MTT (3-(4,5-dimethyl)thiazol-2-yl bromide)-2,5-diphenyltetrazolium, which is reduced by viable cells to produce an insoluble and coloured product;
- Metabolic activity by measuring LDH activity, which is influenced by culture conditions. For this reason, it is widely used in analyses carried out on cultivations conducted in bioreactors.

The number of cell culture passages is directly related to the accumulated number of cell divisions, and will depend on factors such as the plating ratio (e.g. 1:3 means the passage of cells contained in 1 vial to 3 vials of the same size) and cell density (cells/mL or cells/cm^2).

Variations in cell culture technology can be avoided by obtaining the cell lineage from qualified and recognized sources, standardization of the cultivation environment and procedures used, control and testing of different batches of raw materials (e.g. serum and amino acids) and application of Good Cell Culture Practices (GCCP).

GCCP's principles are as follows:

a. Identify and control the relevant factors that can directly affect cell culture in order not to invalidate the results obtained. Key controls: authenticity, genotypic and phenotypic characteristics, microbiological contamination, stability and functional integrity;
b. Guarantee the quality of all materials, methods and their applications in order to maintain the integrity, validity and reproducibility of the results. Key controls: monitoring of cells, reagents and materials, calibration, specification and monitoring of critical equipment that influence cell growth and product formation;
c. Documentation with information necessary for the tracking of materials and methods used. Key controls: the origin of cells and tissues, handling, maintenance and storage;
d. Establishment and maintenance of adequate measures for the individual and environmental protection against potential dangers. Key controls: use of personal protective equipment and appropriate laminar flow cabinets;
e. Compliance with laws and regulations relevant to the area and with ethical principles;
f. To properly train and educate employees in order to promote safety and quality of the work. Key controls: training record, standardization of cultivation techniques and general training on the importance of quality in cell culture.

12.4.10 Development of a Cell Line

The production of recombinant proteins using mammalian cell culture for research and development necessarily begins with obtaining the cell line. The choice of the expression system and the form of expression is generally based on the nature of the protein of interest, the quantity of the final product and the period required for its production. The development of a recombinant cell line is a time-consuming and costly process.

A mammalian cell expression system can be transient for the production of small amounts of proteins in a short time, or stable, in which industrial production occurs for long periods of time and in large quantities.

12.4.10.1 Transient Expression

The gene encoding a protein of interest is introduced into a cell and replicated as an extrachromosomal unit. The factors that determine the expression level of the transfected cell line are the strength of the promoter used to stimulate the expression of the gene encoding the recombinant protein, and the transfection efficiency, which refers to the percentage of cells that have been transfected and express the recombinant DNA. The average production life of the transient expression system is usually limited by toxicity induced by the rapidly multiplying DNA, or the loss of DNA during cell division.

12.4.10.2 Stable Expression

Exogenous DNA needs to be integrated into the host cell genome for stable production of the product. Therefore, in addition to the strength of the promoter and efficiency of transfection, the frequency of DNA integration into the chromosome (number of copies) and the position of integration are important factors in determining total levels of expression.

Basically, there are three methods by which biological products can be expressed in mammalian cells: transfection with a modified bacterial plasmid containing a strong promoter of gene expression in mammals; infection with a recombinant viral vector designed to express high levels of the protein of interest (for example: adenovirus, retrovirus) and infection with viruses of interest, for the production of viral vaccines.

The cell line obtained from Chinese hamster ovaries (CHO) is the most common mammalian cell expression platform for the production of recombinant proteins. The steps for the production of a biopharmaceutical for industrial purposes are:

1. Construction of the expression vector;
   Transfection of the host cell, that is, inserting the vector into the cell cytoplasm so that it is incorporated into the nucleus;
2. Selection of individual and highly expressed clones;
3. Preparation of cell banks for production;
4. Expansion for inoculum production;
5. Growth in bioreactors and biopharmaceutical production;
6. Purification;
7. Formulation and container filling.

12.4.10.3 Construction of Expression Vector

Stable expression of heterologous proteins in mammalian cells is accomplished by vectors that integrate exogenous DNA into the host cell genome. These vectors must have the three main characteristics: (a) the expression must be independent of the integration site in the genome; (b) the level of expression must be correlated with the number of integrated copies of the gene of interest; and (c) the expression needs to be maintained over time (Blaas et al., 2009). Because of these characteristics, the vectors used for mammalian cells are generally plasmids. Therefore, each plasmid vector used for production in recombinant cells must contain sequences with specific functions such as promoter and enhancer that guide the transcription of messenger RNA sequences that help to stabilize and increase the translation of the primary transcript, selection markers. Optional components may include the presence of an antigen epitope that allows the detection/visualization of the expressed gene products, and a reporter gene whose expression can give quantitative indications of transfection or transcriptional activity, for example transferase, luciferase (bioluminescent) or GFP (green fluorescent protein).

12.4.10.4 Promoters

A promoter is a DNA sequence located at the 5′ end that allows the formation of a complex between RNA polymerase II and the exact sequence to start transcription, defining the direction. Enhancers are elements that increase the formation of transcription complexes.

For the expression of recombinant proteins in mammalian cells, the promoter elements of endogenous genes are usually replaced by strong viral promoters or specific combinations for the chosen cell. The two most commonly used are derived from simian virus 40 (SV40) and cytomegalovirus (CMV). However, other sequences are available for cell lines derived from rodents, such as CHO elongation factor (CHEF1-α).

12.4.10.5 Enhancer

Controlled or induced transcription allows for the exact expression of the gene of interest and is often used when the expressed protein is cytotoxic or cytostatic. Inducible promoters are commercially available and capable of inducing high levels of expression. Most are based on bacterial promoters, such as the Lac *E. coli* operon system: 'tet-On/tet-Off' system, an operon that encodes resistance to tetracycline.

12.4.10.6 Elements that Stabilize and Increase the Translation of the Primary Transcript

The original property of most eukaryotic mRNAs is their polyA tail at the 3′ end. The polyA polymerase enzyme adds the polyA tail to the mRNA after releasing the transcription complex. The addition site is marked by the presence of the AAUAAA (polyadenylation signal) sequence in the untranslated region of the mRNA. The role of the poly A tail may be involved with the export of mRNA from the nucleus, prolonging the mRNA half-life in the cytoplasm and enabling a more efficient translation. Therefore, it increases the expression levels of the gene of interest.

12.4.10.7 Selection Markers

Selection markers are genes (mainly bacterial), which establish drug resistance in cell culture. The first category of markers is represented in commercial vectors, for example the aph gene that encodes bacterial aminoglycoside phosphotransferase – which detoxifies the protein synthesis inhibitor drug G418 (neomycin/geneticin) – and phosphotransferase-hygromycin B (hph) – which inhibits hygromycin-B.

The second category of selection genes can amplify expression of the protein of interest. For expression platforms based on CHO cells and NS0 myeloma, the use of genes that express enzymes such as dihydrofolate reductase (DHFR) and glutamine synthetase (GS) have become industry standards. In both cases, selection occurs in the absence of nutrients such as hypoxanthine and thymidine in the case of DHFR and glutamine in the case of GS, thereby preventing the growth of untransformed cells.

Transfection of the DHFR gene in CHO cells deficient (dhfr - CHO) in the activity of this enzyme is the most used method. With this system, expression of the recombinant protein can be increased by exposure of cells to methotrexate (MTX), a substance that blocks DHFR activity and limits, in a dose-dependent manner, a cellular ability to produce the substance tetrahydrofurate (FH4). In the absence of FH4, the primary production pathway for purine and pyrimidine syntheses is inhibited. After 2–3 weeks, most cells die following exposure to MTX, but a small number of cells that have a high level of expression of the DHFR gene can survive. After treatment with MTX, the increase in productivity of a recombinant protein can reach 10–20 times compared to a production system without MTX treatment.

In the system based on transfected NS0 cells, the GS enzyme gene catalyses the production of glutamine from glutamate and ammonia. Thus, the system offers the double advantage of reducing the level of ammonia in the culture medium and providing an amino acid (glutamine) for cell nutrition. An irreversible and specific inhibition of GS can be mediated by the substance methionine sulfoximine (MSX). At concentrations between 10 and 100 mM of MSX, resistant clones can be identified among populations of cells that contain the amplified GS gene and product of interest.

High yields of recombinant proteins can also be achieved using the human cell line PER.C6, obtained from embryonic cells of healthy human retina and immortalized with the adenovirus El gene (Jones et al., 2003). This cell line has been shown to be capable of producing high levels of recombinant protein with a low number of copies of the gene and without the need for amplification protocols. Another advantage of this cell line is the guarantee of protein expression with a human glycosylation profile.

12.4.10.8 Selection of Clones

The levels of expression of recombinant proteins from clones of different cells, including those derived from gene amplification, may vary. As a consequence, the identification of highly producing cell lines is laborious and requires the selection of hundreds of cell lines. Several methods exist for isolating clones, and among the most traditional and commonly used is the method of cloning by limit dilution, which is relatively simple and low cost. In this method, a low cell density suspension is dispensed in microtitre plates to achieve one cell per well. The plate is observed microscopically, and those wells that contain only one cell are marked for further analysis. After cell proliferation, the supernatants of the clones are quantified for the presence of the protein of interest. Clones that have the highest protein titres are subsequently expanded or undergo a further round of serial dilution and screening for the highest producing clone. There are several limitations regarding the use of this method, including the time required to analyse each clone and the number of clones that can be analysed.

The application of selection methods based on flow cytometry and cell separation (FACS) increases the number of cells that can be analysed in less time. Subpopulations and single cells can be isolated even when present at the frequency of 10^{-6} within a population. FACS has become an important instrument in the research development and application of mammalian cell culture methods. However, in some cases, FACS may be limited by the absence of specific antibodies available for recombinant proteins and by the need to optimize the experimental conditions for each cell line. However, several automated methods have been developed in order to select highly producing clones, but these systems are expensive and do select a large number of clones (Browne and Al-Rubeai, 2007). A clone with a specific productivity greater than 10 pg/cell/day can normally be used for the production of recombinant proteins. However, productivities greater than 90 pg/cell/day may be possible with changes in vector technology and in the understanding of parameters that control protein expression (Wurm, 2004).

12.4.10.9 Transfection

Several methods for introducing DNA into mammalian cells have been developed. They can be divided into four classes: chemical, physical, lipid-based lipofection and polymers or dendrimers. The transfection method is chosen according to each cell type.

12.4.10.9.1 Calcium Phosphate Method

This method works with a wide range of cells and was first described in 1973 by Graham et al. In this method, calcium chloride, phosphate buffer and DNA form a co-precipitate that penetrates mammalian cells by endocytosis. In this method, DMSO or glycerol can be used, depending on the type of cell. Although it is a low-cost method, there are disadvantages, as the efficiency of transfection is directly related to the size of the calcium phosphate precipitate and the concentration of DNA. Calcium phosphate crystals form nuclei immediately after adding the reagents and continue to grow rapidly depending on the concentration of calcium chloride and phosphate, pH of the reagents and temperature. Therefore, the method needs to be carefully optimized since small changes in the process or environment can alter the transfection result.

12.4.10.9.2 Electroporation

The electroporation method uses electrical pulses in order to break the voltage gradient that exists across a plasma membrane to create reversible micropores that allow DNA to enter the cell. It is a simple, fast method that involves direct interaction between an electric field and the cell membrane. Therefore, it is a nonspecific method. Parameters such as applied voltage and time need to be determined empirically, taking into account that this method generally decreases post-transfection cell viability.

12.4.10.9.3 Lipofection and Polyfection

Lipofection and polyfection are the most recent and simplest transfection methodologies. They are applicable to a wide variety of cells.

Lipofection consists of the transfer of DNA to cells mediated by cationic lipids. Cationic lipids contain a quaternary amine forming positively charged liposomes, which interact with the negative charge of DNA to form a complex. These complexes can enter the cell through endocytosis or direct fusion with the cell membrane, and DNA is likely to be released into the cytoplasm. The most critical parameters for performing a lipofection are the amount of DNA and lipids used, as well as the ratio between them. Using too little DNA results in low expression, while very large amounts are cytotoxic to cells. In addition, the optimization of the ratio between DNA and lipids is also important, since the final charge of the complex must be necessarily positive to facilitate an interaction with the negative charge of the cell membrane.

Polyfection refers to the transfer of DNA mediated by cationic polymers and dendrimers. Cationic polymers, especially those that are highly branched and spherical dendritic macromolecules, have a high number of primary amine groups on the surface, which interact with DNA forming a complex. This complex protects DNA from degradation before it reaches the cell nucleus.

There are many commercially available lipofection and polyfection reagents, and the most efficient method for a given cell culture system must be determined.

The transfection methods used for transient or stable expression systems are the same; however, in stable expression systems, it is necessary to identify and select cells that have integrated the plasmid DNA into their genome. This is usually accomplished by transfecting a coding sequence for a marker gene into the host cell, along with the expression vector containing the product gene. After the transfection procedure, the cells are cultured under defined conditions that allow the direct selection of those that have incorporated the marker sequence into the genome. Many vectors have been developed in order to guarantee the co-integration of the selection marker and the product-encoding gene. Thus, it is expected that most of the selected cells will also express the gene encoding the desired product.

If the productivity of the cells initially selected falls below expectations, there is a possibility of increasing protein expression by subjecting the transfected cells to repeated exposures with the selection agent. It is possible to analyse the clones individually for increased productivity using 96-well microtitre plate formats.

12.5 PRODUCTION SCALE-UP

The industrial application of mammalian cell culture began in the 1950s, with the production of the polio vaccine in primary monkey kidney cultures. However, it was the first product of commercial interest obtained from the suspension cell culture (interferon in Nawalva cells) that stimulated the adaptation of cell culture to use homogeneous systems in bioreactors that had been used, until then, only for culture of microorganisms. The adaptation of bioreactors was aimed at meeting the needs of mammalian cells, which are more sensitive to mechanical sheer than microbial cells. In the 1960s, microcarriers and disks for anchorage-dependent cells allowed the introduction of bioreactors for large-scale production. In the 1970s, experiments on recombinant DNA and the development of other types of bioreactors (air-lift and hollow fibre) began, in which the idea was to overcome some limitations such as low cell density growth. In the 1980s, the emphasis was on scaling up and producing monoclonal antibodies and other recombinant proteins. By the 1990s, the focus was on improving production yields and scaling efficiency, as well as adapting cell lines for growth in media without serum, protein or animal products and in chemically defined media. Since then, regulatory agencies have been increasingly insistent on removing any and all ingredients from animal sources (cattle, in particular) from the composition of the culture media for the production of therapeutic substances for human use.

An important advance in the use of cell culture technology was the permission by regulatory agencies for the use of continuous cell lines which are capable of growing indefinitely. Cells like BHK, CHO (Chinese hamster ovary cell), myeloma cells (SP2, NSO) and human embryonic kidney cells (HEK) have gradually been accepted for use in cell culture technology. These cells can be grown in suspension and are adaptable to industrial scale (large volumes), and cultivation can be carried out in bioreactors.

12.6 BIOREACTORS AND MAMMALIAN CELL CULTIVATION

There are several models of bioreactors that have been adapted for cell culture, including agitated tank bioreactors, hollow fibre bioreactors and fluidized bed systems. Most of these systems present problems for cell cultivation, such as high costs, product concentration gradients, hydrodynamic forces, cell aggregation and difficulty in sampling, monitoring and control in scale expansion.

Some adaptations and modifications are necessary to meet the requirements of each type of cell and the respective bioproduct, such as oxygen demand parameters, heat transfer, shear sensitivity, sensitivity of cells and target product to process metabolites, current requirements for good manufacturing practices (cGMP) and biosafety requirements (containment levels are typically NB1 and NB2).

Biotechnological processes for obtaining the products of interest use, in most cases, included continuous agitated tank (CSTR) bioreactors, combined with Rushton and/or nautical impellers. Due to the complexity of the culture media of mammalian cells associated with the low specific speed of growth when compared with microorganisms, it is of great importance to ensure aseptic operations in bioreactors with mammalian cells. This is a particularly difficult task in agitated tank bioreactors for the large-scale production, and consequently, the validation of cleaning and sterilization procedures for bioreactors is a mandatory requirement in the production of biopharmaceuticals. Therefore, the repeated use of a bioreactor poses additional challenges for the production of a biopharmaceutical.

The bioreactor provides a closed system where biochemical reactions occur, and its main role is to provide adequate containment conditions for cell growth and product formation. This can be achieved by imitating the cell's original environment, i.e. the original tissue. In addition, other parameters must be controlled such as temperature, continuous supply of oxygen, nutrients and growth factors, and removal of CO_2.

12.6.1 SMALL SCALE CULTURE

In the small-scale culture, most cells are grown in T flasks which have areas ranging from 25 to 225 cm^2. A 175-cm^2 flask yields approximately 10^7 anchorage-dependent cells to 10^8 cells for suspended strains, although the yield depends on the cell used.

Although the different scheduling systems available can be classified in different ways, the most common form of classification is according to the type of cell growth. We can divide the systems into three categories: systems suitable for anchorage-dependent cells, systems suitable for cells that grow in suspension and systems suitable for any of the above.

CHO cells, for example, can adhere to a culture surface in medium containing serum, but also grow in suspension in medium free from serum or proteins. This increases the importance of knowing the essential properties of the cells, as well as defining the culture medium as quickly as possible during the development of the process.

12.6.2 Scaling-Up Problems

Some issues are important during the scale-up process such as:

- *Demand for oxygen*: Mammalian cells need oxygen, but low oxygen solubility in water (about 0.2 mmol/L, at 37°C) can present a significant barrier to the supply of adequate quantities to cells in the large-scale culture. On a small scale, the supply of oxygen to cell cultures is rarely a problem, as the rate of oxygen diffusion through a few millimetres of the culture medium covering the cells is usually sufficient. However, with an increase in the volume of a culture and the concentration of cells, the oxygen supply becomes one of the main problems to be faced during the scale-up.
- *Gradients of nutrients and metabolites*: In static culture, a gradient of secreted nutrients and metabolites tends to form around cells. Large-scale culture becomes inefficient when gradients are formed, and therefore, mixing and homogenization mechanisms must be used, but these can cause problems such as cell disruption and bubble formation. It is recommended to add surfactants to control foaming.
- *Logistics*: Small culture flasks and their contents can be handled manually, moved from one environment to another (e.g. biosafety cabin for the incubator), and can be heated or cooled quickly. As the scale increases, transport and handling become difficult and eventually impossible. For example, the use of autoclaves should be replaced by steam sterilization *in situ* (steam-in-place). Likewise, reaching the correct incubation temperature of the culture medium can take a few minutes on a small scale, but in a large bioreactor, this process can take several hours. These factors have a major impact on the design and operating costs (including downtime) of the large-scale systems.

12.7 ANCHORAGE-DEPENDENT CELL SYSTEMS

12.7.1 Roller Bottles

Roller bottles have been used to grow different types of cells. The inner surface of the bottle or flask is used as a cylindrical growth surface. The cells are introduced with a limited volume of medium and the flask is placed horizontally on equipment that will slowly rotate around an axis parallel to that of the cylindrical surface. During rotations, the cells become attached to the bottle wall and are cyclically immersed in the culture medium. When they are not immersed, there is a thin layer of medium covering the cells, and the rotation speed must be adjusted so that it does not dry out. Oxygen transfer is relatively efficient in this system.

Roller bottles have a surface area of between 490 and 1,800 cm^2, and if an increase in scale is necessary, the number of bottles used is just increased. This makes scaling from the laboratory to production scale relatively simple, as there is no change in the process. However, when a large number of bottles need to be handled, scaling must be carried out with a step automation. Such industrial systems have been used by the vaccine industry. A number of biological drugs have also been produced in facilities using roller bottles, for example recombinant erythropoietin. However, these processes have now been converted to bioreactors, with or without microcarriers.

12.7.2 Stacked-Plate or Multitray Systems

These culture systems are composed of a series of media surfaces stacked in parallel, one on top of the other, in order to increase the area for cultivation. The culture medium and other solutions are added through an access opening and then pumped between the layers. This system does not require agitation, and initial scaling is done by increasing the number of surfaces, which can reach up to 40 surfaces within a single operating unit.

Although the conditions for cultivation in these units are similar to those for culture in other flasks, there are several disadvantages. Unlike a T-bottle or roller culture, there are no means for direct access to the surface of the culture so scraping the cells, for example, is not possible. They are also completely dependent on the integrity of the seals between the various components of the unit. In some models, it is difficult to maintain an atmosphere enriched with CO_2 in order to maintain adequate buffering of the culture medium using the bicarbonate system.

These 'cell factories' have been available for a few decades and have been used for several cell lines and on an industrial scale to manufacture vaccine. As with roller bottles, scaling to adjust to market requirements just involves increasing the number of units.

12.7.3 Microcarriers

Microcarriers were designed as a way to improve the volumetric efficiency (surface area per unit volume) of culture systems for adhered cells. Using these spheres (typical diameter 200 μm), surface areas greater than 30 cm^2/cm^3 of culture medium are easily achieved for batch processes, and higher values can be used for fed-batch or perfusion processes. These values are much higher when compared to T bottle or multitray units.

Initially, DEAE-Sephadex A-50 microcarriers were used, but the density of the positive charges was too high for efficient adherence and cell growth. A reduction in the density of the positive charge has overcome the problem and, currently, negatively charged and amphoteric surfaces are also used. A wide range of microchargers are commercially available. Some have coatings that aim to increase the adhesion of some cell types, while others are made of materials that can be digested by enzymes in order to decrease cell damage after detachment.

Other physical properties are also important to define the use of microcarriers, such as:

- *Density*: The microcarriers need to have enough density so as not to float, but not so dense that they become difficult to maintain in suspension. Values between 1.02 and 1.04 g/cm^3 are the most frequently used.
- *Transparency*: Microcarriers produced from the transparent material allow a microscopic observation of adhered cells. The materials normally used are dextran, cellulose, plastic and gelatin. Some commercial microcarriers are covered by collagen, polystyrene or fibronectin.
- *Porosity*: Most microcarriers are designed in such a way that cells can grow on their surface. Some are designed specifically to be macroporous, so that cells actually multiply within the microcarrier. This can have advantages in terms of protecting cells from damage caused by impacts between microcarriers and shear, although these mechanisms are only significant under high agitation speeds. However, removal of cells from the beads by trypsinization may be more difficult.
- *Diameter*: A large number of spheres/cm^3 is necessary, both to ensure that the suspension is homogeneous and to obtain the necessary volumetric efficiency. Each sphere must have the ability to load several hundred cells, so the diameters are in the range of 150–230 μm, and must be as uniform as possible to provide the greatest homogeneity of cells in the spheres.

Obviously, microchargers should not be toxic to cells. Likewise, the monomeric material from which they are synthesized, and any other substances (such as surface coatings) should not inhibit the cell growth. Although most microcarrier granules are spherical, there are hexagonal or cylindrical polystyrene forms on the market.

One of the advantages of using microcarriers is that culture of adherent cells can be performed in the same type of equipment that is used for suspension culture, with few modifications. Scaling and control gain benefits with the use of this system, which has led to the widespread use of microcarrier, particularly in vaccine production, in which this technology was first applied more than 20 years ago.

12.7.4 Bed Bioreactors

In bed bioreactors, cell culture is conducted hydraulically and with immobilization. We can classify these bioreactors as packaged or fixed bed bioreactors, and fluidized bed bioreactors.

Fixed bed bioreactor has high-density microcarriers, which constitute the fixed bed. A typical fixed bed bioreactor consists of a cylindrical chamber filled with microcarriers, a gas exchanger, a medium storage tank and a pump that circulates the culture medium between the medium storage tank and the bioreactor. The limitations for its application are development of a preferential flow path in the bed, pore blockage and inefficient gas transfer and detachment. The main advantages of this reactor are the low shear stress, the absence of abrasion between particles and the increase in space/time efficiency. For this reason, fixed bed bioreactors (up to 100 L of medium volume) are capable of providing high cell densities in cultures with protein-secreting mammalian cells in suspension, and for anchorage-dependent cells used for cell growth and virus production.

The operation of a fluidized bed bioreactor is carried out so that the flow is upward, causing the bed to expand under high flow rates, and proceeds towards the microcarriers where the cells are attached. This system aims to generate a fluidized bed in order to guarantee a movement of all particles (microcarriers and cells) so as to prevent their sedimentation or flotation.

12.8 SYSTEMS FOR CELLS IN SUSPENSION (OR CELLS ATTACHED TO MICROCARRIERS)

12.8.1 Spinner Bottles

These are cylindrical flasks with devices for stirring, and they vary in capacity from about 100 mL to 36 L. They represent an intermediate level of scaling between T-bottles and bioreactors, but the cost is lower to purchase and use than a small bioreactor of equivalent size. However, there is much less instrumentation and control than is available in a bioreactor. Many different formats are available, with different shaking methods, shaker configurations and sizes.

12.8.2 Shake Flasks

Shake flasks are more correctly called 'Erlenmeyer flasks' and have been used for culturing mammalian cells since the 1950s. These flasks are suitable for a shaker, which mixes the contents and keeps the cells in suspension. This culture method is particularly useful for small to moderate volumes of cells with high oxygen requirements, such as insect cells. Both reusable and disposable bottles are generally available in sizes from 50 mL to 6 L (with up to 2 L of medium per bottle). Ventilated caps can help to increase gas exchange, and scaling can be achieved using multiple bottles.

12.8.3 Culture Bags

Bags suitable to culture cells have been available for many years. The original design was a static system with gas-permeable bags placed in CO_2 incubators, and consequently, the culture volumes were limited to a few litres due to the need to diffuse oxygen through the walls of the bag. A much more suitable system for staggered use was marketed by Wave Biotech, in 1999, and basically consists of a sterile, non-reusable bag to be kept in a device that constantly shakes. The bag is inflated with a suitable mixture of CO_2/air, and then, the medium is introduced into the bag (and heated, if necessary), after which the cells are added. The bag keeps the volume of CO_2/air equal throughout. Shear stresses are very low, but the swing speed must be controlled in order to avoid foaming. The bags are equipped with a filling tube, a collection tube, an inlet filter, an exhaust filter, a sampling opening, a pressure release valve, inputs for in situ pH sensors and dissolved oxygen probes.

The system has been used with mammalian cells in suspension and with microcarriers. With the adjustment of the inclination parameters and the geometry of the bag, scaling is linear in the range of 100 mL–500 L (www.wavesbiotech.com; Pierce and Shabram, 2004). The scaling of this system is also influenced by mechanical, safety and temperature factors.

The main advantages of using culture bags are simplicity, low shear environment and the disposable nature of the bag. Thus, there is no need for validated cleaning, sterilization or maintenance procedures. Even so, in most industrial installations, the system is used to grow seed inoculants for use in more conventional systems, rather than for the large-scale production and obtaining the final product itself.

12.8.4 Single-Use Bioreactors (SUBs)

In recent years, because of the costs and challenges for validating bioprocesses with the safety required for a biopharmaceutical plant, some companies have started to invest in the development of single-use bioreactors (SUBs) (Figure 12.7). These systems represent a new trend of bioreactors used in plants to obtain biopharmaceuticals for application in humans (Mullin, 2016; Eibl et al., 2010). This choice is directly linked to increased flexibility of production plants, reducing investments and operating costs (Challener, 2017).

As with any system, SUBs present the challenges of maintaining a homogeneous physicochemical environment in the bioreactors, without gradients, offering adequate conditions for maintaining cell viability and bioprocess productivity. Many SUBs options are available, each with its advantages and disadvantages, although scalability is often pointed out as one of the biggest limitations (Shukla and Gottschalk, 2013).

FIGURE 12.7 Single-use bioreactor (SUB). (Sartorius (Gottingen, Germany).)

The industry's willingness to use SUBs is currently influenced by the importance of the stage within the production process, the added value of the product and the development and production times. There is also a need for more clarity in understanding the regulatory requirements for the use of SUBs. To date, the FDA has not mentioned anything directly about SUBs, although it has already approved several products developed using these systems.

One of the biggest advantages of SUBs is the flexibility offered to production plants. SUB is the trend for multipurpose plants, that is, the use of the same facility for the production of different biopharmaceuticals, impacting on less time and cost of production, without compromising the quality of the medicine. In multipurpose plants, the main manufacturing bottleneck is the release of the line and validation of cleaning to ensure that there is no transfer and/or cross-contamination from previous batches. The time in which the plant is inactive reduces the number of batches produced in a given period, that is, the downtime reduces productivity.

One of the main contributors to downtime is the preparation of the bioreactor in upstream processes. SUBs save on the time spent preparing the bioreactor for the next batch. For example, on SUB systems, the exchange time is approximately 2 h, including the time required to make all connections, which would be equivalent to a change time of 6–10 h with a stainless steel bioreactor for the same product and 3 weeks for a complete change of the product, due to cleaning, sterilization and validation times in the fixed plant.

The total time for changing the product depends on the model of the manufacturing equipment. In a hybrid system (a combination of SUBs systems in upstream processing and stainless-steel systems in the production process itself), the changeover time is approximately 2 weeks. If the manufacturing line has disposable bioreactors connected to disposable filters and bags using sterile single-use connectors, the total time to exchange between different product lines will take no more than 48 h. Saving time with highly qualified personnel is also an important benefit of SUBs.

Although several relevant advantages are associated with the use of SUBs, there are also some significant challenges, as in the case of plants built specifically to manufacture a specific biopharmaceutical. In such a case, customizing the SUBs to make them suitable for processing specific molecules negates their main advantages of flexible operating.

The lack of homogeneity of the mixture is also one of the critical problems of SUBs, as well as low oxygen transfer coefficients (k_La). As a result, this operation model practically excludes applications in bioprocesses with bacteria and yeasts. Therefore, SUBs are primarily designed only for culturing mammalian cells.

Another critical issue is related to the material SUBs are manufactured from. In many cases, SUBs are produced from plastic derivatives. In addition to concerns about potential leachable and extractable materials, the disposal of such plastic reactors is also now an environmental concern. This is a fundamental challenge, particularly in countries where the disposal of plastics is strictly controlled.

The future development in SUBs technology will be to increase the diversity of processes on a large scale. Central control and automation in the processing of multiple batches using SUBs is another important step in the evolution of this technology, with the greatest focus being on the development of non-invasive sensors *in situ* based on electromagnetic, semiconductor, optical or ultrasonic measurements (Busse et al., 2017). Good single-use sensor technologies already exist. The most common are pH, dissolved oxygen and carbon dioxide sensors based on fluorescence technology.

Another problem common to SUBs is cytotoxicity from components of disposable bags that can leach into the culture medium. For example, bis (2,4-di-tert-butylphenyl) phosphate (bDtBPP) has been found in cell culture media in concentrations harmful to cell growth (Hammond et al., 2014), highlighting a potential safety risk in using SUBs for biopharmaceutical production, which must be managed by the companies that supply this technology.

The processes to manufacture products for cell and gene therapies have largely been conducted in SUBs (Santos et al., 2013; Shah et al., 2016; Schirmaier et al., 2014).

The use of SUBs in continuous processes must be more robust, as the cultivation lasts significantly longer than a fed-batch process, which means that there is a need for high-level detection and analysis to provide a safe production process over a longer period of time. Therefore, understanding how multiple unit operations can be optimized in a continuous environment will also have a major impact on the development of the next generation of SUBs.

12.9 BIOREACTORS

A discussion on bioreactors is provided in more detail in Chapter 10. Here, aspects with specific importance for the cultivation of mammalian cells are discussed.

For the industrial-scale production of mammalian cells and their products, the most widely used method is culture in agitated tanks (stirred tanks) (Figure 12.8); airlift bioreactors are less common. This technology has been used in other biotechnology industries too, and the principles and engineering involved are well understood. For this reason, the first attempts to use the large-scale bioreactors for culturing mammalian cells took advantage of projects that were developed for microbial fermentations. However, it soon became clear that the mass transfer to oxygen, and the characteristics of mammalian cells (Table 12.5), required adaptations to be made to these bioreactors.

Regardless of the scale, stirred tank bioreactors are the most widely used. They are used for inoculum production, screening experiments, for optimization purposes and manufacturing processes.

In airlift bioreactors (Figure 12.9) and bubble columns, mass and heat transfer are largely accomplished by the injection of air or gas through a column or by static gas distributors (diffusers, nozzles, perforated plates, diffuser rings) or dynamic gas distributors (tubes, injectors or ejectors).

While the bubbles of ascending gas injected by columns cause a random homogenization, the circulation of fluids in airlift bioreactors is obtained by the closed circulation of liquids, which allows more efficient mass transfer and improves the flow and homogenization. In comparison with agitated bioreactors, they require less energy expenditure, minimize the need for sterilization (it has no moving parts, axes and mechanical seals) and are easy to scale. However, they present problems when there are large variations in the concentration of biomass, viscosity, surface tension, ionic concentration and foaming.

Stirred tank bioreactors for the culture of mammalian cells can be of volumes close to 20,000 L, while those of airlifts are around 5,000 L.

12.9.1 Hollow Fibre Bioreactors Applied to the Cultivation of Mammalian Cells

Hollow fibre bioreactors allow the cultivation of cells in high density and are an example of a heterogeneous system, since the cells and the culture medium are in separate compartments. The cells adhere to the outer surface, which contains semipermeable fibres of cellulose acetate, and grow in the extracapillary space, while the medium is distributed through the intracapillary space. Nutrients diffuse through fibres, while toxic metabolites diffuse into the extracapillary space and are removed. The unit consists of thousands of fibres arranged in a cylinder or in cartridges. The product accumulates in the extracapillary space and can be collected intermittently. The main advantage of this system is that it can be used with high cell densities, which in turn leads to increased productivity. However, the systems have large diffusion gradients that make scaling difficult. Figure 12.10 shows some types of flasks and bioreactors used in culturing mammalian cells.

TABLE 12.5
Characteristics of the Microbial and Mammalian Cells Culture

Microorganism	Mammalian Cell	Bioreactors Adaptations
Low sensitivity to shear	High sensitivity to shear	Change in the design of the agitator (e.g. naval propeller for better agitation) reducing the rotation rate; removing the deflectors
Low sensitivity to bubbling damage	High sensitivity to bubbling damage	Increased diameter-to-height ratio to maximize surface aeration (although it may decrease the effect with increased culture volume): minimize the size of the gas bubble used by the sprayer: addition of surface-active agent (but foam control)
High viscosity of the medium	Low viscosity of the medium	Efficient mixing can be achieved with rotors operating at low speeds in order to decrease shear: the bottom of more rounded fermenters improves the mixing at lower rotor speeds: magnetic couplings of the motor instead of a mechanical seal to avoid sealing problems
High oxygen demand	Very low oxygen demand	Lower rotation speed should be allowed: reduces the amount (potentially harmful) spraying required

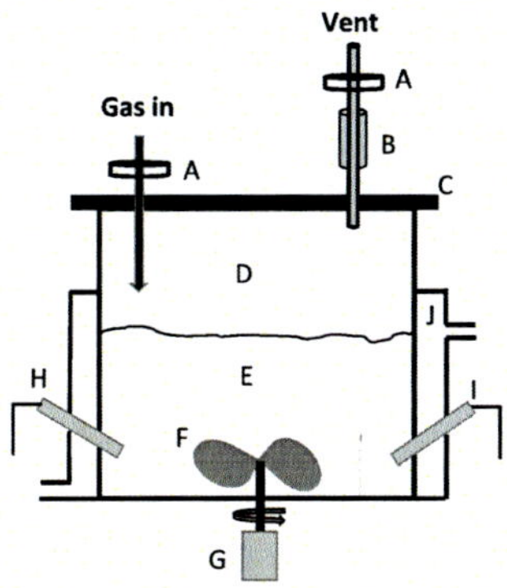

FIGURE 12.8 Diagram of a bioreactor used for mammalian cell culture: (A) air filter, (B) capacitor, (C) removable surface cover, (D) air space between the liquid medium and the surface cover, (E) cell suspension, (F) naval-type propellers, (G) engine, (H) pH sensor, (I) dissolved oxygen (DO) sensor, (J) water jacket.

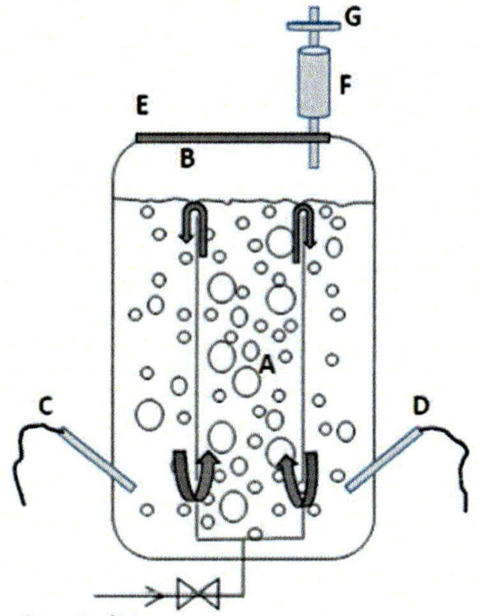

FIGURE 12.9 Simplified diagram of an airlift fermenter: (A) cell suspension, (B) air space between the surface and the cover, (C) pH sensor, (D) dissolved oxygen (DO) sensor, (E) surface cover, (F) condenser, (G) air filter.

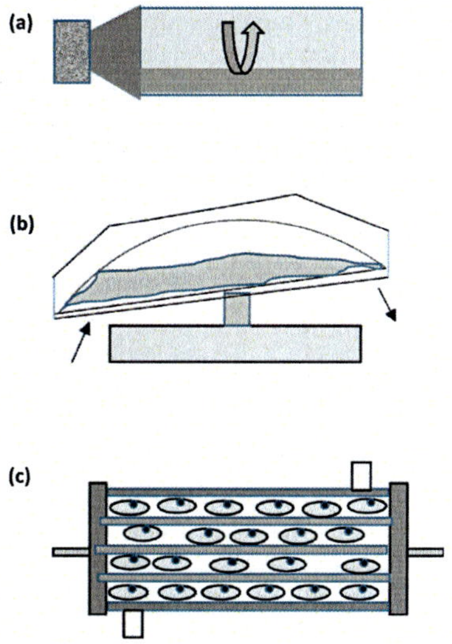

FIGURE 12.10 Basic scheme of some bioreactors for cell culture: (a) roller bottles, (b) bag bioreactor (wave), (c) hollow fibre bioreactor.

12.9.2 Type of Bioprocesses in Cell Culture

Batch or fed-batch processes have been used in cases where a fixed volume of the medium is added to the bioreactor with the cells. Incubation is performed without adding more medium until the end point of the culture is reached, that is, when there is more cell death than growth. However, this process is inefficient due to a limited productivity as there is a depletion of nutrients. In principle, this problem could be easily solved with the addition of fresh medium and/or other nutritive solutions, which led to the development of fed-batch processes, and have been widely implemented in mammalian cell culture. There is also the possibility of continuous removal of the depleted medium and the addition of fresh medium, which generates a process called 'perfusion'. Perfusion processes can be performed with the removal of cells from the system, along with the medium in which the speed of cell removal has to be carefully balanced against the speed of cell growth so as not to deplete the cell population. Another type of perfusion is characterized by a suspension of cells that pass through a cell retention device, where the medium is collected, while the cells are returned to the bioreactor. When the perfusion process is used for the production of extracellular proteins, the mode of cell retention is favoured since it is possible to control the speed of production and the rate of cell growth. Retention devices can be in various forms such as spin filters inside the bioreactors, coupling of filters with tangential flow and the use of continuous flow centrifuges with external cell recirculation.

12.9.3 Cell Metabolism

In order to improve processes that involve cell culture, it is necessary to understand some metabolic pathways. This information can then be used to improve the process, the culture medium and the cell itself. The metabolic pathways of mammalian cells are complex but flexible. The metabolic patterns of mammalian cells are altered when the original homeostatic environment in a tissue of multicellular organisms is changed by stress caused by the *in vitro* environment of the culture system, and by the genetic changes that occur during the immortalization process that establishes a cell line.

This deregulation is characterized by high and inefficient consumption of glucose and glutamine as the main sources of carbon, nitrogen and energy, leading to the generation of unwanted final metabolic products, such as lactate, ammonia and some amino acids (alanine, glutamic acid, etc.), and limiting the performance of mammalian cells in culture.

This metabolic profile is observed in batch culture systems, which are most often used because of their simplicity, in which cells are exposed to high levels of glucose and glutamine in the medium, although this pattern is also found in batch systems fed after a certain period of operation.

Different combined mechanisms result in an unregulated metabolic profile of cells *in vitro*.

Some strategies have been suggested to reduce dysregulation in cell metabolism to obtain more physiologically and metabolically balanced patterns: metabolic engineering, balance and redefinition of the culture medium and engineering design of bioprocesses optimized based on metabolic requirements. The ultimate goal is to generate more efficient cell culture processes. The main metabolic pathways include glycolysis, pentose phosphate pathway, tricarboxylic acid cycle, oxidative phosphorylation and glutaminolysis.

12.9.4 Glucose, Glutamine and Amino Acids as Sources of Energy and Carbon

Culture media contain glucose concentrations in the range of 10–25 mM. As the cytoplasmic membrane is impervious to polar molecules, such as glucose, uptake is obtained by means of transport molecules and is mainly driven by a concentration gradient. Glucose is, above all, a source of carbon and energy, which can be metabolized via glycolysis to pyruvate, which is further oxidized to acetyl-CoA and transformed through the TCA cycle to provide some GTP but also reduced co-factors which are oxidized by the respiratory chain into ATP, water and CO_2. Another important metabolic pathway, which consumes part of the glucose, is the pentose phosphate pathway, which is important for *de novo* synthesis of nucleotides.

Glutamine, present in approximately 1–5 mM in culture media, is an important precursor for the synthesis of purines, pyrimidines, amino sugars and asparagine. It is incorporated into the cell by means of different amino acid transport systems, which can be converted into glutamate and aspartate or metabolized through the TCA cycle. The main metabolite of glutamine transformation is ammonia (about 0.7 mol ammonia/mol glutamine), which is also derived from the chemical breakdown of glutamine to pyrrolidonic carboxylic acid and ammonia. The mechanism of inhibition of cell growth by the presence of ammonia occurs by acidification of the cytoplasm and, therefore, a decrease in intracellular pH.

Amino acids are present in the medium in concentrations ranging from 0.1 to 0.2 mM and are mainly used for protein synthesis. They can be grouped between those that are consumed (arginine, aspartate, cysteine, histidine, isoleucine, leucine, lysine, methionine, alanine, phenylalanine, proline, serine, threonine, tryptophan, tyrosine, valine) and those that are partially produced (alanine, glycine, glutamic acid).

12.9.5 Effects of Lactate and Ammonia

The toxic effects of lactate and ammonia in mammalian cell cultures have been studied in several different cell lines and shown that the tolerance levels for these two products depend on the culture conditions specific to each cell line. Catabolic pathways, especially involving glucose and glutamine, are very similar in different cell types, so the differences between the effects of lactate and ammonia can be related to the different substrate specificities of key enzymes in the main metabolic pathways, and metabolic changes in response to the adverse environment. Understanding these metabolic changes can be essential in the development of more adequate control strategies for bioprocesses (Lao and Toth, 1997).

Studies on the use of nutrients and formation of metabolites have indicated that CHO cells resistant to apoptosis are able to consume available lactate and accumulate little ammonia when they grow in commercially available culture medium (Dorai et al., 2009).

In protein-free culture media with high cell density, cell death by apoptosis occurs in up to 80% cells grown in a fed-batch-type bioreactor, and is induced in response to limitations of nutrients, growth factors, shear, soluble oxygen and accumulation of toxins.

The ability of cell lines resistant to apoptosis to consume lactate, instead of accumulating it, opens the possibility of new culture adaptation strategies to allow growth under conditions of high glucose concentration, as well as after glucose depletion. Therefore, fed-batch cultivation strategies that have been used to limit high toxic levels of lactate and ammonia may not be a problem when cell lines such as these are used.

12.9.6 Role of Oxygen and CO_2 in Cellular Metabolism of Mammalian Cells

Oxygen is an important nutrient for the metabolism of mammalian cells, since it is the final electron acceptor in the mitochondrial respiratory chain and is directly linked to the generation of energy. In addition, oxygen solubility in normal culture media is low, about 0.2 mM, and so the supply of oxygen to cells can be a limiting factor, especially in cultures with high cell concentrations.

Although different cell lines have different ideal values of dissolved oxygen concentration, some general trends are frequently observed, such that the optimum values are in the range of 40%–60% saturation. When the oxygen concentration is below 10% saturation, two trends can be observed: (a) oxygen becomes the limiting substrate, cell growth is reduced and the specific speed of glucose and glutamine consumption is also reduced; and (b) if the dissolved oxygen concentration falls below 0.1%–0.5% of the saturation concentration, cell metabolism is severely affected, specific speeds of glucose and glutamine consumption are very high, and the TCA cycle and the oxidative phosphorylation are inhibited, resulting in increased lactate production.

Carbon dioxide is one of the end products of mammalian cell metabolism. The relevance of CO_2 is related to the regulation of pH because of the equilibrium with bicarbonate $\left(CO_2 + H_2O \Leftrightarrow HCO_3^- + H^+\right)$ in the medium. The CO_2 produced metabolically in bioreactors, observed by increasing the partial pressure of the gas, can lead to the reduced cell viability and other effects related to productivity. Generally, the partial pressure of CO_2 (pCO_2) in the range of 40–50 mmHg is considered ideal, and no deleterious effects on culture are observed (Gray et al., 1996).

12.10 MONITORING AND CONTROL OF MAMMALIAN CELL CULTURE

The monitoring of culture conditions of mammalian cells is similar to those presented in Chapters 7 and 10, in which the same topic is addressed for microbial cells. Strict control and monitoring of parameters such as temperature, pH, dissolved oxygen and CO_2 concentration must be applied to all cultivations used in pharmaceutical biotechnology. However, some aspects should be highlighted in this chapter, as they are especially important for the cultivation of mammalian cells, since non-ideal conditions within the bioreactor can result in a decrease in product formation and a less-efficient process overall. Other parameters such as concentration of metabolites in the culture medium and cell viability can be used to provide information about the stage of the process.

12.10.1 Partial Pressure of O_2 (pO_2)

The solubility of oxygen in aqueous media is low, and its limitation can affect the speed of cell growth and its metabolism. The rate of oxygen consumption depends on the cell line used, growth rates and carbon sources. It is difficult to measure oxygen dissolved in cell culture medium directly, but partial pressure can be used to determine the actual dissolved concentration. At an equilibrium, the partial pressure of oxygen pO_2 in the medium is proportional to the concentration of oxygen (cO_2) in the vapour phase above the medium.

The most common sensors used in monitoring are *in situ* sterilizable galvanic type (potentiometric) electrodes or, more frequently, polarographic-type electrodes (amperometric or Clark). An oxygen-permeable membrane separates the electrode from the culture medium. For both electrode types, the reaction at the cathode (usually platinum) is the reduction of oxygen according to the following reaction:

$$1/2\,O_2 + H_2O + 2e^- \xrightarrow{Pt} 2OH^-$$

In the galvanic-type anode, the reaction is:

$$Pb \rightarrow Pb^{2+} + 2e^-$$

In the polarographic sensor, a constant voltage is applied between the cathode and the anode; the reaction at the anode is:

$$2Ag + 2Cl^- \rightarrow 2AgCl + 2e^-$$

In both types of electrodes, the electrical signal is measured, at steady state, proportional to the oxygen flow in the cathode, which in turn, is proportional to the partial pressure of oxygen in the liquid phase. Two-point calibration is normally performed after *in situ* sterilization, but before inoculation in a culture medium saturated with air and nitrogen.

12.10.2 Partial Pressure of Carbon Dioxide

The concentration of carbon dioxide in the vapour coming out of a bioreactor can be used to calculate the specific respiration rates and cell activity using infrared (IR) analysers (Riley, 2006). The dissolved carbon dioxide can also be measured *in situ* using a chemical fibre optical sensor, but this is not yet used industrially. This type of sensor uses 8-hydroxypyrene-1,3,6-trisulfonic acid trisodium salt (HPTS) to quantify carbon dioxide dissolved in the medium. The protonated and non-protonated forms have different maximum excitations at 396 and 460 nm, which allow a radiometric measurement of pH and carbon dioxide concentrations.

12.10.3 Metabolites and Products

Glucose and the amino acid glutamine are the two main nutrients in mammalian cell cultures, while ammonia and lactate are the two main metabolic by-products which can inhibit cell growth and product formation. So, it is very important to monitor and control the concentrations of both nutrients and by-products in the medium during the culture process.

Biosensors are used to quantify the concentrations of nutrients and derivatives. These are devices that use a specific enzymatic reaction to generate an electrical signal that is proportional to the concentration of the analyte to be quantified. Due to the enzymatic nature of the sensor, only one for each substance should be used. In all cases, the enzymatic reaction should result in a product that can be easily detected, such as hydrogen peroxide. The topic of biosensors is covered in more detail in Chapter 18.

12.10.4 Cell Concentration and Viability

Cell density can be measured using different methods. Offline methods require aseptic sampling and can be as simple as counting cells using a haemocytometer (Neubauer chamber). Trypan blue dye is used to determine the cell viability, which penetrates cells with broken membranes and stains proteins present in the cytosol. More complex automatic counters can also be used to determine the cell density and viability. Indirect methods include the determination of total protein (e.g. Bradford method) and quantification of cell nuclei using a fluorescent DNA dye (DAPI). Other methods for determining the percentage of viable cells within the culture include analysis of MTT (3-(4,5-dimethylthiazol-2yl)-2,5-diphenyl tetrazolium bromide) and determination of LDH.

12.11 FINAL CONSIDERATIONS

Mammalian cell culture technology has become a major field in modern biotechnology, especially in the area of human health, and this is an area of rapid expansion. Fascinating developments achieved in the past decades are impressive examples of an interdisciplinary interplay between medicine, biology and engineering. Cells derived from animals, especially mammals like rodents and humans, have become a major vehicle for producing biologics, a class of medications that includes vaccines, various proteins used in treating cancer, genetic diseases, drugs for cardiovascular, respiratory and immune diseases, and monoclonal antibodies. Tissue engineering or gene therapy opens up challenging new areas. Novel recombinant protein therapeutics have been approved by US Food and Drug Administration (FDA) annually. Advances in cell line development technologies are therefore crucial to support the rapid development of recombinant protein therapeutic products, where improvements in the timeline and the ease of generating high-producing cell lines can contribute to the faster development of biosimilars and innovative products alike. In the case of innovative products, a reduction in time-to-market period for biopharmaceutical manufacturers is also advantageous because it maximizes profit for biologics with the limited period of patent exclusivity. Advances in cell line development technologies centre on improvements in protein expression technologies and new clone screening technologies. Although bioreactors from small (mL range up to 10L) to large scale (up to $20\,m^3$) have been developed over the past decades for mammalian cell culture-based applications, there are a number of complications that make it a difficult process to ensure that the culture is growing under optimal conditions at all times. Complications arising in these systems include the requirement to grow cultures in complex media, the lack of on-line measurements for many of the key substrates, metabolites, and products, the limited and noisy nature of much of the available experimental data and the extremely complex underlying reaction system. In conclusion with the new technologies discussed above, new tools in cell line development can be generated and the process can be further streamlined to facilitate biopharmaceutical drug discovery and development.

REFERENCES

Blaas, L.; Musteanu, M.; Eferl, R.; Bauer, A.; Casanova, E. Bacterial artificial chromosomes improve recombinant protein production in mammalian cells. *BMC Biotechnol.* v.9, 3, 2009.

Brekke, O.; Sandlie, I. Therapeutic antibodies for human diseases at the dawn of the twenty-first century. *Nat Rev Drug Discov.* v.2, 52–62, 2003.

Browne, S. M.; Al-Rubeai, M. Selection methods for high-producing mammalian cell lines. *Trends Biotechnol.* v.25, n.9, 425–432, 2007.

Busse, C.; Biechele, P.; De Vries, I.; Reardon, K. F.; Solle, D.; Scheper, T. Sensors for disposable bioreactors. *Eng Life Sci.* v.17, 940–952, 2017.

Carrel, A.; Burrows, M. T. Cultivation of tissues in vitro and its technique. *J Exp. Med.* v.13, 387–396, 1911.

Challener, C. A. Single-use bioreactors have reached the big time: The decision to use disposable bioreactors is now driven by commercial rather than technological considerations. *BioPharm Int.* v.30, 18–23, 2017.

Cheng, H.-S.; Shen, C.-W.; Wang, S.-R. Effect of storage conditions on detection of mycoplasma in biopharmaceutical products. *In Vitro Cell Dev Biol Anim.* v.43, n.3, 113–119, 2007.

Cruz, H. J.; Moreira, J. L.; Carrondo, M. J. T. Metabolic shifts by nutrient manipulation in continuous cultures of BHK cells. *Biotechnol Bioeng.* v.66, n.2, 104–113, 1999.

Darling, A. Validation of biopharmaceutical purification processes for virus clearance evaluation. *Mol Biotechnol.* v.21, n.1, 57–84, 2002.

Dembowsky, K.; Stadler, P. *Novel therapeutic proteins: Selected case studies.* Wiley-VCH, 2001.

Dorai, H. Kyung, Y. S.; Ellis, D.: Kinney, C., Lin C., Jan D., Moore, G.: Betenbaugh M. J. Expression of anti-apoptosis genes alters lactate metabolism of Chinese hamster ovary cells in culture. *Biotechnol Bioeng.* v.103, n.3, 592–608, 2009. doi:10.1002/bit.22269. PMID: 19241388.

Dutkowski, P.; De Rougemont, O.; Clavien, P. A. Alexis Carrel: Genius, innovator and ideologist. *Am J Transplant.* v.8, n.10, p.1998–2003, 2008.

Eibl, R.; Kaiser, S. C.; Lombriser, R.; Eibl, D. Disposable bioreactors: The current state-of-the-art and recommended applications in biotechnology. *Appl Microbiol Biotechnol.* v.86, n.1, 41–49, 2010. doi:10.1007/s00253-009-2422–9.

Enders, J. F.; Weller, T. H.; Robbins, F. C. Cultivation of the lansing strain of poliomyelitis virus in cultures of various human embryonic tissues. *Science.* v.109, n.2822, 85–87, 1949.

European Medicines Agency. Guideline on similar biological medicinal products containing biotechnology derived proteins as active substance: Non-clinical and clinical issues. 2006. Available at: https://www.ema.europa.eu/en/documents/scientific-guideline/guideline-similar-biological-medicinal-products-containing-biotechnology-derived-proteins-active_en-2.pdf. March 2, 2010.

FDA. Characterization and qualification of cell substrates and other biological materials used in the production of viral vaccines for infectious disease indications (Final Guidance for Industry). 2010.

Freshney, R. *Culture of animal cell - A manual of basic technique.* Wiley–Blackwell, 5th ed. 2005.

Geraghty, R.J. et al. Guidelines for the use of cell lines in biomedical research. *Br J Cancer.* v.111, 1021–1046, 2014.

Graham, F. L.; Van Der Eb, A. J. A new technique for the assay of infectivity of human adenovirus 5 DNA. *Virology.* v.52, 456, 1973.

Gray, D. R. et al. Co2 in large-scale and high-density CHO cell perfusion culture. *Cytotechnology.* v.22, n.1, 65–78, 1996.

Hammond, M.; Marghitoiu, L.; Lee, H.; Perez, L.; Rogers, G.; Nashed-Samuel, Y.; Nunn, H.; Klyne, S. A cytotoxic leachable compound from single-use bioprocess equipment that causes poor cell growth performance. *Biotechnol Prog.* 30, 332–337, 2014.

Harrison, R. G. Observations on the living developing nerve fiber. *Proc Soc Exp Biol Med.* v.4, 140–143, 1907.

Harrison, R. G. The outgrowth of the nerve fiber as a mode of protoplasmic extension. *J Exp Zool.* v.9, 787–846, 1910.

Heldens, J. G. et al. Veterinary vaccine development from an industrial perspective. *Vet J.* v.178, n.1, 7–20, 2008.

Hwang, W. Y. K.; Foote, J. Immunogenicity of engineered antibodies. *Methods.* v.36, n.1, 3–10, 2005.

Jacquemin, M. G.; Saint-Remy, J. M. Factor viii immunogenicity. *Haemophilia.* v.4, n.4, 552–557, 1998.

Jones, D.; Kroos, N.; Anema, R.; van Montfort, B.; Vooys, A.; van der Kraats, S.; van der Helm, E.; Smits, S.; Schouten, J.; Brouwer, K.; Lagerwerf, F.; van Berkel, P.; Opstelten, D. J.; Logtenberg, T.; Bout, A. High-level expression of recombinant IgG in the human cell line per.c6. *Biotechnol Prog.* 19, 163–168, 2003.

Lao, M.-S.; Toth, D. Effects of ammonium and lactate on growth and metabolism of a recombinant Chinese hamster ovary cell culture. *Biotechnol Prog.* v.13, n.5, 688–691, 1997.

Link, T. et al. Bioprocess development for the production of a recombinant mucl fusion protein expressed by cho-k1 cells in protein-free medium. *J Biotechnol.* v.110, n.1, 51–62, 2004.

Monaco, L. et al. Genetic engineering of α2,6-sialyltransferase in recombinant CHO cells and its effects on the sialylation of recombinant interferon-γ. *Cytotechnology.* v.22, n.1, 197–203, 1996.

Mukovozov, I. et al. Factors that contribute to the immmunogenicity of therapeutic recombinant human proteins. *Thromb Haemost.* v.99, n.5, 874–882, 2008.

Mullin, R. Single-use becomes state-of-the-art in bioprocessing. Disposable plastic vessels take the field as companies open big biologics plants. *Chem Eng News* v.94, n.28, 21–22, 2016.

Neusüß, C.; Demelbauer, U.; Pelzing, M. Glycoform characterization of intact erythropoietin by capillary electrophoresis-electrospray-time of flight-mass spectrometry. *Electrophoresis.* v.26, n.7–8, 1442–1450, 2005.

O'Connor, T. P.; Crystal, R. G. Genetic medicines: Treatment strategies for hereditary disorders. *Nat Rev Genet.* v.7, n.4, 261–276, 2006.

Peces, R. et al. Antibodies against recombinant human erythropoietin in a patient with erythropoietin-resistant anemia. *N Engl J Med.* v.335, n.7, 523–524, 1996.

Petricciani, J. C. The acceptability of continuous cell lines: A personal & historical perspective. *Cytotechnology*. v.18, n.1, 9–13, 1995.

Santos, F. F.; Andrade, P. Z.; Silva, C. L.; Cabral, J. M. S. Bioreactor design for clinical-grade expansion of stem cells. *Biotechnol J*. v.8, n.6, 644–654, 2013. doi:10.1002/biot.201200373.

Schernthaner, G. Immunogenicity and allergenic potential of animal and human insulins. *Diabetes Care*. v.16, suppl 3, 155–165, 1993.

Schirmaier, C.; Jossen, V.; Kaiser, S. C.; Jüngerkes, F.; Brill, S.; Agnieszka, S.-N., Siehoff, A.; Van De Bos, C.; Eibl, D.; Eibl, R. Scale-up of adipose tissue-derived mesenchymal stem cell production in stirred single-use bioreactors under low-serum conditions. *Eng Life Sci*. v.14, n.3, 292–303, 2014.

Schwab, R.; Crow, M. K.; Russo, C.; Weksler, M. E. Requirements for T cell activation by OKT3 monoclonal antibody: Role of modulation of T3 molecules and interleukin 1. *J Immunol*. v.135, n.3, 1714–1718, 1985.

Shah, R.; Shah, C.; Park, H. S. Proposed design model of single use bioreactor for mesenchymal stem cells proliferation. *Procedia CIRP*. v.41, 382–386, 2016. doi:10.1016/j.procir.2015.12.127.

Shukla, A. A.; Gottschalk, U. Single-use disposable technologies for biopharmaceutical manufacturing. *Trends Biotechnol*. v.31, n.3, 147–154, 2013.

Simaria, A. S.; Hassan, S.; Varadarajuh, J.; Rowley, J.; Warren, K.; Vanek, P.; Farid, S. S. Allogeneic cell therapy bioprocess economics and optimization: Single-use cell expansion technologies. *Biotechnol Bioeng*. v.111, 69–83, 2014.

Tan, S. Y.; Ponstein, N. Jonas Salk (1914–1995): A vaccine against polio. *Singapore Med J*. v.60, n.1, 9–10, 2019.

Walsh, G. *Pharmaceutical biotechnology: Concepts and applications*. Wiley, 2013.

Walsh, G. Biopharmaceutical benchmarks 2018. *Nat Biotechnol*. v.36, n.12, 1136–1145, 2018.

Weller, T. H. et al. *Growing pathogens in tissue cultures: Fifty years in academic tropical medicine, pediatrics, and virology*. Science History Publications, Canton, MA, 2004.

White, G. C. Hemophilia: An amazing 35-year journey from the depths of HIV to the threshold of cure. *Trans Am Clin Climatol Assoc*. v.121, 61–75, 2010.

WHO Expert Committee on Biological Standardization. WHO Technical Report Series, no. 978, Annex 3, 79–188, 2013.

Witkowski, J. A. Alexis carrel and the mysticism of tissue culture. *Med Hist*. v.23, n.3, 279–296, 1979.

Wrigley, J.D. et al. Cell banking for pharmaceutical research. *Drug Discov Today*. v.19, n.10, 1518–1529, 2014.

Wurm, F. M. Production of recombinant protein therapeutics in cultivated mammalian cells. *Nat Biotech*. v.22, n.11, 1393–1398, 2004.

13 Purification Process of Biomolecules

Adalberto Pessoa Jr. and Beatriz Vahan Kilikian
Universidade de São Paulo

CONTENTS

13.1 INTRODUCTION

In this chapter, unitary operations (also called a defined operation, which encompasses any separate activity during a process – use of a certain type of equipment, application of a methodology, etc.) for the purification of bioproducts are described. An increase in the diversity of biotechnological products has encouraged the development of new and innovative purification processes as well as stimulated the introduction of genetic modifications in production strains during the development of new biomolecules in order to make the purification step more efficient and to improve integration between all of the different stages during process development.

The products of the biotechnology industry are diverse in composition (organic acids, antibiotics, antibodies, polysaccharides, hormones, amino acids, peptides and proteins) and in their location in relation to the producing cell. As a result of this diversity, there are no purification processes of general application. However, conceptually, the process can be divided into four main steps: separation of cells and their fragments (clarification) from the aqueous medium; concentration and/or purification, which comprises the separation of the target molecule, for example, a protein from molecules with significantly different physico-chemical characteristics (water, ions, pigments, nucleic acids, polysaccharides and lipids); purification which frequently comprises separating classes of molecules with some similar physico-chemical characteristics, for example, proteins; and finally operations for the final packaging of the product. In addition, for cell-associated products, it is necessary to disrupt the cells after clarification (Costa-Silva et al., 2018).

The completion of each step does not necessarily comprise a single unit operation. For example, following salt precipitation, dialysis is required to adjust the ionic strength to levels appropriate to a chromatography. On the other hand, there are products (organic acids and industrial enzymes) whose application does not require a high degree of purity, so that chromatographic operations are not necessary. However, in any situation, reducing the number of steps is of fundamental importance for the viability of the process. For example, if each unit operation yields 90% of a product, the application of nine operations will lead to a final yield of only 40%.

The definition of a unit operation in a purification process depends on the use of the target molecule, its physico-chemical characteristics as well as those of the impurities. Products intended for therapeutic and diagnostic use are, of course, those which require a greater degree of purity, and therefore, for these molecules, the purification process is complex. A measure of this complexity is the cost of the purification process in relation to the final cost of the product, which can reach 80%.

13.2 CELL-LIQUID SEPARATION

Purification of biotechnological products is started shortly after cultivation of animal, vegetable or microbial cells from the bioreactor (clarification step), i.e., separation between cells – suspended solids – and liquid medium, from which a clarified liquid results. In industrial processes, the following unitary operations are applied: conventional or tangential filtration and centrifugation.

Conventional filtration is used mainly to suspensions of filamentous fungi, which do not sediment by centrifugation, since they have density very close to the density of the water and cause clogging of the membranes used in tangential filtration. Yeasts, depending on their size between 1 and 8 µm and density 1.05 g/cm³, sediment efficiently by centrifugation. Bacterial suspensions require greater centrifugal forces for clarification of the fermentation broth because they are only 0.1–1.0 µm in size, and often, tangential filtration is used as an alternative.

13.2.1 Filtration

In conventional filtration (also called *dead end* filtration), the cell suspension is fed in a perpendicular direction to the filter media; on the other hand, in tangential filtration (also called microfiltration), the suspension is fed tangentially to the filter membrane. The two types of filtration are performed under pressure, which is applied to the feed medium.

In conventional filtration, continuous deposition of cells and culture medium occurs on the membrane, forming a layer called the *filter cake*, which increases resistance to filtration compared to the resistance caused by the filter membrane alone. The resistance caused by the filter cake to the passage of the cell suspension is directly proportional to the compressibility of the cells and to the pressure applied, since there is synergy between the filtration pressure and compression of the cake (Wheelwright, 1991).

Equation 13.1 is derived from Darcy's law and represents the time required for conventional filtration of a cell suspension (Ladisch, 2001):

$$t = \frac{\mu \alpha' X}{2\Delta P^{(1-s)}} \frac{V^2}{A^2} \qquad (13.1)$$

where μ = suspension viscosity (kg/m·h); α' = constant related to the size and shape of the cells; X = concentration of cells in the suspension (g/L); ΔP = pressure reduction through the bed (N/m²); S = compressibility of the cake (dimensionless ranging from 0 to 1.0); V = volume of filtrate (L); A = filtration area (m²).

The constant α' is specific for a given cell suspension and is determined experimentally as a function of the applied pressure, ΔP. The value of α, is the specific resistance of the cake (m/g) and can be calculated according to Equation 13.2:

$$\alpha = \alpha'(\Delta P)^s \qquad (13.2)$$

For example, in the filtration of a *Streptomyces* suspension at a concentration of 15 g/L, with a viscosity μ of 1.1cp, under ΔP of 6.78×10^4 N/m², the value of α is 2.4×10^{11} cm/g.

Rigid solids are incompressible 'cakes' whose s value (cake compressibility) is zero and, therefore, require significantly lower filtration time compared to 'cakes' made up of microbial cells, whose s value can reach 0.8. Filtering auxiliaries, such as diatomaceous earth or perlite, may be added to the suspension prior to filtration or deposited on the filter medium. The adsorption of the cells onto the soil particles results in reduction in the compressibility of the 'cake' as well as avoids the filter medium clogging as a consequence of the size of the soil particles and their incompressibility (Harrison, 1994; Ladisch 2001).

The effectiveness of filtration auxiliaries is illustrated by a reduction in filtration time in the order of 5–20 times. Although the compressibility (s) is reduced, the viscosity (μ) is increased, which is a variable directly related to the time for a given volume to be filtered. Considering that the pore diameter of conventional filter membrane is in the range of 10–1,000 µm, the use of these filters depends on the use of filtration auxiliaries for the retention of bacteria and fungi, since the size of these microorganisms often is less than 10 µm.

Tangential filtration is the unitary operation most frequently employed for the clarification of microbial suspensions. The tangential flow to the surface of the filter membrane operating at a high linear velocity is between 0.2 and 5 m/s depending on the filter configuration, minimizing accumulation of solids on the membrane surface. Tangential filtration is attractive for its low energy demand although the membrane costs are high (Figure 13.1) (Pyle, 1990).

During the tangential filtration process, there is an increase in cell numbers or solute concentration towards the membrane surface, causing the formation of a gradient of solids, which reduces liquid flow within the suspension. Additionally, a narrowing of the pore diameter of the membrane occurs as a result of the penetration of solutes, a phenomenon called fouling. While the concentration gradient is a reversible phenomenon, fouling is only partially reversible. Increasing the linear velocity of the suspension and the pressure (usual 0.5–2.0 atm) can minimize the effect of concentration gradients, while a variation in the pH of the medium and an increase in the linear velocity can reduce the fouling effect.

The material of the membranes will depend on the medium to be filtered and on the performance of the filtration, both of which can be evaluated experimentally. The choice of membrane material will also depend on the reuse capacity of the membranes after successive cycles of

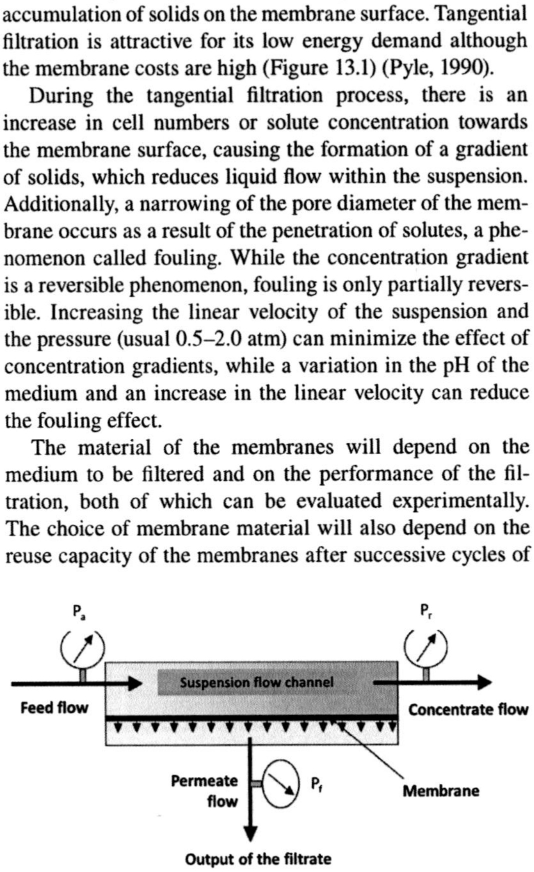

FIGURE 13.1 Scheme of a tangential filtration. Pa, feed pressure; Pr, concentrate pressure; Pf, permeate pressure.

filtration and washing. Regarding the configuration, membranes are generally in the form of flat plates or cylinders, which are then configured into concentric parallel tubes, which are more tolerant to concentration gradients since the filtration operates under turbulent flow (Verrall and Hudson, 1987). The scale up of the tangential filtration process is limited to maximum filter membrane area for a given filter configuration. Since the three pressures indicated in Figure 13.1 are maintained (the pressure in the feed suspension, in the filtrate and in the concentrate) besides the linear velocity of the fed suspension, the filtrate flow is also maintained (Wheelwright, 1991).

13.2.2 Centrifugation

Cells suspended in a liquid medium sediment by the action of gravity when they have a density higher than the density of the liquid. Such sedimentation can be accelerated in equipment in which a centrifugal gravitational field is established. Centrifugation, as well as tangential filtration, is an alternative to conventional filtration because it does not require filtration auxiliaries and allows for aseptic operation.

The sedimentation rate in a centrifugal field (v_c) depends on the particle or cell diameter (d), the difference between the cell density (ρ_c) and the liquid medium density (ρ), the viscosity of the liquid medium (μ), the square of the angular rotation (w) and the radial distance from the center of the centrifuge to the cell (r) (Equation 13.3).

$$v_c = \frac{d(\rho_c - \rho)w^2 r}{18\mu} \tag{13.3}$$

The driving force for natural sedimentation of a cell suspended in a liquid medium is given by the value of g (standard acceleration of gravity). The driving force in a centrifugal field is w^2r. Thus, the ratio of the driving force in a centrifugal field (w^2r) to the standard acceleration of gravity (g) represents a multiple of the latter and is given by Equation 13.4.

$$F_c = \frac{w^2 r}{g} \tag{13.4}$$

A given centrifugation is characterized by the time of application of a certain value of F_c in order to provide a certain degree of clarification. For example, for the separation of yeasts, values of the order of 3,000×g and a few minutes of centrifugation are sufficient to completely sediment the cells.

A simple qualitative criterion for scaling up or simply changing equipment is to maintain the value of the product between F_c and time ($F_c \cdot t$). For example, if 3,000×g for 5 min are sufficient to obtain compact pellet and an acceptably clear supernatant, 1,500×g for 10 min should result in supernatant and sediment with the desired characteristics (Wheelwright, 1991).

Yeast suspensions are efficiently clarified by centrifugation, whereas for bacteria, the smaller cell size requires higher F_c values and, therefore, it is recommended a comparison be made with microfiltration, for performance and cost.

13.3 CELL DISRUPTION

The increase in the demand for intracellular products by the food and pharmaceutical industries has increased the importance to develop suitable processes for cellular disruption, which is a unitary operation that takes place after separation and washing (clarification) of the cells from the culture medium. Criteria used to select a cell disruption technique should consider the next variables: cell size and its shear stress tolerance, temperature control needs, operating time, process yield, energy expenditure, unitary operational cost and investment capital (Harrison, 1994).

Cells enclosed only by cell membranes, such as animal cells and hybridomas, are fragile and readily disrupted under low shear stresses. Consequently, such cells require little energy for their disruption. In addition, they may be broken by simply varying the osmotic pressure of the medium, by adding detergents or by applying low intensity ultrasound. This ease of cell disruption can become a problem in the processing of the medium, since a simple pumping operation can also cause the cells to lyse and the target molecule to be lost. On the other hand, there are cells with a robust wall structure, such as microbial cells, which are difficult to break (Verrall and Hudson, 1987).

The appropriate way to disrupt microbial cell walls depends on the characteristics of the microorganism. Gram-positive bacteria have more rigid walls than Gram-negative bacteria and are, therefore, more difficult to disrupt. However, yeasts and other forms of fungi can be even more difficult to lyse in comparison to bacteria because they have rigid cell walls structures.

Methods of cell disruption can be divided into four classes: mechanical (high-pressure homogenizer, ball mill, French press and ultrasound); non-mechanical or physical (osmotic shock, freezing and thawing, heating; drying), chemical (alkalis, solvents, detergents, acids) and enzymatic (enzymatic lysis or inhibition of cell wall biosynthesis). The cell wall may be fully ruptured or partially permeabilized in order to allow the target molecule to be released into the extracellular medium without cell fragments.

After cell disruption, the cellular homogenate consists of the target molecule, contaminating biomolecules and cellular fragments. Such contaminants are generally undesirable and should be removed by processes such as filtration, centrifugation, precipitation or liquid-liquid extraction. The purification of intracellular products is of higher cost compared to the purification of extracellular products, since the presence of contaminants and cellular fragments requires a greater number of unitary steps. In this sense, molecular biology can contribute to a reduction in the costs of the purification unitary operation of biotechnological products, since it can be applied for the genetic modification of the cell, in such a way that the cell will produce the target biomolecule extracellularly (Harrison, 1994).

Criteria for the selection of a disruption method need to take into account some specific factors, such as yield, specificity, need for temperature control, cost of the unit operation and investment capital.

Enzymes are capable of hydrolyzing walls of microbial cells. Enzymatic lysis methods are suitable for the recovery of biomolecules sensitive to temperature, shear stress or pressures generated by mechanical methods. Disadvantages of using an enzymatic lysis method include the presence of inhibitors, the possibility that the enzyme (which can be expensive) cannot be recovered and used again and resistance to shear stress in the case of enzymatic lysis associated with mechanical disruption. When a certain amount of wall is removed, the internal osmotic pressure of the cell ruptures the cytoplasmic membrane allowing the intracellular contents to be released into the external environment. As in any enzymatic reaction, the study of the efficiency of the enzymes and the determination of the optimal conditions of action are important. As the composition of cell walls varies according to the type of microorganism, the enzymes that can be used for disruption can only be those enzymes specific for the wall substrates. Therefore, in a unitary operation for cell disruption that uses enzymatic lysis, the influence of variables such as pH, temperature, ionic strength, cell and enzyme concentration must be known (Neves et al., 2007).

Yeast cell wall compositions are different from those of bacteria, and therefore, lysis methods are specific to this group of microorganisms. Yeast cell walls have two main layers: the outer layer composed of a protein-mannan complex and the inner layer of glucan. The enzymatic system for the breakdown of yeast is therefore composed of different enzymes such as glucanases, proteases and mannanases, which act synergistically to achieve cell wall lysis, but only two enzymes are essential for complete cell disruption: a protease to degrade the protein-mannan outer layer and a glucanase to degrade the inner layer of glucan.

The resistance of the cell wall of bacteria to disruption varies depending on whether the strains is Gram-positive or Gram-negative, as it was mentioned before. In Gram-positive bacteria, peptidoglycan is present in a greater proportion and is associated with teichoic acids and polysaccharides. On the other hand, Gram-negative bacteria have a double cell wall layer composed of peptidoglycan, proteins, phospholipids, lipoproteins and lipopolysaccharides. The main bacteriolytic enzymes are glycosidases, acetylmuramylalanine amidases, endopeptidases and proteases.

Cell disruption by enzymatic lysis provides many advantages, such as easy control of the pH and temperature of the medium, low capital investment and high specificity for cell wall degradation and, moreover, can be used in association with mechanical or non-mechanical methods. In addition, lytic enzymes can be used to release only biomolecules of interest and, thereby, simplify further purification steps. The main disadvantage is the cost of the lytic enzymes, which often makes its use prohibitive on an industrial scale. Variation in efficiency of enzymatic lysis is dependent on the physiological state of the microorganism, and the need for removing the added enzyme during the purification process is another disadvantage (Verrall and Hudson, 1987).

The efficiency of cell disruption has to take into account the yield of the target biomolecule in its active form. After disruption, proteins can be degraded by proteases, and therefore, it is essential to lower the temperature and add protease inhibitors to minimize their deleterious effects. In addition to the possibility of degradation of the target biomolecule, the release of nucleic acids and structural proteins does occur, which causes an increase in the viscosity of the homogenate. As an example, it has been found that upon disruption of a suspension with 75% (wet mass) of cells, the viscosity of the medium increases by eightfold. Addition of nucleases or proteases may improve rheological characteristics of the medium. pH variation can also be used to reduce the viscosity of the cell homogenate (Ladisch, 2001).

Although there are many specific examples of cell disruption by chemical and enzymatic processes, mechanical methods have been used most widely on an industrial scale. The size and shape of the cells as well as the cell wall structure are determining factors to define the unitary operation to be used for mechanical cell disruption. A high-pressure homogenizer and ball mill are the main equipment used industrially.

Homogenizers consist of pistons designed to apply high pressures to a cell suspension, forcing the suspension through a narrow orifice and onto a surface in the chamber under low pressure. The instantaneous drop of pressure, associated with the impact onto the chamber surface, causes effective cell disruption without damaging proteins. In this type of disruption, larger cells break more easily, and higher pressures increase the breaking efficiency. When the unitary operation is conducted at high pressures, high recovery yields are possible with only one process step. However, multi-step disruptions can also be used to increase process yield (Wheelwright, 1991).

Several operating factors affect the performance of a high-pressure homogenizer: operating pressure, temperature, microorganism growth phase, culture conditions, cell type and cell concentration. In the scale-up of a cell disruption operation using a high-pressure homogenizer, some parameters must remain constant, such as feed velocity, operating pressure and temperature, number of passages through the homogenizer valve, viscosity and cell concentration of the feed.

A ball mill consists of a closed cylindrical chamber (horizontal or vertical), a cooling system and an axis that rotates at high speeds. In the chamber, glass beads and the cell suspension are added. Along the axis of rotation are distributed one or more disks or rods, which rotate at high speed and cause friction between the beads and intact cells, causing cell disruption. Breakage occurs as a function of the shear force applied by the glass beads against the cell wall of the cells. The disruption conditions in such equipment are readily controllable, and the efficiency of the process depends on the type of rupture chamber, the speed and type of agitator, the size of the beads, the bead loading, the cell concentration, the feed rate and the temperature (Pyle, 1990).

Horizontal chambers provide higher disruption efficiency because they can contain larger ball loads and more spheres with smaller diameters. In the vertical chambers, the fluid flow is in the upward direction and, with this, fluidization of the spheres occurs. The speed of the agitator influences the number of contacts between the cells and the beads, and the higher the speed of rotation, the faster the cell disruption. However, the efficiency depends on the size of the cells. Smaller organisms, such as bacteria, need speeds higher than for yeasts. The type of agitator also influences the disruption efficiency. The agitator should be designed to provide maximum kinetic energy transfer to the beads. Although there are a wide variety of agitators, it is not possible to make a direct comparison between the type of agitator and disruption efficiency. The stirrers may be arranged *on* or *off* the central axis, perpendicular or oblique. To aid agitation, they may include grooves, small cuts or holes (Wheelwright, 1991).

When bacteria are to be disrupted, spherical beads (balls) with a small diameter in the order of 0.1 mm are required, whereas yeasts can be disrupted using balls of the order of 0.5 mm in diameter. Balls with smaller diameters tend to provide more efficient disruption as the shear probability is highest with intact cells. In cases, where the target biomolecule is secreted into the periplasmic space, the use of balls with larger diameters is indicated, as these will aid in the release of the biomolecule without the need for total cell disruption.

The balls should occupy between 80% and 85% of the horizontal chamber volume and between 50% and 60% if the chamber is vertical. If the loading volume is too small, there will not be enough collision frequency to provide good disintegration of the cells. If the ball load is too large, the balls will collide with each other and decrease the process efficiency, as well as increasing temperature and power consumption. Cell concentration causes little influence on the efficiency of the lysis; however, the most recommended concentrations range from 30% to 50% (v/v). Lower cell concentrations generate less heat but increase energy consumption per unit cell mass (Harrison, 1994).

The proportion of disrupted cells decreases when the feed flow rate is increased because the time the cells remain in the disrupter is less. The optimal feed flow depends on the speed of the agitator, the ball load, the geometry of the equipment and the properties of the microorganism. Operating temperatures during cell disruption processes need to be controlled to prevent degradation of the target biomolecule. It is recommended that the process be between 5°C and 15°C, and to achieve this, the ball mills should be surrounded by a cooling jacket. In the range of 5°C to 40°C, the effect of temperature on the breaking efficiency is negligible.

To scale up ball mill unitary operations, the following parameters must be kept constant: size of the spheres, volume ratio between the cell suspension and the glass spheres and speed of rotation of the axis or the peripheral speed of the stirrer blades (Wheelwright, 1991).

13.4 CONCENTRATION OF BIOMOLECULES

The concentration step of a solution containing the biomolecule to be purified generally aims to reduce the water content of the clarified medium obtained after removal of the cells. This step is necessary since the biomolecules are diluted in the cultivation medium. Depending on the final application of the biomolecule, the concentrated medium can already be the product available to commercialization, as for example, enzymes used in detergents. When a high degree of purity of the biomolecule is required, reducing the volume of the solution will make subsequent purification operations less costly. In addition to reducing the water content, increasing the concentration of the biomolecule of interest relative to the other molecules will also make the subsequent unitary purification operations easier.

Next, unit operations used to reduce the water content or to increase the concentration of the target molecule will be presented, namely precipitation and ultrafiltration.

13.4.1 Precipitation

During precipitation, a chemical or physical disturbance is promoted into the solution such that proteins, nucleic acids and small metabolites become insoluble forming particles capable of being separated by means of solid-liquid separation operations, which subsequently may be re-solubilized.

Easy scale up of the process and the possibility of continuous operation are an important advantage of precipitation. In addition, several low-cost precipitation agents can be used. The loss of three-dimensional conformation of functional proteins is a disadvantage of this procedure, especially when renaturation is not possible (Wheelwright, 1991).

Assuming that proteins are in solution because of (a) interactions with the solvent and (b) repulsion forces between charges on the surface of the protein and charges in the solution (or even charges on other molecules), then interference in ionic interactions can result in precipitation. The factors responsible for protein precipitation can be divided into two groups: (a) alteration of the composition of the solvent in order to reduce the solubility of the protein, for example, the addition of high concentrations of salts such as ammonium sulphate, organic solvents (ethanol, acetone) or non-ionic polymers, such as polyethylene glycol; and (b) reducing the solubility of the protein by changing its charge by the addition of acids, bases, cationic or anionic precipitants, or by means of direct interactions of the protein with metal ions.

In aqueous solutions, precipitation is promoted by means of increasing (salting-out) or decreasing (salting-in) the salt concentration, addition of organic solvents, polyelectrolytes, non-ionic polymers, a rise in temperature or pH adjustment (Table 13.1). Protein precipitated by salting-out, in general, is not denatured since its activity is recovered after dissolution of the precipitate. In addition, the salts stabilize the proteins against denaturation, proteolysis or bacterial contamination (Pyle, 1990).

The most efficient salts are those with high solubility, which increase the surface tension of the solvent resulting in a lower degree of hydration of hydrophobic zones and, therefore, increase the probability of interactions between these zones. The relative efficiency of neutral salts in *salting-out* was defined by Hofmeister in 1888, who proposed the lyotropic series: $SCN^- > ClO_4^- > NO_3^- > Br^- > Cl^-$ >acetate>citrate> $HPO_4^{-2} > SO_4^{-2} > PO_4^{-3}$ (Verrall and Hudson, 1987).

Unlike *salting-out*, precipitation may also occur when the ionic strength is reduced, for example, less than 0.15 M, which is the typical value inside cells; in this case, precipitation is called *salting-in*. The low interaction of the protein surface charges with those in solution added to insufficient repulsive electrostatic forces between proteins results in the interaction between protein molecules (Harris and Angal, 1995).

TABLE 13.1
Main Methods of Protein Precipitation

Precipitation Agent	Principle	Advantages	Disadvantages
Neutral salts (*salting-out*)	Hydrophobic interactions by reducing the protein hydration layer	Universal use Low cost	Corrosive Ammonia release at alkaline pH
Non-ionic polymers	Exclusion of the aqueous phase protein reducing the amount of water available for protein solvation	Use of small amounts of precipitant	Increased viscosity
Heat	Hydrophobic interactions and interference of water molecules in hydrogen bonds.	Low cost Simple	Risk of denaturation
Polyelectrolytes	Binding with the protein molecule acting as a flocculating agent	Use of small amounts of precipitant	Risk of denaturation
Isoelectric precipitation	Neutralization of the overall protein charge by changing the pH of the medium	Use of small amounts of precipitant	Risk of denaturation
Metal salts	Complex formation	Use of small amounts of precipitant	Risk of denaturation
Organic solvents	Reduction of the dielectric constant of the medium by increasing the intermolecular electrostatic interactions	Recycling facility Easy precipitate removal	Risk of protein denaturation Flammable and explosive

The main effect observed on precipitation by organic solvents is a reduction in the water activity caused by a decrease in the dielectric constant of the medium with a consequent increase in the electrostatic forces of attraction between the molecules of proteins. Precipitated aggregates are formed by a mechanism other than that observed in the *salting-out*, based on interactions between hydrophobic zones. These different mechanisms result in different aggregates, and in precipitation by electrostatic attraction, the precipitates are denser and, therefore, sediment with higher velocity or require less kinetic energy in the centrifugation. In addition, organic solvents promote a density of less than 1.0 g/mL in the medium, and, therefore, the sedimentation of the precipitate formed occurs quickly and does not require centrifugation. Interaction of the organic solvent with internal hydrophobic zones of the protein may result in an irreversible alteration in protein conformation. A reduction in the temperature to values of the order of 0°C or less can minimize this effect.

Precipitation by organic solvents has the advantage that it can be performed under aseptic conditions, and the yield is greater when compared with precipitation using ammonium sulfate following centrifugation. In addition to the antimicrobial properties of solvents, another advantage of solvent precipitation is the possibility of recovering and then recycling the solvent used in the process. Finally, the small volumes of organic solvent (typically around 10% v/v) used for precipitation do not affect the performance of other separation methods, except for hydrophobic interaction chromatography or other adsorption methods that depend on the hydrophobic interactions (Verrall and Hudson, 1987).

Precipitation can be conducted in steps, called fractional precipitation, through which the different solubilities of biomolecules can be exploited. In the first step, generally less soluble proteins are removed, and in the following steps, one or more target biomolecules are precipitated. In fractional precipitation, the concentration of the biomolecule to be precipitated is the main variable.

Precipitation of proteins can be performed in batch, continuous type CSTR (continuous stirred tank reactor) or tubular (plug-flow) reactor.

13.4.2 Tangential Filtration

Tangential filtration when used for concentration and purification is titled *ultrafiltration*, and biomolecules are isolated based on their molar mass. Membranes with a porosity between 1 and 500 nm are used, which allow the passage of water molecules. *Nanofiltration* membranes have a smaller pore size, capable of selectively excluding molecules with molar mass between 300 and 2,000 Da, which includes polysaccharides, antibiotics and small proteins as well as water (Harris and Angal, 1995).

Membrane geometry can be as flat-plate, spiral or cylindrical configurations and can be constructed applied into tubular, capillary or hollow fiber filters. Tangential filtration for concentration purpose is most commonly performed using tubular and capillary filter designs (Harrison, 1994).

In industry, tangential filtration units are arranged so that filtration can be performed on a continuous basis. One possible arrangement is that a liquid medium can be divided and passed through several filters simultaneously, with each filter performing a different size exclusion. The filtrate can then be passed to a subsequent filter, therefore processing the entire medium in a much shorter time period than would have been achieved by sequential passage through individual filters (Wheelwright, 1991).

In addition to increasing the concentration, tangential filtration can be used for protein fractionation resulting in a purification (Ladisch, 2001).

13.4.3 Liquid-Liquid Extraction

Extraction of antibiotics and organic acids in two-phase immiscible liquid systems consisting of an aqueous phase and an organic solvent is used in order to purify these biomolecules. For proteins, however, such systems are not adequate because of the sensitivity of these molecules to denaturation by the organic solvents (Albertsson, 1986).

Proteins can be purified in systems composed of two immiscible aqueous phases (ATPS – aqueous two-phase systems) due to a differential partition of the target molecule and impurities between the liquid phases. The high-water content, 75% to 80% w/w, ensures the preservation of biological properties of proteins. In such systems, the target molecule and the impurities are separated because of their different solubilities in the liquid phases. The surface properties of proteins, such as electric charge and hydrophobicity, as well as molar mass are important factors. The extraction by ATPS has been applied to the purification of products obtained from animal, plant and microbial cells, extraction of viruses, organelles and nucleic acids and for the purification of enzymes (Hatti-Kaul, 2000).

Aqueous two-phase systems are formed by the addition of polymers and/or surfactants to the solution from which the biomolecule will be concentrated or purified. Some common systems include polyethylene glycol (PEG)/dextran (Dx); PEG/potassium phosphate, PEG/magnesium sulfate and PEG/sodium citrate. PEG/salt systems are widely used because they allow for rapid phase separation, and they are cheap and are high selective for the separation of different molecules. Other systems consisting of polymers such as PEG and polyacrylate offer similar advantages to systems using salts (Johansson et al., 2008).

ATPS can be represented in an equilibrium curve, the so-called phase diagram, in which the ordinate represents the mass composition of the molecule having the highest concentration in the upper phase (lower density phase, for example PEG), and the abscissa represents the composition of the molecule with the highest concentration in the lower phase (higher density phase, for example, salt or dextran). Compositions represented by points above the equilibrium curve lead to the formation of two phases and below the curve, to a single phase. The formation of an ATPS, therefore, depends on the concentration of the system components. The bibliography presents phase diagrams for several ATPSs. However, since the diagrams are specific to each type of system and condition (pH, temperature and molar mass of the polymers), sometimes the phase diagrams have to be determined experimentally. The partition of biomolecules between the two phases is governed by the condition of lower chemical potential or greater solubility, that is, the biomolecule will be at the greatest concentration in the phase in which its chemical potential is lower (Albertsson, 1986).

The partition coefficient, K, is determined to evaluate the extraction (Equation 13.5). This coefficient is given by the ratio between the concentrations of a given molecule in the upper and lower phases at equilibrium. When partition coefficients for the molecule of interest and for the other molecules are significantly different, then there is possibility for purification.

$$K = \frac{C_{si}}{C_{Ii}} \tag{13.5}$$

where: C_{si} = concentration of solute i in the upper phase

C_{Ii} = concentration of solute i in the lower phase

Proteins are distributed between the phases according to solubility and physicochemical characteristics of the proteins (hydrophobicity and surface charge) and of the solution (pH and ionic strength) (Hatti-Kaul, 2000).

Clarification for removal of cells and cell debris can be performed in an ATPS. Considering that centrifugation requires high F_c (centrifugation force) for sedimentation of small solids, ATPS can be advantageously applied since the number of unit operations of the process can be reduced because this extraction step can clarify the medium and fractionate proteins simultaneously, in some cases (Johansson et al., 2012; Mazzola et al., 2006).

To scale-up the process successfully, the partition coefficient values, K, in the large-scale operation must be identical to those optimized on a laboratory scale. In this case, the compositions and proportions of the phases, as well as the homogeneity of the system will be the same (Albertsson, 1986).

There are several advantages of ATPS including possibility of continuous operation at large scale and room temperature; maintenance of the proteins in solution by means of polymers, surfactants or salts that protect the proteins from denaturation and possibility of elimination of some steps from the purification process for intracellular molecules (Hatti-Kaul, 2000).

13.5 CHROMATOGRAPHIC PROCESSES

In chromatographic processes, some solutes of a liquid medium – proteins, peptides, antibodies – are specifically adsorbed onto a bed of porous material for further specific remotion by action of a mobile liquid phase – the eluent – to result in purification. The physical configuration requires a stationary phase (the bed of porous material or the so-called matrix), which is packed inside a column through which the mobile phase is pumped.

Chromatographic operations aim to isolate and purify the metabolite of interest in relation to the contaminants leading to a purity of the metabolite appropriate for the intended application. In liquid chromatography, the bioproducts are retained in the bed of the porous material by means of chemical or physical adsorption or molecular exclusion. The stationary phase may consist of porous silica, synthetic organic polymers or carbohydrate polymers in the form of spherical particles of approximately 100 μm, which are embedded in solvent, thereby constituting the major part of the stationary phase (~90%). The subsequent gradual removal of the different solutes occurs by the action of an eluting liquid phase (i.e. the mobile phase), with the differential migration of the solutes between the stationary and mobile phases, resulting in the separation of the different components of the medium.

13.5.1 Molecular Exclusion

One of the most effective methods for separating biomolecules of different molar mass is by size exclusion chromatography (SEC).

The molecules are separated as a function of size, between the mobile phase and the stationary phase of defined porosity. A mixture of proteins dissolved in a solution flows either by gravity or with the aid of pumps through a column filled with microscopic spheres consisting of hydrated porous polymeric material. The porous of the stationary phase has a fractionation range, which means that molecules within this range of molar mass can be separated (Ladisch, 2001).

In the case of a sample containing a mixture of molecules bigger and smaller in size relative to the pores of the stationary phase, the smaller molecules can penetrate all the pores of the matrix and then move slowly through the column, having access to the phase inside the pores and that existing between the particles. Thus, larger molecules will be excluded from the stationary phase first, with smaller molecules last to leaving the column. Molecules of intermediate size may have partial penetration in the stationary phase entering some, but not all, of the pores, thus giving shorter times for elution than the smaller molecules. In this way, the molecules will be eluted according to their sizes, passing through the column at time speeds. The difference in time proteins flow through the column is thus related to the fraction of pores accessible to solutes (Ladisch, 2001).

Molecular exclusion chromatography is the simplest and the mildest of all chromatographic techniques and can be employed in two ways, for the separation of groups of molecules within a certain range of molar mass or for high resolution in the fractionation of biomolecules of different molar mass. One possibility for purification is to separate the sample components into two populations according to their size range. This strategy can be used for the removal of contaminants of high or low molar mass or in the de-salting and exchange of buffer solutions. It can also be applied in high resolution fractionation where components of the sample are separated according to differences in molar masses and can be used to isolate one or more components to separate aggregate monomers, to determine molar mass or to perform analysis of molar mass distribution.

13.5.2 Ion Exchange

Ion exchange is very widely employed for the purification of proteins because it is simple and easy to scale-up and has high resolution and adsorptive capacity.

In ion exchange chromatography, the first step is to reversibly adsorb the electrically charged solutes to groups with opposite charges immobilized onto a solid matrix. The adsorbed solutes are subsequently eluted and replaced by other ions with the same type of charge, but with higher affinity for the stationary phase. The different degrees of electrostatic affinity between the stationary phase and the mobile phase ions govern this type of chromatography (Harris and Angal, 1995).

The basic principle of ion exchange chromatography is based on competition between the ions of interest and contaminants for charged groups of the matrix. Protein molecules have, on their surface, groups with positive and negative charges. The positive charges come mainly from the aminoacids histidine, lysine, arginine and the terminal amines. Negative charges are the result of the presence of aspartic and glutamic acids and of terminal carboxylic groups. The net charge of a protein depends on the ratio of its positive and negative charges to the pH value of the solution. The pH at which the number of positive charges is equal to that of negative charges is called the isoelectric point (pI). Above the pI, proteins have a negative net charge while below, the net charge is positive (Ladisch, 2001).

The separation of proteins is done as a function of the differences in the equilibrium between the ions of the mobile phase and the ions of the stationary phase. For effective ion exchange purification, the stationary phase must be capable of binding to proteins that are positively or negatively charged. Ion exchange matrices containing positively charged groups are called anion exchangers and adsorb proteins with negative net charge. The so-called cation exchangers are negatively charged and adsorb positively charged proteins. The counter ions, also called substitution ions, are ions of low molar mass that bind to the stationary phase or soluble proteins in the mobile phase. In order for protein to bind to the stationary phase, the counter ions must be dissociated. Na^+ and H^+ cations are counter ions commonly found in cation exchangers and, Cl^- and OH^- anions are the most commonly used in anion exchanger. These ions can be classified according to the interacting forces with their respective counter ion groups. As an example, it has been found that chloride (Cl^-) ions will replace hydroxide ions (OH^-) in an anion exchanger. Therefore, prior to use, the ion exchanger should be conditioned with a counter-ion suitable for the desired application.

The purpose of the ion exchange purification process is to adsorb the target protein (or contaminants) to the matrix, with subsequent elution. Once the capacity of the matrix has been determined, the pH and the ionic strength required for most efficient adsorption and elution of a protein have to be defined experimentally. It is necessary to define the most efficient operating parameter for ion exchange chromatography because the adsorption step is a batch process. Protein fractionation by batch process is simple and ideal for treatment of large volumes. In this case, the elution of the biomolecule may also be done in a batch process.

Elution of the chromatographic column can be done using the buffer used for preparing the column. In this case, the adsorbed proteins on the ion exchange matrix will be eluted as the ionic strength increases, or by inclusion of new ionic species or by a change in the pH. It is common to conduct the elution by increasing the concentration of a salt, such as NaCl, thereby increasing competition and reducing the interaction between the exchanger group and the biomolecules to be eluted. Elution of a protein can be conducted in several ways. If the object is to concentrate the protein, the elution can be done with a small volume of eluent. Another form of elution is when the separation of a protein occurs due to differences in the speed with which all the components present in the sample migrate through the column. Stepwise elution is used to purify and concentrate biomolecules. It is divided into steps and, in such a way that at each step there is a change in pH and ionic strength, thereby the target biomolecule is eluted at a given single step and in a small volume. In gradient elution, the ionic strength or pH of the eluent varies continuously as a function of time. This causes, depending on the ionic strength, the proteins to be eluted sequentially depending on the force of their interaction with the matrix. The gradients are obtained by blending the buffers, such that the concentration of the salt in the eluent is increased. This mixture can be made by devices programmed to provide a predefined gradient and with good reproducibility. Gradient elution with increasing pH can be employed with cation exchangers to make the proteins less positively charged and therefore, more easily desorbed from the matrix, whereas a gradient where the pH is reduced can be used with anion exchangers since the adsorbed proteins become less negative.

The use of the column in a novel purification process requires regeneration, i.e. rebalancing with the eluent and removal of the contaminants bound to the matrix. The balance is achieved by passing an eluent volume equal to five to ten times its capacity through the column. Thus, the exchange of ions occurs with the elimination of those contained in the previously used eluents (Wheelwright, 1991; Ladisch, 2001).

13.5.3 Hydrophobic Interaction

Hydrophobic interaction is defined as the tendency of aliphatic groups (or other apolar structures) to associate when present in an aqueous medium. In hydrophobic interaction chromatography, HIC, protein molecules in saline are adsorbed onto a hydrophobic support and then eluted. HIC is an excellent complement to ion exchange and molecular exclusion.

Proteins, although soluble in water, have different amounts of hydrophobic groups, which gives to them their hydrophobic property. The hydrophobic property of a protein can be increased artificially by adding salts to the solution. Hydrophobic interaction chromatography explores the hidrophobicity of proteins by placing them in saline solutions of different concentrations to achieve separation, since hydrophobic interactions increase with increasing salt concentration (Janson and Låås, 1978). Adsorption requires the presence of salting-out ions such as sodium chloride or ammonium sulfate which decrease the availability of water molecules while increasing surface tension and hydrophobic interactions. Consequently, at high salt concentrations, most proteins can be adsorbed by hydrophobic groups into the adsorbent matrix. The efficacy of HIC is reduced by the presence of hydrophobic contaminants. The method is based on the hydrophobic interaction (or the association) between hydrophobic ligands and proteins immobilized on the solid support. Binding is achieved by the attachment of short chain (butyl, octyl, phenyl) hydrophobic groups onto the surface of a solid support by means of arms or spacers.

The use of hydrophobic interaction chromatography is ideal when conducted immediately after salt precipitation, where the ionic strength of the medium will favour hydrophobic interactions. In a purification process using HIC, where the target protein is eluted in a gradient of decreasing ionic strength, ion exchange chromatography may subsequently be used with little need for buffer change. The chance of achieving good separation using a hydrophobic column is considerable, because of the abundant capacity of the column to retain proteins and, the fact that the adsorption is carried out in high concentrations of salt. The main parameters when projecting a HIC are the type of arm or spacer, the type of the hydrophobic short chain bind onto the matrix (butyl, octyl, phenyl), the type of matrix, the type and salt concentration, the pH, the temperature and eventually, additives.

13.5.4 Affinity Chromatography

Affinity chromatography is based mainly on the biological or functional properties of the interacting species: the protein to be separated and the stationary phase (Harrison, 1994).

Chromatographic affinity is a separation technique that depends on highly specific interactions between pairs of biological materials: enzyme-substrate; enzyme-inhibitor; antigen-antibody. One of the components of this interaction (called a linker) is immobilized on an insoluble support, porous matrix, while the other component is selectively adsorbed onto the previously immobilized linker. The adsorbed component may be eluted with a solution that weakens the interactions between the two components. In principle, this technique makes it possible to separate a protein from a complex biological mixture based on the recognition and binding of the target molecule to the specific linkers.

This application can be carried out at any stage of the purification process but comes at a high financial cost. The use of this technique is recommended only after removal or reduction of contaminants by cheaper methods. The purification is of high resolution and the recovery of the active material is generally high, since affinity chromatography presents many advantages: high specificity, purification of proteins from biological mixtures in only one step, separation of native forms from denatured forms of the same protein and removal of small amounts of the protein of interest from a large amount of other contaminating proteins (Verrall and Hudson, 1987).

At the adsorption stage, the sample containing the protein of interest comes into contact with the adsorbent and the desired protein binds reversibly to the immobilized linker. During the adsorption stage some (or all) of the contaminants may diffuse into the matrix pores and depending on the type of matrix and linker, may be adsorbed specifically by the linkers, or there maybe non-specific binding onto the surface of the matrix. In the washing step, the concentration of the contaminants present within the porous particles of the adsorbent is reduced. The adsorbent is then recovered by dissociation of the adsorbed-binding complex (elution stage), and finally the adsorbent is regenerated by contact with the initial buffer (regeneration stage).

Elution of the material bound to the selective stationary phase in affinity chromatography is generally accomplished by the use of specific agents. This step is of fundamental importance so that a good separation of the desired protein is achieved. Elution requires complete dissociation of the adsorbate-adsorbent complex. The two elution methods that have been most widely used are selective and non-selective procedures. The selective method uses the natural properties of specific interactions between the proteins, whereas the non-selective methods use denaturing of proteins or alterations of pH and temperature. Normally, the method used is that of non-selective elution. This method alters the physical properties of the adsorbent with the use of an appropriate eluent, such that the strength of the bond between the linker and the adsorbed protein is reduced, promoting dissociation of the adsorbed linker complex. Another form of selective elution involves the use of a solution containing high concentration of free linker, which may be the same linker that is bound onto the inner surface of the matrix pores. This eluent has significant affinity for the adsorbed protein and competition occurs between the soluble and immobilized linkers, such that if the soluble linker is in excess then the adsorbed will shift almost exclusively to the soluble phase and will be separated from the adsorbent particles. The protein is separated from the soluble linker by exploiting the difference in molar mass of the two species.

13.5.5 Scale Up

The primary purpose of large-scale purification is to reproduce process performance (resolution, time and yield) that has been developed and optimized on a smaller scale. However, scaling up is essential to produce sufficient quantities to meet consumer market demand (Wheelwright, 1991).

Initially, it is necessary to define what exactly a small or large-scale purification process means. Laboratory-scale purification processes yield micrograms or milligrams of bioproduct. At pilot scale these quantities are from milligrams to grams and in large-scale purification the required amount can vary from grams to kilograms as it depends on the end use of the bioproduct.

The scaling up of some chromatographic processes is similar, such as ion exchange, molecular exclusion, affinity and hydrophobic interaction. The stationary phase and the degree of packaging used on a laboratory scale should be retained for scale-up. Column height, linear feed velocity (volumetric flow divided by column cross-sectional area), concentration of the molecule to be purified and solutions or buffers used should be identical to those used during laboratory scale operations. Thus, increasing column capacity for processing of larger volumes of medium is obtained by increasing column diameter. Increasing the scale of a chromatographic process in practice means increasing the diameter of the column in such a way as to accommodate the additional volume of sample to be purified. The enlargement of the width may cause changes in stationary phase packaging due to the fact that a large amount of adsorbent material will be far from the wall of the column, which assists in maintaining the packing of the bed. The result may be deformation, especially in the central part of the bed, which will induce preferential flow of the liquid through this central area, altering the resolution of the chromatographic process. It is for this reason frequently industrial columns have bed height in the order of 30 cm and diameter of around 1 m. However, there are cases where the diameter of the column reaches 2 m. In general, the highest volumes of chromatographic columns are between 700 to 2,000 L as is the case in the purification of whey proteins or human plasma albumin. If greater production is desired, it is recommended to increase the number of columns (Wheelwright, 1991).

13.5.6 Expanded Bed Adsorption

Expanded bed adsorption (EBA) is a technique based on the fluidization of the stationary phase. Although EBA has been available since the 1970s for the purification of proteins, it was not widely used at that time mainly due to technical difficulties such as limitations with the physical properties of the adsorbent matrices. However, there is now increasing interest to use EBA to purify proteins from solutions (whether the solutions contain particulate matter, or not), since it allows integration of the clarification and purification steps.

The conventional arrangement of the adsorbent in a packaged bed requires feeding of a media free from suspended particles, i.e., a pre-clarified media, otherwise the matrix will be clogged. Besides, even when clarified media are applied, the packaged bed causes slow diffusion rendering low productivities. Alternatively, the adsorbent medium may be suspended (stirred, fluidized bed or expanded reactors) which, together with the use of adsorbents which reduce the diffusion time of the liquid, can accelerate the process time. In addition, it is also possible to capture proteins from media with intact cells. In this case, the number of steps in the process and the possibility of losing activity of the target molecule can also be minimized. Further clarification can be performed at the same time as the given biomolecule is purified.

For example, when centrifugation is used for the removal of suspended cells, further microfiltration is required to obtain a medium suitable for packed bed chromatography, because even after centrifugation suspended particles still remain. If this medium is fed to an EBA chromatographic column there is a possibility of loss of fluid flow due to clogging. Additional clarification operations generally result in higher costs and increased overall process times, especially when it comes to media where cells have been disrupted, because this media has a high viscosity (Harrison, 1994).

The possibility of applying chromatographic separations to media containing cell fragments is important in the production of proteins for therapeutic and diagnostic use and, in reducing the action of proteases because the process time is reduced.

Fluidized bed or expanded bed reactors are tubular (such as packaged bed reactors). The adsorbent material is suspended due to the use of solid adsorbents that have a certain particle size distribution and, with a density greater than that of the liquid. The liquid is applied at a rate so that the bed is expanded, and the solid particles occupy a larger volume than the resting volume.

13.6 FINAL TREATMENT

The degree of purity required for a biotechnological product depends on its final application. The simple drying of cultured microorganisms for the production of cellular protein can be sufficient for commercialization. Impure or partially purified enzyme broths can be used as catalysts in industrial chemical conversions, for example, in the production of fructose syrup using the enzyme glucose isomerase. However, final purification is required for most biotechnological products, especially those for pharmaceutical use. In this case, the products must be pure, dry, crystalline or amorphous. To do so, the biomolecules must undergo some final treatments such as crystallization or lyophilization.

Lyophilization is the process of removing a solvent, usually water, by sublimation. The material is frozen and then subjected to low pressure for sublimation of free water. During freezing the water turns into ice and the solutes are concentrated. As a result, the physico-chemical properties (pH, ionic strength, viscosity, freezing point and surface tension) of the non-frozen phase change significantly. Lyophilized materials are dry powders which retain the biological activities of the material for much longer when compared to preservation in aqueous solution. For this reason, many commercial proteins are available in lyophilized form. However, if lyophilization is not adequately planned, denaturation of enzymes may occur. Although it is a widely used technique for the conservation of many biological materials, there are several factors involved in lyophilization that must be considered to ensure good quality material is preserved.

Crystallization is the process of aggregating crystal molecules into supersaturated homogeneous solutions commonly employed in the final stage of protein purification processes, particularly enzymes. Crystallization is very important in many biotechnology industries, since it allows for stable storage because the molecules are immobilized. Crystallized protein solutions contaminated by proteases have their activity preserved, since the proteases also crystallize. After crystallization, the product can be recovered by filtration or centrifugation followed by drying (Verrall and Hudson, 1987).

13.7 PURIFICATION OF MONOCLONAL ANTIBODIES

Monoclonal antibodies are a class of therapeutic proteins of major market significance because they have applications in the treatment of cancer and autoimmune, infectious and cardiovascular diseases. Monoclonal antibodies are immunoglobulins that act on different functions in the human immune system, and immunoglobulin G (IgG) is the most important and most widely produced. The same generic purification process for IgG has been used for decades and employs specific adsorption of the antibody to protein A, which is a cell wall protein of the Gram-positive bacterium *Staphylococcus aureus*. This method is widely used because it results in high yield and purity.

Purification is based on a domain in protein A (called the B domain) which has high affinity for the Fc region of IgG at high pH, but under acidic conditions the interaction between the two molecules breaks down. During the interaction of IgG antibody to protein A in a chromatography column, simultaneous elimination of host cell proteins, DNA, contaminants of viral origin and other impurities occurs. There is also a reduction in the volume of the medium and concomitant concentration of the antibody.

13.8 ENDOTOXIN REMOVAL

In many processes for the purification of biotechnological products, it is necessary to eliminate specific impurities, for example, in the case of molecules that cause harm to the health of users, even though the degree of purity required by the regulatory agencies has been reached in the purification process. Certain proteins considered endotoxins that are part of the cell membrane composition of bacteria, cause fever, that is, they are pyrogens. In Gram-negative bacteria, these endotoxins are lipopolysaccharides, LPS.

Many biomolecules of biotechnological interest are produced using *Escherichia coli* as a heterologous host. *E. coli* is a Gram-negative bacterium and so produces an endotoxin which can aggregate to the product. Limits considered acceptable by the United States Pharmacopeia (USP) and the Food and Drug Administration (FDA) are defined as a function of the dose of drug to be administered to the patient. For example, the endotoxin limit allowed in the water for injection is 0.25 EU/mL and in water for nebulizers is 0.5 EU/mL.

Removal of the endotoxins can be done by affinity chromatography (with L-histidine, poly-L-lysine and polymethyl L-glutamate and Polymyxin-B), size exclusion chromatography (the size of a basic endotoxin subunit is between 10 and 20 kDa), ultrafiltration (requires prior disaggregation with surfactants or chelating agents), hydrophobic interaction chromatography and ion exchange chromatography. More information about LPS can be seen in Chapter 14.

13.9 YIELD AND PURITY

In purification processes, it is necessary to perform routines of analyses to determine the recovery and loss at each stage of the process. Often, in the case of biomolecules, it is essential to determine the degree of purity and, sometimes, the absence of specific contaminants (Collins et al., 1990).

The fraction recovered of the target molecule in each stage of the process requires the measurement of its concentration and the volume of medium at each stage. The fraction multiplied by 100 represents the yield, η, achieved in the isolation of the target molecule at a certain stage of the process (Equation 13.6). P represents the degree of purity achieved in the step. The determination of P requires a prior definition of purity in the specific situation, one of the most common being the relationship between the concentration of the target molecule and the total protein concentration in the system (Equation 13.7) (Ladisch, 2001).

$$\eta = \frac{C_{Xn} \times V_n}{C_{X0} \times V_0} \times 100 \qquad (13.6)$$

$$P = \frac{C_X}{C_T} \qquad (13.7)$$

In Equation 13.6, C_{X_n} represents the target molecule concentration in step n of the purification process, and C_{X0} represents the concentration of the same molecule in the starting medium. V_0 and V_n respectively represent the initial medium volume and the volume of medium in step n.

To determine purity based on Equation 13.7, C_X is the concentration of the target molecule and C_T is the concentration of all molecules. Often, C represents the concentration of proteins, since these constitute the majority of biomolecules of interest – peptides, enzymes, antigens, antibodies and hormones – and also, the main impurities. The variable P represents, therefore, the concentration fraction of the target molecule, relative to the concentration of the molecules comprising the impurities. If C is expressed in terms of mass, P will be the fraction of the mass of proteins (i.e. the target protein), which represents purity.

It is important to determine the yield and the increase of purity that each step confers in order to evaluate the impact of these variables in the complete process and, eventually, the substitution of a given unit operation for another. The increase in purity is given by the increase in the value of P, here called AP. Equation 13.8 defines AP for a given process step in relation to the previous step, where P_n is the purity of the target molecule in stage n and, P_{n-1} is the purity of the same molecule in the previous stage. Similarly, the increase in purity of the whole process is determined by substituting P_{n-1} for P_0 (purity in the starting medium) in Equation 13.8 (Ladisch, 2001).

$$AP = \frac{P_n}{P_{n-1}} \quad (13.8)$$

There are several routinely used methods to determine the concentration of proteins, each based on different fundamental principles and each with specific limitations, especially concerning interfering molecules, for example: Lowry-Folin-Ciocalteau method, Bradford method, copper alkaline biuret-reagent method, absorption of UV rays at 280 nm (aromatic amino acids) or at 205–220 nm (peptides) and bicinchonic acid method (Collins et al., 1990).

Considering that each of these methods is based on different principles, the results obtained cannot be directly compared. In addition, even if a single methodology is adopted, the result obtained will only express the true protein concentration if the calibration curve is determined with a protein solution of the same composition as the target solution (Harris and Angal, 1995).

In determining the yield of a given process step, η, or the purity, P, achieved for the target product, it is necessary to quantify the target molecule to enable the application of Equations 13.6–13.8. Often the target molecule is a protein, however, other kind of molecules will generally be present making it impossible to determine the yield and degree of purity achieved. Indirect quantification is obtained relatively easily when the target molecule has specific biological activity, for example, enzymatic or antigenic activity.

Determining a qualitative degree of purity in a mixture of proteins is widely practiced with the use of electrophoresis, which consists of the separating the proteins by the action of an electric field that forces the movement of the electrically charged molecules through a gel (Collins et al., 1990).

13.10 IDENTIFICATION OF BIOMOLECULES

The quantification of a biomolecule does not guarantee its biological effectiveness; therefore, other methods must be used for the purpose of determining structure and biological activity.

The molar mass is indicative of correct formation of a protein, without undesired truncations or aggregations. The molar mass of a given protein can be estimated by analytical molecular exclusion chromatography, especially for proteins with two or more subunits since, unlike electrophoresis, the subunits are not separated in the preparation of the sample. It is also possible to perform this measurement by means of capillary electrophoresis and gel plate electrophoresis in which molecules of known molar masses are used as a reference band. In molecular exclusion it is possible to verify subunit disintegration, in such a way electrophoresis and molecular exclusion are powerful complimentary tools in the identification of biomolecules based on molar mass.

The hydrophobicity of biomolecules can be confirmed by reverse phase chromatography (RP-HPLC). In capillary electrophoresis (EC), electrophoretic mobility is determined, which can be related to the density of the charged surface of proteins, which is a characteristic property of the three-dimensional conformation of a protein. Mass spectrometry (MS) can determine the chemical structure of molecules, for example, the primary sequence of amino acids in a polypeptide. The correct identity of the target molecule can also be indirectly inferred by using the Western Blotting technique, as this technique is based on biological activity (interaction between a peptide and a specific antibody).

The identity of the purified organic target molecule can ultimately be determined using high resolution nuclear magnetic resonance (*NMR*), in particular proton (*^{1}H-NMR*) and carbon-13 (*^{13}C-NMR*) techniques. By quantifying specific intermolecular interactions between the target molecule and micelles or membranes, the three-dimensional structure of the biomolecule in solution or solid state can be determined. The tertiary structure of proteins is determined by *NMR*, which is an important complementary analysis to chemical methods. Mass spectrometry (*MS*) and nuclear magnetic resonance (NMR) can also identify post-translational modifications such as phosphorylation, sulphation and glycosylation of proteins.

13.11 TRENDS IN PROCESSES APPLIED TO THE PURIFICATION OF BIOMOLECULES

The challenge of purification processes is the treatment of increasing volumes of media containing high value-added biomolecules. Monoclonal antibodies are noteworthy for their applications in the treatment of cancer and diseases related to the immune system such as rheumatoid arthritis, in view of the growth of the elderly population.

Since specific adsorption to protein A is the most expensive step in the purification of IgG monoclonal antibodies, the trend is now towards using multimodal type chromatographic operations that exploit multiple types of interactions between the stationary phase and the mobile phase. These multiple approaches can not only replace the specific requirement for adsorption to protein A, but can differentiate between heterogeneous molecules, for example, glycosylation patterns common to humanized antibodies. Another trend is adoption of continuous processes with many advantages over batch regimes namely: reduction of costs, decrease in processing time, increase in product quality and flexibility in the operation of the processing plant so that more than one product can be produced at any given time. The possibility of obtaining better quality bioproducts compared to batch unit operations is a particular attraction of continuous operations.

In the unit operations of filtration, centrifugation, tangential filtration, precipitation and extraction in two-phase aqueous systems, it is possible to apply a continuous regime.

However, in chromatographic operations the separation efficiency is associated to the discontinuous mode and the long retention times of the molecules caused by the continuous regime can cause aggregations and denaturations of proteins resulting in loss of the target molecule. The possibility of increasing the separation rate of the molecules using a continuous regime must address how such losses can be minimized. Simulated Mobile Bed Chromatography is a continuous operation of adsorptive chromatographic processes applied to the separation of enantiomers in the pharmaceutical industry.

A new trend in chromatographic operations called CPC, *Centrifugal Partition Chromatography* or LLC, *Liquid-Liquid Chromatography* utilizes two liquid mobile phases, one stationary phase because of the centrifugal force applied on it and the other in movement. Separation of the target molecule is achieved on the stationary liquid phase. The CPC system is currently used on a preparative scale for process development and validation of analytical, with real prospects of future industrial scale operation.

Single-use technology (SUT) has been widely adopted by the pharmaceutical industry with disposable devices now largely replacing traditional stainless-steel equipment, especially for small volume and high value-added products. A switch to SUT has largely benefited the industry by reducing health and safety risks, a reduction of capital and energy investments and increased flexibility in the operation of the process. It is a consolidated technique for animal cells, handling and storage of liquids, but it is more challenging for large-scale biomolecule purification processes, especially for chromatographic resins and tangential filtration membranes, the cost of which prohibits single use.

13.12 FINAL CONSIDERATIONS

Increased demand for the development of new, viable and efficient processes for the purification of biotechnological products on an industrial scale has been occurring in recent decades. Although traditional unit operations are still routinely used, new technologies are emerging, with the main objective of reducing process time, increasing purity and yield and minimizing the overall cost. Considering that the bioproduct purification operations are among the main challenges of the biotechnological processes, mainly because it represents a high proportion of the product cost, new studies, both on bench and industrial scale, are continuously carried out. Great challenges are posed to operate efficient scaling, such as guarantees that the structures of biomolecules will remain identical to when they were in the laboratory scale and that there is technical and economic viability of the production process. The wide variety of molecules present in a medium at the end of the process of cultivation and production of the biomolecule of interest will make the development of the purification process, including the appropriate analytical routines, a unique situation.

REFERENCES

Albertsson, Per-Åke. *Partition of cell particles and macromolecules*. 3. ed. New York: Willey Interscience, 1986.

Collins, C. H.; Braga, G. L.; Bonato, P. S. *Introdução a métodos cromatográficos*. Campinas: Editora da Unicamp, 1990, 279p.

Costa-Silva, T. et al. Microbial cell disruption methods for efficient release of enzyme L-asparaginase. *Prep. Biochem. Biotechnol.* 1:707–717, 2018.

Harris, E. L. V.; Angal, S. *Protein purification applications: a practical approach*. Oxford: IR Press, 1995.

Harrison, R. G. *Protein purification process*. New York: Marcel Dekker, 1994.

Hatti-Kaul, R. *Aqueous two-phase systems. Methods and protocols*. Totowa: Humana Press, 2000.

Janson, J-C.; Låås, T. Hydrophobic interaction chromatography on phenyl- and octyl-sepharose CL-4B. In: Roger, E. (Ed.). *Chromatography of synthetic and biological macromolecules*. Chichester: Ellis Horwood, 1978.

Johansson, H., Matos, T., Luz, J., Feitosa, E., Oliveira, C. C., Pessoa, A., Büllow, L., Tjerneld, F. Plasmid DNA partitioning and separation using poly(ethyleneglycol)/ poly(acrylate)/salt aqueous two-phase systems. *J. Chromatogr. A*. 1233:30–35, 2012.

Johansson, H. O., Ishii, M., Minaguti, M., Feitosa, E. Penna, T. C. V., Pessoa Jr. A. Separation and partitioning of Green Fluorescent Protein from Escherichia coli homogenate in poly(ethylene glycol)/sodium-poly(acrylate) aqueous two-phase systems. *Sep. Purif. Technol.* 62(1): 166–174, 2008.

Ladisch, M. R. *Bioseparations engineering: Principles, practice, and economics*. New York: John Wiley & Sons, 2001

Mazzola, P. G.; Lam, H.; Kavoosi, M.; Haynes, C. A.; Pessoa, A.; Penna, T. C. V.; Wang, D. I. C.: Blankschtein, D. Affinity-tagged green fluorescent protein (GFP) extraction from clarified E. coli cell lysate using a two-phase aqueous micellar system. *Bioeng. Biotechnol.* 93(5):998–1004, 2006.

Neves, L. C. M.; Pessoa, Jr, A.; Vitolo, M. Fed-batch production of glucose-6-phosphate dehydrogenase using recombinant Saccharomyces cerevisiae. *Appl. Biochem. Biotechnol.* 136:711–720, 2007.

Pyle, D. L. *Separations for biotechnology*. London: Elsevier Science Publishers, 1990.

Verrall, M. S.; Hudson, M. J. *Separations for biotechnology*. Chichester: Ellis Horwood, 1987.

Wheelwright, S. M. *Protein purification: Design and scale up of downstream processing*. 1. ed. Munich: Hanser Publishers, 1991. 228páginas. ISBN: 3446157034, 9783446157033.

14 Lipopolysaccharides

Methods of Quantification and Removal from Biotechnological Products

André M. Lopes
University of Campinas

Jorge Fernando Brandão Pereira
University of Coimbra

João V. D. Molino
Ronin Institute

Felipe R. Lourenço and Adalberto Pessoa Jr.
Universidade de São Paulo

CONTENTS

14.1 INTRODUCTION

Lipopolysaccharides (LPS) or endotoxins are pyrogenic contaminants found in microorganisms such as Gram-negative bacteria. LPS are the most common molecules and make up about 75% of the outer membrane of these microorganisms. In general, its molecule is constituted of three parts: Lipid A, oligosaccharide core and O-antigen polysaccharide (Raetz and Whitfield, 2002; Falchi et al., 2018; Boratyński and Szermer-Olearnik, 2017).

In biotechnological processes, *Escherichia coli* is one of the main heterologous expression systems used, and some pros and cons have been reported in this sense (Lopes et al., 2018; Teixeira-Pinto et al., 2019). The pros of using microorganisms are: (a) fast and easy transformation employing exogenous DNA; (b) growth in rich and complex media from cheap and readily available raw materials; (c) high cell density culture; and (d) fast growth kinetics. Nevertheless, when the target biomolecule is expressed intracellularly, a further cell disruption step is mandatory for the process. For this reason, downstream processing can raise the cost, as well as reduce the overall yield. Furthermore, cell disruption steps can result in LPS molecules being released into the cell lysate broth (Correa et al., 2017; Falchi et al., 2018; Gyurova et al., 2017).

A single *E. coli* bacterium has around 2×10^6 of these molecules, and although LPS are attached in the outer layer of the cell membrane, they are constantly released into the environment. The release of endotoxin is a concern because contamination of intravenous medicines or during a bacterial infection, for this reason, the side effects from exposure to LPS can be potent. Despite these dangers, a lot of issues related to bacterial endotoxins in human health are left unanswered (Boratyński and Szermer-Olearnik, 2017; Petsch and Anspach, 2000), and these molecules can be considered one of the most exciting and interesting natural biomolecules. In terms of medical approaches, the removal of this contaminant from aqueous solutions, mainly from parenteral solutions, is a challenging and a technically critical topic (Petsch and Anspach, 2000).

In the pharmaceutical industry, different alternatives are adopted for the manufacture of LPS-free products. However, removal of LPS from some products can be complicated, particularly when the LPS molecule is associated with the biomolecule (e.g., protein, enzyme, peptide, amongst others) or product of interest. Particularly, endotoxins are very stable molecules, under conditions of 180°C for 30 min and extreme pH (Hirayama and Sakata, 2002; Petsch and Anspach, 2000). Different approaches have been employed for the LPS removal, e.g., adsorption membranes; ultrafiltration; chromatography processes (i.e., hydrophobic interaction, ion exchange, amongst others). These techniques have different levels of success in removing the LPS, which mainly depends on the physicochemical characteristics of the target molecule (Lin *et al.*, 2005).

A limit for acceptable concentrations of LPS in pharmaceutical preparations has been established by regulatory agencies and is considered decisive in the release of a product for human or animal use. Currently, the limits considered acceptable by the Food and Drug Administration (FDA) (Munson, 1985) for bacterial endotoxins are for pharmaceutical and biological products 5.0 endotoxin units (EU)/kg, radiolabels 2.5 EU/kg, large-volume parenterals 0.5 EU/mL, water for injection 0.25 EU/mL, intrathecal drugs 0.2 EU/mL, products for pharmaceutical use up to 200 EU/unit and intrathecal pharmaceutical products 0.06 EU/unit. The techniques approved by the FDA for LPS detection are testing for pyrogens (elevation of temperature) using rabbits (Williams, 2004) and the *Limulus* Amebocyte Lysate (LAL) test through gel formation (Cooper et al., 1972). The LAL test is the most used technique for detecting LPS in pharmaceuticals during the manufacturing process. In recent years, new methodologies have been developed that use the LAL for both detection and quantification of LPS in solution.

In this chapter, we focus on discussing important considerations regarding this interesting molecule. Thus, the state of the art of LPS characteristics and properties, aspects as effects, detection and quantification, as well as emphasizing its removal from biotechnological products are discussed here.

14.2 LPS CHARACTERISTICS AND PROPERTIES

Any process for quantification and/or removal of a specific biomolecule from complex biotechnological media, broths or formulations should consider the particularities of the target biomolecule. Thus, in this section, a brief analysis of the main characteristics and properties of LPS is described. The functions and mechanism of action of LPS are discussed in Section 14.3.

The lipopolysaccharide layer of Gram-negative bacteria is composed of glycolipids and lipoglycans. A specific part of the LPS known as Lipid A is embedded in the membrane of vegetative cells, but when these cells die, this antigen is released and stimulates the cell-mediated response of the innate immune system of mammals. In addition, the presence of LPS is vital for bacterial growth and viability, particularly in response to environmental stress (Raetz and Whitfield, 2002). Chemically, LPS are mostly amphiphilic in nature and are composed of three different parts shown in Figure 14.1: (a) a lipid moiety known as Lipid A (or endotoxin); (b) an oligosaccharide (OS) core; and (c) a polysaccharidic chain, called the O-antigen, O-side or O-specific chain (Raetz and Whitfield, 2002; Caroff and Karibian, 2003). Lipid A is hydrophobic, while the polysaccharide portion (OS+O-antigen) is hydrophilic in nature, affording opposite physicochemical properties to the LPS (Obeng et al., 2017).

LPS molecules are, in general, classified according the presence and some specific characteristics of the OS and O-antigen portions (Figure 14.1), viz.: (a) Smooth(S)-type LPS composed of the three different parts (Lipid A, OS core and O-antigen); (b) Semi-Rough(SR)-type LPS composed of only one O-chain subunit; and (c) Rough(R)-type LPS composed only of Lipid A and some parts of the OS core (Caroff and Karibian, 2003). The most common LPS structure is the S-type where all the three portions are covalently linked and is found particularly in wild-type strains (Obeng et al., 2017). Although this is the best-known classification of LPS, some Gram-negative bacteria, particularly mucosal pathogens, have a particular type of LPS in their membranes, called lipo-oligosaccharides (LOS) where the O-antigen group possess an oligosaccharide chain attached to the OS inner core (Gnauck et al., 2016).

Lipid A is a glucosamine-based phospholipid composed of four to seven acyl units and two phosphorylated glucosamines (which are negatively charged) (Raetz and Whitfield, 2002; Raviv, 2018). It binds the LPS structure to the outer layer of the cell membrane and is covalently linked to the proximal (inner) moiety of the OS core (Raetz and Whitfield, 2002; Raviv, 2018). The Lipid A portion is the most preserved part of the LPS and has a hexagonal arrangement (Raetz et al., 2007; Magalhães et al., 2007); however, some non-stoichiometric modifications of the Lipid A can occur but to a far less extent than those modifications which occur in both the OS and O-antigen chains (Raetz et al., 2007; Klein et al., 2011). The endotoxin bioactivity is mostly related with the Lipid A portion of the LPS structure (Raetz and Whitfield, 2002; Raetz et al., 2007; Kabanov and Prokhorenko, 2010).

The OS core is composed of 8–12 often branched sugar units divided between the 'inner' and 'outer' core (Gemma et al., 2016). Specifically, the inner core is constituted of mainly 3-deoxy-D-manno-2-octulosonic acid (KDO) and L-glycero-D-manno-heptose (Hep) linked to the Lipid A (Gnauck et al., 2016). It has additional negatively charged phosphate groups which are relevant for providing the integrity of the outer membrane (Raetz and Whitfield, 2002; Caroff and Karibian, 2003). The negative charge

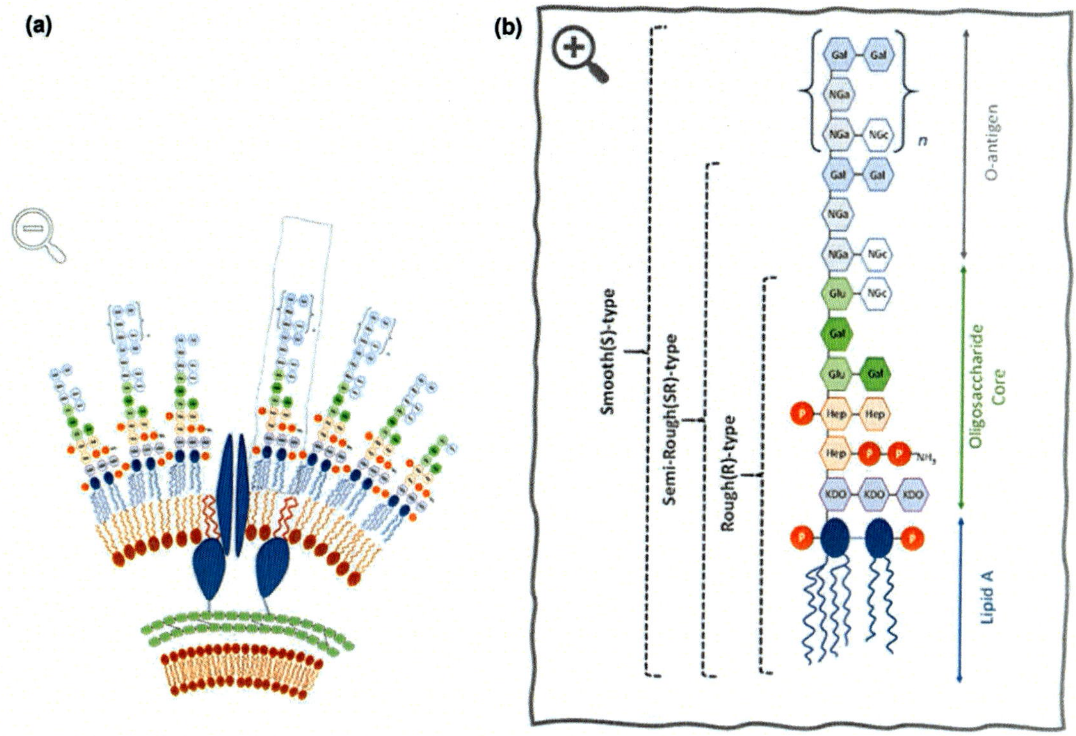

FIGURE 14.1 The outer membrane of Gram-negative bacteria (a) and an endotoxin (lipopolysaccharide – LPS) molecule with its structures showed in detail (b). (Based on Caroff, M., Karibian, D., *Carbohydr. Res.*, 338, 2431–2447, 2003, doi: 10.1016/j.carres.2003.07.010.)

of the LPS is also caused by partial phosphorylation, not only of the OS inner core but of Lipid A. The outer core is mainly composed of hexoses and has a more variable structure (in terms of the constituent saccharides), forming the attachment site to the O-antigen group (if presented in the LPS molecule) (Raetz and Whitfield, 2002).

The third region of the LPS molecules, the O-side chain (mostly known as the O-antigen), is a polysaccharidic chain made of repetitive subunits of liner or branched monosaccharides, which can differ in number, position and stereochemistry of the O-glycosidic links or even by the presence or absence of non-saccharide moieties (i.e., CO-acetyl groups and amino acids). O-antigens are widely diverse in terms of structure and, although they can be found as homopolymers (i.e., a single monosaccharide unit), they are frequently heteropolymers. The structure of the O-side chain in different bacteria is very variable and can be further complicated if different Lipid A-OS-O-antigen structural combinations are found in the bacterial membrane (Raetz and Whitfield, 2002). O-antigen is believed to have a protective role, helping bacteria resist antimicrobials and other environmental stresses (Raetz and Whitfield, 2002; Obeng et al., 2017), but it can also determine the serology type of many bacterial cells (Caroff and Karibian, 2003).

In this section, we highlighted the main characteristics and properties of LPS, summarizing some of the aspects of each portion of the molecule. However, it is important to note that these molecules are highly heterogeneous, even within the same bacterial culture, genus or species, mainly due to changes in the hydrophilic portions of the LPS, i.e., OS core and O-antigen. For example, only evaluating differences in polysaccharide chain length, it is possible to find LPS molecules with very low molecular weight, around 2.5 kDa in the case of endotoxin without an O-antigen chain, or quite a significant molecular weight such as the 70 kDa in the case of very long O-antigen, although the common endotoxin molecular weight range is between 10 and 20 kDa (Magalhães et al., 2007). Therefore, the specific structural properties of LPS, its heterogeneity and abundance (i.e., concentration) in the bacterial outer membrane, will reflect its function and mechanism in a bacterium and, consequently, will influence bacterial-environmental interactions (Gnauck et al., 2016), which are discussed in the following sections.

14.3 LPS FUNCTIONS AND MECHANISMS OF ACTION

Due to the high concentration of LPS in the bacterial membrane, it is highly exposed to the surrounding environment, acting as a defence for bacterial cells (Caroff and Karibian, 2003). However, it is important to highlight that vegetative cells are not able to secret LPS molecules to the extracellular environment, since LPS is an endotoxin and only released from the outer cell membrane under particular conditions, such as cell death, cell division or cell disrupting processes, e.g., the action of certain bactericidal antibiotics (Caroff and Karibian, 2003). Interestingly, the toxicity of LPS is dependent on the relative concentration that is liberated from the cell membrane and into a host, i.e., at low concentration LPS can act as immune stimulator, shrinking tumours, stimulating tissue factor production in mononuclear and endothelial cells (Raetz and Whitfield, 2002); at high concentration LPS causes lethal septic shock (Raetz and Whitfield, 2002; Caroff and Karibian, 2003; Dixon and Darveau, 2005). The functions and mechanisms of action of LPS are organism-specific, i.e., these are different in the producing Gram-negative bacterium compared to when LPS is interacting with human or plant hosts.

Additionally, LPS functions as a protective barrier and has a highly ordered structure due to the following interactions: (a) electrostatic interactions between magnesium ions of the outer membrane and phosphate groups of Lipid A; (b) hydrogen-bonding between the OS saccharides and the hydrophilic regions of adjacent LPS molecules; and (c) hydrophobic interactions between Lipid A chains (Page, 2012; Gnauck et al., 2016). There are wide structural variations in Lipid A (e.g., different acylation patterns) which are required to keep the structural integrity of the bacteria cell under specific growth conditions, such as different pH and temperatures, or the presence antimicrobial agents (Raetz and Whitfield, 2002; Gnauck et al., 2016). The outer core OS and O-antigen are more involved in shielding effects on the cell, particularly O-antigen chains function to limit the access of foreigner molecules to the outer membrane (Page, 2012; Raetz and Whitfield, 2002; Gnauck et al., 2016).

The endotoxicity of the LPS is dose-dependent. At low concentrations, LPS are immunologically significant to mammalian cells. On the other hand, at high concentrations, these molecules trigger multiple events, causing organ failure, septic shock and even death (Chen et al., 2007). The endotoxicity of the Lipid A portion to mammalian cells is due to the formation of LPS complexes with lipopolysaccharide-binding protein (LBP), which activate the human pattern recognition receptor tool-like receptor 4 (TLR4) when linked with Lipid A component (Beutler and Rietschel, 2003). The Lipid A moiety is then responsible for induction of the cell mediate innate immune response, e.g., neutrophils, macrophages, monocytes and dendritic cells (Mack et al., 2014), with a variable endotoxic effect dependent on its structure (Brandenburg and Wiese, 2004).

14.4 MAIN TECHNIQUES FOR LPS QUANTIFICATION IN BIOTECHNOLOGICAL PRODUCTS

The detection and quantification of bacterial endotoxins in pharmaceuticals and other biologicals can be carried out using *in vivo* or *in vitro* methods. The *in vivo* methods are limited because these rely on the use of animal models, measuring on these the pyrogenic effect of LPS. *In vitro* methods are based on the coagulation activity of LPS to *Limulus* amoebocyte lysate (i.e., LAL assay). This assay can be semi-quantitative or quantitative, depending on the type of LAL used.

14.4.1 LPS Test in Rabbits

The pyrogen test was developed to limit to an acceptable level the risks of febrile reactions in patients following intravenous (or intrathecal) administration of a pharmaceutical product. The test includes measuring a rise in rabbit temperature after intravenous administration of a test solution (USP, 2017; ANVISA, 2010).

The LPS test in rabbits was first introduced into the United States Pharmacopoeia (USP XII) in 1942. Despite a lack of knowledge about the different types of endotoxins and their chemical composition, the pyrogenic effects of endotoxins were well recognized (Dare and Mogey, 1954). In the years following its introduction into the USP, the LPS test in rabbits has undergone changes regarding the number of animals required to validate the test, the number of repeat tests allowed and an improvement in the analytical conditions under which the test is performed. These changes were aimed at minimizing biological variability and, consequently, improving the test performance (Braun and Klein, 1960; Benesova, 1970).

Rabbits were chosen as the test species because the thermoregulatory system of these animals most closely resembled humans (Dare and Mogey, 1954). The test was developed for products that can be tolerated by rabbits (limit dose up to 10 mL/kg) administered intravenously for up to 10 min. There is no preference or recommendation as to the sex of rabbits used. However, adult animals with a minimum weight of 1.5 kg are required. Different species of rabbit can be used, such as Dutch, Himalayan or Polish (Bangham, 1979). In order to avoid false-negative results, it is important that each animal that will be used in an *in vivo* pyrogen test has been tested for a febrile response. In this context, the use of LPS as a pyrogenic substance is not recommended due to the issue of natural or cross-immunization, as well as causing temporary tolerance. Alternatively, other pyrogenic substances, such as ribonucleic acid sodium salt and ricin, may be used to test the febrile response of the animals (Benesova, 1970).

There is also a concern about false-positive responses when using the *in vivo* test. In other words, a febrile response can be observed despite of the absence of LPS or other pyrogenic substance in the tested sample. Usually, LPS promotes a temperature rise 60–90 min after the tested sample has been injected, while in the case of a false-positive response, the temperature rises 3 h or more after injection of the tested sample. This difference in the temperature elevation profile is important to differentiate a response due to the presence of LPS from a false-positive result (Pearson, 1985).

Animals can be re-used in this test, theoretically until their death, provided that there are 2–3-days intervals between tests in cases of a non-pyrogenic response or 2–3 weeks in cases where a pyrogenic response occurs (Benesova, 1970). The rabbits should be maintained under proper conditions regarding the temperature and relative humidity, periodic air renewal, adequate lighting, acoustic barriers and protection against insects and rodents.

All materials used during the LPS test in rabbits, such as glassware used in the preparation of samples, syringes and needles used for intravenous injection of the tested sample, must be previously depyrogenized. Usually, it is recommended to use thermal depyrogenation processes. Moreover, each group of animals receiving an intravenous injection of the tested sample must be observed over a period of 3 h. During this period, the animals usually remain in cervical restraint, in a comfortable position. Temperature monitoring during the test is performed by rectal insertion of a thermometer or thermocouple at a depth of 6–7.5 cm (ANVISA, 2010; USP, 2017). It is recommended that, prior to the test, the animals should be acclimatized, and the physiological temperature control be evaluated. Temperature oscillations between two consecutive determinations above 0.3°C may increase the risk of false-positive results (Bangham, 1979; Pearson, 1985).

In principle, all injectable products as well as implantable materials and devices used in parenteral therapy should be subject to quality assessment for LPS contamination. Large volumes parenterals are at greatest risk from contamination, as are intrathecal products. When injected by the intrathecal route, an endotoxin is 1,000 times more potent than if it were introduced intravenously, which may lead to false-positive results. For this reason, products intended to be used intrathecally should be tested using an *in vitro* test (Pinto *et al.*, 2015).

Usually, a dose of 10 mL/kg is used to test a large volume injectable or liquid extractor of polymeric materials. For small volume parenterals, it is recommended that a higher dose be used than normal in therapy, in order to decrease the risk of false-negative results. It is important to ensure the test solution is isotonic by addition of pyrogenic-free sodium chloride, as well as preheating the test solution at a temperature between 37 and 38°C. The test solution should be injected into the rabbit at a rate of between 4 and 6 mL/min so that the dose is administered within 4 min (Bangham, 1979).

The LPS test in rabbits does not apply to products containing drugs that act on the thermoregulatory system, such as antipyretics. In these cases, *in vitro* testing should be used, since the presence of endotoxins in the product, although not causing a febrile condition, may trigger other toxic reactions which would pose a risk to the patient (Braun and Klein, 1960).

It is recommended that the supply of the feed be suspended on the day of the test, however access to water may be free. The animals must be weighed, placed in a container, and their body temperature monitored prior to the test. Each sample should be tested in a group of three animals, and body temperature checked periodically after the sample is administered (Pinto et al., 2015).

A batch can be approved if the temperature of none of the test rabbits rises more than or equal to 0.5°C. If the sample fails the test, then the batch should be re-tested using five different animals. In this case, up to three animals may develop a rise in body temperature greater than or equal to 0.5°C and the sum of temperature increases for the eight animals tested should not exceed 3.3°C for the sample to be approved.

14.4.2 LPS Test – Gel Clot

Researchers observed that horseshoe crab (*Limulus polyphemus*) blood coagulated when in interaction with marine bacteria, causing the death of these animals (Cooper, 2008; Williams, 2004). Specifically, amoebocyte lysates from the *L. polyphemus* (i.e., LAL) are now understood to solidify into a gel in the presence of LPS molecules. The gelation level is directly proportional to the LPS concentration. These studies served as a basis for the progress of the LAL assay as we know it today (Cooper, 2008).

After the discovery that amoebocytes were involved in blood coagulation of *L. polyphemus*, other species of horseshoe-shaped crabs were studied, in which coagulation factors were also found. The coagulation process is comparable to the blood coagulation cascade in mammals, with gel formation mediated by the sequential activation of three proteases: factor B and C, and the proenzyme of coagulation. A schematic of the enzymatic cascade involved in gel formation is shown in Figure 14.2 (Williams, 2004; Pinto et al., 2015).

The enzymatic reaction is dependent on the activation of high molecular weight enzyme by the endotoxin, which gelatinates low molecular weight coagulable proteins. The coagulation enzyme was isolated from the amebocyte lysate and further studies determined that activity was Ca^{2+} dependent. The coagulant protein was called coagulogen. Despite the high specificity of the LAL test, some cellulose derivatives called β-glucans may cause interference to the test (Cooper, 1990; Cooper et al., 1997).

The LAL test is prepared from the blood of the crabs. Usually 0.125% N-ethylmaleimide in a 3% sodium chloride solution is used as the coagulant. The mixture is centrifuged and the supernatant containing haemocyanin is discarded. The amoebocytes are washed with 3% sodium chloride to remove the anticoagulant, and the cells lysed by osmotic shock with the addition of distilled non-pyrogenic water. Alternatively, other cell lysis methods may be employed. The product obtained is lyophilized and must be kept refrigerated for up to 3 years (Williams, 2004; Pearson, 1985).

LAL reagents are standardized for sensitivity (λ) to bacterial endotoxins from *E. coli* (EC-5), which provides a measurement of endotoxin potency expressed in EU/mL. The LAL test for endotoxin determination shows high variability, which can be reduced by treatment with chloroform and the addition of bivalent ions may be used to increase test sensitivity (Williams, 2004; Cooper, 2008).

The simplest and most widely used method is based on endpoint gelation. The endpoint gelation LAL test is usually used as a borderline test, which indicates whether the amount of endotoxin present is above or below the LAL sensitivity. Alternatively, the endpoint gelation test can be used as a semiquantitative test, which provides a measurement of endotoxin from several sample dilutions (Weary and Baker, 1977).

In this test, equal volumes of test solution and LAL reagent (0.1 mL) are transferred to test tubes (10 × 75 mm).

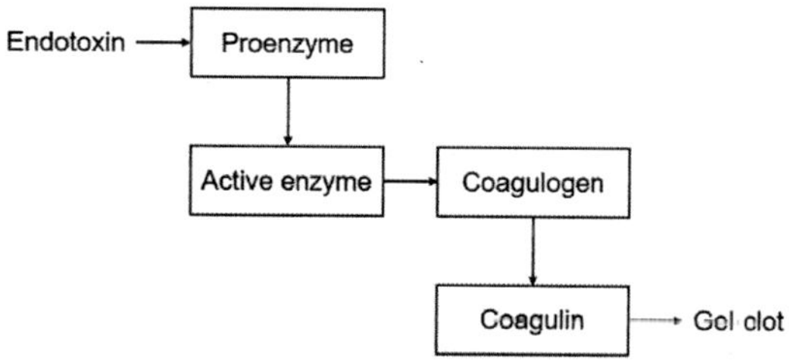

FIGURE 14.2 Schematic of the enzymatic cascade leading to the endpoint gelation of the *Limulus* amoebocyte lysate (LAL) test.

The mixture is gently homogenized and incubated at 37°C for 60 min. It is recommended that the incubation takes place in a water bath because of greater thermal homogeneity in comparison to bacteriological incubators with air circulation. During the incubation period, the test tubes should not be handled to avoid interference with the gelation process (Weary and Baker, 1977; ANVISA, 2010; USP, 2017).

The reading of the result proceeds by carefully inverting each test tube at 180° to verify gel formation (Figure 14.3). Gel that remains firm during the inversion of the tube indicates the presence of endotoxin above the sensitivity limit of the test. No gel formation indicates that the amount of endotoxin is below of the sensitivity limit of the test. Positive and negative controls are required to confirm results obtained using the endpoint gelation LAL test (ANVISA, 2010; USP, 2017).

Validation of this test involves verification of the LAL sensitivity (λ) and determination that possible interfering effects (inhibitory or exacerbating of the reaction) from substances present in the sample are absent (Pearson, 1985; Pinto et al., 2015).

Usually, commercially available LAL reagents have nominal sensitivity between 0.06 and 0.25 EU/mL. The sensitivity of LAL is assessed using a series of dilutions from bacterial endotoxin at concentrations: 0.25λ, 0.5λ, λ, 2λ, and 4λ. All dilutions should be performed using water and other non-pyrogenic materials. From the results obtained (gel formation or not), the endpoints are determined, and the sensitivity is calculated as the geometric mean of the endpoints. To be considered suitable for use, the LAL reagent should have a geometric mean between 0.5λ and 2λ (ANVISA, 2010; USP, 2017).

Substances present in the sample may inhibit or exacerbate the gel formation reaction, which may lead to a false-negative or false-positive result. For this reason, the interference potential of the sample in the LAL test should be evaluated as part of the validation of the test. Determining sample interference entails contaminating the test solution with concentrations of: 0.25λ, 0.5λ, λ, 2λ, and 4λ. The test should be performed in parallel with the same concentrations of endotoxin diluted in non-pyrogenic water. It is considered that the sample does not interfere with the LAL test if the geometric mean of the endpoints is between 0.5λ and 2λ.

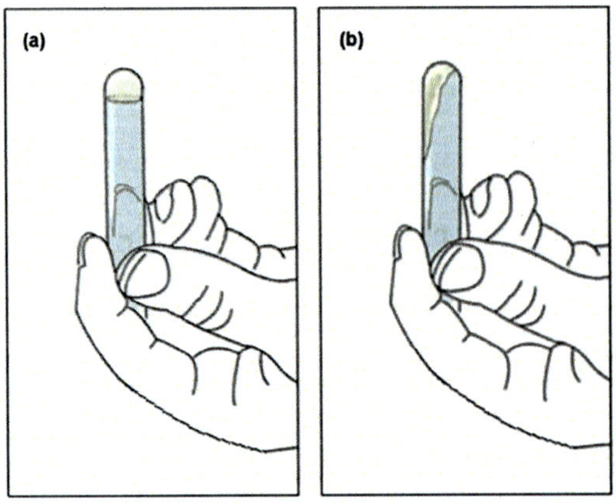

FIGURE 14.3 The endpoint gelation of *Limulus* amoebocyte lysate (LAL) test indicating positive (a) and negative response (b).

In the LAL test, if the sample interferes with the method, it is possible to dilute the sample to reduce interference, provided that the maximum valid dilution (MVD) is used. The MVD is calculated based on the concentration of the active principle in the sample (mg/mL), the endotoxin limit for the active principle (EU/mg) and the sensitivity. When the product is diluted beyond the MDV value, it is possible that the LAL reagent will not detect the presence of endotoxins because it is in a concentration below test sensitivity. In this case, it is not possible to guarantee that the sample is within the endotoxin limit, which invalidates the LAL test (Weary and Baker, 1977; Williams, 2004).

In addition to the MDV requirement, sample interferences can be eliminated with heating (degradation of the interfering substance without bacterial endotoxin degradation), pH adjustment (samples with very acid or alkaline pH may inactivate the enzymes present in the LAL reagent) or addition of a substance that neutralizes the interference of the sample i.e., the addition of ethylenediaminetetraacetic acid (EDTA) can chelate bivalent metals necessary for gel formation (Lourenço et al., 2012; Cooper, 1990).

14.4.3 LPS Test – Chromogenic and Turbidimetric Kinetics

Although the end-point gelation LAL test is the most used and has been widely adopted as a reference test, other methods to detect and quantify bacterial endotoxins based on activation of the LAL enzyme cascade may be used. An increase in optical density due to turbidity caused by the coagulation reaction (turbidimetric assay) or colour produced by enzymatic cleavage of a chromogenic substrate (chromogenic assay) can both be used as an alternative to the final gel point LAL test (Cooper, 2008; Williams, 2004).

The turbidimetric assay allows for a better quantitative measurement of bacterial endotoxins regarding to the endpoint LAL test. This assay (Figure 14.4) is based on increased turbidity as a function of increasing endotoxin concentration due to coagulogen precipitation in the lysate (Pinto et al., 2015).

Mixtures of LAL reagent and test solution are prepared in equal volumes (usually 0.1 mL). After gentle homogenization, the tubes (or microtiter plate) are incubated at 37°C for the recommended time and optical density readings (340 nm) then taken. Usually, the LPS content is calculated from the optical density readings of the test sample employing a standard calibration curve with three to five concentrations of bacterial endotoxin in the range of 0.05 to 50 EU/mL. Positive sample controls should be prepared containing an amount of endotoxin corresponding to the mean concentration of the curve. Negative controls should also be included to ensure the validity of the result obtained (ANVISA, 2010; USP, 2017).

Alternatively, quantification of bacterial endotoxin can be performed in a turbidimetric kinetic assay, in which the optical density measurements are evaluated as a function of reaction time and endotoxin concentration. That is, the assay is not related to turbidity with the endotoxin concentration but the time required for a certain increase in turbidity to

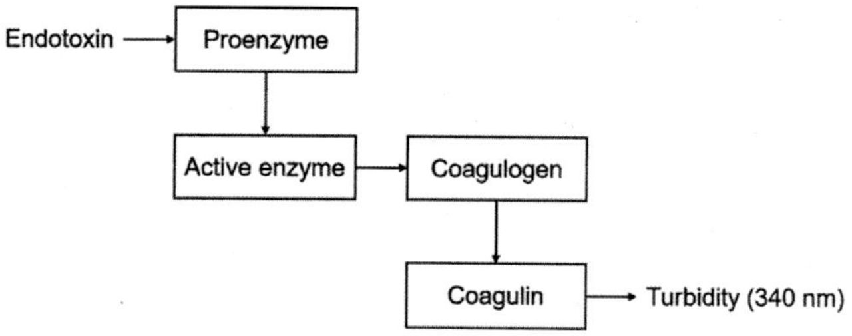

FIGURE 14.4 Schematic of the enzymatic cascade causing turbidity which can be used in the end-point or kinetic turbidimetric *Limulus* amoebocyte lysate (LAL) test.

occur as a function of the endotoxin concentration, i.e., the higher the endotoxin concentration, the lower the reaction time. A standard curve is obtained by plotting the reaction time as a function of the endotoxin concentrations and can be used to determine the amount of endotoxin present in the sample (Cooper, 2008; Williams, 2004).

The chromogenic method is based on cleavage of a chromogenic substrate as a function of the amidase action of the activated coagulation enzyme. The cleavage of the chromogenic substrate leads to the formation of *p*-nitroaniline, a yellow-coloured substance, the intensity of which is directly proportional to the concentration of bacterial endotoxin (Figure 14.5). Thus, it is possible to quantify the amidase activity induced by the bacterial endotoxin formed by *p*-nitroaniline, quantified by absorbance at 405 nm (Haishima et al., 2003; Ostronoff and Lourenço, 2015).

In this method, equal volume of test solution and LAL reagent (0.1 mL) are mixed and preincubated for 8 min at 37°C. The substrate solution (usually 0.5 mL) is then added, and the mixture is incubated for 3 min. The reaction is quenched with glacial acetic acid in water (25% v/v) and the optical density is measured at 405 nm. The LPS concentration in the tested sample is calculated from the optical density readings using a standard calibration curve in the range of 0.05–50 EU/mL. Positive and negative controls may also be included to ensure the validity of the result obtained (ANVISA, 2010; USP, 2017).

When endotoxin is quantified using a chromogenic kinetic assay, the optical density measurements are evaluated as a function of reaction time and endotoxin concentration. The reaction time indicates the time required for detectable levels of *p*-nitroaniline to be released. Similar to the turbidimetric kinetic assay, the higher bacterial endotoxin concentration, the lower reaction time is required. A standard curve is obtained by plotting the reaction time against bacterial endotoxin concentrations in the range of 0.01–100 EU/mL (Cooper, 2008).

14.4.4 New Methods for LPS Quantification

Studies have now shown that in addition to pyrogenic effects, bacterial endotoxins have the potential to induce inflammatory and other immunological responses. These occur due to the stimulation of blood monocytes, which release cytokines which are endogenous pyrogens. Based on these studies, an *in vitro* human pyrogen assay has been developed, which can use whole blood as the biological reagent (Ding et al., 2001).

The analysis of bacterial endotoxins in water for health purposes (i.e., water intended for haemodialysis) is extremely important. A new endotoxin test kit, the Portable Test System (PTS), was developed specially to be used to quantify LPS in water. The PTS consists of an acrylic cartridge containing lyophilized LAL reagent that promotes a chromogenic reaction proportional to the LPS concentration in the tested sample solution. The results of the chromogenic reaction are compared with those obtained with the calibration curve stored in the equipment memory. The test result is obtained in 15 min and is expressed in EU/mL. Recent studies have demonstrated the applicability of PTS to endotoxin detection in radiopharmaceuticals (Pinto et al., 2015).

14.5 TECHNIQUES USED TO REMOVE LPS FROM BIOTECHNOLOGICAL PRODUCTS

Approximately 20% of approved biopharmaceuticals are produced using *E. coli* as a heterologous host (Walsh, 2014). Besides biopharmaceutical production, LPS from Gram-negative bacteria are a concern in any pharmaceutical process, especially for parenteral application. In this regard, the production of plasmid DNA vaccines as prospective bioproducts has been widely described in scientific literature, although its development is still in clinical trials, and LPS removal strategies are being explored (Prather et al., 2003).

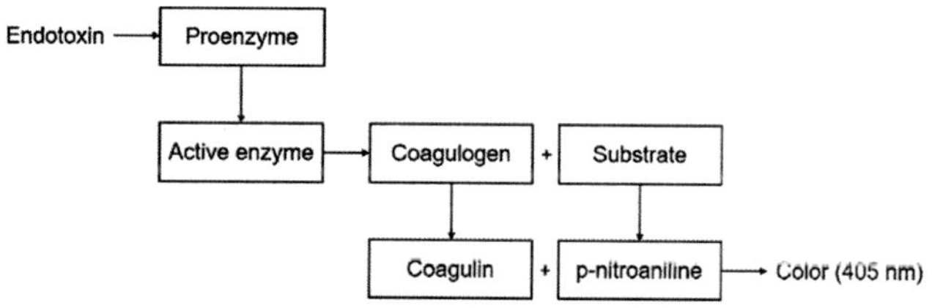

FIGURE 14.5 Schematic of the enzymatic cascade involved in the release of *p*-nitroaniline for end-point or kinetic chromogenic *Limulus* amoebocyte lysate (LAL) tests.

LPS constitutes 5–10% of the total dry weight of Gram-negative bacteria (Rezania et al., 2011), this implies a high amount of LPS is likely to be contaminating the target product. A lab-scale bioreactor can reach over 100 g of dry cell weight/L in the production of a biopharmaceutical, with an equivalent of 5–10 g/L of LPS contaminating the desired product, burdening downstream steps. Lower dry cell quantities are often achieved in large scale production, nevertheless, LPS is still produced in the range of the expressed protein (from hundreds of mg to g/L) (Choi et al., 2006). Figure 14.6 depicts the proportion of each major component types in *E. coli*.

Any downstream process must reduce by several logarithms the amount of contaminant LPS to reach pharmaceutical grade and achieve the safe levels determined by regulatory organizations for each market. For example, the United States Pharmacopeia sets the threshold to 5 USP-EU/kg of body weight for human pyrogenic dose of endotoxin for any parenteral route of administration other than intrathecal (USP, 2015). This is approximately 0.5 ng of *E. coli* endotoxin per kg of body weight of humans (Hochuli, 1988). Bioprocesses can generate materials with high proportion of LPS, with values in the range of 50–100 μg/g of dry weight cells (Rezania et al., 2011).

Downstream approaches must remove *E. coli*-derived LPS, while retaining the desired product with high recovery rates. Like any downstream approach, this is obtained by exploiting the properties of the mixed molecules. LPS are complex amphiphilic molecules with a molecular weight in the few dozens of kDa (Sweadner *et al.*, 1977) and as highlighted are constituted of three parts (i.e., O-antigen, core oligosaccharide and Lipid A). These parts vary in composition in different strains of the same species and amongst species. Still, they often aggregate and form structures with apparent molecular weights greater than 10 kDa (Sweadner et al., 1977). Due to their amphiphilic properties and aggregate size, LPS can be removed through methods that exploit these two main properties:

- Size exclusion (e.g., ultrafiltration);
- Surface properties, such as electrostatic/hydrophobic interactions (e.g., affinity and ion-exchange chromatography, and charged membrane/depth filtration).

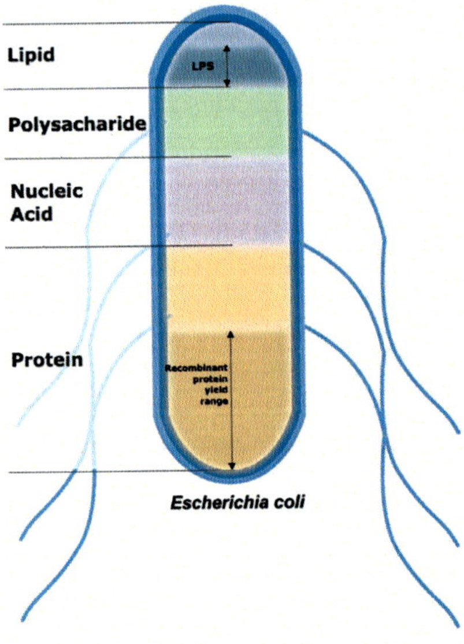

FIGURE 14.6 Schematic representation of major biomolecules in *Escherichia coli* and the proportions to dry weight of cells. Recombinant protein yield varies depending on the expression system.

The operation of any of these techniques depends on the properties of the target product. Ultrafiltration is used, e.g., to remove endotoxins during the manufacture of water for injection, due to the size difference between water molecules and LPS. Products such as recombinant proteins represent a different challenge since the size of many proteins resembles LPS in magnitude (>10 kDa). All things considered, the size of LPS restricts its application to products with a small size, such as small peptides, and solutions employed in manufacturing lines (Sweadner et al., 1977). Fortunately, for proteins, the usual chromatographic steps employed during purification steps to remove other impurities also remove LPS contaminants to a certain extent (Hirayama and Sakata, 2002) and can be set to efficiently remove LPS. Besides filtration and traditional chromatographic columns, LPS can be removed by employing liquid-liquid strategies, such as the aqueous two-phase micellar system (Teixeira-Pinto et al., 2019). Also, a variety of affinity ligands are reported to possess endotoxin specificity, like histamine, histidine and polymyxin B (Anspach and Hilbeck, 1995), deoxycholic acid (Anspach et al., 1999) and dimethylamine ligands (Yuan et al., 2005). Endotoxins are known to bind to several proteins with biopharmaceuticals applications, which increase the complexity of LPS removal (Ongkudon et al., 2012). The use of surfactant is required when endotoxins bind to the target protein surface, leading to an extra step to surfactant removal, and can be used in combination with filtration strategies (Jang et al., 2009).

Accordingly, the removal of LPS related with bacterial expression systems is a complex problem that requires a tailor-made process to achieve the purity required in the final product, ranging from liquid-liquid strategies to liquid-solid phase separations/purifications. In the following sections, we will discuss some of these strategies.

14.5.1 Chromatographic Techniques

Chromatography is commonly used in the purification of several compounds and is widely used in the manufacturing of biopharmaceuticals. Purification relies on the differential affinities of substances used for the mobile and stationary phases through which they pass. This means that to separate the substances in a mixture, some physicochemical properties must be exploited. In the purification of proteins from LPS, it is common to use positively charged surfaces (e.g., cation-exchange chromatography) that bind the LPS to the matrix and allow proteins to be eluted, or conversely, a negatively charged surface that captures the proteins and allows LPS to be eluted (Ongkudon et al., 2012). Anion-exchange chromatography is commonly used as a downstream process and can separate a target molecule from different contaminants simultaneously (i.e., LPS and contaminating proteins). The final chromatographic step can be optimized for maximum protein recovery and contaminant removal by adjusting chromatography parameters including bead size, surface charge, running buffer pH and composition (Fanali et al., 2017). Other chromatographic techniques also used include affinity-base (Fraseur and Kinzer-Ursem, 2018), size exclusion (Batas et al., 1999), hydrophobic interaction (Iuliano et al., 2002) and on a smaller scale, paramagnetic particle chromatography (White, 2003).

14.5.1.1 Affinity Chromatography

This type of chromatography is established on the specific interaction of molecules with immobilized matrices, for example, immobilized antibodies (Hubbard et al., 1987), protein A (Grodzki and Berenstein, 2009), site-specific ligands for enzymes (Blumberg et al., 1970), immobilized dye (Zayed et al., 2016) and recombinant tags, such as immobilized metal affinity ions – IMAC (Sulkowski, 1985). The reversible and specific binding of target molecules to the matrix makes this method one of the most selective, versatile and often complex chromatographic approaches available. The method was developed in the early 1900s (Hage and Matsuda, 2015) to purify α-amylase (Starkenstein, 1910). The specific interaction between the matrix and target molecule is the first step in this chromatography, which allows retention of the target molecule and elution of impurities, such as LPS. Following elution of impurities, a release strategy is used which exploits the reversible nature of target-matrix binding. Elution is achieved either by applying different buffer compositions to enhance release, or by the addition of an agent to bind the product or compete for the affinity ligand (Hage and Matsuda, 2015). Care should be taken when choosing affinity chromatography as a purification strategy, since LPS can also have affinity for the matrix ligand. For example, a common recombinant protein purification strategy is to clone DNA sequences which add a poly histidine tag (His-tag) to the N or C-terminal when the target protein is translated. A chain of six to ten histidine residues have affinity for bivalent metal ions which are immobilized onto a support to capture the tagged protein during purification. LPS carries a negative charge and so will also interact with the immobilized cations, which will trap both the LPS molecules and target protein. Use of a thrombin cleavage site between the protein and His-tag allows the protein to be eluted, while LPS and the His-tag are retained on the solid support (Mack et al., 2014). Although this strategy works, it adds another step to the purification process because the thrombin must be cleaved. Indeed, the direct use of IMAC columns can result in high LPS content after chromatography (i.e., 34,000 EU/mg). Nevertheless, simple strategies, such as the addition of surfactants, can greatly reduce LPS contamination. For example, Reichelt et al. (2006) verified that the addition of amounts, as small as 0.1% wt of non-ionic surfactant (*viz.* Triton X-114 or TX-114) in the washing buffer, can reduce by 99% the initial LPS contamination. Still, the system is required to be kept at 4°C, supposedly to avoid the formation of large micelles by TX-114. Another affinity chromatography strategy is to capture the contaminant in the first step and elute the product with affinity ligands. An example is the use of polymyxin B, a cationic polypeptide that binds to the Lipid A piece (Morrison and Jacobs, 1976) and has been used for LPS removal in different solutions (Teramoto et al., 2002; Morrison and Jacobs, 1976; Sweadner et al., 1977), for the production of proteins (Karplus et al., 1987) and plasmid DNA (Montbriand and Malone, 1996), and as a treatment for patients with bacterial infection (Teramoto et al., 2002). For protein purification, polymyxin can achieve high yields, while removing LPS molecules by several folds. For example, Karplus et al. (1987) demonstrated that contaminated bovine catalase solution could be purified, with small enzyme loss (~11%), while removing LPS contamination by 1,000-fold. Nevertheless, the process is long, due to the slow binding kinetics, and negatively charged proteins cannot be recovered simultaneously with polymyxin B (Ongkudon et al., 2012). Another strategy to specifically remove LPS is the immobilization of anti-O antigen antibody to a solid matrix. Goldbaum et al. (1994) adopted this strategy to study the removal of LPS from *Brucella* spp., since polymyxin B is ineffective against this type of LPS. They were able to immobilize the antibody onto Sepharose 4B resin and reduce LPS 1,000-fold from a cytoplasmic preparation of *Brucella abortus*.

14.5.1.2 Ion-Exchange Chromatography

Any biomolecule with an exposed surface charge can be purified by ion-exchange. The adoption of this technology in bioprocesses is a reflection of the versatility and the improvements observed in the past decades, in terms of increase in the flow rate obtained. The initial soft nature of the chromatography medium, restricted the flow to 1–2 cm/h, leading to a process several-days long, but velocities up to 500 cm/h are obtainable today increasing the application of the technique in industrial processes (Jungbauer and Hahn, 2009). Furthermore, the use of monoliths as chromatographic media has resulted in 5 min separation procedures (Jungbauer and Hahn, 2004). Therefore, ion-exchange chromatography is a suitable method for proteins, polynucleotides, such as pDNA and RNA, and other biomacromolecules such as LPS, since all present an ionic charge. The charge is exploited using either of two strategies: in a cation-exchange resin, in which the solid phase is negatively charged, or in an anion-exchange resin, which is positively charged. LPS possess an ionic core with phosphate groups, negatively charged above a pH of 2.1, meaning under common biomolecule purification conditions, it will be negatively charged. Nevertheless, for protein purification, surface charges vary depending on the isoelectric point (pI) of the proteins and pH of the buffer. For example, the bulk of *E. coli* proteins possess a pI of 3.0 to 6.5 (Lopez-Campistrous et al., 2005), and would be negatively charged at neutral pH. The charge properties of the biomolecules affect the choice of purification strategy, and due to the negative charge of LPS, cation-exchange methods have been used for LPS removal (Lin et al., 2005). For example, a recombinant basic fibroblast growth factor was purified from *E. coli* by applying a sequential process, involving cation-exchange and affinity chromatography strategies. After the cation-exchange chromatography, a 98% reduction in LPS content was observed, and this further improved with affinity chromatography. Anion-exchange chromatography used to remove LPS is generally limited to protein-free solutions, since LPS will also bind in the solid phase. On the other hand, if the protein is stable at pH below its pI, it is possible to use anion-exchange chromatography. LPS will bind to the resin because it is negatively charged, while the positively charged biomolecule will pass through, separating protein from the contaminant (Hou and Zaniewski, 1990). The purification of pDNA is compromised in this approach to remove LPS, since both have negative surface charges.

14.5.1.3 Hydrophobic Interaction Chromatography

LPS molecules also possess a lipid portion that can be exploited to remove LPS. Hydrophobic interaction-based chromatography (HIC) is a technique that exploits hydrophobic domains in biomolecules to bind to hydrophobic ligands on solid matrices character. In particular, HIC is used to remove aggregate species of a protein product, which possess different hydrophobic properties from the monomer (McCue, 2009) or for plasmid DNA purification (Diogo et al., 2000). The process is carried out in an aqueous solution with high salt concentration to control solvent polarity, since water is a poor solvent for non-polar molecules and therefore promotes protein hydrophobic ligand interactions. Alteration of hydrophobic interactions by ions is well known (Duarte et al., 2015), with series of ions strongly or weakly influencing the interactions. The process is generally initiated with a protein solution containing high concentrations of a salt, such as ammonium sulphate and sodium chloride, and the protein bound to the hydrophobic ligand. Elution is obtained by reducing the salt concentration, thereby reducing hydrophobic interactions between the resin and the biomolecule, releasing the product. Each protein presents some degree of affinity for the resin and is released at a different salt concentration. Nevertheless, HIC is not generally used for LPS removal in proteins solutions but is applicable for plasmid purification. Purification is obtained by the differences in hydrophobicity between pDNA and impurities (Diogo et al., 2000). Interestingly, although other nucleic acids are generally present, their removal is possible, since RNA and denatured genomic DNA have exposed hydrophobic parts due to their single-stranded nature. The exposed nitrogenous base has a higher hydrophobicity than double-stranded molecules, within which these regions are shielded, as in pDNA (Freitas et al., 2009). The LPS is also removed, since the Lipid A portion of the molecule interacts with the hydrophobic domain of the resin. Therefore, pDNA tends to be eluted first, while the contaminants are retained and subsequently removed in a low salt buffer elution. This method can remove high amounts of LPS (>98%) in a single step (Freitas et al., 2009). The salt composition of the running buffers can affect the binding pattern of pDNA and impurities to the hydrophobic resins, which can be exploited to separate even similar molecules, such as open circle and supercoiled plasmids. Freitas et al. (2009) reported a successful separation of both isoforms, using a phenyl-sepharose resin and high sodium citrate concentrations (≥1.2 M), with the added benefit of a lower-cost and environmentally friendly buffer, compared to more commonly used ammonium sulphate. Although not commonly used for protein purification, HIC can be used for the objective of removing LPS from protein extracts. Wilson et al. (2001) showed the removal of tightly bound LPS molecules from inclusion bodies proteins. The strategy was to add a combination of guanidine hydrochloride and ammonium sulphate salts at the solubility limit for endotoxins and the target biomolecule. The separation between LPS and the product was obtained with 0.7 M ammonium sulphate and 4 M guanidine HCl, with endotoxin binding to the resin, while the product was obtained in the flow-through. This strategy removed 3.7 logs of LPS, equivalent to 1.1×10^6 to 200 EU/mg (Wilson et al., 2001).

14.5.2 Membrane-Based Filtration

Membrane-based filtration is a process that uses the hydraulic diameters of solutes to separate molecules, readily allowing the flow of solvents and small molecules but blocking the passage of large particles. Nevertheless, due to distinct hydrophobic and hydrophilic portions, LPS tends to aggregate, forming micelle structures with the lipid portion inside the aggregate. The micelles can have different sizes, anywhere between 300 kDa and 1 MDa, and depending on buffer conditions, it can generate smaller aggregates with 10 kDa (Sweadner et al., 1977). The different sizes can be exploited to separate LPS from the product of interest, such as proteins. The filtration would require a set of filters to remove small LPS monomers and another for the aggregates (Figure 14.7).

Still, compared to chromatography, the low cost and scalability are attractive, but problems, such as fouling, during filtration hinders its application in protein purification (Larive et al., 1999). In fact, membrane-based filtrations are preferred in intermediate and final steps in a downstream process, such as protein concentration and buffer exchange, since there is less fouling (Kurnik et al., 1995) at these stages, because the concentration of LPS is already low. Direct filtration, using 10 kDa ultrafiltration membranes, is useful for the preparation of other human products, such as the preparation of water for injection (Grandics et al., 2000), and small molecules, such as antibiotics and active pharmaceutical ingredients, achieving values of LPS lower than 0.03 EU/mg (De Mas et al., 2015).

14.5.3 Ultrafiltration

Ultrafiltration membranes retain molecules within a hydraulic diameter of 5–150 nm, molecules with approximate molecular weights of 1–1,000 kDa, which is basically the range of most proteins, nucleic acids and endotoxins.

Besides the pore size, ultrafiltration parameters can influence LPS removal, especially from protein solutions, since LPS is known to bind to proteins (Elass-Rochard et al., 1995). The buffer solution used influences LPS removal efficiency during ultrafiltration, even though this technique is based on size differences. This occurs because of functional groups in the LPS molecule, and the interaction of LPS molecules with each other and with other molecules biomolecules. This means that solutions with components that disrupt or influence LPS molecules can alter filtration efficiency and the overall purification. Two main approaches include: disturbing LPS interaction with the target protein,

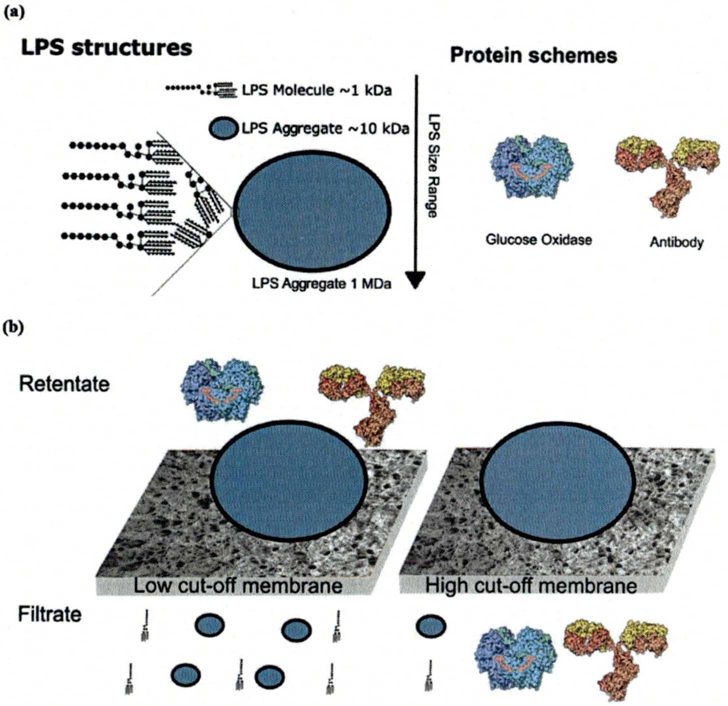

FIGURE 14.7 Schematic representation of lipopolysaccharide (LPS) structures and filtration approaches to remove each form of LPS. Scheme with different LPS structures and the representation of a naturally occurring protein (glucose oxidase) and recombinantly expressed protein (antibody) (a). Filters with low-cut-off (small pore sizes) and high cut-off membranes can be used to remove smaller forms of LPS structures and larger aggregates, respectively (b). Protein images were generated with Illustrate (Goodsell, D. S., Autin, L., & Olson, A. J. *Structure*, 27(11), 1716–1720.e1, 2019, doi: 10.1016/j.str.2019.08.011), using data from PDB ID: 1GPE (Wohlfahrt, G., Witt, S., Hendle, J., Schomburg, D., Kalisz, H. M., & Hecht, H. J. *Acta Crystallographica Section D: Biological Crystallography*, 55(5), 969–977, 1999, doi: 0.1107/S0907444999003431) and PDB ID: 1IGT (Harris, L. J., Larson, S. B., Hasel, K. W., & McPherson, A. *Biochemistry*, 36(7), 1581–1597, 1997, doi: 10.1021/bi962514+), hosted in PDB (http://www.rcsb.org) (Berman, H.M., Westbrook, J., Feng, Z., Gilliland, G., Bhat, T.N., Weissig, H., Shindyalov, I.N., Bourne, P.E., *Nucleic Acids Res.*, 28, 235–242, 2000).

important for the removal of residual LPS from the final product, and increasing self-aggregation of LPS molecules, impacting its filtration profile. In the LPS-protein disruption strategy, surfactants are added to the solution to interact with LPS molecules. The surfactant and LPS can form large micelles that are withheld by the filter while the protein passes through. For example, Jang et al. (2009) studied the influence of different surfactants in LPS removal using a filter with a 100 kDa cut-off. In some of the conditions tested, a removal of more than 99% was observed using 1.0% taurodeoxycholate. Another approach is to promote LPS self-aggregation by cations. LPS molecules possess negative phosphate groups that can be bridged by divalent cations, such as Ca^{2+} and Mg^{2+} (Schindler and Osborn, 1979). Li and Luo (1998) demonstrated that this strategy can be employed to remove LPS bound to proteins by aggregating them with Ca^{2+}. They used a system with two sequential filtration steps. The first step removed the already aggregated LPS molecules by filtration with a membrane with a 300 kDa cut-off. Subsequently, they induced LPS aggregation by adding Ca^{2+} to the filtered solution, forming new LPS aggregates that were retained in the next 300 kDa filtration step, while the studied protein, haemoglobin, passed through the filter (Li and Luo, 1998).

14.5.4 Microfiltration

Microfiltration is a system usually used as a pre-treatment before ultrafiltration to remove particles in the range of about 0.1–10 μm (Baker, 2012). Due to the size of LPS aggregates, part of the LPS molecules is separated by applying this filter. For example, Jang et al. (2009) reported a 33-fold removal of LPS by filtration through a 0.2 μm pore size filter. Ligands such as deoxycholate, histidine, diethylethanolamine (DEAE), polymyxin B, poly-L-lysine (PLL), and poly(ethyleneimine) (PEI) can be immobilized by the addition of cationic functional groups onto the surface of the microfilter membranes (Petsch et al., 1998; Anspach, 2001). Such strategies can enhance LPS binding, for example, Petsch et al. (1998) and Anspach (2001) demonstrated that by immobilizing DEAE to a membrane, they achieved a 20,000-fold greater affinity for LPS than by treating the membrane with bovine serum albumin (BSA) alone.

14.5.5 Depth Filtration

Depth filters are a type of filter composed of a porous medium that has a larger trapping surface for adsorbing material, as well as promoting an extended exposure time for the liquid (Datta and Redner, 1998). This high surface area can be positively charged and has been effectively used to LPS removal from sugar, water and saline solutions (Gerba and Hou, 1985). For example, Hou et al. (1980) showed that a positively charged depth filter was capable to remove more than 99.7% of LPS from a saline solution, containing 10,000 pg/mL of LPS. Use of depth filtration to remove LPS from solutions is limited because proteins also adsorb, thus increasing product loss (Yigzaw et al., 2006).

14.5.6 Novel Methods for LPS Removal

New approaches to remove LPS from pharmaceutical products involve either a combination of existing techniques (De Sousa et al., 2017) or targeting unusual parts of the LPS molecule (Vagenende et al., 2013). Some approaches that are highly selectivity for LPS do not require equilibration steps and are straightforward to use, for example, affinity ligands. Ding et al. (2001), demonstrated the use of a synthetic amphipathic cationic peptide called S3D isolated from the horseshoe crab *L. polyphemus* as an affinity ligand to remove LPS. The immobilized peptide yielded affinity constant values similar to those of antibodies ($K_D = 10^{-6}$–10^{-9} M) and could remove LPS by 20,000-fold from low LPS-concentration solutions. This peptide is manufactured as a recombinant protein, and hence, it is costly and unstable during harsh column sanitization procedures. Another recent approach used nanoparticles to remove LPS from aqueous solutions. Donnell et al. (2016) synthetized nanoparticles using a polymer (i.e., poly(ε-caprolactone) – PCL). PCL is a biodegradable polymer, approved by the FDA and is easily prepared in a one-step phase separation process, giving a removal efficiency per area of 1.49×10^6 EU/cm^2 (Donnell et al., 2016), 6.7-fold more superior than a commercial microporous membrane (Wu et al., 2000). Furthermore, it offered a faster processing time, around 5 min, compared to reverse osmosis systems which can take anywhere between 10 and 40 days (Mokhtar and Naoyuki, 2012). Nevertheless, its use with protein solutions remains to be tested and non-specific protein interactions with the nanoparticles needs evaluating to determine its effectiveness for protein purification.

Expanded bed chromatography is a technique that possesses interesting advantages over packed bed chromatography, due to its ability to handle feedstocks with particulate material. De Sousa et al. (2017) exploited this technique in purifying a *Leishmania* antigen, produced in *E. coli* with a His-tag, from the crude lysate using TX-114 in the washing steps. Expanded bed chromatography using an IMAC resin and a TX-114 wash step could efficiently remove LPS. The use of low concentrations of TX-114 (0.1% wt) was enough to reduce the LPS content by 4 logs, while recovering the target protein in a single step.

Chromatography for endotoxin removal has involved the use of different solid phase absorbents, including anion-exchange, hydrophobic, biological affinity and mixed mode materials. Anion-exchange can be used to bind the negatively charged LPS core as previously discussed. A recent method combined the charge interactions with size fractionation (NIAN et al., 2013) in a technique called anion-exchange in void-exclusion mode – VEAX (Figure 14.8).

Positively charged proteins are repelled from the charged particle pores and pass through the void volume of the column (the region around the particles). This exclusion from the pores allows the larger positively charged proteins to elute first, while smaller positively charge proteins can enter and exit the pores and have a longer elution time from the column. The negatively charged LPS molecules will flow slowly through the column, for this reason these contaminants can be removed from the positively charged biomolecules. Nian et al. (2013) showed 99.7% of LPS removal using a simple HEPES [4-(2-Hydroxyethyl)piperazine-1-ethanesulfonic acid, N-(2-Hydroxyethyl)piperazine-N′-(2-ethanesulfonic acid)] buffer (50 mM HEPES, 50 mM NaCl, pH 7.0). An interesting advancement of this technology is the capacity to separate non-equilibrated samples which is a prerequisite for product recovery in anion-exchange chromatography. This approach is compatible with the use of dissociative agents, that can disrupt LPS interaction with some proteins. This strategy allows the dissociated proteins to flow through the void space, while LPS and the dissociating agent are separated because of their longer elution profiles.

The removal of LPS by interaction with regions of the endotoxin molecule has to-date exploited only the hydrophobic properties of Lipid A and electrostatic interactions with the negative core. The carbohydrates from the O-antigen, despite being a major component on the molecule, have still not been exploited in any removal strategy. The rich carbohydrate nature would allow binding to this part of the molecule using a strategy based on hydrogen bonding, nevertheless identifying a solid phase for this approach is challenging. In this sense, allantoin crystals have been demonstrated as a solid phase for LPS removal, by binding to the carbohydrate rich O-antigen. Allantoin is a metabolic intermediary, produced by several organisms, including humans, which converts uric acid to allantoin (Johnson et al., 2009). Therefore, toxicity concerns are reduced, and it is already used as an additive to human products such as drugs, shampoos, toothpastes and skin care products (Thornfeldt, 2005). This molecule can form crystals with high affinity for endotoxins ($K_a = 10^{10}$ M^{-1} under physiological conditions), superior to specific biologic agents, such as antiendotoxic peptides and endotoxin-binding proteins (Vagenende et al., 2013) (Figure 14.9).

This high affinity has allowed a 3- to 4-log endotoxin reduction while recovering both acidic and basic proteins at high percentages (>95% of protein recovery). Allantoin is compatible with commonly used dissociative agents and buffer components, for example, 20% glycerol, 20% ethanol, 20% dimethyl sulfoxide, 10% detergent, 2 M NaCl and even 6 M guanidine, while still retaining the ability to effectively remove endotoxins (Vagenende et al., 2013).

14.5.6.1 Aqueous Two-Phase Systems

Liquid-liquid extractions employing polymers, surfactants and salts have been performed in several kinds of aqueous two-phase systems (ATPS) for extraction/separation of large target biomolecules, as well as eliminating contamination as an alternative method for LPS removal (Iqbal et al., 2016; Magalhães et al., 2007; Teixeira-Pinto et al., 2019). ATPS uses amphiphilic molecules (surfactants or copolymers)

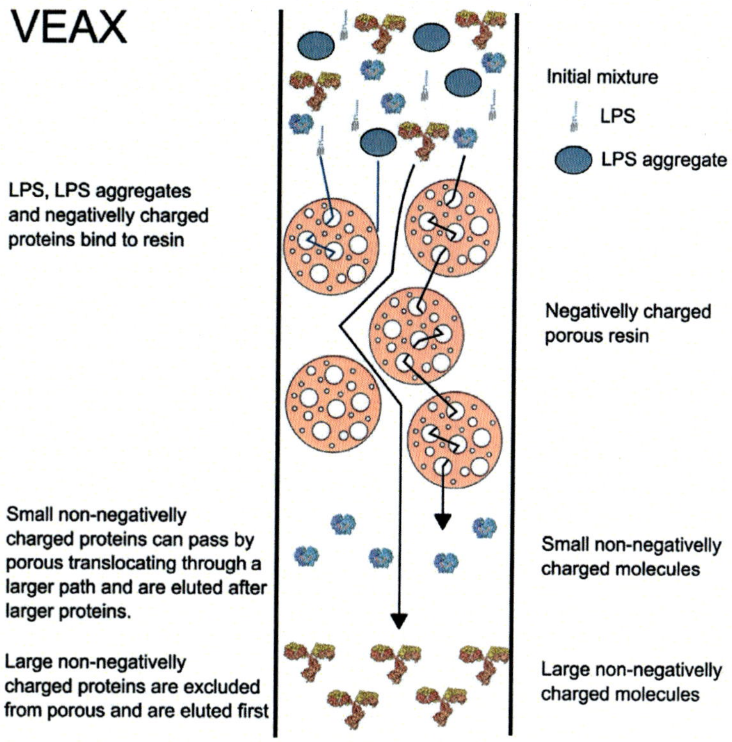

FIGURE 14.8 Volume exclusion anion-exchange (VEAX). A solution containing the protein of interest and contaminants is loaded in the column. The negatively charged components are bound to the resin (LPS and acidic proteins), while the non-negatively charged proteins are separated by size exclusion chromatography. Protein images were generated with Illustrate (Goodsell, D. S., Autin, L., & Olson, A. J. *Structure*, 27(11), 1716–1720.e1, 2019, doi: 10.1016/j.str.2019.08.011), using data from PDB ID: 1GPE (Wohlfahrt, G., Witt, S., Hendle, J., Schomburg, D., Kalisz, H. M., & Hecht, H. J. *Acta Crystallographica Section D: Biological Crystallography*, 55(5), 969–977, 1999, doi: 0.1107/S0907444999003431) and PDB ID: 1IGT (Harris, L. J., Larson, S. B., Hasel, K. W., & McPherson, A. *Biochemistry*, 36(7), 1581–1597, 1997, doi: 10.1021/bi962514+), hosted in PDB (http://www.rcsb.org) (Berman, H.M., Westbrook, J., Feng, Z., Gilliland, G., Bhat, T.N., Weissig, H., Shindyalov, I.N., Bourne, P.E., *Nucleic Acids Res.*, 28, 235–242, 2000).

under certain conditions (e.g., pH, temperature, and above the critical micellar concentration – CMC) and can form nanostructures (cylindrical or spherical micelles or vesicles) in aqueous solution. As result, these nanostructures can be split into micelle-rich and micelle-poor phases; these two phases are immiscible. In this case, this type of ATPS is named aqueous two-phase micellar systems – ATPMS (Liu et al., 1996). ATPMS can be considered low cost, simple and due to the large amount of water used, it can provide a mild environment for target biomolecules purification. Furthermore, the scale-up of these processes make these interesting for industrial applications (Molino et al., 2014), as well as contaminant removal strategies.

In aqueous solution, due to the amphipathic characteristics of LPS molecules, they can form different supramolecular aggregates such as micelles and vesicles. The addition of surfactants in aqueous solution in the presence of LPS allows interactions between the divalent cations and phosphate groups, as well as interactions between nonpolar lipid chains which are favoured, allowing the formation of mixed micelles, i.e., a combination of surfactant and LPS monomers in the aqueous environment. In ATPMS, these mixed micellar aggregates are thus partitioned to the micelle-rich phase, and as a result, the concentration of LPS removed for this phase is very high, which explains the good efficiency of this technique (Anspach, 2001; Lopes et al., 2013; Richter et al., 2011). In this sense, several groups have demonstrated the application of these processes for LPS removal (Boratyński and Szermer-Olearnik, 2017; Lopes et al., 2010, 2011, 2013, 2018; Rozkov et al., 2008; Schädlich et al., 2009; Teixeira-Pinto et al., 2019; Teodorowicz et al., 2017; Zhang et al., 2015).

Alternatively, ATPS can be used as promising platforms to produce enhanced biomolecules therapeutics, such as *E. coli*-based cell-free protein synthesis (CFPS), providing a special possibility to remove LPS prior to protein expression, due to the absence of live cells and its membrane-less environment. Pre-expression LPS removal from CFPS reagents could simplify downstream processing, potentially enabling on-demand production of unique biomolecules therapeutics. In this sense, Wilding et al. (2018) studied a strategy for eliminating *E. coli* endotoxins from cell lysates using a TX-114 two-phase extraction. This process was efficient, eliminating 95% of endotoxins after three steps while maintaining ≈30% protein synthesis capacity.

Remarkably, a single purification step employing $C_{10}E_4$/McIlvaine buffer+Li_2SO_4-based ATPMS, Teixeira-Pinto et al. (2019) obtained 97% recovery of green fluorescent protein (GFP) with a partitioning coefficient (K) of 9.85 with 98% of LPS removed (initial concentration from cell lysate

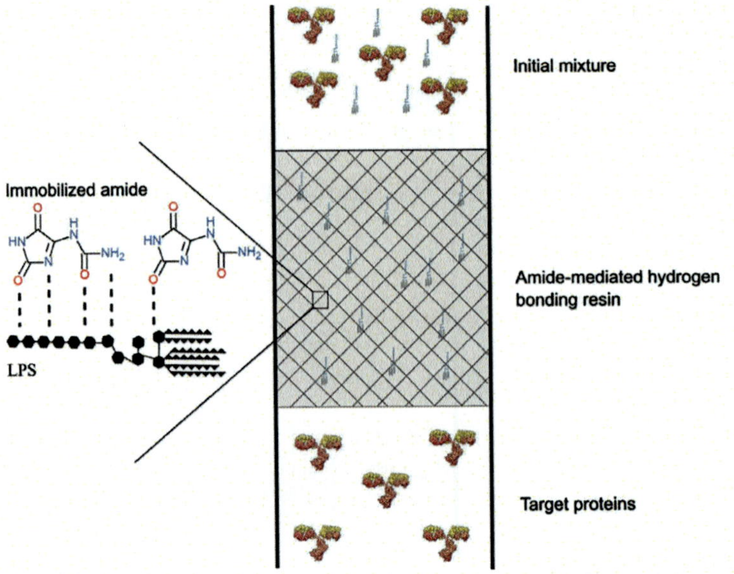

FIGURE 14.9 LPS removal from target proteins applying amide compounds. Amides immobilized in resins capture LPS contaminants forming hydrogen bond with LPS molecules. Protein images were generated with Illustrate (Goodsell, D. S., Autin, L., & Olson, A. J. *Structure*, 27(11), 1716-1720.e1, 2019, doi:10.1016/j.str.2019.08.011), using data from PDB ID: 1IGT (Harris, L. J., Larson, S. B., Hasel, K. W., & McPherson, A. *Biochemistry*, 36(7), 1581–1597, 1997, doi: 10.1021/bi962514+), hosted in PDB (http://www.rcsb.org) (Berman, H.M., Westbrook, J., Feng, Z., Gilliland, G., Bhat, T.N., Weissig, H., Shindyalov, I.N., Bourne, P.E., *Nucleic Acids Res.*, 28, 235–242, 2000).

of 2×10^7 EU/mL). This ATPMS condition could purify 12-fold the target biomolecule (GFP) from undesirable metabolites (i.e., contaminant proteins) present in the cell homogenate. These good values can be explained because both LPS aggregates (i.e., mixed micellar aggregates of LPS+$C_{10}E_4$ monomers) and contaminant proteins gave a greater hydrophobic character, and for this reason migrated preferentially into the $C_{10}E_4$ (micelle)-rich phase.

Table 14.1 summarizes some interesting approaches for the elimination of bacterial LPS applying surfactants or polymers.

Overall, surfactant-based ATPMS is very efficient for the removal of LPS, and because of the addition of adjuvants (i.e., electrolytes), these systems allow phases to separate at very mild temperatures, e.g., ≈14°C with the addition of 0.25 M Li_2SO_4, according to Teixeira-Pinto et al. (2019), Table 14.1. Although, this technique can be employed to extract thermosensitive biopharmaceuticals, some factors should be considered when employing ATPMS for LPS removal. For example, Teodorowicz et al. (2017) verified that the efficiency for the removal of LPS molecules did not increase with additional steps of TX-114-based ATPMS. The formation of protein–endotoxin complexes can explain this difficulty, as described before for different types of biomolecules (Chaby, 2004). The characteristics of the LPS are important, being essential to also know the physico-chemical characteristics of the target biomolecule to be purified, in order to anticipate possible interaction effects between its and the LPS molecules. Finally, other factors such as the cost of surfactants and/or electrolytes used to induce the phase separation, the conditions used (e.g., pH, temperature, number of steps), the viscosity of the surfactants, as well as purification steps to remove traces of surfactants in the sample containing the bioproduct should be also considered before using this technique for LPS removal.

14.6 FINAL CONSIDERATIONS

In recent years, all the major worldwide pharmaceutical regulatory agencies have increasingly required demonstration of methods for detection and quantification of LPS. LPS analysis is one of the main assays used in the quality control for production of injectables because of the risk LPS poses to human health. Different techniques have been tested, but the introduction of *Limulus* amebocyte lysate technique with improvements in the sensitivity of the method has meant this assay is now widely used in the quality control of parenteral and biological products, as well as medical devices for human and animal use. Regarding the current techniques employed for LPS removal, changes in biopharmaceutical formulations (e.g., nanotechnology systems) will require critical revision of existing methods, so that formulations and extraction methods are compatible and effective for LPS removal. Although biomolecules purification/separation/extraction methodologies may decrease the LPS concentration below regulatory requirement levels, a complete guarantee of their removal is utmost difficulty. Cost-effective and efficient LPS removal is possibly one of the most challenging and complex steps in manufacturing biopharmaceutical formulations. Although the development of innovative methods has been reported in recent years, further investigation is needed in this area including the possibility of studies on genetic modification of Gram-negative bacteria as heterologous hosts without LPS expression in their outer membrane.

TABLE 14.1
Summary of Biomolecules' Partitioning (*K*), Recovery (REC_{bio}) and Bacterial LPS Removal (REM_{LPS}) Parameters Achieved by Different Liquid-Liquid Extraction (Using ATPMS or ATPS)

Surfactants-Based Systems	*K*	REC_{bio}		REM_{LPS}	References
2% (v/v) TX-114 for 10 min at 37°C + centrifugation for 20 min at 20,000×*g* at 37°C	-	-		99.9% for β-lactoglobulin 99.2% for soy protein extract	Teodorowicz et al. (2017)
1% (v/v) TX-114 + centrifugation for 10 min at 9,000×*g* at room temperature	-	95%		>98%	Zhang et al. (2015)
1 wt% TX-114 + anion-exchange chromatography and ultrafiltration	-	58% (at the end of four purification steps)		99% (at the end of four cycles with TX-114)	Rozkov et al. (2008)
1 wt% TX-114 for 5 min at 37°C + centrifugation for 1 min at 16,000×*g*	-	86% (after first purification step)	83% (after second purification step)	>99%	Schädlich et al. (2009)
8 wt% TX-114 at 33°C	5.0	-		99.1%	Lopes et al. (2010)
4 wt% TX-114 at 48°C–60°C	5.3-15	96%–118%		99.9%	Lopes et al. (2011)
8 wt% TX-114 at 37.3°C	2.9	-		99.9%	Lopes et al. (2013)
2 wt% $C_{10}E_4$/buffer + 0.25 M Li_2SO_4 at 13.75°C	9.8	97%		98%	Teixeira-Pinto et al. (2019)
Polymers-Based Systems					
12 wt% PEG + 12 wt% NaPA + 0.25 M Li_2SO_4	28.3	83% (PEG-rich phase)		13% (NaPA-rich phase)	Lopes et al. (2018)

-, results not presented or calculated by the reseaerchers.
PEG, polyethylene glycol; NaPA, sodium polyacrylate; TX-114, Triton X-114.

REFERENCES

Anspach, F.B. Endotoxin removal by affinity sorbents. *Journal of Biochemical and Biophysical Methods*, v.49, n.1–3, 665–681, 2001.

Anspach, F.B; Birger, F.; Birger, A.; Dagmar, P.; Wolf-Dieter, D. Model studies on the mechanism of endotoxin adsorption on flat-sheet microfiltration membrane adsorbers. *The Canadian Journal of Chemical Engineering*, v.77, n.5, 921–930, 1999.

Anspach, F.B.; Hilbeck, O. Removal of endotoxins by affinity sorbents. *Journal of Chromatography A*, v.711, n.1, 81–92, 1995.

ANVISA. Agência Nacional de Vigilância Sanitária. Brasília, Brazil: Farmacopeia Brasileira, 2010.

Baker, R. W. Microfiltration. in Baker R. W. *Membrane Technology and Applications* (3rd ed) Chichester: John Wiley and Sons Ltd, 2012, 303–324. doi:10.1002/9781118359686.ch7.

Bangham, D.R. Relevance and standardization in pyrogen tests. *Journal de Pharmacie de Belgique*, v.34, n.3, 134–136, 1979.

Batas, B.; Schiraldi, C.; Chaudhuri, J. B. Inclusion body purification and protein refolding using microfiltration and size exclusion chromatography. *Journal of Biotechnology*, v.68, n.2–3, 149–158, 1999.

Benesova, O. Repeated use of rabbits for the pharmacopeial pyrogenicity test. *Ceskoslovenska farmacie*, v.19, n.6, 207–209, 1970.

Berman, H.M.; Westbrook, J.; Feng, Z.; Gilliland, G.; Bhat, T.N.; Weissig, H.; Shindyalov, I.N.; Bourne, P.E. RCSB protein data bank: Structural biology views for basic and applied research. *Nucleic Acids Research*, v.28, 235–242, 2000.

Beutler, B.; Rietschel, E.T. Innate immune sensing and its roots: The story of endotoxin. *Nature Reviews Immunology*, v.3, 169–176, 2003. doi: 10.1038/nri1004.

Blumberg, S.; Schechter, I.; Berger, A. The purification of papain by affinity chromatography. *European Journal of Biochemistry/FEBS*, v.15, n.1, 97–102, 1970.

Boratyński, J.; Szermer-Olearnik, B. Endotoxin removal from Escherichia coli bacterial lysate using a biphasic liquid system, in: Holst, O. *Microbial Toxins: Methods and Protocols*. New York: Springer, 2017, 107–112. doi:10.1007/978-1-4939-6958-6_10.

Brandenburg, K.; Wiese, A. Endotoxins: Relationships between structure, function, and activity. *Current Topics in Medicinal Chemistry*, v.4, 1127–46, 2004. doi: 10.2174/1568026043388213.

Braun, H.A.; Klein, V.V. Some problems of pyrogen testing. *PDA Bulletin of Parenteral Drug Association*, v.14, 9–15, 1960.

Caroff, M.; Karibian, D. Structure of bacterial lipopolysaccharides. *Carbohydrate Research*, v.338, 2431–2447, 2003. doi: 10.1016/j.carres.2003.07.010.

Chaby, R. Lipopolysaccharide-binding molecules: Transporters, blockers and sensors. *Cellular and Molecular Life Sciences*, v.61, n.14, 1697–713, 2004.

Chen, X.; Howe, J.; Andrä, J.; Rössle, M.; Richter, W.; Da Silva, A.P.G.; Krensky, A.M.; Clayberger, C.; Brandenburg, K. Biophysical analysis of the interaction of granulysin-derived peptides with enterobacterial endotoxins. *Biochimica et biophysica acta*, v.1768, 2421–2431, 2007. doi: 10.1016/j.bbamem.2007.05.001.

Choi, J.H.; Keum, K.C.; Lee, S.Y. Production of recombinant proteins by high cell density culture of Escherichia coli. *Chemical Engineering Science*, v.61, n.3, 876–885, 2006.

Cooper, J.F. Bacterial endotoxins test, in: Prince, R. *Microbiology in Pharmaceutical Manufacturing*. Bethesda, MD: DHI, 2008, 245–274.

Cooper, J.F. Resolving LAL test interferences. *Journal of Parenteral Science and Technology*, v.44, n.1, 13–15, 1990.

Cooper, J.F.; Hochstein, H.D.; Seligmann, Jr, E.B. The Limulus test for endotoxin (Pyrogen) in radiopharmaceuticals and biologicals. *Bulletin of the Parenteral Drug Association*, v.26, n.4, 153–162, 1972.

Cooper, J.F.; Weary, M.E.; Jordan, F.T. The impact of non-endotoxin LAL-reactive materials on limulus amebocyte lysate analyses. *PDA Journal of Pharmaceutical Science and Technology*, v.51, n.1, 2–6, 1997.

Correa, W.; Brandenburg, K.; Zähringer, U.; Ravuri, K.; Khan, T.; Von Wintzingerode, F. Biophysical analysis of lipopolysaccharide formulations for an understanding of the low endotoxin recovery (LER) phenomenon. *International Journal of Molecular Sciences*, v.18, 2737, 2017.

Dare, J.G.; Mogey, G.A. Rabbit responses to human threshold doses of a bacterial pyrogen. *Journal of Pharmacy and Pharmacology*, v.6, n.5, 325–332, 1954.

Datta, S.; Redner, S. Gradient clogging in depth filtration. *Physical Review E*, v.58, n.2, R1203–1206, 1998.

De Mas, N.; Kientzler, D.C.; Kleindienst, D. Endotoxin removal from a small-molecule aqueous drug substance using ultrafiltration: A case study. *Organic Process Research & Development*, v.19, n.9, 1293–1298, 2015.

De Sousa, Jr, F.C.; Ribeiro, V.T.; Chibério, A.S.; Da Mata Costa, L.P.; De Araújo Padilha, C.E.; Martins, D.R.A.; Dos Santos, E.S. Simultaneous recombinant 503 antigen recovery and endotoxin removal from E. coli M15 homogenate using expanded bed adsorption chromatography. *Separation Science and Technology*, v.52, n.11, 1869–1875, 2017.

Ding, J.L.; Zhu, Y.; Ho, B. High-performance affinity capture-removal of bacterial pyrogen from solutions. *Journal of Chromatography. B, Biomedical Sciences and Applications*, v.759, n.2, 237–246, 2001.

Diogo, M.M.; Queiroz, J.A.; Monteiro, G.A.; Martins, S.A.; Ferreira, G.N.; Prazeres, D.M. Purification of a cystic fibrosis plasmid vector for gene therapy using hydrophobic interaction chromatography. *Biotechnology and Bioengineering*, v.68, n.5, 576–583, 2000.

Dixon, D.R.; Darveau, R.P. Lipopolysaccharide heterogeneity: Innate host responses to bacterial modification of lipid A structure. *Journal of Dental Research*, v.84, 584–595, 2005. doi: 10.1177/154405910508400702.

Donnell, M.L.; Lyon, A.J.; Mormile, M.R.; Barua, S. Endotoxin hitchhiking on polymer nanoparticles. *Nanotechnology*, v.27, n.28, 285601, 2016.

Duarte, A.W.F.; Lopes, A.M.; Molino, J.V.D.; Pessoa, A.; Sette, L.D. Liquid–liquid extraction of lipase produced by psychrotrophic yeast Leucosporidium scottii L117 using aqueous two-phase systems. *Separation & Purification Technology*, v.156, 215–225, 2015.

Elass-Rochard, E.; Roseanu, A.; Legrand, D.; Trif, M.; Salmon, V.; Motas, C.; Montreuil, J.; Spik, G. Lactoferrin-lipopolysaccharide interaction: Involvement of the 28–34 loop region of human lactoferrin in the high-affinity binding to Escherichia coli 055B5 lipopolysaccharide. *Biochemical Journal*, v.312, n.Pt.3, 839–845, 1995.

Falchi, F.A.; Maccagni, E.A.; Puccio, S.; Peano, C.; De Castro, C.; Palmigiano, A.; Garozzo, D.; Martorana, A.M.; Polissi, A.; Dehò, G.; Sperandeo, P. Mutation and suppressor analysis of the essential lipopolysaccharide transport protein LptA reveals strategies to overcome severe outer membrane permeability defects in Escherichia coli. *Journal of Bacteriology*, v.200, e00487–17, 2018.

Fanali, S.; Haddad, P.R.; Poole, C.; Riekkola, M.L. *Liquid Chromatography: Fundamentals and Instrumentation*. New York: Elsevier, 2017.

Fraseur, J.G.; Kinzer-Ursem, T.L. Next generation calmodulin affinity purification: Clickable calmodulin facilitates improved protein purification. *PLoS One*, v.13, n.6, p.e0197120, 2018.

Freitas, S.S.; Santos, J.A.L.; Prazeres, D.M.F. Plasmid purification by hydrophobic interaction chromatography using sodium citrate in the mobile phase. *Separation & Purification Technology*, v.65, n.1, 95–104, 2009.

Gemma, S.; Molteni, M.; Rossetti, C. Lipopolysaccharides in cyanobacteria: A brief overview. *Advances in Microbiology*, v.6, 391–397, 2016. doi: 10.4236/aim.2016.65038.

Gerba, C.P.; Hou, K. Endotoxin removal by charge-modified filters. *Applied and Environmental Microbiology*, v.50, n.6, 1375–1377, 1985.

Gnauck, A.; Lentle, R.G.; Kruger, M.C. The characteristics and function of bacterial lipopolysaccharides and their endotoxic potential in humans. *International Reviews of Immunology*, v.35, 189–218, 2016. doi: 10.3109/08830185.2015.1087518.

Goldbaum, F.A.; Rubbi, C.P.; Fossati, C.A. Removal of LPS from a Brucella cytoplasmic fraction by affinity chromatography with an anti-LPS monoclonal antibody as immunosorbent. *Journal of Medical Microbiology*, v.40, n.3, 174–178, 1994.

Goodsell, D. S., Autin, L., Olson, A. J. Illustrate: Software for Biomolecular Illustration. *Structure*, v.27, n.11, 1716–1720. e1, 2019. doi:10.1016/j.str.2019.08.011.

Grandics, P.; Szathmary, S.; Hegyi, E. Method for removing pyrogens from dialysate. USPO 6,106,723. Patent, filed February 10, 1997, and issued August 22, 2000.

Grodzki, A.C.; Berenstein, E. Antibody purification: Affinity chromatography – protein A and protein G sepharose. *Methods in Molecular Biology*, v.588, 33–41, 2009.

Gyurova, A.Y.; Stoyanov, S.V.; Mileva, E. Capture of LPS traces in aqueous solutions by tectomers of four-antennary oligoglycines. *Colloids and Surfaces A: Physicochemical and Engineering Aspects*, v.520, 914–921, 2017.

Hage, D.S.; Matsuda, R. Affinity chromatography: A historical perspective. *Methods in Molecular Biology*, v.1286, 1–19, 2015.

Haishima, Y.; Hasegawa, C.; Yagami, T.; Tsuchiya, T.; Matsuda, R.; Hayashi, Y. Estimation of uncertainty in kinetic-colorimetric assay of bacterial endotoxins. *Journal of Pharmaceutical and Biomedical Analysis*, v.32, 495–503, 2003.

Harris, L. J., Larson, S. B., Hasel, K. W., McPherson, A. Refined structure of an intact IgG2a monoclonal antibody. *Biochemistry*, v.36, n.7, 1581–1597, 1997. doi:10.1021/bi962514+.

Hirayama, C.; Sakata, M. Chromatographic removal of endotoxin from protein solutions by polymer particles. *Journal of Chromatography. B, Analytical Technologies in the Biomedical and Life Sciences*, v.781, n.1–2, 419–432, 2002.

Hochuli, E. Large-scale chromatography of recombinant proteins. *Journal of Chromatography. A*, v.444, 293–302, 1988.

Hou, K., Gerba, C. P., Goyal, S. M., & Zerda, K. S. Capture of latex beads, bacteria, endotoxin, and viruses by charge-modified filters. *Applied and Environmental Microbiology*, 40(5), 892–896, 1980.

Hou, K.C.; Zaniewski, R. Endotoxin removal by anion-exchange polymeric matrix. *Biotechnology and Applied Biochemistry*, v.12, n.3, 315–324, 1990.

Hubbard, H.L.; Eller, T.D.; Mais, D.E.; Halushka, P.V.; Baker, I. A.; Blair, R.H.; Vrbanac, J.J.; Knapp, D. R. Extraction of thromboxane B2 from urine using an immobilized antibody column for subsequent analysis by gas chromatography-mass spectrometry. *Prostaglandins*, v.33, n.2, 149–160, 1987.

Iqbal, M.; Tao, Y.; Xie, S.; Zhu, Y.; Chen, D.; Wang, X.; Hussain, H.I. Aqueous two-phase system (ATPS): An overview and advances in its applications. *Biological Procedures Online*, v.18, n.1, 18, 2016.

Iuliano, S.; Fisher, J.R.; Chen, M.; Kelly, W.J. Rapid analysis of a plasmid by hydrophobic-interaction chromatography with a non-porous resin. *Journal of Chromatography A*, v.972, n.1, 77–86, 2002.

Jang, H.; Kim, H.-S.; Moon, S.C.; Lee, Y.-R.; Yu, K.Y.; Lee, B.K.; Youn, H.Z. Effects of protein concentration and detergent on endotoxin reduction by ultrafiltration. *BMB Reports*, v.42, n.7, 462–466, 2009.

Johnson, R.J.; Sautin, Y.Y.; Oliver, W.J.; Roncal, C.; Mu, W.; Sanchez-Lozada, L.G.; Rodriguez-Iturbe, B.; Nakagawa, T.; Benner, S.A. Lessons from comparative physiology: Could uric acid represent a physiologic alarm signal gone awry in western society? *Journal of Comparative Physiology. B, Biochemical, Systemic, and Environmental Physiology*, v.179, n.1, 67–76, 2009.

Jungbauer, A.; Hahn, R. Ion-Exchange Chromatography. *Methods in Enzymology*, v.463, 349–71, 2009.

Jungbauer, A.; Hahn, R. Monoliths for fast bioseparation and bioconversion and their applications in biotechnology. *Journal of Separation Science*, v.27, n.10–11, 767–778, 2004.

Kabanov, D.S.; Prokhorenko, I.R. Structural analysis of lipopolysaccharides from Gram-negative bacteria. *Biochemistry*, v.75, 383–404, 2010. doi: 10.1134/S0006297910040012.

Karplus, T.E.; Ulevitch, R.J.; Wilson, C.B. A new method for reduction of endotoxin contamination from protein solutions. *Journal of Immunological Methods*, v.105, n.2, 211–220, 1987.

Klein, G.; Lindner, B.; Brade, H.; Raina, S. Molecular basis of lipopolysaccharide heterogeneity in Escherichia coli: Envelope stress-responsive regulators control the incorporation of glycoforms with a third 3-deoxy-α-D-manno-oct-2-ulosonic acid and rhamnose. *Journal of Biological Chemistry*, v.286, 42787–42807, 2011. doi: 10.1074/jbc.M111.291799.

Kurnik, R.T.; Yu, A.W.; Blank, G.S.; Burton, A.R.; Smith, D.; Athalye, A.M.; Van Reis, R. Buffer exchange using size exclusion chromatography, countercurrent dialysis, and tangential flow filtration: Models, development, and industrial application. *Biotechnology and Bioengineering*, v.45, n.2, 149–57, 1995.

Larive, C.K.; Lunte, S.M.; Zhong, M.; Perkins, M.D.; Wilson, G.S.; Gokulrangan, G.; Williams, T. Separation and analysis of peptides and proteins. *Analytical Chemistry*, v.71, n.12, 389R–423R, 1999.

Li, L.; Luo, R.G. Use of Ca+2 to re-aggregate lipopolysaccharide (LPS) in hemoglobin solutions and the subsequent removal of endotoxin by ultrafiltration. *Biotechnology Techniques*, v.12, n.2, 119–22, 1998.

Lin, M.F.; Williams, C.; Murray, M.V.; Ropp, P.A. Removal of lipopolysaccharides from protein–lipopolysaccharide complexes by nonflammable solvents. *Journal of Chromatography B*, v.816, n.1–2, 167–174, 2005.

Liu, C.L.; Nikas, Y.J.; Blankschtein, D. Novel bioseparations using two-phase aqueous micellar systems. *Biotechnology and Bioengineering*, v.52, n.2, 185–192, 1996.

Lopes, A.M.; Magalhães, P.O.; Mazzola, P.G.; Rangel-Yagui, C.O.; Carvalho, J.C.; Penna, T.C.; Pessoa, Jr, A. Green fluorescent protein extraction and LPS removal from Escherichia coli fermentation medium using aqueous two-phase micellar system. *Separation and Purification Technology*, v.81, n.3, 339–346, 2011.

Lopes, A.M.; Magalhães, P.O.; Mazzola, P.G.; Rangel-Yagui, C.O.; Carvalho, J.C.; Penna, T.C.; Pessoa, Jr, A. LPS removal from an E. coli fermentation broth using aqueous two-phase micellar system. *Biotechnology Progress*, v.26, n.6, 1644–1653, 2010.

Lopes, A.M.; Molino, J.V.D.; Santos-Ebinuma, V.C.; Pessoa, Jr, A.; Valentini, S.R.; Pereira, J.F.B. Effect of electrolytes as adjuvants in GFP and LPS partitioning on aqueous two-phase systems 1. *Polymer-Polymer Systems. Separation and Purification Technology*, v.206, 39–49, 2018.

Lopes, A.M.; Santos-Ebinuma, V.C.; Lencastre-Novaes, L.C.; Molino, J.V.D.; Barbosa, L.R.S.; Pessoa, A.; Rangel-Yagui, C.O. LPS–protein aggregation influences protein partitioning in aqueous two-phase micellar systems. *Applied Microbiology and Biotechnology*, v.97, n.14, 6201–6209, 2013.

Lopez-Campistrous, A.; Semchuk, P.; Burke, L.; Palmer-Stone, T.; Brokx, S.J.; Broderick, G.; Ellison, M.J. Localization, annotation, and comparison of the Escherichia coli K-12 proteome under two states of growth. *Molecular & Cellular Proteomics*, v.4, n.8, 1205–1209, 2005.

Lourenço, F.R.; Botelho, T.S.; Pinto, T.J.A. How pH, temperature, ad time of incubation affect false-positive responses and uncertainty of the LAL Gel-Clot Test. *PDA Journal of Pharmaceutical Sciences and Technology*, v.66, 542–546, 2012.

Mack, L.; Brill, B.; Delis, N.; Groner, B. Endotoxin depletion of recombinant protein preparations through their preferential binding to histidine tags. *Analytical Biochemistry*, v.466, 83–88, 2014. doi: 10.1016/j.ab.2014.08.020.

Magalhães, P.O.; Lopes, A.M.; Mazzola, P.G.; Rangel-Yagui, C.; Penna, T.C.V.; Pessoa, A. Methods of endotoxin removal from biological preparations: A review. *Journal of Pharmacy & Pharmaceutical Sciences*, v.10, 388–404, 2007.

McCue, J.T. Chapter 25: Theory and use of hydrophobic interaction chromatography in protein purification applications, in: Burgess R. R. & Deutscher M. P. *Methods in Enzymology*, Guide to Protein Purification. 2nd ed. San Diego, CA: Academic Press. 2009, 405–414. doi:10.1016/S0076-6879(09)63025-1.

Mokhtar, G.; Naoyuki, F. Microfiltration, nano-filtration and reverse osmosis for the removal of toxins (LPS endotoxins) from wastewater. *Journal of Membrane Science & Technology*, v.2, n.3, 118, 2012. doi:10.4172/2155-9589.1000118.

Molino, J.V.; Feitosa, V.A.; Lencastre-Novaes, L.C.; Santos-Ebinuma, V.C.; Lopes, A.M.; Jozala, A.F.; Marques, D.A.V.; Malpiedi, L.P.; Pessoa, Jr, A. Biomolecules extracted by ATPS: Practical examples. *Revista Mexicana de Ingeniería Química*, v.13, n.2, 359–377, 2014.

Montbriand, P.M.; Malone, R.W. Improved method for the removal of endotoxin from DNA. *Journal of Biotechnology*, v.44, n.1–3, 43–46, 1996.

Morrison, D.C.; Jacobs, D.M. Binding of polymyxin B to the lipid A portion of bacterial lipopolysaccharides. *Immunochemistry*, v.13, n.10, 813–18, 1976.

Munson, T.E. Guideline for validation of the LAL test as an end-product endotoxin test for human and biological drug products. *Progress in Clinical and Biological Research*, v.189, 211–220, 1985.

Nian, R.; Chuah, C.; Lee, J.; Gan, H.T.; Latiff, S.M.A.; Lee, W.Y.; Vagenende, V.; Gagnon, P. Void exclusion of antibodies by grafted-ligand porous particle anion exchangers. *Journal of Chromatography A*, v.1282, 127–132, 2013.

Obeng, E.M.; Dullah, E.C.; Abdul Razak, N.S.; Danquah, M.K.; Budiman, C.; Ongkudon, C.M. Elucidating endotoxin-biomolecule interactions with FRET: Extending the frontiers of their supramolecular complexation. *Journal of Biological Methods*, v.4, 71, 2017. doi: 10.14440/jbm.2017.172.

Ongkudon, C.M.; Chew, J.H.; Liu, B.; Danquah, M.K. Chromatographic removal of endotoxins: a bioprocess engineer's perspective. *ISRN Chromatography*, v.2012, 1–9, 2012.

Ongkudon, C.M.; Ho, J.; Danquah, M.K. Mitigating the looming vaccine crisis: Production and delivery of plasmid-based vaccines. *Critical Reviews in Biotechnology*, v.31, n.1, 32–52, 2011.

Ostronoff, C.S.; Lourenço, F.R. Measurement uncertainty of chromogenic LAL assays: Reaction time and proportion of endotoxin and LAL reagent affect release of p-nitroaniline. *Journal of AOAC International*, v.98, n.1, 51–55, 2015.

Page, M.G.P. The role of the outer membrane of gram-negative bacteria in antibiotic resistance: Ajax' shield or Achilles' heel? in Coates, A. R. M. *Handbook of Experimental Pharmacology*, Berlin, Heidelberg: Springer, v.211, 67–86, 2012. doi: 10.1007/978-3-642-28951-4-5.

Pearson, F.C. *Pyrogens: Endotoxins, LAL testing and depyrogenation*. New York: Marcel Dekker, 1985.

Petsch, D.; Anspach, F.B. Endotoxin removal from protein solutions. *Journal of Biotechnology*, v.76, 97–119, 2000.

Petsch, D.; Rantze, E.; Anspach, F.B. Selective adsorption of endotoxin inside a polycationic network of flat-sheet microfiltration membranes. *Journal of Molecular Recognition*, v.11, n.1–6, 222–230, 1998.

Pinto, T.J.A.; Kaneko, T.M.; Pinto, A.F. *Controle biológico de qualidade de produtos farmacêuticos, correlatos e cosméticos*. 4th ed. Barueri: Manole, 2015.

Prather, K.J.; Sagar, S.; Murphy, J.; Chartrain, M. Industrial scale production of plasmid DNA for vaccine and gene therapy: Plasmid design, production, and purification. *Enzyme and Microbial Technology*, v.33, n.7, 865–883, 2003.

Raetz, C.R.H.; Reynolds, C.M.; Trent, M.S.; Bishop, R.E. Lipid A modification systems in gram-negative bacteria. *Annual Review of Biochemistry*, v.76, 295–329, 2007. doi: 10.1146/annurev.biochem.76.010307.145803.

Raetz, C.R.H.; Whitfield, C. Lipopolysaccharide endotoxins. *Annual Review of Biochemistry*, v.71, 635–700, 2002. doi: 10.1146/annurev.biochem.71.110601.135414.

Raviv, U. Interacting bacteria surfaces. *Biophysical Journal*, v.114, 1515–1517, 2018. doi: 10.1016/j.bpj.2018.03.010.

Reichelt, P.; Schwarz, C.; Donzeau, M. Single step protocol to purify recombinant proteins with low endotoxin contents. *Protein Expression and Purification*, v.46, n.2, 483–488, 2006.

Rezania, S.; Amirmozaffari, N.; Tabarraei, B.; Jeddi-Tehrani, M.; Zarei, O.; Alizadeh, R.; Masjedian, F.; Zarnani, A.H. Extraction, purification and characterization of lipopolysaccharide from Escherichia coli and Salmonella typhi. *Avicenna Journal of Medical Biotechnology*, v.3, n.1, 3–9, 2011.

Richter, W.; Vogel, V.; Howe, J.; Steiniger, F.; Brauser, A.; Koch, M.H.; Brandenburg, K. Morphology, size distribution, and aggregate structure of lipopolysaccharide and lipid A dispersions from enterobacterial origin. *Innate immunity*, v.17, n.5, 427–438, 2011.

Rozkov, A.; Larsson, B.; Gillström, S.; Björnestedt, R.; Schmidt, S.R. Large-scale production of endotoxin-free plasmids for transient expression in mammalian cell culture. *Biotechnology and bioengineering*, 99(3), 557–566, 2008.

Schädlich, L.; Senger, T.; Kirschning, C. J.; Müller, M.; Gissmann, L. Refining HPV 16 L1 purification from E. coli: reducing endotoxin contaminations and their impact on immunogenicity. *Vaccine*, v.27, n.10, 1511–1522, 2009.

Schindler, M.; Osborn, M.J. Interaction of divalent cations and polymyxin B with lipopolysaccharide. *Biochemistry*, v.18, n.20, 4425–4430, 1979.

Starkenstein, E. Ferment action and the influence upon it of neutral salts. *Biochemische Zeitschrift*, v.24, n.210, 1910.

Sulkowski, E. 1985. Purification of proteins by IMAC. *Trends in Biotechnology*, v.3, n.1, 1–7, 1985.

Sweadner, K.J.; Forte, M.; Nelsen, L.L. Filtration removal of endotoxin (pyrogens) in solution in different states of aggregation. *Applied and Environmental Microbiology*, v.34, n.4, 382–385, 1977.

Teixeira-Pinto, R.G.R.; Molino, J.V.D.; Santos-Ebinuma, V.C.; Pessoa JR, A.; Valentini, S.R.; Pereira, J.F.B.; Lopes, A.M. Effect of electrolytes as adjuvants in GFP and LPS partitioning on aqueous two-phase systems: 2. Nonionic micellar systems. *Separation and Purification Technology*, v.210, 69–79, 2019.

Teodorowicz, M.; Perdijk, O.; Verhoek, I.; Govers, C.; Savelkoul, H.F.; Tang, Y.; Wichers, H.; Broersen, K. Optimized Triton X-114 assisted lipopolysaccharide (LPS) removal method reveals the immunomodulatory effect of food proteins. *PLoS One*, v.12, n.3, p. e0173778, 2017.

Teramoto, K.; Nakamoto, Y.; Kunitomo, T.; Shoji, H.; Tani, T.; Hanazawa, K.; Kodama, M. Removal of endotoxin in blood by polymyxin B immobilized polystyrene-derivative fiber. *Therapeutic Apheresis*, v.6, n.2, 103–108, 2002.

Thornfeldt, C. Cosmeceuticals containing herbs: Fact, fiction, and future. *Dermatologic Surgery*, v.31, 873–881, 2005.

USP. Bacterial endotoxins test, in: *United States Pharmacopeia 39*. Rockville, MD: U.S. Pharmacopeial Convention, 2015, 161–167.

USP. *United States Pharmacopeia*. Rockville, MD: United States Pharmacopeia Convention, 2017.

Vagenende, V.; Ching, T.J.; Chua, R.J.; Thirumoorthi, N.; Gagnon, P. Amide-mediated hydrogen bonding at organic crystal/water interfaces enables selective endotoxin binding with picomolar affinity. *ACS Applied Materials & Interfaces*, v.5, n.10, 4472–4478, 2013.

Walsh, G. Biopharmaceutical benchmarks. *Nature Biotechnology*, v.32, n.10, 992–1000, 2014.

Weary, M.; Baker, B. Utilization of the limulus amebocyte lysate test for pyrogen testing large volume parenterals, administration sets, and medical devices. *Bulletin of the Parenteral Drug Association*, v.31, n.3, 127, 1977.

White, D. Automated purification of transfection-grade plasmid DNA using wizard MagneSil Tfx system. *Journal of the Association for Laboratory Automation*, v.8, n.4, 50–53, 2003.

Wilding, K.M.; Hunt, J.P.; Wilkerson, J.W.; Funk, P.J.; Swensen, R.L.; Carver, W.C.; Christian, M.L.; Bundy, B.C. Endotoxin-free E. coli-based cell-free protein synthesis: Pre-expression endotoxin removal approaches for on-demand cancer therapeutic production. *Biotechnology Journal*, 1800271, 2018.

Williams, K.L. *Endotoxin: Relevance and Control in Parenteral Manufacturing*, Jimenez, L. (Ed.). New York: Marcel Dekker, 2004, 183–249.

Wilson, M. J., Haggart, C. L., Gallagher, S. P., & Walsh, D. (2001). Removal of tightly bound endotoxin from biological products. *Journal of Biotechnology*, 88(1), 67–75. https://doi.org/10.1016/S0168-1656(01)00256-5

Wohlfahrt, G., Witt, S., Hendle, J., Schomburg, D., Kalisz, H. M., & Hecht, H. J. (1999). 1.8 and 1.9 Å resolution structures of the Penicillium amagasakiense and Aspergillus niger glucose oxidases as a basis for modelling substrate complexes. *Acta Crystallographica Section D: Biological Crystallography*, 55(5), 969–977. https://doi.org/10.1107/S0907444999003431

Wu, X.; Kinsey, J.L.; Ishee, M.; Konstantin, P.; Shertok, J.; Yang, Y. Charged membrane (WIPO 2000069549A1). World Patent, 2000. https://patents.google.com/patent/WO2000069549A1/en.

Yigzaw, Y.; Piper, R.; Tran, M.; Shukla, A.A. Exploitation of the adsorptive properties of depth filters for host cell protein removal during monoclonal antibody purification. *Biotechnology Progress*, v.22, n.1, 288–296, 2006.

Yuan, Z.; Yu, M.; Li, J.; Hou, G.; Wang, H. Endotoxin adsorbent using dimethylamine ligands. *Biomaterials*, v.26, n.15, 2741–2747, 2005.

Zayed, A.; Muffler, K.; Hahn, T.; Rupp, S.; Finkelmeier, D.; Burger-Kentischer, A.; Ulber, R. Physicochemical and biological characterization of fucoidan from fucus vesiculosus purified by dye affinity chromatography. *Marine Drugs*, v.14, n.4, 2016. https://doi.org/10.3390/md14040079.

Zhang, J.; Zhu, C.; Fan, D.; Ma, X.; Mi, Y.; Xue, W. A Two-step protocol to remove endotoxins from human-like collagen. *Separation Science and Technology*, v.50, n.7, 993–1001, 2015.

15 Enzymes
The Catalytic Proteins

Michele Vitolo
Universidade de São Paulo

CONTENTS

15.1 INTRODUCTION

The evolution of enzyme technology can be divided in four phases, i.e., empiric, descriptive, quantitative and applied with planning.

The empiric phase – from 4000/3000 BC up to the end of the 18th century – involved the use of enzymes in high-scale processes (tanning, cheese production, for instance) without any knowledge on what kind of agent would be responsible for the process.

The descriptive phase occurred throughout the 19th century and corresponded to a period in which enzyme activity began to be studied with more attention and methodology. For example, it was observed that aqueous extracts of poultry could decompose animal meat; aqueous extracts of cereals could decompose starch such as ptyalin (present in human saliva), which was capable of liquefying pieces of bread; aqueous extract of baker's yeast was capable of converting sucrose into inverted sugar syrup.

The quantitative phase, which refers to the establishment of mathematical models for quantifying enzyme activity, began at the end of 19th century, with the first quantitative model – known as Michaelis-Menten's model – established in 1913 and improved by Briggs and Haldane in 1926. In the same year, Sumner crystallized urease and concluded that it was a protein. The protein character of enzymes became clear during the 20th century as new enzymes were discovered, described and chemically characterized.

The applied with planning phase, in spite of its beginning in the first decade of the 20th century – with the use of amylase and invertase for the hydrolysis of starch and sucrose, respectively – had renaissance during the 1940s when new enzymes appeared in the market and basic knowledge about them was improving.

It must be borne out that all known enzymes are of protein in nature, but not all protein has catalytic capability. Recently, it was discovered that some ribonucleic acids, the so-called ribozymes, also have catalytic capacity. Their role is circumvented to the modification of some kinds of mRNA before translation of proteins at the ribosome. Ribozymes can be employed in gene therapy (Purich, 2010).

Like all proteins, enzymes are polymers of amino acids linked through peptic bonds. The interactions amongst chemical groups present as side chains of the amino acids that constitute the enzyme can lead the protein to acquire higher order tertiary structure, which allows the enzyme to interact with some other molecules (substrates, co-factors and alike), whose structures are modified to allow the biochemical reaction to proceed.

15.2 ENZYME SPECIFICITY

In enzyme-catalyzed reactions, enzymes, in general, specifically accept one substance as the substrate. Even when an enzyme catalyzes the conversion of more than one substrate, the reaction rates are different. For instance, glucoamylase catalyzes the breakdown of maltose at maximum rate whereas the activity of this enzyme on nigerose and isomaltose is only 7% and 4%, respectively (Kulp, 1975).

Substrate specificity is linked to a region of the enzyme molecule called the active site. The substrate – a reagent which will be modified by the enzyme – fits into the active site, and it is transformed into another compound (called the product). There are two theories explaining enzyme-substrate interactions. One of them – the Key-Lock theory proposed by Fisher at the beginning of the 20th century– states that the substrate (key) enters the active site (lock) following its modification. This theory requires that two conditions must be met, i.e., complementarily between the substrate and the active site and that the substrate and the active site must have compatible polarity and size. The other theory proposed by Koshland during the 1960s is called the Theory of Induced Fitting – proposes that the structure of the active site would be a region of the enzyme molecule susceptible to conformational modification as the substrate nears it, so facilitating the interaction between them. Essentially, both theories only differ on the particular characteristics attributed to the active site, i.e., rigid envisaged by Fisher or flexible as pictured by Koshland.

Nowadays, it is considered that the active site is organized into two sub-regions. At one of them, the substrate fitting occurs (the so-called bonding site), and in the other, the chemical transformation of the substrate occurs (the so-called catalytic site).

Using the example already cited above, glucoamylase (also known as amyloglucosidase) can act on disaccharides isomers (maltose, nigerose and isomaltose), but maltose is preferred as a substrate, because maltose has a higher affinity for the enzyme bonding site than the other isomers.

15.3 ENZYME ACTIVITY

The ability to quantify enzyme activity is essential if an enzyme is to be used in a cost-effective way in an industrial process or other analytical methods.

Focusing on industrial processes, some considerations must be made in order to evaluate the suitability using enzymes as catalysts in specific reactions. The basic considerations should be: (a) the amount of enzyme necessary; (b) the duration of the reaction; (c) the initial substrate concentration necessary; (d) the reaction conditions; and (e) the cost involved.

Insofar as the decision for using an enzyme is taken and knowing that the enzyme per se has a cost, it usually must be purchased at market prices – a peremptory evaluation of the impact the cost of enzyme will have on the overall process ought to be made. In general, the effective cost of an enzymatic process derivates from the ratio between the additional cost due to the enzyme and the improvement of the process yield and/or the aggregate value of the final product, taking into consideration the economic cost of the overall process.

Consequently, the success of using an enzyme is implied by optimizing the triad: amount of enzyme necessary, operational conditions and reaction yield.

Theoretically, since enzyme catalysts are regenerated at the end of a reaction, only a minimum amount of enzyme should be required to catalyze any amount of substrate. However, in reality, there is an optimal and finite correlation between the amount of catalyst, its initial activity and the amount of substrate that can be converted. The duration of industrial processes can vary from minutes to hours. For example, the action of proteases on malted barley with the aim of increasing the amount of nitrogen into the mash to be used in beer brewing takes no more than 60 min, whereas the saccharification of liquefied starch by glucoamylase requires 3 h duration, while lactose hydrolysis of milk whey by β-galactosidase (lactase) takes about as long as 20 h (Godfrey and West, 1996). Enzyme technologies aim to increase a reaction rate while taking into consideration restrictions imposed by the processing conditions and scale, which of course must be justified from an economic point of view.

15.3.1 Quantification of Enzyme Activity

Let's consider the equation as follows:

$$E + S \underset{k_2}{\overset{k_1}{\rightleftharpoons}} ES \xrightarrow[k_3]{} E + P$$

where: E = Enzyme concentration; S = Substrate concentration; ES = Complex Enzyme-Substrate concentration; P = Product concentration; k_1 = First order rate constant (t^{-1}); k_2 = Second order rate constant ($M^{-1} \cdot t^{-1}$); and k_3 = First order rate constant (t^{-1}).

From the equation proposed, it should be noted that before the substrate is converted to a product, the enzyme and the substrate forms an intermediate complex (ES). The formation of the complex constitutes an obligatory step for any enzyme-catalyzed reaction. Depending on the reaction conditions, the ES can either be converted to the product at a rate defined by the rate constant k_3, i.e., the reaction was effective; or freeing the substrate (S), step controlled by the rate constant k_2, in which the reaction did not occur effectively.

Generally, it can be considered that an enzyme catalyzed reaction proceeds through three different steps (Figure 15.1). During the first step – which is observed as soon as the enzyme (E) and substrate (S) are mixed into the reaction medium – the enzyme-substrate (ES) complex is formed and accumulates. In this first step, there is no formation of product. The existence of this complex was predicted by Brown in 1892 and corroborated by Henry in 1902. The idea was used in 1913 by Michaelis-Menten to establish the first mathematical model for evaluating enzyme activity, which in turn, was modified and complemented by Briggs and Haldane in 1926. The experimental identification of the ES complex occurred in 1936. In the second step – a period during which the concentration of the ES complex remains constant in the reaction medium – the substrate concentration diminishes whereas the product concentration increases quickly. In the third step – the concentration of the ES complex in the reaction medium is not constant – the consumption of substrate and formation of product occur more slowly.

Quantitative evaluation of an enzyme activity is made considering the events of the second and third steps.

Now, let's consider the second step.

The starting point will be the evaluation of how the initial substrate concentration (S) varies against time when a fixed amount of the enzyme (E_o) is used. The amount of product (P) formed during the total period of the reaction is measured and can be plotted in a graphic such as $(P) = f(t)$ (Figure 15.2).

From the linear part of the curves, obtained by plotting $(P) = f(t)$, the tangents can be calculated, where each one represents the reaction rate catalyzed by the enzyme (v_1, v_2, v_3, … v_n) against the corresponding initial concentrations of substrate (S_1, S_2, S_3, … S_n) (Table 15.1).

The data from Table 15.1 can then be plotted as $v = f(S)$ to give a profile that generally can be represented as a rectangular hyperbole (Figure 15.3).

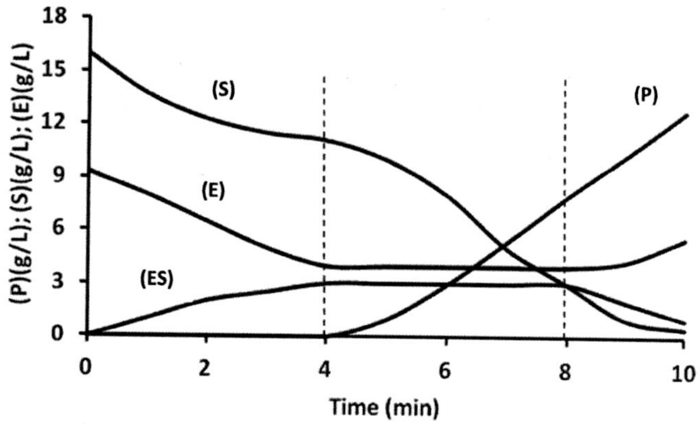

FIGURE 15.1 Variation of enzyme (E), substrate (S) and product (P) concentration along the reaction time. First phase ($0 \le t < 4$), second phase ($4 \le t < 8$) and third phase ($8 \le t < 10$).

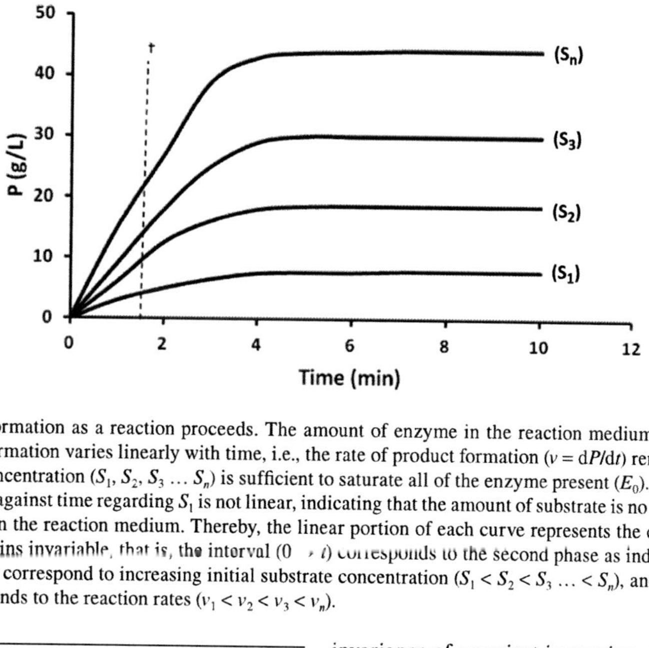

FIGURE 15.2 Product formation as a reaction proceeds. The amount of enzyme in the reaction medium is constant (E_0). At time interval ($0 \rightarrow t$), product formation varies linearly with time, i.e., the rate of product formation ($v = dP/dt$) remains constant, indicating that the initial substrate concentration ($S_1, S_2, S_3 \ldots S_n$) is sufficient to saturate all of the enzyme present (E_0). Over time (t), the correlation between consumption against time regarding S_1 is not linear, indicating that the amount of substrate is no more sufficient to saturate all of the enzyme present in the reaction medium. Thereby, the linear portion of each curve represents the condition under which the concentration of (ES) remains invariable, that is, the interval ($0 \rightarrow t$) corresponds to the second phase as indicated in Figure 15.1. The curves shown in this graph correspond to increasing initial substrate concentration ($S_1 < S_2 < S_3 \ldots < S_n$), and the tangent of the linear part of each curve corresponds to the reaction rates ($v_1 < v_2 < v_3 < v_n$).

TABLE 15.1
Correlation between the Initial Substrate Concentration and the Reaction Rates at Interval $0 \rightarrow t$

Initial Substrate Concentration (g/L)	Reaction Rate (g/L·min)
S_1	v_1
S_2	v_2
S_3	v_3
S_n	v_n

As already mentioned, plotting each pair (v, S) in a Cartesian coordinate system produces a graph shown in Figure 15.3, i.e., v and S correlate through a hyperbolic function, whose asymptote tends to a maximum value of v (the so-called V_{max}, one of the fundamental kinetic constants, that is used to describe quantitative enzyme catalysis). The asymptotic part of the curve, represented by the invariance of v against increasing substrate concentration, would represent enzyme saturation by the amount of substrate in the reaction medium. In other words, all enzyme molecules present in the reaction medium are bound to the substrate, i.e., the ES intermediate predominates (another fundamental component of the enzyme activity) (Purich, 2010).

Decomposition of (ES $\rightarrow E + P$) as the determining step for the reaction can be expressed by Equation 15.1:

$$(-dS/dt) = v = k_3 \cdot (ES) \quad (15.1)$$

At any moment during the reaction there is:

$$E_0 = (E) + (ES) \quad (15.2)$$

When substrate is in excess (S), the reaction will shift totally to the right side (that is, to the direction of product formation) so that all the enzymes will be bound to the substrate

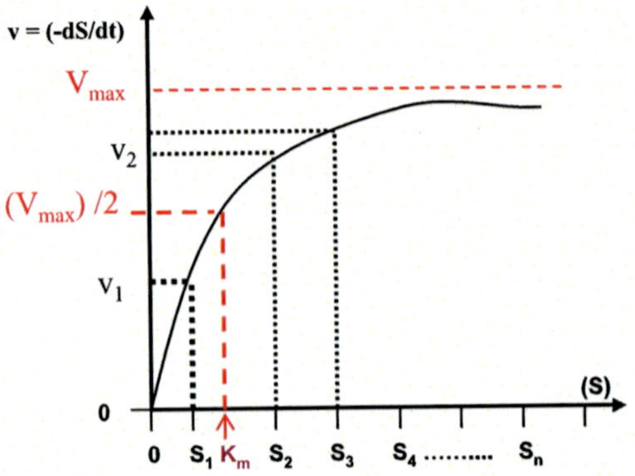

FIGURE 15.3 Reaction rate ($v = -dS/dt$) against the initial substrate concentration. The curve (a rectangular hyperbole) shows that from S_2, the reaction rate does not increase linearly with the substrate concentration. Over S_4, the reaction rate becomes invariant as the substrate concentration increases. Therefore the reaction reached the highest reaction rate (V_{max}). At this point, the enzyme is saturated by substrate, i.e., the complex (ES) is the dominant enzyme form in the reaction medium (second phase of catalysis). In the figure, K_M is also indicated – which correspond to the substrate concentration under which the reaction rate is half of the V_{max}. The kinetic constants K_M and V_{max} at fixed reaction conditions are characteristics of an enzyme.

(that is, at each moment of the reaction there is no enzyme molecule into the reaction medium that is not bound to a substrate molecule). Then, $(E) = 0$.

So, Equation 15.2 becomes

$$E_0 = (\mathrm{ES}) \tag{15.3}$$

Substituting Equation 15.3 in Equation 15.1:

$$v = k_3 \cdot E_0 \tag{15.4}$$

From Equation 15.4, it can be concluded that when the enzyme becomes saturated, the reaction is described as 'pseudo zero order', that is, the kinetic is independent of the substrate concentration. Undoubtedly, analyzing Equation 15.4, it becomes clear that the reaction rate is directly proportional to the total concentration of enzyme present in the reaction medium.

When the enzyme reaches saturation the rate constant (k_3) is called the *turnover number* and is symbolized as k_{cat}. The *turnover number* refers to how many times one enzyme molecule participates in one catalytic cycle at a particular time interval.

At saturation, the product ($k_{cat} \cdot E_0$) corresponds to the maximum rate of the reaction catalyzed by the enzyme (V_{max}). Then,

$$v = V_{max} \tag{15.5}$$

Equation 15.5 can be written as:

$$-dS/dt = V_{max} \tag{15.6}$$

Rearranging and integrating Equation 15.6 results in Equation 15.7:

$$(S) = (S_o) - V_{max} \cdot t \tag{15.7}$$

Thereby, when the enzyme is saturated with the substrate, the concentration of substrate present in the reaction medium diminishes linearly with time. This is a consequence of the reaction proceeding at a constant and maximum rate. Under this condition, the enzyme activity is standardized.

At low substrate concentration, it can be seen from Figure 15.3 **(from 0 up to S_1)** that v varies linearly with S; then the equation can be written as

$$v = (-dS/dt) = k' \cdot S \tag{15.8}$$

in which k' is a first order constant reaction rate.

Rearranging Equation 15.8 followed by integration gives

$$\ln S = \ln S_o - k' \cdot t \tag{15.9}$$

From Equation 15.9, it is clear that the concentration of substrate present in the reaction medium – **in an unsaturated condition and at low substrate concentration** – diminishes exponentially with time.

From Figure 15.3, it can be noted that the reaction rate varies from first order (substrate concentration < S_1) to 'pseudo zero order' (substrate concentration > S_4). To quantify the interval $S_1 \rightarrow S_4$, we have to consider that at the second step (see Figure 15.1), the **(ES)** must be constant. This means that

$$d(\mathrm{ES})/dt = 0 \tag{15.10}$$

Thereby, Equation 15.10 can be interpreted as follows:

Rate of ES formation = rate of ES decomposition, or in mathematical terms

$$k_1(E)(S) = k_2(\mathrm{ES}) + k_3(\mathrm{ES}) \tag{15.11}$$

Rearranging Equation 15.11 gives:

$$(E) = [(k_2 + k_3)/k_1] \times [(\mathrm{ES})/(S)] \tag{15.12}$$

Designating the term $[(k_2+k_3)/k_1] = K_M$, then Equation 15.12 becomes:

$$(E) = [K_M \cdot (\mathrm{ES})] \div (S) \tag{15.13}$$

Substituting Equation 15.13 into Equation 15.2 (this is possible since the reaction conditions preserve the full catalytic capability of the enzyme) gives

$$E_o = \{[K_M(\mathrm{ES})/(S)] + (\mathrm{ES})\} \text{ or } (\mathrm{ES}) = \{[S \cdot E_o] \div [S + K_M]\} \tag{15.14}$$

Substituting Equation 15.14 in Equation 15.1 then gives

$$v = [k_3 \cdot E_o \cdot (S)] \div [(S) + K_M] \text{ or } v = V_{max}(S)/[(S) + K_M] \tag{15.15}$$

Equation 15.15 is a complete description of the hyperbolic curve presented in Figure 15.3. That is, during the

second step of the reaction catalyzed by the enzyme (Figure 15.1), it is possible to know the reaction rate for a given substrate concentration at any time during the reaction.

In Equation 15.15, the so-called kinetic constants V_{max} and K_M appear. These constants characterize an enzyme when the catalysis is carried out under defined reaction conditions (pH, temperature, agitation, reagents concentrations, etc.).

Equation 15.15 can be written as follows:

$$1/v = (1/V_{max}) + 1/S \cdot (K_M/V_{max}) \qquad (15.16)$$

Plotting (1/*v*) against (1/*S*) allows numerical values of the kinetic constants to be obtained.

Three characteristics regarding K_M must be remembered: (a) when $K_M = (S)$, i.e., numerically equal to the substrate concentration, Equation 15.15 gives $v = V_{max}/2$; (b) K_M can be used as a reference to set the initial substrate concentration – when (*S*) is at least 100 times higher than K_M the reaction occurs at substrate saturation; when (*S*) is at least 100 times lower than K_M the reaction still occurs, but the enzyme is not saturated by substrate; and (c) K_M is a characteristic of the enzyme under definite reaction conditions.

Let's analyze the third step of a reaction catalyzed by an enzyme. This is the step when the (ES) is not constant (Figure 15.1).

In order to study this step, let's rewrite Equation 15.15 as follows:

$$v = -(\mathrm{d}s/\mathrm{d}t) = V_{max}/\left[1 + (K_M/S)\right] \qquad (15.17)$$

Rearranging:

$$-(\mathrm{d}s)\cdot\left[1 + (K_M/S)\right] = V_{max}\cdot \mathrm{d}t \qquad (15.18)$$

Integrating:

$$t = \left[(S_o - S) - K_M \cdot \ln(S/S_o)\right] \div V_{max} \qquad (15.19)$$

Equation 15.19 correlates to substrate consumption against time, throughout the duration of the reaction.

Equation 15.19 can be modified as follows:

Defining (*Y*) as the conversion factor:

$$Y = (S_o - S)/S_o \qquad (15.20)$$

Rearranging Equation 15.20 gives

$$S = S_o(1 - Y) \qquad (15.21)$$

Substituting Equation 15.21 into Equation 15.19 results in

$$t = \left[Y \cdot S_o - K_M \cdot \ln(1 - Y)\right] \div V_{max} \qquad (15.22)$$

The very importance of Equation 15.22 is due to the fact that the conversion can be estimated from a previous time point.

15.3.2 Expression of Enzyme Activity

Enzyme activity can be expressed in different ways. Nevertheless, the Biochemistry International Commission considers

a. One unit (U) of any enzyme is the amount that catalyzes the conversion of 1 µmol of substrate per minute, under defined reaction conditions.
b. One katal (k_{at}) is the amount of enzyme that catalyzes the conversion of 1 mmol of substrate per second under defined reaction conditions.
c. The ratio U/mg of protein is called the specific activity.
d. The ratio U/µmol of enzyme is called the molecular activity.

15.3.3 Factors Affecting Enzyme Activity

Roughly speaking, an enzyme macromolecule can be divided in two parts, i.e., a microenvironment in which the active site is located and other domains which are not directly involved in the catalysis. It is obvious that optimal enzyme activity occurs when the reaction conditions do not significantly affect the macromolecule structure. The reaction conditions to be considered are those related to physical-chemical factors (pH, temperature, ionic strength, water activity, etc.), chemical factors (activators, inhibitors, stabilizers and deactivators) and physical factors (pressure and shear forces).

15.3.3.1 Physical-Chemical Factors

15.3.3.1.1 pH

This factor has a generalized action on all biochemical reactions, influencing the reaction rate, equilibrium shifting, ionization and dissolution of the reagents. In the case of enzymes, the effect on the stability of the macromolecular structure and kinetic constants (V_{max} and K_M) must also be considered. Enzyme activity is affected by pH because hydrogen ions influence the ionization state of amino acids that constitute the structure of the enzyme. A bell-shaped curve results when pH versus enzyme activity is plotted on a graph, indicating the gradual pH effect on the macromolecule structure (Tomotani and Vitolo, 2004; Purich, 2010). Combining graphs showing curves of pH versus activity and pH versus stability, it is possible to determine the pH interval in which protein denaturation occurs.

15.3.3.1.2 Temperature

The temperature acts on chemical reactions in a number of ways including the solubility of reagents and the maximum reaction rate. In reactions catalyzed by enzymes, temperature influences the enzyme stability and kinetic constants too. The Van't Hoff's law – which states that the reaction rate doubles with every 10°C increase in reaction temperature – does not apply for enzymes because an increase of 10°C could augment the reaction rate more than twice (Godfrey and West, 1996).

Establishing an optimal temperature for enzyme-catalyzed reactions is difficult because the macromolecular structure of the enzyme is activated and deactivated simultaneously during catalysis. So, the coexistence of activation/deactivation at any period of time during the reaction at a fixed temperature is a significant factor affecting enzyme stability.

The effect of temperature on the shelf-life of a commercially available enzyme is also important to consider because the same enzyme maintained at different temperatures can lose activity at different rates. Introducing modifications into the molecular structure of an enzyme, either by chemical modification (protein engineering) or by genetic modification of the organism that produces the enzyme (molecular biology), in order to increase or decrease thermal stability is an important goal for manufacturers of industrial enzymes.

15.3.3.1.3 Others

Examples of other factors are: (a) Ionic strength – related to the concentration of ions present in the reaction medium – which can affect the solubility and/or the ionization of chemical groups of amino acids side chains of the protein; (b) The buffer composition – related to different water soluble substances, which have buffering capability at the same pH - can diminish the catalytic performance of an enzyme (Vitolo, 2020); and (c) Water activity, i.e., the amount of water present in the reaction medium can affect the activity and catalytic mechanism of the enzyme. The activity of an enzyme could be severely reduced when the water activity medium is low, or the enzyme mechanism may change from hydrolysis to synthesis.

15.3.3.2 Chemical Factors

Chemical factors act differently to the physical and/or physical-chemical factors on specific molecular domains of an enzyme, such as the active site.

15.3.3.2.1 Activators/Deactivators

An activator is a compound (or sometimes it can be a simple molecular ion) which increases enzyme activity. It can insert permanently into the molecular structure of an enzyme (in this case it is called a prosthetic group), or it can be dissolved into the medium, becoming linked to the enzyme molecule in the presence of the substrate at the beginning of catalysis – in such a case, if the activator is an organic substance, it is called a coenzyme. Amongst the six classes of enzymes, the oxidoreductases – which always catalyze oxidation and reduction reactions – require a cofactor, either a prosthetic group (FAD linked to the glucose oxidase) or a coenzyme (NADP in the case of glucose-6-phosphate dehydrogenase) (Tomotani et al., 2005; Tomotani and Vitolo, 2008).

15.3.3.2.2 Stabilizers

A liquid preparation of an enzyme can be stabilized against the effects of temperature if the substrate is also present in the liquid. Moreover, stabilization of an enzyme in solution would also be attained by using a modified substrate (for example, amylases can be stabilized with hydrolyzed starch and proteases with hydrolyzed peptides). Sometimes, for enzymes that catalyze a one-step substrate to product reaction ($S \rightarrow P$), the product could also be used as a stabilizer. A metallic ion can act either as an activator or stabilizer (for example, inclusion of calcium ions into a preparation can stabilize α-amylase against temperature and pH).

15.3.3.2.3 Inhibitors

Inhibitors are specific compounds that, even at low concentration, can diminish the reaction rate of an enzyme catalyzed reaction. Inhibitors can act on the active site or on auxiliary sites without destabilizing the tertiary and/or quaternary structure of the protein molecule; contrarily, chemical reagents (alkalis, acids, salts, urea, detergents, amongst others) can damage the protein structure (Purich, 2010).

From an industrial point of view, it is important to be aware that the presence of any inhibitor will reduce the catalytic efficiency of an enzyme. So, introduction of any substance that might act as an inhibitor during the manufacturing process should be avoided. The main types of inhibitors are: (a) ***Irreversible***: these link to the enzyme molecule through covalent bonds, causing complete inactivation of the enzyme; and (b) ***Reversible***: these link to some part of enzyme molecule by non-covalent bonds, so the effects on the enzyme can be reversed by using an appropriate concentration of the substrate. There are three types of reversible inhibitors: *competitive* – in this case the substrate and inhibitor compete for the active site of the enzyme molecule. Depending on the relative concentrations between the substrate and inhibitor, inhibition can be reduced or eliminated. In other words, the inhibitor and substrate exclude each other; *non-competitive* – here, the inhibitor and substrate do not exclude mutually, because each is linked to different domains of the enzyme molecule – and *uncompetitive* in which the inhibitor only links to the enzyme when the substrate-enzyme **[ES]** complex is present in the reaction medium. Non-competitive inhibition can also occur when a metallic ion is introduced into the enzyme molecule, and the ion becomes susceptible to oxidation or reduction. Glucose oxidase, for instance, has Fe^{+2} ion in its molecular structure, which can be oxidized to Fe^{+3} by hydrogen peroxide at a concentration over 1.22 mM, reducing the catalytic activity of the enzyme (Tomotani et al., 2005).

15.3.3.3 Physical Factors

Two physical factors that cause adversely effects on enzyme activity deserve special mention. These are shear forces – resulting from mechanical agitation of the reaction medium – and the internal pressure of the reactor generated during catalysis.

15.3.4 Thermodynamics of Enzyme Catalysis

The enzyme – like any catalyst – facilitates a reaction to proceed. In essence, it reduces the energy necessary for the reactants to reach the transition state. Nevertheless, in the case of enzyme catalysis, the reaction mechanism always involves formation of the complex enzyme substrate (ES), which is a stable form of the enzyme. Moreover, the energetic level of (ES) is equivalent to that of the free enzyme (E) plus product (P), a condition in which the reaction would stop with the formation of the complex. However, what is commonly observed is that the reaction proceeds yielding product and free enzyme. This can be explained by taking into account a summation of all the instabilities that occur inside the complex (ES), due to different molecular characteristics of the substrate and enzyme. Hence, tensions of different natures (electrostatic and hydrophobic, amongst others) between the chemical groups of both molecules lead to instability of the complex which drives the reaction forward to completion. Besides, the overall entropy of the system diminishes as the complex (ES) accumulates in the reaction medium (remember that the complex is a more organized structure than the separated enzyme and substrate), further exacerbating internal chemical instabilities which favour a drive towards reaction completion (Purich, 2010).

15.4 FINAL CONSIDERATIONS

Enzymes are catalytic proteins capable to convert a compound (substrate) to another (product) at high reaction rate. In enzyme-catalyzed reactions, enzymes, in general, specifically accept only one substrate. Even when an enzyme catalyzes the conversion of more than one substrate, the reaction rates are different. Substrate specificity is linked to a region of the enzyme molecule called the active site, which can be envisaged as a rigid or flexible structure depending on the particular reaction considered. Currently, it is considered that the active site is organized into two sub-regions. At one of them, the substrate fitting occurs (the so-called bonding site), and in the other, the chemical transformation of the substrate occurs (the so-called catalytic site). Generally, it can be considered that an enzyme-catalyzed reaction proceeds through three different steps: first step (period in which the enzyme and substrate are mixed into the reaction medium. The enzyme-substrate complex is formed and accumulates); second step (a period during which the concentration of the ES complex remains constant in the reaction medium, the substrate concentration diminishes, and the product concentration increases); third step (the concentration of the ES complex in the reaction medium is not constant; the consumption of substrate and formation of product occur more slowly). The fundamental aspect is that the catalysis rate and substrate concentration follow a hyperbolic correlation. The enzyme activity can be affected by factors of different nature, i.e., physical-chemical factors (pH, temperature, ionic strength, water activity, etc.), chemical factors (activators, inhibitors –irreversible or reversible (competitive, non-competitive and uncompetitive) –, stabilizers, and deactivators) and physical factors (pressure and shear forces). The ability to quantify enzyme activity is essential if an enzyme is to be used in a cost-effective way in an industrial process or other analytical methods. Focusing on industrial processes, some considerations must be made in order to evaluate the suitability using enzymes as catalysts in specific reactions. The basic considerations should be: (a) the amount of enzyme necessary; (b) the duration of the reaction; (c) the initial substrate concentration necessary; (d) the reaction conditions; and (e) the cost involved. In general, the effective cost of an enzymatic process derivates from the ratio between the additional cost due to the enzyme and the improvement of the process yield and/or the aggregate value of the final product, taking into consideration the economic cost of the overall process.

The enzyme technology has been in constant development due mainly to the discovery of new sources of enzymes, modification of the molecular structure (protein engineering), genetic modification of traditional sources of enzymes (microbial and plants, mainly), improvement of industrial processes and downstream protocols.

REFERENCES

Godfrey, T.; West, S. *Industrial enzymology*. 2. ed. New York: MacMillan Publishers Ltd, 1996, 609 p.

Kulp, K. Carbohydrases. In: Reed, G. (Ed.). *Enzymes in food processing*. New York: Academic Press, Inc., 1975, 53–122.

Purich, D. L. *Enzyme kinetics: Catalysis and control*. London: Elsevier, 2010, 892 p.

Tomotani, E. J.; Das Neves, L. C.; Vitolo, M. Oxidation of glucose to gluconic acid by glucose oxidase in membrane bioreactor. *Applied Biochemistry and Biotechnology*, v. 121–124, 149–162, 2005.

Tomotani, E. J.; Vitolo, M. Immobilization of nicotinamide nucleotides on anionic polystyrene divinylbenzene beads. In: *Brazilian seminary on enzyme technology VIII* (Book of Abstracts). Rio de Janeiro: Federal University of Rio de Janeiro, 2008, 233–234.

Tomotani, E. J.; Vitolo, M. Screening of DOWEX(R) anion exchange resins for invertase immobilization. *Applied Biochemistry and Biotechnology*, v. 113–116, 145–159, 2004.

Vitolo, M. Effect of freezing on invertase activity of Saccharomyces cerevisiae cell. *World Journal of Pharmaceutical Research*, v.9, n.6, 120–131, 2020.

16 Enzymes as Drugs and Medicines

Carlota de Oliveira Rangel-Yagui and Michele Vitolo
Universidade de São Paulo

CONTENTS

16.1 INTRODUCTION

Enzymes are natural catalysts, usually proteins, capable of rapidly converting a substance into another by means of lowering the activation energy of the reaction involved. Invertase, a glycoprotein from yeast, is an example of enzyme; it catalyzes the hydrolysis of sucrose in glucose and fructose almost instantaneously (Addezio et al., 2014). Enzymes play a central role in biotechnology as key molecules for cell life and, therefore, bioprocesses, as well as for all biotechnology tools such as recombinant DNA and cell fusion (Figure 16.1).

The source of an enzyme influences its availability, cost and degree of purity. Common sources of enzymes are animals (e.g., pepsin, trypsin and chymosin), plants (e.g., papain, bromelain and ficin) and microorganisms (e.g., pectinases and amylases). In general, animal and/or vegetable enzymes present bacterial isoforms. **Amylolytic enzymes** that catalyze the hydrolysis of starch, for example, can be found in *Bacillus subtilis*, *Aspergillus oryzae*, *A. niger*, *A. flavus* and *A. awamori*; ***glucose oxidase** that catalyzes glucose oxidation to gluconic acid is found in A. niger, Penicillium amagasakiense* and *P. notatum*; ***lactase** that catalyzes lactose hydrolysis can be produced by Saccharomyces fragilis* and *Zygosaccharomyces lactis*; **lipases** catalyze the hydrolysis of triglycerides and can be found in *Saccharomyces fragilis* and *Zygosaccharomyces lactis*; and **proteases**, responsible for protein hydrolysis, are also produced by *B. subtilis*, *A. oryzae*, *A. flavus*, *Endothia parasítica* and *Mucor pusillus*.

Microbial enzymes are preferred for large scale biotechnological applications owing to the higher growth rate of microorganisms, generating large quantities of enzyme in short times, easy scale up and strain manipulation for over-expression (Meghwanshi et al., 2020). Additionally, they are produced by bioprocess with rigorous control of all steps of production and final preparations are more consistent regarding purity and enzyme activity. The importance of enzymes in pharmaceutical industry can be exemplified by the use of penicillin amidase to the production of tons of 6-aminopenicillanic acid (an intermediate compound for the synthesis of semi-synthetic penicillins).

One aspect that deserves attention is the genetic improvement of the enzyme producing strain based on recombinant DNA technology. Industrial enzymology was certainly one of the first sectors of economy to benefit from this biotechnology tool with the concept 'one gene one enzyme' and the possibility of overexpressing this one gene in recombinant strains. With the advance in enzyme production, the market for biocatalysts went from millions of dollars (in the 1980s) to billions of dollars (in the 21st century). Nowadays, almost all enzymes in household cleaning products are produced by genetically modified microorganisms (Demain, 2007).

In general, microorganisms can be good producers of more than one type of enzyme, making it possible to tune the process to the preferential production of one enzyme specifically. Therefore, companies may adapt industrial production

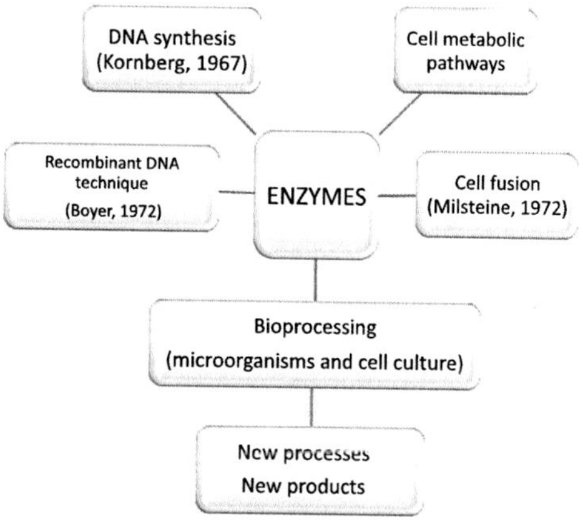

FIGURE 16.1 Enzymes in the biotechnological context.

to market needs and fluctuations. One should keep in mind that this strategy can bring additional concern regarding collateral effects of enzymatic preparations owing to contaminant enzymes produced by the strain, what may impart the final product. A perfect understanding of the enzymatic preparation characteristics and purity is of paramount importance for both producers and consumers. For food products, in particular, it must be informed if the enzyme is from a genetically modified organism. In this case, specific licenses from the regulatory agencies (such as FDA and EFSA) are usually required for the enzyme commercialization as well as for the final food product that will be considered a 'new food'.

Enzymes have been used for centuries in several industrial processes such as tannery, bakery, dairy and beer production. Depending on the process, they can be used isolated (e.g., animal proteases and lipases in tannery and dairy industries) or as part of the whole cell system in bioprocesses (e.g., bakery, beer and wine production and antibiotics production).

Whilst the role of microorganisms in fermentation processes was well described by Pasteur in 1876, the enzymatic mechanisms, on the other hand, have gained a better understanding only at the beginning of the 20th century. At this time, amylase and invertase started to be produced in large scale. More specifically, quantification of enzymatic activity has gained impulse in 1913 and extended throughout this century. Nonetheless, enzymatic catalysis is still an expanding field, especially with the more recent findings of other biomolecules with catalytic activity, as well as the possible development of artificial enzymes. As examples of novel biocatalysts, one can cite ribonucleic acid, modified catalytic antibodies, synthetic peptides and chemically modified proteins. In addition, enzyme immobilization allows the recovery of the catalyst and/or operation in continuous process, facilitating industrial applications.

A general classification refers to industrial and special enzymes (Freedonia®, 2002). Industrial enzymes are used in the production of food, animal food, household cleaning products, textile products, tanning, biofuel, cellulose and paper. Special enzymes, on the other hand, are used for therapeutic purposes (anticoagulant, antivirus, antitumour, amongst others), diagnostic (e.g., polymerases, nucleases), research and fine chemistry (Figure 16.2).

The diverse use of enzymes has generated an established and profitable market of increasing growth. Regarding industrial enzymes, world market in 2004 was about US $2.3 billion with an estimate of 5.7% growth for the following 10 years (Freedonia®, 2005). For special enzymes, it reached US $1.3 billion (biopharmaceuticals corresponding to 38%) with an estimate of US $3 billion for 2014 (Table 16.1). A significant part of the increase in the market for special enzymes is related to enzyme replacement therapy, *i.e.*, the use of enzymes to alleviate the symptoms of the genetic diseases associated with suppressed expression of these enzymes. As examples, one can cite glucocerebrosidase and alpha-galactosidase, used to treat Gauche and Fabry syndromes, respectively. According to a review presenting data from the *Global Market Insights* and *Research and Markets*, the Global industrial enzymes market is expected to grow at a compound annual growth rate of 6.8% during 2019–2024. The global market for proteases, for example, is expected to value around USD 2 billion by 2024, owing to therapeutic applications such as boosting the immune system, preventing inflammatory bowel diseases, and treating skin burns and stomach ulcers, but also to applications in the animal feed segment. Lipases are expected to achieve the increase in sale by around 6.8% by 2024, and the use for the treatment of obesity, is contributing to the increase (Meghwanshi et al., 2020).

It is clear the importance of enzymes and the increasing application in the pharmaceutical field. It is not within the scope of this book to exhaust this topic and, for a deeper understanding, we suggest the readers some classic work on enzymes such as Godfrey and Reichelt (1983), Gerhartz (1990), Godfrey and West (1996), Lauwers and Scharpé (1997), Buchholz et al. (2012) and Devasena (2012).

16.2 ENZYMES IN MEDICINES

Considering the presence of enzymes in animal and plant extracts, the use of these molecules as therapeutic agents is millenary. Rational and planned use of enzymes, nonetheless, started in the 1990s, with the use of proteases (pepsin, papain, trypsin, chymotrypsin) as wound debriding and digestive auxiliary, and streptokinase to treat acute myocardial infarct. More recently, enzymes such as glucocerebrosidase,

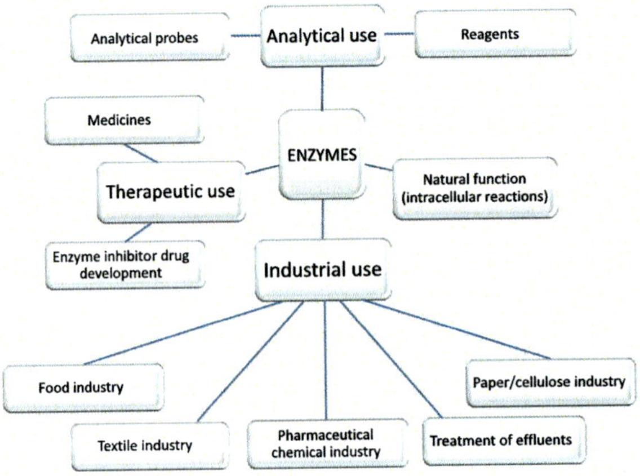

FIGURE 16.2 General uses of enzymes.

TABLE 16.1
Worldwide Commercialization of Special Enzymes (US $ Million)

	Year			
Types of enzymes	1994	2004	2009[a]	2014[a]
Pharmaceutic	205	520	880	1,410
Diagnosis	150	300	400	580
Research	200	420	600	805
Fine chemistry	50	100	180	210
TOTAL	**605**	**1,340**	**2,060**	**3,005**

Source: Freedonia®, *World Enzymes to 2009*, Freedonia Group Incorporated, 336 p., 2005.

[a] Estimative.

α-galactosidase and acid α-glucosidase has gained importance as enzyme replacement therapy in patients afflicted with metabolic disorders (Kang and Stevens, 2009). Popularization of enzymes in therapy increased together with the increase in isolation and purification from diverse sources, therefore providing required amounts and degrees of purity for this field.

Enzymes find several applications in the pharmaceutical field, such as in protocols for the synthesis of chemical drugs (e.g., intermediates production, synthesis of chiral compounds), as biopharmaceuticals (e.g., hyaluronidase as anesthetic adjuvant), as well as possible targets of chemical drugs to be investigated. Cosmetology and clinical analysis are also pharmaceutically related areas that present important examples of the use of enzymes.

For pharmaceutical application, an enzyme should ideally present (a) low K_M; (b) no need of exogenous cofactors; (c) irreversibility of the catalyzed reaction under physiological conditions; (d) high activity and stability under physiological pH; (e) preserved activity in the blood; (f) low blood clearance; (g) no inhibition by its products and/or substances of biological fluids; and (h) possibility to be obtained from non-pathogenic endotoxin-free microorganisms or established animal cell culture. Unfortunately, not rarely one or more of these criteria cannot be fulfilled. To exemplify, still today many recombinant enzymes are expressed in *E. coli*, an endotoxin producer microorganism. Additionally, enzymes must fulfil strict quality control criteria of purity, stability and kinetic properties for human and animal use. Bioavailability and side effects, in particular immunogenicity owing to the exogenous protein character, are also of paramount importance. Therapeutic enzymes should reach the target with preserved catalytic activity and without significant immune response triggering.

In general enzymes are employed as thrombolytic agents (streptokinase, urokinase and plasminogen tissue activator, for example), wound healing agents (fibrinolysin, chymotrypsin, amongst others), anti-cancer agents (L-asparaginase) and anti-inflammatory agents (for example, papain, collagenase, trypsin, bromelain and superoxide dismutase). Therapeutic enzymes can be found in commercial formulations (in pharmacies and drugstores) or for strict use in hospitals when collateral effects are severe or they require special administration such as perfusion (urokinase, streptokinase and L-asparaginase for example). Table 16.2 presents the enzymes currently in use worldwide.

TABLE 16.2
Approved Therapeutic Enzymes (Yari et al., 2017)

Enzyme Group	Enzyme Name	Biotechnological Features
Enzyme replacement therapies	Pancreatic enzymes	Non-recombinant
	Adenosine deaminase	Non-recombinant PEGylated
	Alkaline phosphatase	Recombinant fusion protein
	β-glucocerebrosidase	Non-recombinant (aglucerase)
	α-galactosidase	Recombinant (velaglucerase, taligrucerase)
	α-glucosidase	Recombinant-one point mutation (imiglucerase)
	Lysosomal acid lipase	Recombinant
	Tripeptidyl peptidase-1	Recombinant
	α-L-iduronidase	Recombinant
	Iduronate-2-sulfatase Galactosamine-6-sulfatase Arylsulfatase	Recombinant
		Recombinant
Enzymes in cancer therapy	Asparaginase	Non-recombinant and PEGylated
	Urate oxidase	Recombinant
	Carboxypeptidase	Recombinant
Fibrinolytic enzymes	Streptokinase	Non-recombinant
	Tissue plasminogen activator	Recombinant (alteplase)
		Recombinant - deletion mutation (reteplase)
		Recombinant - three-point mutations (tenecteplase)
Topical enzymes	Collagenase	Non-recombinant
	Deoxyribonuclease	Recombinant
	Microplasmin	Recombinant
	Hyaluronidase	Non-recombinant and recombinant

16.2.1 Enzymes Bioavailability

Enzymes are complex biomacromolecules and, therefore, therapeutic success depends on the enzyme source and immunogenicity, cross-reactivity, administration route (enteral or parenteral) and the ability to transpose the body barriers. The oral route is the preferred one for medicines administration. However, the gastrointestinal tract is a challenge for any protein drug owing to the high molecular mass that difficult the transit in the intestinal epithelium and the susceptibility to hydrolytic enzymes. Digestive system is composed by the oral cavity, esophagi, stomach and intestine, which presents a large superficial area and therefore is the preferential absorption compartment. Therefore, the intestinal epithelium is the main barrier for the absorption of protein drugs in general, including enzymes. Before reaching the intestine, protein drugs have to resist the gastric juice, characterized by excessive acidity and the presence of proteolytic enzymes responsible for the degradation of ingested proteins (trypsin, α-chymotrypsin, elastase and carboxypeptidase).

To overcome acidic and proteolytic degradation, usually drugs are formulated with polymeric covers that undergo degradation only in the intestine owing to the alkaline character (pancreatic and mucolytic juices increases the intestinal lumen pH). Once released in the intestinal lumen, protein drugs must cross the intestinal wall to reach blood circulation. This barrier is mainly composed of (Demeester, 1997):

a. A water film at the external surface of the intestinal mucosa - transport across this layer is by passive diffusion and the resistance is reduced by the intestinal motility. Nonetheless, for molecules above 250 Da, the decrease in diffusion coefficient is proportional to the cubic root of the molar mass.
b. Intestinal mucosa – a viscoelastic gel composed by glycoproteins and water that becomes elastic upon stress. It is certainly an important barrier for protein drug absorption.
c. pH gradient (6.0–7.0) between the lumen and the mucosa – not an important barrier for protein mobility.
d. Glycocalix – a mucopolysaccharide net disposed over absorptive cells with transport and/or catalytic function. Transport of therapeutic proteins is not facilitated by the glycocalix.
e. Membrane microvilosities – a 10 nm pleated membrane that provides large superficial area for absorption. The microvilli are contractile and generate fluidic movements that contribute to molecules absorption.
f. Cell interior – once in the cell cytoplasm, the protein drug might be degraded by lysosomes.
g. Basal membrane of the cell – has to be crossed by the protein drug to reach intracellular space.
h. Intracellular space – rich in blood capillaries that must be crossed by the protein drug to reach blood circulation.

Once in blood, the protein drug must resist liver metabolism and finally reach the target organ. Alternatively, it might reach the lymphatic system via lymphatic ducts at the microvillus, avoiding first pass metabolism in the liver. Membranous cells surround the lymphoid follicles; they present less lysosomal content, a smaller glycocalyx and are rich in vesicles and short microfibrils at the apical portion of the plasmatic membrane. The main function of this layer is antigen recognition and presentation to lymphocytes at the intercellular space, stimulating the immune system. This mechanism of molecules transport across the intestinal epithelium can be explored for enzymes administered by enteral route.

As mentioned before, parenteral routes including intravenous, sublingual and rectal administration can also be used for enzymes administration. Once in blood circulation, enzymes need to reach the target organ by crossing the epithelial layer of the vases by endocytosis, transcytosis or passive penetration via epithelial cells junctions. The microvascular net is responsible for the major part of micro and macromolecules exchange between blood and interstitial fluid. Depending on the tissue/organ, endothelial cells of capillaries are classified as continuous (heart and most of the blood vases), fenestrated (visceral organs) or sinusoids (liver and hematopoietic tissue). The latter are characterized by large intercellular gaps (up to 30–40 μm) and therefore not only water and small molecules can cross freely, but also proteins and even blood cells. Continuous capillaries, on the contrary, allow only water and ions to pass through their intercellular space. For fenestrated capillaries, transport is limited by the dimensions of fenestrations (up to 80 nm); only small molecules and limited amounts of protein can diffuse. Heterogeneity of endothelial cells of capillaries can be found in the same tissue/organ; strong anionic sites for example can be found over small gaps. Tissue physiology and physical forces define the organization and complexity of intercellular junctions of capillaries.

16.2.1.1 Transport across Endothelium

The transport of proteins across the endothelium is a complex process governed by at least three forces: plasmatic driving force, physicochemical properties of the protein and characteristics and properties of the endothelial cells (Antohe and Poznansky, 1997). As mentioned, the cytoplasmic membrane is highly permeable to gases, lipophilic molecules and small hydrophilic molecules (up to 10 Å in diameter) via paracellular transport. Proteins, on the other hand, are internalized by endocytosis or transcytosis. Regarding the physicochemical properties of the protein, transport across endothelium depends on size, shape and superficial charge. Transendothelial transport might be unspecific or specific, *i.e.*, involving the protein binding to membrane receptors or specific sites at the surface of the transport vesicles (receptor-mediated transcytosis).

Endothelial permeability might be regulated by physiological mechanisms, including response to mediators (histamine, serotonin, thrombin and leukotrienes) that bind to specific receptors resulting in an increase in intracellular calcium, contraction of the cytoskeleton and, therefore, the opening of intercellular junctions. Additionally, transcytosis mediated by insulin and transferrin receptors was identified in CNS capillaries.

16.2.2 Therapeutic Enzyme Delivery

Enzymes offer several advantages as drugs such as specificity and low toxicity. However, many challenges are also associated with protein drugs in general. To be used as a drug, proteins need high degree of purity and stability at the pharmaceutical formulation and physiological conditions. Additionally, immunogenicity and inflammatory potential are important concerns. Therefore, nanotechnology has been gaining importance as a toll for enzyme delivery. Safe and effective therapeutic enzyme preparations can be obtained with the help of this tool, and nanodelivery systems usually stabilize protein drugs against denaturation by enzymatic digestion, increasing their potential application in the pharmaceutical field (Pachioni-Vasconcelos et al., 2016).

The type of nanocarrier to deliver proteins has to be carefully chosen with regard to the drug incorporation process and loading capacity, since proteins are usually large and hydrophilic. The pH and thermal protein instability have also to be considered when choosing the encapsulation process. Route of administration is equally important, since proteins are generally unstable in the gastrointestinal tract and present low mucosal permeability. As a consequence, oral bioavailability is usually low and erratic, making the parenteral route the first choice. Even when administered subcutaneously or intramuscularly, systemic bioavailability is often low and variable. Some of the nanotechnological options for protein drugs delivery are discussed in Chapter 19.

Enzyme-albumin conjugates are also an interesting therapeutic alternative. Albumin is the most abundant protein in plasma and works as a transport protein, with half-life of 70h. Poznansky and Antohe (1997) have demonstrated that in the presence of excess albumin, enzymes such as uricase, catalase, superoxide dismutase and L-asparaginase present decreased immunogenicity and protease degradation, and consequently increased half-life. This result is in agreement with enzyme immobilization in solid supports (Fágáin, 2003) with the advantage of the solubility of the enzyme albumin complex. Immunogenicity reduction is associated with allosteric neutralization of antigenic sites at the enzyme surface. Nonetheless, site specificity is not achieved with enzyme-albumin complexes unless a site-directing group is added to the complex.

16.2.3 Acylation of Enzymes

Several enzymes of pharmaceutical interest are classified as serine proteases, *i.e.*, they present a serine residue at the active site and catalyze protein hydrolysis. The β-hydroxy group of serine covalently binds to the substrate molecule, and in general, the imidazolium ring of a histidine residue acts as a proton donor/acceptor (Markwardt, 1997). The peptide (amide) bond cleavage by serine proteases occurs in three steps:

$$\textbf{Step 1}\left[\text{E} + \text{S} \underset{k_1}{\overset{k_{-1}}{\rightleftarrows}} \text{E-S}\right];$$

$$\textbf{Step 2}\left[\text{E-S} \xrightarrow{k_2} \text{P}_1 + \text{E-S}'\right];$$

$$\textbf{Step 3}\left[\text{E-S}' \xrightarrow{k_3} \text{P}_2 + \text{E}\right]$$

where: E=free enzyme; S=substrate (protein or peptide); E-S=1st enzyme-substrate complex; E-S'= 2nd enzyme-substrate complex; P_1 and P_2=reaction products [if S is a protein, the products are peptides); k_1, k_{-1}, k_2 and k_3 are the velocity constants of the reactions involved in the catalytic mechanism.

It is well established that the global velocity of enzymes is controlled by step 3 and (k_3) is the lowest of the constants involved in the catalysis. In the second step, lowering pH and/or temperature can even interrupt the reaction. Serine proteases can be stabilized by means of acylation of the serine residue at the catalytic site with low molecular weight esters. For the therapeutic use of serine proteases, they can be administered in the acylated form, as prodrugs. The active enzyme will be generated *in situ* by means of serine deacetylation at the slightly alkaline blood pH. This strategy allows slow activation and avoids the protease inhibition by normal plasma constituents, significantly increasing the protease plasma half-life. Table 16.3 presents the serine proteases approved as biological drugs.

16.2.4 Molecular Modelling of Enzymes

The biological function of an enzyme is strictly correlated with its tridimensional structure that, in turn, depends on the protein stability in the medium where it will act as a

TABLE 16.3
Serine Proteases Approved as Biological Drugs by the Food and Drug Administration

Use	Protease	Indications	Year of FDA Approval
Thrombolysis	Urokinase (u-PA)	Thrombus, catheter clearing	1978
	t-PA (alteplase)	Acute myocardial infarction (AMI), stroke, catheter clearing	1987 (AMI) 1996 (stroke) 2002 (catheter clearing)
	Reteplase (retavase)	AMI	1996
	TNK-tPA (tenecteplase)	Myocardial infarction	2000
Procoagulant	FIX	Haemophilia B	1990
	FVIIa	Haemophilia A and B	1999
	Thrombin	Bleeding	2006
Sepsis	Activated protein C, (drotrecogin alfa)	Sepsis, septic shock	2001
Digestion	Pancrelipase	Exocrine pancreatic insufficiency	2009

Source: Adapted from Craik, C.S., Page, M., Madison, E.L., *The Biochemical Journal*, 435(1), 1–16, 2011.

drug. Molecular modelling refers to theoretical and computational methods used to model or mimic the behaviour of molecules. Several molecular modelling methods are available; usually, they consider atoms as the smallest individual unit (molecular mechanics approach) or explicitly model electrons of each atom (quantum chemistry approach). Therefore, molecular modelling is an important tool to investigate the structure, dynamics, surface properties and thermodynamics of enzymes and proteins in general. It helps in the elucidation of an enzyme mechanism of action, interaction with the body constituents as well as the planning of structural modifications of this enzyme to improve pharmacokinetics and pharmacodynamics. At the end, it might result in higher efficacy of the enzyme drug and lower cost for the treatment.

Structural information of proteins can be experimentally obtained from X-ray crystallography (a technique that allows precise spatial localization of the atoms of the protein based on its crystal analysis) and NMR spectroscopy (it does not require the protein crystal but larger samples). Based on the previously solved structure of a protein, molecular modelling tools can be used to solve the structure of a related protein (Vriend et al., 1997). Homology degree is important on the success of this strategy and, usually, it is accepted that 80% or higher corresponds to high homology between proteins, 50%–70% corresponds to medium and below 50% we have low degree of homology. In general, the structure of a protein can be precisely inferred when homology models consider proteins with a high degree of homology.

16.2.5 Important Aspects of Therapeutic Enzymes

The activity of therapeutic enzymes has to follow standards, and therefore, it should be investigated under optimal conditions, with and adequate substrate at saturated concentrations thorough the reaction course. The substrate should be as specific as possible, and reaction conditions should be as close as possible to the enzyme physiological conditions, what can be a difficult task. The fibrin plaque adhered method to detect plasmin and plasminogen activators, for example, is difficult to reproduce amongst laboratories. From the physicochemical viewpoint, it is hard to describe a fibrin cloth with plasmin inhibitors adsorbed. Usually, when enzymatic reactions take place at surfaces with non-uniform porosity and absorption capacity, reproducibility is poor. For proteases, already highlighted as therapeutic proteins, some degree of difficulty can be present when natural substrates (proteins) are used in quantification methods. In this case, a standard assay might be run in parallel with a standard substrate, such as albumin (Lauwers, 1997).

Many therapeutic enzymes are dosed using physiological substrates and/or in the presence of biopolymers (proteins, polysaccharides, emulsions etc.), what can generate a series of problems and difficult result interpretation. For proteases, for example, hydrolysis might happen at different protein residues and with different velocities. Some cleavable regions become accessible only after an initial degree of hydrolysis and some of the reaction products are released only after denaturation. Therefore, enzymatic mechanism becomes complex and only apparent kinetic parameters are obtained. Experimental results are strongly influenced by reaction conditions. Bromelain and pancreatic protease, for example, are dosed with casein as a substrate and activity is proportional to the concentration of soluble peptides resulting from casein hydrolysis. Any protein contamination on casein will influence the results.

Lipases represent a particular group of esterases since they act specifically on lipophilic esters, at the oil/water interface. Physicochemical properties of the emulsified substrate determine the kinetic parameters. In this sense, emulsions with the same oil/water ratio can present different sizes of emulsion drop, resulting in different activity measurements. Cofactors and detergents at the emulsion oil/water interface influence the lipolysis velocity and therefore the determination of enzyme activity. Specifically, surfactants lower the interfacial tension and may facilitate the lipase contact with substrate. Colipase, on the other hand, binds to a non-catalytic domain of lipase stabilizing an active conformation and increasing the hydrophobicity of its binding site (Verger, et al., 1999).

Lysozyme, an antimicrobial enzyme, catalyzes the hydrolysis of 1,4-beta-linkages between N-acetylmuramic acid and N-acetyl-D-glucosamine residues in peptidoglycan, the major component of gram-positive bacterial cell wall (Manchenko, 1994). Enzymatic dosage of lysozyme is based on the turbidity measurement generated by the lysis of a *Micrococcus lysodeikticus* suspension. Therefore, to have a precise and reproducible turbidity measurement, cell suspension must be standardized. Characteristics of the suspension (bacterial aggregate), variation between batches (the cell wall width depends on the growing phase at which the sample was collected), pH and ionic strength of the buffer employed (giving the electrostatic interaction between lysozyme and negatively charged proteoglycan) significantly affect the rate of cell lysis ((Lauwers, 1997).

Enzyme formulations for digestive disorders often contain cellulase (from *A. niger*) that is an enzymatic complex constituted by exo β-1,4 glucanases, β-glucosidase and cellobiohydrolase, amongst other carbohydrases. Activity measurements based on the concentration of reducing sugars formed are not sensitive for dosing endocellulase. These enzymes catalyze the random hydrolysis of cellulose molecules generating low amounts of reducing sugars. Therefore, determination of the substrate solution viscosity change is the most appropriate method for dosing endocellulase activity. Conversely, the exocellulase activity generates appreciable amounts of reducing sugars. Nowadays, there is a well-standardized viscosimetric determination of endocellulase activity by using hydroxyethylcellulose as a substrate (a polymer with macromolecular structure well defined). Thereby, a laser device that allows precise determination of the substrate chain length reduction follows the number of glycoside bonds disrupted during hydrolysis.

Highly soluble synthetic substrates of well-defined molecular structure were developed based on the elucidation of enzyme specificity. Based on these substrates, reproducible assays were developed allowing the determination of the enzyme turnover number. Large variety of specific synthetic substrates is found in the market. They present a domain specifically recognized by the enzyme from which results a derivative easily quantified through conventional analytical methods (titration, spectrometry and viscosity, amongst others).

However, natural macromolecular substrates – the so-called physiological substrates – contain a greater number of binding sub-sites than the synthetic ones and are usually more selective for specific enzymes. Synthetic substrates give less information on the structural characteristic of the enzymes binding sites. For instance, some substrates (chromogenic tripeptides) act as inhibitors of serine proteases (Lauwers, 1997). The expression of enzyme activity in terms of unit per molar concentration is possible if the mechanism of reaction and the rate equation are known.

16.3 ENZYMES IN CLINICAL ANALYSIS AND COSMETICS

16.3.1 Clinical Analysis

Enzymes in clinical analysis are employed in three different manners. First, as chemical reagents for dosing specific substances present in biological fluids such as glucose (peroxidase and glucose oxidase), urea (urease associated with glutamate dehydrogenase), cholesterol (peroxidase, cholesterol esterase and cholesterol oxidase), ammonia (glutamate dehydrogenase), inorganic phosphate (sucrose phosphorylase), steroids (β-glucuronidase), creatinine (creatininase) and ethanol (alcohol dehydrogenase).

Second, detection of enzymes in biological fluids as markers for the diagnostic of several diseases. Abnormal enzyme activities present in blood would be useful for identifying illnesses such as prostate cancer and Gaucher's syndrome (acid phosphatase), liver abnormality (alanine aminotransferase and γ-glutamyl transferase), excess of aldosterone (rennin), myocardial infarction and testicular cancer (lactate dehydrogenase), cardiac and muscular pains (creatinine kinase). In all cases, analytical results should be associated to clinical observations.

Third, enzymes linked to antigens or antibodies are employed in the so-called enzyme-immunoassays (EIA), highly sensitive and specific tests that enable the precise determination of an antigen or antibody present in biological fluids. Details on this technique are presented in Chapter 18.

16.3.2 Cosmetics

Enzymes in cosmetology can be used either directly on the skin – to attain a desired effect such as elimination of wrinkles – or indirectly (an enzyme naturally present in the skin is a target for activating or inhibiting substances).

Examples of enzymes used directly are: (a) *proteases*: used in cleansing preparations for denture cleaning and skin peeling; it can be associated with lipase for eliminating acne and scurf; (b) *superoxide-dismutase and catalase*: elimination of free radicals responsible for skin aging; (c) *glucoamylase and glucose oxidase*: used in dentifrices for avoiding the formation of dental plaque; (d) *lactase, peroxidase and uricase*: used for hair dying; (e) *hyaluronidase and lipase*: used in cellulitis treatment; (f) *alkaline phosphatase*: as fibroblast stimulator; (g) *lipase, protease and amylase*: components of shampoos for removing the scalp oiliness; (h) *tyrosinase*: as component of sun tanners.

As examples of enzymes as targets for activators or inhibitors substances, one can mention (a) *metalloproteases*: the inhibition of these enzymes aims to reduce the hydrolysis of collagen and elastin, avoiding skin harshness; (b) *5-α-reductase*: its inhibition avoid the excessive action of sebaceous glands; (c) *tyrosinase*: slowing down its activity helps skin pigmentation; (d) *dopa oxidase*: its activation stimulates melanin production and, therefore, skin tanning; (e) *lipase*: its inhibition by ethyl citrate or ethyl lactate helps to prevent the bad smell associated to transpiration.

Cosmetic products must be adequately formulated and administered as creams, lotions, mouthwashes, dentifrices, soaps (liquid or solid), sprays etc. Therefore, the enzymes incorporated in these formulations must (a) be active along all shell-life of the product; (b) be stable in the presence of all components of the formulation, especially surfactants; and (c) not result in allergic reactions (Soto-Mera et al., 2000).

16.4 FINAL CONSIDERATIONS

One of the most desired properties of a molecule as a drug is selectivity, and enzymes are naturally designed with this purpose. These biomolecules are so attractive that catalytic activity has been even adapted to other biomolecules such as antibodies, in the form of ABzymes, and also nanostructures (nanozymes). Several enzymes find application as drugs, and the advances in molecular biology and nanotechnology have been contributing significantly to overcome the natural challenges of enzymes as drugs and medicines. In this chapter, we presented the main features of enzymes as drugs and medicines, as well as some specific enzymes worth mentioning giving the importance in therapy. Certainly, novel therapeutic enzymes will arise in the near future, especially for enzyme replacement therapy, as well as novel alternatives to improve the therapeutic potential of this class of molecules. Nonetheless, we should keep in mind that basic concepts of biochemistry and enzymatic catalysis are the ground floor for a successful development of a novel therapeutic enzyme alternative.

REFERENCES

Addezio, F.; Yoriyaz, E.J.; Cantarella, M.; Vitolo, M. Sucrose hydrolysis by invertase using a membrane reactor: effect of membrane cut-off on enzyme performance. *Brazilian Journal of Pharmaceutical Sciences*, v.50, 257–259, 2014.

Antohe, F.; Poznansky, M.J. The endothelium monolayer: barrier to enzyme and drug delivery. In: Lauwers, A.; Scharpé, S. (Eds.). *Pharmaceutical enzymes*. New York: Marcel Dekker Inc., 1997, 21–38.

Buchholz, K.; Kasche, V.; Bornscheuer, T.V. *Biocatalysts and enzyme technology*. Weinheim, Germany: Wiley-VCH, 2012, 583p.

Craik, C.S.; Page, M.; Madison, E.L. Proteases as therapeutics. *The Biochemical Journal*, v.435, n.1, 1–16, 2011.

Demain, A.L. The business of biotechnology. *Industrial Biotechnology*, v.3, n.3, 269–283, 2007.

Demeester, J. Biopharmaceutical aspects of enzyme adsorption. In: Lauwers, A.; Scharpé, S. (Eds.). *Pharmaceutical enzymes*. New York: Marcel Dekker Inc., 1997, 1–21.

Devasena, T. *Enzymology*. Oxford: Oxford University Press, 2012, 588p.

Fágáin, C. Enzyme stabilization – recent experimental progress. *Enzyme Microbial Technology*, v.33, n.2–3, 137–149, 2003.

Freedonia®. Enzymes: Specialty & Industrial, Forecasts to 2006 & 2011 (Study #1574), 2002.

Freedonia®. World Enzymes to 2009, Freedonia Group Incorporated, 336 p, 2005.

Gerhartz, W. *Enzymes in industry: Production and applications*. Weinheim, Germany: Wiley-VCH, 1990, 516 p.

Godfrey, T.; Reichelt, J. *Industrial enzymology: The application of enzymes in industry*. Surrey: MacMillan Press, 1983, 582 p.

Godfrey, T.; West, S. *Industrial enzymology*. 2. ed. London: MacMillan Press, 1996, 609 p.

Kang, T.S.; Stevens, R.C. Structural aspects of therapeutic enzymes to treat metabolic disorders. *Human Mutation*, v.30, n.12, 1591–1610, 2009.

Lauwers, A. Good standardization practices. In: Lauwers, A.; Scharpé, S. (Eds.). *Pharmaceutical enzymes*. New York: Marcel Dekker Inc., 1997, 335–341.

Lauwers, A.; Scharpé, S. *Pharmaceutical enzymes*. New York: Marcel Dekker, 1997, 401 p.

Manchenko, G.P. Lysozyme. *Handbook of detection of enzymes on electrophoretic gels*. Boca Raton: CRC-Press, 1994, 223.

Markwardt, F. Acyl enzymes as prodrugs. In: Lauwers, A.; Scharpé, S. (Eds.). *Pharmaceutical enzymes*. New York: Marcel Dekker Inc., 1997, 53–72.

Meghwanshi, G.K.; Kaur, N.; Verma, S.; Dabi, N.K.; Vashishtha, A.; Charan, P.D.; Purohit, P.; Bhandari, H.S.; Bhojak, N.; Kumar, R. Enzymes for pharmaceutical and therapeutic applications. *Biotechnology and Applied Biochemistry*, v.67, 586–601, 2020, https://doi.org/10.1002/bab.1919.

Pachioni-Vasconcelos, J.A.; Lopes, A.M.; Apolinário, A.C.; Valenzuela-Oses, J.K.; Costa, J.S.; Nascimento, L.O.; Pessoa Jr, A.; Barbosa, L.R.S.; Rangel-Yagui, C.O. Nanostructures for protein drug delivery. *Biomaterials Science*, v.4, n.2, 205–218, 2016.

Poznansky, M.J.; Antohe, F. Enzyme-albumin conjugates: Advantages and problems with targeting. In: Lauwers, A.; Scharpé, S. (Eds.). *Pharmaceutical enzymes*. New York: Marcel Dekker, Inc., 1997. 39–52.

Soto-Mera, M.T.; Lopéz-Rico, M.R.; Filgueira, J.F.; Villamil, E.; Cidrás, R. Occupational allergy to papain. *Allergy*, v.55, n.10, 983–984, 2000.

Verger, R.; Van Tilbeurgh, H.; Cambillan, C.; Bezzine, S.; Carriere, F. Colipase: Structure and interaction with pancreatic lipase. *Biochimica Biophysica Acta*, v.1441, n.2–3, 173–184, 1999.

Vriend, G.; Hendriks, D.; Scharpé, S. Molecular modeling of pharmaceutically important targets. In: Lauwers, A.; Scharpé, S. (Eds.). *Pharmaceutical enzymes*. New York: Marcel Dekker Inc., 1997, 73–105.

Yari, M.; Ghoshoon, M.B.; Vakili, B.; Ghasemi, Y. Therapeutic enzymes: Applications and approaches to pharmacological improvement. *Current Pharmaceutical Biotechnology*, v.18, 531–540, 2017.

17 Aspects of the Immobilization Technique

Michele Vitolo
Universidade de São Paulo

CONTENTS

17.1 INTRODUCTION

The cell is a very small structure (its volume is about 10^{-12}mL) containing a great number of molecules which, inevitably, must interact between themselves. Moreover, the majority of known enzymes are linked to internal cell membranes. So inside a cell, enzymes still can act at high reaction rates, in spite of their interaction with insoluble structures. Taking this fact as a start point, researchers have been trying to reproduce this condition by using inert materials upon which enzymes can be bound using physical and/or chemical methods.

An immobilization technique was attempted for the first time in 1916 by Nelson and Griffin with the enzyme invertase being immobilized onto an activated charcoal support, but it wasn't until the 1960s before research into enzyme immobilization gained traction. Immobilization techniques have been developed so intensely that there are now several commercially available immobilized enzyme preparations. Immobilized glucose isomerase (Sweetzyme®) – just as an example – is used in the conversion of glucose from hydrolyzed starch into fructose syrup.

Besides enzymes, cells, organelles and medicines can now be immobilized too.

17.2 TYPES OF IMMOBILIZATION

17.2.1 ENTRAPMENT

Entrapment, which is extremely useful for immobilizing cells and high molecular weight (MW) enzymes, consists of separating the biocatalyst from the reaction medium through a semi-permeable membrane. The membrane must have adequate porosity in order to entrap the biocatalyst and leave free diffusion for molecules of low MW.

The main advantages of entrapment are immobilizing enzymes, cells or organelles; easy confection with the semi-permeable membrane; the material necessary to make the support is commercially available and affordable; the biocatalyst is protected against hydrolases (mainly proteases) and/or high MW inhibitors and the immobilization procedure does not cause significant damage to the macromolecular structure of the biocatalyst. However, two main disadvantages are that the support material usually cannot be recycled, and the method is not appropriate for enzymes which act on high MW substrates.

Immobilization of the biocatalyst, particularly enzymes, is typically carried out by one of three different methods, i.e., cross-linked matrices, encapsulation and microencapsulation.

17.2.1.1 Cross-Linked Matrices

Here, enzymes are entrapped within the interstitial spaces of cross-linked polymers. Entrapment is accomplished by cross-linking of the polymer after addition of the enzyme. The most commonly used polymer for cross-linking is polyacrylamide gel, which is formed by the reaction between acrylamide and N, N′-methylenebisacrylamide.

Cross-linking in the proceeding example is achieved by dissolving the enzyme in a buffer containing acrylamide (monomer), bisacrylamide (cross-linking reagent) and ammonium persulphate (or a mixture of hydrogen peroxide and ferrous sulphate). A polyacrylamide gel results after exposing the mixture to ultraviolet radiation (Cantarella, Alfani, and Cantarella, 1998).

Polymerization can have two types of damaging effects on the enzyme. First, monomer and free radicals can react with chemical groups in the active site of the enzyme, which are essential for the catalysis. Second, the enzyme could denature because of the heat generated during the polymerization reaction. Alfani et al. (1988) circumvented the overheating by conducting the polymerization between two glass plates sealed with a silicone strip, which is then submerged into a water bath at 5°C. The submersion is at a maximum 5 cm depth of water so that UV radiation can still penetrate through the water and glass and cause the cross-linking reaction to occur. An alternative method would be to perform the reaction in an emulsion of water and surfactant, with the later absorbing the heat generated. In this case, the water drops' size would determine the size of gel particles formed (Cantarella et al., 1996).

Another method that can reduce the HEAT generated during polymerization would be to substitute the monomer with an oligomer (polyethylene glycol dimetacrylate, for instance) susceptible to polymerization by UV radiation. Thereby, the gel formed is chemically inert and mechanically stable. Functional chemical groups can be inserted into the gel. A method based on this idea is the use of uretan derivatives as oligomers, which when mixed with water liberates carbonic gas and polyurethane foam. If the enzyme is added before the foaming, enzyme molecules become entrapped inside the interstices of the matrix. The final characteristic of the support can also be varied by choosing an uretan derivative most suitable for the application. The main advantages for using oligomers are (a) simple entrapment and smooth reaction conditions; (b) no enzyme inhibition; (c) control of matrix structure according to the oligomer MW; and (d) physical-chemical characteristics such as ionization capability and hydrophilic and lipophilic equilibrium can be changed during oligomer synthesis, and before addition of the enzyme (Cantarella et al., 1996).

17.2.1.2 Encapsulation

Several natural polysaccharides such as alginate and k-caragena will, under certain conditions, gelatinize and become useful supports for encapsulation.

Sodium alginate is a linear heteropolymer of D-manuronic and L-guluronic acids that is soluble in water. When an enzyme is dissolved, or cells or organelles are suspended in a solution of sodium alginate, then in the presence of a polyvalent cation (Ca^{2+}, Ba^{2+} amongst others), the solution becomes gelatinous, and the biocatalyst is encapsulated inside spheres which develop (Vitolo, 2019a, b). During the encapsulation procedure, special attention must be taken in order to avoid inclusion of ion quenching substances such as phosphate or EDTA, which have the capacity to disrupt the gel structure causing the biocatalyst to leach into the reaction medium. Encapsulation is a widely used immobilization method due to its technical simplicity, soft gelatinizing conditions and the availability of an abundant variety of low-cost alginates on the market (Arruda and Vitolo, 1999).

k-Caragena is another water-soluble polysaccharide obtained from seaweed, which gelatinizes in the presence of K^+ or NH_4^+ ions. If a biocatalyst is dissolved into a solution of k-caragena before gelatinization occurs, the biocatalyst remains entrapped within the gel. Additional gel hardening can also be achieved by addition of glutaraldehyde or hexamethylenediamine, which can also contribute to stabilization of the biocatalyst.

17.2.1.3 Microencapsulation

Enzymes can be immobilized into microcapsules formed by polymeric membranes, which are permeable to low MW substances and impermeable to the biocatalyst. Basically, there are two types of microencapsulation methods. One of these consists of preparing an emulsion by mixing an aqueous solution of the enzyme with an organic solution containing a polymer, such as cellulose nitrate, ethylcellulose, polystyrene or polyvinyl acetate. A second solvent is then added to the mixture in order to precipitate the polymer at the interface between a micro-drop of water which contains the enzyme and the organic phase containing the polymer. Another method consists of emulsifying the enzyme in a water immiscible solvent containing a hydrophilic monomer. By adding a hydrophobic monomer, the polymerization occurs at the immiscible interface between the water droplets and the solvent. This method of microcapsules has been used to produce nylon, polyester, polyurea and polyurethane.

The semipermeable membrane can protect the enzyme against antibodies and hydrolytic enzymes. The enzyme can also be further stabilized by cross-linking with a bifunctional reagent before being microencapsulated.

Microencapsulation is a versatile immobilization procedure because it allows simultaneous entrapment of different enzymes and cofactors because they are linked to hydrophilic polymers.

When the microcapsules are composed of lipid layers in an aqueous medium, the microcapsules are called liposomes. The lipid layer is typically formed by mixing egg lecithin, cholesterol and a lipid; while the enzyme is dissolved in the aqueous solvent. Due to the selective permeation of the liposome, an enzyme entrapped within the liposome is protected against potential inhibitors – such as the β-D-glucosidase, which is not inhibited by copper ion (Gerhartz, 1990). Adapting the composition of the lipid layer allows specific immunoglobulin to be introduced into the membrane so that the liposome can be targeted to specific tissues or regions of the body.

17.2.2 Bonding Formation

The side chains of amino acids that form an enzyme macromolecule have hydrophilic and/or hydrophobic chemical groups with different reactivity. Moreover, the macromolecule itself has a hydrophobic domain composed of amino acids that favour linking the enzyme with an inert support through weak bonds (absorption), simple or cross-linking bonds, which is adequate for immobilizing cells and enzymes.

17.2.2.1 Adsorption

Adsorption can be considered as the adhesion of an enzyme to the surface of a carrier that has not been specifically modified to allow for covalent attachment (Yoriyaz and Vitolo, 2014).

Adsorption is the first option when immobilization is planned, because it is cheap and easily accomplished. It must be remembered that the first immobilization attempted was the adsorption of invertase onto active charcoal (Vitolo, 2020a, b). Another example is absorption of cellulase onto cellulose for glucose production (Yoriyaz and Vitolo, 2014).

The first significant industrial application of an immobilized system was aminoacylase adsorbed onto DEAE-Sephadex for the separation of racemic mixtures of amino acids obtained by chemical synthesis. The enzyme-Sephadex interaction was made between the amino groups of the support and the carboxylic groups of the enzyme. A fixed-bed type reactor was used and consisted of a column filled with a DEAE-Sephadex aqueous suspension. After complete sedimentation of the Sephadex, a buffered solution (phosphate, pH 7.0) containing aminoacylase was introduced, and the enzyme remained adsorbed onto the bed. The column was then washed repeatedly with deionized water and the acyl D and L mixture of amino acids then added onto the top of the column. The L-amino acid could be crystallized of the fractions recovered from the column. The column maintained 60% of its initial activity for 32 days of continuous operation at 50°C and pH 7.0. The reactor activity could be restored by adding fresh aminoacylase buffered solution. The performance of this process could be measured by the fact that the support did not lose its adsorbing capability and there was no structural deterioration, which would lead to the enzyme detaching during the reaction along 2 years of continuous use (Godfrey and West, 1996).

Although this type of immobilization is easy to accomplish, the mechanisms involved in forming the union between the support and enzyme are complex and involve several kinds of non-covalent bonds which form simultaneously between the support and the enzyme, i.e., hydrogen bond, Van der Waals force, dipole-dipole interaction, electrostatic force and hydrophobic bonds, among others. These weak bonds are easily disrupted by changes in pH, ionic strength and the medium solvent, as well as the reaction temperature. The weakness of the interaction between the support and enzyme is the main factor restricting wider application of this kind of immobilization. The stability of the support/enzyme during use depends on how similar are the conditions of adsorption and the reaction medium. If electrostatic forces are responsible for the interaction between the support and enzyme, then any changes in pH and/or ionic strength of the reaction medium could cause the enzyme to detach from the support. In order to optimize the interaction between the support and enzyme, it is necessary to use the greatest surface area possible when using a porous support or, if the support has low porosity than it must be formed by the smallest of granules. For example, acid phosphatase adsorbed onto CM-cellulose can be stably maintained using a 2 M acetate-acetic acid buffer (pH 5.0). In the case of a porous support, enzyme desorption can be avoided if the porous diameter is on average at least twice the diameter of the enzyme molecule. It must be remembered that enzyme molecules adsorbed onto a porous support are not uniformly distributed across the entire surface, and that the support chosen must allow for the enzyme molecules to become distributed as near as to the surface as possible, while being inserted into the support to prevent against the action of denaturing substances (Yoriyaz and Vitolo, 2014).

The relative ratios of enzyme to support required in order to attain the maximal adsorption must be optimized. In principle, it would be reasonable to suppose that the higher the enzyme concentration, then more of the support that becomes saturated, if the amount of support remains constant. Often, this is observed. However, if the support has been saturated with enzyme, the interaction between the substrate and enzyme is dependent on the diffusion of substrate to the enzyme active site, and the reaction will preferentially occur with those enzyme molecules located at the support surface. As consequence, the overall efficiency of enzyme catalysis diminishes.

The catalytic performance of an enzyme can diminish as a consequence of the interaction between the support and enzyme molecules because: (a) of rearrangement of the molecular structure of the enzyme during adsorption due to the interactions between the protein amino acid side chains and the support. This possibility will be magnified if the enzyme molecule has a quaternary level of arrangement, i.e., the molecule is formed by two or more peptides; (b) the interaction of support with a prosthetic group linked to the enzyme molecule (indispensable for the enzyme catalysis); and (c) the enzyme is adsorbed on the support by an amino acid located at the active site. Moreover, an increase in solvent temperature could improve the interaction between the enzyme and support by increasing the diffusion of the enzyme into the support material, provided this raise in temperature does not affect the stability of the enzyme.

Adsorption can be improved either by modification of the enzyme structure or the structure of the support. For example, krill's chitin treated with CS_2 has higher adsorbing capability than untreated chitin (Barros and Vitolo, 1992). This is because there is about a 30% conversion of amino acids in the chitin to negative charged dithiocarbamide derivatives, which have the capability on interacting with amine, guanidine, indol, imidazol and tiol groups of amino acid side chains of an enzyme. Whereas the amino groups of untreated chitin might only make bonds with carboxylic groups of an enzyme, for example, as in the adsorption of soy β-amylase onto phenylborate-agarose (Gerhartz, 1990).

Hydrophobic interactions could be improved by introducing chemical hydrophobic groups such as methyl-4-phenyl-butirimide into a protein molecule.

The stability of the adsorbed enzyme can be improved through additional cross-linking made with glutaraldehyde. For example, Rucka and Turkiewicz (1989) stabilized lipase in a politetrafluoroethylene membrane by using glutaraldehyde as the cross-linking agent. The authors also demonstrated that the procedure was simple, cheap and non-toxic. Glioxal and formaldehyde can also be used too. Moreover, the compounds cited also have antimicrobial properties which prevent contamination of the immobilized system during use. This type of immobilization is considered of mixed type in which one fraction of the protein is adsorbed and the remaining fraction is linked onto the support through covalent bonds.

Finally, it should be noted that several materials have been employed as adsorbent supports, including sandy alumina, amberlite CG-50, bentonite, phosphate-calcium gel, active charcoal, CM-cellulose, CM-sephadex, collagen, DEAE-cellulose, DEAE-Sephadex, controlled porous glass, silica gel, chitin, quitosana and agarose.

17.2.2.2 Covalent Binding

Binding between an insoluble support and an enzyme is established amongst the chemical groups of the support and those of the amino acid side chains of the enzyme. The amino acids involved in the bounding must not to be located at the active site of the enzyme.

The main chemical groups of amino acids involved in the linking are lysine ε-amino, cysteine sulphydril, aspartic acid β-carboxyl, glutamic acid γ-carbonyl, tyrosine hydroxyl, serine hydroxyl and threonine hydroxyl. Carboxyl and hydroxyl amino groups are the preferential targets for linking because these are the most abundant in proteins.

In general, this immobilization occurs in two steps. First, the support is activated with an appropriate reagent; second, after removing any excess activator, the enzyme and the activated support are placed together. The mixture is left at room temperature for a determined period of time in order for covalent bonds between the enzyme molecules and support to form.

Immobilization by covalent linking has advantages such as strong bonding between enzyme and support; easy interaction between enzyme and substrate (due to the superficial location of the enzyme molecules); improved thermal stability and resistance to the shearing force (thereby enlarging the choice of a reactor).

Nevertheless, this type of immobilization also has disadvantages, for example, the enzyme can become damaged due to the kind of reagents employed; conformational stress imposed upon the molecular structure of the enzyme resulting from the union with the support material could make insertion of the substrate into the active site difficult; difficulty on establishing the best immobilization conditions; inadequate for immobilizing cells; difficulty in recovering the support material after the system has lost its catalytic capability.

However, this method is useful when the enzyme is expensive and is thermally labile, such as enzymes typically employed in analytical methods.

The covalent binding between enzyme and support can be achieved using several organic synthesis pathways for example, diazotization, condensation, alquilation, reaction with cyanogen bromide, silane or azide derivatives).

17.2.2.3 Cross-Linking

Enzymes can be immobilized by cross-linking directly using functional cross-linking agents, which possess two identical functional groups, for example, glutaraldehyde, bis-diazobenzidine-2,2′-disulfonic acid, 4,4′ difluoro-3,3′dinitrodiphenyl-sulfone, toluene-2,4-diisothio-cyanate or hexamethylene-diisocyanate.

The cross-linking agent can aggregate the enzyme macromolecules forming either insoluble or soluble aggregates. When the aggregation results only from the interaction between enzyme molecules and cross-linking agent, the immobilization is called 'pure cross-linking'. Nevertheless, it is quite common to carry out cross-linking in the presence of a selected support in order to obtain a more stable immobilized enzyme system. In this case, the functional cross-linking agent establishes preferential bridges between the enzyme and the support, rather than between enzyme to enzyme. For example, diazobenzidine forms bonds between phenyl and amino groups belonging to both an enzyme and the support, respectively.

The addition of a cross-linking agent can be made to an enzyme already adsorbed onto the support, or onto the free enzyme followed by adsorption onto the support. Moreover, it is also possible to mix the support and the cross-linking agent first, followed by the enzyme (Pessela et al., 2008).

Another possibility can be cross-linking the enzyme with a polymer which is soluble under the enzyme reaction conditions but which becomes insoluble when a reaction parameter such as pH is changed. Takeuchi and Makino (1987) immobilized cellulase by cross-linking with poli-L-glutamic acid which is soluble at pH 4.5 but insoluble at pH 3.0.

17.2.3 Supports

Many solid or gelatinous materials can be used as supports for immobilization. The main requirements for a material to be used as a support are: (a) the physical characteristics are adequate for use in reactors; (b) maintaining the chemical and mechanical stability under the reaction conditions; (c) to have chemical groups adequate for linking the biocatalyst; (d) to be composed of particles of different diameters in order to attain an equilibrium between reduction of diffusion effects and adequate operation of the reactor, particularly when the reactor is of a fixed bed type; (e) having a porosity compatible with the biocatalyst to be immobilized; (f) to be resistant against microorganisms; (g) to have thermal stability; and (h) to be easily regenerated.

As previously mentioned, the diversity and characteristics of a support material can be extremely varied, beyond just being either inorganic or organic in chemical nature. Morphologically, the support can be porous or not porous. A porous support type, in turn, can be solid with defined porosity (pores of the same diameter and distributed homogeneously over the support surface) or irregular porosity (porous having different diameters), as well as semi-permeable reticulated, membranous or copolymerized gels. The physical characteristic of the support material can be as a powder, fibre or membrane amongst others.

The main advantage of a non-porous support material is the deposition of enzyme molecules along the external surface of the particles, which facilitate the interaction between the enzyme and the enzyme substrate. Nevertheless, the small surface area is the main disadvantage of this kind of support. However, this unfavourable aspect can be circumvented if the support particles have a quite small diameter. It must be borne out that support materials composed of small particles can become useless if the immobilized system is intended to be used in a fixed bed reactor because after a short period of operation, the bed becomes impermeable to the flux of substrate solution or in a fluidized bed reactor where floatation of the particles makes retention of the enzymes within the particles difficult. However, these problems do not occur in membrane type reactors.

A porous support has a greater surface area where enzyme molecules can become distributed either over the external or internal porous surfaces. Charged chemical groups located inside the pores can facilitate catalysis depending on the degree of ionization of the substrate under the reaction conditions. However, this aspect can become useful when optimizing the catalysis if the enzyme and reactor characteristic are also adequately evaluated.

It must be remembered that when a substrate is moving inside porous particles, this diffusion can make interaction with the catalyst difficult. In such a case, the substrate MW becomes a decisive factor that determines the yield of the reaction. Nevertheless, the inner location of the enzyme molecules could offer extra protection against the turbulence of the reaction medium inside the reactor.

Another point to be considered is the physical stability of the support material, i.e., rigid (inorganic) or elastic (organic). Rigid material has advantages such as resistance to deformation when packed in a column type reactor, protection conferred to the enzyme molecules against shear forces, avoiding damage to the tertiary and/or quaternary structure of the enzyme molecules. Elastic material can be manufactured as a small width membrane having a great surface area thus avoiding poor diffusion problems. Nevertheless, due to its flexibility, ridged materials can deform during use which can affect the conformational structure of the enzyme.

Finally, the growing interest in biotransformations has stimulated the increasing commercial availability and diversity of supports in general. Covalent binding immobilization supports that are currently available include kaolin activated with glutaraldehyde (Biofix®), derivatised polyacrilamide (Enzacryl®), controlled porous glass (Glycophase®), kieselgur activated with glutaraldehyde (Macrosorb®) and Sepharose® activated with CNBr.

The choice of the immobilization method will depend on two factors: (a) the type of biocatalyst; and (b) the envisaged use of the immobilized system.

General rules for choosing the support material and immobilization method do not exist and can only be established empirically. However, many of the empirics can be circumvented insofar as the protein engineering that is being developed or optimized (Pessela et al., 2008).

17.2.4 Effects Caused by Immobilization

17.2.4.1 Steric and Conformational Effects

When an enzyme is linked to a support, the enzyme can undergo conformational changes that diminish its catalytic performance. The way in which an enzyme and support interact can cause the active site domain of the enzyme to become less accessible to the substrate (Steric hindrance), leading to reduced catalytic efficiency. These steric and conformational effects are difficult to evaluate.

17.2.4.2 Diffusion and Mass Transport Effects

When an enzyme is immobilized onto or inside a solid support, the substrate must diffuse from the reaction medium to the enzyme active site. So, when the diffusion rate of the substrate is lower than the transformation rate of the enzyme, the overall reaction rate observed is lower than that expected for a given enzyme concentration, because the enzyme is not fully saturated with the substrate.

This phenomenon can be expressed by the effectiveness factor (f). This factor is defined as the ratio between the observed (v') and standard (v) rates:

$$f = v'/v \quad (17.1)$$

If the enzyme kinetics follows the equation:

$$v = \left[V_{max} \cdot (S)\right] \div \left[K_M + (S)\right] \quad (17.2)$$

Substituting Equation 17.2 into Equation 17.1 gives:

$$v' = f\left\{\left[V_{max} \cdot (S)\right] \div \left[K_M + (S)\right]\right\} \quad (17.3)$$

Equation 17.3 can then be rearranged as follows:

$$\left(1/v'\right) = \left[K_M/(S) \cdot f \cdot V_{max}\right] + \left(1/f \cdot V_{max}\right) \quad (17.4)$$

Equation 17.4 can be linear or non-linear because 'f' is a function of the substrate concentration available for catalysis.

There are two types of diffusion process:

a. *External diffusion*: this occurs when the substrate moves from the reaction medium to the catalysis surface. Thereby, the substrate molecules must cross a fluid layer during which substrate to product transformation does not occur. According to Fick's law, only a linear gradient of substrate concentration would be established throughout the fluid layer. The external diffusion effect can easily be minimized by increasing the agitation of the reaction medium.
b. *Internal diffusion*: this occurs when the substrate moves inside the pore of the supporting material. In this case, catalysis and diffusion occur simultaneously, so that a linear substrate gradient concentration does not occur throughout the immobilized system.

17.2.4.3 Effects of Microenvironment

When an enzyme is immobilized, it enters a different spatial environment, especially if the matrix or support has a net charge. This new spatial environment can affect kinetic parameter values (V_{max}, K_M).

Microenvironment effects are dependent on the physical and chemical composition of the support material and can promote an unequal distribution of substrate, cofactors and product between the region surrounding the immobilized enzyme and the external solution. Generally, this is due to the influence of electrostatic and/or hydrophobic interactions between the support and low molecular weight chemical species present in the external solution.

17.2.4.3.1 Partition

The kinetic behaviour of an enzyme linked to a charged support differs from the kinetics of the enzyme if the enzyme were in the solution, even in the absence of diffusion effect. This can be attributed to the concentration of chemical species (substrate, product, ions etc.) near the enzyme which can be different from that in the external solution, due to electrostatic interactions between those chemical species and the charges present in the support. Partition is an example of a microenvironment effect (Figure 17.1).

Let's consider the case in which the chemical specie is the hydrogen ion (H_3O^+). In order to obtain a mathematical model, let's take into account the following: $^*\mu$=electrochemical potential of the microenvironment; $^*\mu_1$=electrochemical potential of the bulk solution; μ=chemical potential of the microenvironment; μ_1=chemical potential of the bulk solution; z=charge of the chemical species; ψ=electrostatic charge of the support; K=Boltzmann constant; T=absolute temperature; a=activity of the chemical species; μ^o=standard chemical potential of the charged chemical species (it is a characteristic parameter of a particular chemical species, independent on the region in which the species is located, i.e., near the enzyme or in the bulk solution).

When the system reaches equilibrium:

$$^*\mu = {}^*\mu_1 \tag{17.5}$$

$$\text{But } {}^*\mu = \mu + z\psi \tag{17.6}$$

$$^*\mu_1 = \mu_1 \tag{17.7}$$

$$\mu = \mu^o + K \cdot T \cdot \ln a \tag{17.8}$$

$$\mu_1 = \mu^o + K \cdot T \cdot \ln a_1 \tag{17.9}$$

Substituting Equations 17.6 and 17.7 into 17.5 gives:

$$\mu + z \cdot \psi = \mu_1 \tag{17.10}$$

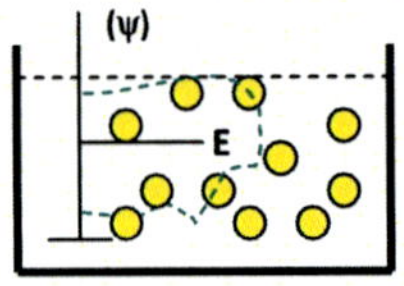

FIGURE 17.1 Partition effect. The dashed line (green) represents the border between the region in which the enzyme molecules (E) are quite near to the charged support (ψ) [microenvironment] and the bulk of solution. The yellow circles represent the chemical species (hydrogen ion and/or substrate ionized) distributed between the bulk solution and microenvironment.

Substituting Equation 17.8 and 17.9 into 17.10 and considering $z=1$ (the hydrogen ion has a charge of one):

$$-\ln a - (-\ln a_1) = \psi / K \cdot T \text{ or } \mathrm{pH} - \mathrm{pH}_1 = (\psi \cdot 0.43)/K \cdot T \tag{17.11}$$

From Equation 17.11, it can be observed that due to the charges of the support material, the pH of the microenvironment near the enzyme differs from that of the bulk solution (pH_1). Drawing pH versus activity plots for soluble and immobilized enzyme, it can be noted that the profiles of the curves are not coincident.

From Equation 17.11 it can be observed: (a) *anionic support* ($\psi<0$): as $\psi/K{\cdot}T<0$ than $pH<pH_1$; and (b) *cationic support* ($\psi>0$): as $\psi/K{\cdot}T>0$ than $pH>pH_1$.

Let's consider the case of a charged substrate.

Taking into account the same considerations made for the hydrogen ion, the resulting equation is:

$$S = S_1 \cdot (e)^{(-z\psi/K \cdot T)} \tag{17.12}$$

where S=substrate concentration in the microenvironment; S_1=substrate concentration in the bulk solution; z=electric charge of the substrate.

If the enzyme kinetics follows the equation:

$$v' = (V_{max} \cdot S)/(K_M + S) \tag{17.13}$$

Where v'=the immobilized enzyme activity.

Substituting Equation 17.12 into 17.13 gives:

$$v' = \left[V_{max} \cdot S_1 \cdot (e)^{(-z\psi/K \cdot T)}\right] \div \left[K_M + S_1 \cdot (e)^{(-z\psi/K \cdot T)}\right] \tag{17.14}$$

If $S_1=K_M{\cdot}(e)^{(z\psi/K{\cdot}T)}$ then $v'=(V_{max}/2)$. So, the substrate concentration in the bulk solution (S_1) which leads to $V_{max}/2$ corresponds to an apparent Michaelis constant (K'_M) related to K_M (when the enzyme is not immobilized) by the equation:

$$K'_M = K_M \cdot (e)^{(z\psi/K \cdot T)} \tag{17.15}$$

Equation 17.15 shows clearly that the electrostatic potential of the support material interferes directly with the K_M and on the overall catalytic performance of the enzyme.

The partition effect can be used to favour enzyme activity if the charges of both substrate and support are opposite. Attraction favours the accumulation of ionized substrate molecules into the microenvironment, in which they are transformed to product by the enzyme.

17.2.4.4 Advantages and Disadvantages of the Immobilization Technique

The main advantages of the immobilization technique are in enzyme recovery, increase in enzyme stability and use in a continuous process. When immobilization is successful, the three points cited above occur simultaneously, and the application of the immobilized system on an industrial scale can in some cases lead to a reduction of about 50% in overall process costs (Powell, 1996).

As immobilization is an artificial technique – insofar as the enzyme and the support do not have any natural affinity – immobilization also has its disadvantages such as unpredictable interactions amongst the chemical groups of both enzyme and support which could result, for instance, in only a fraction of enzyme molecules becoming immobilized relative to the support material, and in a configuration that completely prevents substrate access to the enzyme active site. A general method for immobilization does not exist, which means that the most optimal pairing between enzyme and support material must be optimized empirically. The success of covalent binding immobilization depends on the availability of enzyme in high purity. The unpredictable nature of the enzyme to support interaction – in the sense that any amino acid of the protein, including amino acids of the active site domain, can be bound to the support – can exacerbate steric and conformational effects leading to a loss in enzyme activity (Table 17.1).

17.2.4.5 Applications

Immobilized enzymes are employed either in bioreactors or analytical methods.

There are a large variety of analytical methods used nowadays, and so only two, enzyme electrodes and enzyme immunoassay, will be described here.

17.2.4.5.1 Enzyme Electrodes

An enzyme electrode is a device formed in two parts, an electrochemical sensor and an enzyme immobilized onto the sensor surface (Figure 17.2). An electrochemical sensor can be amperometric or potentiometric, depending on the kind of compound to be analysed and the enzyme employed. Immobilization of the enzyme can be made using any of the immobilization methods already cited.

Generally speaking, the procedure for using an enzyme electrode begins with the electrode being placed into a sample that contains the compound to be determined (which is the substrate of the enzyme linked to the sensor surface). The substrate then diffuses across the semi-permeable membrane of the sensor surface (the membrane acts to stabilize, fix and protect the enzyme) reaching the enzyme active site, where a product is formed. The concentration of the product is measured through the electrochemical sensor. Two widely used examples of enzyme electrodes are

a. *Glucose/glucose oxidase (GO) electrode*: the reaction measured by this electrode is

$$\text{Glucose} + O_2 + H_2O + \text{GO} \rightarrow \text{Gluconic acid} + H_2O_2 + \text{GO}$$

TABLE 17.1
Variation in Kinetic Parameters for Yeast Invertase Immobilized onto Several Supports and by Different Immobilization Methods

a. Parameter	Calcium Alginate (Entrapment)	b. DOWEX 1X8-50 (Adsorption)	c. Krill Chitin (Covalent Bonding)	d. PED[a] (Covalent Bonding)	e. Enzyme (Soluble)
K_M (mM)	7.20	29.1	52.2	32.2	28.0
E_a (kJ/mol)	24.4	40.5	51.6	42.4	31.6
ΔH (kJ/mol)	21.9	37.9	49.0	39.8	30.0
pH	4.6	4.0	4.0	4.6	4.6
T (°C)	60	55	60	70	56
IC (%)	50.0	87.0	40.0	90.0	-

PED, low density polyethylene; IC, immobilization coefficient.

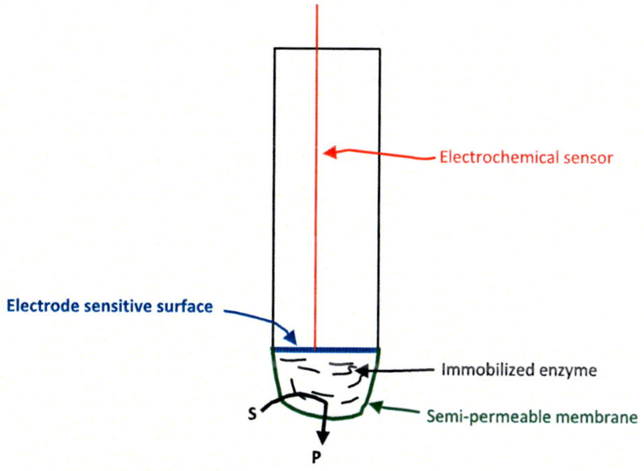

FIGURE 17.2 Generic sketch of an enzyme electrode.

The electrochemical sensor can be either amperometric (for measuring oxygen consumed or hydrogen peroxide formed) or potentiometric (for measuring pH change by formation of gluconic acid).

b. *Urea/urease electrode*: the reaction measured by this electrode is

$$\text{Urea} + 2H_2O + H^+ + \text{urease} \rightarrow 2NH_4^+ + HCO_3^- + \text{urease}$$

The electrochemical sensor is potentiometric, measuring either decrease in H^+ or increase in NH_4^+.

Undoubtedly, the invention of enzyme electrodes is one of the greatest applications of enzyme immobilization. In spite of great advancements in electronics and protein engineering, there are several problems still to be solved regarding enzyme electrodes, i.e., high cost of the pure enzymes required, difficulty in handling cofactors required for some enzymes, sensors are not available for all substances that one would wish to measure, short half-life, high sensitivity of potentiometric sensors to buffer solutions and monovalent cations and the poor performance of amperometric electrode at low oxygen conditions.

17.2.4.5.2 Enzyme Immunoassay

This technique is based on antigen-antibody interaction in which the enzyme is used as a marker. The test becomes quantitative due to the distribution of enzyme between the triad (antibody-enzyme-antigen) and the free form of the enzyme (antigen-enzyme or antibody-enzyme). This method is similar to other immunoassays (radio immunoassay, immune fluorescence etc.) but with the advantages of: (a) sensitive and specific assay; (b) extended shelf-life of the reagents; (c) simple manipulation; (d) fast assay times; (e) the possibility for performing several assays simultaneously; and (f) automation.

Enzymes commonly used include malate dehydrogenase (EC.1.1.1.37), glucose-6-phosphate dehydrogenase (EC.1.1.1.49), glucose oxidase (EC.1.1.3.4) and peroxidase (EC.1.11.1.7) amongst others.

These enzymes must have the following characteristic to be amenable for widespread application: (a) available in high purity; (b) high specific activity; (c) be stable under assay conditions, with a long storage shelf-life; (d) high solubility; (e) measurement of enzyme activity must be simple, fast and sensitive; (f) absent from biological fluids to be tested; (g) substrate and inhibitors also should be absent from the biological fluids; and (h) retain catalytic activity after immobilization.

There are basically two types of enzyme immunoassay, i.e., ELISA (Enzyme-linked Immunosorbent Assay) and EMIT (Enzyme Multiplied Immunoassay Technique). The ELISA always requires an insoluble step whereas EMIT is carried out with all reagents in the soluble form.

The mechanisms involved in ELISA and EMIT tests are quite variable, so that only two of them are presented below, just as examples:

a. ELISA (here the enzyme-marked antibody (Ac*) is in excess):

$$Ac^* + Ag \rightarrow Ac^* - Ag + Ac^* \rightarrow$$

$$Ag{-\!\!|} \rightarrow Ac^* - Ag{-\!\!|} + Ac^* - Ag$$

The mechanism has three steps: (a) interaction of all antigens present in the sample with the marked antibody (Ac*); (b) separation of all free marked antibodies by the immobilized antigen on a polystyrene surface $(Ag{-\!\!|})$ resulting $(Ac^* - Ag{-\!\!|})$; and (c) activity measurement of the soluble complex $(Ac^* - Ag)$. The antigen concentration in the original sample is calculated by the difference in the activities between Ac* (originally in excess) and $Ac^* - Ag$ (Goldberg, 2005).

b. EMIT (in the case of measuring hapten [H]):

The first assay based on this mechanism was used to determine morphine using lysozyme (EC.3.2.1.17) marked with a hapten (a chemical analogue of morphine). The modified enzyme, when put in contact with the morphine antibody, becomes inactivated. When the sample containing morphine is introduced, the antibody binds with all morphine molecules present in the sample, releasing an equal number of modified lysozyme molecules (Figure 17.3). The lysozyme activity is proportional to the concentration of morphine in the sample.

17.3 FUNDAMENTALS OF ENZYME REACTORS

A bioreactor is a conventional chemical reactor adapted for operating using biocatalysts (cells, enzymes and organelles).

When a bioreactor is designed, it must be taken into account if the biocatalyst to be employed will be cells or an

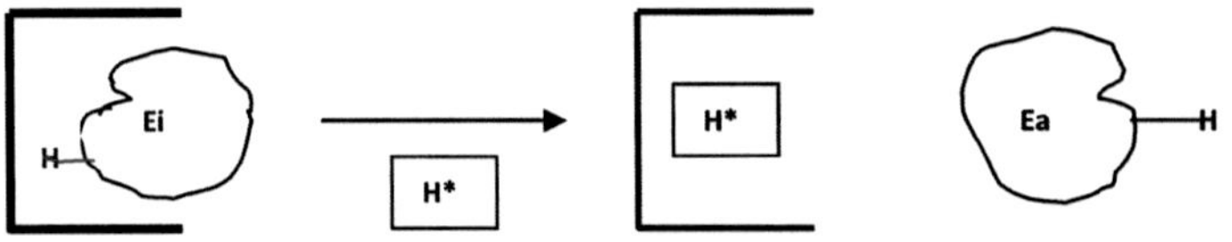

FIGURE 17.3 Generic sketch of the EMIT technique. The enzyme containing the hapten molecule [H] – which is a structural analogue of the molecule whose concentration will be measured [H*] – becomes bound to the specific antibody (Ac) for [H] and is inactivated (Ei). However, when a sample containing [H*] is added to the medium – the affinity of [H*] for the antibody is greater than that of the hapten [H], the enzyme is freed and becomes catalytically active (E_a). Thereby, the enzyme activity measurement is proportional to the [H*] present in the sample, insofar as the stoichiometry is 1:1.

enzyme extracted from the cells. In any case, the decision will be based on the following considerations: (a) formation of byproducts by the cells. Sometimes, the byproducts are not easily separated and will contaminate the desired final product. Moreover, a significant amount of substrate could be diverted for byproduct synthesis, leading to a loss in product yield during the bioprocess; (b) consumption of the product by the cell. If the cell metabolizes the product, the final yield will significantly be diminished; (c) cost for isolating, purifying and immobilizing the enzyme; (d) the origin of the enzyme: extracellular (low cost and available in huge amount) or intracellular (high cost and available in low amount); and (e) necessity of a cofactor for enzyme activity. For example, enzymes that require cofactors such as β-nicotinamide-adenine-dinucleotides are not used in industrial scale due to the high cost of the cofactor, which is sometimes more expensive than the enzyme. (Tomotani and Vitolo, 2008a, b).

Nevertheless, if it is concluded that the use of an enzyme is appropriate, then the next evaluation will be based on which form of the enzyme is best employed, i.e., soluble or immobilized. The following criteria will be taken into account: (a) will the catalytic capability of the enzyme be altered after immobilization. For instance, in Table 17.1 are shown some of the variations in kinetic parameters of invertase immobilized on different supports and by several immobilization methods; (b) the type of process in which the enzyme will be used. For instance, in bakery - α-amylase, pentosanase, amongst others – and in meat tenderizing (papain, bromelain), the enzymes must be soluble; (c) the operational stability of both forms of the enzyme. Thanks to the advances on enzyme engineering, it is nowadays possible to obtain highly stable immobilized enzyme systems (Pessela, et al., 2008); and (d) the biogenic origin of the enzyme either extracellular (used either in a soluble or an immobilized form) or intracellular (used only in an immobilized form). As a general rule, intracellular enzymes are – normally expensive – and are used in an immobilized form on an industrial scale, whereas extracellular enzymes can be used in a soluble form, except in some particular processes in which a continuous reactor is economically more interesting to industry. For example, extracellular enzymes such as glucoamylase, papain and quimosin can be used in immobilized forms for syrup production (starch saccharification), beer production (avoiding beer chill proofing at low temperature) and cheese production (conversion of k-casein into p-k-casein) (Nagodawithana and Reed, 1993).

17.3.1 Types of Enzyme Reactors

Enzyme reactors has been divided in two main types, i.e., discontinuous (including the fed-batch type) and continuous (Neves and Vitolo, 2020). Continuous reactors, in turn, can be fluidized bed or stationary bed types (Figure 17.4).

Discontinuous reactors in which the residence time is the same for reagents, products and catalyst (soluble or immobilized) have been used when the biocatalyst is cheap or has a short half-life during the process (Yoriyaz and Vitolo, 2014). Though quite simple in handling, modelling and adaptable for multiple uses (as a storage tank, a mixer, a decanting tank, amongst others), discontinuous reactors fail to protect the biocatalyst (an enzyme for example) against the inhibition, which is inevitably caused by excess of substrate or product in the reaction medium. To circumvent the latter problem, the so-called fed-batch process can be used, in which the substrate is added into the reactor in a step-wise manner so that the substrate concentration remains below a level that would otherwise be inhibitory. Moreover, in the fed-batch process the filling of the reactor and reaction occur simultaneously. If well planned, complete filling of the reactor would match the completion of the reaction. This procedure eliminates the so-called 'dead-time' associated with batch reactor types, which is the period of time required to completely fill the

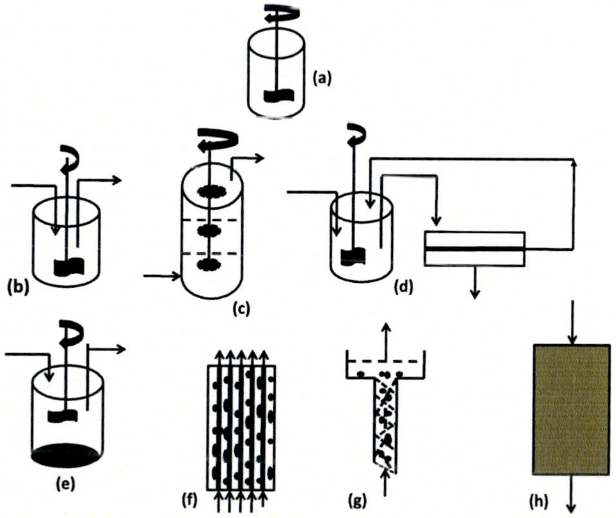

FIGURE 17.4 Types of bioreactors: Batch (a), continuous stirred tank reactor (CSTR) (b), multi-stage continuous reactor (c), bimodular membrane reactor (d), unimodular membrane reactor (e), hollow fibre reactor (f), fluidized bed reactor (g) and packed bed reactor (h).

reactor with the substrate solution. The dead-time in high volume batch reactor is huge.

Continuous types of bioreactor resulted from the development of the immobilization technique, which was effectively introduced on an industrial scale from the 1970s. As biocatalysts are linked to inert and insoluble supports by chemical or physical methods, continuous bioreactors have a variety of different designs to accommodate these supports such as embedded column with an immobilized system (e.g. plug flow reactor) or a column containing the immobilized system held in constant suspension (e.g. a fluidized bed reactor). In the latter type of reactor, the immobilized system is maintained in suspension by using mechanical pumps.

Another mode for maintaining the immobilized system in suspension would be the use of a discontinuous reactor where input and output pipelines are adapted. This reactor is called a continuous stirred tank reactor (CSTR). Fluidized bed reactors have the advantage of not presenting problems such as the formation of pH, temperature, substrate and product gradients inside the column.

In general, a CSTR is the first choice in reactor type for developing a new process because of operational flexibility (for example, the option of working with long periods of agitation) and utilization (these reactors are not designed to a specific process).

Amongst the different types of continuous reactors that deserve special mention are those that use a semi-permeable membrane. A Membrane Reactor (MR) can be designed either by adapting a membrane into an existing CSTR or packing several tubular membranes into a column (a so-called hollow fibre reactor, HFR). The HFR has mainly been employed for treatment of residual water, desalinization of sea water and in enzyme catalysis (in this case, the enzyme solution is disposed outside the hollow fibres and the substrate solution flows inside the fibres) (Romero et al., 2004; Yoriyaz and Vitolo, 2014; Nakla et al., 2006).

Membrane reactors – except the HFR – can be classified in two types by considering how the membrane and the biocatalyst are arranged. In one type, the biocatalyst is not linked to the membrane surface (in this case the membrane acts as a separation surface); whereas in the other type of reactor design, the biocatalyst is bound onto the membrane surface (the membrane, in this case, acts as both catalysis and separation surface) by entrapment, deposition in jelly form or linked through chemical interaction (adsorption, covalent bonding, for example).

There are two possible membrane/reactor arrangements, i.e., the membrane is located in a cassette coupled in series with the CSTR (MR-bimodular), or the membrane is adapted on the base of the CSTR (MR-unimodular; ultrafiltration cell type reactor).

The membrane reactor is a viable alternative to traditional reactors when one is using an immobilized enzyme (PFR, CSTR-TRADitional and fluidized bed reactor), because the catalyst need not necessarily be in an immobilized form as in CSTR-TRAD. However, a MR presents the same operational versatility as a CSTR-TRAD. Moreover the MR, in principle, allows integrating in one step the conversion, separation and concentration of the product, and recovery of the catalyst. These aspects can afford high productivity and reduction in costs when the process is scaled up (Giorno and Drioli, 2000). A MR has the following advantages: homogeneous catalysis; absence of diffusion, steric and conformational limitations; high activity by volume unit; possibility of working under aseptic conditions; constant productivity and the possibility of using multi-enzyme systems. Besides, the MR is quite useful on a laboratory scale for continuous monitoring of enzyme activity against inactivation effects caused by pH, temperature and other inhibitors. The operational conditions established on a laboratory scale are easily reproduced during process scale up (Tomotani and Vitolo, 2007b).

The crossing of solutes through a membrane can occur either by diffusion (the separation is based on the concentration gradient established between the upper and lower surfaces of the membrane) or by convection (establishment of temperature or pressure gradients across the membrane). The most commonly employed membrane reactors are of a convection type (Curcio et al., 2002). Moreover, the input of flux into a MR and the crossing of this flux through the membrane can occur in parallel (both perpendicularly to the membrane surface) or perpendicular to each other (the flux input and crossing through the membrane surface are tangential and perpendicular, respectively). The latter pattern is the most widely used in a MR because it avoids membrane polarization (Vitolo, 2020a).

In actual fact, the cost of the membrane does not affect the overall operational cost of a MR because the membranes are quite stable and can be regenerated several times. Different types of membranes are commercially available and are suitable for different applications. There are microfiltration (pore diameter: 0.1–0.5 μm), ultrafiltration (pore diameter: 0.001–0.1 μm) and nanofiltration (width between layers lower than 2 nm) membranes, which in turn have hydrophilic, hydrophobic, neutral or ionic characteristic. Membranes are made from several different types of materials including polysulphone, cellulose, cellulose acetate and polytetrafluorethylene, amongst others (Shin and Kang, 2003).

Undoubtedly, MR designs such as the CSTR (uni or bimodular) are employed in several processes. According to Yoriyaz and Vitolo (2014) these processes have yielded a large variety of products such as cyclodextrins, fructooligosaccharides, catechol, casein hydrolysate, fructose, ester synthesis and enzymatic hydrolysis of lactose from milk and whey. In addition, gluconic acid has been produced from the oxidation of glucose by glucose oxidase either soluble or immobilized on anionic resin (Neves and Vitolo, 2007; Tomotani and Vitolo, 2007a; Taraboulsi et al., 2012; Silva et al., 2011).

The choice of bioreactor for a particular process will be influenced by some of the following considerations: (a) mode of operation: discontinuous (less expensive and with multiple operational options) or continuous (more expensive and designed for a particular process); (b) catalyst cost compared with the overall process cost; (c) stability of enzyme during the process; (d) ability for effective control of pH and temperature; (e) maintaining the amount of substrate in the reaction medium below a concentration that would become inhibitory; (f) characteristics of the raw material used as substrate (for example, a continuous PFR would be inadequate if the substrate solution was composed of solid particles in suspension); and (g) ease of replacing the exhausted enzyme immobilized system as quickly as possible without interrupting the process (for example, the CSTR is quite versatile for this purpose).

17.3.2 Enzyme Reactor Kinetics

In order to quantify the performance of a reactor, equations that describe the conversion of substrate to product during the reaction must be formulated (Blanch and Clark, 1996; Dunn et al., 1992; Isunza and De Hoyos, 2004).

As each type of reactor has its own process equation (Figure 17.4), deduction of the equation for a batch reactor will only be presented here (Figure 17.5).

The following premises will be taken: one enzyme, ideal flux inside the reactor, an irreversible reaction, absence of inhibitors and enzyme that obey the Briggs-Haldane's model.

Let's considering the equations:

$$v = (V_{max} \cdot S)/(K_M + S) \tag{17.16}$$

$$(1/v) = (K_M/V_{max}) \cdot (1/S) + (1/V_{max}) \tag{17.17}$$

$$R_E = R_A - R_S - R_C \tag{17.18}$$

where:

v=reaction rate (substrate consumed/time. L);

V_{max}=maximum rate at a fixed temperature (substrate consumed/time. L);

K_M=Michaelis-Menten constant (mol/L);

S=concentration of the substrate present into the reaction medium (mol/L);

R_A=mass rate of substrate input (kg/h);

R_S=mass rate of substrate output (kg/h);

R_C=substrate consumption due to the reaction (kg/h);

R_E=variation of the substrate concentration into the reaction medium (kg/h).

Let's consider the following definitions:

m=initial mass of substrate (kg);

M=total mass inside the reactor (kg);

V_B=volume of the reacting mixture inside the reactor (L);

x'=conversion factor regarding the total mass inside the reactor (M);

x=conversion factor regarding the initial mass of substrate (m).

The conversion factors would be rewritten as follows:

$$x' = (\text{mass of } S \text{ consumed}/M) \tag{17.19}$$

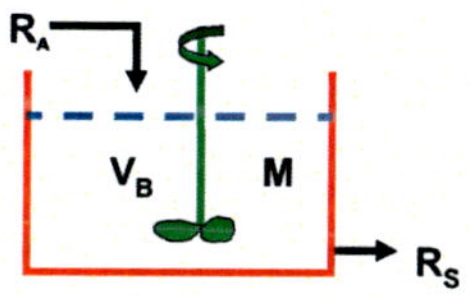

FIGURE 17.5 Generic sketch of a batch reactor.

$$x = (\text{mass of } S \text{ consumed}/m) \tag{17.20}$$

As for a batch reactor R_A=R_S=0, then Equation 17.18 becomes:

$$-R_C = R_E \tag{17.21}$$

Remembering that:

$$R_C = v \cdot V_B \tag{17.22}$$

$$R_E = -M \cdot (\mathrm{d}x'/\mathrm{d}t) \tag{17.23}$$

Substituting Equations 17.22 and 17.23 in Equation 17.21:

$$v \cdot V_B = M \cdot (\mathrm{d}x'/\mathrm{d}t) \tag{17.24}$$

Integrating Equation 17.24:

$$t = \left(\frac{M}{V_B}\right) \cdot \int_0^{x'} \frac{\mathrm{d}x'}{v} \tag{17.25}$$

Combining Equations 17.19 and 17.20, both written as differentials:

$$M \cdot \mathrm{d}x' = m \cdot \mathrm{d}x$$

Rearranging

$$\mathrm{d}x' = (m/M) \cdot \mathrm{d}x \tag{17.26}$$

Substituting Equation 17.26 into Equation 17.25,

$$t = S_0 \cdot \int_0^{x} \mathrm{d}x/v \tag{17.27}$$

Where S_0=initial substrate concentration (m/V_B).

Rewriting Equation 17.20 as follows,

$$x = (S_0 - S)/S_0 \quad \text{or} \quad S = S_0 \cdot (1 - x) \tag{17.28}$$

Substituting Equations 17.17 and 17.28 in 17.27:

$$t = \left(\frac{K_M}{V_{max}}\right) \cdot \int_0^{x} \frac{\mathrm{d}x}{1-x} + \left(\frac{S_0}{V_{max}}\right) \cdot \int_0^{x} \mathrm{d}x$$

Finally,

$$t = (x \cdot S_0/V_{max}) - (K_M/V_{max}) \cdot \ln(1-x) \tag{17.29}$$

Equation 17.29 is called the process equation related to an enzyme batch reactor, in which the total reaction time (t) is correlated to the conversion factor (x).

17.3.3 Operation of Enzyme Reactors

The application of a process equation – such as Equation 17.29 – on an industrial scale depends on the operator capability in circumventing some of the more commonly encountered operational problems like these:

a. *Diffusion* effects, which can cause a reduction in the per cent of substrate converted into product because the movement of molecules (enzyme, substrate and product) inside the reaction medium becomes restricted. Reducing the probability that an enzyme and substrate will interact leads to less ES (complex enzyme-substrate) being formed and, as a consequence, the overall reaction rate is reduced.
b. *Back-mixing*, a phenomenon observed in fixed or fluidized bed reactors operating continuously and with an immobilized enzyme, causing imperfect flux of the fluid inside the reactor. The perturbation on the flux profile throughout the reactor results from attrition between the fluid layer (solution of substrate, for instance) and the particles of the support material used in the immobilized system. Depending on the intensity of back-mixing, whirlpools form around the particles of the support leading to less ES being formed. Often, back-mixing and diffusion occur simultaneously when immobilized enzymes are used in reactors, as a consequence the overall reaction rate is again reduced.
c. *Formation of a temperature gradient inside the reactor*: This phenomenon has been observed in bed fixed reactors, in which the heat must flow by conduction from the reactor wall up to the centre of the bed. As the materials used to build the reactor (stainless steel 316, generally) and the support for immobilization (organic or inorganic polymers) are quite different; the thermal transference from the reactor wall, which is in direct contact with the heating jacket, up to all points of the bed is not uniform particularly during the first hours of continuous operation. Thereby, substrate conversion into product remains below the real catalytic capability of the immobilized enzyme, insofar as a fraction of enzyme molecules are catalyzing the reaction at the optimal reaction temperature (mainly those near the internal surface of reactor's wall) whereas the remainder of the enzyme molecules are not (those located more profoundly within the bed). Under these conditions, the operational reactor regimen remains transient for a period of time, the duration of which will depend on how different the materials that form the immobilization support and the equipment used to manufacture the reactor vessel are. When the support and equipment materials are thermally incompatible, overheating in some parts of the bed can occur (mainly those parts of the bed near the internal face of the reactor wall) causing enzyme molecules to become denatured. Yield loss under these conditions is permanent and cannot be recovered. It must be borne out that the conversion yield during a transient regimen is lower than that once a steady-state regimen has been achieved (which of course is the operating goal of a continuous regimen).
d. *Creating a pH gradient*: The appearance of a pH gradient is also observed in fixed bed reactors but is not as serious as the formation of a temperature gradient. When an immobilized enzyme is exposed to a pH gradient, the catalytic potential of the enzyme may become reduced, but the enzyme does not suffer irreversible denaturation. To minimize this effect, it is usually sufficient to leave the immobilized enzyme in a buffer at the desired pH before the enzyme is introduced inside the reactor. Thereby, when the reactor is fed with the buffered substrate solution, probably all of the ionic groups of the enzyme molecule will be adequately charged for the catalysis.
e. *Variations in internal reactor pressure*: This is a critical aspect when a high-volume fixed bed reactor is operated. The pressure difference between the top and the bottom of the bed depends on column height, substrate solution flux rate across the bed, viscosity of substrate solution and embedding degree of the bed. At first glance, it seems obvious to use a support composed of small particles (less than 45 mesh, US-sieve standard). However, small particles have the tendency to become tightly pack so that the pressure difference between the extremities of the column becomes so high that powerful pumps for maintaining the flux of substrate solution throughout the bed are required.
f. *Loss of reactor performance during the reaction*: This can occur due to the **escape of the enzyme** from the reactor (disintegration of the support and/or detachment of the enzyme from the support), **ineffective enzyme-substrate interaction** (irregular flux pattern inside the reactor, enzyme particles covered by a polymer impermeable layer, for example), **loss of enzyme activity** (inhibition, microbial degradation, amongst others), **loss of product** (mainly by microbial degradation) or **formation of a pH gradient** throughout the reactor.

Finally, it must be remembered that the goal for operating an enzyme reactor is to achieve a reduction in the overall cost of the process. The general equation which correlates the total production (P_t) and process time (t) is:

$$P_t = F \cdot \mathrm{d}t \qquad (17.30)$$

Where F=feeding rate.

In order to solve the above equation it is necessary to know the relation between substrate conversion and decay of the enzyme catalytic power.

If enzyme decay follows an exponential function such as:

$$F = F_i \cdot \mathrm{e}^{(-t \cdot \ln 2/t^*)} \qquad (17.31)$$

Where F_i=initial feeding rate; t^*=half-time of immobilized enzyme.

Substituting Equation 17.31 into Equation 17.30 and integrating,

$$P_t = \left(F_i \cdot t^*/\ln 2\right) \cdot \left[1 - \mathrm{e}^{(-\,tp \cdot \ln 2/t^*)}\right] \qquad (17.32)$$

Where tp=production time.

In industrial practice, knowing that the immobilized enzyme obeys a decaying law and that after a period of continuous operation the process must be forcibly stopped to replace the immobilized system is not an auspicious fact in terms of process yield.

A controlled increase of temperature – taking into account the high thermal stability of immobilized enzyme – and/or the use of reactors in series would be an adequate approach that can circumvent this problem. Increasing the reaction temperature during the process – of course between the limits of temperature stability of the immobilized enzyme – allows the initial reaction rate to be maintained in spite of the loss in catalytic activity as the process proceeds. Regarding to the number of reactors it must be taking into account the duration of time each step in the process will take reactor filling, feed beginning, passage from a transient to stationary regimen, loss of enzyme activity (once the operating temperature is reached) and emptying/cleansing of the reactor. According to Pitcher et al. (1975), it is possible to calculate the ideal number of reactors considering the half-life of the immobilized system. These strategies have been well developed for several industrial processes such as the production of high fructose syrup using glucose isomerase and saccharification of starch to obtain glucose syrup using glucoamylase (Nagodawithana and Reed, 1993; Powell, 1996; Godfrey and West, 1996; Antrim, Lloyd, and Auterinen, 1989).

17.4 FINAL CONSIDERATIONS

Immobilization is a technique in which a biological material (enzymes, cells, organelles and medicines) is coupled to an inert support aiming the catalyst reutilization and run continuous biotechnological processes. There are different types of immobilization methods, such as entrapment (cross-linked matrices, encapsulation and microencapsulation) and bonding formation (adsorption, covalent binding and cross-linking). The choice of the immobilization method will depend on two factors, that is, the type of biocatalyst and the envisaged use for the immobilized system.

Moreover, hundreds different types of supports are available for immobilization. They can be organic (chitin, cellulose, alginate etc.) or inorganic (charcoal, bentonite, silica etc.) chemicals, porous/not porous and solid/gelatinous material. The use envisaged for an immobilized enzyme will determine the choice of the support regarding its physical and chemical characteristics. The growing interest in biotransformations has stimulated the increasing commercial availability and diversity of supports in general.

As the immobilization is an artificial technique, several effects on the catalytic activity are observed, i.e., steric, conformational, diffusion, mass transport and related to the microenvironment (gradients of pH, temperature and substrate over the bulk solution).

Since the sixties of the last century – when the immobilization technique was invented – a lot of applications were observed in industry (mainly, continuous enzymatic processes), chemical and biomedical analysis. Undoubtedly, enzyme electrodes and enzyme immunoassays (ELISA and EMIT) are the most striking applications of immobilized systems in biomedicine. In industry, the immobilization allowed to develop a series of bioreactors adapted for safe handling of biological materials. Bioreactors have been divided in two main types, i.e., discontinuous (including the fed-batch type) and continuous (including the fluidized bed, stationary bed, continuous stirred tank reactor, and membrane reactor).

General rules for choosing the immobilization method and support material does not exist and can only be established empirically. On one hand, these handicaps can introduce limitations to the use of immobilized systems, mainly in terms of research costs, insofar as the most suitable support-method pair can be found by chance. On the other hand, having no predetermined rules then any inert available material can be used as the support for immobilizing a biocatalyst through any method leading to a new product. This is quite common in developing new industrial processes (for example, substituting the fermentation by the enzymatic process in the gluconic acid production) as well as immunoassays suitable for detecting and quantifying new pathogens during human or animal epidemic caused by viruses, microorganisms, parasites etc.

REFERENCES

Alfani, F.; Cantarella, M.; Cantarella, L.; Vitolo, M. Biological upgrading of wastes from sucrose processing. *Annals of the New York Academy of Sciences*, v. 542, p. 346–350, 1988.

Antrim, R. L.; Lloyd, N. E.; Auterinen, A. L. New isomerization technology for high fructose syrup production. *Starch/Starke*, v. 41, p. 155–160, 1989.

Arruda, L. M. O.; Vitolo, M. Characterization of invertase entrapped into calcium alginate beads. *Applied Biochemistry and Biotechnology*, v. 81, p. 23–33, 1999.

Barros, D. P.; Vitolo, M. Sucrose hydrolysis by invertase immobilized on chitin. *Lebensmittel Wissenschaft und Technologie*, v. 25, n. 3, p. 240–243, 1992.

Blanch, H. W.; Clark, S. D. *Biochemical engineering*. New York: Marcel Dekker, Inc., 1996, 702p.

Cantarella, L.; Alfani, F.; Cantarella, M. Stability in organic solvent mixtures of β-glucosidase and β-fructofuranosidase as dry powder and entrapped in poly-HEMA. *Annals of New York Academy of Sciences*, v. 864, p. 219–223, 1998.

Cantarella, M.; Alfani, F.; Cantarella, L.; Gallifuoco, A. Entrapment of enzymes and cells in poly (2-hydroxylethyl methacrylate) supports. In: Bickerstaff, G. F. (Ed.). *Methods in biotechnology: immobilization of enzymes and cells*. v.1. New York: Humana Press Inc., 1996, p. 67–76.

Curcio, S.; Calabró, V.; Iorio, G. Monitoring and control of TMP and feed flow rate pulsatile operations during ultrafiltration in a membrane module. *Desalination*, v. 146, p. 217–222, 2002.

Dunn, I. J.; Heinzle, E.; Ingham, J.; Prenosil, J. E. *Biological reaction engineering*. Weinheim: VCH, 1992, 374p.

Gerhartz, W. *Enzymes in industry: production and applications*. Weinheim: VCH, 1990, 321p.

Giorno, L.; Drioli, E. Biocatalytic membrane reactors: applications and perspectives. *Trends Biotechnology*, v. 18, n. 8, p. 339–349, 2000.

Godfrey, T.; West, S. *Industrial enzymology*. 2nd ed. New York: MacMillan Publishers Ltd., 1996, 610p.

Goldberg, D. M. Clinical enzymology: an autobiographical history. *Clinica Chimica Acta*, v. 357, n. 2, 93–112, 2005.

Isunza, F. L.; De Hoyos, S. F. Models of immobilized enzyme fixed bed reactors. In: Ochoa, S. H. (Ed.). *Reactores enzimáticos*. Ciudad de México: Universidad Autonoma Metropolitana, 2004, 37–92.

Nagodawithana, T.; Reed, G. *Enzymes in food processing*. 3rd ed. San Diego: Academic Press Inc., 1993, 480p.

Nakla, G.; Lugowski, A.; Patel, J.; Rivert, V. Combined biological and membrane treatment of food-processing wastewater to achieve dry-ditch criteria: pilot and full-scale performance. *Bioresource Technology*, v. 97, 1–14, 2006.

Neves, L. C. M.; Vitolo, M. Continuous production of gluconic acid and fructose using membrane bioreactor. *World Journal of Pharmacy and Pharmaceutical Sciences*, v.9, n.10, 423–440, 2020.

Neves, L. C. M.; Vitolo, M. Use of glucose oxidase in a membrane reactor for gluconic acid production. *Applied Biochemistry and Biotechnology*, v. 136–140, 161–170, 2007.

Pessela, C. B.; Mateo, C.; Carrascosa, A.; Lafuente, R. F.; Guisan, J. M. Strategies for fast, simple and cheap covalent immobilization of industrial enzymes. In: *Brazilian seminary on enzyme technology VIII* (Book of Abstracts), Rio de Janeiro: UFRJ, 42–43, 13 a 15 ago, 2008.

Pitcher Jr, W. H. Design and operation of immobilized enzyme reactors. In: Messing, R. A. (Ed.). *Immobilized enzymes for industrial reactors*. New York: Academic Press, 1975, 151–199.

Powell, L. W. Immobilized enzymes. In: Godfrey, T.; West, S. (Ed.). *Industrial enzymology*. 2nd ed. New York: Stockton Press, 1996, 265–272.

Romero, L.; Chávez, R.; Huerta-Ochoa, S.; Prado, A. Enzyme membrane bioreactors: applications. In: Ochoa, S. H. (Ed.). *Reactores enzimáticos*. Ciudad de México: Universidad Autonoma Metropolitana, 2004, 93–124.

Rucka, M.; Turkiewicz, B. Hydrolysis of sunflower oil by means of hydrophobic membrane with lipolytic activity. *Biotechnology Letters*, v. 11, n. 3, 167–172, 1989.

Shin, H. S.; Kang, S. T. Characteristics and fates of soluble microbial products in ceramic membrane bioreactor at various sludge retention times. *Water Research*, v. 37, n. 1, 121–127, 2003.

Silva, A. R.; Tomotani, E. J.; Vitolo, M. Invertase, glucose oxidase and catalase for converting sucrose to fructose and gluconic acid through batch and membrane-continuous reactors. *Brazilian Journal of Pharmaceutical Sciences*, v. 47, n. 2, 399–407, 2011.

Takeuchi, T.; Makino, K. Cellulase immobilized on poly-L-glutamic acid. *Biotechnology and Bioengineering*, v. 29, n. 2, 160–164, 1987.

Taraboulsi Jr, F. A., Andreotti, D. Z., Tomotani, E. J., Vitolo, M. Oxidation of glucose to gluconic acid and regeneration of NADP catalyzed by glucose oxidase and dehydrogenases. *20th European Biomass Conference and Exhibition*. Milan: Eta-Florence, 2012, 1972–1975.

Tomotani, E. J.; Vitolo, M. Immobilized glucose oxidase as a catalyst to the conversion of glucose into gluconic acid using a membrane reactor. *Enzyme and Microbial Technology*, v. 40, 1020–1025, 2007a.

Tomotani, E. J.; Vitolo, M. Production of high-fructose syrup using immobilized invertase in a membrane reactor. *Journal of Food Engineering*, v. 80, 662–667, 2007b.

Tomotani, E. J.; Vitolo, M. Screening of DOWEX(R) anion exchange resins for invertase immobilization. *Applied Biochemistry and Biotechnology*, v. 113–116, 145–159, 2008a.

Tomotani, E. J.; Vitolo, M. Immobilization of nicotinamide nucleotides on anionic polystyrene divinylbenzene beads. In: *Brazilian Seminary on enzyme technology VIII* (Book of Abstracts). Rio de Janeiro: UFRJ, 2008b, 233–234.

Vitolo, M. Calcium-alginate beads as carriers for biocatalyst encapsulation. *World Journal of Pharmacy and Pharmaceutical Sciences*, v. 8, n. 10, 1–17, 2019a.

Vitolo, M. Encapsulation of invertase in barium alginate beads. *World Journal of Pharmaceutical Research*, v. 8, n. 10, 222–233, 2019b.

Vitolo, M. Adsorption of invertase on anionic resin. *European Journal of Pharmaceutical and Medical Research*, v. 7, n. 7, 113–117, 2020a.

Vitolo, M. Membrane reactor applied to enzyme conversion of sucrose and glucose. *World Journal of Pharmacy and Pharmaceutical Sciences*, v. 9, n. 9, 43–70, 2020b.

Yoriyaz, E. J.; Vitolo, M. Feasibility on using the membrane reactor for sucrose hydrolysis. *World Journal of Pharmacy and Pharmaceutical Sciences*, v. 3, n. 12, 66–72, 2014.

18 Biomolecules in Analytical Methods

João Henrique Picado Madalena Santos
University of São Paulo; Institute for Technological Research

Carlos Miguel Nóbrega Mendonça
University of São Paulo; University of Aveiro

Francislene Andréia Hasmann
University of Amazon

Adalberto Pessoa Jr.
Universidade de São Paulo

CONTENTS

18.1 INTRODUCTION

Blood tests as well as urine and feces are extremely important and challenging classic example in the field of pharmaceutical sciences. These tests show important physiological and pathological conditions, as for example, about cancer metastasis, infection and immune status. The complex composition of human, or animal, fluids, like blood, urine and feces usually required several sample preparation steps to purify the targeted analytes, such as cell lysis, dilution, centrifugation and/or staining. Also, well-trained technicians and modern bench-top instruments are fundamental for reliable results (Kuan and Huang, 2020). Then, analytical methods that are sufficiently innovative, robust and accurate, which demonstrate potential for societal impact, are increasingly needed such as molecular methodologies; biosensors; bioengineering; drug development; pharmaceutical and clinical analysis; microfluidics; nanotechnology; and omics studies (proteomics, metabolomics, transcriptomics or glycomics); environmental, agricultural and food science; beverages; liquid fuels; neuroscience; biochemical; forensic analysis; and industrial process. This chapter summarizes several of the available chemical and electrochemical strategies towards the sensing of biomolecules, along the side with the novel trends in biosensing for clinical diagnostics.

18.2 ENZYMES AS DIAGNOSTIC TOOLS

The largest market for analytical enzymes is the *in vitro* clinical diagnostics industry. This market, which includes non-enzymatic tests, is estimated to move billions of dollars annually. The sector is composed of several segments such as: clinical chemistry, immunoassays, microbiology, serology, haematology, cytology, amongst others (Wohlgemuth et al., 2015). Enzyme production technologies for industrial and diagnostic use are differentiated. The former has a low requirement with regard to purity and a need for higher concentrations, whilst in the case of enzymes applied in diagnosis, lower concentrations are required but a higher level of purity. In this sense, the industrial process of producing enzymes applied in clinical diagnosis generally requires a greater number of purification operation units (i.e., especially including chromatographic processes and cross-flow ultrafiltration, both high resolution techniques), significantly increasing the production cost of these enzymes. The high levels of purity are associated with the high specificity of the enzyme by the substrate that is required, and the presence of other contaminating enzymes is unacceptable, as it can lead to false results. The main field of use of enzymes for diagnosis is in the medical area. Many kits, analyzers and biosensors have been modified and/or adapted for use in other areas such as the food manufacturing and environmental management. However, the available assays and market size are quite small when comparing to the medical area.

The detection and follow-up of a wide range of diseases is now highly demanding for the routine examination of clinical samples and other related assays. These typically require analytical methodologies that demand both expertise and time to collect the necessary number of specimens for clinical testing. Enzymes that are used for the detection/diagnosis or prognosis of disease conditions are referred to as 'diagnostic enzymes' (Singh et al., 2019). The substrate specificity and quantitative activity make enzymes preferred in diagnosis and can therefore be used as a diagnostic tool for disease detection. The diagnosis of the serum level of certain enzymes has been used as an indicator of cellular damage that results in the release of intracellular components into the blood stream. Hence, when a medical doctor reveals that a person needs to undergo a neurological enzyme assay, the purpose is to ascertain whether or not there is brain damage. Commonly assayed enzymes for the diagnosis of various diseases are alkaline phosphatase, amylase, hexokinase, aminotransferases (i.e., alanine aminotransferase and aspartate aminotransferase), dehydrogenases (i.e., glucose-6-phosphatase dehydrogenase and lactate dehydrogenase), cyclooxygenase, glucose oxidase, tartrate resistant acid phosphatase and so forth (Table 18.1). Several additional enzymes are also engaged in the medical, both human and veterinary, for the clinical diagnosis of diverse illnesses.

TABLE 18.1
List of the Most Used Enzymes for Clinical Diagnosis and Their Associated Diseases

Enzyme	Disorder/Disease
Acid phosphatase	Malaria
Alanine aminotransferase	Hepatocellular damage
	Hepatitis B and C
Alkaline phosphatase	Chronic kidney disease
	Paget disease or rickets/osteomalacia
	Type II diabetes
	Obstructed liver
Amylase	Pancreatitis
	Myocardial infarction
Aspartate aminotransferase	Hepatic diseases
	Dental disorder
	Liver fibrosis
Butyrylcholinesterase	Schizophrenia
	Alzheimer's disease
	Parkinson's disease
Cathepsin-D	Renal cell carcinoma
	Dental disorder
	Breast cancer
	Rheumatoid arthritis
Creatine kinase	Myocardial damage
	Neuroleptic malignant syndrome
Cysteine cathepsins	Premalignant lesions in colon, thyroid, brain, liver, breast and prostate
Gamma glutamyl transferase	Cardiovascular mortality
Gelatinase-B	Gastric cancer
	Vascular dementia
	Rheumatoid arthritis
	Malignant gliomas
Glycogen phosphorylase-BB	Myocardial infarction
Glucose-6-phosphatase	Gierke disease
	Hypoglycaemia
Glucose-6-phosphatase Dehydrogenase	Gastric cancer
Lactate dehydrogenase	Pyroptosis
	Necrosis
	Breast cancer
Leukocyte esterase	Periprosthetic joint infection
	Urinary tract infection
	Bacterial peritonitis
	Ascitic fluid infection
Lipase	Acute pancreatitis
	Skin disorders
Lysozyme	Rheumatoid arthritis
	Tuberculous meningitis
	Tuberculous pericarditis
Prostatic acid phosphatase	Prostate cancer
Sorbitol dehydrogenase	Prostate cancer
	Colorectal neoplasms
Tartrate-resistant acid Phosphatase	Osteoarthritis
Tartrate-resistant acid phosphatise-5b	Giant cell tumour
	Bone metastases

Source: Adapted from Ram, S., Singh, T., Ashish, S., Enzymes as diagnostic tools, in: *Biomass, Biofuels, Biomass: Advances in Enzyme Technology*, Elsevier, Amsterdam, 2019, 225–271, https://doi.org/10.1016/B978-0-444-64114-4.00009-1.

18.2.1 Diagnosis of Gastric Disorders

There are several gastric disorders that can affect people to varying degrees of severity (malabsorption, Crohn's disease, celiac disease, non-specific disorders, etc.). Because the severity of symptoms can be so variable, these conditions are not always considered by healthcare providers as a high priority for diagnosis and treatment, but many patients must still live with the unpleasant consequences of these disorders (vomiting, diarrhoea and weight loss amongst others) perhaps for many years without having any definitive diagnosis ever being made. Current diagnostic methods (endoscopy, intestinal and stomach biopsies, amongst others) are invasive and uncomfortable to the patient, and the results are often inconclusive. Thus, the development of simple enzyme-based methods that cause less discomfort to the patient and provide more definitive diagnoses are important.

In the cases of disorders of absorption, since the 1970s it has been possible to use enzymes as diagnostic tools. These disorders cause patients to absorb some sugars abnormally. Thus, tests for sugar permeability can be used to detect these disorders. Generally, a monosaccharide (or a polyol) and a disaccharide are used in the test, for instance, mannitol and cellobiose (or lactulose). This test is very simple, fast, and inexpensive. The concentrations of mannitol and cellobiose present in urine collected from patients 5 h after drinking a hypertonic beverage containing cellobiose (or lactulose), mannitol and sucrose can be measured using mannitol dehydrogenase (MDH), cellobiase (β-glucosidase) and lactase-β-dehydrogenase, which are enzymes that are all commercially available.

Figure 18.1 shows the reaction of mannitol in the presence of MDH, which is measured indirectly at 339 nm by the amount of NADH formed (Dooley, 1992; Makristathis et al., 2004; Hoepffner et al., 2006).

18.2.2 Neonatal Screening

Neonatal screening consists of a variety of laboratory tests performed on newborns for the early detection of inherited diseases of metabolism and other asymptomatic pathologies. Most of these diseases, if detected between the 3rd and 30th day of life, can be successfully treated or efficiently controlled, without severe long-term consequences to the health of the newborn child (Detolve, 2016).

Neonatal screening was initially only available for the diagnosis of phenylketonuria (PKU) and congenital hypothyroidism (CH). Later, screening for cystic fibrosis, sickle disease and haemoglobinopathies also became routinely available (Breveglieri et al., 2019). There are also many different versions for tests which detect enzymatic disorders related to the metabolism of carbohydrates, fatty acids, and organic acids.

Phenylketonuria is a recessive, autosomal inherited disease in metabolism of the essential amino-acid phenylalanine. This disorder was first described by a Norwegian physician Ivar Asbjørn Følling in 1934, who observed that two mentally retarded siblings both produced a foul-smelling urine which contained elevated levels of phenylpyruvate and phenylalanine. The patients lack the enzyme phenylalanine hydroxylase (PAH), which converts excess phenylalanine into tyrosine and, as a consequence, phenylalanine metabolism is shunted to an alternative route, which results in the biosynthesis of phenylpyruvate (Figure 18.2) (Dooley, 1992).

Neonatal screening for PKU was initially performed by using the Guthrie test, a microbiological procedure for semi-quantitative analysis of phenylalanine in the blood. This test has now been replaced by a faster and much simpler enzymatic-colorimetric method. Diagnostic kits have been commercially available since 1995 and measure production of NADH by reaction with iodine-nitro-toluene through a colorimetric assay, which is directly proportional to the concentration of phenylalanine present in the sample (Borrajo, 2016).

18.2.3 Cancer Diagnosis

Cancer is defined as a disease in which a group of cells grows abnormally, resulting in their uncontrolled growth and proliferation. In addition, 90% of cancer-related mortality is due to the process called metastasis. Cancer can occur in many different parts of the body, including the lungs, breast, colon, prostate, brain, mouth or even blood. Most cancers are initially recognized either because of the appearance of signs or symptoms or through screening. Neither of these leads to a definitive diagnosis, which requires the examination of a tissue sample by a pathologist. People with suspected cancer are investigated with medical tests. Detecting cancer, or cancer diagnosis, always entails detailed examination. These medical exams commonly include blood tests, X-rays, CT (contrast) scans and endoscopy. During the state of illness, the level of certain enzymes in the bloodstream becomes abnormal, which further acts as a biomarker of cancer prognosis (Singh et al., 2019). A few enzymes that are involved in the detection of cancer are mentioned as follows: acid phosphatases (ACP), cathepsin D, cysteine cathepsins, cyclooxygenase-2 (COX-2), dehydrogenases (sorbitol dehydrogenase and lactate dehydrogenase) and tartrate-resistant acid phosphatase, as it can be seen at Table 18.1.

For example, to detect prostate cancer more than one diagnostic enzyme can be ascertained, such as sorbitol dehydrogenase, prostatic acid phosphatase, amongst others. Sorbitol dehydrogenase (SDH, E.C. 1.1.1.14) catalyzes the reversible oxidation-reduction between the polyhydric alcohol D-sorbitol and D-fructose using NAD^+/NADH as a coenzyme. This enzyme is located primarily in the cytoplasm and mitochondria of the human liver, kidney, and seminal vesicles. In case of prostate cancer, an abnormal serum concentration of SDH is detected (Szabó et al., 2010), and the same occurs for precancerous colorectal neoplasms (Uzozie et al., 2014). Moreover, an enhanced level of SDH can be observed during acute liver damage and parenchymal hepatic diseases.

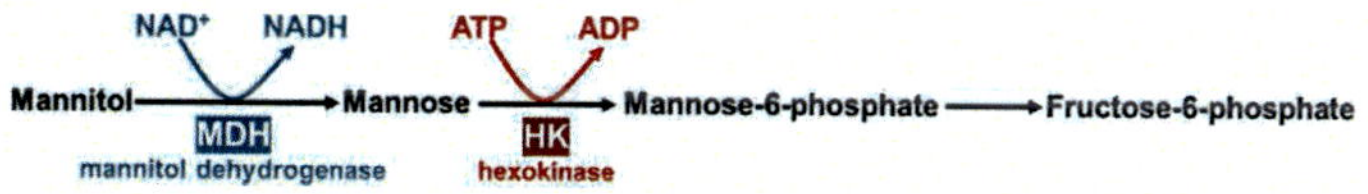

FIGURE 18.1 Schematic representation of the conversion of mannitol into mannose with the consequent formation of NADH, used as an indirect determination of the activity of mannitol dehydrogenase (MDH).

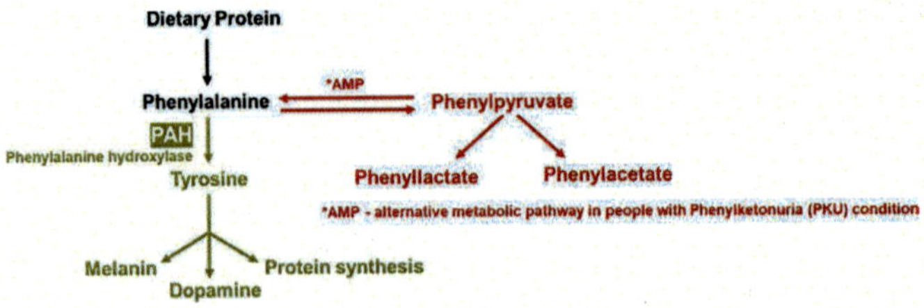

FIGURE 18.2 Hydroxylation of phenylalanine catalyzed by phenylalanine hydroxylase and shunt products when this enzyme is absent (PKU condition) (Dooley, 1992).

18.2.4 Enzyme Detection of Micronutrients, Microorganisms and Cholesterol

Micronutrients are elements that are essential to maintain many cellular and multicellular processes but which are required in only very small quantities. Excess of these micronutrients is toxic or even lethal to cells. New techniques have provided precise determinations of micronutrients present at extremely low concentrations in body fluids. Sodium and potassium, for example, can now be determined with the use of microelectrodes. Enzymes, however, provide cheaper alternative analytical approaches. Several enzymes are activated or inhibited by ions and this provides the principles for the assay design. Some examples: glucokinase, pyruvate kinase and β-galactosidase require either magnesium, potassium or sodium as a co-factor. Enzyme activity is, therefore, dependent on the concentration of the co-factor present. Hence, the concentrations of these micronutrients in a sample will be directly proportional to the specific activity of the enzyme. Conversely, the concentration aluminium can be measured indirectly from the specific activity of the enzyme dihydropteridine reductase, which is inhibited by this metal.

Although high serum cholesterol levels are well-recognized by the lay public as a risk indicator for heart attack or stroke, cholesterol as a biomarker is not in itself a reliable diagnostic predictor of cardiovascular function. Thus, it is necessary to employ techniques that allow the detection of other types of lipoprotein present in serum. Commercial methods that are available use phospholipases to selectively precipitate lipoprotein fractions, which have different densities: HDL (high density lipoprotein); LDL (low density lipoprotein) and VLDL (very low-density lipoprotein) (Aloulou et al., 2018).

The application of biomolecules in diagnostic kits is very broad and, every day, new technologies are being developed and commercialized. A good example is the enzyme amplification method, through which it has become possible in colorimetric based detection of substances at very low concentrations (pg or ng/mL). This technology has been used to predict pregnancy by the detection of estradiol levels and has replaced very widely used immunoassays, since the enzymatic technique is a more rapid alternative (Ro et al., 1988).

Also, in the food industry, luciferase has been used to determine the presence of microorganisms on equipment and work surfaces. This assay is based on the relationship between the concentration of ATP (from microbial cells) and the formation of light and pyrophosphate catalyzed by this enzyme, causing a luminescence on the machinery and surfaces being assessed for contamination. The equipment used to quantify the light emitted from the reaction is called luminometer (Taylor et al., 1998).

18.3 TEST STRIPS

Diagnosis test strips are an interesting example of how patients with chronic medical conditions can be easily and frequently monitored. For example, these systems are available for rapid determination of glucose, cholesterol and triglycerides in either a chemical pathology laboratory, in a doctor's office or a patient's home. The diagnostic strip market is dominated by blood glucose and urine tests. It is estimated that in the United States alone, 1 million patients frequently use test strip to monitor blood glucose levels (glyco-strip) (Godfrey and West, 1996; Whiting et al., 2005).

Strip-based systems are also widely available commercially as a point-of-care and over-the-counter self-testing kits for the determination and control of serum cholesterol levels. In such systems, the enzymes glucose oxidase, cholesterol oxidase and peroxidase are employed to generate a semi-quantitative colour reaction. The results are accurate according to criteria set by the National Cholesterol Education Program (NCEP - is a program managed by the National Heart, Lung and Blood Institute, a division of the National Institutes of Health, and its goal is to reduce increased cardiovascular disease rates due to hypercholesterolaemia in the United States) and are amenable to routine testing in a conventional laboratory (Pugia, 2000).

Diagnostic strips have also been evaluated as a rapid and efficient method for the diagnosis of meningitis. In these systems, proteins, glucose and cells can all be detected in cerebrospinal fluid. Bacterial meningitis has a high mortality rate unless treated quickly with antibiotics. These strip methods are especially useful for rapid diagnosis, especially in cases where it is difficult to obtain a sufficient volume of cerebrospinal fluid to perform routine analysis (cytobacteria) or in locations where there is no immediate laboratory facilities available (Azoulay et al., 2000).

18.4 BIOSENSORS

A biosensor can be defined as a semiconductor detection device that brings together a living organism or products derived from biological systems (enzymes, antibodies, DNA, etc.) and a transducer that provides the indication or signal (a form of recognition) of a substance to be determined (Box 18.1) (Eggins, 2007).

BOX 18.1 BIOSENSORS

According to the International Union of Pure and Applied Chemistry (IUPAC), a biosensor is defined as 'an independently integrated receptor transducer device, which is capable of providing selective quantitative or semi-quantitative analytical information using a biological recognition element'. The features of biosensors comprise the bio-recognition unit and the transduction mechanism from biological signals to measurable electronic signals, e.g., colour, current, voltage, capacitance, light intensity, wavelength, and phase.

Major parameters to assess the performances of biosensors include the following: high sensitivity, quick response or real-time analysis and diagnosis, low consumption of sample volume, high throughput and ease of operation. Biosensors may be low cost and have the convenience of being produced as portable devices or as a disposable portion of a piece of equipment.

The objective of these devices is to simplify analytical methods, with the possibility that these devices be taken outside of a laboratory and used at the point of testing, e.g., at a hospital bedside or factory production line. In recent years, the number of different kinds of biosensor available on the market has increased, also allowing for a reduction in the size (or volume) of the samples being tested and a saving in time and cost of each analysis. Research into biosensors has had a major impact on laboratories and the commercial sector, with special emphasis in clinical diagnostics. Other recent achievements that have contributed to an improvement in the design of biosensors and an expansion in the number of areas biosensor-based analyses can now be applied has been the development of proteins engineered specifically for application, thus making possible the industrial scale manufacture of highly reproducible biosensors (Vaughan and Guilbault, 2006).

Biosensors are inherently unstable because they use biological systems that function outside of the natural environment of biological. Efforts to develop more stable biosensors that ensure reliable and reproducible results after long-term storage and over extended operating times has therefore been the main focus and major innovative driver in biosensor design. The development of biosensors requires multidisciplinary efforts and knowledge from areas such as biotechnology, chemistry, physics, signal processing and instrumentation (Grieshaber et al., 2008).

A typical biosensor consists of (a) the biological element – responsible for the recognition of a given parameter or experimental variable to be determined, for example, the concentration of a substance (carbohydrates and alcohols, for example), cofactors, antigens, hormones and metals, amongst others); (b) the transducer, such as an electrode, that acts as the communication signal that results from a change in a certain property in the vicinity of the biological agent; (c) an amplifier which increases the signal generated in the transducer; and (d) a signal processor and (e) a display (analog or digital). Figure 18.3 shows a schematic diagram of a biosensor (Luong, Male, and Glennon, 2008).

18.4.1 General Characteristics

A biosensor must present some prerequisites to have commercial application; most important is selectivity that will guarantee reliable and reproducible results. Selectivity is determined by the affinity between the biological element and the substance to be analysed. Speed of detection is also an important factor to consider, as well as the stability and reusability of the biosensor (Karunakaran, Rajkumar, and Bhargava, 2015). The thermostability and long term of stability of biosensors can be enhanced through bioconjugation of the biological element, for example biocatalyst, with a conventional polymer, such as polyethylene glycol (PEG) (Santos et al., 2019).

Most traditional analyzers utilize a spectrophotometric methodology, in which the reaction generates a chromophore that can be a colorimetric, fluorescent or luminescent indicator, amongst others. Biosensors based on chemical methods are based upon electrochemical principals, as in the case of the pH (potentiometric) electrode and the oxygen (amperometric) electrode. Biosensors are not restricted to the categories described, since any variable that can be measured in a reaction can be used to generate the signal.

Biosensors based on bio-affinity measure binding of the substance to be analysed to a biological material, which is immobilized. This binding causes a conformational change in the immobilized biomaterial resulting in a physical alteration to the immobilization medium causing a signal (a charge, temperature or colour) to be generated. In an enzymatic (or metabolic) biosensor, the recognition of the substrate by the immobilized receptor (enzymes or other) is followed by conversion of the substrate into the corresponding product, which is then is detected and measured.

Efficient detection cannot always be obtained by simple contact between the surface of the electrode and the compound to be analysed, it is sometimes necessary to use mediators that promote the transport of electrons between

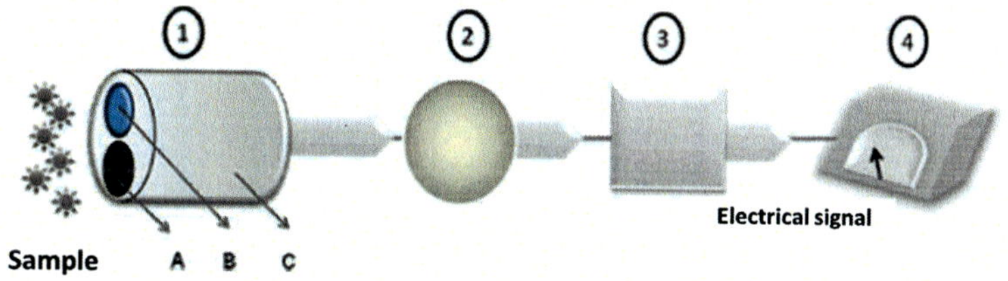

FIGURE 18.3 General scheme of a biosensor: transducer (1), working electrode (A), organic matrix with immobilized enzyme (B), reference electrode (C), amplifier (2), signal processor (3) and analogue/digital panel (4).

the enzyme and the electrode. Generally, the mediator should be a compound of low molecular weight and with a pair of electrons in close proximity to the prosthetic group of the enzyme (Grieshaber et al., 2008; Ronkainen, Halsall, and Heineman, 2010).

Biosensors can be classified according to the recognition method and type of immobilization of the biomolecule. The biomolecule may be entrapped between the transducer and a membrane, covalently attached or adsorbed onto the surface of the transducer. In the absence of oxygen or other reagents, it is necessary to immobilize a co-substrate. The bio-component and the mediator can also be co-immobilized directly into the medium, so that when the transistor captures and amplifies the signal produced by changes to the surface properties of the immobilized biomaterial, a secondary electronic signal is generated.

There are several transduction systems in use and the most important ones are:

i. *Electrochemicals*: measurement of electric current at constant electric potential (amperometric), or measurement of electrical potential at zero electric current (potentiometric).
ii. *Electrical*: conductivity measurement.
iii. *Optics*: measure of chemiluminescence, fluorescence or absorbance.
iv. *Thermal*: calorimetric measurement.
v. *Piezoelectric*: mass measurement employing crystalline quartz microbalances, or surface acoustic waves.

Biosensors have been applied to a wide variety of applications in the pharmaceutical, environmental, bioprocess, safety and defence industries, but it is in clinical diagnoses, that biosensors have gained their greatest commercial success (Kirsch et al., 2013; Luong, Male, and Glennon, 2008).

18.4.2 Electrochemical Detectors

18.4.2.1 Amperometric Biosensors

In an amperometric detector, the detected current is proportional to the concentration of the current generating species. In biosensors that use such a detection method, an electrical potential difference is applied to the electrode, which causes the chemical reaction of the species to be detected and the current is generated. The current intensity will be directly proportional to the concentration so that detection limits can be in the micromolar range. The major advantage of biosensors employing amperometric methods is the possibility of analysis of strongly stained or turbid samples, such as blood, without pretreatment. In addition, there is a wide range of substrate-enzyme systems which allow modification of one or both redox components, for example, biosensors that measure levels of glucose, lactate, pyruvate, urea, 1-alanine, phenylalanine and cholesterol (Mirsky, 2004).

The detection system generally consists of three electrodes: (a) the electrode in which the reaction occurs; (b) the reference electrode (that regulates the potential in the first electrode); and (c) a counter electrode that provides the electrical current to the electrode (Grieshaber et al., 2008).

The choice of the electrical potential to be employed is determined by the likely presence of interferences (baseline deviation) and by the detection limits required (Singh et al., 2004).

Particularly attractive for biosensor applications are reactions involving oxidase enzymes, such as glucose oxidase, that can be monitored by measuring oxygen consumption or product formation (hydrogen peroxide), using a reference electrode (Ag/AgCl). The signal generated by the presence of hydrogen peroxide in the medium can be measured at +650 mV and can be represented by Equation 18.1 (Singh et al., 2004).

$$\begin{aligned} \text{Substrate} + O_2 &\rightarrow \text{Product} + H_2O_2 \\ H_2O_2 &\rightarrow O_2 + 2H^+ + 2e \end{aligned} \tag{18.1}$$

Dehydrogenases, on the other hand, make a poor choice as recognition agents in biosensors, and, in practice, few enzyme electrodes based on these enzymes have been developed. This is because of the high polarization potential (+800 mV Ag/AgCl) required for the oxidation of NADH to NAD^+ by dehydrogenases. These high potentials cause degradation of the electrode surface because of reactions with free radicals generated electrochemically (Jena and Raj, 2006).

Biosensors for measuring cholesterol are interesting because they have high stability, respond rapidly and are inexpensive. They are generally constructed by immobilizing the enzyme cholesterol oxidase (which catalyzes the oxidation of cholesterol generating hydrogen peroxide) onto the surface of an electrode. When hydrogen peroxide comes into contact with the surface of the electrode, the current generated by the enzymatic reaction is proportional to the concentration of cholesterol in the sample, which can be using a calibration curve. Figure 18.4 represents the reactions involved in measuring cholesterol by this enzyme biosensor process (Ram et al., 2001).

A class of biosensors that have attracted great attention are those pre-loaded with antibodies coupled to a system for electrochemical detection, since these allow design of immunosensors sensitive to several substances. Table 18.2 shows the main characteristics of some enzyme electrodes (Mădăraş et al., 1997).

18.4.2.2 Potentiometric Biosensors

In a potentiometric device, a selective electrode is used to identify and determine the concentration of an ion of interest. Many enzymes catalyzed reactions are acid-base reactions that involve proton transfer between the reactants. The extent of proton transfer can be predicted by the pK_a of the reactants and can be monitored by changes in the pH of the reaction medium. Several enzymes catalyzed reactions can be monitored by potentiometric biosensors and include determination of phenylalanine (with NH_4^+ dosage), peroxide (iodide dosage), penicillin and urea (H^+ dosing).

The first potentiometric biosensor that determined the concentration of urea was composed of urease immobilized in polyacrylamide gel and a selective ammonium ion electrode. The biosensor was sensitive enough to detect changes in urea concentration of the order of 5×10^{-5} M, with a

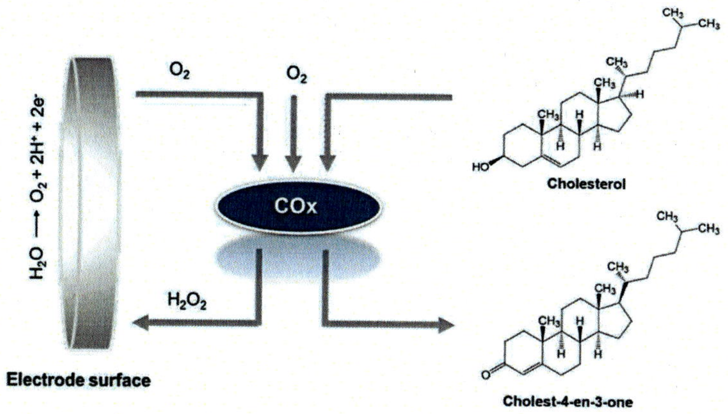

FIGURE 18.4 Schematic representation of the reaction in an amperometric cholesterol monitoring biosensor.

TABLE 18.2
Characteristics of Enzymatic Electrodes

Substance	Enzyme	Linearity	Storage Period	Response Time
Glucose[a]	Glucose oxidase	2 μmol/L 3 mmol/L	>3 months	10 s
Alcohol	Alcohol oxidase	Up to 5 mg%	120 days	2 min
Urea[b]	Urease	1 μmol/L	-	-
Creatinin[c]	Creatin phosphokinase	30 μmol/L	>3 months	1 min
L-aminoacids	L-aminoacid oxidase	1 μmol/L	10–12 days	1 min

[a] In whole blood samples.
[b] In plasma samples.
[c] In serum samples.

response time of 35 s and the biosensor itself had an average lifetime of 14 days. Currently, many ammonia gas sensor electrodes are used for the determination of urea, since the gas generated by the enzymatic hydrolysis of urea diffuses through a membrane until the partial pressure of the gas is equal on both sides. This pressure is proportional to the enzymatic activity. Within these biosensors, there is a reference solution (ammonium chloride) in which diffused ammonia hydrolyses (Equation 18.2) (Werkmeister, Koide, and Nickel, 2016):

$$CO(NH_2)_2 + 3H_2O \rightarrow CO_2 + 2NH_4^+ + 2OH^- \quad (18.2)$$

Other biosensors used for the determination of urea are described in the literature (Soldatkin et al., 2003; Yang and Zhang, 2013), amongst these are those that use carbon dioxide with a reference solution of sodium bicarbonate, where the potential varies with the pH change of the internal solution and is proportional to the initial concentration of urea. The great advantage of these electrodes is in a reduction of interferents. However, the response time is longer, especially for very dilute samples.

18.4.2.3 Optical Detectors

Optical or photometric biosensors have a wide potential for use because of their sensitivity and simplicity, especially those employing optical fibres. Optical measurements can be based on variations in the index of refraction, fluorescence, chemiluminescence or absorbance (Karunakaran, Rajkumar, and Bhargava, 2015; Myszka, 1999).

Optical biosensor, in general, consists of an enzyme linked to a chromophore, which is attached to a membrane. The pH change generated by the enzymatic reaction changes the colour of the chromophore/membrane complex. The transducer system consists of a simple photodiode coupled to the system. These biosensors do not require a reference sensor signal because there is a comparison signal that is generated by the light source used in the sample. For example, when Beer-Lambert law for the absorbance of the indicator chromophore is applied, the signal in the transducer will be directly proportional to the concentration of the analysed substance. The success of the technique depends on the ability of the indicator chromophore to specifically and reversibly bind to the substance of interest.

An example of the use of these biosensors is the colorimetric detection of aromatic compounds. In these, the peroxide generated in the reaction of the analyte with the enzyme peroxidase can be easily detected. Amongst the compounds that have been analysed by this system are heterocyclic compounds such as 4-aminoantipyrine coupled to a condensation reaction with the phenol to give a quinoneimine-like dye.

Optical biosensors have gained increasing interest with the development and miniaturization of optical fibres and optoelectronic transducers, which allow the detection of non-electronic remote signals produced by the reaction with the compound to be analysed. An example of such a biosensor is used in the determination of lactate, which measures the variation of molecular oxygen as a consequence of oxygen binding to a fluorescent dye. A decrease in the concentration of oxygen is reflected by an increase in the fluorescence of the dye placed onto the optical fibre. An increase in lactate concentration reduces the oxygen concentration, which decreases the coalescing bond between the oxygen and the dye, promoting an increase in the fluorescence signal. In this system, the enzyme lactate monooxygenase is used.

18.4.2.4 Thermal Detectors

Calorimetric biosensors use a device capable of recording temperature differences produced by a biochemical reaction. Generally, a linear response of the temperature as a function of analyte concentration is obtained, with temperature variations of the order of 0.0001°C. This heat exchange is monitored by an appropriate thermometric transducer.

In order to avoid thermal fluctuations from the environment interfering with the measurements, these devices are placed within isolated boxes. The enzymatic reaction occurs in a small reactor and, the sample to be analysed passes through a heat exchanger to be stabilized. The measurement is made with the aid of thermistors and, the type of resistor used to measure temperature changes is based on variations in electrical resistance (Karunakaran, Rajkumar, and Bhargava, 2015).

The main advantage of a calorimetric biosensor is in its use in analysing concentrated solutions and, consequently, there are a wide range of applications in biotechnology and medicine. However, the temperature must be constantly maintained in the system, which may lead to the denaturation of enzymes, thus affecting the functioning of the biosensor (Zhang and Tadigadapa, 2004).

18.4.2.5 Piezoelectric Detectors

Piezoelectricity is defined as a property which the crystals possess to generate a voltage in response to an external vibration. The effect generated is reversible and thus all piezoelectric crystals vibrate in the presence of an electric field at a frequency that depends on the type of crystal, its thickness and the type of crystal cut (Alves et al. 2011). This frequency has the ability to absorb or desorb molecules on its surface (Luong and Guilbault, 1991). The frequency variation is proportional to the mass variation of the adsorbed material and, such variations can be determined with the aid of electronic circuits. Therefore, these types of biosensors contain enzymes immobilized onto piezoelectric crystals that act as transducers of the electrical signal produced at the interface (Alves et al. 2011).

Humidity can interfere with the functioning of these biosensors and, therefore, its application in the determination of analytes in solution is limited. However, piezoelectric biosensors are relatively inexpensive, small and capable of rapid response. Piezoelectric biosensors are used in the detection of gaseous formaldehyde, using dehydrogenase immobilized onto a quartz crystal (Alves et al., 2011; Luong and Guilbault, 1991; Vaughan and Guilbault, 2006).

18.4.2.6 Immunodetection Biosensors

Immunodetection biosensors can be designed for direct binding of an antigen to an immobilized antibody, generally in the transducer, and a second antibody bound to an antigen previously attached to the first antibody (Figure 18.5). It is common for the second antibody to have been previously conjugated to an immunosensor enzyme by means of transducers (optical or amperometric). The activity expressed by the enzyme is dependent on the concentration of the enzyme-labelled antibody, which is bound to the antigen, and this in turn is bound to the first antibody immobilized on the transducer. Immunosensors are one of the most sensitive methods capable of detecting antigens at concentrations of picograms, both in liquid and gas phases (Honda et al., 2005).

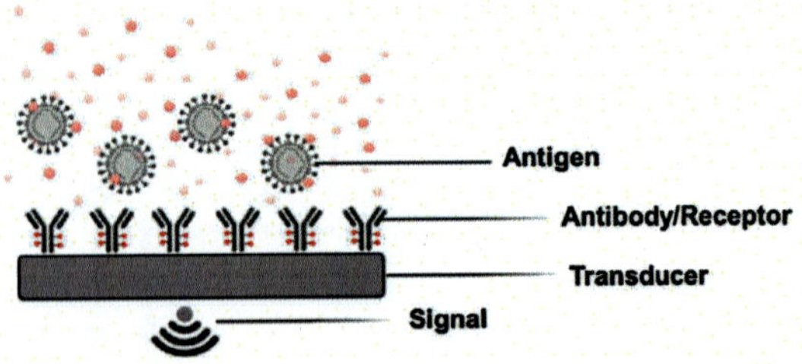

FIGURE 18.5 Schematic representation of an immunosensor.

Significant attention has been given to the development of rapid and safe methods for the diagnosis of tuberculosis, since traditional methods may require 4–8 weeks for a definitive result. The use of immunosensors for detecting mycobacterial antigens in biological fluids could bring great advantages for successful clinical outcome, since the time to diagnosis would be shortened (Díaz-González et al., 2005).

The main problems of these biosensors are related to non-specific binding and irreversible binding of antigen to antibody, reducing the active surface of the transducer and sensitivity in consecutive measurements.

18.5 NOVEL TRENDS IN BIOSENSING TECHNOLOGY

18.5.1 Microfluidic Integrated Biosensors: Towards Lab-on-a-Chip

Microfluidic can be defined as the science and technology of systems that process or manipulate small amounts (10^{-9} to 10^{-18} L) of fluids, using channels with dimensions of tens to hundreds of micrometers (Wang et al., 2020). This technology allows the miniaturization of biosensors, with faster, controlled, and reproducible biosensing activity. Advantages such as high heat and mass transfer of reagents, low reagent and energy consumption, safety, portability of microreactors, and so on, enable a plethora of new possibilities for biosensors devices (Luka et al., 2015). Microfluidic devices can be fabricated in distinct materials such as polymers, glasses, silicon, metals and ceramics, being the most common polydimethylsiloxane (PDMS). Due to the small size of micro-systems, a single microfluidic biosensor can perform extensive/detailed analysis including continuous sampling, sample separation and mixing, pre-concentration and treatment. Furthermore, these microfluidic biosensors offer improved analytical performance, high throughput, real-time detection and high-speed reaction rates, making detection adaptable to point of care (POC) applications (Wang et al., 2020). The operation of microfluidic integrated biosensors is depicted in Figure 18.6, being the most common biological recognition elements exhibited, such as: antibodies, enzymes and aptamers. There are three major types of microfluidic integrated biosensors available, namely: droplet-based microfluidics, digital microfluidic-based biosensor and continuous microfluidic-based biosensor. From the aforementioned types of devices, the droplet-based microfluidics is the most employed system to be integrated with the biosensor technology. The digital microfluidics (DMF) is applied for the manipulation of liquid samples as microdroplets; it is compatible with a wide range of detection methods (i.e., electrochemical and optical) and avoids the use of syringe pumps, valves and channels, unlike the continuous microfluidic-based biosensors (Luka et al., 2015).

Overall, the integration of biosensors with the microfluidic devices creates a compelling analytical tool that will be an innovative step towards the home-testing approach which will benefit both developing and developed countries (Luka et al., 2015).

18.5.2 Wearable Biosensors for Healthcare Monitoring

One of the brand-new technologies in the field of health is wearable biosensors. Biosensors hold considerable promise for wearable applications due to their high specificity, speed, portability, low cost and low power requirements. These novel devices are arousing considerable interest due to their potential to provide continuous, real-time physiological information through dynamic, non-invasive measurements of biochemical markers in biofluidics such as sweat, tears, saliva and interstitial fluid (ISF). The prospective utility of wearable biosensors is obvious from the promptly escalating rate of recently reported proof-of-concept findings. Several of these devices are under clinical assessment; however, successful translation to the commercial market has been challenging due to the need of large-scale validation studies, the necessary device regulatory approvals and final marketing paths (Kim et al., 2019). Even so, it is possible to see in Table 18.3 the great versatility of developed devices that are either under development (i.e., running clinical trials) or already FDA approved and currently in commercialization.

In this next generation of biosensors, body fluids are sampled in a non-invasive manner for biosensing aims, meaning that they can be speedily accessed without disrupting the outermost protecting layers of the body's skin (the stratum corneum) and without contacting with blood (avoid the use of hypodermic needle and syringe). As such, non-invasive sensing procedures pose minimal risk of harm or infection and are generally more user friendly (Sempionatto et al., 2017). In this sense, wearable biosensors have already been widely applied to a variety of head-to-toe application sites, targeting an array of important analytes in proof-of-concept demonstrations (Lee et al., 2016). Wearable monitoring devices can provide understandings into dynamic biochemical processes in these biofluids by allowing continuous, real-time examining of biomarkers, related to a wearer's *in situ* health condition (Kim et al., 2015, 2019; Imani et al., 2016). Such real-time monitoring can give information on health and wellness, boost the management of chronic diseases (such as prediabetes, diabetes, and associated complications, amongst others) and alert the user or medical professionals of irregular or unexpected situations (Gao et al., 2016).

In this way, these new devices show a high practicality and can be used in everyday life without causing any inconvenience. The study of new nanomaterials and smart materials (materials that respond to external stimuli) has been applied for the development of wearable devices with better performance, that provide the necessary flexibility and stretchability (Kim et al., 2017; Jeerapan et al., 2016). Continuous multidisciplinary development of new biosensing technologies will allow the development of new devices that non-invasively monitor a wide range of biomarkers with a larger global market and should be the future of real-time clinical diagnosis (Senior, 2014).

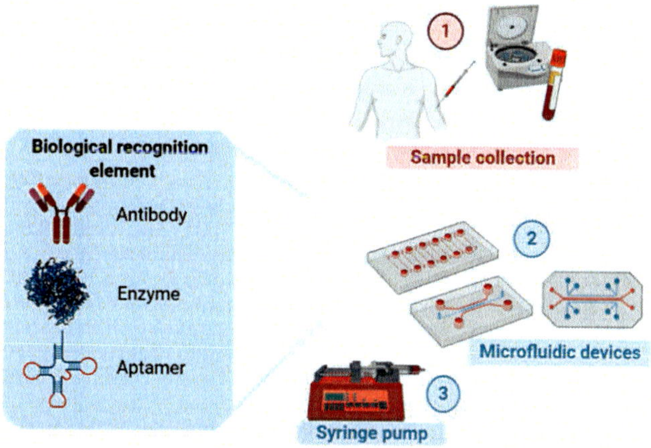

FIGURE 18.6 Schematic representation of continuous microfluidic-based biosensors for the detection of body metabolites using antibodies, enzymes and aptamers as biological recognition elements. The fundamental procedure involves three steps: (1) sample collection from the patient (a low-volume sample is needed), (2) development of a specific microfluidic device (specific geometry, material, microchannels size and other features) and (3) pumping of the collected sample into the microchannels of the continuous microfluidic-based biosensor for clinical diagnostic.

TABLE 18.3
Selected Examples of Commercial Non-Invasive or Slightly Invasive Biosensors

Product, Company	Analyte and Sample	Wearable Platform	Monitoring Mechanism
Smart contact lens, Google and Novartis	Glucose in tears	Contact lens	Electrochemistry
GlucoWatch, Cygnus Inc.	Glucose in ISF	Watch type	Electrochemistry
BioMKR, Prediktor Medical	Blood glucose	Wrist strap similar to a smart watch	Near infrared spectroscopy, bioimpedance
GlucoWise, MediWise	Blood glucose	Finger clip	Radio frequency
Freestyle Libre, Abbott	Glucose in ISF	Patch	Electrochemistry
Dexcom G6 CGM, Dexcom	Glucose in ISF	Patch	Electrochemistry
GlucoTrack, Integrity Applications	Blood glucose	Finger clip	Ultrasonic, electromagnetic, thermal waves
Eversense, Senseonics	ISF glucose	Subcutaneous small stick implant	Fluorescence
NovioSense tear glucose sensor, NovioSense	Tear glucose	Small stick (spiral type) placed under the lower eyelid	Electrochemistry

Source: Adapted from Kim, J., Campbell, A.S., De Ávila, B.E., et al., *Nat. Biotechnol.*, 37, 389–406, 2019.

18.6 FINAL CONSIDERATIONS

The growing number of various diseases and the increase of environmental contamination are the causes for the development of novel methods for their detection. Bioanalytical methods are one of the hot topics in the scientific literature as they can provide valuable information regarding portable or potentially portable devices (such as microfluidic devices and wearable biosensors), field tests and intelligent devices that can be used in different areas of the pharmaceutical field. It can provide a vital tool and important bioinformatics for the early prevention, diagnosis and treatment of diseases. Specificity and practicability of the method in real samples have to be verified, indicating its potential applications. In addition, a reliable, high selectivity, low detection limits, portable and high-throughput method for the detection of biomolecules is highly desirable. The development of a sensitive, low-cost and general sensing platform for the analysis of a biomolecule and, eventually, its mutation is important for early disease screening and offer a powerful tool for biomedical research and clinical diagnosis. Therefore, new methods, simple and fast, for biomolecule quantification are highly advantageous and must be developed and widely used.

REFERENCES

Aloulou, A.; Rahier, R.; Arhab, Y.; Noiriel, A.; Abousalham, A. Lipases and Phospholipases. In: Sandoval, G. (Ed.), *Methods in Molecular Biology*. New York: Humana Press Inc., 2018, vol. 1835, 69–105. https://doi.org/10.1007/978-1-4939-8672-9_3.

Alves, W.; Alves, W.; Sousa, C. P.; Kogikoski Jr S.; Silva, R. F.; Amaral, H.; Liberato, M. S.; Oliveira Jr V. X.; Martins, T.; Takahashi, P. M. Biosensors based on biological nanostructures. In: Serra, P.A. (Ed.), *New Perspectives in Biosensors Technology and Applications*. IntechOpen, 2011. https://doi.org/10.5772/16260.

Azoulay, E.; Fartoukh, M.; Galliot, R.; Baud, F.; Simonneau, G.; Le Gall, J. R.; Schlemmer, B.; Chevret, S. Rapid diagnosis of infectious pleural effusions by use of reagent strips. *Clinical Infection Diseases*, vol. 31, 914–919, 2000.

Borrajo, G. J. C. Newborn screening for phenylketonuria: Latin American consensus guidelines. *Journal of Inborn Errors of Metabolism & Screening*, vol. 4, 1–5, 2016. https://doi.org/10.1177/2326409816682764. Available at: https://us.sagepub.com/en-us/nam/open-access-at-sage. Accessed on: 4 Oct. 2020.

Breveglieri, G.; D'Aversa, E.; Cosenza, L. C.; Boutou, E.; Balassopoulou, A.; Voskaridou, E.; Gambari, R.; Borgatti, M. Detection of the sickle hemoglobin allele using a surface plasmon resonance based biosensor. *Sensors and Actuators, B: Chemical*, vol. 296, 126604, 2019. https://doi.org/10.1016/j.snb.2019.05.081.

Detolve, G. R. Newborn screening. In: Rubin, I. L.; Merrick, J.; Greydanus, D. E.; Patel, D. R. (Eds.), *Health Care for People with Intellectual and Developmental Disabilities across the Lifespan*. Cham: Springer International Publishing, 2016, vol. 31, 631–640. https://doi.org/10.1007/978-3-319-18096-0_55. Available at: /https://www.springer.com/gp/book/9783319180953. Accessed on: 4 Oct. 2020.

Díaz-González, M.; González-García, M. B.; Costa-Garcia, A. Immunosensor for Mycobacterium tuberculosis on screen-printed carbon electrodes. *Biosensors and Bioelectronics*, vol. 20, 2035–2043, 2005.

Dooley, K. C. PKU screening method using phenylaianine dehydrogenase. *Clinical Biochemical*, vol. 25, 271–275, 1992.

Eggins, B. *Chemical Sensors and Biosensors*. 1st edition. New York: Wiley, 2007. https://doi.org/10.1002/9780470511305.

Gao, W.; Emaminejad, S.; Nyein, H.; Challa, S.; Chen, K.; Peck, A.; Fahad, H. M.; Ota, H.; Shiraki, H.; Kiriya, D.; Lien, D. H.; Brooks, G. A.; Davis, R. W.; Javey, A. Fully integrated wearable sensor arrays for multiplexed in situ perspiration analysis. *Nature*, vol. 529, 509–514, 2016.

Godfrey, T.; West, S. I. *Introduction to Industrial Enzymology*. 2nd edition. New York: MacMillan Publishers Ltd., 1996.

Grieshaber, D.; Mackenzie, R.; Vörös, J.; Reimhult, E. Electrochemical biosensors - Sensor principles and architectures. *Sensors*, vol. 8, no. 3, 1400–1458, 7 Mar. 2008. https://doi.org/10.3390/s80314000. Available at: https://www.ncbi.nlm.nih.gov/pmc/articles/PMC3663003/#:~:text=Typical%20recognition%20elements%20used%20in,suppress%20any%20non%2Dspecific%20interaction. Accessed on: 4 Oct. 2020.

Hoepffner, N.; Shastri, Y. M.; Hanisch, E.; Rosch, W.; Mossner, J.; Caspary, W. F.; Stein, J. Comparative evaluation of a new bedside faecal occult blood test in a prospective multicentre study. *Alimentary Pharmacology & Therapeutics*, vol. 23, 145–154, 2006.

Honda N.; Inaba, M.; Katagiri, T.; Shoji, S.; Sato, H.; Homma, T.; Osaka, T.; Saito, M.; Mizuno, J.; Wada, J. High efficiency electrochemical immuno sensors using 3D comb electrodes. *Biosensors and Bioelectronics*, vol. 20, 2306–2309, 2005.

Imani, S. Bandodkar, A.; Mohan, A.; Kumar, R.; Yu, S.; Wang, J.; Mercier, P.P. A wearable chemical-electrophysiological hybrid biosensing system for real-time health and fitness monitoring. *Nature Communications*, vol. 7, 11650, 2016.

Jeerapan, I.; Sempionatto, J. R.; Pavinatto, A.; You, J. M.; Wang, J. Stretchable biofuel cells as wearable textile-based self-powered sensors. *Journal of Materials Chemistry* A, vol. 4, 18342–18353, 2016.

Jena, B. K.; Raj, C. R. Electrochemical biosensor based on integrated assembly of dehydrogenase enzymes and gold nanoparticles. *Analytical Chemistry*, vol. 78, no. 18, 6332–6339, 15 Sep. 2006. https://doi.org/10.1021/ac052143f. Available at: https://pubs.acs.org/doi/abs/10.1021/ac052143f. Accessed on: 4 Oct. 2020.

Karunakaran, C.; Rajkumar, R.; Bhargava, K. Introduction to biosensors. In: Karunakaran, C.; Bhargava, K.; Benjamin, R. (eds.), *Biosensors and Bioelectronics*. New York: Elsevier, 2015, 1–68. https://doi.org/10.1016/B978-0-12-803100-1.00001-3.

Kim, J.; Campbell, A. S.; De Ávila, B. E.; Wang, J. Wearable biosensors for healthcare monitoring. *Nature Biotechnology*, vol. 37, 389–406, 2019.

Kim, J.; Kumar, R.; Bandodkar, A. J.; Wang, J. Advanced materials for printed wearable electrochemical devices: A review. *Advanced Electronic Materials*, vol. 3, 1600260, 2017.

Kim, J. et al. Wearable salivary uric acid mouthguard biosensor with integrated wireless electronics. *Biosensors and Bioelectronics*, vol. 74, 1061–1068, 2015.

Kirsch, J.; Siltanen, C.; Zhou, Q.; Revzin, A.; Simonian, A. Biosensor technology: Recent advances in threat agent detection and medicine. *Chemical Society Reviews*, vol. 42, no. 22, 8733, 2013. https://doi.org/10.1039/c3cs60141b. Available at: http://xlink.rsc.org/?DOI=c3cs60141b.

Kuan, D.; Huang, N. Recent advancements in microfluidics that integrate electrical sensors for whole blood analysis. *Analytical Methods*, no. 26, 3318–3332, 2020. https://doi.org/10.1039/D0AY00413H.

Lee, H.; Choi, T. K.; Lee, Y. B.; Cho, H. R.; Ghaffari, R.; Wang, L.; Choi, H. J.; Chung, T. D.; Lu, N.; Hyeon, T.; Choi, S. H., Kim, D. H. A graphene-based electrochemical device with thermoresponsive microneedles for diabetes monitoring and therapy. *Nature Nanotechnology*, vol. 11, 566–572, 2016.

Luka, G.; Ahmadi, A.; Najjaran, H.; Alocilja, E.; DeRosa, M.; Wolthers, K.; Malki, A.; Aziz, H.; Althani, A., Hoorfar, M. Microfluidics integrated biosensors: a leading technology towards lab-on-a-chip and sensing applications. *Sensors (Basel)*, vol. 15, no. 12, 30011–30031, 2015.

Luong, J. H. T; Guilbault, G. G. Analytical applications of piezoelectric crystal biosensors. In: Blum, L. J.; Coulet, P. R. (Eds.), *Biosensors, Principles and Applications*. New York: Dekker, 1991.

Luong, J. H. T.; Male, K. B.; Glennon, J. D. Biosensor technology: Technology push versus market pull. *Biotechnology Advances*, vol. 26, no. 5, 492–500, 2008. https://doi.org/10.1016/j.biotechadv.2008.05.007.

Mădăraş, M. B.; Spokane, R. B.; Johnson, J. M.; Woodward, J. R. Glutamine biosensors for biotechnology applications, with suppression of the endogenous glutamate signal. *Analytical Chemistry*, vol. 69, no. 18, 3674–3678, 15 Sep. 1997. https://doi.org/10.1021/ac970173f. Available at: https://pubmed.ncbi.nlm.nih.gov/9302872/. Accessed on: 4 Oct. 2020.

Makristathis, A.; Hirschl, A. M.; Lehours, P.; Mégraud, F. Diagnosis of Helicobacter pylori infection. *Helicobacter*, vol. 9, 7–14, 2004.

Mirsky, V. M. Ultrahin electrochemical chemo and biosensors: Technology and performance. In: Wolfbeis, O. S. (Ed.). *Chemical Sensors and Biosensors*. Sensors Series. vol. 2. New York: Springer, 2004.

Myszka, D. G. Improving biosensor analysis. *Journal of Molecular Recognition*, vol. 12, no. 5, 279–284, 1999. https://doi.org/10.1002/(SICI)1099-1352(199909/10)12:5<279::AID-JMR473>3.0.CO;2-3.

Pugia, J. M. Technology behind diagnostic reagent strips. *Laboratory Medicine*, vol. 31, 1–5, 2000.

Ram, M. K.; Bertoncello, P.; Ding, H.; Paddeu, S.; Nicolini, C. Cholesterol biosensors prepared by layer-by-layer technique. *Biosensors and Bioelectronics*, vol. 16, 849–856, 2001.

Ro, P.-C.; Chan, S. Y. W.; Grace, W. K.; Tang, M. B. Diagnosis of early pregnancy by enzyme immunoassay of Schwangerschafts-protein 1. *Fertility and Sterility*, vol. 49, no. 1, 76–80, 1988. https://doi.org/10.1016/S0015-0282(16)59652-6.

Ronkainen, N. J.; Halsall, H. B.; Heineman, W. R. Electrochemical biosensors. *Chemical Society Reviews*, vol. 39, no. 5, 1747–1763, 2010. https://doi.org/10.1039/b714449k.

Santos, J.H.P.M.; Carretero, G.; Ventura, S.P.M.; Coverti, A.; Rangel-Yagui, C.O. PEGylation as an efficient tool to enhance cytochrome c thermostability: a kinetic and thermodynamic study. *Journal of Materials Chemistry* B, vol. 7, 4432–4439, 2019.

Sempionatto, J. R.; Nakagawa, T.; Pavinatto, A.; Mensah S. T.; Imani S.; Mercier P.; Wang J. Eyeglasses based wireless electrolyte and metabolite sensor platform. *Lab Chip* vol. 17, 1834–1842, 2017.

Senior, M. Novartis signs up for Google smart lens. *Naure Biotechnology*, vol. 32, 856, 2014.

Singh, S.; Chaubey, A.; Malhotra, B. D. Amperometric cholesterol biosensor based on immobilized cholesterol esterase and cholesterol oxidase on conducting polypyrrole films. Analytica Chimica Acta, vol. 502, 229–234, 2004.

Singh, R.; Singh, T.; Singh, A. Enzymes as diagnostic tools. In: *Biomass, Biofuels, Biomass: Advances in Enzyme Technology*. Amsterdam: Elsevier, 2019, 225–271. https://doi.org/10.1016/B978-0-444-64114-4.00009-1.

Soldatkin, A. P.; Montoriol, J.; Sant, W.; Martelet, C.; Jaffrezic-Renault, N. A novel urea sensitive biosensor with extended dynamic range based on recombinant urease and ISFETs. *Biosensors and Bioelectronics*, vol. 19, no. 2, 131–135, 15 Nov. 2003. https://doi.org/10.1016/S0956-5663(03)00175-1. Available at: https://linkinghub.elsevier.com/retrieve/pii/S0956566303001751. Accessed on: 4 Oct. 2020.

Vaughan, R.D., Guilbault, G.G. Piezoelectric Immunosensors. In: Janshoff, A., Steinem, C. (Eds) *Piezoelectric Sensors. Springer Series on Chemical Sensors and Biosensors (Methods and Applications)*, vol 5. NewYork: Springer, 2006, 237–280.

Szabó, Z.; Hämäläinen, J.; Loikkanen, I.; Moilanen, A.M.; Hirvikoski, P.; Väisänen, T.; Paavonen, T.K.; Vaarala, M.H. Sorbitol dehydrogenase expression is regulated by androgens in the human prostate. *Oncology Reports*, vol. 23, no. 5, 1233–1239, 2010. https://doi.org/10.3892/or_00000755.

Taylor, A. L.; Kudlow, B. A.; Marrs, K. L.; Gruenert, D. C.; Guggino, W. B.; Schwiebert, E. M. Bioluminescence detection of ATP release mechanisms in epithelia. *American Journal of Physiology Cellular and Physiology*, vol. 275, 1391–1406, 1998.

Uzozie, A.; Nanni, P.; Staiano, T.; Grossmann, J.; Barkow-Oesterreicher, S.; Shay, J.W.; Tiwari, A.; Buffoli, F.; Laczko, E.; Marra, G. Sorbitol dehydrogenase overexpression and other aspects of dysregulated protein expression in human precancerous colorectal neoplasms: A quantitative proteomics study. *Molecular and Cellular Proteomics*, vol. 13, no. 5, 1198–1218, 2014. https://doi.org/10.1074/mcp.M113.035105.

Wang, J.; Ren, Y.; Zhang, B. Application of microfluidics in biosensors. In: Ren, Y. (Ed.), *Advances in Microfluidic Technologies for Energy and Environmental Applications*.

London: IntechOpen, 20 May 2020. https://doi.org/10.5772/intechopen.91929. Available from: https://www.intechopen.com/books/advances-in-microfluidic-technologies-for-energy-and-environmental-applications/application-of-microfluidics-in-biosensors.

Werkmeister, F. X.; Koide, T.; Nickel, B. A. Ammonia sensing for enzymatic urea detection using organic field effect transistors and a semipermeable membrane. *Journal of Materials Chemistry B*, vol. 4, no. 1, 162–168, 17 Dec. 2016. https://doi.org/10.1039/c5tb02025e. Available at: www.rsc.org/MaterialsB. Accessed on: 4 Oct. 2020.

Whiting, P.; Westwood, M.; Watt, I.; Cooper, J.; Kleijnen, J. Rapid tests and urine sampling techniques for the diagnosis of urinary tract infection (UTI) in children under five years: A systematic review. *BMC Pediatrics*, vol. 5, no. 4, 1–13, 5 Apr. 2005. https://doi.org/10.1186/1471-2431-5-4. Available at: https://pubmed.ncbi.nlm.nih.gov/15811182/. Accessed on: 4 Oct. 2020.

Wohlgemuth, R.; Plazl, I.; Žnidaršič-Plazl, P.; Gernaey, K. V.; Woodley, J. M. Microscale technology and biocatalytic processes: Opportunities and challenges for synthesis. *Trends in Biotechnology*, vol. 33, no. 5, 302–314, 2015. https://doi.org/10.1016/j.tibtech.2015.02.010.

Yang, Z.; Zhang, C. Single-enzyme nanoparticles based urea biosensor. *Sensors and Actuators, B: Chemical*, vol. 188, 313–317, 2013. https://doi.org/10.1016/j.snb.2013.07.004.

Zhang, Y.; Tadigadapa, S. Calorimetric biosensors with integrated microfluidic channels. *Biosensors and Bioelectronics*, vol. 19, no. 12, 1733–1743, 15 Jul. 2004. https://doi.org/10.1016/j.bios.2004.01.009.

19 Nanotechnology and Biopharmaceuticals

Alexsandra Conceição Apolinário
University of São Paulo

Valker Araujo Feitosa
University of São Paulo and Institute for Technological Research

João Henrique Picado Madalena Santos and Carlota de Oliveira Rangel-Yagui
University of São Paulo

CONTENTS

19.1 INTRODUCTION

Nanotechnology comprises a multidisciplinary science dedicated to create, explore and manipulate a broad array of materials and/or systems in a nanometer size scale, i.e. with at least one dimension ranging from approximately 1 to 1,000 nm. For pharmaceutical and medical applications, however, nanotechnology usually refers to systems of 5 to approximately 200 nm.

The U.S. Food and Drug Administration (FDA) considers nanotechnology products as materials, ingredients and other substances, including drugs, biologicals, medical devices, cosmetics, foods, dietary supplements, which are engineered with at least one dimension in the nanoscale, usually presenting physicochemical properties and/or biological effects attributable to its dimensions (FDA, 2017). Similarly, the European Medicines Agency (EMA) defines nanotechnology as the use of tiny structures – less than 1,000 nm across, which are designed to have specific properties (EMA, 2019). To illustrate the concept of the nanoscale range, Figure 19.1 presents the biological systems and both organic and inorganic nanostructures ranging from ~10 nm (e.g. proteins), ~50–150 nm (e.g. viral particles), up to 1,000 nm (e.g. polymeric nanoparticles). Many of these nanostructures will be described in this chapter.

Nanotechnology has gained pivotal roles in biology and medicine. A significant development was observed in the past decades; as a result, nowadays nanostructures can be fabricated, characterized and modified for extensive life science applications (Mirza and Siddiqui, 2014). In this context, nanobiotechnology emerges as a science which converges nanosystems and nanomaterials with biological systems.

19.2 NANOBIOTECHNOLOGY FOR BIOPHARMACEUTICALS

Biopharmaceuticals such as peptides, therapeutic proteins, enzymes, monoclonal antibodies, nucleic acids, antibody–drug conjugates, hormones and vaccines have structural complexity (e.g. high molecular mass) and are susceptible to degradation (e.g. oral delivery of peptides and proteins is challenged by gastrointestinal degradation), which brings additional challenges for the development of pharmaceutical formulations. Biopharmaceuticals may also have

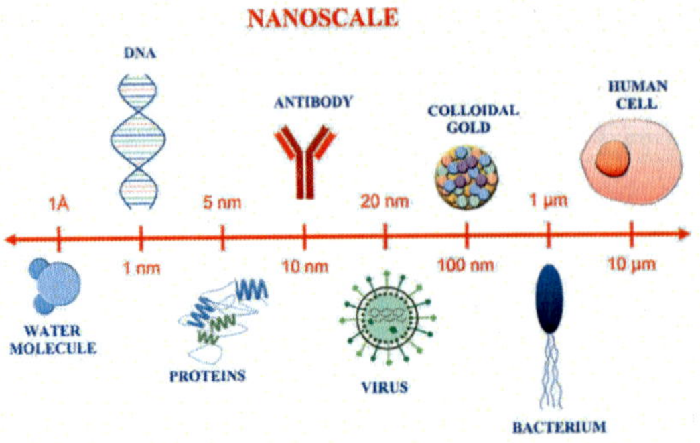

FIGURE 19.1 Illustration of nanoscale comparing small molecules, macromolecules and cells. Water molecule is illustrated as a small molecule, but non-biological drugs can also be cited, which are in the angstrom range, i.e. 0.1 nm. Biological structures can vary in ranges from 1 nm up to sizes higher than 1,000 nm; for example, smaller nanostructures are DNA (<5 nm) and proteins (~5–10 nm) such as antibodies, but nanostructures such as virus are higher (~100 nm). Unicellular organisms such as bacteria, yeasts and human cells are in the range of micrometres. Nanocarriers of biopharmaceuticals, e.g. proteins, can be in the feasible range of size to accommodate the macromolecules, as well as achieve pharmacological effects either through cell uptake or long circulation in bloodstream. The size of the nanocarriers can be adjusted depending on the application. Overall, according to FDA, nanocarriers can range from 1 to 1,000 nm. (Reproduced with adaptations from: https://www.scielo.br/pdf/qn/v43n2/0100-4042-qn-43-02-0212.pdf, Apolinário, A.C., Salata, G.C., Bianco, A.F.R., Fukumori, C., Lopes, L.B., *Química Nova*, 43, 212–225, 2020, https://www.scielo.br/scielo.php?script=sci_abstract&pid=S0100-40422020000200212&lng=en&nrm=iso, an Open Access Article distributed under the terms of the Creative Commons Attribution License.)

reduced permeation across biological barriers, complicating the delivery to specific tissues and/or intracellular targets. Furthermore, some may suffer fast elimination due to renal filtration (up to 50 kDa), fast enzymatic degradation (e.g. through plasmatic proteases), uptake by the mononuclear phagocytic system, as well as accumulation in non-targeted organs and tissues such as liver and spleen. Therapy with biopharmaceuticals usually involves recurrent injections, which is often not economical, has poor patient adherence and sometimes impossible owing to toxicity (Torchilin, 2011; Mitragotri, Burke and Langer, 2014; Orive et al., 2004).

In this way, the pharmaceutical industry has a demand for formulations of biopharmaceuticals with improved stability, effective absorption, maintenance of therapeutic concentrations within the target tissue and long-term release of the drugs (Orive et al., 2004). Nanobiotechnology offers alternatives to meet tissue/cellular penetration. Attributes like target specificity, extended half-life and/or cellular uptake can result from biopharmaceuticals' encapsulation in nanostructures, as well as from the covalent attachment to polymers, a so-called bioconjugation process. For instance, owing to their subcellular and submicron size, nanomedicines can penetrate tissues through the epithelial fenestrations of fine capillaries and are generally taken up efficiently by cells (Panyam and Labhasetwar, 2003).

Since the late 1980s, the enhanced permeability and retention (EPR) effect is described as responsible for the accumulation of nanostructures and macromolecules at much higher concentrations in tumour tissues than in healthy tissues and organs. This effect is observed for macromolecules and nanostructures of apparent molecular sizes larger than 50 kDa and is related to the increased permeability of the tumour vasculature owing to larger interendothelial gaps generated by a rapid and disorganized angiogenesis. These leaky spots, together with poor lymphatic drainage, are described as responsible for the passive diffusion and accumulation of larger molecules/structures in solid tumours (De Lázaro and Mooney, 2020). This concept has been a central paradigm supporting the formulation of nanomedicines to treat solid tumours. Nonetheless, recent investigations on mouse models and human tumour models suggest that up to 97% of nanoparticles penetrate tumours by an active process through endothelial cells (Sindhwani et al. 2020).

Cellular uptake is needed for therapeutic applications of small interfering RNA (siRNA), mRNA, oligonucleotides or plasmid DNA (pDNA), and avoiding the degradative endolysosomal compartment of the cytoplasm is crucial (Panyam and Labhasetwar, 2003). The low permeability of cell membranes to macromolecules represents another drawback for the development of formulations of biopharmaceuticals (Torchilin and Lukyanov, 2003). Some nanocarriers not only are capable of passive permeation into tissues but can also be designed to facilitate the cell penetration and endosomal escape depending on their size, surface charge and surface functionalization. Cell-penetrating peptides (CPPs), for example, can be used for functionalization (Box 19.1). Stability, targeting specificity and cargo release kinetics may also be optimized (Vercauteren et al., 2012; Gu et al., 2011).

BOX 19.1 CELL-PENETRATING PEPTIDES (CPPs)

CPPs promote the cell uptake of therapeutic proteins, peptides and nucleic acids. These small peptides (approximately 30 amino acids) can be either obtained from natural proteins or synthesized. Some proteins present intrinsic biological functions and have the ability to enter into mammalian cells, so they can be employed to enhance the internalization of a variety of biomolecules, including plasmid DNA, siRNA, oligonucleotides, peptide–nucleic acid, peptides, proteins and nanocarriers. CPPs effects rely on the number of positively charged amino acids; i.e., approximately 8–15 positively charged residues usually result in high cell penetration capacity. The specific pathway of CPPs across the plasma membrane is still a matter of debate, but it is accepted that a multimechanism process relies on some issues, including the amino acid sequence, concentration, hydrophobicity and net charge; the payload; cell type; and temperature (Koren and Torchilin, 2012; Munyendo et al., 2012; Bolhassani, Jafarzade and Mardani, 2017; Zhu and Jin, 2018).

Target specificity is particularly important for immunotherapy. Sipuleucel-T (Provenge®), for example, is an antigen-presenting cell (APC) complex loaded with a fusion protein (PA2024) of the prostatic acid phosphatase antigen linked to an immunostimulatory granulocyte–macrophage colony-stimulating factor. The FDA approved it in 2010 for the treatment of metastatic castration-resistant prostate cancer (Anassi and Ndefo, 2011; Wang and Wang, 2012). In 2018, the FDA approved the first-of-its kind targeted RNA-based therapy (Onpattro®) to treat polyneuropathy caused by hereditary transthyretin-mediated amyloidosis (hATTR) in adults. In this medicine, lipid nanocarrier is used to deliver the drug directly into the liver, by infusion.

Nanoformulations protect proteins from proteolysis and subsequent rapid clearance from the bloodstream, modulating their half-lives (Yadav, Kumari and Yadav, 2011; Sleep, Cameron and Evans, 2013). Nanocarriers based on polymeric matrix lead to a slow diffusion and controlled drug release free from the burst effect, increasing the drug apparent half-life and reducing both the systemic side effects associated with high drug concentrations and the frequency of drug administration (Pakulska et al. 2016). An example is Eligard®, a lyophilized drug delivery system commercially available for acetate leuprolide, which is a synthetic nonapeptide analog of gonadotropin-releasing hormone indicated for the therapy of advanced prostate cancer. Eligard® is administered subcutaneously via an implanted depot delivery system (Atrigel®) composed of the biodegradable polymer poly(*D*, *L*-lactide-*co*-glycolide) (PLGA) dispersed in *N*-methyl-2-pyrrolidone. It releases the drug in a controlled manner over defined intervals of 1, 3, 4 or 6 months (Sartor, 2003; Tombal and Berges, 2005). Although this example refers to a synthetic peptide, similar nanoformulations can be used for peptides from biotechnological routes.

The choice of nanocarrier depends on the therapeutic goal (e.g. controlled release, cellular uptake), drug characteristics (e.g. solubility, molecular weight) as well as other drawbacks that must be overcome (e.g. short half-life, immunogenicity, aggregation with serum proteins and enzymatic degradation). Biocompatibility, biodegradability and some physical–chemical aspects regarding stability are important to develop a successful nanomedicine formulation. Some of the main types of nanocarriers used for biological drugs/products delivery are briefly described in this chapter.

19.3 LIPOSOMES

Liposomes are the classic example of nanocarriers used by the pharmaceutical industry. Liposomes are spherical vesicular nanostructures formed by one (unilamellar) or concentric lipid bilayers (multilamellar) enclosing an aqueous phase (the core). Vesicles can be regarded as closed lamellar bilayers, and therefore, the geometry of the phospholipids, or any amphiphilic molecules in general, is important and can be related to the aggregate shape by the critical packing parameter (cpp), according to Equation 19.1 originally proposed by J. Israelachvili (Bueno, Oliveira and Rangel-Yagui, 2018).

$$\mathrm{cpp} = \frac{v}{a_0 \cdot l_c} \qquad (19.1)$$

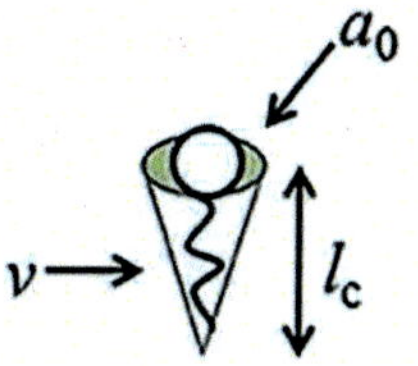

in which cpp = critical packing parameter, v = volume of the hydrophobic tail, a_0 = area of the cross section of the polar head group and l_c = length of the hydrophobic chain(s) of the surfactant.

The following structure progression is observed with an increase in cpp value: micellar → hexagonal → bilayers. When $0.5 < \mathrm{cpp} < 1$, vesicles (liposomes) are predominantly formed.

Liposomes can entrap hydrophilic drugs in their internal aqueous compartment, and hydrophobic drugs can be embedded into the membrane. Liposomal systems have advanced over the years since the first preparations in the 1960s (Figure 19.2). Characteristics like size, charge and surface properties can be varied simply by the selection of an appropriate phospholipid, by the addition of other components such as cholesterol and/or by the method preparation (Torchilin, 2005).

Currently, stealth PEGylated liposome technology (e.g. Doxil®), non-PEGylated liposomes (e.g. Myocet®) and DepoFoam™ technology are available on the market. Figure 19.3 illustrates the DepoFoam™ technology, composed of microscopic spheroids (3–30 μm) with a granular structure and single-layered lipid particles composed of several multivesicular liposomes with internal aqueous chambers that encapsulate the drug (Olusanya et al. 2018).

Several techniques have been reported for liposomes preparation, such as thin-film hydration followed by sonication or extrusion to trigger unilamellar vesicles, microfluidization, an ethanol injection method, a proliposome–liposome method and reverse-phase evaporation. However, scale-up is a challenge for several of these methods (e.g. sonication does not ensure the uniformity of size for liposomes). High-pressure extrusion by means of a homogenizer or microfluidization has been shown to be scalable. Extrusion by using membranes of nanometric pore size is also common in liposomes preparation on an industrial scale (Charcosset et al. 2015). Lipex® extruder, for example, is a 10-mL industrial unit equipped with supplementary 100- and 800-mL units (Figure 19.4).

Liposomes composed of cationic lipids are most widely used for *in vitro* nucleic acid transfection (in molecular biology techniques). Some products are commercially available for this purpose, e.g. Lipofectin®, Transfectam®, LipofACE™, LipofectAMINE™ and LipoTAXI™. In these systems, encapsulation is greatly improved due to the electrostatic interaction between negatively charged plasmids and positively charged lipids, resulting in more complex assembles called lipoplexes. Another advantage of these systems is that since the external leaflet of biological membranes is usually negatively charged, DNA/RNA are efficiently delivered into the cells (Del Pozo-Rodríguez et al., 2010).

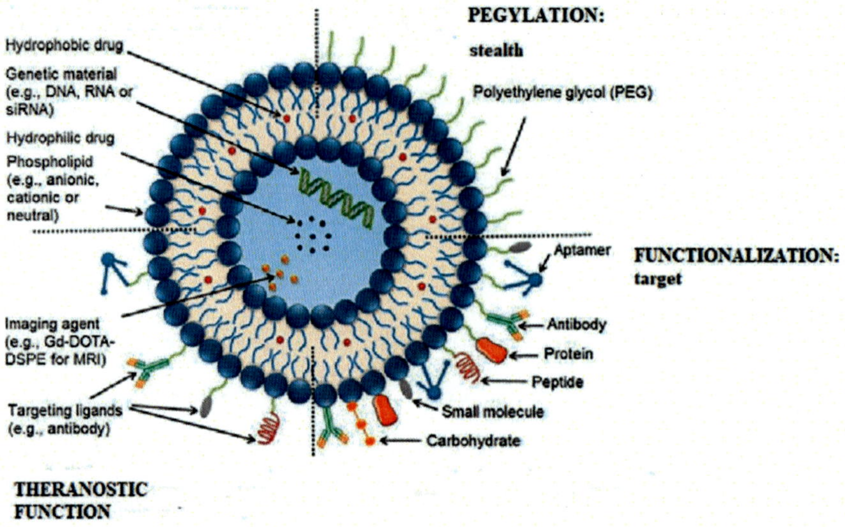

FIGURE 19.2 Representation of conventional and functionalized unilamellar liposomes. In the top-left section, it is depicted conventional liposome and the different payloads, such as small molecules, e.g. hydrophobic drugs entrapped into the lipidic bilayer and hydrophilic drugs encapsulated into the aqueous core. It is not illustrated, but macromolecules like proteins and peptides can be loaded either bilayer or core of vesicles, relying on chemical proprieties. PEGylated liposomes formed from the lipid containing PEG molecules (in green) are illustrated in the top-right section. In bottom sections, targeted liposomes are illustrated for both theranostic (at the left section) and therapy (at the right section); in these cases, liposome surface presents ligands such as antibodies, peptides and carbohydrates to target specific sites, e.g. a receptor expressed in cells of a solid tumour. (Reproduced from: Olusanya, T.O.B., Haj Ahmad, R.R., Ibegbu, D.M., Smith, J.R., Elkordy, A.A., *Molecules*, 24443, 1–17, 2018, https://www.mdpi.com/1420-3049/23/4/907, an Open Access Article distributed under the terms of the Creative Commons Attribution License.)

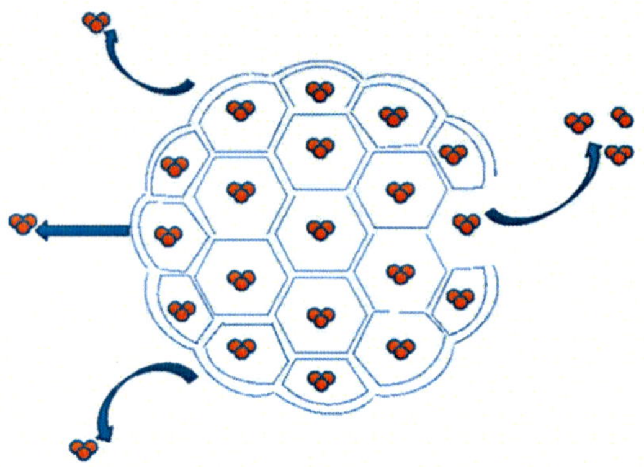

FIGURE 19.3 Representation of multivesicular DepoFoam™ technology, which consists of multivesicular liposomes formed from a unique structure of multiple non-concentric networks of lipid membranes forming aqueous chambers with diameters of 0.2–5 µm. DepoFoam™ provides a sustained release of drugs encapsulated into the aqueous compartment. Orange small spheres represent the drugs, and the blue arrows illustrate the drug release. Reproduced with adaptations from: Bulbake, U., Doppalapudi, S., Kommineni, N., Khan, W., *Pharmaceutics*, 9, 2–17, 2017, https://www.mdpi.com/1999-4923/9/2/12, an Open Access Article distributed under the terms of the Creative Commons Attribution License.)

Derivations of liposomes of particular importance in the context of biopharmaceuticals are the virosomes, phospholipid vesicles incorporating virus-derived proteins to allow fusion with target cells. In clinically designed virosomes, the viral genetic material is substituted with biopharmaceuticals, including genes or vaccination agents (Kalra et al., 2013). Biopharmaceuticals can be incorporated in the aqueous cavity of virosomes or antigens can be coupled to lipid anchors so they will aggregate with the phospholipids of the bilayer. The most common virosomes are based on influenza virus, i.e. composed of spiked neuraminidase and hemagglutinin into liposomes (Almeida et al., 1975). Nonetheless, virosomes can also be made from other viruses such as Sendai, Epstein–Barr, Sindbis, Semliki Forest, herpes simplex and human immunodeficiency virus. These nanostructures are prepared by the solubilization of viral membranes by detergents having no denaturing tendency. After solubilization, the viral nucleocapsid is removed by ultracentrifugation, and the viral membrane is reconstituted with the envelope proteins (Stgmann et al., 1987; Bron et al., 1994).

19.4 POLYMERSOMES

Analogous to liposomes are polymersomes (Figure 19.5b), which are vesicles first described in 1999 and composed of high molecular weight amphiphilic block copolymers. The copolymers are composed of distinct hydrophilic and hydrophobic blocks; when the hydrophilic block is poly(ethylene glycol) (PEG), polymersomes are considered stealth. The geometry of amphiphilic block copolymers also dictates the final aggregate shape in water (micelles or vesicles). Nonetheless, usually a simplification of the critical packing parameter (cpp), known as the 'hydrophilic fraction' (*f*), is used for amphiphilic block

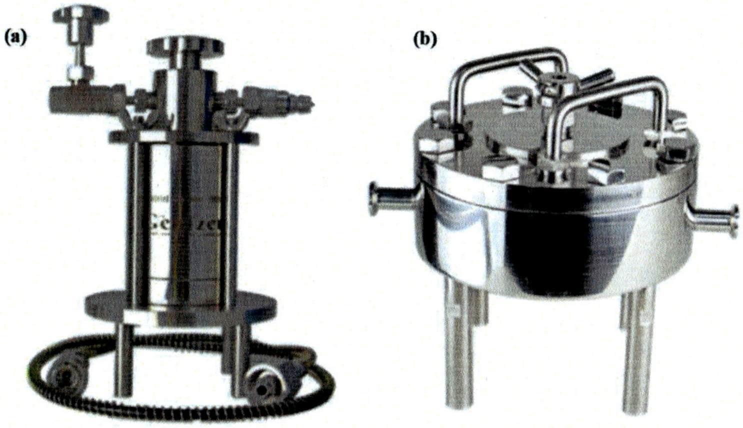

FIGURE 19.4 Lipex® extruders used to execute the one-step technique extrusion, which produces liposomes by utilizing a constant-pressure force and forcing the aqueous suspensions of lipidic vesicles through polycarbonate filters of a predefined pore size, leading to achieve homogeneous populations of liposomes. The maximum volumes and operating pressure range from 10 to 10,000 mL and 3,000–6,000 psi, respectively, for Jacketed Liposome Extruder shown in (a). For extruder model presented in (b), the volume varies from 1.5 to 800 mL and the maximum operating pressure is 850 psi. The side wall of the thermobarrel is jacketed to offer temperature control of the extruder at the maximum value of 80°C for both. (Courtesy of Genizer™.)

copolymers. Generally, polymersomes formation is favoured when f is 10%–40%. At f ~40%–55%, cylindrical micelles tend to form, and at f ~50%–70%, spherical micelles are predominantly formed (Discher and Ahmed, 2006; Pachioni-Vasconcelos et al., 2016).

Like liposomes, polymersomes are usually spherical, enclosed vesicles, which can cargo hydrophilic and/or hydrophobic drugs. Unlike lipid-based vesicles, the increased length and conformational freedom of polymer chains enhance the toughness and reduce the permeability of polymersomes (Discher et al., 1999; Messager et al., 2014). Additionally, recent advances in polymer chemistry have allowed the design of several functional membranes in polymersomes (e.g. pH responsive, temperature responsive, permeable).

Polymersomes are promising nanocarriers for protein drug delivery. Nonetheless, formulations of protein drugs nanoencapsulated in polymersomes are still under development, and to date, none have been approved for clinical use. Currently, a company in Singapore is developing novel veterinary vaccines using artificial cell membrane (ACM) technology based on polymersomes to stabilize and deliver functional, folded viral and bacterial cell-surface antigens to induce a strong immune reaction.

For therapeutic enzymes, some proof-of-concept formulations have been reported using polymersomes as nanoreactors (Joseph et al., 2017; Blackman et al., 2018). In this context, substrates can enter the nanostructure through the membrane, while products either accumulate inside the nanoreactor or are released into the bulk medium (Renggli et al., 2011). Using this approach, a therapeutic enzyme would not need to be released (Blackman et al., 2018) (Figure 19.6).

At a laboratory scale, polymersomes can be produced by the same techniques used for liposomes. Additionally, specific features of the copolymer can be used to facilitate the aggregation into polymersomes, such as a switch in temperature or pH (Apolinário et al., 2017).

19.5 POLYMERIC MICELLES

Polymeric micelles (Figure 19.5a) are colloidal nanoparticles with a core–shell structure that spontaneously and reproducibly form in aqueous environments by amphiphilic molecules (e.g. copolymers and PEGylated phospholipids) above a certain concentration known as 'the critical micelle concentration' (CMC) (Movassaghian, Merkel and Torchilin, 2015). Thermodynamically, polymeric micelles are obtained from the self-assembly of amphiphilic copolymers that organize themselves such that the hydrophobic blocks of the copolymer molecule are assembled mutually and directed away from the aqueous medium, while the hydrophilic blocks are kept in close contact with water. These structures exist in a dynamic equilibrium, with a constant exchange of copolymer unimers in the aqueous environment with those present in the micelles (Banerjee and Onyuksel, 2012).

Since the 1980s, extensive studies have been carried out on the properties of micelles, highlighting micelles as promising alternatives for multiple pharmaceutical applications due to several unique advantages over other delivery systems, including (a) ease of production and sterilization; (b) nanosize range (i.e. typically <100 nm) and narrow size distribution; (c) high static and dynamic structural stability; (d) long-term circulation time; (e) ability to incorporate large amounts of drug; and (f) potential to be fine-tuned for specific therapeutic applications (Movassaghian et al., 2015).

Polymeric micelles have been used for entrapment and delivery of therapeutic agents (e.g. small-molecule drugs, genes, peptides and proteins), as well as for diagnostic applications. In general, hydrophobic drugs can be solubilized in the micellar core, while water-soluble biomolecules are thought to exist in the micellar hydrophilic palisade layer (Banerjee and Onyuksel, 2012).

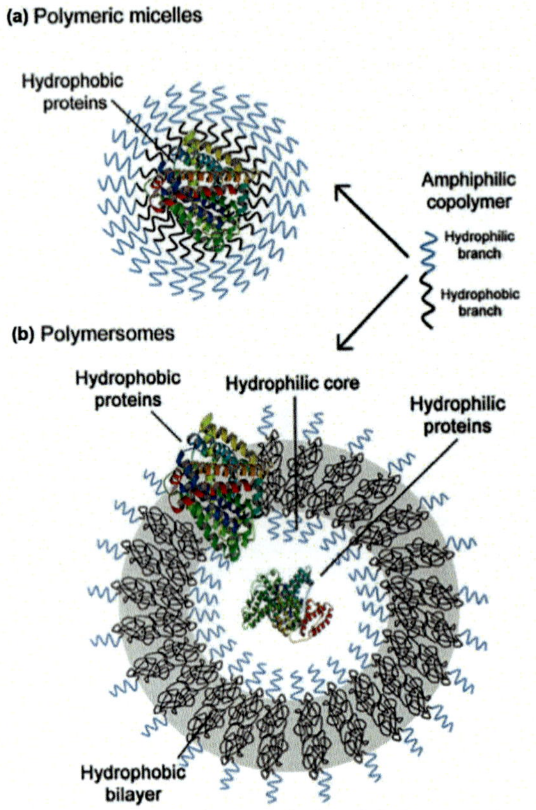

FIGURE 19.5 Schematic representation of a cross section of (a) polymeric micelles and (b) polymeric vesicles or polymersomes; both nanostructures are formed from the self-assembly of amphiphilic copolymers. In this example, it is represented as a diblock copolymer that consists of hydrophilic block (in blue) covalently linked to hydrophobic block (in black). It is illustrated the insertion of hydrophobic proteins into micellar core and bilayer of polymersomes and the loading of hydrophilic proteins into aqueous core of vesicles. (Reproduced from Pachioni-Vasconcelos, J.E.A., Lopes, A.M., Apolinário, A.C., Valenzuela-Oses, J.K., Costa, J.S., Nascimento, L.E.O., Pesso, A., Barbosa, L.R., Rangel-Yagui, C.O., *Biomater Sci*, 4, 205–218, 2016, with permission of The Royal Society of Chemistry.)

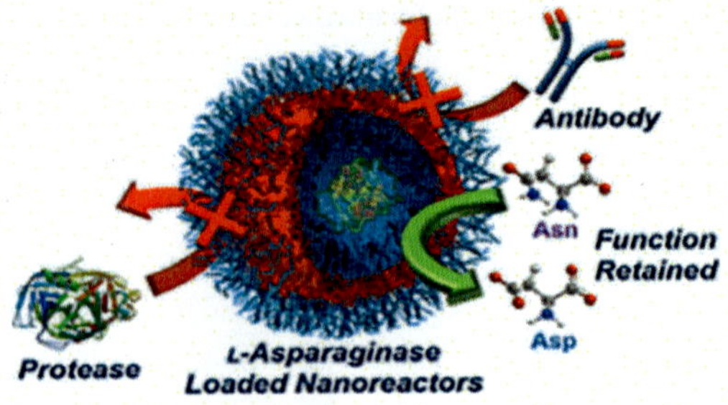

FIGURE 19.6 Representation of a polymersome acting as a nanoreactor for enzyme *L*-asparaginase, a biopharmaceutical that is key in therapeutic scheme for acute lymphoblastic leukaemia by depleting the free asparagine, which is essential for leukemic cells, but not for the healthy cells. In this case, the vesicle is formed by a permeable bilayer, which can allow the penetration of substrate L-asparagine (L-Asn). The hydrolysis of amino acid occurs inside the polymersome, so the enzyme is kept encapsulated, leading to improved proteolytic stability and decreased antibody recognition. (Adapted with permission from Blackman, L.D., Varlas, S., Arno, M.C., Houston, Z.H., Fletcher, N.L., Thurecht, K.J., Hasan, M., Gibson, M.I., O'Reilly, R.K., *ACS Cent Sci*, 4, 718–723, 2018, https://cdn-pubs.acs.org/doi/full/10.1021/acscentsci.8b00168. Further permissions related to the material excerpted should be directed to the ACS.)

19.6 POLYMERIC NANOPARTICLES: NANOCAPSULES AND NANOSPHERES

Polymeric nanocapsules (Figure 19.7) are nanocarriers in which the drug is confined to a reservoir, or within a core enclosed by a polymer membrane or coating. The core can be aqueous or oily, surrounded by a single layer of polymer. Polymeric nanospheres, on the other hand, are matrix-type colloidal particles in which drugs are dissolved, entrapped, encapsulated, chemically bound or adsorbed to the constituent polymer matrix (Letchford and Burt, 2007).

Polymers usually employed for drug delivery applications include poly(*N*-(2-hydroxypropyl) methacrylamide) (PHPMA), poly(oligoethylene glycol methyl ether methacrylate) (POEGMA), PLGA, poly(glutamic acid) (PGA), poly(*N*-isopropyl acrylamide) (PNIPAM), poly(*N, N'*-diethyl acrylamide) (PDEAM), polystyrene and PEG. Currently, microspheres of PLGA are used for parenteral delivery of

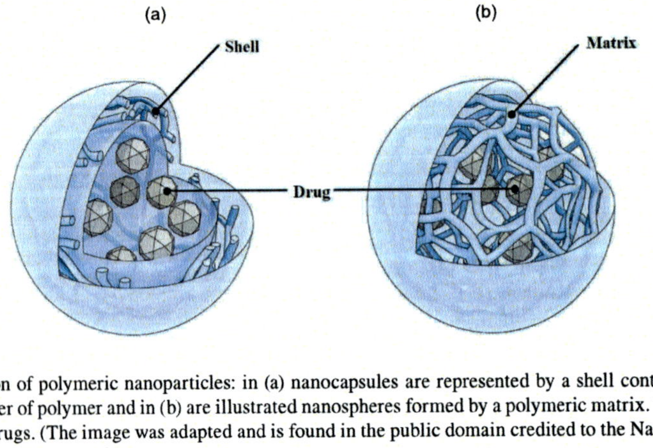

FIGURE 19.7 Illustration of polymeric nanoparticles: in (a) nanocapsules are represented by a shell containing an oily liquid core surrounded by a single layer of polymer and in (b) are illustrated nanospheres formed by a polymeric matrix. The yellow structures correspond to encapsulated drugs. (The image was adapted and is found in the public domain credited to the National Institutes of Health/Department of Health and Human Services.)

analogues of the luteinizing hormone-releasing hormone (LHRH) (e.g. Lupron Depot® and Decapeptyl Depot®), human growth hormone (Nutropin Depot®) and octreotide acetate (Sandostatin LAR®) (Almeida and Souto, 2007).

The definition of polymeric nanocarriers goes beyond the nanostructures described above and frequently is extended to polymer–protein drug conjugates like PEGylated proteins (Section 19.9). Nonetheless, the classification of polymer–protein drug conjugates as nanostructures is controversial, and several authors exclude these.

19.7 NANOEMULSIONS AND MICROEMULSIONS

Microemulsions are thermodynamically stable isotropic liquids formed by mixing oil, water and surfactants, which form a variety of different systems depending on their composition and environmental conditions (particularly temperature). Nanoemulsions, on the other hand, are conventional emulsions of two immiscible liquids in which one of the liquids is dispersed as nanometric droplets (Figure 19.8). Nanoemulsions are thermodynamically unstable, but kinetically stable. Both micro- and nanoemulsions can be in the sub-100 nm size range (McClements, 2012).

Nanoemulsions were first licensed for a biological product in 1997, for an influenza vaccine (Fluad®) containing both the antigen and *N*-acetyl-l-alanyl-d-isoglutanimyl-l-alanine-2-(1,2-dipalmitoyl), which is the synthetic form of muramyl dipeptide, a natural component of bacterial cell wall that can trigger innate immunity. This formulation of oil droplets of about 160 nm had Tween 80 and Span 85 (nonionic surfactants) as stabilizing agents (O'Hagan et al., 2012). Nano- and microemulsions are constantly investigated for drug delivery, including biopharmaceuticals, and several approved excipients are available.

Regarding micro- and nanoemulsions preparation, high-pressure homogenization, including Microfluidizer® processor (Figure 19.9a), and high-amplitude ultrasonic processing (Figure 19.9b) are currently the leading methods

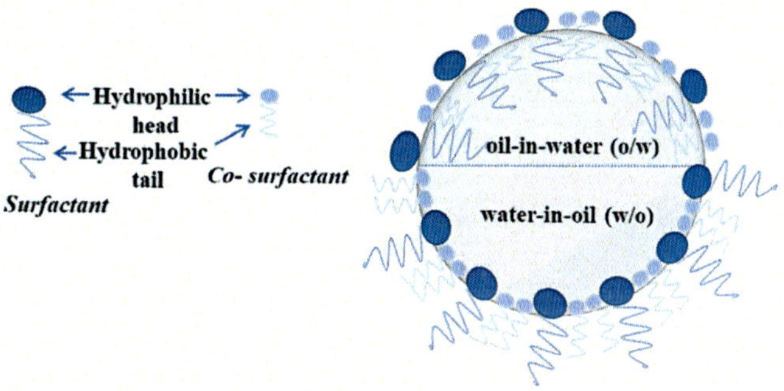

FIGURE 19.8 Representation of micro- and nanoemulsion structures. Oil (or organic phase), surfactants, co-surfactants, which are amphiphilic molecules presenting a polar head and an apolar tail, and water (or aqueous phase) might form microemulsions and nanoemulsions. Microemulsions and nanoemulsions are thermodynamically and kinetically stable, respectively. The surface with the highest tension will be concave and consequently will encompass the other surface of the liquid, making the internal phase of the system that can be oily, i.e. oil-in-water (o/w), or aqueous, i.e. water-in-oil (w/o). The dimensions represented in this illustration do not reflect the actual size for both nanostructures, as microemulsions have droplets of smaller sizes than nanoemulsions despite the name suggesting otherwise.

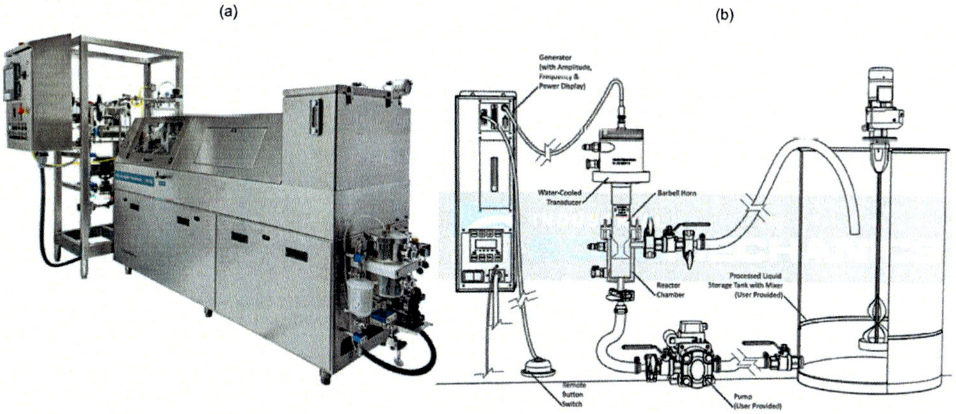

FIGURE 19.9 (a) Microfluidizer® processor with a multi-slotted fixed-geometry interaction chamber able to process pressures up to 689, 1,379 or 2,068 bar. This technology generates uniform droplets and particle size reduction, leading to stable nanoemulsions and liposomes. (b) Industrial-scale ultrasonic processors most commonly supplied with a 3,000-W ultrasonic generator, water-cooled transducer and flow-through reactor chamber (flow cell). It can operate at extremely high amplitudes. (Courtesy of Microfluidics International Corporation (https://www.microfluidicscorp.com/) and Industrial Sonomechanics, LLC (https://www.sonomechanics.com/3000_w_industrial-scale_processor/.)

to produce these systems, with high-amplitude ultrasonic processing resulting in nanoemulsions of superior quality (O'Hagan et al., 2012).

19.8 SOLID LIPID NANOPARTICLES AND NANOSTRUCTURED LIPID CARRIERS

Solid–lipid nanoparticles (SLNs) are obtained from oil-in-water emulsions with a solid–lipid matrix that melts at body temperature (Figure 19.10). Similar to other self-aggregated systems, stabilization is guaranteed by surfactants and the use of high-pressure homogenization methods (Müller, Mäder and Gohla, 2000).

SLNs are stable, biodegradable and simple to scale up manufacture. The hydrophobic matrix of SLN does not seem suitable to encapsulate hydrophilic proteins. However, hydrophilic peptides (such as calcitonin, insulin, LHRH, protein antigens) and model proteins (such as bovine serum albumin and lysozyme) have been incorporated into SLN formulations (Almeida and Souto, 2007).

Like liposomes, positively charged SLNs have been reported as alternative nonviral transfection systems. They are usually composed of cationic peptides, cationic polymers and/or cationic lipids, which form DNA–lipid complexes by the direct mixing of the positively charged lipids with the negatively charged DNA (Del Pozo-Rodríguez et al., 2010).

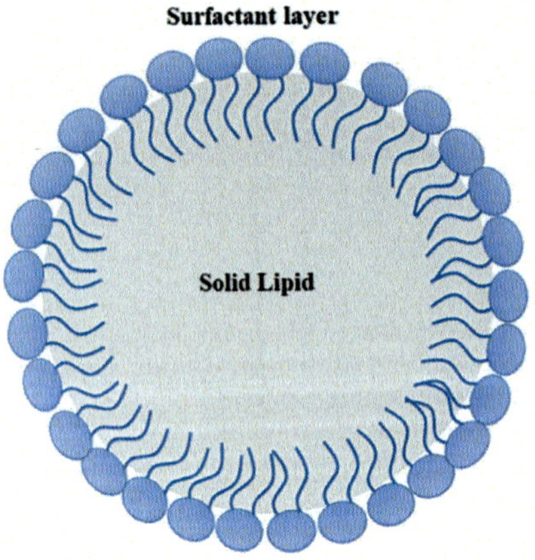

FIGURE 19.10 Representation of solid–lipid nanoparticles with a lipidic solid core.

Nanostructured lipid carriers (NLCs) are a second generation of lipid nanoparticles prepared from a blend of solid and liquid lipids that solidify below 40°C. Consequently, NLCs are more flexible for drug-release modulations, avoiding premature cargo release. Additionally, the loading capacity of NLCs is usually higher, due to a less-ordered lipid matrix that is formed. Biopharmaceuticals such as calcitonin have been loaded to NLCs for oral delivery, and recently, these nanocarriers have also been suggested as promising vehicles for antigen delivery (Martins et al. 2007; Courant et al., 2017).

19.9 PEGYLATION

In the late 1970s, Professor Frank Davis and his colleagues covalently attached methoxy-poly(ethylene glycol) (mPEG) to two proteins (albumin and catalase) using cyanuric chloride as the activating agent (Jevsevar, Kunstelj and Porekar, 2010). Their studies showed that the covalent attachment of PEG markedly improved the stability of proteins. Chemical conjugation of polymers to protein drugs may bring several benefits such as increased solubility, reduced immunogenicity, increased plasma residence time *in vivo* and overall improved pharmacokinetics. PEG is a biocompatible polymer, has low immunogenicity and toxicity, is soluble in water and organic solvents and is easily eliminated from the body, making it the polymer of first choice for bioconjugation (Santos et al., 2018). In addition, PEG is among the few synthetic polymers considered safe for internal administration by the regulatory agencies such as the FDA and EMA.

Currently, PEGylation is well established and used industrially in the production of biopharmaceuticals (Figure 19.11). This technique can be applied to peptides, proteins (such as enzymes and fragments of antibodies), nucleotides and even small organic molecules (Jevsevar, Kunstelj and Porekar, 2010; Li et al., 2013).

The design of a PEGylation bioprocess involves several stages: (a) reaction design, (b) separation of PEGylated conjugates, (c) mass and structure analysis, (d) bioactivity analysis and (e) selection of appropriate PEGylated conjugates (González-Valdez, Rito-Palomares and Benavides, 2012). Regarding the reaction design, the first-generation PEGylation refers to nonspecific random conjugations, generally promoted by the attachment of PEG to the ε-amino groups of lysines and *N*-terminal amino acid (Santos

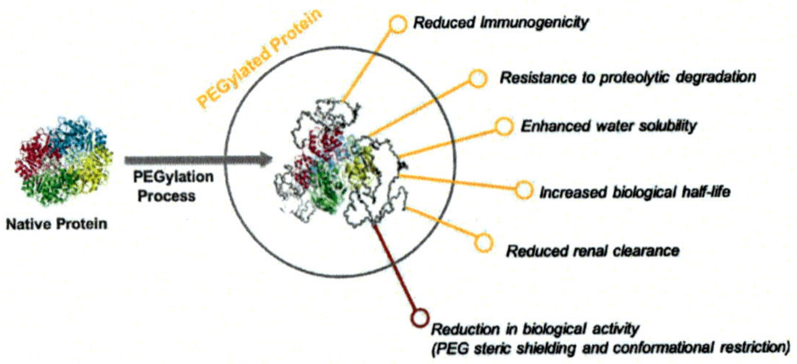

FIGURE 19.11 Effects of PEGylation process on protein properties.

TABLE 19.1
PEGylated Protein Drugs Approved in the United States of America and/or Europe

Brand Name, Generic Name	Company	PEGylation Site	PEG Size (Da)	Indication(s)	Year of Approval
Adagen®, pegademase bovine	Enzon	Lysines	5,000	Severe combined immunodeficiency diseases (SCID)	1990
Oncaspar®, pegaspargase	Enzon	Lysines	5,000	Acute lymphoblastic leukaemia (ALL)	1994
PEG-INTRON®, PEG-interferon alpha 2b	Schering-Plough	Histidine, cysteine, lysines, serine, tyrosine, histidine	12,000	Hepatitis C	2000
PEGASYS®, PEGinterferon alpha 2a	Hoffman-La Roche	Lysines	40,000	Hepatitis C	2001
Neulasta®, pegfilgrastim	Amgen	*N*-terminal	20,000	Neutropenia	2002
Somavert®, pegvisomant	Pfizer	Lysines, *N*-terminal	5,000	Acromegaly	2003
Macugen®, pegaptanib	Pfizer	Lysines	40,000	Age-related macular degeneration	2004
Mircera®, PEG-epoetin beta	Hoffman-La Roche	Lysines	30,000	Anaemia associated with chronic renal failure	2007
Cimzia®, certolizumab pegol	UCB	*C*-terminal	40,000	Rheumatoid arthritis, Crohn's disease, axial spondyloarthritis, psoriatic arthritis	2008
Krystexxa®, pegloticase	Savient	Lysines	10,000	Chronic gout	2010
Sylatron™, PEGinterferon alpha 2b	Merck	Histidine, cysteine, lysines, serine, tyrosine	12,000	Melanoma	2011
Omontys®, peginesatide	Affymax/ Takeda	Lysines	40,000	Anaemia associated with chronic kidney disease	2012
Plegridy®, peginterferon beta-1	Biogen	*N*-terminal	20,000	Relapsing multiple sclerosis	2014
Adynovate®, PEGylated antihaemophilic factor	Baxalta	Lysines	20,000	Haemophilia A	2015
Rebinyn®, coagulation factor IX glycoPEGylated	Novo Nordisk	*N*-linked glycans	40,000	Haemophilia B	2017
Jivi®, antihaemophilic factor PEGylated-aucl	Bayer	Cysteine	60,000 (Branched PEG)	Haemophilia A	2018
Palynziq®, PEGylated phenylalanine ammonia lyase	BioMarin	Lysines	20,000	Phenylketonuria	2018
Revcovi®, PEGylated adenosine deaminase	Leadiant Biosciences	Lysines	5,000	SCID	2018
Asparlas® (calaspargase pegol-mknl)	Servier	Lysines	5,000	Acute lymphoblastic leukaemia (ALL)	2018
Jivi® (antihaemophilic factor PEGylated-aucl)	Bayer	Cysteine (recombinant)	60,000	Haemophilia A	2018
Esperoct® (antihaemophilic factor, glycopegylated-exei)	Novo Nordisk	Glyco-PEGylation	40,000	Haemophilia A	2019

et al., 2018). Higher polydispersity, intrinsic to this type of PEGylation, makes it challenging to reproduce drug batches and can contribute to the poor clinical performance of the PEGylated biopharmaceuticals conjugate. Furthermore, the first-generation reactions mainly use the linear PEG derivatives with lower molecular weight (12 kDa or less). Notwithstanding these disadvantages, two first-generation PEGylated drugs are still in use today: Adagen (adenosine deaminase) and Oncaspar (*L*-asparaginase), for the treatment of severe combined immunodeficiency disease and acute lymphoblastic leukaemia, respectively. Second-generation PEGylation chemistry refers to PEG derivatives that can be linked to specific sites at the protein surface (i.e. PEG-acetaldehyde and PEG-propionaldehyde). Site-specific PEGylation can minimize the loss of biological activity and enhance the batch-to-batch control. Third-generation PEGylation is a recent innovation to balance effectiveness and prolonged circulation half-life by using a prodrug approach, i.e. by incorporating cleavable PEG groups (Gong, Leroux and Gauthier, 2015). It also refers to the introduction of specific PEGylation sites on proteins using molecular biology, for example Ambrx's ReCODE and EuCODE technologies. With an influx of new techniques, assays to track and characterize the released PEG molecules are increasingly requested by regulatory agencies.

To date, several PEGylated proteins, peptides, antibody fragments and oligonucleotides have been approved by the FDA and are on the market (Table 19.1). Most of these formulations are PEGylated proteins, with an estimated total market value of over US $ 8 billion per year (Ginn et al., 2014).

19.10 CHARACTERIZATION TECHNIQUES

Nanostructures require specific techniques to analyse particle/aggregate size, shape, as well as to guarantee colloidal stability. Some of the most relevant analytical methods for the characterization of nanostructures are dynamic light scattering (DLS), zeta potential (ZP), nanoparticle tracking analysis (NTA), microscopy and chromatography.

19.10.1 Dynamic Light Scattering (DLS)

In the field of nanobiotechnology, one of the most useful characterization techniques is DLS because it allows measurements of the mean hydrodynamic diameter and polydispersity index of the nanostructures. DLS, also known as 'photon correlation spectroscopy' or 'quasi-elastic light scattering', is based on the illumination of a suspension of nanoparticles or large molecules (such as proteins and polymers) in Brownian motion by a laser beam. It is noninvasive and requires a minimal sample preparation, and no pre-experimental calibration is needed (Bhattacharjee, 2016). The measurement of light scattered by the nanoparticles/molecules (assumed to be spherical) provides the translational diffusion coefficient (D_f), which is converted into a hydrodynamic diameter (D_h) using the Stokes–Einstein equation.

$$D_f = \frac{k_B \cdot T}{6 \cdot \pi \cdot \eta \cdot R_h} \quad (19.2)$$

in which D_f = diffusion coefficient, k_B = Boltzmann constant ($1.38064852 \times 10^{-23}$ J/K), T = temperature, π = mathematical constant (3.14159), η = viscosity and R_h = hydrodynamic radius.

The DLS also provides the polydispersity index (PdI), a parameter that estimates the size variation based on the intensity of light scattered by the different populations of nanoparticles in a system. In general, systems presenting PdI ≤ 0.1 are considered to be highly monodisperse, whereas PdI > 0.4 are observed in highly polydisperse systems. Values greater than 0.7 indicate that the sample has a very broad size distribution and is probably not suitable to be analysed by this technique. Values between 0.1 and 0.4 are considered moderate and usually adequate for drug delivery purposes, but should be analysed on a case-by-case basis depending on the intended application.

19.10.2 Nanoparticle Tracking Analysis (NTA)

Another tool suitable for the analysis of particles in the nanometre size range, as well as for the characterization of protein aggregates, is NTA. This technique combines laser light scattering microscopy with a charge-coupled device camera, allowing visualization and recording of nanoparticles in solution. It can track individual nanoparticles moving by Brownian motion and then converts the signal into particle size according to the Stokes–Einstein equation. Additionally, it provides an approximate nanoparticle concentration. The difference between DLS and NTA is the low peak resolution for DLS; i.e., it can only resolve nanoparticles that differ in size at least by a factor of three (Filipe, Hawe and Jiskoot, 2010).

19.10.3 Zeta Potential (ZP)

ZP estimates the magnitude of the electrostatic repulsion/attraction between particles; it is related to stability and can explain permeation and/or uptake of drugs from nanocarriers (Baspinar and Borchert, 2012; Voigt et al., 2014). Specifically, the ZP corresponds to the potential difference between the medium in which the nanoparticles are dispersed and the stationary layer of fluid surrounding the dispersed nanoparticle and can be correlated with the particle charge (Figure 19.12).

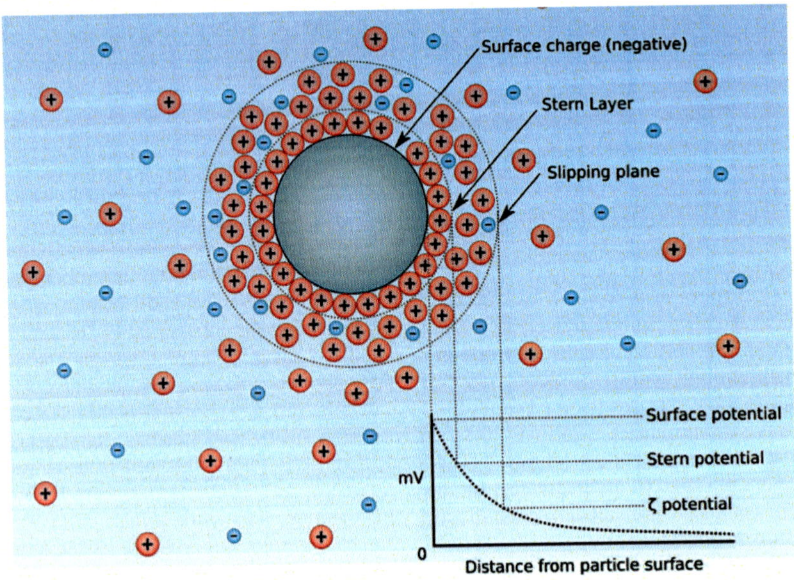

FIGURE 19.12 Diagram showing the zeta potential (ζ) as a function of distance from the charged surface of a nanoparticle in a dispersion medium. (Reproduced from Wikimedia, Diagram of Zeta potential and slipping, 2019, an Open Access File distributed under the terms of the Creative Commons Attribution License.)

The ZP is a key indicator of colloidal stability (ability to resist sedimentation or particle aggregation) since it correlates the degree of electrostatic repulsion between similarly charged particles in dispersion. For molecules and particles that are small enough, high ZP values indicate the stability. When the potential is small, attractive forces may exceed the electrostatic repulsion, and particle aggregation and sedimentation may occur. Usually, it is assumed that colloidal systems with high absolute values of ZP (approximately 30 mV) are electrically stable, while colloids with low ZP values tend to coagulate or flocculate (Vogel et al. 2017).

Techniques for ZP measurement based on electrokinetic phenomena, such as electrophoretic mobility (U_E) measurements, are largely used and are usually provided in the same light scattering instruments applied to determine the hydrodynamic size (Sze et al., 2003; Delgado et al., 2007). The electrophoretic mobility is proportional to the electric field and involves a relative movement between rigid and mobile parts of the double layer. The conversion of electrophoretic mobility to ZP can be performed by Henry's equation (Hsu et al., 2014; Pfeiffer et al., 2014).

$$U_E = \frac{2 \cdot \varepsilon \cdot \zeta \cdot f(\kappa a)}{3 \cdot \eta} \tag{19.3}$$

in which U_E = electrophoretic mobility, ζ = zeta potential (ZP), ε = dielectric constant, η = viscosity and $f(\kappa a)$ = Henry's function.

19.10.4 Microscopy

Microscopy is an important tool to evaluate the shape and surface of nanostructures. Transmission electron microscopy (TEM), for example, is most frequently used for vesicular nanostructures like liposomes and polymersomes, while scanning electron microscopic (SEM) techniques are more suitable for nanospheres, since SEM provides images from the scanned surface of the nanostructures. Atomic force microscopy (AFM) is more appropriate for surface analysis and can give details on surface porosity. More recently, cryo-TEM has been used to analyse vesicular nanostructures, since it allows a direct visualization in a frozen-hydrated state (Gaumet et al., 2008; Kuntsche, Horst and Bunjes, 2011).

19.11 NANOTOXICITY

Nanotoxicology is a branch of bionanoscience that deals with the study of the toxicity of nanomaterials (Ai et al., 2011). The term 'nanotoxicology' was first introduced in the 2010s; it reflects the unique properties of nanostructures and suggests that their interactions with organisms may be unpredictable (Shvedova, Pietroiusti and Kagan, 2016).

Two factors are important with regard to nanoparticle toxicity: (a) size and (b) surface chemistry composition. A reduction in the size of nanostructures results in an increase in particle surface area. Therefore, more chemical molecules may attach to this surface, which will enhance reactivity and potential toxic effects (Ai et al. 2011). Besides size and surface chemistry, other parameters need to be considered, such as (a) shape, (b) ZP, (c) purity, (d) crystallinity, (e) dosage, (f) aqueous solubility/stability, (g) free radicals production, (h) cellular uptake and (i) intracellular persistence (Ai et al., 2011; Siegrist et al., 2018).

Recently, the Nanotoxicological Classification System (NCS) was proposed, combining both particles' size and degree of biodegradability (i.e. persistency in the body). Accordingly, nanomaterials can be divided into four levels of increasing toxicity, namely class I (size above 100 nm and biodegradable – lower risk), class II (size above 100 nm and non-biodegradable – intermediate risk), class III (size below 100 nm and biodegradable – intermediate risk) and class IV (size below 100 nm and non-biodegradable – higher risk) (Keck and Müller, 2013).

19.12 REGULATORY ASPECTS

New or altered physicochemical properties at the nanoscale must be carefully monitored to assess how these changes might influence the product safety and clinical effectiveness. The regulatory agencies of the EU (EMA), USA (FDA) and Japan (PMDA/MHLW) have been working together since 2009 to discuss characterization, formulation development, manufacturing and safety evaluation of nanomaterial-containing drug products for human use, in an attempt to harmonize common perspectives (Sainz et al. 2015). Several guidelines have been published as a result of these joint efforts.

Some of the issues that must be considered when developing a nanostructured drug formulation include route of administration, clinical indication, structural complexity and maturity of the technology regarding manufacturing processes and analytical techniques. The nanomaterial should be properly described, including technical information on size, charge, morphology, composition, complexation, as well as the profile of interaction with biological systems (i.e. bioavailability, distribution, biodegradation, accumulation).

Taking into consideration parameters of quality by design (QbD), the critical quality attributes (CQAs) of the nanomaterial should be employed to determine the potential impact on the product performance (Sadrieh and Tyner, 2010). The following CQAs should be included and measured for all nanomaterials incorporated into medicines: (a) chemical composition, (b) average particle size, (c) particle size distribution, (d) shape and morphology and (e) physicochemical stability. Additional CQAs may also be required, depending on the route of administration, indication and patient population: (a) *in vitro* release, (b) capacity of drug loading, (c) surface properties, (d) core–shell structure/lamellarity, (e) coating properties, (f) nanoparticle concentration, (g) ratio between encapsulated (or bioconjugated) drug vs free drug; (h) impurities, (i) sterility and endotoxin levels and (j) crystalline form. Some other important aspects to take into account for the regulation of nanobiological drugs are standard methods currently available for characterization, e.g. ISO 22412:2017 and ASTM E2859-11(2017), and the environmental impact of nanomaterials, to determine whether the drug product can be qualified as a safe product in terms of Environmental Assessment (EA).

19.13 FINAL CONSIDERATIONS

Nanotechnology and biotechnology are intrinsically related, and several biological structures are comprised in the nanoscale. More recently, the term 'nanobiotechnology' has gained attention and can be broadly understood as the application of nanotechnology in the life sciences. As we saw in this chapter, nanobiotechnology offers several alternatives for biopharmaceuticals to meet tissue/cellular penetration, target specificity and extended half-life. Strategies as incorporation into nanostructures and bioconjugation are becoming frequent, and we already see products in the market. Nonetheless, toxicological and regulatory aspects are still evolving, and we expect this to be a constant movement, since novel alternatives will certainly arise in the next years.

REFERENCES

Ai, J.; Biazar, E.; Jafarpour, M.; Montazeri, M.; Majdi, A.; Aminifard, S.; Zafari, M.; Akbari, H.R.; Rad, H.G. Nanotoxicology and nanoparticle safety in biomedical designs. *Int J Nanomed*, v. 6, 1117–1127, 2011.

Almeida, A.J.; Souto, E. Solid lipid nanoparticles as a drug delivery system for peptides and proteins. *Adv Drug Deliv Rev*, v. 59, 478–490, 2007.

Almeida, J.D.; Brand, C.M.; Edwards, D.C.; Heath, T.D. Formation of virosomes from influenza subunits and liposomes. *Lancet*, v. 2, 899–901, 1975.

Anassi, E.; Ndefo, U.A. Sipuleucel-T (provenge) injection: the first immunotherapy agent (vaccine) for hormone-refractory prostate cancer. *Pharmacy and Therapeutics*, v. 36, 197–202, 2011.

Apolinário, A.C.; Pachioni-Vasconcelos, J.D.A.; Pessoa, Jr, A.; Rangel-Yagui, C.O. Polymersomes versus liposomes: The "Magic Bullet" Evolution. *Química Nova*, v. 40, 810–817, 2017. https://www.scielo.br/scielo.php?pid=S0100-40422017000700810&script=sci_abstract.

Apolinário, A.C.; Salata, G.C.; Bianco, A.F.R.; Fukumori, C.; Lopes, L.B. Opening the Pandora's box of nanomedicine: There is indeed 'plenty of room at the bottom'. *Química Nova*, v. 43, 212–225, 2020. https://www.scielo.br/scielo.php?script=sci_abstract&pid=S0100-40422020000200212&lng=en&nrm=iso.

Banerjee, A.; Onyuksel, H. Peptide delivery using phospholipid micelles. *Wiley Interdiscip Rev Nanomed Nanobiotechnol*, v. 4, 562–74, 2012.

Baspinar, Y.; Borchert, H.H. Penetration and release studies of positively and negatively charged nanoemulsions: Is there a benefit of the positive charge? *Int J Pharm*, v. 430, 247–252, 2012.

Bhattacharjee, S. DLS and zeta potential: What they are and what they are not? *J Control Release*, v. 235, 337–351, 2016.

Blackman, L.D.; Varlas, S.; Arno, M.C.; Houston, Z.H.; Fletcher, N.L.; Thurecht, K.J.; Hasan, M.; Gibson, M.I.; O'Reilly, R.K. Confinement of therapeutic enzymes in selectively permeable polymer vesicles by polymerization-induced self-assembly (PISA) reduces antibody binding and proteolytic susceptibility. *ACS Cent Sci*, v. 4, 718–723, 2018.

Bolhassani, A.; Jafarzade, B.S.; Mardani, G. In vitro and in vivo delivery of therapeutic proteins using cell penetrating peptides. *Peptides*, v. 87, 50–63, 2017.

Bron, R.; Ortiz, A.; Wilschut, J. Cellular cytoplasmic delivery of a polypeptide toxin by reconstituted influenza-virus envelopes (virosomes). *Biochem*, v. 33, 9110–9117, 1994.

Bueno, C.Z.; Oliveira, C.A.; Rangel-Yagui, C.O. Polymeric and liposomal nanomaterials. In *Nanobiomaterials: Nanostructured Materials for Biomedical Applications*, ed. R. Narayan. Woodhead Publishing, 2018.

Bulbake, U.; Doppalapudi, S.; Kommineni, N.; Khan, W. Liposomal formulations in clinical use: An updated review. *Pharmaceutics*, v. 9, 2–17, 2017.

Charcosset, C.; Juban, A.; Valour, J.-P.; Urbaniak, S.; Fessi, H. Preparation of liposomes at large scale using the ethanol injection method: Effect of scale-up and injection devices. *Chem Eng Res Des*, v. 94, 508–515, 2015.

Courant, T.; Bayon, E.; Reynaud-Dougier, H.L.; Villiers, C.; Menneteau, M.; Marche, P.N.; Navarro, F.P. Tailoring nanostructured lipid carriers for the delivery of protein antigens: Physicochemical properties versus immunogenicity studies. *Biomaterials*, v. 136, 29–42, 2017.

De Lázaro, I.; Mooney, D.J. A nanoparticle's pathway into tumours. *Nat Mater*. v. 19, 486–487, 2020.

Del Pozo-Rodríguez, A.; Delgado, D.; Solinís, M.A.; Pedraz, J.L.; Echevarría, E.; Rodríguez, J.M.; Gascón, A.R. Solid lipid nanoparticles as potential tools for gene therapy: In vivo protein expression after intravenous administration. *Int J Pharm*, v. 385, 157–162, 2010.

Delgado, A.V.; González-Caballero, F.; Hunter, R.J.; Koopal, L.K.; Lyklema, J. Measurement and interpretation of electrokinetic phenomena. *J Colloid Interface Sci*, v. 309, 194–224, 2007.

Discher, D.E.; Ahmed, F. Polymersomes. *Annu Rev Biomed Eng*, v. 8, 323–341, 2006.

EMA. http://www.ema.europa.eu (access in January 2019).

Discher, B.M.; Won, Y.Y.; Ege D.S.; Lee, J.C.; Bates, F.S.; Discher, D.E.; Hammer, D.A. Polymersomes: tough vesicles made from diblock copolymers. *Science*, v. 284, 1143–1146, 1999.

FDA. Drug Products, Including Biological Products, that Contain Nanomaterials: Guidance for Industry. Food and Drug Administration, 2017.

Filipe, V.; Hawe, A.; Jiskoot, W. Critical evaluation of Nanoparticle Tracking Analysis (NTA) by NanoSight for the measurement of nanoparticles and protein aggregates. *Pharm Res*, v. 27, 796–810, 2010.

Gaumet, M.; Vargas, A.; Gurny, R.; Delie, F. Nanoparticles for drug delivery: The need for precision in reporting particle size parameters. *Eur J Pharm Biopharm*, v. 69, 1–9, 2008.

Ginn, C.; Khalili, H.; Lever, R.; Brocchini, S. PEGylation and its impact on the design of new protein-based medicines. *Future Med Chem*, v. 6, 1829–1846, 2014.

Gong, Y.; Leroux, J.-C.; Gauthier, M.A. Releasable conjugation of polymers to proteins. *Bioconjugate Chem*, v. 26, 1172–1181, 2015.

González-Valdez, J.; Rito-Palomares, M.; Benavides, J. Advances and trends in the design, analysis, and characterization of polymer-protein conjugates for "PEGylaided" bioprocesses. *Anal Bioanal Chem*, v. 403, 2225–2235, 2012.

Gu, Z.; Biswas, A.; Zhao, M.; Tang, Y. Tailoring nanocarriers for intracellular protein delivery. *Chem Soc Rev*, v. 40, 3638–3655, 2011.

Hsu, C.; Lee, D.-J.; Hsu, J.-P.; Wang, N.; Tseng, S. Electrophoresis of pH-regulated particles in the presence of multiple ionic species. *AIChE J*, v. 60, 451–458, 2014.

Jevsevar, S.; Kunstelj, M.; Porekar, V.G. PEGylation of therapeutic proteins. *Biotechnol J*, v. 5, 113–28, 2010.

Joseph, A.; Contini, C.; Cecchin, D.; Nyberg, S.; Ruiz-Perez, L.; Gaitzsch, J.; Fullstone, G.; Tian, X.; Azizi, J.; Preston, J.; Volpe, G.; Battaglia, G. Chemotactic synthetic vesicles: Design and applications in blood-brain barrier crossing. *Sci Adv*, v. 3, e1700362, 2017.

Kalra, N.; Dhanya, V.; Saini, V.; Jeyabalan, G. Virosomes: As a drug delivery carrier. *Am J Adv Drug Deliv*, v. 1, 29–35, 2013.

Keck, C.M.; Müller, R.H. Nanotoxicological classification system (NCS): A guide for the risk-benefit assessment of nanoparticulate drug delivery systems. *Eur J Pharm Biopharm*, v. 84, 445–448, 2013.

Koren, E.; Torchilin, V.P. Cell-penetrating peptides: Breaking through to the other side. *Trends Mol Med*, v. 18, 385–393, 2012.

Kuntsche, J.; Horst, J.C.; Bunjes, H. Cryogenic transmission electron microscopy (cryo-TEM) for studying the morphology of colloidal drug delivery systems. *Int J Pharm*, v. 417, 120–137, 2011.

Letchford, K.; Burt, H. A review of the formation and classification of amphiphilic block copolymer nanoparticulate structures: Micelles, nanospheres, nanocapsules and polymersomes. *Eur J Pharm Biopharm*, v. 65, 259–269, 2007.

Li, W.; Zhan, P.; Clercq, E.D.; Lou, H.; Liu, X. Current drug research on PEGylation with small molecular agents. *Prog Poly Sci*, v. 38, 421–444, 2013.

Martins, S.; Sarmento, B.; Ferreira, D.C.; Souto, E.B. Lipid-based colloidal carriers for peptide and protein delivery: liposomes versus lipid nanoparticles. *Int J Nanomed*, v. 2, 595–607, 2007.

McClements, D.J. Nanoemulsions versus microemulsions: terminology, differences, and similarities. *Soft Matter*, v. 8, 1719–1729, 2012.

Messager, L.; Gaitzsch, J.; Chierico, L.; Battaglia, G. Novel aspects of encapsulation and delivery using polymersomes. *Curr Opin Pharmacol*, v. 18, 104–11, 2014.

Mirza, A.Z.; Siddiqui, F.A. Nanomedicine and drug delivery: A mini review. *Int Nano Lett*, v. 4, 94, 2014.

Mitragotri, S.; Burke, P.A.; Langer, R. Overcoming the challenges in administering biopharmaceuticals: Formulation and delivery strategies. *Nat Rev Drug Discov*, v. 13, 655–672, 2014.

Movassaghian, S.; Merkel, O.M.; Torchilin, V.P. Applications of polymer micelles for imaging and drug delivery. *Wiley Interdiscip Rev Nanomed Nanobiotechnol*, v. 7, 691–707, 2015.

Müller, R.H.; Mäder, K.; Gohla, S. Solid lipid nanoparticles (SLN) for controlled drug delivery – A review of the state of the art. *Eur J Pharm Biopharm*, v. 50, 161–177, 2000.

Munyendo, W.L.; Lv, H.; Benza-Ingoula, H.; Baraza, L.D.; Zhou, J. Cell penetrating peptides in the delivery of biopharmaceuticals. *Biomolecules*, v. 2, 187–202, 2012.

O'Hagan, D.T.; Ott, G.S.; De Gregorio, E.; Seubert, A. The mechanism of action of MF59- an innately attractive adjuvant formulation. *Vaccine*, v. 30, 4341–4348, 2012.

Olusanya, T.O.B.; Haj Ahmad, R.R.; Ibegbu, D.M.; Smith, J.R.; Elkordy, A. Liposomal drug delivery systems and anticancer drugs. *Molecules*, v. 24443, 1–17, 2018.

Orive, G.; Gascón, A.R.; Hernández, R.M; Domínguez-Gil, A.; Pedraz, J.L. Techniques: New approaches to the delivery of biopharmaceuticals. *Trends Pharmacol Sci*, v. 25, 382–387, 2004.

Pachioni-Vasconcelos, J.E.A.; Lopes, A.M.; Apolinário, A.C.; Valenzuela-Oses, J.K.; Costa, J. S.; Nascimento, L.E.O.; Pesso, A.; Barbosa, L.R.; Rangel-Yagui, C.O. Nanostructures for protein drug delivery. *Biomater Sci*, v. 4, 205–218, 2016.

Pakulska, M.M.; Donaghue, I.E.; Obermeyer, J.M.; Tuladhar, A.; Mclaughlin, C.K.; Shendruk, T.N.; Shoichet, M.S. Encapsulation-free controlled release: Electrostatic adsorption eliminates the need for protein encapsulation in PLGA nanoparticles. *Sci Adv*, v. 2, e1600519, 2016.

Panyam, J.; Labhasetwar, V. Biodegradable nanoparticles for drug and gene delivery to cells and tissue. *Adv Drug Deliv Rev*, v. 55, 329–347, 2003.

Pfeiffer, C.; Rehbock, C.; Hühn, D.; Carrillo-Carrion, C.; De Aberasturi, D.J.; Merk, V.; Barcikowski, S.; Parak, W.J. Interaction of colloidal nanoparticles with their local environment: the (ionic) nanoenvironment around nanoparticles is different from bulk and determines the physico-chemical properties of the nanoparticles. *J R Soc Interface*, v. 11, 20130931, 2014.

Renggli, K.; Baumann, P.; Langowska, K.; Onaca, O.; Bruns, N.; Meier, W. Nanoreactors: Selective and responsive nanoreactors. *Adv Funct Mater*, v. 21, 1206–1205, 2011.

Sadrieh, N.; Tyner, K.M. Nanotechnology and therapeutic delivery: a drug regulation perspective. *Ther Deliv*, v. 1, 83–89, 2010.

Sainz, V.; Conniot, J.; Matos, A.I.; Peres, C.; Zupancic, E.; Moura, L.; Silva, L.C.; Florindo, H.F.; Gaspar, R.S. Regulatory aspects on nanomedicines. *Biochem Biophys Res Commun*, v. 468, 504–510, 2015.

Santos, J.H.P.M.; Torres-Obreque, K.M.; Meneguetti, G.P.; Amaro, B.P.; Rangel-Yagui, C.O. Protein PEGylation for the design of biobetters: From reaction to purification processes. *Braz J Pharm Sci*, v. 54, e01009, 2018.

Sartor, O. Eligard: Leuprolide acetate in a novel sustained-release delivery system. *Urology*, v. 61, 25–31, 2003.

Shvedova, A.; Pietroiusti, A.; Kagan, V. Nanotoxicology ten years later: Lights and shadows. *Toxicol Appl Pharmacol*, v. 299, 1–2, 2016.

Siegrist, S.; Cörek, E.; Detampel, P.; Sandström, J.; Wick, P.; Huwyler, J. Preclinical hazard evaluation strategy for nanomedicines. *Nanotoxicology*, v. 13, 73–99, 2018.

Sindhwani, S.; Syed, A.M.; Ngai, J; Kingston, B.R.; Maiorino, L.; Rothschild, J.; Macmillan, P.; Zhang, Y.; Rajesh, N.U.; Hoang, T.; Wu, J.L.Y.; Wilhelm, S.; Zilman, A.; Gadde, S.; Sulaiman, A.; Ouyang, B.; Lin, Z.; Wang, L.; Egeblad, M.; Chan, W.C.W. The entry of nanoparticles into solid tumours. *Nat Mater*, v. 19, 566–575, 2020.

Sleep, D.; Cameron, J.; Evans, L.R. Albumin as a versatile platform for drug half-life extension. *Biochim Biophys Acta*, v. 1830, 5526–5534, 2013.

Stgmann, T.; Morselt, H.W.M.; Booy, F.P.; Van Breemen, J.F.L.; Scherphof, G.; Wilschut, J. *EMBO J*, v. 6, 2651–2659, 1987.

Sze, A.; Erickson, D.; Ren, L.; Li, D. Zeta-potential measurement using the Smoluchowski equation and the slope of the current-time relationship in electroosmotic flow. *J Colloid Interface Sci*, v. 261, 402–410, 2003.

Tombal, B.; Berges, R. How good do current LHRH agonists control testosterone? Can this be improved with Eligard®? *Eur Urol Suppl*, v. 4, 30–36, 2005.

Torchilin, V.P. Recent advances with liposomes as pharmaceutical carriers. *Nat Rev Drug Discov*, v. 4, 145–160, 2005.

Torchilin, V.P. Tumor delivery of macromolecular drugs based on the EPR effect. *Adv Drug Deliv Rev*, v. 63, 131–135, 2011.

Torchilin, V.P.; Lukyanov, A.N. Peptide and protein drug delivery to and into tumors: Challenges and solutions. *Drug Discov Today*, v. 8, 259–266, 2003.

Vercauteren, D.; Rejman, J.; Martens, T.F.; Demeester, J.; De Smedt, S.C.; Braeckmans, K. On the cellular processing of non-viral nanomedicines for nucleic acid delivery: Mechanisms and methods. *J Control Release*, v. 161, 566–581, 2012.

Vogel, R., Pal, A.K.; Jambhrunkar, S.; Patel, P.; Thakur, S.S.; Reátegui, E.; Parekh, H.S.; Saá, P.; Stassinopoulos, A.; Broom, M.F. High-resolution single particle Zeta potential characterisation of biological nanoparticles using tunable resistive pulse sensing. *Sci Rep*, v. 7, 17479, 2017.

Voigt, N.; Henrich-Noack, P.; Kockentiedt, S.; Hintz, W.; Tomas, J.; Sabel, B.A. Surfactants, not size or Zeta-potential influence blood-brain barrier passage of polymeric nanoparticles. *Eur J Pharm Biopharm*, v. 87, 19–29, 2014.

Wang, H.Y.; Wang, R.F. Enhancing cancer immunotherapy by intracellular delivery of cell-penetrating peptides and stimulation of pattern-recognition receptor signaling. *Adv Immunol*, v. 114, 151–176, 2012.

Wikimedia. 2019. Diagram of Zeta potential and slipping plane. https://commons.wikimedia.org/wiki/File:Diagram_of_zeta_potential_and_slipping_planeV2.svg

Yadav, S.C.; Kumari, A.; Yadav, R. Development of peptide and protein nanotherapeutics by nanoencapsulation and nanobioconjugation. *Peptides*, v. 32, 173–187, 2011.

Zhu, P.; Jin, L. Cell penetrating peptides: A promising tool for the cellular uptake of macromolecular drugs. *Curr Protein Pept Sci*, v. 19, 211–220, 2018.

20 Biosafety Applied in Pharmaceutical and Biotechnological Processes

Camila Manoel Crnkovic
Universidade de São Paulo

Celso Pereira Caricati
Instituto Butantan

Ricardo Pinheiro de Souza Oliveira and Marco Antonio Stephano
Universidade de São Paulo

CONTENTS

20.1 INTRODUCTION

There are four theoretical and practical concepts to consider when talking about biosafety: the environment, risk, accident and ethics. These concepts are interconnected in the definition of biosafety (Figure 20.1).

One cannot assess risks for human and animal health without considering the environment. Risk assessment, in turn, depends on anticipating potential accidents. If these concepts are trivialized in view of human social behaviour and ethics is extrapolated to benefit individuals or corporations, laws and regulations become necessary

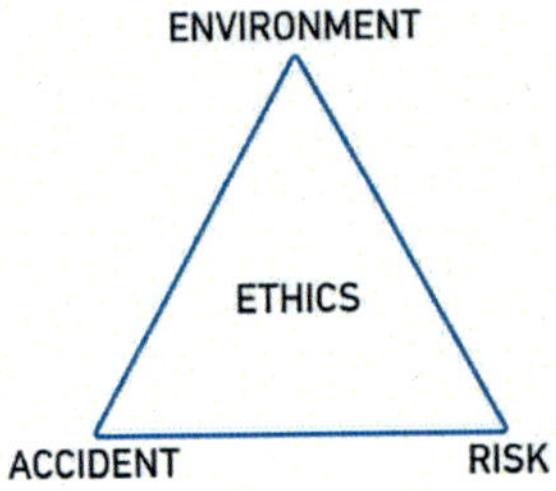

FIGURE 20.1 Interconnection of ethics, environment, risk and accident in defining biosafety.

to prevent accidents for which risks have already been identified.

Prior to defining biosafety, we must first define accident, risk and ethics:

- *Accident*: Fortuitous and harmful event, usually regrettable, unfortunate.
- *Risk*: The probability that a dangerous event will occur.
- *Ethics*: The branch of philosophy that deals with the moral principles that discipline, motivate, distort or guide human behaviour towards values, norms, prescriptions and pressures present in any society.

Biosafety comprises the practices used to contain organisms aiming to reduce risks and avoid accidents, primarily those affecting the environment and, consequently, that may compromise human and animal health. Such procedures require standardized behavioural patterns known as Good Practices (e.g. Good Laboratory Practices, Good Manufacturing Practices, Good Clinical Practices).

Good Practices (GxP) are general guidelines of conduct applied to ensure the quality, safety, consistency, and reliability of products and services. These guidelines entail principles that are applicable to certain activities, but do not necessarily specify how a certain task may be executed. Such details are defined on a case-by-case basis, according to the risks that are inherent to each activity.

Various official conferences have been held and conventions were established to ensure that such guidelines are applied to organism containment plans. These conventions were critical for the development of biotechnology, as they proposed regulations for exploiting biological diversity and for genetic engineering.

One of the first conferences in this regard was the United Nations Conference on the Human Environment that took place in Stockholm in 1972. The environment and development were among the major topics of this convention. It was recognized that both industrialized and developing economies contribute to environmental issues. The Stockholm Conference encouraged various countries to revise environmental policies, create agencies and monitor environmental conditions.

In 1992, during the Earth Summit held in Rio de Janeiro (RIO ECO-92), further discussions addressed recombinant DNA technology and the environment, as well as the need for environmental impact studies to support the controlled release of transgenic plants into the environment. This conference resulted in the Convention on Biological Diversity (CBD), a multilateral treaty that established goals for biodiversity conservation and sustainable use of genetic resources.

The CBD was supplemented by the Cartagena and Nagoya Protocols. The Cartagena Protocol on Biosafety addresses the transboundary movements of living modified organisms (LMOs) as an environmental protection measure. Drafted since 1995 by the Open-ended Ad Hoc Working Group on Biosafety, the Cartagena Protocol entered into force only in 2003. The Nagoya Protocol, in turn, governs the 'Access to Genetic Resources and the Fair and Equitable Sharing of Benefits Arising from their Utilization'. In force since 2014, the Nagoya Protocol contributed towards implementing the objectives of the CBD regarding the sustainable use of biodiversity.

In addition to these environmental protection mechanisms, one must always reflect on potential accidents with pathogenic microorganisms, whether genetically modified or not.

The history of laboratory accidents holds several examples related to biosafety. Systematic studies in this regard started in 1941 with a publication by Eddie and Meyer, where the authors describe 74 cases of laboratory-acquired brucellosis in the United States. Eight years later, Sulkin and Pike (1949) compiled information on 222 viral infections acquired in laboratories. Pike later summarized and analysed a large set of 3,921 cases of laboratory-associated infections (Pike, 1976).

When considering industrial applications, it is also important to anticipate risks associated with the large-scale production of microorganisms. Large scale refers to culture volumes above 10 L in liquid media or 15 kg in solid media.

20.2 DEFINITIONS OF BIOSAFETY

Biosafety has been defined in various ways. The World Health Organization (WHO) defines laboratory biosafety as:

> The containment principles, technologies and practices that are implemented to prevent unintentional exposure to pathogens and toxins, or their accidental release.
>
> *WHO (2004)*

This definition focuses on mechanisms of risk prevention, including both organisms and their toxic chemicals as potentially harmful agents.

According to the Food and Agriculture Organization of the United Nations (FAO), biosafety aims:

> To prevent, manage, minimize or eliminate hazards to human health and security and to protect the environment from biological agents and organisms used in research and trade.
>
> *FAO (2011)*

Beyond prevention, the FAO includes risk management, minimization and elimination as the intended goals of biosafety measures. It also explicitly states that such measures apply to both human health and environment protection.

The Biosafety Clearing-House (BCH) of the CBD provides the following:

> In its broad sense, the term biosafety refers to the protection of human health and the environment from potential

> harm due to biological agents. Under the Convention on Biological Diversity (CBD), and more specifically under the Cartagena Protocol on Biosafety, the term biosafety essentially refers to safety procedures aimed at regulating, managing or controlling the risks associated with the use and release of LMOs [living modified organisms] resulting from biotechnology which are likely to have adverse environmental impacts that could affect the conservation and sustainable use of biological diversity, taking also into account risks to human health.
>
> *CBD (2014)*

The BCH was established by the Cartagena Protocol on Biosafety as a mechanism used to integrate and facilitate the exchange of information regarding LMOs among countries and agencies. While, in the broad definition, biosafety concerns all biological agents, the CBD and the Cartagena Protocol apply this concept specifically to LMOs, thus emphasizing its importance towards biotechnology and its industrial application.

20.3 CLASSIFICATION OF MICROORGANISMS BY RISK GROUP

From a biosafety standpoint, microorganisms can be divided into the two main groups: nonpathogenic and pathogenic microorganisms.

Nonpathogenic microorganisms do not cause disease to humans, animals or plants. They may belong to the so-called normal microbiota and participate in an organism's health and survival. Nonpathogenic microorganisms can be handled in Biosafety Level 1 (BSL-1) facilities, complying with the most basic laboratory safety and containment measures.

Pathogenic microorganisms are known to cause disease. They are classified according to their ability to cause infection, their pathogeny, transmission, morbidity, mortality, epidemic, potential alien effect and according to whether there are treatment options for the disease they cause.

Pathogenic microorganisms should be handled in Biosafety Level 2, 3 or 4 (BSL-2, 3 or 4) facilities depending on their specific characteristics. BSLs are described in Section 20.4.

Infection is the capacity of a pathogenic agent to enter and colonize an organism until it presents the first symptoms. For example, it takes on average 25–40 days for the first symptoms of rabies to appear after the virus infects a mammal. It takes 24–72 h for the influenza virus, 8–16 h for the Ebola virus, 12–24 h for enteropathogenic *Escherichia coli* and 3–7 days for *Bacillus anthracis*.

Pathogeny (sometimes referred to as pathogenesis) corresponds to the severity of symptoms caused by a certain microorganism. The Ebola virus causes severe haemorrhagic fever, leading to death within a matter of hours due to the severity of symptoms. There is no specific treatment against the rabies virus once it spreads to the central nervous system, resulting in poor prognosis. An infection by enteropathogenic *Escherichia coli* may have a more favourable prognosis depending on the host's immune status, but in immunocompromised individuals, infection can lead to death. *Brucella abortus* may lead to infertility in both men and women.

The transmission of a pathogen may be aggravated depending on their route of entry into the host. Natural barriers may reduce or increase the ability of infection, depending on how effective they are against a pathogen. Microorganisms transmitted by the inhalation route are often more harmful than those transmitted orally, since they can be transmitted more easily between people and animals. Pathogens that are transmitted by direct inoculation often show lower incidence of disease than those transmitted orally.

Morbidity is the relative number of infected individuals, animals or plants expressed as a proportion of the total population that was exposed to the pathogen.

Mortality is the relative number of individuals, animals or plants that died due to a certain disease or condition.

Epidemic is the outbreak of a disease to a large number of people in a certain region in a short period of time. It can be endemic if regionally restricted, or pandemic if widespread across continental regions.

The alien effect is the result of a potential accident that releases in the environment any organism that does not exist in that region. For example, handling Ebola virus in the United Kingdom or working with foot-and-mouth virus in North America poses such risk.

With regard to treatment, it can be preventive, therapeutic or palliative. A preventive treatment aims to improve an individual's or animal's immunity. It can be an active measure (e.g. vaccination) or a passive measure (e.g. transfer of specific antibodies). A therapeutic treatment relies on utilizing pharmaceutical or biopharmaceutical drugs that act directly on the pathogen, such as antibiotics, antivirals and monoclonal antibodies. A palliative treatment aims to mitigate the pathophysiological effects caused by the pathogen without directly acting upon it. Examples include the administration of anti-inflammatory, analgesic, anticonvulsant, antipyretic and antiemetic drugs.

The WHO classifies infective microorganisms into four risk groups:

> 1. ***Risk Group 1*** (no or low individual and community risk): a microorganism that is unlikely to cause human or animal disease.
> 2. ***Risk Group 2*** (moderate individual risk, low community risk): a pathogen that can cause human or animal disease but is unlikely to be a serious hazard to laboratory workers, the community, livestock or the environment. Laboratory exposures may cause serious infection, but effective treatment and preventive measures are available and the risk of spread of infection is limited.
> 3. ***Risk Group 3*** (high individual risk, low community risk): a pathogen that usually causes serious human or animal disease but does not ordinarily spread from one infected individual to another. Effective treatment and preventive measures are available.
> 4. ***Risk Group 4*** (high individual and community risk): a pathogen that usually causes serious human or animal disease and that can be readily transmitted from one individual to another, directly or indirectly. Effective treatment and preventive measures are not usually available.
>
> *WHO (2004)*

TABLE 20.1
Selected Pathogens Classified by Risk Group

Risk Group	Pathogens
1	*Lactobacillus* sp., asporogenic *Bacillus subtilis*, *Saccharomyces cerevisiae* (yeast)
2	*Helicobacter pylori*, *Salmonella* spp., *Staphylococcus aureus*, *Streptococcus* spp., *Vibrio cholera*, *Candida albicans*, *Leishmania* sp., *Plasmodium* sp., *Schistosoma* sp., *Toxoplasma* sp., herpes virus, rubella virus, Dengue virus, Epstein–Barr virus, Influenza viruses (types A, B, and C), papilloma viruses
3	*Bacillus anthracis*, *Brucella* spp., *Mycobacterium tuberculosis*, *Yersinia pestis* (black plague), yellow fever virus, human immunodeficiency virus (HIV), Middle East respiratory syndrome coronavirus (MERS-CoV), SARS-associated coronavirus (SARS-CoV and SARS-CoV-2[a])
4	Ebola virus, Marburg virus, Lassa virus

[a] SARS-CoV-2 risk group classification according to NIH (2020), Government of Canada (2020) and European Union (2020).

Examples of pathogens classified in each risk group can be found in Table 20.1 (NIH, 2019).

Importantly, risk groups often relate, but do not always correlate with BSLs. To determine the required BSL for a certain pathogen, a risk assessment study must be carried out.

20.4 BIOSAFETY LEVELS

To ensure safe and effective laboratorial activities, BSLs are classified into four categories: BSL-1, BSL-2, BSL-3 and BSL-4. These levels comprise containment precautions, laboratory practices and protection equipment categorized in an increasing order of complexity. BSLs apply to laboratories developing clinical analyses, research, production, diagnosis or teaching activities utilizing biological agents.

A brief summary of each BSL is presented below (CDC, 2009; WHO, 2004):

- *BSL-1*: As the lowest level, it is recommended for work involving low-risk biological agents that are nonpathogenic to healthy adults and present a minimal hazard to laboratory workers and to the environment. Procedures can be executed on an open bench with basic personal protective equipment (PPE), such as gloves, eye protection and a laboratory coat.
- *BSL-2*: This level allows work with biological agents that pose moderate risk and can cause mild disease. Appropriate PPE is required, including face shield as needed. Procedures involving potentially hazardous splashes or aerosols must be carried out in a biological safety cabinet (BSC).
- *BSL-3*: For work with biological agents that are either native or exotic and pose a significant risk, such as lethal diseases that are transmitted by the inhalation route. BSL-3 facilities require controlled access and directional airflow, and personnel must wear special PPE. All hazardous work must be performed within a BSC.
- *BSL-4*: As the highest level, it is recommended for work with highly dangerous biological agents transmitted by the inhalation route that cause lethal diseases for which there are no available vaccines or treatments. All work involving the biological agent must be performed within an appropriate BSC. Personnel must be highly trained and wear proper PPE. BLS-4 facilities can be designed as a cabinet laboratory or a suit laboratory. The latter requires the use of a protective suit that has air-supplied and is under a constant positive pressure.

The cost of construction and maintenance of containment laboratories vary according to the BSL. Table 20.2 presents the cost of construction (US$) per m^2 and cost of maintenance per m^2 per year for the different BSLs.

Costs in Table 20.2 include the HVAC (heating, ventilation and air-conditioning) system, specific autoclaves and cabinets required for the different BSLs, and special air supply systems for BSL-4 laboratories for the complete isolation of the operators. Costs do not include the construction or maintenance of adjacent areas. Costs in Table 20.2 may vary according to the origin and quality of the equipment to be installed, such as autoclaves, BSCs, HVAC system, water supply system, effluent treatment system, power supply and electric systems, safety systems, decontamination systems and incinerators.

20.4.1 Biosafety Level 1 (BSL-1)

BSL-1 represents the most basic biocontainment level. It requires the application of standard microbial practices with no specific recommendations for protection equipment as the work can be performed on an open bench. It can be used for work involving biological agents classified in Risk Group 1 that are well characterized and are not associated with disease in healthy adults or animals. Professionals working in BSL-1 facilities should follow the minimum requirements described below and have a proper biosafety training. Special containment equipment or devices, such as BSC and autoclaves, although welcome, are not mandatory.

20.4.1.1 Laboratory Practices for BSL-1

Laboratory access should be controlled following the institutional guidelines. When infectious agents are present, a

TABLE 20.2
Construction (US$/m^2) and Maintenance (US$/m^2/year) Costs for Containment Laboratories of Different Biosafety Levels

	Construction Costs (US$/m^2)	Maintenance Costs (US$/m^2/year)
BSL-1	3,000.00	800.00
BSL-2	7,000.00	1,100.00
BSL-3	12,000.00	1,900.00
BSL-4 (cabinet laboratory)	25,000.00	3,000.00
BSL-4 (suit laboratory)	40,000.00	7,000.00

biohazard symbol should be displayed at the entrance to the laboratory. It may also include the laboratory supervisor's name and contact information.

Handwashing is required after manipulating biological agents and before leaving the laboratory.

Pipetting should be carried out using appropriate devices. Mouth pipetting is always prohibited.

It is not allowed to store food inside the laboratory, except when these are research objects. Likewise, eating, drinking, smoking, applying cosmetics and handling contact lenses are not allowed in the laboratory.

Reagents and materials must be stored in designated locations within the laboratory. Benchtop and work surfaces must be decontaminated after work and in case of a spill or contamination with biological agents or potentially infectious biological material.

Sharp materials and needles must be handled carefully. Needles should never be bent, sheared, recapped, removed or manipulated by hand before being disposed. There are special puncture-resistant containers for the disposal of sharps and needles. These containers must be correctly identified and located inside the laboratory area. Likewise, broken glassware should be disposed in specific containers. The removal of broken glassware should never be performed by hand, but using forceps, a broom and a dustpan and brush instead.

Biological materials should be properly decontaminated before discharged or disposed of, according to the institutional hazardous waste management policies.

Operation manuals, contingency plans and emergency plans must be established and shared with the laboratory personnel. The laboratory supervisor must ensure that laboratory users understand these guidelines. A first-aid kit should be available in case of a potential accident.

Additionally, a surveillance programme as well as proper medical evaluation and treatment should be provided as needed.

A routine of arthropod and rodent control should be implemented and maintained to ensure laboratory areas remain pest-free.

20.4.1.2 Safety Equipment for BSL-1

PPE should be worn as needed. These pieces of equipment include a laboratory coat or apron, gloves, eye protection and facemasks. Wearing laboratory clothing is prohibited outside the laboratory area. Open-toed footwear is also prohibited inside the laboratory.

Safety glasses should be worn when performing experiments that can generate hazardous aerosols. They are also required for personnel who wear contact lenses in the laboratory.

Although not mandatory, the use of BSCs and autoclaves is desirable.

The laboratory must have an emergency eye wash station. Emergency showers should be located either inside the laboratory or at an easily accessible location.

20.4.1.3 BSL-1 Laboratory Facility

The laboratory should have doors for access control and to separate it from the rest of the building. It is recommended that doors have vision panels and are, preferably, self-closing. Safety systems should consider emergency showers, eyewash stations, electrical emergencies and fire.

The water supply system should provide sufficient water for laboratory activities and a reservoir for firefighting operations. Drinking water and water for laboratory systems must be separate.

Laboratory areas should have ample spaces to allow laboratory work, cleaning and maintenance. It should also be easy to clean and decontaminate. Carpets, rugs and curtains are not appropriate. A sink for washing hands must be available.

The use of sun shades, glazing or blinds is recommended to control harsh sunlight in the laboratory. Windows that can be opened should be protected with insect screens.

Laboratory benches and furniture should be designed to support estimated loads. Benchtops should be impermeable to liquids and resistant to heat, organic solvents and other chemicals. Chairs and seats used in the laboratory should be covered with impermeable material that allows for easy cleaning and decontamination.

The laboratory should contain a space for storing laboratory clothing and other PPE. Personal items should be stored outside the laboratory.

Designated spaces should be provided in the laboratory for storing frequently used supplies and substances. It is recommended that areas for the long-term and/or high-volume storage be provided outside of the laboratory, but in an easily accessible location (Figure 20.2).

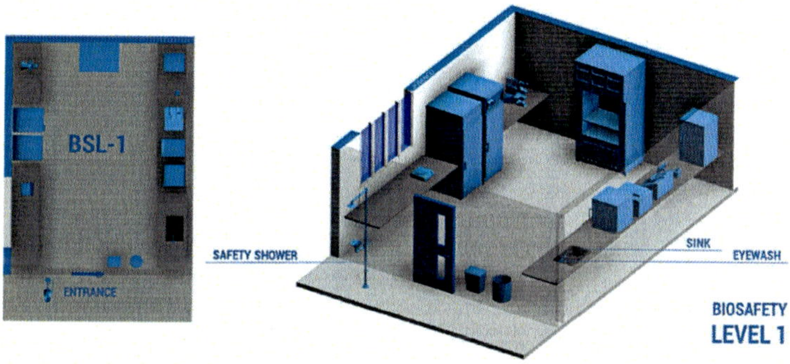

FIGURE 20.2 Layout of a Biosafety Level 1 (BSL-1) laboratory. (Graphics kindly provided by Júlio Franco, Eng., M.Sc.)

20.4.2 Biosafety Level 2 (BSL-2)

This level is required for work with biological agents that pose a moderate risk to laboratory workers and to the environment, often classified in Risk Group 2. Personnel should be properly trained to work with contained biological agents and supervised by an experienced professional with expertise in handling potentially pathogenic biological materials. All activities that may generate hazardous aerosols or spills must be performed in a BSC.

20.4.2.1 Laboratory Practices for BSL-2

Standard practices described for BSL-1 apply to BSL-2. Additional practices are described below.

Displaying the international biohazard sign is mandatory at the entrance of laboratories where there is work involving microorganisms classified in Risk Group 2 or higher. Along with this sign, it is recommended to identify the microorganisms present in the laboratory, and to include required PPE and recommended immunizations, the laboratory supervisor's name and contact information.

Laboratory doors should remain closed while work is being carried out. They must be locked outside of working hours.

Personnel should be provided with medical evaluations and proper immunization according to the pathogenic agents being used in the laboratory. An additional measure includes storing serum samples from personnel.

Susceptible individuals, such as those who are immunocompromised and immunosuppressed, should never perform potentially hazardous laboratory activities.

The development and implementation of a laboratory biosafety manual is recommended. It should mainly refer to the pathogenic agents frequently used in the laboratory and must be made available to all personnel.

The laboratory supervisor is responsible to ensure that laboratory workers fully understand laboratory practices before working with pathological biological agents. Annual trainings and refresher trainings in laboratory biosafety are recommended.

Laboratory equipment should be regularly decontaminated. Decontamination should also be performed after spills or contamination with potentially pathogenic agents and materials. Accidents that can lead to exposure to biological agents should be immediately addressed following the laboratory biosafety manual guidelines and the laboratory supervisor informed.

All activities that can generate hazardous aerosols or spills must be carried out in a BSC or appropriate containment device.

20.4.2.2 Safety Equipment for BSL-2

Personnel are required to wear an appropriate PPE, as described for BSL-1.

Disposable gloves should never be washed or reused. They must not be worn outside the laboratory.

Laboratory clothing (e.g. laboratory coats, smocks, gowns) should be removed before leaving the work area. These should be deposited in designated containers for decontamination prior to reuse or disposal.

The use of a BSC (Class I or II) in addition to an appropriate PPE is required when manipulating pathogenic biological agents in procedures that can create aerosols, splashes or spills. Examples of such procedures include pipetting, centrifugation, agitation, sonication, opening vials that contain the agent, intranasal inoculation in animals and collecting infected tissue from animals or eggs.

An autoclave must be available in an easily accessible location to allow for decontamination of materials and waste prior to use or disposal.

20.4.2.3 BSL-2 Laboratory Facility

BSL-2 laboratory facility should meet the criteria for BSL-1 facilities plus the following items. When BSL-2 criteria are not compatible with the BSL-1 standards, the requirement for the highest level of containment will prevail.

The laboratory should be equipped with self-closing doors and a lock system, since access is restricted to properly trained professionals.

A sink for washing hands must be available. It may be operated manually, using feet or elbow, or automatically.

Since air fluctuations may interfere with the proper operations of a BSC, these should be installed away from doors, ventilation windows or busy locations.

Exhaust air from a Class II BSC can be recirculated back into the laboratory after passing through HEPA filters that must be tested and certified annually. Alternatively, BSCs can be connected to the central exhaust air system. Exhaust air should be discharged above laboratory facilities and neighbouring buildings, away from housing, air currents and neighbouring ventilation systems.

There must be an eyewash station in the laboratory in an easily accessible location.

When planning new facilities, it is best to consider a ventilation system that provides a single-direction air, with no recirculation to areas inside the building (Figure 20.3).

20.4.3 Biosafety Level 3 (BSL-3)

BSL-3 applies to laboratories where work involves biological agents that can cause potentially lethal diseases in humans or animals and are transmitted by inhalation. These agents are often classified in Risk Group 3.

In addition to the requirements for BSL-1 and BSL-2, a BSL-3 facility requires safety devices and a laboratory infrastructure that are more efficient in containing the pathological agents.

Personnel must be provided with specific training for the manipulation of infectious agents and biological materials and must be supervised by the responsible professional.

All work involving the manipulation of infectious agents must be developed in a BSC. BLS-3 laboratories should be registered with official health authorities.

20.4.3.1 Laboratory Practices for BSL-3

BSL-3 requires a rigorous application of microbiological practices and criteria established for BSL-2 plus the mandatory use of Class II or III BSCs. Additional practices are described below.

All professionals in the laboratory must be fully aware of potential hazards and risks present in the environment. Access should be rigorously controlled to allow only professionals that will perform or assist activities involving

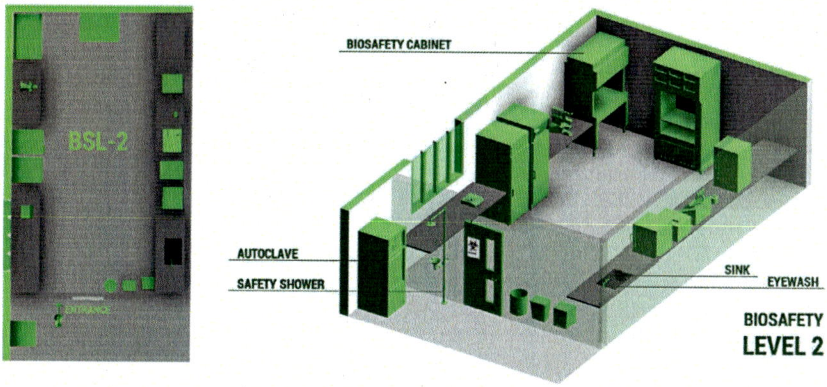

FIGURE 20.3 Layout of a Biosafety Level 2 (BSL-2) laboratory. (Graphics kindly provided by Júlio Franco, Eng., M.Sc.)

the manipulation of pathogenic agents. Depending on the risk posed by the biological agent, it is recommended that at least two professionals enter the facility. Susceptible individuals, such as immunocompromised and immunosuppressed, should never enter the facility.

Laboratory personnel should receive a proper biosafety training, be informed of the potential risks, required precautions to avoid or minimize exposure to the biological agents, and actions to take in case of exposure. Periodic refresher trainings should be provided, and specific training should be given in case of changes in regulations or guidelines.

All technical and administrative procedures must be described in a biosafety manual that should be available and accessible to all personnel.

Personnel should be provided with periodic medical evaluations and immunizations according to the pathogenic agents being manipulated, or present in the laboratory. Serum samples from all professionals, but especially from those who are directly exposed to the biological agents, must be collected and stored for future reference. Additional samples may be collected periodically depending on the specific guidelines.

All work involving the manipulation of infectious agents must be developed within a BSC inside a containment facility.

PPE is not allowed outside the laboratory. These must be decontaminated before reuse or disposal. Similarly, all materials used in the laboratory must be decontaminated before reuse and the waste must be sterilized before being disposed or removed from the laboratory.

HEPA filters and pre-filters from BSCs and ventilation systems must be securely stored in hermetically sealed containers after they are removed for decontamination by sterilization.

Incidents or accidents that may result in exposure to infectious agents or materials should be immediately reported to the laboratory supervisor or responsible professional. They should be rapidly addressed according to the laboratory guidelines, and professionals should be provided with medical examinations, surveillance and treatment. Written records should be kept regarding the incident and the measures that were taken.

The responsible professional must ensure that all BSL-3 instruments and installations are documented, operational parameters are verified and equipment is properly functioning before the laboratory procedures begin. The facility should be inspected annually and instruments, including safety equipment, must be periodically verified regarding their operation, calibration and efficiency according to the manufacturer's specifications or Good Laboratory Practices.

20.4.3.2 Safety Equipment for BSL-3

All procedures involving infectious materials, such as infectious cultures, clinical samples or environmental samples containing hazardous agents (often classified in Risk Group 3), must be performed in a Class II or III BSC. Additionally, the appropriate combination of PPE and physical containment devices must be established.

An autoclave should be available inside the BSL-3 containment laboratory.

Appropriate safety clothing is mandatory, as well as face protection, gloves and shoe covers. Professionals wearing contact lenses should wear additional eye protection. When work involves infected animals, face, eye and respiratory protection equipment pieces are required.

Gloves must be changed when compromised or when necessary. The use of two pairs of gloves is recommended to avoid accidental exposure to the infectious agents in case the gloves tear. When work with the infectious agents has been completed, gloves must be removed and disposed with other biohazardous waste. Disposable gloves should not be washed or reused.

Handwashing protocols must be rigorously followed after any procedure is completed in a BSL-3 facility.

20.4.3.3 BSL-3 Laboratory Facility

BSL-3 laboratory facilities should meet the criteria for BSL-2 facilities plus the following items. When BSL-3 criteria are not compatible with BSL-2 standards, the requirements for the highest level of containment will prevail.

Laboratory access must be restricted. Access should be through two interlocking and self-closing doors, ensuring that only one door is open at a time. Between the two doors, there must be an anteroom with differential pressure for changing laboratory clothing and other PPE.

A sink for washing hands should be provided next to the exit door. It should be automatically operated or have hands-free controls.

An autoclave should be available in the containment laboratory to allow for the sterilization of materials and decontamination of biohazardous waste.

An independent ventilation system is required to ensure a directional airflow into the laboratory area. It draws air from lower-risk to higher-risk areas within the facility. A safety system should monitor that a proper directional airflow is maintained.

Exhaust air must not recirculate into other areas of the building. Laboratory exhaust air (other than from BSCs) should pass through a HEPA filter prior to being recirculated into the laboratory, or being discharged above the laboratory facilities and neighbouring buildings, away from housing, air currents and other ventilation systems. HEPA filters should be installed in such a way that allows for their decontamination and verification.

Exhaust air from Class III BSCs should be directly connected to the exhaust system to avoid interference with the air balance of the cabinets or the building.

Instruments that may generate hazardous aerosols should be installed under fan hoods, or other equipment that exhausts air through the HEPA filter.

Safety showers, eyewashes and sinks with hands-free operation must be available in the laboratory and adjacent areas.

It is recommended that furniture is modular and adaptable to facilitate portability.

Floor surfaces should be sealed and impermeable. Ceilings and walls should be sealed and finished with smooth surfaces. Door and window frames should be easy to clean and decontaminate. The laboratory area should be able to be sealed for decontamination.

Water supplies should be protected to avoid backflow. Vacuum lines should be fitted with HEPA filters or chemical traps. Alternatively, portable vacuum pumps with no connection to the exterior and equipped with HEPA filters may be used.

A backup power system should support alarms and the emergency system as well as essential equipment, such as BSCs, autoclaves, freezers, refrigerators, incubators, and the ventilation and exhaust air systems (Figure 20.4).

20.4.4 Biosafety Level 4 (BSL-4)

This is the maximum containment level, recommended for work with hazardous and exotic agents that are highly transmissible, often by inhalation and for which there are no approved prophylactic and therapeutic options.

There are two primary containment systems for BLS-4 laboratories:

a. *Cabinet laboratories*: Laboratories where biological agents are manipulated in Class III BSCs.
b. *Suit laboratories*: Laboratories where biological agents are manipulated by professionals wearing air-supplied suits that are positively pressurized.

BSL-4 laboratories must be authorized by the competent national authorities and must be inspected by the health authorities.

20.4.4.1 Laboratory Practices for BSL-4

BSL-4 requires a rigorous application of the microbiological practices and criteria established for BSL-3, plus the mandatory use of a Class II B2 BSC combined with air-supplied protective suit (suit laboratory) or a Class III BSC (cabinet laboratory). Additional practices are described below.

Laboratory access should be restricted to professionals that are directly involved in laboratory activities. Access should be securely controlled and documented by a recording system or logbook where all information on date/time of people entering and leaving the facility is kept. Doors should remain locked.

Laboratory workers must be warned of the potential risks and be instructed on safety measures. A laboratory-specific biosafety manual should be available and easily accessible to all personnel.

The laboratory supervisor is responsible to ensure that personnel (a) are proficient in the practices required for work with biological agents in Risk Group 4; (b) receive proper training in practices and procedures that are specific to the BSL-4 laboratory; and (c) receive updated information and additional training in case of any changes in procedures or guidelines.

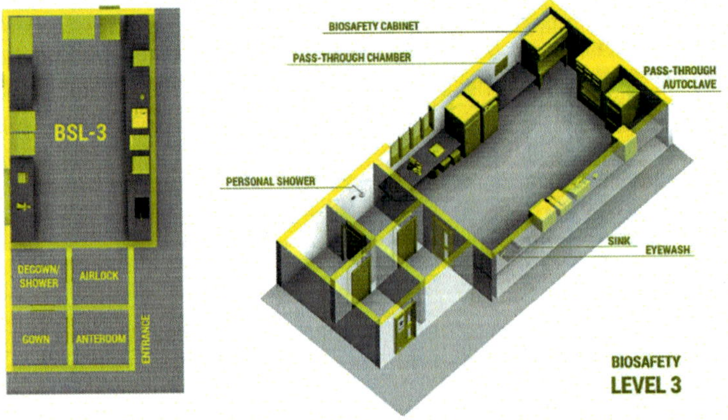

FIGURE 20.4 Layout of a Biosafety Level 3 (BSL-3) laboratory. (Graphics kindly provided by Júlio Franco, Eng., M.Sc.)

Personnel should be provided with periodic medical examinations and immunizations according to the pathogenic agents being manipulated or present in the laboratory. Serum samples from all professionals, but especially from those who are directly exposed to the biological agents, must be collected and stored for reference. Additional samples may be collected periodically according to laboratory-specific guidelines.

Written records should be kept on laboratory incidents and accidents. They should be evaluated and addressed according to the biosafety manual or specific guidelines. Considering the risk of potential contamination in case of an accident, there should be a designated area for the isolation and primary medical care of personnel.

On entering the laboratory, personnel must change clothes and shoes. Upon leaving, they must take a shower, except in case of an emergency. Used laboratory clothes should be decontaminated prior to disposal or laundering. On leaving a suit laboratory, professionals wearing a protective suite must undergo a chemical decontamination.

Materials that are not associated with the work of the laboratory should not be allowed inside the facility. Supplies and materials must be decontaminated before entering the BSL-4 laboratory. There should be an autoclave or fumigation chamber with secure, interlocked double doors to allow for material decontamination and transfer.

Materials and waste must be properly decontaminated or sterilized before removed from the BLS-4 facility. The exceptions are biological materials that must remain viable or intact.

In this case, the viable biological agent must be packed in a sturdy and sealed primary container, which should be placed in an additional sturdy and sealed secondary container. Package should be decontaminated by immersion in a disinfectant tank, passing through a fumigation chamber or other systems validated for this purpose. Biological materials that enter or leave the facility should be properly documented with all data required for identification and tracking.

20.4.4.2 Safety Equipment for BSL-4

In cabinet laboratories, activities involving the manipulation of hazardous agents are performed in Class III BSCs. In suit laboratories, activities involving the hazardous agents are performed in Class II B2 BSCs by professionals wearing a full-body proactive suit with a dedicated air supply. The protective suit is under a constant positive pressure and is equipped with a life support system protected by the HEPA filters. The life support system should include a breathing air compressor, safety alarms and emergency air tanks.

20.4.4.3 BSL-4 Laboratory Facility

BSL-4 laboratory facilities should meet the criteria for BSL-3 facilities plus the following items. When BSL-4 criteria are not compatible with BSL-3 standards, the requirements for the highest level of containment will prevail.

The BLS-4 laboratory should be located in a separate building, or in an area within the building that is completely isolated. It has specific construction features to prevent hazardous biological agents escaping into the environment. The laboratory may be designed for either of the two containment models (cabinet or suit laboratory), or for a combination of both. For the latter, the building should meet the requirements of both models.

The various rooms in the facility (dirty and clean changing areas, personal shower and chemical shower in the case of suit laboratories) should be connected in such a way as to ensure a sequential passage.

Access should be controlled by a rigorous safety system, such as iris or fingerprint scanners.

Entrance and exit of personal should be through a changing facility with differential pressure and equipped with automatic, interlocking double doors. Materials and samples should enter the facility through a pass-through system. Waste should only be removed from the facility after decontamination by autoclaving.

Doors should be able to be sealed for decontamination. Walls, floor and ceiling in the containment laboratory should be structured to form a sealed shell to allow for fumigation procedures.

Containment areas should have a controlled and dedicated air system providing a directional airflow from the lower- to higher-risk zones, ensuring a differential pressure is always maintained. Air system should be constantly monitored. A safety alarm system should indicate inconsistencies. Exhaust air should never recirculate to other areas of the building. Exhaust air should pass through two sequential HEPA filters prior to being discharged above the laboratory facility and neighbouring buildings, away from housing, air currents and other ventilation systems.

HEPA filters should be certified according to the manufacturer's instructions. HEPA filter casings should allow for in-place decontamination and replacement.

Gas lines should be equipped with filters or other validated systems to prevent the backflow. The vacuum system should be equipped with two sequential HEPA filters.

Liquid effluents, e.g. from sinks, autoclaves and drains, must be sterilized by heat treatment prior to discharge into the sewerage system.

The facility should have communication systems to allow for emergency communications between the laboratory and the outside, including support areas and the building's technical support.

There must be an emergency power source connected to the lighting circuits, alarms, entry/exit control, communication systems, BSCs, life support systems and other essential equipment. Emergency, life support and other essential systems must be periodically tested.

All BSL-4 laboratories must have pass-through autoclaves with interlocked double doors for the decontamination of materials and waste. If connected to the outside, the walls around the autoclave must be well sealed with proper materials.

In cabinet laboratories, double-door autoclaves must be directly connected to the Class III BSC for the decontamination of materials and waste. Exhaust air from a Class III BSC must pass through two sequential HEPA filters.

In suit laboratories, it is recommended that activities be concentrated on a single floor. Containment areas should be properly sized and arranged to allow for the transit of people wearing air-supplied protective suits, minimizing the risk of accidents.

Benchtops should have monolithic surfaces, with no grooves or gaps, and be fixed and sealed. They should be impermeable and resistant to heat, organic solvents and other chemicals used for decontamination.

Furniture should require a minimum maintenance and be resistant to gases, chemicals and moderate heat (Figure 20.5).

20.5 LABORATORY ACCIDENTS

Around 20% of laboratory accidents involving infectious material have a known cause. They are often associated with the manipulation of hazardous microorganism with sharps and needles. The remaining 80% have unknown causes, with clinical symptoms being presented days, weeks or even years (e.g. oncovirus) after manipulation of the biological agent. The disruption of protection barriers, due to improper PPE or issues in the laboratory containment system, can lead to exposure to aerosols containing pathological agents, causing imperceptible contaminations.

Major causes of laboratory accidents involving infectious material, from most to least frequent, are (Pike, 1979; Sewell, 2000):

a. Splashes, spills and sprays;
b. Manipulation of needles and syringes;
c. Accidents with broken glass or sharps;
d. Animal bites and scratches;
e. Mouth pipetting;
f. Accidents with centrifuge.

20.5.1 Historical Accidents

- *Marburg virus*: This virus was first reported in 1967, when 37 people fell ill in Marburg and Frankfurt in Germany and in Belgrade, Serbia. It was associated with infected *Cercopithecus aethiops* macaques imported from Uganda for use in the development of polio vaccines. It was the first filovirus to be discovered.
- *Lymphocytic choriomeningitis virus*: Researchers in London were infected after manipulating contaminated hamsters. Everyone who was in contact with the animals showed symptoms of the disease between 15 and 30 days after leaving the laboratory.
- *Oncovirus*: Two researchers from a renowned laboratory in Paris developed a deadly form of lymphoma after working with oncovirus.
- *Foot-and-mouth disease virus*: A world reference laboratory for foot-and-mouth disease vaccines accidentally released the virus due to a defective valve in their manufacturing plant. Animals that were kilometres away from the laboratory were found with symptoms of the disease. Genome sequencing analysis identified the strain and revealed the source of the virus (ANDERSON, 2008).

20.5.2 Accident Prevention

20.5.2.1 Proper Laboratory Supervision

Laboratory supervision is an administrative task that includes designing experiments and defining procedures that involve the manipulation of pathogenic agent, planning tasks with qualified and trained personnel, directing all laboratory activities in their technical and administrative aspects, delegating tasks accordingly, management of activities, and comparing plan vs. actual results.

With this, the probability of an accident occurring is minimized, since monitoring of activities is ongoing.

20.5.2.2 Training and Awareness

Proper training in the manipulation of pathogenic agents is required for all new researchers and employees. It is important to ensure awareness of the risks posed to the own researcher, their colleagues and the environment.

20.5.2.3 Continued Education

One of the biggest challenges in biosafety trainings consists of maintaining awareness on the inherent risks posed by the manipulation of pathogenic agents. Overconfidence and disregard for danger are often shown after long periods with no training sessions, leading to a higher chance of accidents

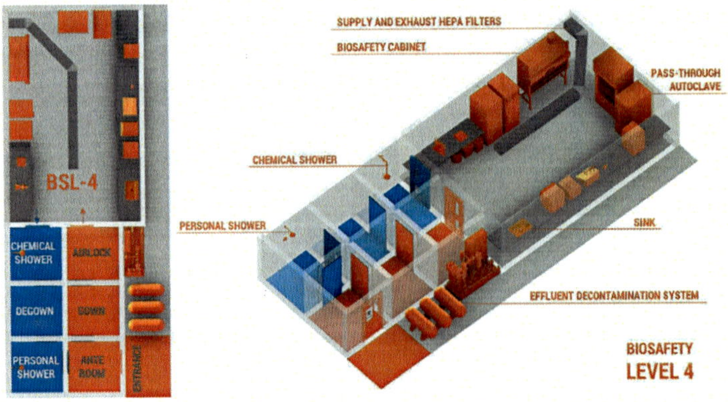

FIGURE 20.5 Layout of a Biosafety Level 4 (BSL-4) laboratory. (Graphics kindly provided by Júlio Franco, Eng., M.Sc.)

occurring. Therefore, it is important that people working in containment facilities receive periodic refresher trainings on the risks associated with the activities that they perform, and the microorganisms that they manipulate. The recommended frequency of refresher training sessions varies according to the BSL: at least annually for BLS-2, at least twice a year for BSL-3 and at least four times a year for BSL-4 facilities.

20.5.2.4 Proper Laboratory Techniques

The use of inappropriate techniques increases the risk of accidents when manipulating biological agents. Examples include the use of Bunsen burners inside BSCs (which change the air temperature and disrupt the airflow and efficiency of HEPA filters) and the use of autoclaves not validated for decontamination.

20.5.2.5 Proper Laboratory Equipment

Using proper equipment in the correct way reduces the chances of a laboratory accident from happening. The use of non-compatible adapters and accessories increases the chances of an accident.

20.5.2.6 Laboratory Organization

Laboratories should be audited with regard to their organization and safety record. The storage of pathogenic agents and chemicals must follow the pre-established guidelines and official regulations.

20.5.2.7 Vaccination

Laboratory personnel should be protected by proper vaccination where possible (e.g. diphtheria, tetanus, polio, hepatitis, influenza). In cases where animals are handled, personnel should be vaccinated against rabies.

20.6 ACTIVITY PLANNING

Every containment facility, for research or production of biotechnological products, must have a detailed activity and maintenance plan. Planning should be organized by specific programmes, such as the:

a. Engineering programme, which encompasses preventive actions for the facility supply systems. These include:
   1. The HVAC system, representing the air inlet, treatment, distribution and exhaustion systems;
   2. Hydraulic systems, encompassing water inlet, treatment and distribution, as well as the effluent treatment system;
   3. The steam generation system, compressed air and the supply of other gases.

b. Biological safety programme, which involves the acquisition, storage and distribution of biological agents, such as cells, biological fluids, tissues, organs or anatomical specimens (WHO, 2006).
   1. A biological product requires a detailed documentation regarding its acquisition. The donation, acquisition or purchase of any biological agents must follow ethical principles of use. In case of donation, the source, preparation and transport of the material should be documented according to official regulations.
   2. Documentation regarding the storage of a biological product should specify the responsibilities of those who have access to the material and how it may be removed from the facility or manipulated in the laboratory. The use of standard operating procedures helps minimize the risk of accidents.
   3. The distribution of biological products, from a biosafety standpoint, includes the proper identification and packaging of the material for transportation (WHO, 2019). It is important to document and track specific information about the biological agents, such as source, species, strain, size, volume, replications, and macroscopic and microscopic morphologies. The approval of immediate superiors should also be documented.

c. Animal safety programme, which considers various aspects related to the animal species, risk of zoonoses, genetically modified organisms (GMOs) and manipulation of chemical and biological agents.
   1. *Animal species*: Under stress, animals display different mechanisms of defence, such as biting (with or without the injection of venom), scratching and kicking. These defence mechanisms may be avoided by using proper containment methods and techniques specific for each species. Accidents are often due to the improper use of methods and lack of PPE.
   2. Zoonoses are infectious diseases that are transmitted from animals to humans. Noteworthy, rodents and lagomorphs (e.g. rabbits) can transmit more than 23 different diseases to humans, including some lethal infections (e.g. pasteurellosis, salmonellosis, and tularaemia).
   3. *Genetically modified organisms (GMO)*: Genetic manipulation of laboratory animals has become a common practice. Work includes gene deletions, modifications and the insertion of genes from a different species (transgenic manipulations). Various aspects and the consequences of such modifications if animals are released to the environment remain unknown, for example those related to the transfer of genes to the offspring or resistance/susceptibility to diseases. Therefore, containment systems must ensure an environmental protection.
   4. The manipulation of chemical agents associated with animals should be carefully performed to ensure a proper testing of the biotechnological products in toxicity, immunogenicity and potency assays.
   5. The manipulation of biological agents associated with animals represents an additional risk factor. The required BSL is determined by the characteristics of the biological agent. The animal should be kept under appropriate containment conditions.

TABLE 20.3
Selected Examples of Collective and Personal Protective Equipment

Collective protective equipment (CPE)	Autoclaves, BSCs, emergency shower, eyewashes, handwashing sink, chemical absorbents and spill kits, chemical showers
Personal protective equipment (PPE)	Laboratory coats, aprons, gowns, coveralls, gloves, safety glasses, goggles, facemasks, face shields, respirators, protective suits

d. Chemical safety programme, which considers the use of chemicals along with biological agents. The use of hazardous chemical substances increases the risks posed by biological agents. This is due to the fact that such chemicals can interfere with the body's primary defence by irritating mucous membranes and causing chemical burns to the skin, thus facilitating the penetration of infectious agents. If laboratory procedures involve the use of chemicals associated with biological agents, it is important to consider an increased risk for the operator, which can be mitigated by the use of proper safety equipment.

e. *Plant safety programme*: In addition to potential hazards to humans and animals, wild and genetically modified plants may affect environmental balance, leading to changes in ecosystems that may only be detected in over long periods of time.

20.7 SAFETY EQUIPMENT

See Table 20.3.

20.8 FINAL CONSIDERATIONS

Biosafety is an essential concept underlying all processes in the pharmaceutical and biotechnological industries. Aiming to minimize risks and avoid accidents, its ethical application ensures the well-being of humans, animals and the environment.

Specific guidelines and regulations apply to different BSLs. They describe the practices, equipment and installations required for work with hazardous biological agents of different risk groups.

To ensure the best application of biosafety measures, laboratory personnel must be properly trained, supervised and equipped to deal with the challenges present in containment facilities.

REFERENCES

Anderson, I. Foot and mouth disease 2007: A review and lessons learned. London, 2008. Available at: https://www.gov.uk/government/publications/foot-and-mouth-disease-2007-a-review-and-lessons-learned (Accessed: 10/11/2020).

CBD - Convention on Biological Diversity. Training manual on risk assessment of living modified organisms in the context of the Cartagena protocol on biosafety. Pyeongchang, 2014. Available at: https://bch.cbd.int/cpb_art15/training.shtml (Accessed: 10/11/2020).

CDC - Centers for Disease Control and Prevention. *Biosafety in Microbiological and Biomedical Laboratories (BMBL)*. 5th Edition. Atlanta, 2009. Available at: https://www.cdc.gov/labs/BMBL.html (Accessed: 10/11/2020).

Eddie, B; Meyer, K. F. Laboratory infections due to Brucella. *Journal of Infectious Diseases*, 1941, 68(1), 24–32. DOI: 10.1093/infdis/68.1.24.

European Union. Commission Directive (EU) 2020/739 of 3 June 2020 amending Annex III to Directive 2000/54/EC of the European Parliament and of the Council as regards the inclusion of SARS-CoV-2 in the list of biological agents known to infect humans and amending Commission Directive (EU) 2019/1833. *Official Journal of the European Union*, 2020, 175, 4.6.2020, 11–14. Available at: http://data.europa.eu/eli/dir/2020/739/oj (Accessed: 10/11/2020).

FAO - Food and Agriculture Organization of the United Nations. Biosafety resource book. Rome, 2011. Available at: http://www.fao.org/3/i1905e/i1905e00.htm (Accessed: 10/11/2020).

Government of Canada. ePATHogen - Risk group database. Available at https://health.canada.ca/en/epathogen (Accessed: 10/11/2020).

NIH - National Institutes of Health. NIH guidelines for research involving recombinant or synthetic nucleic acid molecules (NIH Guidelines). Bethesda, 2019. Available at: https://osp.od.nih.gov/wp-content/uploads/NIH_Guidelines.pdf (Accessed: 07/22/2020).

NIH - National Institutes of Health. Interim laboratory biosafety guidance for research with SARS-CoV-2 and IBC requirements under the NIH guidelines. Available at: https://osp.od.nih.gov/biotechnology/interim-lab-biosafety-guidance-for-research-with-sars-cov-2/ (Accessed: 07/22/2020).

Pike, R. M. Laboratory-associated infections: summary and analysis of 3921 cases. *Health Laboratory Science*, 1976, 13(2), 105–14.

Pike, R. M. Laboratory-associated infections: incidence, fatalities, causes, and prevention. *Annual Review of Microbiology*, 1979, 33(1), 41–66. DOI: 10.1146/annurev.mi.33.100179.000353.

Sewell, D. D. Laboratory-acquired infections. *Clinical Microbiology Newsletter*, 2000, 22(10), 73–77. DOI: 10.1016/S0196-4399(00)80009-4.

Sulkin, S. E.; Pike, R. M. Viral infections contracted in the laboratory. *New England Journal of Medicine*, 1949, 241(5), 205–213. DOI: 10.1056/NEJM194908042410505.

WHO - World Health Organization. *Laboratory Biosafety Manual*. 3rd Edition. Geneva, 2004. Available at: https://www.who.int/csr/resources/publications/biosafety/WHO_CDS_CSR_LYO_2004_11/en/ (Accessed: 10/11/2020).

WHO - World Health Organization. Biorisk management: Laboratory biosecurity guidance. Geneva, 2006. Available at: https://www.who.int/csr/resources/publications/biosafety/WHO_CDS_EPR_2006_6/en/ (Accessed: 10/11/2020).

WHO - World Health Organization. Guidance on regulations for the transport of infectious substances 2019–2020. Geneva, 2019. Available at: https://www.who.int/ihr/publications/WHO-WHE-CPI-2019.20/en/ (Accessed: 10/11/2020).

21 Pharmaceutical Quality System for Biotechnology Products

Laura de Oliveira Nascimento
University of Campinas

Francisco Vitor Santos da Silva and Marco Antonio Stephano
Universidade de São Paulo

CONTENTS

21.1 INTRODUCTION

Patients and caregivers assume that the consumption of medicines will result in the performance described by their manufacturers, maintaining efficacy and safety over their shelf lives. These assumptions imply that the production occurred in a manner that ensures quality – its lack can result in reduced pharmacological efficacy and physiological side effects. Quality does have a broader aspect since patients and society also seek ethical, environmental, social and marketing values. However, this chapter focuses on quality aspects that can be directly measured on process and product to assure efficacy, safety and stability of the product physicochemical profile.

The efficacy of biopharmaceuticals is not straightforward and requires a broader view. As an example, the vaccine against human papillomavirus (HPV) can protect women against cervical cancer, with two major brands of this vaccine on the market. However, one contains four serotypes and 98% efficiency, while the other has two serotypes and 89% efficiency (Pandhi and Sonthalia, 2011). But this quantitative difference might disappear depending on the vaccinated population because immunization efficiencies may change significantly in heterogeneous groups.

Safety relates to the occurrence and severity of side effects, which can come from the inherent properties of the active ingredient, inadequate prescription/use or product quality deviations. For example, some immunomodulatory biologics do increase the infection susceptibility of patients; infected patients undergoing these treatments, together or not with immunosuppressive pathologies, may die from infections that would not threaten life in normal conditions, such as histoplasmosis (Olson et al., 2011). This adverse effect was related to the intrinsic immunomodulatory properties of the studied drugs, which were worsened by their use without prescreening of patients for fungal infections. In this case, even a 'high-quality produced' biological medicine would provoke a deadly adverse event, considering that production process adjustments can only mitigate quality deviations of a predefined product.

Pharmaceutical specificities generate a demanding regulatory framework for pharmaceuticals, which results in a significant fraction of the production costs in the biopharmaceutical industry. The necessity for stringent regulations is backed up by historical drawbacks of insufficient pharmaceutical industry oversight. Therefore, current rules permeate each step of the product lifecycle, including drug and process development, market approval, manufacturing, end product testing and follow-up. These steps will be further discussed in this chapter.

Another aspect comes from the globalized market, which leads to a common view of pharmaceutical quality and harmonized rules to obtain it. The most prominent initiative to construct a global regulatory view pertains to the International Council for Harmonization of Technical Requirements for Pharmaceuticals for Human Use (ICH), founded by regulatory agencies and industry associations from Europe, USA and Japan. Currently, several other countries are members or observers of ICH, with the pharmaceutical world progressively converging to its view. Guidelines and articles published by this foundation regulate or support decisions for a variety of institutions, including the United States Pharmacopeia (USP) and the Pharmaceutical Inspection Co-operation Scheme (PIC/S). Both are complimentary in quality terms: USP focuses on drug and dosage form requirements, describing the analytical test parameters and targets to be pursued, whereas PIC/S focuses on harmonizing Good Manufacturing Practices (GMP) and inspection procedures worldwide. PIC/S has already 52 participating authorities, especially in Europe and North America, with a great potential for expansion. Besides enhancing global quality standards, harmonizing regulatory inspections decreases duplicate inspections from regulatory agency members, with the resulting decrease in regulatory costs and speeding up of registration processes (PIC/s, 2020). The concepts discussed here will be related to the ICH and PIC/S views.

21.2 PHARMACEUTICAL QUALITY SYSTEM (PQS)

Considering the regulatory framework, quality assurance of medicines requires a robust pharmaceutical quality system (PQS), which, as stated by the guideline ICH Q10: assures that the desired product quality is routinely met, suitable process performance is achieved, the set of controls are appropriate, improvement opportunities are identified and evaluated, and the body of knowledge is continually expanded. PQS covers the entire lifecycle of the product, from pharmaceutical development to product discontinuation, but needs to be combined with regional GMP (ICH, 2008).

Firstly, the PQS of a given company must be based on a manual describing the quality policy, planning, communication and review, together with management responsibility assignments for each activity. Quality, as defined by ICH Q7, is a responsibility of everyone involved in manufacturing; this view can be broadened here, considering that product quality is every personnel matter. However, quality assurance and quality control should form units or a combined unit independent of production, with its unit structure defined in the manual.

The quality manual must at least contain the following elements (ICH, 2008):

a. A monitoring system for process performance and product quality, which aims to maintain a state of control.
b. A Corrective Action/Preventative Action (CAPA) system for process and product continuous improvement and understanding.
c. A change management system to analyse and implement changes in a rational, timely and effective manner.
d. A management review plan of the quality system.

The monitoring system (a) conception requires other ICH guidelines to define and maintain the 'state of control'. Control ranges and targets come from the product development (Figure 21.1), followed by analytical (ICH Q2) and process validation inputs (ICH Q11) (ICH, 2012). Design concepts of manufacturing workflows and respective critical parameters will be detailed in Section 21.3. The system may also include regular internal audits (ICH Q7) to ensure GMP practices, periodical quality reviews and regular personnel training.

The CAPA system (b) describes the actions upon detected deviations in routine process/product monitoring, audits (non-conformances), annual product reviews, complaints, etc. The corrective actions prevent recurrence, and the preventive ones are taken to prevent occurrence (ICH Q10). These actions should be proportional to the deviation risk for patient safety, defined in root cause and risk analysis (ICH, 2005). Risk management became essential with increased process complexity – it puts 'first things first' – otherwise, the parameters for a strict control and action would scale up and reach incompatible levels of financial and personnel efforts. This approach will permeate all elements of the PQS to facilitate a scientific assessment of the pharmaceutical quality.

The change management system (c) evaluates the proposed changes, such as new excipient suppliers, in relation to current process and product understanding. As for CAPA, the level of effort and formality should be proportional to the level of risk. When a change is evaluated, approved and implemented, its impact should be monitored to assure the state of control.

Last, the management review of the quality system (d) ensures its continuing effectiveness in maintaining product quality and process performance over the product lifecycle. The evaluations require experienced personnel with leadership roles, possibly organized in quality councils and company complexity considerations.

Considering that this global unified view is driven by the Quality-by-Design approach, a deeper understanding of this matter becomes essential to gain process and product understanding.

21.3 QUALITY-BY-DESIGN IN BIOTECHNOLOGY PRODUCT AND PROCESS DEVELOPMENT

21.3.1 Historical Context

Pharmaceutical companies have brought most medicines to market using a traditional process development paradigm, or Quality-by-Testing (QbT) (Kenett and Kenett, 2008; Sommeregger et al., 2017). According to this paradigm, the manufacturer initially identifies a product and then designs

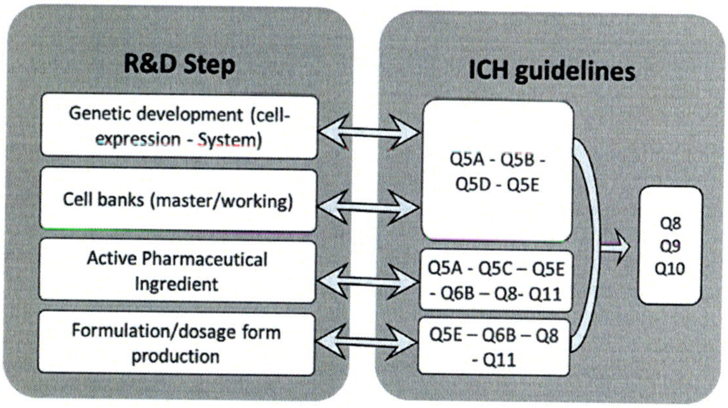

FIGURE 21.1 R&D steps of a biotechnology product and their connection to ICH quality guidelines (second row of white squares), to support the framework of a Pharmaceutical Quality System (ICH Q10). All development and process/product changes are encouraged to be driven by quality risk management (ICH, 2005). Q5 guidelines approach the quality of biotechnological: Q5A(R1) Viral Safety Evaluation, Q5B – Analysis of the Expression Construct in Cell Lines Q5C – Stability testing, Q5D – Derivation and Characterization of Cell Substrates, Q5E – Comparability of Biotechnological/Biological Products Subject to Changes in their Manufacturing Process. Q6B – Specifications: Test Procedure and Acceptance Criteria for Biotechnological/Biological. Q8(R2) – Pharmaceutical Development Q11 – Development and Manufacture of Drug Substances. (Chatterjee, B., April 16, 2018, Considerations for Biologic Drug Substance and Drug Product Testing, Outsourced Pharma..)

its production process within narrow limits, which (hopefully) guarantee consistency of the drug quality profile. Despite its historical success, this approach has limited the potential for improvement in several established production processes and has resulted in products that are vulnerable to small perturbations in process control parameters, which has, in some instances, led to recalls and drug shortages.

In the early 1990s, the FDA (the U.S. Federal Drug and Food Administration) altered its policy from regulating single products to seeing the whole pharmaceutical industry. This change included the creation of several new requirements for the industry, which resulted in a substantial increase in the Chemical Manufacturing Control (CMC) reviewing process time. Since CMC approval is required before the implementation of technological innovations in the production process, manufacturers have become cautious of investing in process development and the regulatory agencies have been constantly overwhelmed by the growing number of process change applications filed every year (Sangshetti et al., 2017). This shift has stifled technological development in manufacturing processes, although the products themselves have undergone a remarkable evolution in the past decades (Elliott et al., 2013).

To fulfil the demand for a more modern pharmaceutical development and oversight philosophy, the FDA started, in 2002, *The Pharmaceutical Current Good Manufacturing Practices (CGMPs) for the 21st Century: A Risk-Based Approach* initiative (FDA, 2004c). The purpose of this initiative was to simplify the reviewing process of manufacturing changes, to support the pharmaceutical industry to introduce innovation in the manufacturing processes, and to facilitate an alignment between innovations and the existing regulatory framework (FDA, 2004b). Amongst other proposals, this document suggested the adoption of Quality-by-Design (QbD) and Process Analytical Technology (PAT) in manufacturing processes (FDA, 2004c).

The QbD philosophy was initially popularized by J. M. Juran and, since the early 1990s, has been widely adopted by the automobile industry. QbD prioritizes a 'bottom-up' approach for product design, which sees the requirements of the customer as the cornerstone of process development. Juran defined quality as being both the presence and reliability of product attributes, which lead to customer's satisfaction. Juran believed that quality should be designed into the manufacturing process instead of being tested in the final products – hence, Quality-by-Design (Juran, 1992).

The EMA (the European Medicines Agency) and the PMDA (the Japanese Pharmaceutical and Medical Devices Agency) have worked together to align their respective regulatory frameworks with the FDA's 21st-century CGMP initiative. Guidelines Q8 through Q12 of ICH summarize these unified recommendations.

The first official recommendation to describe QbD-based pharmaceutical development was the ICH guideline Q8, published in 2004, revised in 2008 (Q8(R2)) and adopted in 2009, which contains descriptions and examples of the most relevant QbD concepts in drug formulation development (ICH, 2009). The next ICH guideline to be published was Q9 (ICH, 2005), which describes the necessary elements of Quality Risk Management plans. Afterwards, in June 2008, guideline Q10 for PQSs was adopted (ICH, 2008). In May 2011, ICH guideline Q11 achieved step 4 and thus becomes the official recommendation for the implementation of QbD in the development of active pharmacological substances, including biologicals (ICH, 2012). More recently, the ICH issued guideline Q12 (ICH, 2019). This guideline expands upon the necessary aspects of the Product Lifecycle Management Plan, initially put forward in guideline Q10.

In June 2018, the ICH listed the forthcoming guidelines (ICH, 2018). ICH guideline Q13 will describe the quality requirements for continuous pharmaceutical manufacturing, a greatly anticipated progress for the industry because of the potential to reduce equipment scales and idle times, as well as to accelerate processing rates and resulting increase in product shelf life (Benyahia, Lakerveld, and Barton, 2012; Mascia et al., 2013; Santos da Silva and Seidel-Morgenstern, 2016; Kim et al., 2017). Moreover, ICH guideline Q2 (Analytical Validation) will be revised and extended to become guideline Q2(R2)/Q14, to be better aligned with QbD principles, which should incorporate the elements of Analytical Quality-by-Design or AQbD (ICH, 2018).

The adoption of QbD by pharmaceutical companies seems to be gaining adherents, as exemplified by Januvia®, which became the first medicine to be approved within the FDA's new framework in 2006 (Woodley, 2018). Since then, a total of 143 medicines that utilized elements of QbD during development have received marketing authorization from the EMA (see Figure 21.2a) (EMA, 2020). From these, only 24 claimed a Design Space, i.e. the possibility of altering operating conditions and product specifications within certain limits without the prior need of approval from the regulatory agencies (see Figure 21.2b). The disparity between these two numbers can either indicate the difficulty in obtaining a Design Space approval or indicate the unwillingness of manufacturers to pursue it. The trend is, however, encouraging as the number of medicines approved within the new framework continues to grow (Figure 21.2).

The first case study of a QbD-based biopharmaceutical development focused on the manufacturing process of a-Mab, a humanized IgG1 antibody used in the treatment of indolent non-Hodgkin's lymphoma in adults. Drug development was undertaken as a joint initiative by some of the major companies in the field (Abbott, Amgen, Eli Lilly & Co., Genentech, GlaxoSmithKline, MedImmune and Pfizer), who collaborated with the regulatory agencies to develop a comprehensive case study for process development of similar products in future (CMC Biotech Working Group, 2009). This case study was published in 2009, and in 2013, the FDA approved the first biopharmaceutical to claim a Design Space within the QbD paradigm, Gazyvaro® (obinutuzumab, Roche AG, Switzerland) (Sommeregger et al., 2017; Committee for Medicinal Products for Human Use (CHMP), 2014). Currently, there are three more biopharmaceuticals approved by the EMA with a Design Space:

- Tecentriq® (atezolizumab, Roche AG, Switzerland) (Committee for Medicinal Products for Human Use (CHMP), 2017);

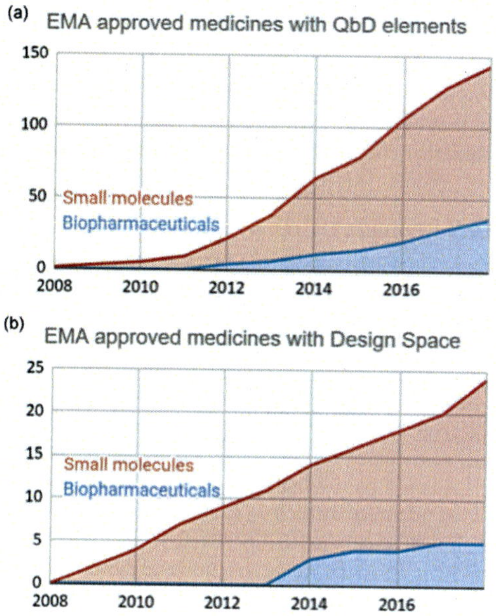

FIGURE 21.2 (a) The total number of medicines approved by the EMA where elements of QbD have been used during drug development organized by year. (b) The total number of medicines approved by the EMA where a Design Space, i.e. flexible operation, has been claimed. (Data sourced from EMA.)

- Imrestor® (pegbovigrastim for veterinary use, Eli Lilly & Co. Ltd., UK) (Committee for Medicinal Products for Veterinary Use, 2015);
- Gardasil 9® (HPV vaccine, Sanofi Pasteur, France) (Committee for Medicinal Products for Human Use (CHMP), 2015).

21.3.2 QbD-based Development

As shown in Figure 21.3a, in the traditional drug development paradigm (or Quality-by-Testing), the process development usually starts by defining the active compound, followed by iterations of redesign and adjustment of its manufacturing process until the final product meets all of the safety and efficacy specifications required by the medicines regulator (Figure 21.3a). On the other hand, as shown in Figure 21.3b, within the QbD paradigm, the starting point of the development process are the customers' needs, which form the prospective Quality Target Product Profile (or QTPP) (ICH, 2009). The QTPP defines a very specific application, as well as the desired safety and efficacy of a product, and not only the active substance (ICH, 2012). The QTPP should be translated into measurable surrogate parameters defined as critical quality attributes (CQA) that guide process development and risk management (ICH, 2009).

To start with the end in mind is the spirit of QbD-based product development. This goal-oriented philosophy of product development is not only in line with Juran's perspective, but also with the most successful strategies in a variety of fields, ranging from business management to military warfare. Therefore, simply taken at the face value, QbD already stands as a more reasonable approach for drug development, compared to its traditional QbT counterpart. Additionally, one would expect that a product rationally designed would better respond to market expectations and thus lead to a more robust business model (Elliott et al., 2013).

As will be described in more detail later, the manufacturer typically decides the QTPP based on experience with the product. This effort can, however, be significantly simplified when a rich literature describing the

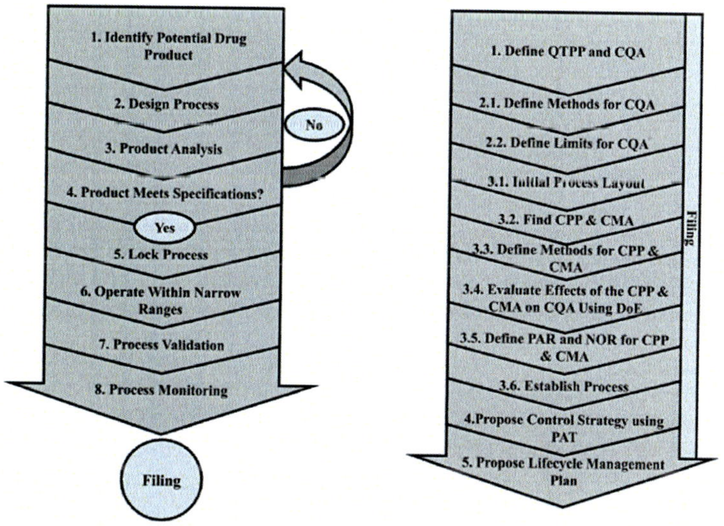

FIGURE 21.3 Comparison between the traditional process development paradigm (or Quality-by-Testing, QbT) (a) and Quality-by-Design (QbD) (b).

product's pharmacokinetics, pharmacodynamics, potential side effects, and main issues is available (ICH, 2009). Consequently, the application of QbD principles in the process development of biosimilar drugs and biobetters is very favourable (Kenett and Kenett, 2008; Vulto and Jaquez, 2017; Brumano et al., 2019; Benyahia et al., 2020a).

Within the traditional QbT paradigm, once the pharmaceutical product has successfully met specifications, the regulatory agency approves the product, and the manufacturer fixes the production process operating parameters to guarantee the product consistency. On the downside, this approach demotivates companies from investing in process improvement, since even the smallest changes in the process need to be approved by regulatory agencies, based on a thorough demonstration that the proposed changes will not affect the final product characteristics. In addition to this, each CMC supplement application costs over 250,000 USD (Elliott et al., 2013), which, according to the FDA (FDA, 2004c), has led manufacturers not to investigate the mechanisms correlating the process variables and the product quality profile. This has resulted in a poor understanding of the causes of product variability, which may have negative effects on population health. Consequently, within the new QbD paradigm, regulatory agencies have sought to facilitate the review process of manufacturing improvements on the condition that manufacturers can properly demonstrate a sufficient level of production process understanding that justifies the proposed changes. This should, in the long run, reduce regulatory compliance costs (Elliott et al., 2013).

Another significant difference between the QbT and QbD paradigms is that in QbD, data and process mathematical models gathered during the developmental and scale-up phases are an integral part of the filing process with the regulatory agencies (Benyahia et al., 2020b). Consequently, the regulatory agencies can afterwards use the well-validated process models (and their supporting data) to evaluate whether a degree of process understanding has been demonstrated by the manufacturer to justify the proposed alterations without the need of process revalidation. With this, the regulatory agencies expect to encourage companies to pursue a continuous improvement of manufacturing processes, motivated by the prospect of reducing production costs, waste and product variability more strongly (Elliott et al., 2013). Conversely, in the traditional QbT approach, only data concerning the final version of the manufacturing process is used for filing.

A major shortcoming of the traditional QbT process development approach is the absence of the requirement for a reliable risk management policy by which the manufacturer could proactively assess and mitigate the effects of process changes on the final product quality profile. With the addition of this recommendation, the regulatory agencies expect to reduce shortages caused by the loss of product batches, which can harm the public financially as well as healthwise. Besides the obvious economic advantage of losing less of the final product, the manufacturers would also profit from a swifter product release process. The long-term goal of the regulatory agencies is to achieve the real-time release of pharmaceutical products, i.e. the release of product batches based on the process data alone, instead of the traditional QbT approach of extensive final product testing before release (Sommeregger et al., 2017). To achieve this ambitious goal, regulatory authorities are recommending the development and implementation of widespread PAT to production lines (FDA, 2004b). PAT refers to advanced analytical tools that will allow on-, at- and in-line measurement of process variables, such as host cell gene expression levels, the concentration of secondary metabolites and other variables that could help predict the process outcomes and control product variability (FDA, 2004b).

21.3.2.1 Product Design Space

As shown in Figure 21.3b, the first step of process development within the QbD paradigm is the definition of the QTPP of the prospective drug product, which should consider its intended use, route of administration, dosage form/design/strength, container closure system, pharmacokinetic characteristics and drug product quality criteria (e.g. assay, stability, purity, sterility) (ICH, 2009). As previously mentioned, in the case of biosimilar and biobetter drugs, the QTPP is based on the literature data and prior knowledge about the reference medicine (Zalai, Dietzsch, and Herwig, 2013). As an example, Eli Lilly & Co. Ltd. (UK) defined the QTPP of Imrestor® (pegbovigrastim, a pegylated bovine granulocyte colony-stimulating factor) in terms of its stability, pharmacodynamics, pharmacokinetics, toxicology, immunogenicity, user (human) safety and target animal (bovine) tolerance (Committee for Medicinal Products for Veterinary Use, 2015). Each quality attribute must contain a target outcome and a justification for the chosen specification, generally linked with safety and pharmacological efficacy assurance.

After the QTPP is defined, the CQAs can be selected from a range of quantifiable characteristics of the final product, and typically include physical, chemical, biochemical, and microbiological properties (ICH, 2009). The selection of CQAs forms the prospective Analytical Target Profile of the drug, a concept that is expected to be officially incorporated into the coming ICH guideline Q2(R2)/Q14 (ICH, 2018).

The first step of the CQA selection process is the identification of potential connections between the product attributes and its quality profile using risk assessment tools like Ishikawa (fishbone) diagrams and statistical correlations (ICH, 2009). These factors are then classified using Risk, Ranking and Filtering (RRF) strategies as, for example, the failure mode and effects analysis (FMEA).

The A-Mab and Gazyvaro® (Roche AG, Switzerland) dossiers provide insightful examples of how to implement the risk assessment into the CQA selection process (Committee for Medicinal Products for Human Use (CHMP), 2014; CMC Biotech Working Group, 2009). In both instances, more than one parallel scoring system was used to define the CQAs:

1. *Prior knowledge*: The products were monoclonal antibodies, with a vast pre-existing literature in the field; therefore, product specifications of similar products already on the market were automatically considered CQAs.
2. *Criticality* = *Impact* × *Uncertainty*: The impact scores of the quality attributes were defined based

on their effect on four major aspects of the QTPPs: biological activity, pharmacokinetics, immunogenicity and safety. Uncertainty scores were defined based on the level of knowledge pertaining to the effect of varying each given attribute on the four QTPP aspects. The result of this system was the creation of four parallel CQA lists.
3. *Criticality = Severity × Likelihood*: The severity scores were defined much in the same way as the impact in the previous system. The likelihood, however, was defined as the probability of each given attribute to fall outside of specifications, which led to a different framework to evaluate criticality and, consequently, a different CQA list.

The attributes found to be critical in each one of these categories were grouped together to make the CQA assortment as comprehensive as possible. The authors considered this strategy to be sufficiently conservative to offset the novelty of the QbD approach (Committee for Medicinal Products for Human Use (CHMP) 2014; CMC Biotech Working Group, 2009). However, this method should be used with parsimony; otherwise, attribute criticality may become irrelevant due to the many competing quality attributes to track.

The CQAs determined for Gazyvaro® (Roche AG, Switzerland) based on this methodology were divided into seven categories (Committee for Medicinal Products for Human Use (CHMP), 2014):

- *Presence of product variants*: Size-related, charge-related, oxidation-related, glycosylation-related and structural variants.
- *Contamination with process-related impurities*: Host cell protein, host cell DNA and leached protein A.
- *Contamination with raw materials*: Cell culture and purification components (nutrients, trace elements, salts, buffers, etc.).
- *Contamination with leachables*: Filters, packaging materials, etc.
- *Contamination with adventitious agents*: Viruses, bacteria, endotoxins.
- *Drug substance and drug product composition and strength*: Protein content, osmolality, pH, appearance (colour, opalescence, clarity), L-histidine content, trehalose and poloxamer 188 content.
- *Drug product (formulation) characteristics*: Sub-visible and visible particles, extractable volume and sterility.

Using a similar strategy to the one employed by Roche AG (Switzerland), Eli Lilly & Co. Ltd. (UK) found the CQAs of Imrestor® (pegbovigrastim) were protein aggregation, oxidation, carbamylation, disulphide bond scrambling, acetyl-hydrazide formation, residual solvent contamination and bioactivity (Committee for Medicinal Products for Veterinary Use, 2015).

The frameworks listed above demonstrate that there is no single or correct way to rank quality attributes, which shows the importance of prior reflection on the best strategy for each scenario (Benyahia et al., 2020b).

It is also possible to refine the CQA selection process by intentionally varying product attributes and assessing their combined effects on safety and efficacy using experimental design techniques, for example Plackett–Burman and Taguchi. However, when designing such experiments, it is important to include wide ranges for each attribute, even if the normal operation of the manufacturing process is perfectly able to control these within narrow margins. Such caution is necessary to prevent neglecting quality attributes that may be very critical when out of specification (OOS), but not on normal operating scenarios. Ideally, the characterized region of the product Design Space should be vastly larger than its normal operating region in order to prevent process operation to wander dangerously close to uncharted territory (Elliott et al., 2013).

Within the QbD paradigm, it is also essential to develop a mechanistic understanding of these correlations in parallel to the statistical studies, to better ground the assignment of criticality to certain product attributes. Moreover, in the case of biosimilar products, it is important that the CQAs found match as closely as possible those of the originator/innovator drug (Vulto and Jaquez, 2017).

Following the definition of the CQAs, the analytical methodologies for their evaluation must be established and evaluated, as described in Section 21.5 (Kenett and Kenett, 2008). Then comes the definition of CQA acceptable ranges, which are set in a multidimensional space that defines the Product Design Space (ICH, 2009). This is opposite to the traditional QbT paradigm, which utilizes one-dimensional product specifications. As will be further explored in the next section, multidimensional spaces (both for the product and for the process specifications) result in a more flexible operation, since the regulatory agencies allow the movement within an approved Design Space without the need for CMC preapproval (ICH, 2009).

21.3.2.2 Process Design Space

One of the guiding principles of the QbD philosophy is that *the product is the process* (Vulto and Jaquez, 2017; Rekhi et al., 2015). Accordingly, after the establishment of QTPP and CQAs, the manufacturing process development may begin. A layout of the production process is then proposed, followed by the identification of the critical process parameters (CPP) and critical material attributes (CMA). CPP include the physicochemical variables of the unit operations, such as pH, temperature, metabolite concentrations, optical density, conductivity, etc., which can be kept within predefined ranges by automated in-process controls. CMA, on the other hand, range from solvents, starting materials, excipients, packaging materials, buffers, culture media, stock solutions to chromatographic resins and filter membranes, amongst many others (ICH, 2005).

As shown in Figure 21.3b, before criticality assessment may begin, the necessary analytical methods need to be established. These include sensors for the unit operations, physicochemical parameters, which belong to PAT (FDA, 2004b), as well as laboratory analytical routines to evaluate the raw material quality aspects. Afterwards, the criticality of process parameters and material attributes is assessed in the same way as that of the product's quality attributes;

i.e., it starts with RRF methods, followed by sensitivity analyses, and ends with the proposing mechanistic models that relate these to the CQAs.

ICH Guideline Q9 suggests some RRF methods for QbD process development, including fault tree analysis (FTA), FMEA, hazard analysis and critical control points (HACCP), hazard operability analysis (HAZOP) and other statistical tools to support them (ICH, 2005). Normally, a risk assessment should start with the unit operations having the greatest potential to negatively affect the QTPP, and then progress back towards the ones least likely to do so. This ranking of unit operations according to their criticality is itself an RRF but, in practice, it might be difficult to organize unit operations in this way. An easier and commonly employed strategy is to focus on the final operations of the downstream steps for greater scrutiny first, as these are the ones closest to the patient. Then, progress to the upstream steps, since small perturbations in these can be corrected by the later processing steps. Furthermore, it is wise to study and validate different alternatives for each unit operation, since once a process Design Space that spans different upstream and downstream options has been approved by the regulatory agency, it becomes possible to switch between these without the need for resubmission (Elliott et al., 2013).

After the initial risk assessment, the combined effects of altering process parameters and raw materials are evaluated using experimental design methods (Design of Experiments, or DoE) such as the ones cited in the previous section. This study has a combined purpose of assigning criticality to process parameters and generating initial mathematical models of the production process. The mathematical models obtained in the DoE methodology are usually empirical surface response functions and tend to be quite reliable at predicting process outcomes due to their statistical nature. Additionally, these surface response models allow the inference of CQA sensitivity to the parameter variation (i.e. sensitivity analysis) and hint at the possible underlying causal relations (Sommeregger et al., 2017).

Empirical models are, however, by their own nature hard to extrapolate to untested conditions as well as to different process scales. Given that ICH Guideline Q8(R2) incentivizes dimensionless process Design Space validation, it is advantageous for the manufacturer to develop more sophisticated mechanistic models that explain the connections between process parameters and CQAs.

Once in possession of an adequate mechanistic computational model of the manufacturing process, it is possible to feedback information to the risk analyses using uncertainty and sensitivity analyses *in silico*, both of which can be carried out using ensemble modelling by Monte Carlo simulations. An important tool for risk-based control (ICH, 2009), uncertainty analysis, is the estimation of the probability that CQAs will fall outside of the specified ranges and can serve as the basis for *in silico* estimations of the Process Capability Indexes (Cpk[b]) (Sangshetti et al., 2017).

The Monte Carlo method is named because of the exhaustive process of randomly assigning values to variables in a complex system and evaluating outcomes, which is analogous to what happens in casinos. Although randomly generated, the input variables follow the predefined probability distributions, resulting in the outcomes also following probability distributions. Since mathematical models of systems – in this case of a biopharmaceutical manufacturing process – may be comprised of complex nonlinear equations, often requiring numerical solutions, outcome probability distributions can become quite intractable to estimate analytically. Monte Carlo simulations offer, therefore, an attractive, albeit computationally demanding, alternative. Monte Carlo simulations result in an improved understanding of process variability when compared to the more basic alternatives of point estimation and what-if scenarios. They are, therefore, useful tools in the QbD process development paradigm.

Once all the sources of CQA variability originating in the process operation have been identified, their impact on the final product quality has been estimated and mechanistic/mathematical models explaining these causal relations have been validated, the next step is the definition of the process Design Space itself. The process Design Space incorporates the proven acceptable ranges (PAR) and normal operating ranges (NOR) of process operating parameters, as well as the CMAs, expressed as a multidimensional object that considers interactions among variables (ICH, 2012). This is in a stark contrast to the traditional one-dimensional way in which regulatory agencies have historically required the definition of acceptable ranges for process control variables and material attributes and, as demonstrated in Figure 21.4, may in due course result in a more flexible process operation.

Once the regulatory agency has approved the process Design Space, operation within its specifications is not considered a change and, therefore, does not require resubmission and approval (ICH, 2012). Furthermore, the process Design Space includes not only physicochemical process parameters but also the evolution profile of time-dependent variables in the case of complex operations such as fed-batch cultivation, chromatographic separation, lyophilization, precipitation, etc. Furthermore, the process Design Space may be defined individually for each unit operation or spanning the entire manufacturing process. When approving a Design Space for the entire production process, the regulatory agencies create more flexibility for post-approval changes that do not influence outputs (ICH, 2009).

As briefly mentioned before, the regulatory agencies also advise the process Design Space to be defined in terms of scale-independent variables (i.e. dimensionless), such as the aeration number, Reynolds number and power number. The advantage for the manufacturers, in this case, is that scale changes within the validated process model range can be performed without the need for post-approval resubmission (ICH, 2009; CMC Biotech Working Group, 2009).

21.3.2.3 Control Strategy

The next stage of the QbD process development cycle (Figure 21.3b) is proposing a risk-based control strategy for the production process (Lakerveld et al., 2013, 2015). This is in line with the QbD principles, in the sense that their overarching goal is to grant more autonomy to drug-producing companies on the condition that a deeper scientific understanding of the production process is demonstrated. Therefore, a risk-based control strategy must be based on risk knowledge and reliable process models (ICH, 2009, 2012).

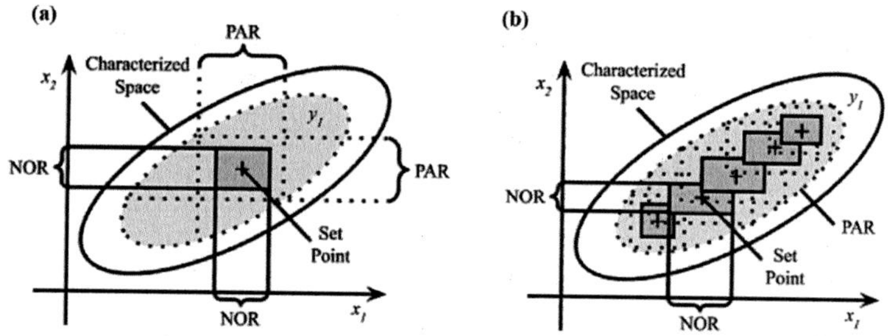

FIGURE 21.4 Comparison between one-dimensional process operating limits (a) and the multidimensional process Design Space recommended by QbD (b). This example considers, for clarity sake, only two process control variables (x_1 and x_2). However, the same conclusions could be drawn when a larger number is considered. As shown in (a), when only one-dimensional NOR and PAR are defined, the operating region is restricted to a rectangle centred on the setpoint, leaving out all the other potential setpoints shown in (b). Therefore, by defining the process Design Space as a multidimensional object, the implementation of QbD principles results in a more flexible process operation. (Rathore, A.S., Winkle, H., *Nature Biotechnology*, 27, 2009, https://doi.org/10.1038/nbt0109-26; Stockdale, G.W., Cheng, A., *Quality Technology & Quantitative Management*, 6(4), 391–408, 2009, https://doi.org/10.1080/16843703.2009.11673206.; ICH, 2012, Development and Manufacture of Drug Substances (Chemical Entities and Biotechnological/Biological Entities) Q11, in: *International Conference on Harmonization of Technical Requirements for Registration of Pharmaceuticals for Human Use*, https://database.ich.org/sites/default/files/Q11%20Guideline.pdf.)

An integral part of a risk-based control strategy is PAT, a concept that was put forward by the FDA (2004b) with the purpose of motivating the manufacturers to obtain a deeper understanding of process variables. PAT specifically refers to the collection of multivariate analytical methods, which allows the measurement and control of the often neglected variables. Since the regulatory agencies have emphasized the importance of method reliability, there is now an increasing interest in the development of sophisticated analytical methodologies such as HPLC (high-performance liquid chromatography), circular dichroism spectroscopy, mass spectrometry, capillary electrophoresis (CE), Raman spectroscopy and a myriad of hyphenated techniques (Vulto and Jaquez, 2017; Sangshetti et al., 2017). These techniques can generate a richer data to help to understand many of the bioprocess unit operations, especially in the case of cell culture, a process notoriously difficult to control (Vulto and Jaquez, 2017; Sommeregger et al., 2017).

The advent of PAT allows the measurement and control of product attributes (such as molecular weight, amino acid sequence, methionine oxidation, denaturation, aggregation etc.) and process-related variables (such as secondary metabolite concentration, host cell gene expression levels, plasmid concentration etc.), which may eventually result in improved quality assurance protocols (Vulto and Jaquez, 2017). An interesting prospect is the use of these data as a basis for more sophisticated computational models of the manufacturing process, which can include gene expression and metabolic networks to allow fine-tuning of cell culture parameters (Sommeregger et al., 2017).

A risk-based control strategy should include the control of the input material attributes (such as the raw materials cited in the previous section), product specifications and unit operations. This strategy should also include a monitoring programme aimed at the continuous validation and extension of the process models using PAT (ICH, 2009).

The long-term goal of the regulatory agencies is to move towards the real-time release, i.e. CQA assessment and control during processing (ICH, 2009). By shifting quality control upstream and supporting the development of feed-forward, i.e. proactive, process controls, the regulatory agencies seek to reduce post-processing waiting times and simplify the final product testing (Schlindwein and Gibson, 2018). The advantage for the manufacturer would be a reduction in recalls and batch losses, as well as an increase in the shelf life of the drug product. The regulatory agencies, therefore, expect to reduce drug short supplies and to provide 'fresher' biopharmaceuticals to the public.

21.3.2.4 Lifecycle Management Plan

The QbD philosophy entails the continuous improvement of both the drug substance and its manufacturing process based on the accumulation of knowledge pertaining to production process performance, effectiveness of the control strategy and post-approval clinical experience. According to ICH guidelines Q10, Q11, and Q12, there should be a prior strategy for how further knowledge can be gathered and applied for a continuous improvement (ICH, 2008, 2012, 2019). The product lifecycle phases listed in the ICH guideline Q10 include pharmaceutical development, technology transfer to other manufacturing sites or companies, commercial production, process improvement and discontinuation (ICH, 2008).

It is important at this point to draw a distinction between knowledge and data. While the continuous gathering of process and product data is fundamental for the expansion of knowledge, it is not on its own enough to guarantee the latter. Knowledge is actionable data, which paves the way for decision-making. Therefore, a knowledge management system should include not only the model for data acquisition but also the model for data analysis, storage and dissemination (ICH, 2008).

One of the many ways in which new process data can be incorporated into process knowledge is the continuous refinement and revalidation of its computational models, be it in the form of recalculating mean parameter values or reassessing statistical significance (Elliott et al., 2013), both of which can be achieved within a Bayesian statistical framework (Kenett and Kenett, 2008; Peterson, 2004). Additionally, larger data sets (either in the form of larger samples or as more diverse physicochemical data) can be used for proposing new mechanistic models for unit operations.

In the QbD framework, product specifications can be moved within the approved product Design Space without the need for preapproval by the regulatory agencies. Therefore, the way in which these changes may be implemented and documented should also be part of the Lifecycle Management Plan. To justify and, more importantly, motivate such changes, data concerning consumer response to the drug product should be gathered.

It is important to remember that in the context of the pharmaceutical industry, consumers are not only patients but also medical staff, healthcare providers, insurance companies as well as the regulatory agencies. Thus, post-approval decision-making might be driven by the market reaction from all these stakeholders (Schlindwein and Gibson, 2018). When it comes to patients, for instance, following the population-wide response to treatment, including adverse reactions as well as unanticipated interactions with other medications, may substantiate the need for altering the drug product specifications. Most often, however, the decisions regarding treatment course are taken not by the patients themselves, but by the medical staff responsible for their care. Therefore, it is also essential to keep an open communication channel with the medical community in order to collect information about their potential concerns regarding treatment risks, limitations and cost benefits, which might justify post-approval movement within the product Design Space. On a higher level, treatment coverage is predefined by insurance companies, which, depending on the country, might be public, private, or, as it is most often the case, a mixture of both. Treatment limitations such as cost-effectiveness and public perception might play a significant role in coverage selection, which, in turn, have major effects on drug marketability and may warrant some level of product specification change post-marketing authorization.

21.4 PRACTICAL NOTES ON PROCESS DEVELOPMENT

The first practical step of process design for a biological product is the choice of the expression system (Table 21.1 and Section 21.4.1). Once selected, laboratory production takes place, followed by scaling to pilot quantities. Since there are many ways to achieve a high-quality product, developing and justifying process steps may be more important than the actual choice. The first scaling goes up to enough production for animal safety studies, in general; before human trials, the final dosage form design and GMP-validated process should be accomplished. Successful clinical trials, together with safety data from animal studies and manufacturing documentation, constitute the backbone of register files. Due to differences in registering requirements among countries, the subject will not be directly explored here, except for related topics attesting quality, such as validation.

21.4.1 Expression Systems

Each expression system demands adaptations of process parameters, or even a change in sequence or type of operation units, directly impacting the process development. Bacteria, yeasts, fungi and mammalian cells can be used, but the questions that need to be answered are: 'Which is the best expression system?' 'Which one is economically viable?' and 'What is the production capacity of my expression system?'

21.4.1.1 Bacterial Systems

- *Advantages*: Well established, with several proteins already expressed in bacteria; grows fast at relatively low growth medium costs and high yield (g/L).
- *Disadvantages*: Unsuitable for post-translational modifications, possible protein loss under overproduction, inefficient secretion, possible phage contamination, generally require the presence of tags for simpler protein purification, often require

TABLE 21.1
Timeline for the Development of a Biological Pharmaceutical Product

Development phase	YEARS							
	01	02	03	04	05	06	07	08
Cell biology and clonage	■							
non-GMP process development	■	■						
Bank cell creation		■						
Pilot scale (patial GMP)		■	■					
Process development			■					
non-clinical studies (mainly toxicology)			■	■				
Formulation development		■	■					
Scale up (complete GMP with validation)			■	■	■	■		
Clinical Trial (Phase I, II and III)				■	■	■		
Preparation of registration report						■	■	
Regulatory office analyzes							■	■
Commercial release								▲

the use of antibiotics for the selection of bacteria carrying the cloned plasmid. Since tags are immunogenic and antibiotics cause resistant bacterial selection, both are not recommended for biopharmaceutical production.

21.4.1.2 Filamentous Fungi and Yeast Systems

- *Advantages*: Well-known genetic characteristics and mechanisms of expression, simple gene manipulation, good industrial application, better post-translational machinery than *E. coli*, no endotoxins and reduced risk of adventitious agents.
- *Disadvantages*: Production of endogenous proteases, glycosylation pattern is different from other eukaryotic cells, with a high concentration of mannose. Due to the presence of mannose receptors on the membrane of macrophages, this expression system is ideal for vaccines, but it is often not ideal for biopharmaceuticals because it increases immunogenicity.

21.4.1.3 Mammalian Cell Systems

- *Advantages*: Better post-translational glycosylation, extracellular expression of proteins, optimal for more complex structures.
- *Disadvantages*: Costly, difficult to handle in large volumes, easily contaminated and may contain viruses and prions.

21.4.2 Where Does Biopharmaceutical Scale-Up Start?

- Map the process or operation, including subsequent procedures and concurrent operations. Besides the importance of step sequence, it is common to ignore subsystems such as the production of steam, purified water, water for injectables, etc. Academic research generally focuses on upstream and downstream steps, undervaluing cleaning cycles and product packaging, which are important unit operations for the clinical study phase. Process design should follow ICH Q11 to adequately assess GMP from the beginning. Most concepts of this guideline were discussed in the QbD session (ICH, 2012).
- *Select each step of the process by unit operation*: Selecting the unit operation or each stage of the process and trying to reproduce laboratory results will lead to a reduction in time lost with production adjustments. The intention is for upstream and downstream unit operations to be replicated from the bench to the pilot scale, and from the pilot to the industrial scale. Some authors suggest that scale-up should follow a factor of 10: for example, if the laboratory scale studied a 10-L production, the pilot should have up to 100 L and the production scale up to 1,000 L. In practice, this is challenging because it depends on the unit operations, which are not always reproducible.
- *Conduct a risk assessment on the selected process or operation*: Losing or contaminating the product is critical in a biopharmaceutical production process, so a scale-up should map the risks, rank and change process for risk mitigation if needed. Noteworthy, a process Design Space, if defined in pilot or laboratory scales, could need adjustments when applied to the industrial scale. An important note for risk assessment is the complexity of assessing the risk uncertainty in FMEA or other tools for risk structure (e.g. FMEA items are ranked from 1 to 10 and multiplied between them, generating one risk value for each attribute). With that in mind, FDA stated that a 'simple' classification with low-medium-high scales would be enough to rank risks and support criticality decisions in defining attributes.
- *Collect and evaluate data from existing processes and operations against the Critical Process Parameters (CPPs)*: Analysing the data obtained during process scaling is not an easy task. Application of statistical methods is the credibility basis of the process. Thus, statistical analyses are the main tool for determining the CPPs, as detailed in Section 21.3.2.2. It is important to highlight the need for geometrical, thermal and mechanical similarities among equipment at different scales for the successful scale-up. Otherwise, critical parameters will not be scalable. Chemical and biochemical similarities are also desirable but will depend on how cells grow and produce at each scale.
- *Determine additional data requirements to better understand the process*: When we begin the laboratory study of a biopharmaceutical, the main requirements are productivity, biological activity and purity. However, when scaling up, we must think about the analytical methods of quality and process control, the safety and efficacy of the product, as well as all the quality parameters necessary in current national and international legislations.

21.5 GOOD MANUFACTURING PRACTICES (GMP)

The process development belongs to a greater frame that affects quality in manufacturing. Although each country may have a local GMP, the global market progressively changes its rules, resulting in the PIC/S initiative, for instance, as described before. The ICH Q7 covers the API manufacturing, including biotechnology-derived drugs, and will be used to guide the principles of this section. Logically, the specific requirements of GMP go through constant updates to reflect innovation and harmonization, but the sections (main areas of concern) remain practically unaltered. Since production includes much more than the unit operations, the guide starts with quality management, planning and review. In line with ICH Q10, the approach highlights the importance of attribute responsibilities for each activity, describes who can release/evaluate and act upon materials and the product being manufactured and plan the reviews, including the audits. Then, it specifies personnel requirements, such as proper training (regular, documented,

TABLE 21.2
Sections of ICH Q7 – Good Manufacturing Practices (GMP), Divided by Major Areas of Concern (ICH, 2000)

Data and Quality Monitoring	Operational API Obtention	Operational Packing/Distribution/Outsourcing
Quality management	Buildings and facilities	Storage and distribution
Personnel	Material management	Agents, brokers, traders, distributors, repackers and relabellers
Documentation and records	Process equipment	Contract manufacturers
Validation	Specific guidance for APIs manufactured by cell culture/fermentation	Packaging and identification labelling of APIs and intermediates
Change control	APIs for use in clinical trials	
Complaints and recalls	Rejection and re-use of materials	
Production and in-process controls		

adequate for the function) and personnel hygiene for operator and product safety. Buildings and facilities constitute another section focused on direction to prevent mix-ups or contamination. It includes utilities, water and infrastructure requirements. To better demonstrate the scope of GMP, Table 21.2 contains all the respective sections.

21.5.1 Validation – Production Process

One of the essential tools for quality assurance of manufacturing relates to consistency, named validation. ICH Q7 defines process validation as a 'documented evidence that the process, operated within established parameters, can perform effectively and reproducibly to produce an intermediate or API meeting its predetermined specifications and quality attributes'. This item will then follow this guideline, PIC/s and WHO recommendations (PIC/s, 2007; WHO, 2005).

Prior to validation, equipment and systems must be qualified, according to the following activities:

- *Design or Design Qualification (QD)*: Evidence (documents) that facilities, support systems, utilities, equipment and processes have been designed in accordance with GMP requirements.
- *Installation Qualification (IQ)*: Ensurance that facilities used in production processes (including equipment, infrastructure, measuring instruments, utilities and manufacturing areas) and/or computer systems are adequate for its end and correctly installed as described in their specifications.
- *Operational Qualification (QO)*: Ensurance that, under specified conditions, the system or subsystem operates as expected, in all operational ranges considered. All equipment used in the performance of the tests shall be identified and calibrated before use.
- *Performance Qualification (QP)*: Ensurance that the equipment or system performs in a consistent and reproducible manner, according to the defined parameters and specifications, in a certain period.

All the items above are achieved by a well and verifiable set of documented operations.

To conduct any process validation, one must manufacture at least three complete and consecutive cycles of production. These include, in addition to the process, an equipment clean-up validation for the next cycle. A periodic review of validated systems will check if systems are still operating as expected, or revalidation must be performed.

Prospective validation: The preferred type by the regulations, it is done during the product development stage, generally based on a risk analysis on the basis of experience to evaluate what can be critical.

Concurrent validation: Accepted when few batches have been produced, or a validated process has been modified, it happens in parallel to the routine production of commercial products.

Retrospective validation: Based on the analysis of past production experience, provided that the composition, equipment and workflows remain the same. It should include all batches produced in the respective period, regardless of failures.

Cleaning validation: Documented evidence demonstrating that cleaning procedures remove waste at predetermined levels of acceptance, considering factors such as batch size, dosage, toxicological data, solubility and area of contact of the equipment with the product.

Validation of computer systems: Assurance that a computerized system used to analyse, control or recording data is performed as expected, and that the processed data meets the predetermined specifications.

21.5.2 Process Validation in Bioreactors (Upstream)

The validation of an upstream process for the cultivation of bacteria, yeast, fungus or mammalian cells has practically the same basic variables: agitation, addition of gases, pH, temperature, cell growth and growth medium components (substrates, nutrients, etc.), which vary according to the cell type. Then, the outputs consist of cell growth rate, substrate consumption, product production and mass transfer, amongst others. The risk of microbial contamination must be mitigated in every step, such as working with closed systems and vessels or, when needed, working in a controlled environment with open vessels (e.g. biosafety cabinet).

Process validation in bioreactors comes after the setup of input parameters to obtain the desired outputs. Data analysis of cycles is based on statistical tools such as mean and standard deviation comparisons, together with control charts to evaluate the maintenance of values between upper and lower confidence intervals and specifications. It is very important to use at least two or three output parameters (e.g. cell growth rate, production rate and substrate consumption rate) to attest consistency.

All batches or process cycles should be monitored; these results should be within the confidence intervals obtained in the validation, and OOS happens when results are outside the confidence limits. All reports of noncompliance must be conclusive, including corrective maintenance and/or preventive maintenance of the equipment, as well as operational errors. After 1 year of production and analysis, all data are updated and their result should be compared with the previous year by Fisher's F test, Student's t-test or comparable statistical methods. The statistical choice will depend on the adopted criteria (variance, difference between medians, etc.).

21.5.3 Validation of Extraction and Purification Processes (Downstream)

Extraction and purification steps based on centrifugation, tangential filtration, chromatography and other methods follow the principles of process validation described Section 21.5.2. As expected, each operation unit will have its own set of parameters to be evaluated (centrifugation – time, force × *g*, sedimentation velocity, chromatography – retention time, pressure, flow, volume of application, volume obtained, number of theoretical plates, resolution and peak symmetry, tangential filtration – pressure differential, flow, filtration, holding capacity). However, extraction and purification methods based on solubility differences demand a greater effort. The solutions used for precipitation or phase separation also need to be prepared under a validated manner. For example, a saline solution must comply with predetermined values of pH, conductivity, dissolved solids, ion concentration (Na, Cl, K, PO_4, SO_4, NH4, CO_3), etc. The statistical methods are the same as those used in the previous section and should always be performed in three consecutive process cycles.

21.5.4 Validation – Viral Safety

The risk of viral contamination is a common feature of all biological products that production involves the use of animal or human material. Viral contamination of a biological nature may arise from materials, for example animal cell banks, human blood, human or animal tissues, or as adventitious agents introduced by the production process. For this chapter, as in ICH Q5A, the term 'virus' excludes nonconventional bovine spongiform encephalopathy and scrapie.

Human plasma and plasma products are used for the prophylactic and therapeutic treatment of human diseases. Prior to the adoption of viral screening and inactivation techniques, blood products often transmitted viruses to patients. The wake-up case consisted of hepatitis A virus (HAV) infection of haemophiliacs upon Factor VIII administration in Europe. The rise of human immunodeficiency virus (HIV) infections increased the concern. Although HIV rapidly inactivates outside the human body, there is a need to prove the efficiency of its removal during the manufacturing process. For viral safety, methods and tests are required to be of statistically proven efficiency and accuracy for detecting a certain amount of virus, i.e. establishing the limits of viral particle detection and quantification in the feasibility assays. Methods other than those provided by regulations must demonstrate reproducibility, repeatability and linearity (EMA, 1996).

Analogous to human-derived products, animal-derived products may carry diseases that are transmitted from animals to humans. To illustrate, in equine plasma, we can find the hendra virus or equine morbidity virus, equine adenovirus and the equine herpes virus.

Viral removal treatments may be physical, chemical or the combination of both. Regardless of the treatment, the process should decrease the concentration of viable infectious particles by an order of four logarithmic fold changes (WHO, 2004).

21.5.4.1 Physical Methods

The two main physical treatments for reducing viral particles in biopharmaceuticals or biological products are heat treatment and filtration. When it comes to thermal removal, pasteurization (60°C for 10h) of a 5% bovine albumin solution has proved to be inefficient at reducing or eliminating HAV (human hepatitis A virus) especially in the presence of high concentrations of sucrose (WHO, 2004). Treatment of the freeze-dried Factor VIII at 100°C for 30min showed to be efficient at removing HIV-1, VSV (vesicular stomatitis virus), BVDV (viral bovine diarrhoea virus), and HAV viruses. However, by this method, PSR (pseudorabies virus), SV40 (simian virus 40) and BPV (bovine papillomavirus) viruses maintained a residual infectivity. SV40 and BPV were only inactivated after a 2-h incubation at 100°C. An intermediate treatment at 80°C for 72h of the lyophilized Factor VII showed an efficiency in the reduction of cell infectivity when this factor had a moisture content greater than or equal to 0.8%, and little efficiency when the humidity was lower than 0.8%, with the same effect for porcine parvovirus virus (Roberts et al., 2007).

The use of nanofiltration was also able to remove viruses during the production process of a 7% human immunoglobulin solution. The method of tangential filtration with membranes of 75 and 35nm cut-offs and two steps was efficient for most of the viruses studied, among these: bovine diarrhoea virus, simian virus 40, feline calicivirus (CAL), murine encephalomyocarditis virus (CME) and HAV. However, it was not effective for the so-called small viruses, such as porcine virus and bovine parvovirus (Burnouf and Radosevich, 2003).

21.5.4.2 Chemical Methods

Several chemicals can be used to inactivate adventitious viruses from biological products, such as detergents (TNBP and Tween), sodium caprylate, caprylic acid and acidic solutions (pH 4). Treatment with detergent has several advantages, since the concentration used is relatively low and consequently not toxic to patients, but there have been cases of type I hypersensitivity to some products (Dichtelmüller, 2009).

A 0.3% concentration of TNBP with 1% Tween 80 was able to neutralize Sindbis virus in intravenous immunoglobulin and coagulation factor VIII preparations after 60 min of reaction. In turn, treatment with sodium caprylate was 100% efficient (reduction of viral infectivity of more than four logarithms in base 10) at 12 mM concentration for 1 h at 25°C for HIV-1, BVDV and PRV. The addition of 25 g/L. acid in an intravenous immunoglobulin solution was able to inactivate HIV-1, BVDV, PRS and Sindbis virus. The most interesting observation was that by gradually increasing the concentration of caprylic acid in the solution, the PRS virus became more resistant. At last, treating an intravenous immunoglobulin solution at pH 4.25 for 40 h was 100% effective at inactivating HIV-1, BVDV, PSR, SV40 and EMC viruses. Treatment for 16 h was ineffective against BVDV virus (Li, 2019; Burnouf et al., 2004; Caricati et al., 2013).

The validation process is obtained by deliberately adding (retro-contamination) a material with a known concentration of virus at various stages of production. Its removal or inactivation will be measured during the subsequent step or steps. This will identify production steps that are effective in reducing the level of virus and provide an estimate of the overall process capability to eliminate the viral infectivity. An effective method is one that can reduce four log units at base 10 of viral concentration, equivalent to 10,000 viral particles. According to the recommendations of experts, there is no need to validate each individual step of the manufacturing process for viral safety. During the process validation, only the viral inactivation or removal steps are studied.

Donor screening practices and rigorous testing of blood donations for virus markers have significantly improved the safety of plasmas and other blood derivatives. However, no matter how rigorous, the screening processes cannot exclude the possibility of a 'window period' of detection of infectious agents in plasma, causing the donations to contaminate a plasma pool to be used in the production of blood products. This is the reason why a wide range of highly effective methodologies have been incorporated into the manufacturing processes of plasma derivatives.

Even with these processes that have dramatically increased the safety of end products, it is necessary to deploy a validation of viral inactivation to ensure the safety of plasma derivatives. Although there is no known transmission of any infectious agent by means of antivenoms, theoretical concerns about infection transmitted by equidae and biangulados to humans have contributed to a further increase in the safety of products derived from these species. Natural reservoirs of infectious agents have recently been found in animals, and the inherent risk of disease is emerging. Examples of such infections from animals include HIV, Ebola virus, Hantaan, Lassa, Nipah and other paramyxoviruses, equine morbillivirus and probably acute coronavirus respiratory syndrome (Abbot and Cyranoski, 2003).

21.5.5 Validation – Analytical Methods

The validation of the analytical methodology constitutes an essential and initial activity of a well-structured quality assurance programme, being a critical factor in the validation of the production process. There is no process validation before the analytical method is validated. ICH dedicated one guideline solely to Analytical Validation (ICH Q2), highlighting its importance. Although an ongoing revision and new guideline on the subject will come soon, to align with QbD principles and more analytical techniques, the concept of validation parameters will not change drastically.

According to the USP, validation of analytical methods 'is the process by which it is established by laboratory studies that the performed characteristics of the method satisfy the requirements for the analytical applications practiced'.

21.5.5.1 Reference Standard

The recovery of drugs and their metabolites from biological samples (matrices) is evaluated by matrix spiking with reference standards at known concentrations, followed by their analytical determination. It is desirable that biological matrices be certified or at least traceable. The degree of purity of the reference standard may influence the quality of the result. Therefore, certified reference standards with a high degree of purity are required; if possible, the reference standard should be the same molecular form as the analyte. When this is not possible, a stable chemical form (base or free acid, salt or ester) of known purity may be used. Three types of reference standards can be used:

1. Certified reference standard.
2. Reference standards from trustable companies in the respective field.
3. Standards synthesized on demand for the client by an analytical laboratory or other institution.

The origin, batch number, expiration date, certificates of analysis and any other documents which are essential for ascertaining the purity, identity and origin of the standard should be available.

21.5.5.2 Method Development

The fundamental parameters for any validation of analytical methods are accuracy, linearity, precision, range, specificity, sensitivity, reproducibility and stability. The measurements of each analyte in the biological matrix must be validated. The stability of the analyte in fortified samples of the biological matrix must also be determined. A typical methodological validation design for bioanalytical methods includes the following determinations:

a. *Linearity*: It is the ability of an analytical methodology to demonstrate that the measurements obtained are directly proportional to the analyte concentration in the sample, within a specified range. Generally, this property is demonstrated by the analysis of at least five samples in different concentrations, followed by the linear regression of the data. Some assays, like immune response analysis, will need a mathematical transformation to follow a linear response.

b. *Specificity*: It is the ability of the analytical method to differentiate and quantify the analyte in the presence of other components of the sample. Some substances may interfere with the selectivity, such as some components of the biological matrix, degradation products, excipients, products used for adulteration and other medications used in animals.
c. *Accuracy*: The closeness of agreement between the value obtained by the measurement and the expected reference value.
d. *Precision*: The closeness of agreement between the values obtained by measurements of multiple sampling from the same starting sample.
e. *Range*: The working interval within the upper and lower quantification limits of an analytical method. In general, it comes from the linearity study and is linked to the application of the method.
f. *Limit of detection*: The smallest amount of analyte detected in a sample by the method realized under specified experimental conditions, but not necessarily quantified.
g. *Limit of quantification (determination)*: The smallest amount of analyte determined in a sample by the method realized under specified experimental conditions, with adequate accuracy and precision.
h. *Robustness*: The robustness of an analytical method determines its ability to withstand small and deliberate variations in analytical parameters. It indicates the confidence interval during normal use and might be assessed by experimental designs such as shown in Table 21.3.

$$\frac{(s+t+u+v)}{4}=\frac{(4A)}{4}=A\frac{(w+x+y+z)}{4}=\frac{4a}{4}=a$$

H_0 = no significant difference in *test t of Student* between A and a ($A = a$)

H_1 = significant difference in *test t of Student* between A and a ($A \neq a$)

i. *Stability of the analyte in spiked samples*: The stability of the drug in a biological sample (matrix) is linked to its storage conditions. These procedures should evaluate the steps such as sample collection and handling, short- and long-term storage, refreezing tests, stock solution stability and post-preparation stability.

If possible and when applicable, other tests can complement the assay validation, e.g. interference of foreign material in the sample to be analysed, type of processing that was sampled prior to the test and so on. The purpose of the test will define the parameters to be evaluated, as shown in Tables 21.4 and 21.5.

This information applies to:

a. Analytical techniques using gas chromatographic (GC) or HPLC methods.
b. Non-chromatographic methods provided they offer an acceptable selectivity (e.g. titration, UV-VIS spectrophotometry).
c. Immunological or microbiological tests provided the degree of variability usually associated with these techniques is observed.

21.5.6 Validation – Biological Assays

There are three broad categories of bioassays that are commonly used for biological products: binding assays, cell-based assays and animal assays.

Binding assays, e.g. immunoassays, involve the binding of two or more molecules. They are used to monitor a molecule during purification and equipment cleaning validation steps. Binding assays are generally not considered acceptable as potency assay due to the presence of a molecule is not necessarily an indication of biological activity.

Cell assays measure the product response in specific cells: agglutination, cell lysis, cell fusion or generation of a detectable specific chemical. These assays may be more variable than binding assays and should be performed with caution to ensure the consistent results. Cell-based assays are often used for potency assays. The results of binding assays vary typically between 5% and 20%. Variation in the results from cell and animal assays may range up to 50%.

Animal testing is more difficult and involves maintenance and handling of the animals. They are time-consuming, expensive, with highly variable results, and are usually only carried out for the release of the final product.

TABLE 21.3
Robustness Test Matrix Model Based on the Youden & Steiner Test

Variable	Analysis							
A, a	A	A	A	A	a	a	A	a
B, b	B	B	b	b	B	B	B	b
C, c	C	c	C	c	CC	c	C	cD
D, d	D	D	d	d	d	d	D	D
E, e	E	e	E	e	e	E	E	E
F, f	F	f	f	F	F	f	f	F
G, g	G	g	g	G	g	G	G	g
Results	s	t	u	v	w	x	y	z

TABLE 21.4
Types of Tests Classified Validation Parameters

Category	Purpose of Test
I	Quantitative tests for the determination of the active principle in pharmaceuticals or raw materials
II	Quantitative tests or limit tests for the determination of impurities and degradation products in pharmaceutical products and raw materials
III	Performance tests (e.g. dissolution, release of the asset)
IV	Identification tests

TABLE 21.5
Parameters Applicable to Validation According to the Analytical Category to be Used

Parameter	Category I	Category II Quantitative	Category II Limit Assay	Category III	Category IV
Specificity	Yes	Yes	Yes	[a]	Yes
Linearity	Yes	Yes	No	[a]	No
Interval	Yes	Yes	[a]	[a]	No
Precision repeatability	Yes	Yes	No	Yes	No
Intermediate	[b]	[b]	No	[b]	No
Detection limit	No	No	Yes	[a]	No
Limit of quantification	No	Yes	No	[a]	No
Accuracy	Yes	Yes	[a]	[a]	No
Robustness	Yes	Yes	Yes	No	No

[a] May be required, depending on the nature of the specific test.
[b] If there is evidence of reproducibility, no proof of accuracy is required.

The biological response of a suitable species to a biological product in test is usually compared with a reference product or to group controls that were not inoculated, so its activity can be measured by comparison. These tests are used for pyrogen tests, general safety tests and potency tests.

Complex assays may involve more than one test category, for example test for the inhibition of cytotoxicity of the diphtheria toxoid in Vero cell, having the following steps: animal immunization, diphtheria toxin production, Vero cell toxicity assay, animal assay to produce antibodies against diphtheria toxin, binding assay for the inhibition of serum cytotoxicity and toxin neutralization.

The most frequently used statistical test for the validation of biological animal tests is the homogeneity test: by the study of mean and variance, at least three consecutive analyses with three replicates, by two different operators. The results will be compared for each repetition (mean and standard deviation) and for the distribution of the three results (analysis of variance). This method does not apply to *in vitro* potency tests. The test is part of the robustness test and the interlaboratory test, once it is validated by the parameters described above.

21.5.7 Change Control

A change control must be made to any changes of equipment, system, process, or procedures because they may also change the outcomes that were already validated. The change control operation should be formalized, with control happen upon planning and a change proposal, including the rationale and expected impact on function, operation or performance of the process or system that will suffer the change, besides indirect impacts on other processes/procedures/utilities. The formalization format should be part of a Quality Assurance SOP or included in the Validation Master Plan. The proposal comes from the responsible for the process but must be approved by the change control team. If the process changes, new validations must be carried out; approved equipment/systems/tests/processes should be also formally reviewed and approved by means of a change control procedure.

21.6 ANALYTICAL METHODS FOR LABORATORY QUALITY CONTROL

The biotechnology industry can utilize a large array of methods to develop products and control their quality attributes. Specific methods and protocols are published in pharmacopeias, such as USP, including the regulatory limits and desired outcomes for the results. These compendiums relate to the current local legislation (USP – USA) but are also accepted by other countries. An exception is the International Pharmacopoeia, published by WHO and available to any WHO member to adopt it in its local legislation. Here, the focus is on the foundation for the most common methods, based on the Brazilian Pharmacopeia, but it is not exhaustive (Anvisa, 2019). Product-specific characteristics can be found in the pharmacopeias.

21.6.1 Physicochemical Tests

21.6.1.1 High-Performance Liquid Chromatography (HPLC)

Most of pharmaceutical analyses are based on the partition separation method, especially HPLC. This technique separates, detects and quantifies the active ingredient and related impurities or degradation products present on the sample. Molecule separation occurs between a flowing liquid mobile phase and an immiscible solid stationary phase, with separation occurring inside a pre-packed cylindrical column.

The analysis happens in a HPLC system: at first, individual solutions or premixed mobile phase and column are connected. The mobile phase is then pumped inside the chromatographic system to wash and equilibrate the column. When equilibration ends, the sample dissolved in the mobile phase or other solvent is injected into the chromatographic system with a syringe or automatic sampler of vials. Analysis starts at a predetermined flow rate and mobile-phase mixture programming (isocratic when fixed ratios of components, gradient mode for ratio changes over time). At the end of the column, a detector identifies and quantifies the eluted molecules, and an integrator or register converts the signals to peaks and valleys over a baseline (chromatogram) in printed or electronic forms.

Separation mechanisms vary according to the stationary phase and include partition, adsorption, ion exchange, molecule size and affinity properties. The affinity of the analyte for the stationary phase influences its retention time inside the column and can be modulated by the polarity of the mobile phase. A 'normal-phase' chromatography uses polar stationary and nonpolar mobile phases, whereas 'reverse-phase' process goes with nonpolar stationary and polar mobile phases.

Nonpolar stationary phases are mainly based on organic molecules linked to silica particles of 3–10 μm. The functional groups vary and include octyl, phenyl, octadecyl, cyanopropyl and polar nitrile groups. The amount of free silanol groups influences the separation and shape of peaks, so some columns are coated to reduce their effects. Most of these silica columns were designed to work in a 2–8 pH range. Columns made of porous graphite or polymeric materials, such as styrenodivinylbenzene, perform well over a broader pH range.

Besides the chosen separation mechanism, composition of the phases, particle diameter of the column, flow of mobile phase and analysis temperature also influence the separation outcomes. For example, smaller particles can provide faster and better transfer of substances between the liquid and solid phases.

As for detection, the most common for HPLC are spectrophotometric detectors, assembled as a flow cell that receives light (tungsten or mercury for visible, deuterium for UV) to be absorbed by chromophores of the moving molecules. The absorbance is recorded by the detection system, usually in a single wavelength. Diode array detectors (DADs) can detect multiple wavelengths in the same analysis and improve peak identity and a concomitant quantification of substances that absorb in different regions of the spectrum.

Alternatives to UV/Vis are refractive index detectors, fluorometric detectors (sample fluorophores or fluorescent derivatives by chemical reaction), potentiometric/voltammetric/electrochemical detectors (oxidable or reducible analytes, for highly selective and sensitive measurements), conductivity detectors and mass spectrometry detectors. Mass spectrometry detectors deliver high selectivity due to single molecule mass detection by simple quadrupole design (MS) or tandem (MS/MS). The most common sources of ionization are electrospray and chemical ionization.

21.6.1.2 Gel Electrophoresis

Polyacrylamide gel electrophoresis is another separation-based technique used mainly to evaluate the pool of proteins of a sample, which gives an idea of purification efficiency. Polyacrylamide gel sieves molecules through a three-dimensional porous network. The matrix results from bisacrylamide and polyacrylamide chains cross-linked in a chemical reaction catalysed by ammonium persulfate (PSA) and tetramethylethylenediamine (TEMED).

Charged particles of an electrolytic solution migrate through a porous gel matrix to the electrode of opposite charge. The smaller the pore size, the higher its concentration in acrylamide and the slower the molecule migration rate. Protein properties also account for electrophoretic mobility, such as size and pKa. Migration rate also depends on size, shape and nature of the modified charge of the particles. Since charge depends on ionization, concentration and buffer pH will affect the migration. Equipment effects on the migration include running temperature and electric field intensity.

Polyacrylamide gel electrophoresis under denaturing conditions using sodium dodecyl sulphate (SDS-PAGE) is the most used electrophoresis technique to evaluate the pharmaceutical quality of proteins. Heating combined with the SDS detergent denatures the sample; opened protein chains then bind to the negative SDS and gain a constant negative charge/mass ratio, which is independent of the protein or peptide sequence. This independence leads to the migration of SDS–polypeptide complexes throughout the gel with mobilities mainly related to the size of the polypeptide. Therefore, the comparison of relative mobility (RM) of samples with the ones from molecular weight markers gives an estimate of sample molecular weight. Noteworthy, SDS binds in a different way in the presence of glycolic groups, so the apparent molecular weight of glycosylated proteins may not correlate with the real molecular weight.

The characterization of protein mixtures generally happens in batch-mode buffer systems, which include two continuous gels: a (top gel) stacking gel and a separating or resolution gel (bottom gel). These two gels have different porous size, pH and ionic strength. When the electric field is applied, the sample proteins migrate to the front between the stacking and resolving gels, become concentrated in a small volume and then migrate in a slower rate into the resolving gel, which has smaller pores. The molecular sieving separates SDS–protein complexes about their respective molecular masses.

21.6.1.2.1 Reducing Conditions

Polypeptide chains often form the stable three-dimensional structures because of disulphide bonds between the chains. These disulphide bonds can be disrupted using reducing conditions so that individual polypeptides can be separated from the larger protein complex. Complete denaturation and dissociation of chains can be achieved by the treatment with 2-mercaptoethanol or dithiothreitol (DTT), which causes unfolding of these structures. Therefore, the molecular mass of protein subunits can be determined by comparison with the migration pattern of molecular weight standards.

21.6.1.2.2 Non-Reducing Conditions

For some analyses, a complete dissociation of a protein into its constitutive peptide subunits is undesirable. The absence of reducing agents maintains the covalent disulphide bonds and preserves the oligomeric conformation of the protein. Consequently, the migration velocity is slower for SDS–oligomer complexes than for polypeptide subunits, and SDS may not bind to these oligomers in a constant mass ratio. These phenomena impair the accurate molecular mass determination since proteins can present different conformations, even when they possess the same molecular weight.

21.6.1.2.3 Detection of Proteins and Evaluation of Molecular Weight

The most commonly used method for staining proteins in gel electrophoresis is Coomassie Blue, which works in the 10–100 µg protein concentration range. Alternatively, for more diluted samples (in the 10–100 ng range), silver nitrate is more often used. All stages for staining gels are performed at room temperature, and gentle agitation is done by an orbital movement using an appropriate equipment.

The stained protein migration is measured with respect to its distance to the dye front and generally accounted as RM. The RMs of protein standards are correlated with their molecular weights to create a logarithmic graph and generate a standard curve by the linear regression method. Therefore, unknown proteins have their molecular weight calculated upon their RMs if they fall on the linear part of the curve.

21.6.1.2.4 Quantitative Determination of Impurities

If a monograph specifies a protein impurity, one must obtain a standard solution corresponding to its allowed limit (e.g. 5%, dilute the standard to this concentration before application) to apply in the gel and compare by normalization with the analysed sample (its comparison by an integrating densitometer).

21.6.1.3 NMR – Nuclear Magnetic Resonance

For almost half a century, the determination of three-dimensional (3D) structures of proteins has been achieved using the methods of crystallography and X-ray diffraction. However, about 20 years ago, nuclear magnetic resonance (NMR), which already occupied a prominent place in structural studies of small molecules, also began to be used for the determination of the 3D structure of peptides and proteins in solution, independently of crystallographic data. This was made possible by significant advances in instrumentation and methodology of this technique, including the application of high-intensity homogeneous magnetic fields, new pulse techniques and the introduction of more advanced computational methods that generate and interpret NMR data of macromolecules.

Currently, more than 2,000 3D protein structures determined by NMR have their atomic coordinates deposited in the Brookhaven Protein Data Bank, and this number is growing rapidly. Significant progress has also been made in the structural determination of other biopolymers by NMR, as in the case of nucleic acids and oligosaccharides previously obtained only by X-ray diffraction.

With the advent of the post-genomic era, the methods for obtaining 3D structures of biomolecules were never so much demanded. Traditionally, this work has been carried out with almost 'artisanal' characteristics, due to the high complexity of the task, the time needed to obtain only one structure (several months) and the high costs of infrastructure and equipment maintenance. For these and other reasons, the use of NMR in biochemistry is still restricted to a few excellence groups around the world and its theoretical underpinnings remain obscure to most of those who have some interest in the latest breakthroughs of this new era of scientific discovery.

Spectroscopy is the study of the interaction of electromagnetic radiation (EMR) with matter. This radiation is a wave with two components, one electric and the other magnetic, which oscillate perpendicular to each other, both of which are also perpendicular to the direction of propagation of light (Figure 21.5).

The interaction between EMR and matter can occur in two ways: by its electrical component or by its magnetic component. One of the fundamental differences between NMR and other forms of spectroscopy lies in the fact that in NMR, this interaction occurs with the magnetic field of EMR and not with the electric field, as is the case, for example, infrared or ultraviolet.

EMR of interest in chemical analysis ranges from gamma (high-energy) rays to radio waves (low energy). For each type of spectroscopy, a type of excitation is required, and for each one, there is a definite amount of energy; i.e. these phenomena are quantized. This means that a certain frequency and characteristic radiation is absorbed for a given transition.

NMR is in the region of radio waves (radio frequencies). In principle, it can be said that NMR is another form of absorption spectroscopy. In a magnetic field, under certain conditions, a sample can absorb EMR in the radiofrequency (rf) region – this absorption is governed by the characteristics of the sample. This absorption is a function of certain nuclei present in the molecule, which are sensitive to the applied radiation, and therefore, these nuclei are the object of study for the understanding of NMR spectra.

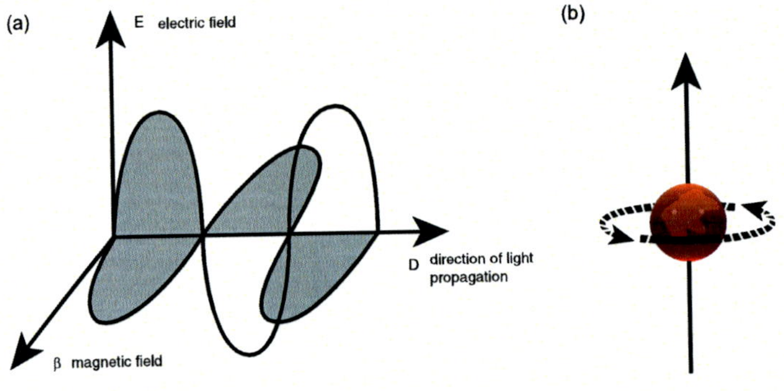

FIGURE 21.5 The electromagnetic radiation (a) and magnetic moment μ generated from angular momentum (b).

The physical quantity involved in NMR is nuclear spin. The concept of nuclear spin comes from quantum mechanics, not having an equivalent concept in classical mechanics. This concept is fundamental to the understanding of the phenomenon and can be understood as a property that certain nuclei are present. These nuclei, by virtue of their nuclear configuration, assume a behavioural characteristic of angular momentum capable of generating a magnetic moment.

The magnetic moment μ can be described as number of spin I, whose values are calculated by quantum mechanics and can be 0, 1/2, 1, 3/2 etc. As it is known, nuclei that present both atomic masses and even atomic numbers do not have spin, and consequently, the spin number is equal to zero. This is the case of 12C, 16O, 32S etc. (Table 21.6). These nuclei have no associated angular momentum and thus do not exhibit magnetic properties, which means there will be no detectable NMR signals. Nuclei with spin $I \neq 0$ are, in principle, detectable by NMR. The detection of these nuclei is related to the experimental conditions.

21.6.1.4 Capillary Electrophoresis

CE consists of an electrophoresis applied with potential difference in a capillary tube. The analysis is based on the migration of a solute carried out inside capillary tubes by an electrolyte solution under an electric field. CE separates molecules based on their electrophoretic mobility, isoelectric point, molecular size, and phase partition. Two factors are responsible for the separation in the CE: the movement of the solute in the capillary as a function of the applied electric field (E) and the electrophoretic speed (electrolytic flow due to the charge on the surface of the capillary wall).

Electrophoretic mobility (μ_{ep}) is correlated with molecule size, shape and electrical charge, but also with solution viscosity, pH, electrolytes or the presence of additives. Electrophoretic speed (V_{ep}) relates to the flow of solute through the electrolyte when a voltage is applied, measured in cm/s and determined by the equation:

$$V_{ep} = \mu_{ep} \times E = \left(\frac{q}{6\pi\eta r}\right)\left(\frac{V}{L}\right)$$

where

μ_{ep} = electrophoretic mobility;
E = applied voltage;
q = effective solute charge;
η = viscosity of the electrolyte;
r = Stoke's radius;

TABLE 21.6
Correlation between Atomic Number (*Z*) and Mass Number (A) with Spin Number (*I*)

I	*A*	*Z*	Examples
Half whole	Odd	Even or odd	$^{1}H(1/2)$, $^{17}O(5/2)$, $^{15}N(1/2)$
Whole	Even	Odd	$^{2}H(1)$, $^{14}N(1)$, $^{10}B(3)$
Zero	Even	Even	$^{12}C(0)$, $^{16}O(0)$, $^{32}S(0)$

V = voltage applied to the system;
L = total capillary length.

An electrolytic flow is generated around the capillary when an electric field is applied. The migration of different solutes present in the sample will happen along the capillary towards the detector, indicating the involvement of an additional force that goes beyond mobility and is independent of the ionic charge. The absence of this additional force would cause negative compounds to migrate against the detector direction and neutral solutes would not migrate.

Electroosmotic flow (EOF) is the additional force that plays an important role in different types of CE, as it directs the solute through the capillary tube. EOF generates pHs above 3 by ionizing silanol groups on the capillary internal wall. In this way, the electrolyte cations are attracted by the negative groups of the silanoate forming an internal layer on the capillary wall. Therefore, a double layer, essentially static, is formed near the capillary surface. There is also a mobile, diffuse layer, and due to the electrical voltage, they move to the cathode together with the hydration water. These two layers, mobile and static, generate a friction plane and consequently an electrical imbalance that corresponds to the potential difference (zeta potential, ζ) that crosses the two layers.

The density of charges on the capillary internal wall is related to the speed of the EOF, which is dependent on the electroosmotic mobility (μ_{eo}). Thus, the calculation of the speed of the electroosmotic flow (V_{eo}) is given by the equation:

$$V_{eo} = \mu_{eo} \times E = \left(\frac{\varepsilon\zeta}{\eta}\right)\cdot\left(\frac{V}{L}\right)$$

where

ε = dielectric constant of the electrolyte;
ζ = zeta potential of the capillary surface;
η = viscosity of the electrolyte;
V = voltage applied to the system;
L = total capillary length.

The electrophoretic and electroosmotic mobilities of a solute go parallel or in opposite directions, depending on the charge of the solute and the velocity of the solute (v), as described by the following equation:

$$V = V_{ep} \pm V_{eo}$$

Depending on whether the mobilities act in opposite directions or in the same direction, the difference or the sum between the two speeds will determine the shape of the CE. For example, in the most common CE, anions migrate in the opposite direction to the EOF; cations will migrate in the same direction as the EOF and their speeds will be greater than the speed of the EOF. Therefore, the separation of cations and anions can occur in the same electrophoretic run, provided that the main condition is that there is a faster speed than the electrophoretic speed.

The function that determines the time (t) ratio for the solute to travel a distance (l) from the capillary inoculation terminal to the capillary detector (effective capillary length) is defined by the equation:

$$t = \frac{l}{V_{ep} \pm V_{eo}} = \frac{l(L)}{V(\mu_{ep} \pm \mu_{eo})}$$

where

l = distance from the capillary injection terminal to the capillary detection window (effective capillary length);
V_{ep} = electrophoretic velocity;
V_{eo} = velocity of electroosmotic flow.

The maintenance of the EOF between different runs is related to the reproducibility of the migration speed for the solutes. Sometimes the reduction or suppression of the EOF is necessary for some special applications. This reduction or suppression is achieved by means of changes in the capillary wall or in the composition, concentration and pH of the electrolyte solution.

According to the intrinsic mobility, each solute of the sample migrates close to the electrolyte as an independent band, after introducing the sample into the capillary. The molecular diffusion of the solute in the capillary (longitudinal diffusion) is the only factor that contributes to the widening of the bands, when the CE is performed under ideal conditions. Thus, the number of theoretical plates (N) is related to the efficiency of the bandwidth, where (N) is determined by the equation:

$$N = \frac{(\mu_{ep} \pm \mu_{eo}) \cdot (Vl)}{2DL}$$

where

D = coefficient of molecular diffusion of the solute in the electrolyte.

The other terms were discussed earlier.

The modifications due to the electrophoretic mobility of the solutes, the increased efficiency of the band of each solute under analysis and the EOF are responsible for the separation of the sample bands. The resolution can be calculated using the equation:

$$Rs = \frac{\sqrt{N}(\mu_{epb} - \mu_{epa})}{4(\mu_{ep} + \mu_{eo})}$$

where

μ_{epa} and μ_{epb} = electrophoretic mobilities of two solutes to be separated;
μ_{eo} = mobility of the electroosmotic flow;
μ_{ep} = mean electrophoretic mobility of the solutes.

21.6.1.4.1 *Equipment*

CE equipment consists of:

- a high-voltage source;
- two electrodes inside electrolyte solution reservoirs, one for the anode and another for the cathode;
- capillary made of fused silica with a detection window that aligns with detectors;
- Injection system for samples by hydrodynamic or electrokinetic action;
- Detector (generally UV, UV-VIS or fluorimetry, but also electrochemical detectors or mass spectrometry);
- temperature control system;
- computerized system to record and interpret the readings.

21.6.1.4.2 *Instrumental Parameters*

Voltage: The applied voltage is proportional to the separation time. However, if it is too high, it causes excessive heat production (Joule effect). This increase in heat is responsible for increasing the bandwidth and resolving the solute due to the formation of viscosity gradients in the electrolyte inside the capillary.

Polarity: Under normal conditions, the polarity of the electrode is based on anode at the entrance and cathode at the exit, and the EOF will move towards the cathode. However, if the polarity of the electrode is reversed, the EOF will be in the opposite direction to the outlet, and there will only be an electromagnetic flow towards the outlet charged solutes with greater electrophoretic mobility.

Temperature: The electrolyte's electrical conductivity and viscosity are the main effects of temperature. When these properties are changed in the electrolyte, there are changes in the migration rate.

Capillary: The internal diameter and length of the capillary tube are the factors that directly influence the analysis parameters, interfering in the efficiency of the separation, in the load capacity and mainly in the total migration time of the solutes. Keeping the voltage constant, the electric current can decrease due to the increase in the effective and total length of the capillary, which will influence the increase in the migration time of the analyte. The dissipation of the heat generated by the electric current (Joule effect) improves as the internal diameter of the capillary decreases, allowing the reduction of the analysis time due to the increase in voltage. The capillary internal diameter also influences the detection limit of the method, which is also influenced by the volume of sample injected and the type of detection system. All these parameters can increase the efficiency of the separation by reducing the capillary internal diameter. Electrophoretic methods should have an efficient washing procedure of the capillary to keep reproducible migration times of the solutes between analyses.

The use of extreme pHs, adsorption of positively charged electrolytes and modification in the composition of electrolytes are the strategies used to prevent the adsorption and the interaction of the sample on the capillary internal wall (mainly with protein samples), which is a limitation of the CE efficiency. The use of polymers in the lining of the capillary inner wall through covalent bonds can prevent the interaction of proteins with the negatively charged silica from the capillary surface. These types of capillaries coated with neutral-hydrophilic, anionic and cationic polymers are now commercially available.

21.6.1.4.3 Parameters of the Electrolytic Solution

- *Nature and concentration of the buffer*: The electrolytes used in the CE must be able to buffer the sample to maintain low mobility and decrease the generation of electric current. Electrophoretic peak distortions are minimized by combining the mobility of the electrolyte ion with the mobility of the solute. The efficiency of the separation, detection and the uniformity of the solute will be directly influenced by the choice of the sample solvent.
- *Electrolyte pH*: The factors directly affected by the pH of the electrolyte are the change in flow and the modification of the solute charge. The separation of proteins in the CE is directly influenced by changes in the pH of the electrolyte in relation to the isoelectric point of peptides and proteins, changing from negative to positive due to the change in the net charge. In general, the increase in the EOF is directly influenced by the increase in the pH of the electrolyte.
- *Organic solvents*: Methanol, acetonitrile and other organic solvents are used mainly to increase the solubility of the solute in the aqueous electrolyte. The addition of organic solvents or other substances present in the electrolyte influences the ionization degree of solutes and can reduce the osmotic flow. Organic additives such as urea and methanol can be used to improve the resolution.

Cyclodextrins of different sizes (α, β and γ), modified with neutral groups (hydroxyalkyl, methyl, ethyl etc.) or ionizable groups (sulfobutyl, aminomethyl, carboxymethyl ether, etc.), are used as additives in the development of methods for an enantiomeric separation. However, others such as polysaccharides, proteins and crown ethers are used as chiral selectors. The chiral selector is responsible for the chiral separation, which is influenced by the temperature of the analysis and the pH of the electrolyte.

21.6.1.5 Spectrophotometric Methods

21.6.1.5.1 Atomic Absorption

The main application of atomic absorption spectrometry in biotechnology is the determination of metals in metalloproteins. Element determination basically consists of four techniques: atomic absorption with flame, graphite furnace, generation of hydrides and generation of cold vapour. The flame and graphite furnaces (like atomizers) can determine around 70 elements, mostly metals. The hydride generation technique can determine arsenic, antimony, selenium, bismuth, tellurium, lead, indium, tin, germanium and thallium. For mercury determination, it is necessary to generate a cold.

The concentration of an analyte by atomic absorption relates to a radiation intensity in a specific wavelength, which depends on the analysed element. The attenuation of the radiation is proportional to the analyte concentration (Lambert–Beer law).

The instrumentation for these types of analysis includes a source of radiation, an atomizer (flame or graphite furnace), a monochromator, a detector and a data acquisition/processing system. Radiation may come from hollow cathode lamps and electrodeless discharge lamps that emit an intense radiation of the same wavelength as that absorbed by the element to be determined. The monochromator separates the chosen wavelength from the radiation; then, the different wavelengths in a diffraction grating reach the detector (photomultiplier) that transforms light into electrical current followed by amplification and interpretation by the acquisition system.

21.6.1.5.2 Flame Atomic Absorption Spectrometry

This type of analysis is performed in a premix chamber where the fuel and the oxidant mix and burn. The sample solution enters by a pneumatic nebulizer that generates an aerosol towards the flame. The flame dissociates and atomizes samples depending on the temperature. A low-temperature flame does not produce neutral atoms, whereas an extremely high temperature forms a lot of ions that may not absorb radiation. The most popular flames are generated by air-acetylene (2,100°C–2,400°C) and acetylene-nitrous oxide (2,650°C–2,850°C). The former is applied to the analysis of elements with low atomization temperatures such as Na, K, Mg, Cd, Zn, Cu, Mn, Co etc. The acetylene-nitrous oxide flame goes for elements with refractory properties, such as Al, V, Ti, Si, U, among others.

21.6.1.5.3 Atomic Absorption Spectrometry with Hydride Generation

This technique consists of generation, transport and atomization of hydrides. It is useful for the determination of volatile hydride-forming elements, most commonly As, Se, Sb, Bi, Ge, Sn, Pb and Te. To generate hydrides, the analyte reacts with a reducing agent ($NaBH_4$), commonly in acid medium. Hydrides reach the quartz cell by transport through an inert drag gas such as argon or nitrogen. A purge must be performed to remove atmospheric gases when elements absorbing at wavelengths of less than 200 nm are in place. Atomization happens in a quartz cell heated electrically or by a burner typical of flame atomization systems, reaching cell temperatures between 850°C and 1,000°C.

21.6.1.5.4 Atomic Absorption Spectrometry with Cold Vapour Generation

This system is applied to mercury determination. The analysis uses the same system of atomic absorption by hydride generation, but with quartz cell lightly heated, since mercury reduction leads to volatile metallic mercury at room temperature. Purges are also unnecessary because Hg analysis wavelength is 253.7 nm, different from the atmospheric gases. To prevent water vapour condensation, an infrared lamp heats the quartz.

21.6.1.5.5 Atomic Absorption Spectrometry with Graphite Furnace

This sensitive technique uses an oven with a graphite tube (3–5 cm length, diameter of 3–8 mm, pyrolytic graphite coating). The furnace is heated by an electric current that passes in a longitudinal or transverse manner. To prevent furnace combustion, inert gas streams, such as argon, are kept externally and internally. Samples are injected by an automated system, and the internal flow pulls out atmospheric air and the vapours generated in drying and pyrolysis steps.

21.6.1.5.6 Mass Spectrometry

Inductively coupled plasma mass spectrometry detects elements by their mass determination, being able to detect in the order of 1 part per trillion. It contains an ionization source (signal generator), a mass separator, an electron multiplier detector, and a data acquisition/processing unit. After ionization, the sample is transported as plasma ions at atmospheric pressure (760 Torr) to the mass separator that operated under vacuum (10^{-6} Torr). The ions pass through an interface of two perforated cones (the sample cone followed by the skimmer cone), and then, they are focused to the mass analyser. This is achieved using an ionic lens or ion lens assembly.

The mass separator is generally a quadrupole, with four equal cylindrical or hyperbolic metal bars as electrodes. Upon the application of direct current (dc) and alternating current (ac), only ions with a given mass/charge ratio (m/z) can be conducted through this unit. Other ions are retained in the electrodes or removed. Ions that reach the end of the quadrupole are converted to electrical signals by dynodes, the Faraday Cup and Channeltron detectors, among others.

The technique suffers spectral interferences depending on the analyte, such as the ones derived from polyatomic, isobaric, double charge ions and refractory oxide ions. Various methods can suppress the interferences, from isotope substitution to software manipulation.

Non-spectral interferences may derive from deposition on the equipment cones, another element easily ionizable in the sample and effect space load. In these cases, correction can be done with an internal standard of mass/charge ratio and ionization potential close to the ones from the analyte. Scandium and rhodium are common internal standards applied for the correction of samples with low and high mass/charge ratio, respectively.

21.6.1.5.7 Ultraviolet (UV), Visible (VIS) and Infrared (IR)

Spectrophotometry relates to the ability of molecules to absorb electromagnetic energy depending on their concentration and structure. It can be classified based upon the frequency range used for the electromagnetic energy emission (waves): ultraviolet, visible and infrared. The wavelength bands of electromagnetic energy used in spectrophotometry are cited in Table 21.7.

TABLE 21.7
Wavelength Range of Interest for Spectrophotometry

Region	Wavelength Range
Ultraviolet (UV)	190–380 nm
Visible (VIS)	380–780 nm
Near-infrared (NIR)	780–2,500 nm (12,800–4,000 cm^{-1})
Medium-infrared (MIR)	4–25 µm (2,500–400 cm^{-1})
Far-infrared	25–300 µm (400–33 cm^{-1})

The total energy of the molecule involves the energy derived from its vibration, rotation and the electronic energy derived from the configuration of its electrons. By absorbing energy, molecules transit to the excited state, which usually happens in stages. In the UV/Vis region, electronic transitions occur in molecule moieties named chromophores: electrons are promoted between molecular orbitals, generally σ and π ligands and non-ligands, to the immediately superior, or π* and σ* anti-ligand energy orbitals. In the medium-infrared region (MIR), transitions of vibrational energy occur, and infrared radiation-induced vibrations comprise stretches and tensions of interatomic bonds and modifications in bond angles.

At last, near-infrared (NIR) spectra relate to radiation absorption by overtones and fundamental vibrational modes of bonds C-H, N-H, O-H, S-H and others. Bands of an NIR spectrum are generally weaker than the bands of the MIR spectrum.

NIR spectrophotometry is applied mainly to quantify and identify molecules from formulations, their polymorphic forms, and disintegration patterns, and therefore, it can be used for development and process control.

Spectra from these techniques can be obtained in different ways. UV/VIS methods rely on transmission, whereas infrared spectra can be obtained by transmission, reflection or even a combination of both modes (transreflection).

- *Transmission mode*: When the radiation passes through the sample, it measures an intensity decrease of the radiation at proper wavelengths. The sample is arranged in the optical beam between the source and the detector. The transmission (T) is then calculated as follows:

$$T = \frac{I}{I_0}$$

where
I_0 = intensity of incident radiation
I = intensity of transmitted radiation.

Transmission spectra can be converted to absorbance:

$$A = \log_{10} \frac{I}{I_0}$$

Diffuse reflection: The sample reflects light and the intensity of this reflection is compared with the one reflected by a reference reflective surface. The unabsorbed radiation is reflected towards the detector.

Attenuated total reflection (ATR): An infrared beam is directed to the sample that reflects it in a chamber with an internal reflection element (high refractive index). This creates an internal evanescent wave that will be attenuated by the sample that absorbs energy in the IR spectrum.

Transreflection: A reflective surface reflects the radiation transmitted through the sample, focusing again on the sample to double the optical path and reach the detector.

21.6.1.5.7.1 Instrumentation Used in Ultraviolet (UV) and Visible (VIS) Regions Equipment consists of a radiation source, wavelength selector, absorption cells (cuvettes) into which the samples are placed which are then positioned between the monochromatic light beam and radiation detector, and a signal processing unit. Its use in pharmaceutical analysis concerns mainly substance content determination, including drugs and contaminants.

The most common sources of radiation used in spectrophotometry that emit in ultraviolet and visible regions are deuterium and tungsten lamps, which radiate from 160 to 380 nm and 320 to 2,500 nm, respectively. Most equipment uses a monochromator or a wavelength filter to detect and measure only the bands of interest. Photometers or colorimeters are simpler instruments that use a filter for wavelength selection and are usually used in the visible region. Spectrophotometers use monochromators for wavelength selection and are operated in the UV/VIS regions.

The sample is placed inside a cuvette that is transparent on at least two sides so that the radiation can pass through the sample during analysis. For analysis in the UV region, quartz cuvettes are required, but when VIS spectroscopy is used, the cuvettes are made from glass or acrylic.

The most common detectors are phototubes, photodiode arrays and charge transfer devices. Phototubes are simple, and based on a photoelectric effect, DADs can follow all wavelengths at once; charge transfer devices have two silicon-integrated circuits where the photogenerated charge goes to an output amplifier.

Instruments can have single or dual beams, but only the dual beam compensates for radiant fluctuations emitted by the source. Multichannel instruments, in turn, use DADs and load transfer devices for a faster reading.

Currently, most spectrophotometers are connected to a microcomputer supported by an appropriate software, which allows the absorption spectra of a substance to be manipulated digitally and to be transferred to the third-party software.

21.6.1.5.7.2 Instrumentation Used in Medium-Infrared (MIR) and Near-Infrared (NIR) Regions The equipment for the MIR and NIR regions (between 750 and 2,500 nm or 13,300–400 cm^{-1}) analysis has a light source, a monochromator or interferometer and a detector. Samples can be analysed in conventional cuvettes or by optical fibres, transmission cells and diffuse reflection accessories. Transmission cells are used to hold the sample and are suitable for spectral acquisition in both diffuse reflection and attenuated total reflection configurations. Currently, infrared (4,000–400 cm^{-1}) spectrophotometers use an interferometer, and the source of the polychromatic radiation is emitted on the bottom of the sample. Spectra are obtained with the aid of the Fourier transformation.

21.6.2 Others (pH, Total Solids, Preservatives, Isotonicity)

In the concept of pH, when $[H^+] = [OH^-]$, the concentration of each of these species is $1.0 \times 10^{-7} M$ at 25°C. Under these conditions, the solution is said to be at neutral pH. pH is defined as the inverse of the logarithm of the H^+ concentration. By this definition, neutral pH is defined as being numerically equal to 7. When $[H^+] < [OH^-]$, the solution will have a pH greater than 7 and is said to be basic or alkaline. When $[H^+] > [OH^-]$, the solution has a pH of less than 7, which is said to be an acidic solution.

By the definition given here, it is possible to establish a numerical pH scale ranging from 1 to 14. Note that when the pH rises by one value, in fact the higher pH solution is ten times more basic, due to the logarithmic nature of the scale used. Two difference values correspond to a difference of 100 times, three values a 1,000 times etc.

How important is the pH of a solution? Many substances have groups that can undergo a protonation, i.e. incorporating one or more protons. In the same way, many substances can undergo deprotonation, i.e. to lose protons. In many cases, the protonation state of a molecule affects its biological activity. An example of this is protonation of several side chains of amino acids that constitute enzymes. Sometimes an amino acid simply does not have a proton that is essential for the catalysis, so that an entire enzyme can lose its biological activity. In addition, the solubility of a protein in an aqueous medium is dependent upon the pH of the solution. Proteins can precipitate at a pH when protonation and deprotonation are balanced, i.e. the isoelectric point. Under these conditions, the degree of dissolution is changed.

The pH of a solution can be measured in various ways. The most sensitive method is to use a pH electrode, which is an electrochemical device that measures the concentration of H^+ in a solution. The electrode is partially immersed in the test solution and produces an electrical current proportional to the H^+ concentration, which is then converted to a numerical value. For less-sensitive readings, pH tapes or indicator solutions may be used. The indicator solutions change colour at the so-called turning point, having a certain colour below that pH value and another above. pH tapes use the same principle, but in general use combinations of indicators for a more accurate pH measurement. It is also possible to establish a scale of pOH, like that of pH. However, this scale is not commonly used.

The determination of solids that are soluble in biological materials is a widely used measure in the processing and preservation of protein solutions when evaluating:

- solution stability;
- development of standard solutions;
- quality of prepared solutions.

The concentration of soluble solids in a sample is measured using a refractometric method, which is simple and fast, and has a good degree of precision. It uses, as a principle, the index of refraction of a pure substance, which is constant for a given temperature and pressure condition. For example, the concentration of solids soluble in sand can be determined because the known refractive index of water at 20°C is 1.3330. The presence of soluble solids in the water causes the refractive index of the water to change. Knowing the refractive index of the aqueous solution, it is possible to determine the amount of solute present. This property is used to then determine the concentration of soluble solids of solutions of proteins or buffered solutions.

The most used equipment is called an Abbe refractometer. Its use is simple and can only be calibrated with distilled water before any determinations. There are tables that allow corrections to be made due to the influence of temperature. The reading is performed as follows: the liquid sample without the presence of solids, particles or other insoluble substances is placed on the prism of the apparatus through which a beam of light will pass. The result is read immediately by a scale on the device. The reading value is expressed as a percentage of soluble solids or Brix.

Many protein solutions are suitable media for the growth of bacteria and fungi. Often in concentrated or multidose solutions, it is necessary to add preservatives that function as bactericidal or bacteriostatic, preventing the contamination of these solutions. The mostly used substances for this purpose are thimerosal, phenol and cresol. The use of thimerosal is being phased out because its base is mercury.

Tonicity is another parameter to consider, used to describe the response of cells immersed in an external solution. There are three classifications of tonicity that one solution may have over another: hypertonic, hypotonic and isotonic. Protein solutions for parenteral must be isotonic to avoid haemolysis (cytolysis of red blood cells).

21.6.3 In vitro Tests

21.6.3.1 Cytotoxicity

Cytotoxicity testing allows the toxic or mutagenic effects of a test sample to be investigated using cell or tissue cultures. This testing is of extreme importance during the development of pharmaceutical products intended for human or animal use. Three categories of cytotoxicity tests are commonly used: extract test, direct contact test and indirect contact test. The choice of which ones should be used will depend on sample nature, intended site of use and expected activity. The extent of any cytotoxic effect can be evaluated after a predetermined exposure time. Various endpoints can be used to determine cytotoxicity, such as:

- measurements of cell growth;
- measurements based on cell damage, like leakage of intracellular proteins;
- measurements of specific aspects of cell metabolism, like mitochondrial activity;
- evaluation of morphological changes, such as visual inspection by microscopy.

21.6.3.2 Isoenzymes

The sensitive isoenzyme assay can detect cell cross-contamination and verify the origin of the production cells. Isoenzymes are defined by a group of multiple molecular forms of the same enzyme that occurs in one species because of the presence of more than one encoding gene. Isoenzyme analysis is the most direct and rapid way to genotypically assess many loci in many individuals (WHO, 2013). The basic principle of the technique lies in the use of electrophoresis and the differential migration of molecules with different loads and sizes in starch gel, and in the visualization of the enzymatic product by histochemical methods.

Because it is a codominant marker, all alleles are evidenced in the gel, except the nulls. Thus, it is assumed that differences in the mobility of isoenzymes in an electric field are due to differences in the DNA sequences that encode such enzymes. And, if the band patterns of two individuals differ, these differences are genetically based and inheritable. This allows us to readily estimate parameters (such as genotype and gene frequencies, observed and expected heterozygosity) and test if the population is in Hardy–Weinberg (HW) equilibrium, the proportion of polymorphic loci and inbreeding coefficients, and can also test the distance–isolation models (Araújo et al., 2019).

21.6.3.3 Microbiological Tests

21.6.3.3.1 Fungi and Bacteria

Assays for the microbiological control of biopharmaceuticals are the same as any other injectable product. Even if the product is administered orally, such as some vaccines (polio and rotavirus), they must be kept sterile to maintain stability and avoid the ingestion of possible toxic degradation products. These tests should be performed in clean rooms using good microbiological practices that avoid cross-contamination, since different practices between laboratories can lead to different assay results (FDA, 2004a).

The choice of the culture medium depends on the pharmacopeia monograph being followed, but common culture media are thioglycolate and soy casein. All microbiological tests are performed in comparison with positive controls, which use standard strains such as *Staphylococcus aureus* (Gram-positive cocci), *Bacillus subtilis* (endospore-forming Gram-positive rod), *Escherichia coli* (Gram-negative rod), *Candida albicans* (yeast) and *Aspergillus niger* (filamentous fungus).

21.6.3.4 Mycoplasma

Mycoplasma species can be introduced into cell lines from the operator, from the medium (e.g. trypsin or serum of porcine origin) and from the cell line itself. This is one of the major problems of contamination in eukaryotic cell culture. Three different assays are used routinely to detect contamination: an immunological method (ELISA or direct immunofluorescence), a genetic method (PCR) and a direct isolation in PPLO broth medium. Generally, five positive controls are used: *Mycoplasma pneumoniae*, *Mycoplasma gallisepticum*, *Mycoplasma orale*, *Ureaplasma urealyticum* and *Acholeplasma laidlawii* (WHO, 2013).

21.6.3.4.1 Adventitious Viruses

This is one of the most difficult and laborious tests to perform because it consists of detecting the presence of viruses in biopharmaceuticals and vaccines derived from mammalian cell cultures. These viruses may originate from the cell culture itself, for example the pig virus circus, which was recently found, in cells used to produce rabies vaccine.

The test is performed by inoculating the biopharmaceutical or vaccine suspension into five different cell lines to observe a cytopathic effect. However, there may be a virus that does not easily adapt to the new cells or does not have a clear cytopathogenic effect to verify the presence of adventitious viruses. Another assay is to inoculate subcutaneously a suspension of the biopharmaceutical or vaccine into at least three species of laboratory animals (mouse, rat and guinea pig). No animals can present pathophysiological features related to infection. This assay is also very subjective since many viruses do not have receptors in cells of laboratory animals and can often present false-positive results. Both assays have proven ineffective for the detection of adventitious viruses, so validation trials of viral clearance or inactivation of biopharmaceuticals and/or viral vaccines are required, except in live virus-containing vaccines such as rotavirus vaccine, vaccine polio (Sabin), yellow fever vaccine, and in gene therapy, in which viruses are used with DNA vector.

21.7 PRECLINICAL SAFETY EVALUATION

According to FDA Guidance for industry S6 preclinical tests, together with *in vitro* ones, define pharmacological and toxicological profiles before the human testing. Regardless of the assay, the essential parameters should be specified: species, age, physiological state, dose/route of administration/treatment regimen, and test material stability under the intended conditions of the animal test. For in vivo preclinical studies, the trials presented in the following sections are recommended. For in-depth details, the reader can consult the OECD guidelines (OECD, 1981–2020).

21.7.1 Acute Toxicity

An acute toxicity test is a study that is performed when an animal is administered with a single dose of the drug, or at fractional doses such that the total administration period does not exceed for 24 h.

Acute toxicity studies should be planned in such a way that it is possible to obtain (a) a case fatality rate (not necessarily with a high level of statistical pressure); (b) the way in which the drug causes death; (c) establish a quantitative relationship between the doses administered and the signs of toxicity, including changes in body weight and food consumption, behavioural, biochemical (blood and urine), haematological and histopathological observations; (d) all animals that die must be autopsied and the other animals must be euthanized and autopsied at the end of the observation period; (e) the ideal observation period should be 14 days, and never less than 7 days; and (f) control groups should be performed (animals injected only with the vehicle and untreated animals).

21.7.2 Subacute Toxicity or Repetitive Doses

In a subacute toxicity test, the medicinal product is administered at regular intervals for a minimum period of 14 days; three spaced dosing levels should be used, the lowest dose corresponding to the highest dose not producing detectable toxic effects after the single dose administration.

Whenever possible, studies should be planned so that changes produced by the drug during treatment can be compared to pre-treatment levels for each animal.

The evaluation should be as broad as possible, including behavioural observations, biochemical profiles (blood and urine), haematological and histopathological changes.

21.7.3 Chronic and Subchronic Toxicity

In a chronic toxicity test, the medicinal product is administered at regular intervals for a minimum period of 90 days.

In the subchronic toxicity test, the medicinal product is administered at regular intervals for a minimum period of 30 days.

For both the chronic toxicity test and the subchronic toxicity test, one should consider the following: (a) three doses should be used, the lowest corresponding to the highest dose that does not produce detectable toxic effects after the single administration; (b) duration of studies (previously described); (c) the evaluation should be as wide as possible, including behavioural observations, biochemical profile (blood and urine), haematological alterations and histopathology; (d) whenever possible, the choice of species for the subchronic and chronic studies should take into account the pharmacokinetic similarity to humans; and (e) if in chronic studies, the drug is added to food or water, it should also be ensured that the drug under these conditions is stable and that its concentrations in the food or water are carefully adjusted in order to keep a constant daily doses in relation to the weight of the animal.

21.7.4 Teratogenicity and Reproductive Disorders

Teratogenicity tests are performed using animals during gestation or fertile periods with doses like those used for chronic toxicity testing, in order to observe any teratogenic effects in foetuses or newborns and/or reproductive disorders in fertile animals.

21.7.5 Immunotoxicity

Animal tests are used to determine the antigenicity of a biological product, i.e. studies to assess the induction of antibodies against the biopharmaceutical, studies of inflammatory or anti-inflammatory activity by the measurement of inflammatory cytokines, and studies of self- and non-self-recognition to assess the induction to autoimmunity.

21.7.6 Pharmacokinetics

PK studies are generally only applied to biopharmaceuticals other than vaccines or hyperimmune sera and blood derivatives where bioavailability is not a requirement for these products.

21.7.7 Carcinogenicity or Oncogenicity

Most carcinogenicity studies are carried out using rodent species, following guidance of the Organization for Economic Co-operation and Development (OECD, 1981–2020). The three main routes of administration used in carcinogenicity studies are oral, dermal and inhalation. This route choice depends on the physical and chemical characteristics of the test substance and the intended route of delivery in humans. The aim of carcinogenicity studies covered by the test guideline is to identify the carcinogenic properties of a chemical, resulting in an increased incidence of neoplasms compared to control groups.

The objectives of carcinogenicity studies include:

- identification of the targeted organs;
- tumour dose characterization: response rate;
- identification of the greatest dose that has no-observed-level-of-adverse-effects (NOAEL) or starting point for the establishment of a benchmark dose (BMD);
- extrapolation of carcinogenic effects to a lower dose than levels of human exposure;
- providing data to test hypotheses about the carcinogenic mechanism if present.

21.7.7.1 *In vitro* Preclinical Studies – Mutagenicity

The recommended mutagenicity tests are mammalian chromosomal aberration test, mammalian cell mutation test and bacterial retromutation test, as discussed below, and all of them follow OECD guidelines.

21.7.7.2 *In Vitro* Mammalian Chromosomal Aberration Test (OECD 473)

The objective of the in vitro chromosome aberration test is to identify the agents that cause structural chromosomal or chromatid aberrations in mammalian cell cultures over a period. Cells must be chosen based on their growth rate, stability of the karyotype and frequency of spontaneous aberrations. These aberrations trigger genetic disorders and consequent disease, such as changes in oncogenes and/or cell tumour suppressor genes that were already related to cancer development. Most chemical mutagenic agents induce chromatid-like aberrations, with less often damages of chromosome-type abnormalities. Although aneuploidy can be observed in these tests, its evaluation must be done by a micronucleus test.

21.7.7.3 *In Vitro* Mammalian Cell Gene Mutation Test (OECD 476)

This test intends to detect mutations of reporter genes induced by chemicals in cultured cell lines. Reporter genes include hypoxanthine-guanine phosphoribosyl transferase (HPRT test) and xanthine-guanine phosphoribosyl transferase transgene (XPRT test). The tests complement each other since they evaluate different mutational events. Common cell lines include CHO, V79, AS52 and CHL (Chinese hamster cells), TK6 (human cells) and L5178Y (mouse cells).

21.7.7.4 Recombinant Bacterial Testing (OECD 471)

The bacterial reverse mutation assay uses amino acids requiring strains of bacteria to detect point mutations. It detects mutations that revert the initial mutation of the strains and restore their ability to produce the amino acids. Noteworthy, in vitro metabolic activation systems do not fully reproduce conditions in vivo in mammals. The test also does not offer direct mutagenic and carcinogenic evaluations of bacterial-derived substances in mammals.

21.8 LEGAL ASPECTS FOR CLINICAL TRIALS

Clinical studies are investigations in humans for the purpose of discovering or verifying the toxicity, pharmacological and other effects of the product, including adverse reactions, with the goal of ascertaining that the product is safe and efficient.

21.8.1 Phases of the Clinical Trial

21.8.1.1 Phase I

Initial assessment in humans (number of individuals 20–100), to evaluate tolerance in healthy volunteers: (a) highest tolerable dose, (b) lower effective dose, (c) dose/effect relationship, (d) duration of effect, (e) side effects and (f) pharmacokinetics in humans (metabolism and bioavailability).

It is the first study in humans, including generally healthy people, for a new active substance or new dosage form/formulation. These studies aim to assess a preliminary safety and pharmacokinetic profile and, where possible, a pharmacodynamic profile.

21.8.1.2 Phase II (Pilot Therapeutic Study)

This is the first controlled study in patients to demonstrate the potential drug effectiveness (number of individuals 100–200), concerning (a) indication of efficacy, (b) confirmation of safety, and (c) bioavailability and bioequivalence of different formulations.

The objectives of the Pilot Therapeutic Study are to demonstrate the activity and establish the short-term safety of the active substance in patients affected by a particular disease. Trials include a small number of people and are often followed up by a questionnaire study. It should also be possible to establish dose–response relationships, to obtain data to endorse extended therapeutic studies.

21.8.1.3 Phase III

In this phase, large-scale international studies are carried out in multiple centres with different populations of patients to demonstrate efficacy and safety (minimum population approximately 800). Multicentre studies aim to (a) expand knowledge of the product; (b) establish the therapeutic profile, such as dose and route of administration, contraindications, and precautionary measures; (c) show the demonstration of therapeutic advantage (e.g. comparison with competitors); (d) pharmacoeconomics and quality of life; and (e) publication and communication strategy (e.g. conferences and workshops).

Expanded Therapeutic Study

These studies are carried out on large and varied groups of patients, aiming to determine (a) the short-/long-term risk/benefit outcome of different formulations of the active substance and (b) the overall relative therapeutic value.

The type and profile of the most frequent adverse reactions, as well as special characteristics of the medicinal product, is explored at this stage, for example clinically relevant interactions, the main factors modifying the effect, such as age, etc.

21.8.1.4 Phase IV

After approval for commercialization of the product, studies are carried out to:

- Detect infrequent or unexpected events (post-marketing surveillance).
- Support marketing decisions.
- Compare the product with its competitors.
- Evaluate new formulations (palatability, ease of ingestion).

Phase IV research should follow the same ethical and scientific standards applied to previous phase research.

21.8.2 Ethical Aspects of Human Trials

The ethics of research includes:

a. free informed consent of the target individuals and the protection of vulnerable and legally incapacitated groups (autonomy). In this sense, research involving human beings must preserve their dignity, their autonomy and respect their vulnerability;
b. weighting between risks and benefits, both current and potential, individual or collective (charity), committing to maximum benefits and minimum damages and risks;
c. guarantee that predictable damages will be avoided (not maleficence);
d. social relevance of the research with significant advantages for the subjects of the research, minimizing the burden for the vulnerable subjects (justice and equity).

Most international agencies have specific GMP requirements for Phase I studies, for example validations of analytical methods must be completed; the clean-up validation procedures must have been approved and good clinical practice procedures should be described and approved. The production of a biopharmaceutical is often carried out in companies commissioned by the producing laboratory, the so-called Contract Manufacturing Organization (CMO).

These companies are defined as the ones offering production services in small quantities to support preclinical trials or at higher volumes when necessary for clinical trial and marketing purposes.

21.9 FINAL CONSIDERATIONS

In this chapter, we gave an overview of the most regulatory-based aspects of quality applied to pharmaceutical biotechnology products. From development to scaled-up production, the whole process needs to be under a PQS that defines how, when and what should be considered in each phase. As expected, although each company or group can define their quality strategies, they must be in accordance with all current legislation, locally and sometimes globally. The tendency to a global quality understanding grows fast and points to universal quality guidelines verified by international groups, aiming to a uniform quality of pharmaceuticals throughout the world. We also introduced the reader to the Quality-by-Design philosophy, which emphasizes process robustness and quality assurance from the earliest stages of bioprocess development. This is fundamental for the present and next generation of scientists, since the FDA, EMA, and ICH are increasingly supporting the pharmaceutical companies to manage risk and to adopt QbD as their development paradigm.

REFERENCES

Abbot, A.; Cyranoski, D. Biologists seek to head off future sources of infection. *Nature*, v. 423, 3, 2003. https://doi.org/10.1038/423003b.

Anvisa. *Brazilian Pharmacopeia*, 6th edition, 2019 (Portuguese).

Benyahia, B.; Brumano, L.P.; Pessoa, Jr, A.; and Santos da Silva, F.V. Chapter 2: Biopharmaceutical Molecules. In: Rodrigues, A.G. (ed.), *New and Future Developments in Microbial Biotechnology and Bioengineering*. New York: Elsevier. 2020a, 31–68. https://doi.org/10.1016/B978–0–444–64301–8.00002-0.

Benyahia, B.; Brumano, L.P.; Pessoa, Jr, A.; and Santos da Silva, F.V. Chapter 3: Biopharmaceutical Development, Production, and Quality. In: Rodrigues, A.G. (ed.), *New and Future Developments in Microbial Biotechnology and Bioengineering*. New York: Elsevier. 2020b, 69–89. https://doi.org/10.1016/B978-0-444-64301-8.00003-2.

Benyahia, B.; Lakerveld, R.; Barton, P.I. A plant-wide dynamic model of a continuous pharmaceutical process. *Industrial & Engineering Chemistry Research*, v. 47, n. 51, 15393–15412, 2012. https://doi.org/10.1021/ie3006319.

Brumano, L.P.; Santos da Silva, F.V.; Costa-Silva, T.A.; Apolinário, A.P.; santos, J.H.P.M.; Kleingesinds, E.K.; Monteiro, G.; Rangel-Yagu, C.O.; Benyahia, B.; and Pessoa, Jr, A. 2019. Development of L-asparaginase biobetters: current research status and review of the desirable quality profiles. *Frontiers in Bioengineering and Biotechnology*, v. 6, 212. https://doi.org/10.3389/fbioe.2018.00212.

Burnouf, T.; Griffiths, E.; Padilla, A.; Seddik, S.; Stephano, M.A.; Gutiérrez, J.M. Assessment of the viral safety of antivenoms fractionated from equine plasma. *Biologicals*, v. 32, n. 3, 115–128, 2004. https://doi.org/10.1016/j.biologicals.2004.07.001.

Burnouf, T.; Radosevich, M. Nanofiltration of plasma-derived biopharmaceutical products. *Haemophilia*, v. 9, n. 1, 24–37, 2003. https://doi.org/10.1046/j.1365-2516.2003.00701.x.

Caricati, C.P.; Oliveira-Nascimento, L.; Yoshida, J.T.; Caricati, A.T.P.; Raw, I.; Stephano, M.A. Safety of snake antivenom immunoglobulins: Efficacy of viral inactivation in a complete downstream process. *Biotechnology Progress*, v. 29, 972–979, 2013. https://doi.org/10.1002/btpr.1818.

Chatterjee, B. April 16, 2018. Considerations for Biologic Drug Substance and Drug Product Testing, *Outsourced Pharma*.

CMC Biotech Working Group. 2009. A-Mab: A Case Study in Bioprocess Development. https://cdn.ymaws.com/www.casss.org/resource/resmgr/imported/A-Mab_Case_Study_Version_2-1.pdf. Last access on July 3rd, 2020.

Committee for Medicinal Products for Human Use (CHMP). 2014. CHMP Assessment Report: Gazyvaro. London. https://www.ema.europa.eu/documents/assessment-report/gazyvaro-epar-public-assessment-report_en.pdf. Last access on July 5th, 2020.

Committee for Medicinal Products for Human Use (CHMP). 2015. Assessment Report: Gardasil 9. London. https://www.ema.europa.eu/en/documents/assessment-report/gardasil-9-epar-public-assessment-report_en.pdf. Last access July 5th, 2020.

Committee for Medicinal Products for Human Use (CHMP). 2017. Assessment Report: Tecentriq. London. https://www.ema.europa.eu/en/medicines/human/EPAR/tecentriq#product-information-section. Last access July 5th, 2020.

Committee for Medicinal Products for Veterinary Use. 2015. CVMP Assessment Report: Imrestor. London. https://www.ema.europa.eu/documents/assessment-report/imrestor-epar-public-assessment-report_en.pdf. Last access July 5th, 2020.

Dichtelmüller, H.O.; Biesert, L.; Fabbrizzi, F.; Gajardo, R.; Gröner, A.; Von Hoegen, I.; Jorquera, J.I.; Kempf, C.; Kreil, T.R.; Pifat, D.; Osheroff, W.; Poelsler, G. Robustness of solvent/detergent treatment of plasma derivatives: A data collection from Plasma Protein Therapeutics Association member companies. *Transfusion*, v. 49, 1931–1943, 2009. https://doi.org/10.1111/j.1537-2995.2009.02222.x.

Elliott, P.; Billingham, S.; Bi, J.; Zhang, H. Quality by design for biopharmaceuticals: A historical review and guide for implementation. *Pharmaceutical Processing*, v. 1, n. 1, 105–122, 2013. https://doi.or/10.4155/PBP.13.6.

EMA. 2020. European Medicines Agency. https://www.ema.europa.eu. Last access July 5th, 2020.

EMA. The European Agency for the Evaluation of Medical Products. Human Medicines Evaluation Unit. Committee for Proprietary Medicinal Products. 14 Feb 1996. Note for Guidance on Virus Validation Studies: The Design, Contribution and Interpretation of Studies Validating the Inactivation and Removal of Viruses. London. https://www.ema.europa.eu/en/documents/scientific-guideline/note-guidance-virus-validation-studies-design-contribution-interpretation-studies-validating_en.pdf. Last access July 5th, 2020.

FDA. Food and Drug Administration. Sept 2004a. Guidance for Industry - Sterile Drug Products Produced by Aseptic Processing. https://www.fda.gov/media/71026/download. Last access July 5th, 2020.

FDA. Food and Drug Administration. 2004b. Guidance for Industry PAT—A Framework for Innovative Pharmaceutical Development, Manufacturing, and Quality Assurance. Rockville, MD. https://www.fda.gov/downloads/drugs/guidances/ucm070305.pdf. Last access July 5th, 2020.

FDA. Food and Drug Administration. 2004c. Pharmaceutical CGMPS for the 21st Century—A Risk-Based Approach. Rockville, MD. https://www.fda.gov/media/77391/download. Last access July 5th, 2020.

ICH. 2000. Good manufacturing practice for active pharmaceutical ingredients Q7. In *International Conference on Harmonization of Technical Requirements for Registration of Pharmaceuticals for Human Use*. https://www.ema.europa.eu/en/ich-q7-good-manufacturing-practice-active-pharmaceutical-ingredients.

ICH. 2005. Quality Risk Management Q9. In *International Conference on Harmonization of Technical Requirements for Registration of Pharmaceuticals for Human Use*. https://database.ich.org/sites/default/files/Q10%20Guideline.pdf.

ICH. 2008. Pharmaceutical Quality System Q10. In *International Conference on Harmonization of Technical Requirements for Registration of Pharmaceuticals for Human Use*. https://database.ich.org/sites/default/files/Q10%20Guideline.pdf.

ICH. 2009. Pharmaceutical Development Q8(R2). In *International Conference on Harmonization of Technical Requirements for Registration of Pharmaceuticals for Human Use*. https://database.ich.org/sites/default/files/Q8%28R2%29%20Guideline.pdf.

ICH. 2012. Development and Manufacture of Drug Substances (Chemical Entities and Biotechnological/Biological Entities) Q11. In *International Conference on Harmonization of Technical Requirements for Registration of Pharmaceuticals for Human Use*. https://database.ich.org/sites/default/files/Q11%20Guideline.pdf.

ICH. 2018. ICH Press Release. In *International Council on Harmonization of Technical Requirements for Registration of Pharmaceuticals for Human Use*, 1–2. Kobe, Japan. https://www.ich.org/pressrelease/meddra-management-committee-kobe-japan-june-2018. Last access July 5th, 2020.

ICH. 2019. Development and Manufacture of Drug Substances (Chemical Entities and Biotechnological/Biological Entities) Q12. In *International Conference on Harmonization of Technical Requirements for Registration of Pharmaceuticals for Human Use*. https://database.ich.org/sites/default/files/Q12_Guideline_Step4_2019_1119.pdf.

Juran, J. M.. 1992. *Juran on Quality by Design: The New Steps for Planning Quality into Goods and Services*. 2nd edition. New York: Free Press. ISBN 0–02–916683–7.

Kenett, R.S.; Kenett, D.A. Quality by design applications in biosimilar pharmaceutical products. *Accreditation and Quality Assurance*, v. 13, n. 12, 681–90, 2008. https://doi.org/10.1007/s00769-008-0459-6.

Kim, K.-M.; Weon Lee, J.; Kim, S.; Santos da Silva, F.V.; Seidel-Morgenstern, A.; and Lee, C.-H. Advanced operating strategies to extend the applications of simulated moving bed chromatography. *Chemical Engineering & Technology*, v. 40, n. 12, 2163–78, 2017. https://doi.org/10.1002/ceat.201700206.

Lakerveld, R.; Benyahia, B.; Braatz, R.D.; Barton, P.I. Model-based design of a plant-wide control strategy for a continuous pharmaceutical plant. *AIChE Journal*, v. 59, n. 10, 3671–3685, 2013. https://doi.org/10.1002/aic.14107.

Lakerveld, R.; Benyahia, B.; Heider, P.L.; Zhang, H.; Wolfe, A.; Testa, C.J.; Ogden, S.; Hersey, D.R.; Mascia, S.; Evans, J.M.B.; Braatz, R.D.; Barton, P.I. The application of an automated control strategy for an integrated continuous pharmaceutical pilot plant. *Organic Process Research & Development*, v. 19, n. 9, 1088–1100, 2015. https://doi.org/10.1021/op500104d.

Li, Y. The application of caprylic acid in downstream processing of monoclonal antibodies. *Protein Expression and Purification*, v. 153, 92–96, 2019. https://doi.org/10.1016/j.pep.2018.09.003.

Mascia, S.; Heider, P.L.; Zhang, H.; Lakerveld, R.; Benyahia, B.; Barton, P.I.; Braatz, R.D.; Cooney, C.L.; Evans, J.M.; Jamison, T.F.; Jensen, K.F.; Myerson, A.S.; Trout, B.L. End-to-end continuous manufacturing of pharmaceuticals: Integrated synthesis, purification, and final dosage formation. *Angewandte Chemie International Edition*, v. 52, n. 47, 12359–12363, 2013. https://doi.org/10.1002/anie.201305429.

OECD. Organization for Economic Co-operation and Development. OECD Guidelines for the Testing of Chemicals, Section 4, Health Effects. 1981–2020. https://doi.org/10.1787/20745788.

Olson, T.C.; Bongartz, T.; Crowson, C.S.; Roberts, G.D.; Orenstein, R.; Matteson, E.L. Histoplasmosis infection in patients with rheumatoid arthritis, 1998–2009. *BMC Infectious Diseases*, v. 11, 145, 2011. Published 2011 May 23. https://doi.org/10.1186/1471-2334-11-145.

Pandhi, D.; Sonthalia, S. Human papilloma virus vaccines: Current scenario. *Indian Journal of Sex Transmitted Diseases and AIDS*, v. 32, n. 2, 75–85, 2011.

Peterson, J.J. A posterior predictive approach to multiple response surface optimization. *Journal of Quality Technology*, v. 36, n. 2, 139–153, 2004. https://doi.org/10.1080/00224065.2004.11980261.

PIC/s. Pharmaceutical Inspection Co-Operation Scheme. Validation Master Plan Installation and Operational Qualification Non-Sterile Process Validation Cleaning Validation. PI 006-3. 2007. http://academy.gmp-compliance.org/guidemgr/files/PICS/PI%20006-3%20Recommendation%20on%20Validation%20Master%20Plan.pdf. Last access July 5th, 2020.

Rathore, A.S.; Winkle, H. Quality by design for biopharmaceuticals. *Nature Biotechnology*, v. 27, no. January, 26, 2009. https://doi.org/10.1038/nbt0109-26.

Rekhi, R.; Smith, J.A.; Arshad, Z.; Roberts, M.; Bountra, C.; Bingham, I.; Brindley, D.A. Decision-support tools for monoclonal antibody and cell therapy bioprocessing: Current landscape and development opportunities. *Bioprocessing International*, v. 13, n. 11, 32–39, 2015. https://bioprocessintl.com/manufacturing/monoclonal-antibodies/decision-support-tools-for-monoclonal-antibody-and-cell-therapy-bioprocessing-current-landscape-and-development-opportunities/. Last access July 5th, 2020.

Roberts, P.L.; Dunkerley, C.; Mcauley, A.; Winkelman, L. Effect of manufacturing process parameters on virus inactivation by dry heat treatment at 80°C in factor VIII. *Vox Sanguinis*. v. 92, n. 1, 56–63, 2007. https://doi.org/10.1111/j.1423-0410.2006.00845.x.

Sangshetti, J.N., Deshpande, M., Zaheer, Z., Shinde, D.B.; Arote, R. Quality by design approach: Regulatory need. *Arabian Journal of Chemistry*, 10, S3412–S3425, 2017. https://doi.org/10.1016/j.arabjc.2014.01.025.

Santos da Silva, F.V.; and Seidel-Morgenstern, A. Evaluation of center-cut separations applying simulated moving bed chromatography with 8 zones. *Journal of Chromatography A*, v. 1456, 123–36, 2016. https://doi.org/10.1016/j.chroma.2016.05.060.

Schlindwein, W.S.; Gibson, M. (eds.). *Pharmaceutical Quality by Design: A Practical Approach. Advances in Pharmaceutical Technology*. 1st ed. New York: John Wiley and Sons Ltd. 2018, 368. ISBN 1118895223, 9781118895221.

Sommeregger, W.; Sissolak, B.; Kandra, K.; Von Stosch, M.; Mayer, M.; Striedner, G. Quality by control: Towards model predictive control of mammalian cell culture bioprocesses. *Biotechnology Journal*, v. 12, n. 7, Special Issue: Biochemical Engineering Science, 2017. https://doi.org/10.1002/biot.201600546.

Stockdale, G.W.; Cheng, A. Finding design space and a reliable operating region using a multivariate Bayesian approach with experimental design. *Quality Technology & Quantitative Management*, v. 6, n. 4, 391–408, 2009. https://doi.org/10.1080/16843703.2009.11673206.

Vulto, A.G.; Jaquez, O.A. The process defines the product: What really matters in biosimilar design and production? *Rheumatology (Oxford, England)*, v. 56, Suppl 4, iv14–iv29, 2017. https://doi.org/10.1093/rheumatology/kex278.

Woodley, J.M. Chapter 3: Innovative process development and production concepts for small-molecule API manufacturing. In: Singh, R.; Zhihong, B.T., *Process Systems Engineering for Pharmaceutical Manufacturing*, Computer Aided Chemical Engineering. New York: Elsevier. 2018, 41, 67–84. https://doi.org/10.1016/B978-0-444-63963-9.00003-8.

World Health Organization. Supplementary Guidelines on Good Manufacturing Practices (GMP): Validation. Working Document QAS/03.055/Rev. 2. 2005. https://www.who.int/medicines/services/expertcommittees/pharmprep/Validation_QAS_055_Rev2combined.pdf. Last access July 5th, 2020.

World Health Organization. Recommendations for the Evaluation of Animal Cell Cultures as Substrates for the Manufacture of Biological Medicinal Products and for the Characterization of Cell Banks. Geneva. 2013. https://www.who.int/biologicals/vaccines/TRS_978_Annex_3.pdf?ua=1. Last accessed July 5th, 2020.

World Health Organization. Expert Committee on Biological Standardization. Guidelines on Viral Inactivation and Removal Procedures Intended to Assure the Viral Safety of Human Blood Plasma. Geneva. 2004, 26–30. https://www.who.int/publications/m/item/WHO-TRS924-Annex4. Last access July 5th, 2020.

Zalai, D., Dietzsch, C.; Herwig, C. Risk-based process development of biosimilars as part of the quality by design paradigm. *PDA Journal of Pharmaceutical Science and Technology*, v. 67, 569–580, 2013. https://doi.org/10.5731/pdajpst.2013.00943.

22 Techno-Economic Evaluation of Biotechnological Processes and Pharmacoeconomic Analysis

Rafael da Gama Ferreira
Intelligen Brasil

Nicola Sgrignoli Fuganholi, Larissa Pereira Brumano, and Adriano Rodrigues Azzoni
Universidade de São Paulo

Sindelia Freitas
Instituto SENAI de Inovação em Biotecnologia

CONTENTS

22.1 INTRODUCTION

The industrial bioprocesses used to produce biopharmaceuticals are complex and costly to design, develop, construct and operate. Perhaps one of the first questions that arise during the design and development of a new biopharmaceutical production facility is how many resources will be needed to build the plant and produce a specific annual amount of bioproduct (Petrides et al., 2014). Here, the definitions of resources include the set of process equipment, utilities, labour and materials, among others (Toumi et al., 2010; Petrides et al., 2014). Important questions such as the total capital to be invested, the estimated manufacturing cost of a product batch, the time to produce a batch or the optimized time between the production of two consecutive batches have also to be answered (Heinzle, Biwer, and Cooney, 2007; Petrides et al., 2014). As the development of the project progresses, so does the level of detail of the process. New issues such as increasing final yield, improving product quality, increasing robustness, reducing the environmental impact of the process and identifying manufacturing bottlenecks should also be specifically addressed (Heinzle, Biwer, and Cooney, 2007; Toumi et al., 2010).

In this context, process simulation is very useful for systematizing and guiding efforts to find answers to the questions posed above. Simulation software packages have been used in the chemical industry since the 1960s, initially for the simulation of continuous processes at the steady state and for some perturbations of this state, such as those found during the plant start-up and shutdown (Toumi et al., 2010; Petrides et al., 2014). However, in the mid-1990s, the first simulators focused specifically on pharmaceutical and biopharmaceutical processes were developed, in which production is usually conducted in the form of discontinuous or semi-continuous processes (Petrides et al., 2014).

A tool capable of simulating discontinuous and semi-discontinuous bioprocesses is SuperPro Designer software, developed by Intelligen (USA) in the mid-1990s. For each unit operation of a process, the simulator has a mathematical model capable of performing mass and energy balances. Based on the mass balances, the equipment is then dimensioned (Petrides et al., 2014). The software also has an extensive database of thermodynamic and cost properties. Nevertheless, it is the responsibility of the researcher to correctly define the operations and structure of the process from the literature information and experimental tests, thus creating a base (or initial) process flow diagram drawn in the simulator. Through the constant exchange of information on the performance of the calculated process *in silico* and the experimental results, continuous refinement and improvement of the simulation of the process happens. In a fast and dynamic way, different process scenarios can be analysed and the economic impact evaluated. Bottlenecks or critical operations in terms of economic impact can be identified, and new solutions can be proposed (Ferreira, Azzoni, and Freitas, 2018). The different scenarios and process configurations can also be analysed in terms of mitigation of the environmental impact caused by the process, which will ultimately be reflected in the cost reduction as well (Ferreira, Azzoni, and Freitas, 2018).

As important as the economic analysis, the development of the process flow diagram in a simulator tends to facilitate the use of other tools essential to the design of a modern biopharmaceutical production plant. This is the case of the quality by design (QbD) philosophy, which has been encouraged by the US Food and Drug Administration (FDA) as an important tool for the development of new pharmaceutical production processes. The QbD philosophy advocates the understanding of productive processes through mathematical and statistical models so that the focus is placed on the robustness of the process and the quality profile of the final product (Rathore, 2018). Its purpose is to replace the traditional philosophy of the development of production processes, according to which a process is developed, validated and operated within narrow margins (Rathore, 2014). Based on an in-depth study of the production process, the QbD philosophy allows the process to be operated in a more flexible way while ensuring the final quality profile of the product in a more robust way (ICH, 2005, 2008, 2009). Ultimately, processes developed within the QbD philosophy also tend to be more economical, since fewer lots are lost during production (Rathore, 2014).

Therefore, the techno-economic analysis through simulation and the QbD tool should be used in a complementary way in the development of new pharmaceutical bioprocesses. By detailing the process flow diagram and identifying the critical unit operations, a rapid reassessment and optimization of the process can be achieved, so that new process configurations and/or conditions can be experimentally validated and tested on a pilot scale. In this way, those involved in simulation and techno-economic analysis should be in constant and close contact with those working in experimental work and quality evaluation, in order to support the development of an economically viable large-scale process under good manufacturing practice conditions, and with the highest quality standards assured for the final product. Figure 22.1 describes the flow of information that can be followed during the development of a new production process initiating with the creation of a baseline design via simulation, which can be constantly improved and validated with the introduction of new information coming from experimental and QbD studies.

22.2 BIOPROCESS DESIGN AND ECONOMICS

22.2.1 Design Basis

The design of a bioprocess using simulation software such as SuperPro Designer (Intelligen, USA) typically begins by determining the amount of product to be manufactured on an annual basis (that is, the **annual production rate**), the basic properties and purity requirements of the product, and the fundamental biological (microbial, cellular or biocatalytic) process capable of generating the desired product. These elementary but important pieces of information will guide all the process designs, and, in particular, will lead to the creation of the first process flow model, which may be called the 'baseline process'.

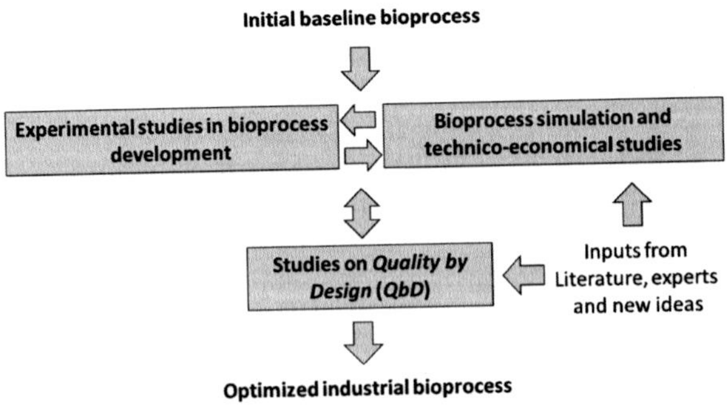

FIGURE 22.1 Expected flow of actions and information during the techno-economic evaluation studies for the development of a large-scale process for the production of biopharmaceuticals. (This figure was inspired and modified based on Figure 1.2 – Integrated development of bioprocesses, presented by Heinzle, E., Biwer, A.P., Cooney, C.L., *Development of Sustainable Bioprocesses Modeling and Assessment*, John Wiley and Sons, Chichester, West Sussex, 2007, 294 p.)

The annual production rate may be estimated by considering the current market size of the product on a national, or even global, scale, and then assessing a reasonable market share goal to be filled by the bioproduct in question. In the case of a new biopharmaceutical, for instance, one may combine data on the total number of patients in the country, dosage and length of treatment, in order to estimate the total annual demand. Then, a fraction (or even the entirety) of this figure can be considered as a target value for the product/process under study. The annual production rate is essential to simulate and determine the number and size of every piece of equipment that will compose the process. In other words, it establishes the process scale.

The basic properties and purity requirements of the product are also crucial for the process design, especially for selecting the type and number of operations of recovery and purification. Among the physical, chemical and biological properties generally relevant for process development, one may consider molecular weight; polar/hydrophobic characteristic; an affinity for cations, anions, metals, epitopes or other ligands; heat and pH stability; resistance to shear stress; solubility in water; and ability to crystallize. For instance, certain bioproducts such as ethanol and n-butanol are quite stable under heat and therefore can be efficiently separated through high-temperature operations, such as distillation, while other bioproducts, such as proteins, are highly thermolabile and thus require gentler separation methods, such as tangential filtration or chromatography. Similarly, the purity requirements of different bioproducts can vary widely depending on their application. For example, a lipase enzyme employed in the formulation of laundry detergent may present significant levels of endotoxins without consequence, as opposed to a therapeutic enzyme such as L-asparaginase, which is used to treat leukaemia patients.

The fundamental biological transformation underpinning a bioprocess is naturally of paramount importance. The type of biological transformation behind the product synthesis not only determines the kind of equipment, operational conditions and raw materials required for producing the bioproduct itself, but also largely determines the equipment, operations and materials that must come before that transformation (the so-called Upstream Section) and after that transformation (the so-called Downstream Section). For instance, if a recombinant monoclonal antibody is to be produced using CHO cells, a sophisticated bioreactor will be required for the cell culture. Moreover, a series of auxiliary bioreactors will be necessary to generate the inoculum for the main bioreactor (those additional bioreactors will form the Upstream Section). Likewise, given that antibody production by CHO cells is extracellular, it will be necessary to separate the cells from the liquid phase containing the antibodies afterwards. Besides, given that the cells may harbour viruses, a step of viral inactivation will also be imperative in the Downstream Section. In sum, the biological transformation has a vast impact on the whole process design.

Due to the importance of the biological transformation step, the equipment, operations and raw materials directly involved in the core biological transformation are together considered as a distinct section of the bioprocess, called 'Bioreaction Section' or, alternately, 'Fermentation Section'. Thus, every bioprocess may be divided into three major parts: Upstream Section, Bioreaction Section and Downstream Section.

22.2.2 Upstream Section

When dealing with biological transformations that involve live cells, it is usually necessary to prepare a fresh and sizable inoculum to start the cell culture. This is especially important when developing large-scale processes, in which sometimes bioreactors as large as 100 m^3 are employed; otherwise, the duration of the cell culture would take so long that it would make the process unfeasible. Having defined the size of the main bioreactor and the volume of culture it will carry, and having determined an adequate inoculum volume to feed the reactor, for example, 10%, it is possible to work backwards and add a seed bioreactor with 10% of the culture volume. This reactor will require the same auxiliary equipment as the primary one, that is, blending tanks, heat sterilizers, air filters, etc. If the seed reactor is also considerably large so that it requires an inoculum volume that cannot be neglected, the procedure is repeated: a smaller seed bioreactor is added ahead of the previous one with, for example, 10% of culture volume. Generally, when the inoculum is small enough to be produced in a laboratory setting, the inoculum preparation can be neglected. In terms of modelling, the same cell growth models applied to the main bioreactor, described in the next section, may also be applied to seed bioreactors; roughly speaking, each seed bioreactor may be simulated in the same way as the main bioreactor.

22.2.3 Bioreaction Section

The Bioreaction Section is the core of the bioprocess, in which the major biological transformation of the process occurs. That transformation may consist of either a (cell-free) enzymatic reaction or a cell culture in which live cells grow and generate the desired product. In the former case, the Bioreaction Section is comprised of a biocatalytic (enzymatic) reactor as well as the tanks containing the enzyme and the substrates that feed the reactor. In the case of cell culture, a bioreactor appropriate for the cultivation of microbial cells or animal cells is used, as well as the equipment necessary to prepare and sterilize the culture medium and airflow that feed the reactor. Moreover, the flowsheet of this section in the simulator may include equipment to sterilize the airflow that *leaves* the bioreactor, for biosafety reasons.

In order to simulate and economically evaluate the bioprocess, it is necessary to quantitatively model the biological transformation that happens in the Bioreaction Section, so that the amount of each substrate consumed and of each product generated can be calculated. Two basic types of mathematical model are ordinarily used: stoichiometric and kinetic models. A stoichiometric model is simply a representation of the biotransformation as a chemical equation. In case of an enzymatic reactor, this is often relatively straightforward; for example, let us consider the process of converting sucrose into glucose and fructose using the enzyme invertase

$$1\,\text{sucrose} + 1\,H_2O \rightarrow 1\,\text{glucose} + 1\,\text{fructose}$$

In this case, all the molar coefficients are equal to one; together with the molar weights of each molecule, we can also write the equation in terms of mass:

$$342.30\,[g]\ \text{sucrose} + 18.02\,[g]\ H_2O \rightarrow$$

$$180.16\,[g]\ \text{glucose} + 180.16\,[g]\ \text{fructose}$$

Note that the invertase enzyme is not actually consumed or produced in the process since it functions as a catalyst.

Associating the stoichiometric equation with yield data (which may be experimental or theoretical), it is possible to calculate the amount of glucose and fructose produced for a given amount of sucrose, or conversely, it is possible to calculate the amount of sucrose required to produce a certain amount of glucose and fructose. For instance, let us assume that we intend to produce 180.16 g of glucose + 180.16 g of fructose and that we wish to know the amount of sucrose necessary to be fed to the reactor. Moreover, let us suppose that laboratory data indicate that the extent of the invertase reaction is equal to 80%. The extent of reaction can be defined as follows:

$$\text{Extent of reaction} = \frac{(\text{moles of glucose produced})}{(\text{moles of total sucrose})}$$

Let x be the unknown amount of sucrose. Substituting the value of the extent of reaction and that of the amount of glucose (180.16 g = 1 mol), we have:

$$0.80 = \frac{1\ \text{mol of glucose}}{x\ \text{moles of total sucrose}}$$

$$x = 427.88\ \text{g of total sucrose}$$

That is, 427.88 g of sucrose would have to be fed to the reactor to produce 180.16 g of glucose and 180.16 g of fructose in this case. Note that a certain amount of sucrose (85.58 g) is not consumed in this case, thus remaining in the solution together with the glucose and the fructose at the end of this bioreaction.

In the case of biotransformation involving live cells, not just one reaction, but thousands of reactions occur in the reactor, inside each cell. Nevertheless, it is often possible to represent this kind of process through a stoichiometric equation as well, employing an empirical formula for the cells in question. The bacterium *Escherichia coli* can be represented by the formula $CH_{1.8}O_{0.5}N_{0.2}$, for example. As such, the aerobic growth of *E. coli* in a medium containing glucose as the carbon source and NH_3 as the nitrogen source may be represented by the following equation:

$$\nu_1 C_6H_{12}O_6 + \nu_2 NH_3 + \nu_3 O_2 \rightarrow \nu_4 CH_{1.8}O_{0.5}N_{0.2}$$

$$+ \nu_5 H_2O + \nu_6 CO_2$$

where the coefficients ν_j function as stoichiometric coefficients of a chemical reaction. Assuming that the yield of biomass (dry cell weight) with respect to glucose is known and equal to 0.50 g of cells/g of glucose, we may rewrite the equation above as:

$$1\,[g] C_6H_{12}O_6 + \nu_2 NH_3 + \nu_3 O_2 \rightarrow 0.5\,[g] CH_{1.8}O_{0.5}N_{0.2}$$

$$+ \nu_5 H_2O + \nu_6 CO_2$$

Four stoichiometric coefficients are unknown in that equation. However, the principle of conservation of atoms asserts that the amounts of carbon, hydrogen, oxygen, and nitrogen have to remain constant in the process. Consequently, four conservation equations can be written, one for each type of atom, e.g. for the carbon atoms:

$$\text{mass of C in glucose} = \text{mass of C in biomass}$$

$$+\text{mass of C in } CO_2$$

Given that the atomic weights of C, H, O and N are 12.01, 1.01, 16.00 and 14.01 g/mol, respectively, we can calculate the empirical molar weight of the biomass as 24.63 g/mol. The molar weights of glucose and CO_2 are 180.16 and 44.01 g/mol, respectively. Thus, we can rewrite the equation above as:

$$1*\left(6*\frac{12.01}{180.16}\right) = 0.5*\left(1*\frac{12.01}{24.63}\right) + \nu_6*\left(1*\frac{12.01}{44.01}\right)[g]$$

$$\nu_6 = 0.57\,g$$

By performing a similar procedure for the other three atoms, we can calculate all the unknown coefficients and finally write the stoichiometric equation as follows:

$$1.00\ C_6H_{12}O_6 + 0.07\ NH_3 + 0.38\ O_2 \rightarrow$$

$$0.5\ CH_{1.8}O_{0.5}N_{0.2} + 0.38\ H_2O + 0.57\ CO_2\ [g]$$

Or, if we multiply the equation above by the molar weight of glucose (180.16 g/mol):

$$180.16\ C_6H_{12}O_6 + 12.46\ NH_3 + 69.10\ O_2 \rightarrow$$

$$90.08\ CH_{1.8}O_{0.5}N_{0.2} + 68.55\ H_2O + 103.08\ CO_2$$

As in the case of the enzymatic reactor, the equation above will allow the simulation software to calculate the amount of each substrate consumed and that of each product generated during the bacterial culture. Besides, the extent of reaction may also be defined to account for an incomplete growth process.

In addition to stoichiometric models, biotransformations can be represented by kinetic models. These models take into account the time dependence of the biological process and can be useful if kinetic data, such as reaction or growth rates, as well as yield data, are available. In the case of enzymatic reactors, a simple and common kinetic model that can be applied to many monosubstrate reactions is that proposed by Michaelis and Menten, which can be represented as follows:

$$E + S \leftrightarrow E\text{-}S \rightarrow E + P$$

where E is the enzyme, S is the substrate and P is the product of the reaction. Considering certain simplifying assumptions, it can be demonstrated that the enzymatic reaction rate is given by the following formula:

$$v = v_{\max} * \frac{[S]}{(K_m + [S])}$$

in which v is the reaction rate, $v_{\max}$ is the maximum reaction rate, $[S]$ is the concentration of the substrate in the solution, and K_m is a constant that depends on the affinity between the enzyme and the substrate in question. Consequently, if the parameters $v_{\max}$ and K_m are known, the equation above can be numerically solved, and thereby, the amount of substrate consumed and that of product generated can be calculated for a given reaction duration t.

A kinetic model very similar to the Michaelis–Menten equation also exists for cell growth, the so-called Monod equation of microbial growth. It describes well the growth of cells such as bacteria, yeast or even animal cells, and consists of the following equations:

$$\mu = \mu_{\max} * \frac{[S]}{(K_s + [S])} \tag{22.1}$$

where μ is the specific growth rate, $\mu_{\max}$ is the maximum specific growth rate, $[S]$ is the concentration of the limiting substrate (often the carbon source), and K_s is a constant that depends on the microorganism and the substrate in question. In this case, the relationships between the reaction rate, the substrate S and the product X (the biomass) are not so obvious; in fact, to describe these relationships, we must develop mass balances of S and X, such as follows:

$$\frac{dX}{dt} = \mu * X \tag{22.2}$$

$$-\frac{dS}{dt} = \frac{1}{Y_{X/S}} * \mu * X \tag{22.3}$$

where $Y_{x/s}$ is another parameter, the yield of biomass on the substrate. Thus, if the three parameters $\mu_{\max}$, K_S and $Y_{x/s}$ are known, Equations 22.1–22.3 can be used to calculate the amounts of S and X over time.

It is worth noting that simulation software specialized in bioprocesses, such as SuperPro Designer, already has these stoichiometric and kinetic models built-in. However, it falls on the user to fill in the reaction components, coefficients and/or kinetic parameters so that the program is able to calculate the amounts of substrates consumed and products generated in the Bioreaction Section.

Besides the reactor, the Bioreaction Section typically requires a blending tank in which the substrates are mixed prior to the introduction into the reactor. In the case of cell cultures, these substrates (the culture medium) also usually go through a heat exchanger to ensure their sterilization before coming into the bioreactor. If the cell culture is aerobic, an air compressor may also be added ahead of the bioreactor, and an air filter may be introduced between the compressor and the reactor so that the reactor will be fed with sterile air (or even oxygen). Finally, another air filter may be added right after the gas outlet of the reactor, so that the gas effluents (primarily CO_2 and N_2) are also sterilized (for biosafety concerns).

22.2.4 Downstream Section

The Downstream Section is usually the most variable one, due to the multitude of products, properties and purity requirements found in the biotechnological industry. Nevertheless, some unit operations are quite common, such as centrifugation, homogenization, cross-flow filtration and chromatography. The selection and ordering of these unit operations usually follow certain rules of thumb, as summarized by Harrison et al. (2015):

1. Remove the most abundant impurities (including water) first;
2. Remove the impurities easiest and cheapest to separate first;
3. Choose the unit operations that exploit the largest differences in properties between the product and its impurities;
4. Select and sequence unit operations with different driving forces.

As a consequence of these principles, the overall structure of the Downstream Section generally follows one of three patterns, depending on whether the product is intracellular, extracellular or enzymatic (cell-free process):

1. *Intracellular product*: Generally, the first step is the harvesting (concentration) of cells, that is the removal of water, typically performed by centrifugation or microfiltration. The following step, cell disruption, is usually carried out in biopharmaceutical processes using a high-pressure homogenizer, which in fact requires a minimal concentration of cells to work properly. Then, the cell debris is separated from the liquid phase containing the product, again using centrifugation or ultrafiltration. Next, the product is usually captured (concentrated) and major impurities are removed, using operations such as ultrafiltration, precipitation, solvent extraction and/or aqueous two-phase liquid extraction. Subsequently, chromatographic steps are employed to remove minor impurities. Finally, the product is subject to polishing steps of purification and further concentration procedures, often related to the product formulation and presentation form; crystallization, ultrafiltration or freeze drying are typical examples.
2. *Extracellular product*: As in the previous case, the first step is to separate the cells from the liquid phase; here, however, the product is present in the liquid phase, so that one can proceed directly to the first operations of product capture (concentration) and purification. The remaining steps follow the same structure as described above.
3. *Enzymatic product (cell-free process)*: In this case, no cell harvesting is necessary; one can proceed directly to the first steps of capture and purification. The remaining steps stay the same as well.

It becomes clear that, in general, the Downstream Section of an intracellular product is generally more extensive and elaborate than that of an extracellular product, and that the latter is generally more extensive and elaborate than that of a cell-free enzymatic product.

In any case, it should be stressed that each unit operation in the Downstream Section will require unique data and assumptions to be modelled and simulated. For instance, let us assume a scenario in which an extracellular product is produced by yeast. In order to remove the yeast cells, we may consider the use of a disc-stack centrifuge or a microfilter. In the case of centrifugation, we may directly assume a certain percentage of solid (cells) removal or provide data on the diameter and density of the cells to a simulator, which will thereby calculate the level of cell removal using a proper sedimentation equation. If, however, microfiltration is selected, then parameters such as the rejection coefficient of each component in the mixture (e.g. RC = 100% for whole cells) and the ratio of feed/retentate must be determined. For detailed explanations and models of downstream operations, we recommend references like Doran (2013) and Harrison et al. (2015).

22.2.5 Process Flow Diagram and Process Simulation

Once the major unit operations that comprise the bioprocess are defined, one can draw a process flow diagram, which is simply a graphical representation of all the unit operations, in order, connected by the appropriate material streams. Examples of process flow diagrams are given in the next section of this chapter. Process flow diagrams can be created in modern process simulators such as SuperPro Designer. Moreover, if the user provides the components and major parameters of each unit operation, such as medium composition and quantities, stoichiometric coefficients, centrifugation parameters, etc., to the simulator, the program is able to perform mass and energy balances throughout the process, so that the correct amounts of every component, in every process stream, become known. In particular, it is possible to discover the amount and concentration of the product in the final stream. It is worth noting that the size of each piece of equipment may be either defined in advance by the user or automatically calculated by the software from the amounts of raw materials and process parameters introduced by the user.

22.2.6 Economic Analysis

Estimating the operating costs, capital costs, profits and other economic information concerning a prospective bioproduct is usually vital when designing a bioprocess. In order to perform such analysis, a process simulator is also extremely helpful, since the software can relate economic data such as costs of raw materials, equipment, labour and waste treatment, with all the process data, such as the amounts of raw materials and waste streams, type and size of each piece of equipment and duration of each operation. In fact, process simulators often have built-in economic data banks and models. However, it is incumbent on the user to check, complete and update those data, e.g. by providing the prices of raw materials that are not available in the program, introducing labour rates and taxes appropriate for the country and industry under consideration, adjusting prices to inflation, etc. Consequently, significant work on data collection is necessary when conducting an economic analysis. Next, we describe the two major elements of economic analysis: cost analysis and profitability analysis.

22.2.7 Cost Analysis

The costs associated with a manufacturing process can be divided into two major categories: capital costs and operating costs. Capital costs are related to the investment necessary to build the facility, including the purchase and installation of equipment, as well as the start-up and validation of the plant. Operating costs account for ongoing production costs such as those of raw materials, labour, utilities, etc. Next, we briefly list the cost components of each category, indicating how they can be calculated or estimated.

Capital Costs

- *Direct fixed capital (DFC)*: Includes the purchase and installation costs of all the equipment employed in the bioprocess, such as reactors, storage tanks, filters, centrifuges and chromatographic columns; all the process piping and electrical installation; and the cost of buildings, engineering and construction. The purchase cost of equipment is quite difficult to come by, given that equipment is often custom-made and that equipment manufacturers do not openly provide their prices. However, some simulators do have built-in models that estimate the cost of equipment based on their size, material of construction and other characteristics. Note that these models will apply to a certain time and place, and therefore may significantly differ from real prices. The other components of the DFC cost can be estimated by multiplying appropriate factors to the equipment purchase cost.
- *Working capital*: Corresponds to the money the company needs to maintain its day-to-day operations. It can be estimated by applying a multiplier to the DFC cost or by calculating the amount of resources (raw materials, utilities, labour, etc.) necessary to run the plant for a short period of time, e.g. 30 days.
- *Start-up and validation costs*: The process of starting up a new plant or production process can be extremely complex and critical, especially when dealing with a novel and intricate process. Consequently, these costs may need to be taken into consideration. They may be estimated as a fraction of the DFC cost.
- *Up-front R&D and royalties cost*: Comprises the costs of R&D and, in the case of using a patent from another corporation or entity, the royalties that are due, before the beginning of production.

Operating Costs

- *Costs of raw materials*: Encompasses the costs of all the chemical reagents used in the process, including the components of culture media, buffer solutions and cleaning solutions.
- *Labour cost*: Comprises the cost of all the personnel related to manufacturing, particularly equipment operators. Moreover, the time dedicated by each operator to each piece of equipment has to be taken into account. Generally, the average labour rate (plus benefits) of an operator is introduced, from which the cost of the managerial and other administrative positions is calculated as a percentage. It is worth noting that labour rates may vary significantly from country to country, and even among different regions within a single country.
- *Utilities cost*: Involves the cost of process water, electricity, cooling and heating agents such as steam and chilled water. Note that the cost of pure water for preparing culture media, chromatographic buffers, etc. is usually allocated in the cost of raw materials.
- Cost of waste treatment and disposal involves the cost of treatment and disposal of aqueous and solid wastes, including hazardous chemicals such as solvents employed in liquid extractions and chromatographic buffers and biohazardous materials such as filtration and centrifugation cakes. Generally, each waste stream is assigned an appropriate cost of treatment and disposal per unit volume/mass, so that the overall cost of waste treatment is calculated by multiplying the total volume/mass of each waste stream by the corresponding cost of treatment and disposal per unit volume/mass.
- *Cost of consumables*: Covers the cost of materials that are subject to fouling or damage over time, and thus have to be periodically discarded and replaced. Typical examples are filtration membranes, chromatographic resins, and disposable reactors. To calculate this cost, it is necessary to determine the unit cost of the consumable and its replacement frequency. Chromatographic resins, in particular, are usually very expensive and may have a considerable impact on the production cost.
- *Cost of laboratory and quality control*: Includes all the costs associated with the off-line analysis of the physical, chemical and biological properties of the final product, as well as those of raw materials and intermediate samples. In general, this component can be estimated as a fraction of the labour cost. In industries with stringent quality and regulatory requirements such as the pharmaceutical industry, this cost can be quite significant.
- *Facility overhead cost*: Comprehends the cost of depreciation and maintenance of the plant, as well as insurance, local taxes and factory expenses that are not directly related to manufacturing, such as accounting, security, cafeteria, etc. It may be estimated as a fraction of the DFC cost.
- *Miscellaneous*: Embraces R&D, validation, marketing and sales activities associated with the bioproduct. If the process employs a patent owned by another company or entity, the royalties that must be paid can also be included in this topic.

An important parameter to economically evaluate a bioprocess is the **unit production cost** of the final product. It is defined as the annual operating cost divided by the annual production rate. As such, it does not directly include capital costs.

22.2.8 Profitability Analysis

The revenue side of the economic analysis depends primarily on the selling price of the bioproduct. This value can be introduced into the process simulator, which then multiplies the selling price by the amount of product manufactured over 1 year, which translates into the **annual revenues** of the project. By relating the costs and revenues, it is then possible to evaluate how feasible and attractive a bioprocess is. Some common parameters that are used to this purpose are as follows:

- *Gross profit*: It is defined as the difference between the annual revenues and the annual operating cost.
- *Gross margin*: It is defined as the gross profit divided by the annual revenues; as such, it is a number between 0 and 1.
- *Net profit*: It is equal to the gross profit minus income taxes; the annual depreciation may be added to the net profit as well.
- *Return on investment (ROI)*: It consists of the net profit divided by the total capital costs.
- *Payback time*: It is the time required for the total capital costs to be counterbalanced by the accumulated annual profits, that is the point from which the investment actually starts to 'make money'. It is defined as the total capital costs divided by the net profit and given in a number of years

22.3 EXAMPLES OF TECHNO-ECONOMIC ANALYSIS

This section presents the examples of the production processes of two different biomolecules with the objective of consolidating the concepts presented in the previous sections. We also try to demonstrate how techno-economic analysis assists the decisions taken during the development of the process and how the interpretation of the simulation results depends, to a large extent, on the final application of the bioproduct.

The two bioprocesses we analyse here are the production of a pharmaceutical protein, the biosynthetic human insulin (BHI), previously presented by Petrides (2018), and the production of an industrial low-cost recombinant enzyme, β-glucosidase (BGL), presented by Ferreira, Azzoni and

TABLE 22.1
Description of the Production Processes of Biosynthetic Human Insulin (BHI) (Petrides, 2018) and a Thermophilic β-Glucosidase (Ferreira, Azzoni, and Freitas, 2018)

	BHI	BGL
Application	Diabetes treatment	Enzymatic hydrolysis of cellulose
Expression platform	*Escherichia coli*	*Escherichia coli*
Seed train	Two seed bioreactors	Two seed bioreactors
Main bioreactor type	Stirred tank	Stirred tank
Operation mode	Batch	Fed-batch
Production bioreactor volume	$50\,m^3$	$100\,m^3$
Fermentation time/batch	18h	30h
Biomass concentration	30 gDCW/L	100 gDCW/L
Downstream processing steps	Three sections (26 unit operations)	One section (4 unit operations)
Overall recovery yield	32%[a]	90%[b]
Batches/year	160	264
Production/batch	10.9kg	334.4kg

a Amount of BHI in the final product stream related to the amount of the proinsulin contained in the chimeric protein (Trp-Lc-Met proinsulin) synthesized by E. coli.
b Amount of BGL obtained in the final product stream related to the BGL amount in the lysate.

Freitas (2018). In both cases, the simulation was performed using the software SuperPro Designer (Intelligen Inc., USA). The main characteristics of each process are summarized in Table 22.1, and they will be used to guide our analysis. However, it is recommended that the papers be consulted in their entirety aiming at a greater understanding and learning about the requirements of each process, which directly affected the results obtained in the techno-economic analysis.

Insulin is a protein that plays a key role in the regulation of carbohydrate metabolism and is the main drug for the treatment of diabetic patients. Currently, recombinant human insulin has been predominantly produced using *Escherichia coli* (*E. coli*) and *Saccharomyces cerevisiae* for therapeutic use in humans (Baeshen et al., 2014). The production process developed by Petrides (2018) employs *E. coli*, and its process flow diagram is presented in Figure 22.2. The process starts with two stages of seed bioreactors

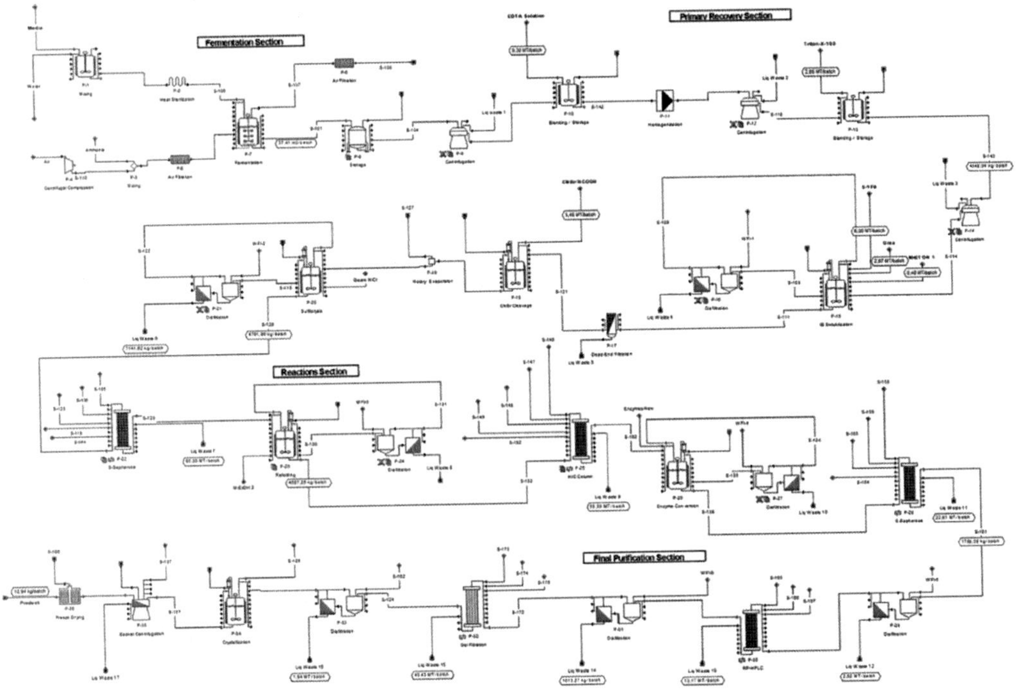

FIGURE 22.2 Production process of biosynthetic human insulin in *Escherichia coli* (Petrides, 2015). (This figure was modified from the Examples folder of SuperPro Designer with the authorization of Dr. Demetris Petrides (Intelligen Inc, USA).)

(not considered in Figure 22.2) to inoculate the 50-m^3 production bioreactor. The bioreactor operation is in batch mode and the fermentation time is 18 h achieving a bacterial cell concentration of 30 g/L (dry cell weight). The product synthesized is a chimeric protein, Trp-LE′-MET-proinsulin, which is expressed intracellularly and corresponds to 80% of the total amount of the inclusion bodies. The author assumed that the inclusion bodies represent 20% of the total dry cell mass. At the end of this step, the cell suspension undergoes a long sequence of downstream steps divided into three sections: primary recovery, reactions and final purification. The downstream processing steps are designed to ensure that active insulin is recovered and that the requirements of purity level and good manufacturing practices for biopharmaceutical products are achieved.

The other case study is the production process of BGL designed by Ferreira, Azzoni and Freitas (2018), whose process flow diagram is presented in Figure 22.3. BGL is an enzyme responsible for the hydrolysis of β-1,4-glucosidic linkages, and it has been recognized as a key enzymatic activity for the breakdown of lignocellulosic biomass into fermentable sugars for biorefinery purposes. This enzyme is an important input in the second-generation ethanol production processes, and therefore, the low cost of the enzyme becomes mandatory in order to avoid increasing the final ethanol cost (Singhania et al., 2013). In this process, *E. coli* is considered an appropriate host since the BGL gene selected was obtained from the genome of a bacterium, *Thermotoga petrophila* (Haq et al., 2012). The cultivation was conducted in a fed-batch mode to achieve high cell densities (ca.100 g/L dry mass) and minimize the formation of acetic acid, which is one of the metabolites primarily responsible for the inhibition of bacterial growth. The recombinant BGL is expressed intracellularly, and therefore, the first step of downstream is cell disruption through a high-pressure homogenizer. The enzyme present in the cell lysate does not require a significant purity for its application. However, the product concentration followed by the buffer exchange is necessary in order to ensure the stabilization of the enzyme.

22.3.1 Comparison of the Insulin and BGL Production Processes

By analysing Figures 22.1 and 22.2, one can easily notice the difference in the complexity of the downstream process stage. While the BGL production process requires 4 unit operations for recovery and purification, the BHI production requires 26 unit operations to ensure both the recovery of the protein with therapeutic efficacy and the high-purity level adequate for a biopharmaceutical product. Based on Table 22.1, the BHI production scale is about two times smaller than that used for BGL production, considering the volumes of the main reactors of each plant. However, the annual production rate is 52 times lower. Although both processes are based in the same microbial platform for expression (*E. coli*), the BHI biosynthesis step, differently from the BGL process, was designed to be conducted in a batch mode, resulting in a low-to-moderate cell density. Furthermore, based on the author's assumptions, the insulin precursor represents only 30% of the chimeric protein (Trp-LE′-MET-proinsulin) synthesized by *E. coli* and the chimeric protein represents 80% of the amount of inclusion bodies accumulated in the cytoplasm. As a recombinant protein expressed intracellularly, the product yield is a consequence of the biomass yield. Therefore, the low production rate could be attributed to the very low overall yield of pure insulin obtained after all the processing steps, including the low yield in the biosynthesis step.

Regarding the cost analysis of the two processes (Table 22.2), the capital invested in the BHI production plant and the annual operating cost are, respectively, around three times and four times higher when compared to the BGL production plant. Moreover, the production of 1 kg of purified crystals of recombinant insulin is around 200 times more expensive to produce than the same amount of recombinant BGL. Despite the high annual production cost per kg of insulin, it is important to take into account that the market price of this biopharmaceutical (including regular insulin and insulin analogs) is in the range of 24,750–100,000 dollars per kg (Gotham et al., 2018) so that the BHI production

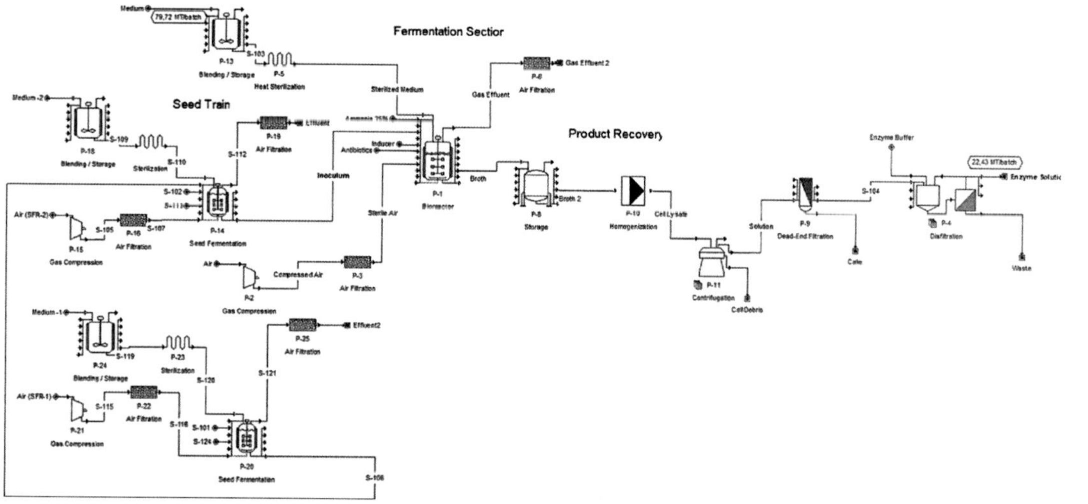

FIGURE 22.3 Production process of recombinant β-glucosidase in *Escherichia coli* (Ferreira, Azzoni, and Freitas, 2018).

TABLE 22.2
Cost Analysis and Breakdown of Annual Production Costs for the Production Process of Biosynthetic Human Insulin (Petrides, 2018) and Recombinant β-Glucosidase (Ferreira, Azzoni, and Freitas, 2018)

	Insulin	BGL
Capital investment ($)	202,527,000	70,783,000
Operating cost ($/year)	110,795,000	27,916,000
Unit production cost ($/kg)	64,630.00	316.21
Annual Operating Cost (AOC) Breakdown per Section		
Upstream	-	11.0%
Fermentation	7.5%	43.0%
Downstream	92.5%	46.0%
Annual Operating Cost (AOC) Breakdown per Item		
Facility dependent	25.60%	45.0%
Consumables	32.60%	23.0%
Labour dependent	7.50%	4.0%
Raw materials	25.7%	25.0%
Laboratory/QC/QA	1.88%	1.0%
Utilities	0.22%	2.0%
Waste treatment/disposal	6.43%	0.0%

process analysed here can be considered profitable at the highest price level per kg of product.

In the case of BGL, the cost obtained in the economic analysis of the baseline scenario was quite high for the cellulase market, and because of this, the authors report a set of alternative scenarios, obtained by simulation, aimed at reducing costs reaching a final value of US $37/kg. Despite the drastic reduction of the production cost of this enzyme obtained in the most optimized scenario, it is still considerable when compared, for example, to the amylase used in the ethanol industry, whose market price is US $25/kg (Lynd et al., 2017).

The distribution of the annual operational costs (AOC) per section is also very different between the two plants (Table 22.2). In the case of BHI, the Downstream Section is responsible for 92.5% of the AOC, whereas in the case of BGL production, this value is 46.0%. Interestingly, the Upstream Section accounts for 11% of the AOC in the BGL production, and it was considered negligible in the insulin production. In fact, the cost associated with seed propagation is commonly neglected in the case of production processes of high-value bioproducts, but as we can see, it can make a considerable economic impact on low-cost bioproducts.

The distribution of AOC per item shows that the portion related to facility-dependent costs is by far the highest cost in the BGL process, followed by the raw materials and consumables in third place. This result is attributed to the cost of the bioreactor, the cost of the carbon source (glucose) and the inducing reagent (IPTG) used in the Fermentation Section, and also due to the micro- and ultrafiltration membranes used in the Downstream Section.

In the case of BHI production, the cost related to consumables is the most relevant followed by raw materials and facility-dependent costs in third place. The high impact of consumables to the AOC is attributed to the chromatographic resins (five different types of chromatography are used in the Downstream Process Section) and the filtration cartridges required to achieve the purity level of the insulin crystals.

Based on the examples discussed, it is possible to identify the technical bottlenecks and the factors that most impact the economic viability of a designed process from the detailed analysis of the data generated from the simulation. By simulating different process scenarios and configurations, such as reducing the number of unit operations, replacing expensive unit operations, replacing or recycling expensive raw materials and reagents, or even using more productive microbial strains, it is possible to improve the understanding about the process economics with the goal to achieve improved process designs.

22.4 BEYOND TECHNO-ECONOMIC ANALYSIS

Beyond the cost of producing a biopharmaceutical, evaluated through techno-economic analysis, the selling price of a medicine depends on other factors, such as competitor strategies, public opinion and political pressures. For health-care systems, the value of a medicine is not considered only by the pharmaceutical price, but also by patient health gains for the money spent. This is the principle of pharmacoeconomics.

Pharmacoeconomics studies proved to be an important tool that can be applied to help strategic and operational decisions regarding pharmaceutical development, production and market insertion. It is also increasingly important to prove not only the product's safety, efficacy and quality (the first, second and third hurdles) but also the clinical effectiveness and cost-effectiveness (the fourth hurdle) for regulatory agencies. The fourth hurdle, which includes pharmacoeconomic analysis, is usually described as what is needed to be done to gain market access and assure success for medical technology, pharmaceutical or biotech product (Paul and Trueman, 2001).

According to the last data reported by the World Health Organization in 2018, the domestic general government health expenditure has been growing, achieving 3.5% of world gross domestic product (GDP) on average. Considering United States of America and United Kingdom, the spends achieved 8.5% and 7.9% of GDP according to the last report (World Health Organization, 2018). In this context, pharmacoeconomics is an important tool for the rational evaluation of health technologies, including biopharmaceuticals, aiming better decision-making in the use of different medicines and drug therapies. This field of economic analysis has raised in the past 20 years, and its application helps to better target the resources employed in health (Garrison, 2013).

22.5 COSTS CONSIDERATIONS IN BIOPHARMACEUTICALS

As discussed in the techno-economic analysis, costs are important for the estimation of resources or inputs that are necessary for the production of a product or service. In this context, biopharmaceutical products (innovative and biosimilar) have important differences in the production process, regulatory pathway, intellectual property protection and marginal costs when compared to small-molecule drugs (Garrison, 2013). The production of biopharmaceuticals has greater complexity when compared to that of traditional synthetic pharmaceuticals, usually resulting in higher costs of therapies. The cost of biological drugs fall outs from the riskiness of the biopharmaceutical development process. Biopharmaceutical industries spend approximately 25% of revenues on research and development, whereas about 20% is invested by the traditional pharmaceutical industry, making the biopharmaceutical industry the most R&D-intensive industry in the US economy (Eldeman, 2004; Lamattina, 2018). Other price considerations exist once biopharmaceutical is on the market, due to the nature of these drugs. Biopharmaceuticals present are usually not so stable as synthetic drugs, frequently needing special handling and adequate administration procedures. Furthermore, many of the biopharmaceutical therapies are target to relatively uncommon diseases or for a small subset of patients with a highly prevalent condition such as asthma disease, significantly raising the costs (Eldeman, 2004).

Considering all the elements involved in the development, production, marketing, distribution and use, the final price of a biopharmaceutical must be carefully defined, and pharmacoeconomics can play an important role. The use of a higher-cost biopharmaceutical product does not necessarily lead to cost offsets from decreased hospital days, physician visits, or other utilization of the health-care system. However, sometimes a higher-price medicine can be considered advantageous if ultimately it enhances the patient's quality of life and daily living, and helps to decrease hospital admissions, as for an example. In this direction, improvements in the process such as the change of an adjuvant, which reflects in higher-quality product, cost offset of the treatment (less adverse effects, hospitalization, etc.) and better patient's quality of life, should be done, even if in a first step, in a techno-economic analysis this improvement impacts significantly in the final medicine price. By performing pharmacoeconomic studies, one can evaluate if a higher-price medicine is advantageous or not, considering the outcomes beyond the shelf price.

22.6 PHARMACOECONOMIC ANALYSIS

The term 'pharmacoeconomics' is defined as the description and analysis of the several costs involved in drug therapy, regarding the health-care systems and the society (Ray, 1979; Trask, 2011). It categorizes, measures and compares the cost and outcomes of pharmaceutical products and other health technologies. Physicians, health managers and other decision-makers can use pharmacoeconomic evaluation to assess and compare the total costs of treatment options and possible outcomes associated, aiming to make use of resources more efficient and effective (Rascati, 2014). Pharmacoeconomic analysis can be posed from a variety of perspectives, such as the societal perspective, the payer perspective, the provider perspective or the patient perspective, reflecting the differing values and incentives of the individuals or institutions conducting or using the analysis (Garrison, 2013).

There are different methods for pharmacoeconomic evaluation, all of them can be used to pharmaceuticals. These methods are cost-minimization analysis (CMA), cost-effectiveness analysis (CEA), cost-benefit analysis (CBA) and cost-utility analysis (CUA).

- *CMA*: By using this method, one can measure and compare input costs, assuming that outcomes are equivalent. As such, there is a limitation of the types of interventions that can be evaluated (Rascati, 2014). Therefore, when comparing medicines with similar composition, dosage and pharmaceutical properties, only the costs of the medicine itself need to be compared, since outcomes should be the same. This method is often used to evaluate brand versus generic drugs or generic drugs made by different companies (Rascati, 2014).
- *CEA*: According to Rascati (2014), CEA is the most common pharmacoeconomic analysis. It measures the cost in monetary value (i.e. dollar) and outcomes in natural health units (i.e. cure, lives saved or blood pressure reduction), which indicate an improvement in health. The outcomes must be measured in the same clinical unit for both products evaluated, making possible their comparison. A graphical depiction of CEA comparison is presented as a cost-effectiveness plane (Figure 22.4). The point where the x- and y-axes cross indicates the costs and effectiveness starting point for the standard comparator.

  For the visual representation of a CEA comparison between a new intervention and an existing standard, a point is placed on the plane comparing the cost (y-axis) and effectiveness (x-axis) (Figure 22.4). That means, comparing to the existing standard that is the central point of the graph, if the new intervention is more or less effective and if it has cost-increasing or cost-saving, the Quadrant I, II, III or IV can be pointed. For example, if the new intervention is more expensive and more effective than the standard one, this point would be at Quadrant I.

  The trade-off between the increase in cost and the increase in benefits must be considered when the point of the new intervention is at Quadrant I. Similarly, if the new intervention is at Quadrant III, the following trade-off would have to be considered whether the costs saved by the alternative compensate its decreased ineffectiveness. In case the point falls in Quadrant II, the new intervention dominates the standard alternative and its use is advantageous in cost and effect. However, if the point is at Quadrant IV, the new intervention is disadvantageous in cost and effect, and therefore is dominated by the existing standard and its use is not recommended (Cohen and Reynolds, 2008; Rascati, 2014).
- *CBA*: It is the most widespread method aiming to compare alternatives, since it creates a common measurement for all costs and benefits, using monetary units, thus allowing to compare several

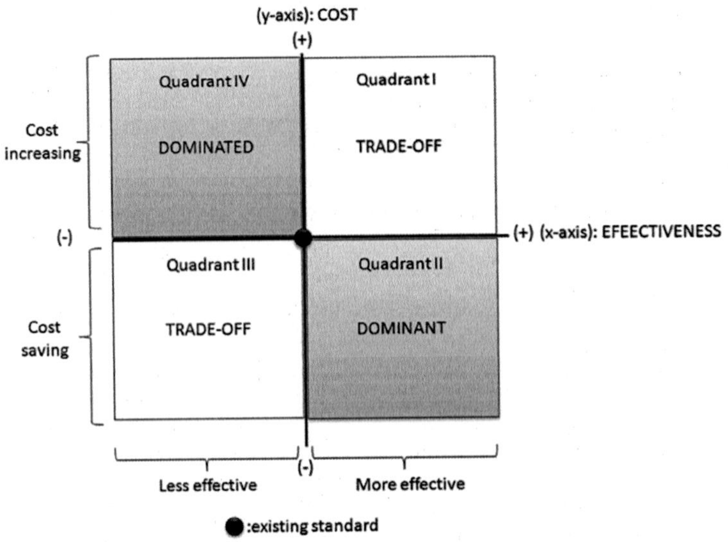

FIGURE 22.4 Cost-effectiveness plane. (This figure was constructed based on the work presented by Cohen, D.J., Reynolds, M.R., *J. Am. Coll. Cardiol.*, 52(25), 2119–26, 2008.

outcomes. The concept of this analysis originated in the United States in the 1930s, based in welfare economics, and then was applied in many fields, including pharmacoeconomics. CBA permits the identification, assessment and also comparison of costs and benefits of the new technology. These are expressed as a ratio (a benefit-to-cost ratio), a net cost or a net benefit (Willians, 2008). A decision would be made choosing the treatment option that presents the highest net benefit or the greatest benefit-to-cost (B:C) ratio (Trask, 2011):

B:C ratio > *1*: The benefits realized by new technology outweigh the cost of providing it.
B:C ratio = *1*: The benefits realized by the new technology are equivalent to the cost of providing it.
B:C ratio < *1*: The cost of the new technology outweighs the benefits realized by it.

The disadvantage of this method is the difficulty of giving economic values on medical outcomes, and there is not a universal agreement on one standard method for their accomplishment (Rascati, 2014).

- *CUA*: This analysis is recommended when the new technology possibly will not alter the life expectancy of the patient but still offer some value beyond reduction or avoidance of side effects or symptoms, increase in the quality of life, etc, that can alter both the quantity and quality of life, significantly (Cohen and Reynold, 2008). This analysis can also evaluate and compare quality, cost and the number of patient-years. The therapeutic outcome is measured in patient-weighted utilities (instead of physical units, used in CEA), and the cost is measured in monetary value. CUA is considered as a subset of CEA in which the outcomes are measured by the quality-adjusted life year (QUALY), a special type of clinical outcome measure. CUA takes patient preferences (utilities) when measuring health consequences. QUALY incorporates quality (morbidity) and quantity (mortality) of life. Other unities as disability-adjusted life years (DALYs) and healthy-years equivalents (HYEs), among others, are less frequently used in this analysis (Rascati, 2014). In economic theory, gold standard methods of directly eliciting utilities from patients are well established, but it is difficult and time-consuming to apply in practice (Cohen and Reynold, 2008). Similarly to CE, CUA is expressed as a ratio (the cost-utility ratio C:U), which can be defined as the cost per QUALY gained (or another health-state utility measurement such as DALY, HYE). The disadvantage of this analysis comes from the high degree of subjectivity of QUALY and other utility measures, causing some disagreement among specialists about which scales should be preferred for measuring utility (Trask, 2011).

22.7 BIOPHARMACEUTICAL PHARMACOECONOMIC STUDY EXAMPLE

A pharmacoeconomic study of a biopharmaceutical that exemplifies the importance of this analysis when an alternative product is evaluated in comparison with a standard one was presented by Kurre et al. (2002). The authors compared the use of a therapeutic enzyme, the PEGylated asparaginase (PEGasparaginase), with native *E. coli* asparaginase, as a therapeutic drug for treating acute lymphoblastic leukaemia. The authors performed a CBA and considered societal costs (lodging, missed workdays, food, babysitter, transportation) and payer (frequency of encounters) cost data. Further payer costs, such as cost per clinic visit, medicine costs and cost per patient day stay, were collected from patients and participating institutions (Kurre et al., 2002). Even though the price of the PEGasparaginase can be much higher than that of native asparaginase (6,572% per unit of enzyme: PEGasparaginase costs $1,183.00 for a 3,750-unit vial, whereas native asparaginase costs $4,800.00 for a 10,000-unit vial – prices from 1998 when the study was performed), considering the reduced number of doses, side effects episodes and clinic visits, the total patient costs of PEGasparaginase are similar to those of native *E. coli* asparaginase.

Despite the expressive higher cost per dose of PEGasparaginase compared with *E. coli* L-asparaginase, the overall costs of both therapies were relatively comparable, because medicine cost represented only a small part of the total costs, being approximately 5% for PEGasparaginase patients and 4% for *E. coli* L-asparaginase patients (Kurre et al., 2002). This is an interesting example of how pharmacoeconomics studies can be applied to aid administrators to take strategic and operational decisions, especially in the context of the recent biobetter[1] developments in the biopharmaceutical industry.

22.8 FINAL CONSIDERATIONS

As discussed above, the creation of a process flow diagram and simulation analysis allows the identification of the critical unit operations and the bioprocess optimization. New process configurations and/or conditions can be experimentally validated and tested on a pilot scale, making possible the development of economically viable large-scale process. In a broader view, pharmacoeconomics also play a very important role in the challenging task of definition of the final price of a biopharmaceutical.

REFERENCES

Baeshen, N.A.; Baeshen, M.N.; Sheikh, A.; Bora, R.S.; Ahmed, M.M.M. Cell factories for insulin production. *Microbial Cell Factories*, v. 13, n. 141, 2014.

Brumano, L.P.; Silva, F.V.S.; Costa-Silva, T-A.; Apolinário, A.C.; Santos, J.H.P.M.; Kleingesinds, E.K.; Monteiro, G.; Rangel-Yagui, C.O.; Benyahia, B.; Pessoa Jr, A. Development of L-asparaginase biobetters: Current research status and review of the desirable quality profiles. *Frontiers in Bioengineering and Biotechnology*, v. 6, 212, 2019.

Cohen, D.J.; Reynolds, M.R. Interpreting the results of cost-effectiveness studies. *Journal of the American College of Cardiology*, v. 52, n. 25, 2119–2126, 2008.

Doran, P.M. *Bioprocess Engineering Principles*. 2nd ed. Oxford: Academic Press, 2013, 919 p.

Eldeman, B. Explaining the cost of biotech therapies. *Biotechnology Healthcare*, v. 1, n. 2, 37–41, 2004.

Ferreira, R.G.; Azzoni, A.R.; Freitas, S. Techno-economic analysis of the industrial production of a low-cost enzyme using E. coli: The case of recombinant β-glucosidase. *Biotechnology for Biofuels*, v. 11, 1–13, 2018.

[1] The term 'biobetter', also called 'biosuperior', refers to new drugs designed from existing peptide or protein-based biopharmaceuticals by improving their properties such as selectivity, stability against degradation and immunogenicity (Brumano et al., 2019).

Garrison Jr., L.P. Pharmacoeconomics and Drug Pricing. In: Ho, R.J.Y.; Gibaldi, M., *Biotechnology and Biopharmaceuticals*. Hoboken, NJ: John Wiley & Sons, 2013, 125–136.

Gotham, D.; Barber, M.J.; Hill, A. Production costs and potential prices for biosimilars of human insulin and insulin analogues. *BMJ Global Healthy*, v. 3, n. 5, 2018.

Harrison, R.G.; Todd, P.W.; Rudge, S.R.; Petrides, D.P. *Bioseparations Science and Engineering*. New York: Oxford University Press, 2015.

Haq, I.U.; Khan, M.A.; Muneer, B.; Hussain, Z.; Afzal, S.; Majeed, S.; Rashid, N.; Javed, M.M.; Ahmad, I. Cloning, characterization and molecular docking of a highly thermostable β-1,4-glucosidase from Thermotogapetrophila. *Biotechnology Letters*, v. 34, n. 9, 1703–1709, 2012.

Heinzle, E.; Biwer, A.P.; Cooney, C.L. *Development of Sustainable Bioprocesses Modeling and Assessment*. Chichester, West Sussex: John Wiley and Sons, 2007, 294 p.

International Conference on Harmonisation of Technical Requirements for Registration of Pharmaceuticals for Human Use (ICH). Harmonised Tripartite Guideline. Pharmaceutical Development Q8(R2). 2009. Acesso em: janeiro de 2019. Disponível em: https://database.ich.org/sites/default/files/Q8_R2_Guideline.pdf.

International Conference on Harmonisation of Technical Requirements for Registration of Pharmaceuticals for Human Use (ICH). Harmonised Tripartite Guideline. Pharmaceutical Quality System Q10. 2008. Acesso em: janeiro de 2019. Disponível em: https://database.ich.org/sites/default/files/Q10_Guideline.pdf.

International Conference on Harmonisation of Technical Requirements for Registration of Pharmaceuticals for Human Use (ICH). Harmonised Tripartite Guideline. Quality Risk Management Q9. 2005. Acesso em: janeiro de 2019. Disponível em: https://database.ich.org/sites/default/files/Q9_Guideline.pdf.

Kurre, H.A.; Ettinger, A.G.; Veenstra, D.L.; Gaynon, P.S.; Franklin, J.; Sencer, S.F.; Reaman, G.H.; Lange, B.J.; Holcenberg, J.S. A pharmacoeconomic analysis of pegaspargase versus native Escherichia coli L-asparaginase for the treatment of children with standard-risk, acute lymphoblastic leukemia: The Children's Cancer Group study (CCG-1962). *Journal of Pediatric Hematology/Oncology*, v. 24, n. 3, 175–181, 2002.

Lamattina, J. Pharma R&D Investments Moderating, But Still High. *Forbes*. Available in: https://www.forbes.com/sites/johnlamattina/2018/06/12/pharma-rd-investments-moderating-but-still-high/#15a18bdb6bc2. Access in 20 nov 2018.

Lynd, L.R.; Liang, X.; Biddy, M.J.; Allee, A.; Cai, H.; Foust, T.; Himmel, M.E.; Laser, M.S.; Wang, M., Wyman, C.E. Cellulosic ethanol: Status and innovation. *Current Opinion in Biotechnology*, v. 45, 202–211, 2017.

Paul, J.E.; Trueman, P. "Fourth hurdle reviews", NICE, and database applications. *Pharmacoepidemiology and Drug Safety*, v. 10, n. 5, 429–438, 2001.

Petrides D.; Carmichael D.; Siletti C.; Koulouris, A. Biopharmaceutical process optimization with simulation and scheduling tools. *Bioengineering*, v. 1, 154–187, 2014.

Petrides, D. Bioprocess Design and Economics. In: Roger G.H.; Paul W.T.; Scott R.R. and Demetri P.P.; *Bioseparations Science and Engineering*. 2. ed.. Oxford University Press (2015), ISBN 978-0-19-539181-7. Avaiable in: https://www.intelligen.com/wp-content/uploads/2020/05/BioProcessDesignAndEconomics_March_2015.pdf. Access in 19 February 2021.

Petrides, D. Biosynthetic Human Insulin Production. Available in: http://www.intelligen.com/demo.html. Access in 17 November 2018.

Rascati, K.L. *Essentials of Pharmacoeconomics*. 2. ed. Baltimore, MD: Lippincott Williams & Wilkins, 2014, 310 p.

Rathore, A.S. QbD/PAT for bioprocessing: Moving from theory to implementation. *Current Opinion in Chemical Engineering*, v. 6, 1–8, 2014.

Rathore, A.S. Quality by design (QbD)-based process development for purification of a biotherapeutic. *Trends in Biotechnology*, v. 34, 358–370, 2018.

Ray, M. Pharmaco-economics. In: Remington, A., *Text Book of Pharmaceutical Science and Practice of Pharmacy*. 19. ed. Philadelphia: Lippincott Williams & Wilkins, 1979, 1931–1933.

Singhaniar, R.; Patel, A.K.; Sukumaran, R.K.; Larroche, C.; Pandey, A. Role and significance of beta-glucosidases in the hydrolysis of cellulose for bioethanol production. *Bioresource Technology*, v. 127, 500–507, 2012.

Toumi, A.; Jurgens, C.; Jungo, C.; Maier, B.A.; Papavasileiou, V.; Petrides, D. Design and optimization of a large scale biopharmaceutical facility using process simulation and scheduling tools. *Pharmaceutical Engineering*, v. 30, n. 2, 1–9, 2010.

Trask, L. Pharmacoeconomics: Principles, Methods, and Applications. In: Dipiro, J.T.; Talbert, R.L.; Yee, G.C.; Matzke, G.R.; Wells, B.G.; Posey, L.; *Pharmacotherapy: A Pathophysiologic Approach*. 8. ed. New York: McGraw-Hill, 2011, 1–3.

Willians, B. Cost-benefit analysis. *Economic & Labour Market Review*, v. 2, n. 12, 67–70, 2008.

World Health Organization. Domestic general government health expenditure (GGHE-D) as a percentage of gross domestic product (GDP). WHO Global Health Expenditure Database, 2018.

23 Perspectives for Pharmaceutical Biotechnology

Michele Vitolo
Universidade de São Paulo

CONTENTS

23.1 INTRODUCTION

Pharmaceutical biotechnology is an important discipline in modern biotechnology and has been growing intensely since the 1980s. The first great advance in pharmaceutical biotechnology came with the production of insulin through fermentation process using genetically modified *Escherichia coli*. The gene encoding insulin was cloned from β-cells of the human pancreas and stably integrated into the *E. coli* genome, which, in turn, then expressed human insulin. This approach offered an alternative supply of insulin which had previously been purified from pig pancreas, which although similar to the human hormone, invariably led to an undesirable immune response in the user.

23.2 SOME NEURODEGENERATIVE DISEASES

Amyotrophic lateral sclerosis (ALS) results from a hardening and healing of motor neuron cell bodies that are located at the side area of the spinal cord (so the term 'lateral sclerosis') and leads to a complete axon degeneration that ultimately results in atrophy and weakening of the skeletal muscles (so the term 'amyotrophic', i.e., muscle deprived of electric pulses crossing the myoneural junction). Besides ALS, other examples of neurodegenerative diseases include Huntington's disease (the damaged neurons belong to the smooth muscle), Parkinson's disease (the damaged neurons are located at the mesencephalon) and Alzheimer's disease (the damaged neurons are located in the brain cortex and hippocampus).

According to Coleman (2005), there are different mechanisms that when acting collectively lead to the degeneration of the body and axon of a motor neuron. Such mechanisms include toxic excitation (when an excess of glutamate that is released into the synapses can cause cell death), axon strangling (deposition of fibrils along the axon preventing the free flux of nutrients), neuroglial toxins (microglia cells and astrocytes, which normally surround motor neurons and provide a homeostatic environment, can produce toxins which damage the neuron), proteases (which digest the cytoskeleton, weakening the cell body of the motor neuron), breakdown of proteasome function (the intracellular organelle that degrades cytoplasmic macromolecules, but when these macromolecules are present in excess, the proteasome fails to function).

Today, there is only one medicine (Riluzole®) available on the market that has been clinically proven to delay the progression of ALS (Taylor et al., 2016).

The future contribution of pharmaceutical biotechnology in the area of neurodegenerative diseases will likely be based on therapeutics derived from the stem cell technology (which when delivered to the spinal cord should lead to the production of growth factors, such as VEGF and IGF-1 (insulin growth factor-I), which in turn would protect motor neurons) and RNA interference techniques (RNAi) (which could inhibit the translation of incorrect RNA, thus avoiding the translation of proteins or their products that damage motor neurons).

23.3 SCHISTOSOMIASIS

Schistosomiasis is a disease endemic in tropical regions and is caused by parasites of the genus *Schistosoma* (*S. mansoni*, *S. japonica* and *S. haematobium*). The disease, also commonly known as bilharzia, was estimated to affect at least 252 million people worldwide in 2015. The adult male and female worms live *in copula* inside the mesenteric veins of the intestine or bladder of humans. A female can lay around 300 eggs per day, and the immune system reacts to the eggs by producing granulomas that entrap the eggs. The granulomas promote collagen to become deposited resulting in fibrosis, which can damage organs and tissues. Schistosomiasis can drastically reduce the person's capacity to work leading to a huge medical cost to the society (McManus and Loukas, 2008).

The astonishing point is that the adult worms are not detected by the immune system, despite the worms living in the bloodstream. According to McManus and Loukas (2008), three aspects can explain this phenomenon; i.e., the immune system is inactivated by inhibitory substances produced by the worms, the worms tegument is poorly immunogenic and the worms sequester host proteins onto the tegument, camouflaging the worm from the immune system. Schistosomiasis can be treated using the drug praziquantel® and by the introduction and maintenance of adequate public sanitation programmes. The latter is fundamental because it eliminates the freshwater snail of the genus *Biomphalaria* (these snails are commonly found in lakes polluted with human excrements and act as the intermediate host for the parasite).

Unfortunately, in developing countries, sanitation is poor and so it is difficult to eliminate this parasite. It is not uncommon for a person to be cured only to become reinfected once they enter a polluted lake to work, e.g. in rice cultivation.

Pharmaceutical biotechnology can contribute to the fight against this infection through the development of vaccines, focusing on specific antigenic proteins present on the adult worm tegument (glutathione-S-transferase and tetraspanines). A vaccine against glutathione-S-transferase of *S. haematobium* is under development (McManus and Loukas, 2008).

23.4 AIDS

Similar to schistosomiasis, pharmaceutical biotechnology may in future provide a vaccine against AIDS – a worldwide endemic disease caused by HIV. Capsid proteins of HIV undergo rapid antigenic variations because of modifications introduced by reverse transcriptase during RNA conversion into the double-stranded DNA. For this reason, it is quite difficult to design a vaccine – a preventive medicine – that would be effective for immunizing people against all of the antigenic variations of HIV that exist. Nevertheless, chemotherapy for HIV infection has been transformed since the introduction of triple antiretroviral cocktail combinations that specifically inhibit several steps in virus proliferation. However, the treatment is lifelong and cannot be stopped because HIV can become dormant within inactive memory T cells (which are spread throughout the bloodstream and lymph glands), as well as in the so-called anatomical reservoirs – such as the central nervous system, gastrointestinal system and genital system – poorly reached by the immune cell system and barriers across which current drugs penetrate poorly.

Hence, the eradication or total cure for HIV infection is still some way off. Nevertheless, pharmaceutical biotechnology offers real perspectives in fighting AIDS, insofar as new therapeutic targets are being identified. NEIL (2008) proposed the following targets: (a) inhibition of infective viral factor (VIF) that inhibits protein A3G (abundant in macrophage and lymphocytes) that will otherwise cause irreversible damage to the newly synthesized viral DNA; (b) inhibition of lens epithelium-derived growth factor (LEDGF), a normal cell protein which is used by the virus to stimulate integrase, another normal cell enzyme, to integrate newly synthesized viral DNA into the cell genome; (c) alteration of infected T-cell chromatin in order to translate proteins that are trafficked to the membrane of the cells and which stimulate the immune system to destroy the infected cells; and (d) inhibition of the viral protein 'U' (V_{pu}), which is fundamental in liberating new virus particles from T cells (Neil, 2008).

Conn et al. (2009) have suggested fighting HIV by using the so-called allosteric drugs, which can reduce the affinity of cytoplasmic membrane receptors for binding HIV. Such drugs would bind to secondary domains of a receptor, leaving the main functional domain of the receptor free for binding the natural activator. It must be borne in mind that allosteric drugs differ fundamentally from non-allosteric drugs (specific inhibitors or activators that bind directly the major functional domain of a receptor; the anti-HIV cocktail is composed of these kinds of drugs). By modulating memory T-cell membrane receptors, it could be possible to prevent HIV latency within this cell reservoir. Celsentri® is an allosteric type of drug licensed for the treatment of HIV and already available on the market.

23.5 CANCER

Cancer represents a group of illnesses characterized by uncontrolled cell multiplication, which, if not detected and treated adequately, is fatal, especially if the cancer metastasizes.

Cancer cells develop either through viral infection or by genetic mutations. These factors prevent the normal cell cycle arrest and cell multiplication proceeds unchecked, resulting in a malignant growth and a situation where some cells become detached from the tumour and migrate (metastasis) to other tissues, forming new tumours (secondary cancers). In addition, it is a widely held suspicion that *Helicobacter pylori* can cause a type of stomach cancer. If this is correct, it would be a rare example of a cancer caused by a bacterium (Bhavsar et al., 2007).

Nowadays it is well established that genetic alterations are responsible for the appearance of a premalignant tissue, which goes on to become malignant when exposed to a prolonged period of the inflammatory process. In such cases, cells of the innate immune system (macrophages and lymphocytes, for instance) would promote carcinogenesis. In other words, genetic alteration would predispose a cell to become malignant, whereas a prolonged exposure to the innate inflammatory response of the immune system would induce this predisposition. This concept is the so-called immune paradox.

The immune paradox could be envisaged as follows: a dendritic cell exposes antigens of the tumour to adaptive T (cytotoxic) and B (antibody producer) cells. Generally, T cells and antibodies neutralize cancerous cells. Sometimes, B cells respond to chemical signals – not identified yet – that direct them to produce specific antibodies, which, in turn, stimulate the innate immune system to produce cells with the capacity to help cancerous cells survive and grow (Visser et al., 2006).

The role of macrophage in coordinating several critical steps in cancer development and metastases has been studied (Lewis and Pollard, 2006; Yang et al., 2018). It is thought that macrophage can digest the membranous envelope enclosing solid tumours, releasing abnormal cells. These cells migrate along a chemotactic gradient (a growth factor – also produced by the macrophage) to the nearest blood vessel in which the cancerous cells penetrate and subsequently spread throughout the body. The macrophage also stimulates angiogenesis within the solid tumour, allowing oxygen to penetrate to cancer cells in the inner portion of the tumour; macrophage may also produce inhibitory proteins against components of the adaptive immune response. There is further evidence that macrophages produce inflammatory compounds (for example, tumour necrosis factor) capable of activating nuclear factor kappa B (NF-kB) in tumour cells. The NF-kB translocates to the cell nucleus where proteins that inhibit apoptosis and promote tissue inflammation and cell proliferation are induced.

Chronic inflammation could also be responsible for other diseases, i.e. **heart, stroke** (macrophages engulf LDL that agglutinates blood cells inside the endothelium of the blood vessel lumen, causing the formation of arterial plaques. Sometimes a plaque leaches from the vessel wall and the contents spread through the body via the bloodstream and can enter narrow arteries, blocking them and resulting in stroke), **diabetes** (cells of the innate immune system and adipose cells when under metabolic stress, produce cytokines – signalling molecules such as the tumour necrosis factor – that interfere with the insulin-dependent mechanism for glucose uptake), **Alzheimer's disease** (microglia cells – 'the macrophages' of the nervous system – produce cytokines and free radicals, which in association with β-amyloid protein lead to the formation of deposits of plaques throughout the brain), **depression and schizophrenia** (patients with these anomalies present with high concentrations of serum interleukin-6 and active protein C, two molecules with high inflammatory capability).

Recognizing the role of chronic inflammation in carcinogenesis can open the possibility that therapeutic protocols based on anti-inflammatory medicines could prevent premalignant cells from becoming cancerous or inhibiting their capability of metastasizing.

The picture drawn above should be completed by including a discussion about cancerous stem cells, which are derived from normal stem cells following DNA mutation. The mutation could arise from exogenous factors (toxic substances, radiation, amongst others) and/or endogenous one (errors occurring during cell division – asymmetric chromosome distribution, poorly positioned crossing-over events or insertion of transposons into DNA to activate silenced oncogenes). The final result would be the deregulation of cell division. This observation is corroborated by the fact that cells belonging to solid tumours are not capable for an indefinite proliferation. Indeed, the tissue of a solid tumour is a heterogeneous mosaic of different cell types, giving the impression that the tumour is a badly formed organ. Hence, a strategy for a complete cure from cancer by destroying all malignant cells in a tumour might be possible using a combination of approaches – through surgical excision, chemotherapy and/or radiotherapy – augmented with methods for systemic cancerous stem cell hunting. If these cells could be detected and a method for eradication be successful, then the cancer would not return.

It must be considered that a solid tumour is not just an aggregation of cells that are continuously dividing in an unsynchronized and deregulated manner, but tumours have a microenvironment composed of several types of immune cells, producing abundant chemical signals and a web of blood vessels forming a distinct circulatory system.

It is well known that tissues and surrounding extracellular matrix produce chemical signals that affect the behaviour of adjacent cells (and stem cells). In the same way, the microenvironment surrounding a tumour strongly influences tumour integrity and likelihood that the tumour will metastasize.

If normally healthy and growing cells are removed from the body and cultivated in the laboratory, these cells will eventually become undifferentiated and ultimately die. Stem cells, on the contrary, need to be cultivated in a medium containing signalling compounds that retain the cells in an undifferentiated form; otherwise, stem cells rapidly differentiate as a preprogrammed behaviour. Inside the body, stem cells are surrounded by special cells which control their multiplication and differentiation capabilities (stroma cells of the conjunctive tissue of bone marrow, for example).

Stem cells seldom migrate from the tissues from where they originated, because stem cells are attached to adjacent tissues by adhesion compounds. When a stem cell differentiates, it is attached to a guard cell that directs the stem cells from the tissues to the tissue where differentiation is completed. It has been suggested that an analogous mechanism also occurs with cancerous stem cells (Chu et al., 2020).

Another aspect for the genesis and development of a malignant tumour is related to the pattern of angiogenesis established inside its structure.

Solid tumours have distributed throughout their structure abnormal blood vessels with abnormal structure (the vessel walls which are highly porous or impermeable, for example) and blood flow (the blood flux inside the vessel can vary from fast moving to stagnant). Moreover, in a same vessel, the flux of blood can change the direction (Ahir et al., 2020).

The chaotic vascular anatomy and function of a solid tumour contribute to local inflammation and the inefficiency of drug treatment. The distribution of a medicine throughout the tumour is not uniform due to the accumulation of fluid around interstitial spaces caused by high vessels' porosity and stagnation of blood inside the vessels (the osmotic pressure between the lumen of the vessel and the interstitial space is zero). Sometimes, the interstitial fluid leaches from the tumour into body cavities spreading tumour cells and abnormal proteins, which, in turn, promote the formation of additional lymph and blood vessels in normal tissues and the nearest lymph nodules. In other words, conditions for the cancer to metastasize are favoured. Moreover, the chaotic distribution of vessels inside the tumour induces zones of low oxygen tension and pH. This unnatural microenvironment reduces the efficiency of immunological cells as well as chemotherapy and radiotherapy. Chemo- and radiotherapy both require oxygen to destroy tumour cells by generating free radicals on the spot. Adjuvant therapies that reverse extremes of pH within solid tumours are a rapidly growing research area in cancer chemotherapy.

Angiogenesis under normal physiological conditions – for example, as part of a wound healing process – is controlled by the complimentary action of signalling molecules (for example, vascular endothelial growth factor, VEGF) and inhibitors (for instance, thrombospondins). Nevertheless, in solid tumours and other chronic diseases – atherosclerosis (fat plaques accumulate inside the blood vessels), diabetic retinopathy, and ageing macular degeneration, amongst others – rapid angiogenesis prevails. Therefore, medicines that are available for use in the clinic are designed to inhibit VEGF (Lucentis®, Macugen® and Avastin®), or to stimulate the anti-angiogenesis molecule trombospondine-1 (Herceptin®).

For some solid tumours, it was observed that the use of antiangiogenesis medicines alone, like Avastin®, did not increase patient survival rates in spite of significantly reducing angiogenesis. However, by using Avastin in combination with chemo- or radiotherapy, patient survival rates significantly increased. The synergy in combination therapy is because VEGF inhibition by Avastin reduces the abnormal vessel formation, so that vessels with a normal anatomy and physiology predominate within the tumour. As a consequence, chemo- or radiotherapy can reach the overall tumour mass, destroying it (Ahir et al., 2020).

According to Dai (2007), there is evidence that heat shock proteins (HSPs) could be useful in fighting cancer, and these proteins are synthesized by cells under stress. Tumour formation can also be envisaged as a stressor, insofar as it is an anomalous tissue inside the body.

There are several HSPs known: HSP-40, HSP-70, HSP-60 (foldase), HSP-90 and HSP-100 (unfoldase). Although these proteins are often produced under stress conditions caused by heating, their expression is not limited to induction by heat. Other stressors that induce HSPs expression include cold, hypoxia, dehydration and nutrient limitation. The HSPs help cells maintain their function against physiological conditions of stress. When stressed, many vital intracellular proteins become destabilized and lose their function. But HSPs generated under stress have the capability of rescuing essential cellular proteins, recycling damaged proteins and restoring cell metabolism to homeostasis. It is interesting to speculate that the practice of physical exercises leads to an increase in body temperature – a stressing condition for cells in general – and, as a consequence, stimulates HSP biosynthesis. A possible benefit to health of HSP biosynthesis, therefore, would be a reduction in intracellular protein damage and stimulation of the innate immune system (Dai, 2007).

HSPs act by binding to other proteins with the aim of inhibiting undesirable interactions with other macromolecules. Moreover, HSPs can also act on several intracellular proteins at the same time, coordinating tertiary structure folding and quaternary protein–protein interactions during times when a cell is under stress. For example, membrane receptors are proteins that are trafficked by HSPs from the cytoplasm to be anchored into the exact position in the cytoplasm membrane. After polypeptide chains have been biosynthesized on ribosomes, the physical–chemical characteristics of the cytoplasmic environment can induce secondary structures formation, but the functional form of the final protein is, in part, determined by binding of HSPs.

The capacity of HSPs to bind to proteins can be useful in oncology. Some anomalous proteins or peptides can also bind to a HSP, which presents these as antigens to the innate immune system, stimulating a strong immune response. However, there are signalling HSPs, such as HSP-90, that bind to proteins in both normal and cancerous cells (HSP-90 is involved in cell proliferation and is also believed to be a factor that influences the life span of cells). Medicines that inhibit HSP-90 (e.g. alvespimycin and tanespimycin that target abnormal cells) have been developed.

To review, it is becoming increasingly clear that cancer results from a complex combination of factors shared by both normal and cancerous cells (angiogenesis, immune response, inflammation, etc.). It is also now clear that there is a metabolic and physiological equilibrium between normal and abnormal cells at which cancer cells seem not to proliferate and develop into tumours destined to metastasize. However, if this equilibrium becomes unbalanced and the reasons for this seem complex and are far from clear, clinical symptoms of cancer develop and a treatment with a combined use of medicines and surgery is required if the patients stand any chance of survival.

Taking into account new knowledge on cancer cell and tumour biology – elucidation of angiogenesis mechanisms, immune and inflammatory responses, amongst others – new approaches to cancer treatment which act as adjunct therapies to existing treatments are now on the market and have been developed because of our knowledge in pharmaceutical biotechnology. Examples of biopharmaceuticals: non-steroidal anti-inflammatory drugs; monoclonal antibodies (e.g. Avastin® – a potent angiogenesis inhibitor – and Rituxan® – an inhibitor of the inflammatory response); and vaccines (e.g. HybriCell® that combine *in vitro* dendritic cells with cancerous cells removed from the patient). Vaccines against cancer, such as HybriCell®, are an excellent example of 'individualized therapy', also called 'personalized medicine', which is becoming an important goal for pharmaceutical biotechnology (Ahir et al., 2020). Another example of individualized therapy is the use of the so-called cell-T-chimeric antigen receptor (CAR-cell T), which does not need the direct intervention of dendritic cells, but only of modified T cells (Sadelain, 2013). The approach consists of modifying T cells taken from a patient and then introducing into these cells a gene capable of expressing a specific antibody on the external surface of the cytoplasm membrane. The antibody is designed to be specific against the CD19 antigen, which is abundantly expressed by malignant cells present in the blood of patients with leukaemia (Maus, 2014). A special mention should be made of the vaccine developed to provide immunity against human papilloma virus (HPV). This vaccine has been widely adopted into the vaccination schedules of many countries and can also offer immunity not just against cervical cancer, but also against HPV-induced cancers of other tissues as well in the mouth, throat and anus (Curran and Glisson, 2019).

Developing individualized patient therapies is gaining increasing attention in the fight against cancer, with the biotechnological development of devices capable of rapidly identifying specific markers to determine if a cell is potentially cancerous or not, of equipment that can detect tumours by whole-body scans and of diagnostic tools that can predict the likely response of a patient to a particular therapy (also known as pharmacogenomics). According to Sequist et al. (2009), a chip called 'chip-CTC' is constructed from a silicon chip coated with antibodies specific for molecules located on epithelial adhesive cancer cells (EpCAM), a chamber within which the chip is submerged in a buffer and a pneumatic pump, has been developed for this purpose.

Individualized therapy depends extensively on the availability of precision and miniaturized diagnostic gadgets that are available at low cost. For example, the cost for identifying each protein associated with a particular type of cancer by taking a drop of blood is approximately $1.00/protein using current chip devices (Gmeiner and Ghosh, 2013).

Generally, a miniaturized gadget for diagnostic purposes is built based on an understanding of metabolism which can be described using a mathematical algorithm. All biochemical events in the human body (related to genome, metabolism and proteome) are coordinated by a network of proteins inside the cell. Moreover, these proteins interact with each other and with other biologically active molecules (hormones, lipids, sugars etc.), producing a metabolic map for each cell type in the body. Any disruption in one of these proteins can cause the appearance of a disease. For example, prostate cells – as with any other cell type – contain a network of proteins which interact with each other to create a metabolic web. Error in a protein such as MAPK8 (associated with cell locomotion) and the sudden expression of a protein such as SDC1 (an abnormal protein) in prostate cells result in these proteins being released into the bloodstream. The capability of detecting these two proteins amongst thousands of proteins present in the blood using the chip technology can now alert the doctor about a possible problem occurring in the prostate. Other prostate proteins such as the prostate specific antigen (PSA), membrane prostate-specific antigen (PSMA) and platelet factor-4 (PF-4) have also been detected in blood samples using a biosensor constructed from graphite electrodes covered with antibodies conjugated to a chemiluminescent compound (Kadimisetty, 2016). Diagnostic devices must be capable of generating the desired information by analysing a great number of biological molecules as easily and as quickly as possible. Gmeiner and Ghosh (2013) described a prototype chip for measuring the concentrations of proteins related to certain kinds of cancer from a drop of blood in 10 min at a cost of $0.10 per protein.

Pharmaceutical biotechnology offers the future possibility that cancer will be converted from a lethal disease to a chronic illness through the development and a combined use of new practical diagnostic devices and therapeutic protocols.

23.6 STEM CELLS

Stem cells – from which medical science hopes to use for regenerating damaged nervous, cardiac and muscular tissues – represent a large field for studies in both biology and pharmaceutical sciences.

Stem cells can be divided into two types, i.e. embryonic stem cells and tissue-specific stem cells.

Embryonic stem cells can differentiate to become any tissue type in the body. If left in cell culture without the addition of any signalling molecules, these cells naturally differentiate into many miscellaneous cell types (neurons, epithelial and endothelial cells, for instance), although it seems that the stem cells have a natural propensity to differentiate into specific tissue types (cardiac tissue, for instance). This was the first indication that each kind of tissue might have a specific stem cell progenitor (Rosenthal, 2003).

There have been some successes using stem cells in cardiology. Adult stem cells extracted from the bone marrow of a patient can be grown in culture, and then, these cells are surgically introduced back into the patient to restore the damaged cardiac tissue. However, the mechanism involved in this tissue regeneration is not well understood. There are two possibilities. One is a fusion between the stem cells and normal cardiac cells, which can lead to the formation of hybrid cells capable of their multiplication and differentiation. Another possibility is that the stem cells when in contact with the cardiac tissue can produce the growth factor locally, inducing specific cardiac stem cells to grow and multiply (Haraguchi et al., 2012).

When muscle fibres are damaged, these must be replaced by new muscle fibres. Chemical signals such as IGF-I and myostatin stimulate muscle stem cells to multiply. Stem cells have been observed to fuse with muscle fibres releasing their nuclei to the muscle fibres, resulting in the regeneration of the muscle. One useful application of stem cell therapy will be the treatment of Duchenne muscular dystrophy, whereby a gene that encodes dystrophin (a protein responsible for the contractile muscular response) (Camera et al., 2016) can be cloned into a patient's stem cell. The stem cell is then cultured and allowed to multiply before being transplanted into the patient's muscles. The genetically modified stem cells will then deliver the gene to replace the faulty gene in the muscle tissue. In this way, stem cell therapy can be developed in future as a cure of genetic hereditary diseases, mainly those resulting from the damage of a single gene. Transplantation of stem cells is already a reality. Using surgical approaches, scientists can identify the individual stem cells responsible for angiogenesis in the cardiac muscle in order to stimulate the formation of a collateral web of coronary vessels nearby, or around the damaged cardiac tissue (Rubanyi, 2015).

A number of salient points need to be addressed if stem cell therapy is to be fully exploited, such as a more reliable source of stem cells, the identification of chemical signals that direct embryonic stem cells to differentiate into specific tissues and a better understanding about a communication between embryonic stem cells and specific tissue stem cells.

The most testing problem for the future success of stem cell technology is to secure a reliable source for stem cells. Many experts advocate that the best source of stem cells should be of embryonic origin. However, the use of embryos that are destroyed during the process of extracting pluripotent cells from a blastocyst raises a series of objections (religious, moral and ethics). Another critical point is to avoid genetic incompatibility between the embryo donor and recipient. However, two alternative ways to obtain stem cells have been successfully developed. First, the genetic reprogramming of human epithelial cells (also called 'induced pluripotent stem cells') is done by inserting a plasmid carrying transcription factor-encoding genes into the nucleus of epithelial cells. When expressed, these transcription factors repress the gene overexpression in the epithelial cells to avoid the formation of cancerous tissue in the recipient. The rejection problem can easily be avoided if the epithelial cells are harvested from the recipient. Second, the nuclei from the unfertilized ova can be substituted by nuclei obtained from the somatic cells of the patient. The ova are allowed to multiply in cell culture from which stem cells can then be separated (Cox and Rizzino, 2010; Rosenthal, 2003).

The development of stem cell cultures *in vitro* – regardless of whether the stem cells are from embryo, specific tissue or genetically reprogrammed origin – will promote advances in genetic and embryology research; production of bioactive molecules; discovery of new drugs; understanding the mechanism of action of drugs; establishing protocols to evaluate toxic and other collateral effects of drugs and bioproducts for human use; identification and characterization of intracellular signalling molecules; and gene therapy.

The real contribution of pharmaceutical biotechnology in this area will be the introduction of methods capable of identifying signalling molecules, as well as methodologies to understand the mechanisms of stimulation and action. In future, it should be possible to stimulate signalling molecules directly within a patient to induce stem cells for tissue repair.

23.7 BIOTECHNOLOGY INTERFACES

During the past 20 years, pharmaceutical biotechnology merged the boundaries between biology and pharmaceutical sciences, and has taken from other scientific disciplines such as informatics, optical physics, nanotechnology, mechanical engineering and electronics.

Optics is an area of physics in which the pharmaceutical biotechnology has made use for various applications. For example, spherical nanoparticles of silicon (diameter of 100 nm) can be coated with a layer of gold (thickness of 10 nm) onto which monoclonal antibodies directed to specific antigens (located in cancerous cells, for example) can be linked. The silicon-gold nanoparticles would be injected intravenously into a patient and be targeted to the tumour because of the specificity of the nanoparticle–antibody conjugate for the antigens on the surface of the tumour. Since the vascular system of tumours is extremely porous (as discussed previously), the nanoparticles cross the epithelial surface and interact directly with the tumour cells. The frequency and wavelength of infrared radiation are translucent to human cells. When an infrared laser is focused onto the tumour, electrons along the gold and silicon interface begin to oscillate, resulting in a local increase in temperature from 37°C to 45°C, under which the tumour cells die (Ozbay, 2006).

Materials and devices that are manufactured on a nanometre scale can interact with cells and molecules conspicuously. Nanometre-scale technology (nanotechnology or nanobiology) has become widely adopted in research and clinical therapy, varying from 10 nm (the average size of antibodies) up to 100 nm (the size of an average virus). Size does limit the clinical use; for example, nanoparticles between 10 and 100 nm are easily eliminated by the kidneys (<10 nm) or have difficulty penetrating the damaged tissue (>100 nm). Nanometre devices have been employed as sensors for detecting bioactive molecules (proteins and DNA, for instance); fixing organ images; targeting specific tissues; and targeting the delivery of drugs to specific regions of the body (Gmeiner and Ghosh, 2013).

The medical–pharmaceutical nanotechnology is advancing by constructing the following devices:

a. *Nanowires*: They are anticipated for use in sensors. A nanowire conductor – being of 10–20 nm in thickness – can be inserted into a channel in which a sample of biological fluid is next introduced. The nanowire is covered with a monoclonal antibody or oligonucleotide that can detect and identify proteins or DNA present in the sample. When the protein or DNA interact with antibody on the sensor surface, the conduction capacity of the nanowire is changed, which can be detected electronically.
b. *Consoles*: These are expected to have sensory applications. For example, strings of nanowire filaments can be coated with DNA sequences and packaged into an appropriate container (the console). When a sample containing an unknown mixture of DNA sequences is introduced into the console, only DNA with complimentary sequences to the DNA of the console will bind. Complimentary DNA-DNA pairing will cause a perturbation on the string, which is translated as an electronic signal for computing purposes. More recently, devices like this have been used for detecting DNA and RNA of bacteria and viruses present in the blood in order to evaluate the type and severity of infection (Rackus, 2015; Kelley, 2014).
c. *Quanta points*: They are projected for use in imaging applications. Nanocrystals of cadmium covered with latex or metal, when exposed to a specific laser radiation, become fluorescent and emit light at several wavelengths and intensities. Antibodies raised to be specific for a given tissue (a tumour, for instance) can be linked to the crystals allowing an image of the tumour to be visualized through printed images. Nowadays, in order to minimize the number and the suffering of experimental animals (such as guinea pigs, mice, monkeys) used during drug testing, more and more imaging devices are being utilized.

   Introducing cancer cells carrying a gene encoding luciferase into experimental animals can be considered another imaging approach. The proliferation of the tumour and metastasis throughout the body will be visualized through the luminescence emitted by the cancer cells. The intensity of the emission can then be related to the growth rate of the tumour. This procedure allows the effect of new drugs to be evaluated in a real time, without the unnecessary period waiting for tumour consolidation, thereby reducing suffering of the animal significantly – demonstration of a reduction in experimental animal suffering is now a requirement in modern ethics of research (Mukerjee, 1997; Tiwari et al., 2020).

d. *Nanolayers*: They are used for imaging and/or carriers for drugs. Nanospheres of silica covered with gold layers coated with a specific antibody-therapeutic drug conjugate are introduced into the bloodstream, which accumulates in the damaged tissue. By directing a laser light to the tissue nanosphere, a temperature gradient is established between the spheres and the tissue, resulting in tissue death. If adequately planned, the nanolayers can absorb or disperse the light in order to give the sharpest of images of an organ, tumour etc.
e. *Nanoparticles*: They are used for drug targeting. Particles composed of several kinds of materials can contain drugs that are delivered to a defined location. These vehicles can be made of a thin layer of lipid (liposome), which passively crosses the walls of tumour blood vessels to the intercellular space of the tissue. Therefore, the drug – often, a chemotherapy drug – is delivered slowly. The FDA has already approved a liposome carrying doxorubicin for the treatment of ovarian cancer and multiple myeloma. More complex nanoparticles have been made containing monoclonal antibodies to target proteins of a particular tumour.

Nanometre particles can be considered as new pharmaceutical formulations made with biocompatible materials. Gmeiner and Ghosh (2013) have described cyclodextrin nanoparticles in which siRNA molecules (therapeutic agent) and transferrin (carrier) are inserted. These particles are introduced into the bloodstream, which after a period of time become anchored onto the tumour, as the tumour blood vessels are abnormal and are permeable to the nanoparticles. When the particles reach the intercellular space of the tumour, they become linked to the cells through transferrin and transferrin–receptor interaction. The particles cross the plasma membrane of the cells by endocytosis. Once inside the cell cytoplasm, the particles decompose freeing the siRNA, which, in turn, binds to complementary mRNA, thus preventing its translation into a vital protein for the cancer cell.

23.8 ADMINISTRATION OF BIOACTIVE MOLECULES

Undoubtedly, the greatest challenge for pharmaceutical biotechnology is the development of pharmaceutical formulations to deliver the arsenal of new drugs to the patients.

For a long time, it has been known that the route of administering drugs to the patient plays a relevant role in the efficiency of the treatment. Moreover, drug administration depends on the pharmacokinetic properties of the drug. In other words, a drug can become toxic if it enters too quickly and reaches a high dose in the bloodstream, or becomes ineffective if absorption is too slow and its concentration is low in the bloodstream. This implicates that both optimal choice of preparation and route of administration (enteric or parenteral) are chosen. Enteric delivery imposes a series of barriers which the drug must cross, i.e. stomach and duodenum acidity, the alkalinity of the intestines, hydrolysis promoted by enzymes, crossing of intestinal wall and, after entering into the bloodstream, resistance to liver metabolism. Such aspects become important when high molecular weight drugs (proteins, for instance) must cross membranes and circumvent the hydrolysis by enzymes naturally present in the body.

At the first glance, drug administration by a parenteral route would seem an attractive alternative to circumvent the problems cited. Nevertheless, delivering a drug by injections causes pain and hematomas in many patients, which undermines their willingness to take the medication. Fortunately, other methods for administering drugs include adhesives on the skin, intradermal implant, injection for slow drug delivery, topical jellies and nasal sprays. The FDA has already approved two new medicines based on the methods cited, i.e. Nutropin-Depot® (injectable microspheres of degradable polymer which deliver human growth hormone for 4 weeks) and Gliadel® (a brain implant for the direct delivery of chemotherapeutic agents) (Langer, 2001; Li et al., 2019).

The skin, the respiratory tract and the intestinal tract are traditional routes for drug delivery. Biomolecules, however, have difficulty in crossing these barriers because they have high molecular weight and have antigenic effects and their structures are often sensitive to ionic strength and pH. Crossing the skin implies that the drug must cross the corneum layer; meanwhile in the intestinal tract, the drug must resist hydrolytic enzymes and have the capability to cross the gut epithelium barrier. In the respiratory tract, a drug must resist phagocytosis by macrophages, which are abundant in the lung alveoli (Langer, 2001).

To circumvent such difficulties, new formulations must be developed. For example, a drug could reach the bloodstream from the gut lumen if it were associated with a polymer capable of adhering to the intestinal wall, or could be linked to natural epithelium carriers. If the respiratory tract was the administration pathway chosen, delivering the drug through a spray formulation would allow to produce an aerosol made up of nanometre particles which would be quickly absorbed through the lung alveoli, thus circumventing the destruction of the drug by macrophages (Langer, 2001).

Administration of biopharmaceutical drugs through the skin is possible by iontophoresis. This approach consists of two small adhesives, having opposite charges, connected to a reservoir containing the drug and to a small battery set under the patient's clothes. Painless electrical pulses are generated, which favour the drug (a protein, for instance) to pass from the epidermis to derma and to the bloodstream. Iontophoresis has been tested for administering hormones such as parathormone (osteoporosis) and gonadotropin (fertilization *in vitro*). Another possibility for the drug administration would be the use of ultrasound for opening small channels throughout the epidermis.

Iontophoresis is becoming a popular method for continuous and controlled drug release – with the aim of maintaining a drug at therapeutically relevant levels in the bloodstream. Advances in this field include microchip implants inside the skin, spinal cord and, in future, perhaps even brain implants that might be available. According to Langer (2001), a microchip is constructed with very small reservoirs covered with a nanometre gold layer, which would be slowly decomposed by a faint electrical charge (1 V maximum) to release the drug at a constant rate.

A more advanced device might in future have the capacity to detect specific chemical signals in the blood, which would allow a controlled delivery of the drug dose. A record of the dose delivered could be collected by an external device, which would then be directed to the doctor and/or hospital computers to monitor the clinical progress of the patient. The goal for this technology would be to administer a biotherapeutic with efficiency, specificity, timing and exact dosage to any part of the human body.

23.9 INDIVIDUALIZED THERAPEUTICS

Pharmaceutical biotechnology has advanced by introducing a new generation of technologies based on the next-generation DNA sequencing. The cost of sequencing an entire human genome is now around US $1,000, and it is hoped that within the coming decade, this cost will further reduce and become accessible by most people. Knowing the genome sequence of a patient will allow doctors to predict how the patient might respond to treatment, taking into account the physiological differences between age, gender, ethnicity etc (Conforti, 2018).

Undoubtedly, exploitation of DNA sequencing by pharmaceutical biotechnology depends on reducing the number of steps to acquire sequence, including miniaturization of sequencing devices and the ability to sequence thousands of molecules simultaneously. Amongst the many sequencing techniques now available, the nanopore technique is increasingly popular. The basic idea is based upon the negative charge of a DNA molecule, so that in an electrical field DNA will move in the direction of the anode. If the electrical field is passed through a 1.5-nm-thick membrane, only the single filamentous DNA molecules will cross. As the single-stranded molecule moves through the pores of the membrane, the nucleotides cause a small and sudden change in electrical conductance (in the order of a picoampere) of the membrane. Different bases (adenine, thymine, guanine and cytosine) cause different changes in membrane conductance, producing a distinct pattern for each DNA sequence. Improvements in this technique could allow for longer sequences to be read so that potentially an entire human genome could be sequenced in 1 day (Shendure et al., 2004).

Individualized medicine considers the human body as a complex dynamic web of molecular interactions. Any deviation in these interactions can be measured and correlated with health or illness. Methodologies are now capable of handling (at a nanometre scale) samples of body fluids and/or small amount of cells removed from diseased tissues, allowing many different biomolecules to be identified quickly, precisely and cheaply.

The benefits of individualization of medicine are being realized by public health systems. For examples, patients with the same type of cancer and who receive an identical radiotherapy and/or chemotherapy treatment protocols often respond differently – some may be cured, whereas other are not. In the case of prostate cancer, it is recognized that some 80% of prostate tumours grow so slowly that the tumours do not cause problems to the patient. However, the other 20% of tumours can grow so fast that metastasis occurs and death is the inevitable prognosis. For this group of patients, treatment must be rapid and aggressive. Conversely, the majority of patients with slow-growing tumours can spare unnecessary surgeries and radiotherapy and/or chemotherapy. The collateral effects of unnecessary surgery include urinary incontinence, sexual impotence and abdominal pain. All these sufferings will be minimized if changes in the metabolic web, as referred above, can be identified.

The circulatory system crisscrosses the body, transporting proteins and thousands of other molecules. Blood is, therefore, an ideal body fluid for analyses to give a panoramic view of the functioning of the entire body at any given moment. The ability to detect an unbalance of certain proteins or messenger RNA molecules in the blood can signal the presence of an illness before clinical symptoms develop. Every organ expresses approximately 50 different messenger RNA molecules at any one time. Translation of each molecule results in specific proteins being released into the bloodstream. The level of each protein will, therefore, be an indication of metabolic efficiency of a particular organ, i.e. functioning of the metabolic web. When this web is disturbed due to illness, protein levels in the blood are altered. Identification of these alterations allows a particular disease to be diagnosed. Hypothetically, it is possible to determine the concentration of about 25 different proteins produced by any given organ from a single sample of blood. Unfortunately, the cost of such a procedure is still prohibitive to most health systems. For example, the cost for determining the blood concentration of PSA is about US $50, whereas a more realistic cost to analyse a single protein from a mixture of many thousands should be around $0.05–0.10 (Heath and Davis, 2008). The perspective for analysing thousands of proteins in a small sample of blood still depends on the development of fluid dynamic and surface chemistry in a micro- and/or nanometre scale, as well as computer programs to rapidly interpret the huge amount of data the analysis would produce.

Developing protocols for disease prevention using either DNA sequencing or protein quantification of biomolecules from small volumes of blood, and then transforming the results into individualized treatment require rapid, sensitive and low-cost methods. The following is an example based on the identification and quantification of antibodies that can be a diagnostic for several illnesses, which are currently under development.

In autoimmune diseases (diabetes type I, Addison syndrome, celiac disease, multiple sclerosis, systemic lupus erythematosus, amongst others), antibodies appear into the bloodstream many years before the symptoms of the illness. The availability of tests capable of detecting these antibodies will give doctors the opportunity to take early measures in order to minimize the severity of the illness. For example, the systematic and slow attack over many years of B and T lymphocytes on the β-cells of the islets of Langerhans leads to type I diabetes. Measuring the concentration of antibodies against insulin, glutamic acid decarboxylase (GAD) and islet antigen-2 (IA-2) in the blood will enable doctors to design the most appropriate therapeutic strategy (Notkins, 2002; Redondo et al., 2018). Moreover, devices that automatically regulate glucose levels in the blood have been intensely studied, and several of these devices have been submitted for approval to the FDA appreciation to undergo Phase I and II clinical trials (Bekiari et al., 2018).

According to Notkins (2002), the detection of autoantibodies can be used as prognostic risk indicators (i.e. to determine the likelihood of a person that might develop a particular disease in future and the cost/benefit ratio for taking preventive therapeutic measures), markers for disease progression (i.e. to anticipate the severity of the illness and the probable rate of progression) and rationalizing clinical trials in humans (knowing individuals under a real risk of developing a rare disease, the number of volunteers needed for clinical trials should be reduced).

Therapeutic protocols to treat autoimmune diseases can be based on stimulating regulator T cells (T-REG), which naturally suppress other *cells* in the immune system, control the immune response to self-antigens and help prevent autoimmune disease. The thymus, a gland in which all T cells mature, will normally remove all abnormal T cells from the circulation. Sometimes thymus awareness fails. In this case, T-REG cells are produced by the body to circumvent thymus failure (Notkins, 2002).

There is the perspective that during the next two to three decades, pharmaceutical biotechnology will provide chips coupled to computers that can detect autoantibodies. For example, a blood sample from a patient will be taken and the serum then separated. The serum will next be dispersed into a plate containing chips coated with several types of self-antigens (antigens known to stimulate an autoimmune reaction). Any autoantibody present in the serum will bind to the corresponding self-antigen causing an electrical pulse on the plate, which will be directed to a computer program to translate the pulse into the required information (Notkins, 2002).

23.10 SYNTHETIC BIOMOLECULES

A new field that is becoming increasingly popular with many new and exciting applications in pharmaceutical biotechnology is synthetic biomolecules.

Examples of synthetic biomolecules are the **bis-amino acids** (compounds composed of two carboxyl and two amino groups which, after polymerization, form structures similar to proteins), **polymeric moulds of biomolecules** (plastic matrices of functional domains of natural biomolecules; for instance, for an enzyme, the functional domain is the active site, whereas for an antibody, it is the antigen-binding site) and **peptide–nucleic acids** (hybrids of protein and DNA or RNA).

A great challenge in developing this field is to create mimetic molecules of high MW that act like natural proteins which can bind to cell receptors, enzyme catalysts, carriers inside body fluids (blood, lymph, for example) amongst others. Proteins are polymers of amino acids biosynthesized on ribosomes from a messenger RNA (mRNA) transcript. After translation of the mRNA is completed, the linear peptide dissociates from the ribosome and forms the tertiary structure required for its biological role. The precise tertiary structure depends on the type and sequence of amino acids along the chain; understanding the specific sequence of amino acids essential for their function has been a huge challenge for protein engineering (Pomsuwan et al., 2006).

According to Pomsuwan et al. (2006), bis-amino acids could be used as monomers of pseudo-peptides capable to become moulds of natural protein. The central idea is not to reproduce the complete protein but only its active domain. The foreseen uses of these pseudo-proteins are drugs, catalysts, components of biosensors, 'nanometre valves' and memory devices for future biological computers ('biocomputers').

As an alternative to bis-amino acids, Borman (2003) proposes the building models of biomolecules using substances (e.g. cyclodextrins) capable of polymerizing around a target molecule, resulting in a mould quite similar to the original structure of the biomolecule ('polymeric molecular mould'; PMM). PMMs will be useful for separating isomers from racemic mixtures, for substituting industrial enzymes insofar as only the active site of the enzyme will be recreated, for reproducing the antibody-binding site of an antigen, which can be used in kits for identifying antibodies and for designing inhibitors for enzymes naturally present in the body. In short, the development of PMMs technology should allow reproducing *in vitro* structural aspects of thousands of biomolecules whose tridimensional images have already been determined by high-resolution spectroscopy (Fromme, 2015; Waldrop, 2014).

Hybrids of protein–nucleic acids (PNAs) have a promising future in the field of synthetic biomolecules because such molecules have the dual capability of storing information (afforded by the nucleic part of the molecule) with the physical–chemical stability of the protein. A PNA is basically a sequence of amino acids intercalated with molecules of thymine (uracil in case of RNA) and cytosine. These nitrogenous bases establish hydrogen bonds – the so-called Hoogsteen bridges (Orgel, 2004) – with adenine and guanine molecules in natural nucleic acids (DNA or RNA), resulting in the association between PNA and a particular nucleic acid. The complex formed between PNA and a natural nucleic acid can be used in various applications. For example, PNA containing a particular sequence of cytosine and uracil (or thymine) can bind by complimentary base pairing to a particular mRNA in order to inhibit binding to a ribosome and inhibit the protein synthesis (responsible for a hereditary illness, for instance). Due to their peculiar chemical nature, PNAs have a long half-life in the cytoplasm, because they are highly resistant to hydrolytic enzymes present inside the cells. Moreover, the affinity of PNAs to DNA can allow a controlled regulation of gene expression similarly to natural regulators such as proteins (transcription effectors) and interference (silencing) RNA. A future application will be the site-specific insertion of PNA into the double-stranded DNA to partially separate the two complementary strands, leaving one strand exposed and onto which RNA polymerase can bind to initiate transcription. Thus, knowledge of specific regions of the human genome can be obtained.

As science is always advancing, so developments in synthetic biology (Lu, 2016) are increasingly focused on the genetic modification of microbial cells for medical–therapeutic benefits. For example, Jusiak (2016) has described the engineering of *E. coli* for the treatment of uncontrolled accumulation of ammonia inside the human body due to a genetic dysfunction in the urea cycle (UCD). The treatment consists of substituting the normal bacterial gene that inhibits the overproduction of arginine by two synthetic genes: one encodes arginine production and the other is a repressor. When the engineered bacterium is delivered into the colon of a patient suffering from UCD dysfunction, the low oxygen tension of the colon inhibits the transcription of the repressor gene, thereby allowing transcriptional activation of the other gene and promoting the synthesis of arginine from the excess of ammonia.

23.11 ELECTRONICS IN PHARMACEUTICAL BIOTECHNOLOGY

There are some well-developed electronic technologies that have been used for many years in medicine, for example, to detect brain waves, to control prostheses and to overcome deafness (implant of a sensor into the cochlea that stimulates the audition nerve with sounds captured by a microphone). Moreover, a wide diversity of electronic sensors are used in equipment to monitor bodily activities (e.g. in intensity therapy units, haemodialysis, etc.), for directing surgical devices through the body (catheters, surgery without a scalpel, etc.) and for conducting clinical prognostic and/or diagnostic tests (polysomnography, nuclear magnetic resonance, electrocardiography, mammography, among others).

The technologies cited above resulted essentially from neurological studies based on the use of electrodes for stimulating and registering the activity of isolated brain cells (neurons). But the manner by which neurons act cooperatively to give us thoughts and behaviours is, still, a great challenge to biology. The big question is how neuroscientists can in future visualize neural circuits in action (Clyne and Miesenböck, 2008).

One exciting new approach combines gene therapy with light – a technique called 'optogenetics' (Clyne and Miesenböck, 2008; Zhang, 2010; Mei and Zhang, 2012).

Neurons transmit information by electrical signals (potential of action). These pulses (with a voltage of about one-tenth of an AA battery) induce brain cells to release neurotransmitter molecules, which in turn either activate or inhibit a neural circuit (Clyne and Miesenböck, 2008).

Optogenetics was developed from the idea that gene manipulation could be used to visualize the specific neural circuits. Through techniques developed for gene therapy, a gene that encodes for a photosensitive protein can be inserted into neural cells; if these cells were functioning in a coordinated fashion in response to a stimulus, then the photosensitive protein would be expressed and could be detected fluorescence at a given wavelength (Clyne and Miesenböck, 2008).

All cells of the organism contain the same genes. What distinguishes one cell from another is the different combination of genes that are either expressed or repressed. For instance, neurons designed for producing dopamine have all genes related to the biosynthesis of the neurotransmitter expressed, whereas the same genes are repressed in non-dopamine-producing neurons. In theory, an inducer for dopamine biosynthesis can be translationally coupled to a gene that produces a photosensitive pigment upon irradiation. If this pigment is expressed, then the inducer for dopamine biosynthesis would be co-expressed and only cells that produce dopamine would be observed in tissue sections.

In spite of its theoretical appeal, optogenetics must circumvent a major practical obstacle in humans, the insertion of a foreign gene – which encodes the photo-induced activator– into the brain. Up until now, gene therapy techniques have still not been able to overcome this challenge and, as a consequence, the regulatory authorities – worried on the risk involved – have yet to give ethical approval for human trials until results of cell- and animal-based experiments provide more convincing efficacy to this approach (Clyne and Miesenböck, 2008).

When the identification of neural circuits has been solved, an interface between brain and machine will next need to be developed from which applications involving pharmaceutical biotechnology, electronics and cybernetics will become a reality. So far, the brain–machine interface to be explored has been the surgical removal of a mouse hippocampus which has been replaced by a silicon chip. According to Kohler et al. (2017), there are some promising results from this hippocampus chip, which appears to be capable of processing received neural signals and then sending reproducible and coherent responses with 90% precision. The hippocampus is the region of the brain where neural circuits are involved in memory – manipulating this region of the brain in a controlled and programmable way can, in future, lead to the development of humanoid robots, although experts in the field admit we are some 50 years from this, at least (Horgan, 2005; Bang et al., 2019). Finally, interfacing between the body's physiology, electronics and cybernetics can be developed to build a mechanical exoskeleton (controlled by brain waves) to help paraplegics recover the ability to walk (Velliste, 2008).

23.12 FINAL CONSIDERATIONS

Pharmaceutical biotechnology has been growing intensely since the 1980s, when the microbial insulin production was made by using genetically modified *Escherichia coli*. This approach offered an alternative supply of insulin which had previously been purified from pig pancreas, which, although similar to the human hormone, invariably led to an undesirable immune response in the user. It must be borne out that during the past 20 years, pharmaceutical biotechnology merged the boundaries between biology and pharmaceutical sciences, and has taken from other scientific disciplines such as informatics, optical physics, nanotechnology, mechanical engineering and electronics.

Nanotechnology, for example, handles material and devices – nanowires, consoles, quanta points, nanolayers and nanoparticles (liposomes) – that are manufactured on a nanometre scale (10–100 nm), which can interact with cells and molecules. Nanometre devices have been employed as sensors for detecting bioactive molecules (proteins and DNA, for instance); fixing organ images; targeting specific tissues; and targeting the delivery of drugs to specific regions of the body.

A greatest challenge for pharmaceutical biotechnology is the development of pharmaceutical formulations to deliver the arsenal of new drugs to the patients. The drug administration can be made by injections, adhesives on the skin, intradermal implant, topical jellies and nasal sprays. Iontophoresis and ultrasound are two innovative techniques that allow administering drugs efficiently through the skin.

Undoubtedly, the pharmaceutical biotechnology success based on the high-quality and complex researches stirs up the human being confidence on the perspective of pain reduction caused by neuropathic and inflammatory diseases. Moreover, it is also expected that cancer (if not completely erased, at least transformed from a lethal disease to a chronic illness), neurodegenerative diseases (based on therapeutics derived from stem cell and RNA_i technology), AIDS (based on allosteric drugs leading to the inhibition of essential growth factors –VIF, LEDGF, protein U etc. –for the virus survival inside the host cell), and hereditary dysfunctions would be subdued.

Certainly, in the next three or four decades, the pharmaceutical biotechnology will promote a significant improvement in the precision of prognosis and diagnosis tests, diversification and improvement in biodrugs production, and the establishment of effective therapeutic protocols for an individualized medicine. All these will be based on the development of chips coupled to computers –detection of autoantibodies, for instance – by using synthetic biomolecules (bis-amino acids, polymeric moulds of biomolecules and peptide–nucleic acids, i.e. hybrids of protein and DNA or RNA) as markers, and by using equipment pieces coupled with electronic sensors – used for monitoring bodily activities (vital signals of patients in ITU, haemodialysis etc.), for directing surgical devices through the body (catheters, surgery without a scalpel etc.) and for conducting clinical prognosis and/or diagnosis tests (polysomnography, nuclear magnetic resonance, electrocardiography, mammography, among others).

REFERENCES

Ahir, B.K.; Engelhard, H.H.; Lakka, S.S. Tumor development and angiogenesis in adult brain tumor: glioblastome. *Molecular Neurobiology*, v.57, n.5, 2461–2478, 2020.

Bang, S.; Jeong, S.; Choi, N.; Kim, N.H. Brain-on-a chip: a history of development and future perspective. *Biomicrofluidics*, v.13, 051301, 2019. doi:10.1063/1.5120555.

Bekiari, E.; Kitsios, K.; Thabit, H. Artificial pancreas treatment for outpatients with type 1 diabetes: systematic review and meta-analysis. *British Medical Journal*, v.361, k1310, 2018. doi: 10.1136/bmj.k1310.

Bhavsar, A.P.; Guttman, J.A.; Finlay, B.B. Manipulation of host-cell pathways by bacterial pathogens. *Nature*, v.449, 827–834, 2007.

Borman, S. Two ways to shape new drugs. *Chemical and Engineering News*, v.81, n.2, 40–41, 2003.

Camera, D.M.; Smiles, W.J.; Hawley, J.A. Exercise-induced skeletal muscle signaling pathways and human athletic performance. *Free Radical Biology and Medicine*, v.98, 131–143, 2016.

Chu, D.T.; Nguyen, T.T.; Tien, N.L.B. Recent progress of stem cells therapy in cancer treatment: molecular mechanisms and potential applications. *Cell*, v.9, n.3, 563–580, 2020.

Clyne, D.J.; Miesenböck, G. Sex-specific control and tuning of the pattern generator for courtship song in Drosophila. *Cell*, v.133, n.2, 354–363, 2008.

Coleman, M. Axon degeneration mechanisms: commonality amid diversity. *Nature Reviews Neuroscience*, v.6, n.11, 889–898, 2005.

Conforti, F. Cancer immunotherapy efficacy and patient's sex: a systematic review and meta-analysis. *Lancet Oncology*, v.19, 737–746, 2018.

Conn, P.J.; Christopoulos, A.; Lindsley, W.C. Allosteric modulators of GPCRs: a novel approach for the treatment of CNS disorders. *Nature Reviews Drug Discovery*, v.8, n.1, 41–54, 2009.

Cox, J.L.; Rizzino, A. Induced pluripotent stem cells: what lies beyond the paradigm shift. *Experimental Biological Medicine*, v.235, n.2, 148–158, 2010.

Curran, M.A.; Glisson, B.S. New hope for therapeutic cancer vaccines in the era of immune checkpoint modulation. *Annual Review of Medicine*, v.70, 409–424, 2019.

Dai, C. Heat shock factor 1 is a powerful multifaceted modifier of carcinogenesis. *Cell*, v.130, n.6, 1005–1018, 2007.

Fromme, P. XFELs open a new era in structural chemical biology. *Nature Chemical Biology*, v.11, n.12, 895–899, 2015.

Gmeiner, W.H.; Ghosh, S. Nanotechnology for cancer treatment. *Nanotechnology Reviews*, v.3, n.2, 233–239, 2013.

Haraguchi, Y.; Shimizu, T.; Yamato, M.; Okano, T. Cell therapy and tissue engineering for cardiovascular disease. *Stem Cells Translational Medicine*, v.1, 136–141, 2012.

Heath, J.R.; Davis, M.E. Nanotechnology and cancer. *Annual Review of medicine*, v.59, 251–265, 2008.

Horgan, J. The forgotten era of brain chips. *Scientific American*, v.293, n.4, 66–73, 2005.

Jusiak, B. Engineering synthetic gene circuits in living cells with CRIPR technology. *Trends in Biotechnology*, v.34, n.7, 535–547, 2016.

Kadimisetty, K. 3D - printed supercapacitor – powered electrochemiluminescent protein immunoarray. *Biosensors and Bioelectronics*, v.77, 188–193, 2016.

Kelley, O.S. Advancing the speed sensitivity and accuracy of biomolecules detection using multi-length-scale engineering. *Nature Nanotechnology*, v.9, n.3, 969–980, 2014.

Kohler, F.; Gkogkidis, A.; Bentler, C. Closed-loop interaction with the cerebral cortex: a review of wireless implant technology. *Brain-Computer Interfaces*, v.4, n.8, 146–154, 2017.

Langer, R. Drug delivery: drugs on target. *Science*, v.293, 58–59, 2001.

Lewis, C.E.; Pollard, W.J. Distinct role of macrophages in different tumor microenvironments. *Cancer Research*, v.66, n.2, 605–612, 2006.

Li, C.; Wang, Y.; Jin, Y. Recent progress in drug delivery. *Acta Pharmaceutica Sinica B*, v.9, n.6, 1145–1162, 2019.

Lu, K.T. The era of synthetic biology and genome engineering: where no man has gone before. *Journal of Molecular Biology*, v.428, n.5, part B, 835–836, 2016.

Maus, M.V. Adaptive immunotherapy for cancer or viruses. *Annual Review of Immunology*, v.32, 189–225, 2014.

McManus, D.P.; Loukas, A. Current status of vaccines for schistosomiasis. *Clinical Microbiology Review*, v.21, n.1, 225–242, 2008.

Mei, Y.; Zhang, F. Molecular tools and approaches for optogenetics. *Biological Psychiatry*, v.71, n.12, 1003–1038, 2012.

Mukerjee, M. Trends in animal research. *Scientific American*, v.276, n.2, 86–93, 1997.

Neil, S.J.D. Tetherin inhibits retrovírus release and is antagonized by HIV-1 Vpu. *Nature*, v.451, 425–430, 2008.

Notkins, A.L. Immunologic and genetic factors in type 1 diabetes. *Journal of Biological Chemistry*, v.277, n.46, 43545–43548, 2002.

Orgel, L.E. Prebiotic chemistry and the origin of the RNA world. *Critical Reviews in biochemistry and Molecular Biology*, v.39, n.2, 99–123, 2004.

Ozbay, E. Plasmonics: merging photonics and electronics at nanoscale dimensions. *Science*, v.311, 189–193, 2006.

Pomsuwan, S.; Bird, G.; Schafmeister, C.E. Flexibility and lengths of bis-peptide nanostructures by electron spin resonance. *Journal of the American Chemical Society*, v.128, n.12, 3876–3877, 2006.

Rackus, G.D. A digital microfluidic device with integrated nanostructured microelectrodes for electrochemical immunoassays. *Lab on a Chip*, v.15, n.18, 3776–3784, 2015.

Redondo, M.J.; Steck, A.K.; Pugliese, A. Genetics of type 1 diabetes. *Pediatric Diabetes*, v.19, n.3, 346–353, 2018.

Rosenthal, N. Prometheus's vulture and the stem-cell promise. *New England Journal of Medicine*, v.349, n.3, 267–274, 2003.

Rubanyi, G.M. Angiogenic gene therapy for refractory angina. *Expert Opinion on Biological Therapy*, v.16, n.3, 303–315, 2015.

Sadelain, M. The basic principles of chimeric antigen receptor design. *Cancer Discovery*, v.3, n.4, 388–398, 2013.

Sequist, L.V.; Nagrath, S.; Toner, M.; Haber, D.A.; Lynch, T.J. The CTC-chip: an exciting new tool to detect circulating tumor cells in lung cancer patients. *Journal of Thoracic Oncology*, v.4, n.3, 281–283, 2009.

Shendure, J.; Mitra, R.D.; Varma, C.; Church, M.G. Advanced sequencing technologies: methods and goals. *Nature Reviews Genetics*, v.5, 335–344, 2004.

Taylor, P.J.; Brown, R.H.; Cleveland, D.W. Decoding ALS: from genes to mechanism. *Nature*, v.539, 197–206, 2016.

Tiwari, D.K.; Tiwari, M.; Jin, T. Near-infrared fluorescent protein and bioluminescence-based probes for high-resolution in vivo optical imaging. *Material Advances*, v.1, 967–987, 2020.

Velliste, M. Cortical control of a prosthetic arm for self feeding. *Nature*, v.453, 1098–1101, 2008.

Visser, K.; Eichten, A.; Coussens, L.M. Paradoxical roles of the immune system during cancer development. *Nature Reviews Cancer*, v.6, n.1, 24–37, 2006.

Waldrop, M.M. X-ray science: the big guns. *Nature*, v.505, 604–606, 2014.

Yang, M.; McKay, D.; Pollard, J.W.; Lewis, C.E. Diverse functions of macrophages in different tumor microenvironments. *Cancer Research*, v.78, n.19, 1–13, 2018.

Zhang, F. Optogenetic interrogation of neural circuits: technology for probing mammalian brain structures. *Nature Protocols*, v.5, n.3, 439–456, 2010.

Index